Local Density Approximations in Quantum Chemistry and Solid State Physics

Local Density Approximations in Quantum Chemistry and Solid State Physics

EDITED BY

Jens Peder Dahl

Technical University of Denmark
Lyngby, Denmark

AND

John Avery

University of Copenhagen
Copenhagen, Denmark

Plenum Press • New York and London

Library of Congress Cataloging in Publication Data

Symposium on Local Density Approximations in Quantum Chemistry and Solid State Theory (1982: University of Copenhagen)
Local density approximations in quantum chemistry and solid state physics.

Includes bibliographical references and index.
1. Quantum chemistry—Congresses. 2. Solid state chemistry—Congresses. 3. Hartree-Fock approximation—Congresses. I. Dahl, Jens Peder. II. Avery, John, . III. Title.
QD462.A1S95 1982 541.2′8 84-3482
ISBN 0-306-41667-0

Proceedings of a Symposium on Local Density Approximations in Quantum Chemistry and Solid State Theory, held June 10–12, 1982, at the University of Copenhagen, Copenhagen, Denmark

Printed in the United States of America

PREFACE

The simplest picture of an atom, a molecule or a solid is the picture of its distribution of charge. It is obtained by specifying the positions of the atomic nuclei and by showing how the density, $\rho(\underline{r})$, of the electronic charge-cloud varies from place to place.

A much more detailed picture is provided by the many-electron wavefunction. This quantity shows not only the arrangement of the electrons with respect to the nuclei, but also the arrangement of the electrons with respect to each other, and it allows the evaluation of the total energy and other properties. The many-electron wavefunction is in principle obtained by solving the many-electron Schrödinger equation for the motion of the interacting electrons under the influence of the nuclei, but in practice the equation is unsolvable, and it is necessary to proceed by methods of approximation. Needless to say, such methods will as a rule depend on the complexity of the system considered.

The present volume is concerned with an approximate scheme of a very general type, in the sense that it is essentially independent of the complexity of the system studied. The scheme reduces the many-electron Schrödinger equation to a one-electron Schrödinger equation, and it assumes that the effective potential in this equation is the sum of the potential from the nuclei, the classical Coulomb potential from the electronic charge-cloud, and a so-called exchange-correlation potential. Moreover, it is assumed that the value of the exchange-correlation potential at a point $\underline{r}$ only depends upon the value of the electron density ρ at that point (or perhaps upon the spin-up and spin-down parts of ρ separately). The resulting one-electron equation must be solved in an iterative fashion such that the scheme becomes self-consistent.

This self-consistent theory, with its built-in local density approximation, incorporates the electron density - and hence the simple picture of an atom, a molecule or a solid - into the calculation in a very direct way. Accordingly, it provides a simple conceptual framework for the discussion of electronic properties, and it has turned out to be a remarkably accurate theory, both in quantum chemistry and solid state physics. Let us briefly comment on the origin and development of the theory, often referred to as a density functional theory.

In quantum chemistry, the main approach used for reducing the many-electron Schrödinger equation to a one-electron equation has been the Hartree-Fock approximation supplemented by configuration interaction. However, if the Hartree-Fock approximation is applied to the electrons in a metallic solid, it leads to a zero density of states at the Fermi surface. Furthermore, configuration interaction calculations to correct the Hartree-Fock approximation are almost impossible to carry out in the case of a conducting solid. Hence, the Hartree-Fock approach is seldom used in solid state theory.

In 1951, J.C. Slater introduced an alternative method for reducing the many-electron Schrödinger equation to an approximate one-electron equation. F. Bloch and P.A.M. Dirac had been able to show that in the case of a uniform electron gas, the energy due to exchange is proportional to the 4/3 power of $\rho(\underline{r})$, and the exchange potential is proportional to the cube root of $\rho(\underline{r})$. This fact led Slater to propose a general approximate exchange potential proportional to the cube root of the local electron density, the approximation being made in the same spirit as the Thomas-Fermi approximation.

In spite of the lack of rigorous theoretical justification for using the Slater potential in cases where the density is non-uniform, it has been used for many years with great success in solid state theory. Because of this not well understood but nevertheless very real success, there has recently been great interest in introducing the Slater potential (and other exchange-correlation potentials derived from electron gas theory) into quantum chemistry.

The first calculations of this kind were made by Slater's coworkers, who developed the multiple-scattering $X\alpha$ method. In the $MSX\alpha$ method, which was modelled after existing solid state methods, each atom in a molecule or cluster is surrounded by a sphere, within which the potential is spherically averaged. In the region between the spheres, the potential is approximated by a constant. Solutions to the effective one-electron Schrödinger equation are found by matching internal and external solutions at the boundaries.

An alternative method developed by later workers avoided the problematical "muffin tin" approximation of the $MSX\alpha$ method. Instead, the effective one-electron Schrödinger equation was solved using a variational LCAO method. Matrix elements of the effective one-electron Hamiltonian were evaluated by Diophantine integration. Careful calculations on small molecules using this method showed that the

Slater potential was able to give results which could compete favourably in accuracy with Hartree-Fock results.

Much interest has been focused on the problem of how to improve the $\rho^{1/3}$ potential. Many-body theory applied to a uniform electron gas has yielded the energy as a function of density with both exchange and correlation effects included, and other terms have been added to correct for the effects of non-uniformity. Work along these lines has led to the proposal of several improved exchange-correlation potentials.

The publication in 1964 of the Hohenberg-Kohn theorem (showing that the ground-state energy of a system of electrons in an external potential is a unique functional of the density) led to a great outburst of interest in density functional theory, and recent work in this field has yielded many important results.

The field of local density approximations and density functional theory is currently a very rapidly developing and lively area of research. Nevertheless, many fundamental questions remain to be answered: For example, why do potentials derived from the theory of a uniform electron gas work at all in cases where the electron density is highly non-uniform? How can these potentials be systematically corrected to account for effects of non-uniformity? What is the best way to treat excited states and transition probabilities? How should relativistic systems be treated? These are only a few of the questions on which attention is now focused.

Because of the excitement and rapid pace of development in the field, we were happy to have had the opportunity to organize a symposium in Copenhagen devoted to local density approximations. The symposium took place on June 10-12, 1982 as a satellite of the 4th International Congress in Quantum Chemistry, which was held in Uppsala a few days later. The present multi-author book includes both lectures presented at the symposium and also invited articles from a number of distinguished experts in the field who were unable to attend the Copenhagen meeting. We are extremely grateful to the distinguished scientists who contributed to this book and thus enabled us to present a coherent and up-to-date picture of the field.

For practical reasons we have divided the book into two parts, Fundamental Theory and Applications, but this division should not be taken too literally. Important contributions to theory may be found in the articles of part two, and likewise important applications may be found in the articles of part one. We have included a few purely experimental papers in order to stress the close connection to experimental work on electron densities. As the last contribution we have included an article by an ab-initioist, who serves as the advocatus diaboli.

We give our warmest thanks to Annalise Andresen, Louise Helweg-Mikkelsen, Janne Larsen, Astrid Schantz, and Katrine Winther-Petersen of Chemistry Department B, the Technical University of Denmark, who typed all of the chapters of this book in photoready form.

November, 1983

John Avery
Jens Peder Dahl

CONTENTS

FUNDAMENTAL THEORY

APPLICATIONS

FUNDAMENTAL THEORY

DENSITY FUNCTIONAL THEORY AND LOCAL POTENTIAL

APPROXIMATIONS FROM MOMENTUM SPACE CONSIDERATIONS

Vedene H. Smith, Jr.

Department of Chemistry
Queen's University
Kingston, Ontario K7L 3N6, Canada

INTRODUCTION

In this article we shall be concerned with three topics. The first of these is the relationship between position and momentum space and among various quantities defined in the respective spaces. We shall consider next the basic theorems of density functional theory from the point of view of quantum mechanics in momentum space. We turn then to consideration of the Schrödinger equation in momentum space and in particular the Hartree-Fock equations. By following the approach of Huo and Lassettre, we shall show that a family of local exchange potentials may be derived starting from either the exchange potential in the Hartree-Fock equations or from the exchange-correlation term in the density functional expression for the energy. These correspond respectively to the Slater and Kohn-Sham-Gaspar approaches to the derivation of the $\rho^{1/3}$ exchange potential, i.e. $\alpha=1$ and $\alpha=2/3$ respectively in the $X\alpha$ exchange potential.

MOMENTUM SPACE

Why momentum space?

Although the momentum space and position space representations of the wavefunction of an N-electron system are related in quantum mechanics via the Dirac-Fourier transformation, our experience leads us to prefer the position space representation, and to ask the question in the title of this section. Experiments using Compton scattering[1] of X-rays, γ-rays and high energy electrons[2] can yield information about $\Pi(\vec{p})$, the probability that an electron in the system under examination has momentum between $\vec{p}$ and $\vec{p} + d\vec{p}$. Similarly the so-called (e,2e) knockout reaction[3,4] provides information about orbital overlap distributions in momentum space. As a result there is an increasingly large amount of experimental information becoming available which can provide a test for theoretical calculations whether made in momentum space or made in position space and transformed to momentum space. Furthermore this experimental data must be interpreted in either momentum space or position space concepts about electronic structure. Because of our bias towards position space comparatively little has been done to discuss bonding and similar matters in momentum space;[5,6] what has been done is relatively unfamiliar to most chemists and physicists. In fact there is often confusion as to the relationship between the momentum density $\Pi(\vec{p})$ and its more familiar position space counterpart $\rho(\vec{r})$.

Relationship among momentum and position space functions

To introduce the relevant quantities, and to see their relationships, we shall consider the simple case of the one-electron atom (hydrogen). Here and throughout this article we shall use atomic units. Then the Schrödinger equation takes the form

$$-1/2\nabla^2\psi(\vec{r}) - Z/r\ \psi(\vec{r}) = E\psi(\vec{r}) \tag{1}$$

in position space and the form

$$-1/2p^2\chi(\vec{p}) - 2(2\pi)^{-2}\int\chi(\vec{p})\,|\vec{p}-\vec{p}'|^{-2}d\vec{p}' = E\chi(\vec{p}) \tag{2}$$

in momentum space. The wavefunctions $\psi(\vec{r})$ and $\chi(\vec{p})$ are related by the Dirac-Fourier transform

$$\chi(\vec{p}) = (2\pi)^{-3/2}\ F_3(\psi(\vec{r})) \tag{3}$$

where F_3 is the N=1 case of the 3N-dimensional Fourier transform

$$F_{3N} = \int\exp[-i\sum_{k=1}^{N}\vec{r}_k\cdot\vec{p}_k]d\vec{r}_1 d\vec{r}_2\ldots d\vec{r}_N\ . \tag{4}$$

By inspection of the pair of equations (1) and (2) we see that in Eq. (1) the kinetic energy operator is non-local while in Eq. (2) the electron-nuclear or external potential term is non-local. This observation will prove to be important in our later discussion.

If we define the density functions

$$\rho(\vec{r}) = |\psi(\vec{r})|^2\ , \tag{5}$$

$$\Pi(\vec{p}) = |\chi(\vec{p})|^2 \tag{6}$$

as well as their respective Fourier transforms, the form factor $F(\vec{p})$,[7]

$$F(\vec{p}) = F_3(\rho(\vec{r})) \tag{7}$$

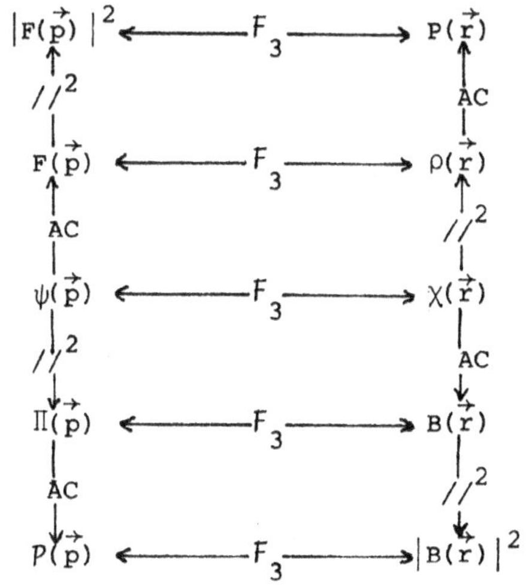

Fig. 1. Relationships among various position and momentum space functions illustrated in the case of a one-electron wavefunction (orbital).

and the internally folded density $B(\vec{r})$,[8-10]

$$B(\vec{r}) = F_3(\Pi(\vec{p})) \tag{8}$$

then we may display their interrelationships in a chart (Fig.1)[9,11] where F_3, $//^2$ and AC denote three-dimensional Dirac-Fourier transformation, modulus squared and "auto-correlation", respectively. As indicated in the chart

$$F(\vec{p}) = \chi^*(\vec{p}) \otimes \chi(\vec{p}) = \int \chi^*(\vec{p}-\vec{p}\,')\chi(\vec{p}\,')d\vec{p}\,' \tag{9}$$

and

$$B(\vec{r}) = \psi^*(\vec{r}) \otimes \psi(\vec{r}) = \int \psi^*(\vec{r}+\vec{r}\,')\psi(\vec{r}\,')d\vec{r}\,' \tag{10}$$

where $\otimes$ denotes the convolution operation and "auto-correlation" is the convolution of a function with its complex conjugate. We observe that in the one-electron case $B(\vec{r})$ is equivalent to an overlap integral.

We have included in the chart for the sake of completeness, the Patterson function or density auto-correlation function $P(\vec{r})$[11]

$$P(\vec{r}) = \int \rho(\vec{r}\,')\rho(\vec{r}\,'+\vec{r})d\vec{r}\,' = F_3(|F(\vec{k})|^2) \tag{11}$$

and a momentum space analogue $\mathcal{P}(\vec{p})$[11]

$$\mathcal{P}(\vec{p}) = \int \Pi(\vec{p}\,')\Pi(\vec{p}+\vec{p}\,')d\vec{p}\,' = F_3(|B(\vec{r})|^2) \tag{12}$$

These concepts may be generalized to the N-electron case through the use of the one-particle reduced density matrix (one-matrix) in position space[7]

$$\Gamma^{(1)}(x_1;x_1') = N\int\Psi^*(x_1',x_2',\ldots x_N') \times\Psi(x_1,x_2,\ldots,x_N)dx_2\ldots dx_N \tag{13}$$

and its momentum space analogue[10]

$$\Gamma^{(1)}(y;\ y_1') = N\int\psi^*(y_1',y_2'\ldots y_N') \times\ \psi(y_1,y_2,\ldots y_N)dy_2\ldots dy_N \tag{14}$$

where $\Psi(x_1,x_2,\ldots x_N)[\Phi(y_1,y_2,\ldots,y_N)]$ is the position [momentum] space representation of the N-electron wavefunction and $x_i = (\vec{r}_i,\sigma_i)$ $[y_i = (\vec{p}_i,\sigma_i)]$ is a combined position [momentum]-spin coordinate. Since spin is irrelevant in the present discussion, we will work with the spinless one-matrices

$$\Gamma^{(1)}(\vec{r}_1;\vec{r}_1') = \int\Gamma^{(1)}(x_1;x_1')\delta(\sigma_1-\sigma_1')d\sigma_1 d\sigma_1' \tag{15}$$

and

$$\Gamma^{(1)}(\vec{p}_1;\vec{p}_1') = \int\Gamma^{(1)}(y_1;y_1')\delta(\sigma_1-\sigma_1')d\sigma_1 d\sigma_1' \ . \tag{16}$$

They are related themselves by a 6-dimensional Dirac-Fourier transformation

$$\Gamma^{(1)}(\vec{p}_1;\vec{p}_1') = (2\pi)^{-3}F_6(\Gamma^{(1)}(\vec{r}_1;\vec{r}_1')) \tag{17}$$

Then

$$\rho(\vec{r}) = \Gamma^{(1)}(\vec{r};\vec{r}) \tag{18}$$

and

$$\Pi(\vec{p}) = \Gamma^{(1)}(\vec{p};\vec{p}) \tag{19}$$

while

$$F(\vec{p}) = \int\Gamma^{(1)}(\vec{p}';\vec{p}'+\vec{p})d\vec{p}' \tag{20}$$

$$F(\vec{p}'-\vec{p}) = \int\Gamma^{(1)}(\vec{p};\vec{p}')d\vec{P} \tag{21}$$

where $\vec{P} = (\vec{p}'+\vec{p})/2$ and $\vec{p} = \vec{p}'-\vec{p}$ are respectively the extracular and intracular coordinates corresponding to $\vec{p}$ and $\vec{p}'$. Similarly

$$B(\vec{r}) = \int\Gamma^{(1)}(\vec{r}';\vec{r}'+\vec{r})d\vec{r} \tag{22}$$

$$B(\vec{r}'-\vec{r}) = \int\Gamma^{(1)}(\vec{r};\vec{r}')d\vec{R} \tag{23}$$

where $R = (\vec{r}'+\vec{r})/2$ and $\vec{r}'-\vec{r}$ are respectively, the extracular and intracular coordinates corresponding to $\vec{r}$ and $\vec{r}'$. Eqs. (21) and (23) make it clear[10] that F and B respectively are the <u>relative</u> or <u>intracular projections</u> of the momentum-space and position-space one-matrices.

We may illustrate the nature of these operations by sketching the $\vec{r},\vec{r}'$ plane on which the one-matrix $\Gamma^{(1)}(\vec{r},\vec{r}')$ is defined. Then $\rho(\vec{r})$ is its value at the points along the line $\vec{r}=\vec{r}'$ (illustrated by

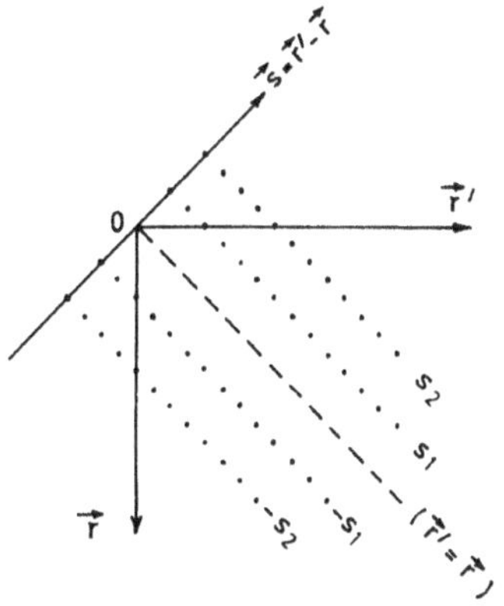

Fig. 2. Illustration of the operations leading to $\rho(\vec{r})$ and $B(\vec{r}'-\vec{r})$.

the dashed line) while $B(\vec{r})$ is the relative or intracular projection of $\Gamma^{(1)}(\vec{r}',\vec{r})$ illustrated by the dotted lines and defined by Eq. (23). Several descriptive names have been attached to $B(\vec{r})$ in the literature besides the intracular projection of the one-matrix. Because of the operation involved in Eq. (22), it has been called the internally folded density.[10] The name reciprocal form factor[9] has been used because it is a form factor in a space reciprocal to the usual one. The designation of characteristic function of the momentum density[12] has an obvious meaning.

With these definitions for the N-electron case, our chart for the one-electron case may be generalized (Fig. 3). Here, the F's, $//^2$ and AC have their usual meanings while R^{N-1} denotes the spin-traced reduction from the N-particle density matrix to $\Gamma^{(1)}$ as defined by Eqs. (13) and (15) in position space and Eqs. (14) and (16) in momentum space and IP denotes the intracular projection defined by Eqs. (21) and (23) or the internal folding defined by Eqs. (20) and (22).

We note that the two charts (Fig. 1 and Fig. 3) are related through the (natural) expansion of the one-matrix in terms of its eigenfunctions, $\psi_i(\vec{r})$, $\chi_i(\vec{p})$ and eigenvalues λ_i, i.e.

$$\Gamma^{(1)}(\vec{r};\vec{r}') = \sum_{i=1}^{\infty} \lambda_i \psi_i(\vec{r})\psi_i^*(\vec{r}') \tag{24}$$

or

$$\Gamma^{(1)}(\vec{p};\vec{p}') = \sum_{i=1}^{\infty} \lambda_i \chi_i(\vec{p})\chi_i^*(\vec{p}') \ . \tag{25}$$

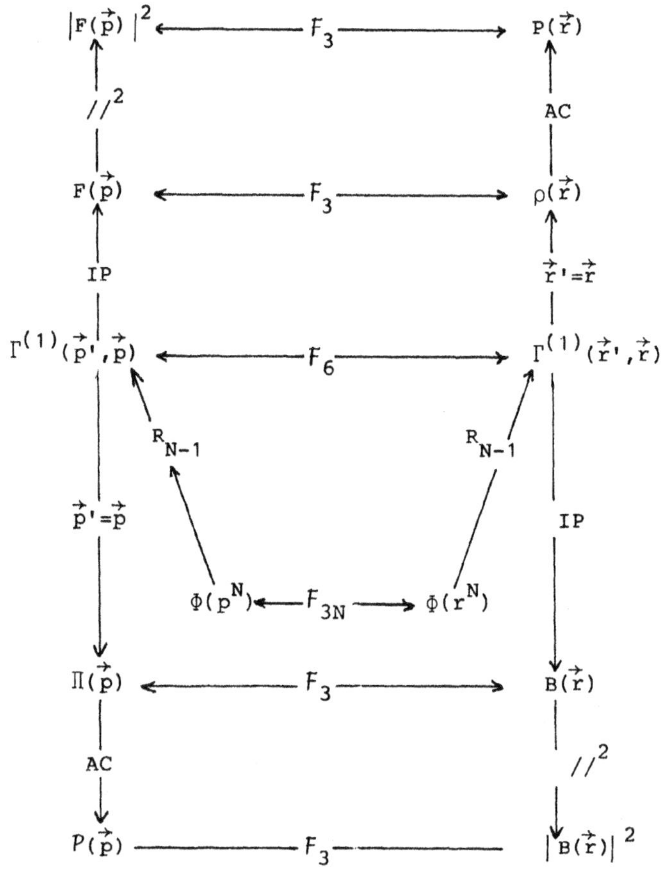

Fig. 3. Relationships among various (single-particle) position and momentum space functions illustrated in the N-electron case.

The single-particle chart (Fig. 1) applies for each value of the index i to these natural orbitals and the quantities defined in terms of them. Insertion of the natural expansions (24) and (25) for the one-matrices leads to natural expansions of $F(\vec{p})$, $\Pi(\vec{p})$, $B(\vec{r})$ and $\rho(\vec{r})$ in terms of $F_i(\vec{p})$, $\Pi_i(\vec{p})$, $B_i(\vec{r})$ and $\rho_i(\vec{r})$, i.e.

$$F(\vec{p}) = \sum_i \lambda_i \; F_i(\vec{p}) \tag{26}$$

$$\ell(\vec{p}) = \sum_i \lambda_i \; \Pi_i(\vec{p}) \tag{27}$$

$$B(\vec{r}) = \sum_i \lambda_i \; B_i(\vec{r}) \tag{28}$$

$$\rho(\vec{r}) = \sum_i \lambda_i \; \rho_i(\vec{r}) \tag{29}$$

For the top and bottom row of the charts, the situation is more complicated, e.g.

$$|F(\vec{p})|^2 = \sum_{i,j} \lambda_i \lambda_j |F_i(\vec{p})|^2 \; |F_j(\vec{p})|^2 \qquad (30)$$

and

$$|B(\vec{r})|^2 = \sum_{i,j} \lambda_i \lambda_j |B_i(\vec{r})|^2 \; |B_j(\vec{r})|^2 \qquad (31)$$

ROUTES TO MOMENTUM SPACE PROPERTIES

From Position Space

As is evident from the preceeding discussion, one procedure to calculate properties corresponding to operators in momentum space, is simply to Dirac-Fourier transform a given position-space wavefunction $\Psi(x^N)$ into its momentum space equivalent $\Phi(y^N)$, i.e.

$$\Phi(y^N) = (2\pi)^{-3N/2} \; F_{3N}(\Psi(x^N)) \qquad (32)$$

and proceed to calculate the desired expectation value in the usual manner. If the wavefunction has been reduced to the one-particle density matrix $\Gamma^{(1)}(x';x)$ then only the 6-dimensional Dirac-Fourier transformation is needed, i.e.

$$\Gamma^{(1)}(y;y') = (2\pi)^{-3} \; F_6(\Gamma^{(1)}(x;x')) \qquad (33)$$

and one-particle operators (both local and non-local) may be evaluated in the normal way.[7,13] For q-particle operators, the 3q-dimensional Dirac-Fourier transformation of the q-particle density matrix is needed.[10]

By Solution of the Schrödinger Equation in Momentum Space

The possibility of solving the Schrödinger equation in momentum space has been considered on occasion through the years[5,14-18] and more often recently.[19-25] To solve it is an attractive possibility for several reasons. To see these we write the equation down in the case of a two-electron system:

$$\begin{aligned}(\tfrac{1}{2}p_1^2 + \tfrac{1}{2}p_2^2 - E)\chi(\vec{p}_1,\vec{p}_2) &= 2(2\pi)^{-2} \int d\vec{q}\; S(\vec{q}) q^{-2} \; [\chi(\vec{p}_1-\vec{q},\vec{p}_2) + \chi(\vec{p}_1,\vec{p}_2 - \vec{q})] \\ &\quad - (2\pi)^{-2} \int d\vec{q}\; q^{-2} \; [\chi(\vec{p}_1-\vec{q},\vec{p}_2+\vec{q})] \qquad (34)\end{aligned}$$

where

$$S(\vec{q}) = \sum_{\alpha=1}^{A} Z_\alpha \exp(i\vec{q}.\vec{R}_\alpha) \qquad (35)$$

and $\vec{R}_\alpha$ is the position vector which locates one of the A nuclei, namely the α'th nucleus of charge Z_α. The first thing we observe by inspection of Eq. (34) is that there are no multicentre singularities as

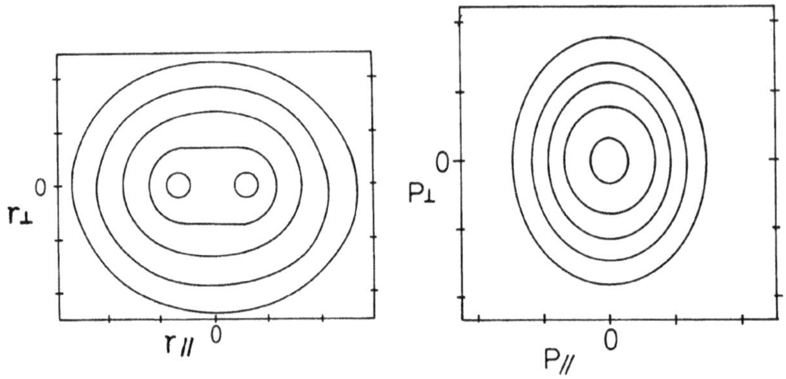

(a) (b)

Fig.4 Comparison of the (a) position density $\rho(\vec{r})$ and (b) momentum density $\Pi(\vec{p})$ for the ground state of molecular hydrogen in the $(r_\perp, r_{//})$ and $(p_\perp, p_{//})$ planes respectively.

there are in position space at each of the nuclear points R_α, and the electron-electron coalescence; the only singularity at $\vec{q} = 0$ is eliminated by the integration over $\vec{q}$ in Eq. (34). The singularities at $r_{12} = 0$ and $\vec{r}_\alpha = 0$ do affect the asymptotic behaviour ($q \to \infty$) of the wavefunction[5,7]. The momentum distribution is single-centred about $\vec{p} = 0$.[13] We sketch the momentum density and position density for the hydrogen molecule in Figure 4. Inspection indicates the single--centredness of $\Pi(\vec{p})$ and the two-centred nature of $\rho(\vec{r})$. In fact $\vec{p} = 0$ is a centre of inversion for $\Pi(p)$, i.e. $\Pi(\vec{p}) = \Pi(-\vec{p})$.[26]

We note further that since the factor $(\frac{1}{2}p_1^2 + \frac{1}{2}p_2^2 - E)$ in Eq. (34) is positive, we may divide both sides of this equation by it and proceed to develop an iterative method for its solution.

Total Energy Expressions Involving Momentum Space Quantities

By use of the one-matrix $\Gamma^{(1)}(\vec{p};\vec{p}')$ and two-matrix $\Gamma^{(2)}(\vec{p}_1,\vec{p}_2;\vec{p}_1',\vec{p}_2')$, we may write the total energy E in the form

$$E = T[\Pi] + U[\Gamma^{(1)}(\vec{p};\vec{p}')] + Eee[\Gamma^{(2)}(\vec{p}_1,\vec{p}_2;\vec{p}_1',\vec{p}_2')] \qquad (36)$$

where the kinetic energy term is:

$$T[\Pi] = 1/2\int p^2 \Pi(\vec{p})d\vec{p} \; , \tag{37}$$

the electron-nuclear term is:

$$U[\Gamma^{(1)}] = -2(2\pi)^{-2}\int d\vec{q}d\vec{p}S(\vec{q})\;\Gamma^{(1)}(\vec{p}-\vec{q};\vec{p}) \tag{38}$$

and the electron-electron term is:

$$Eee[\Gamma^{(2)}] = (2\pi)^{-2}\int d\vec{q}d\vec{p}_1 d\vec{p}_2 q^{-2}\Gamma^{(2)}(\vec{p}_1-\vec{q},\vec{p}_2+\vec{q};\vec{p}_1,\vec{p}_2) \; . \tag{39}$$

We write down for comparison the corresponding expression over position-space quantities[27]

$$E = T[\Gamma^{(1)}(\vec{r};\vec{r}\,')] + U[\rho(\vec{r})] + Eee[\Gamma^{(2)}(\vec{r}_1,\vec{r}_2;\vec{r}_1,\vec{r}_2)] \tag{40}$$

HOHENBERG-KOHN-LEVY THEOREMS AND THE MOMENTUM DENSITY

In the context of our discussion, it is natural to ask whether there are theorems similar to the celebrated theorems of density functional theory, i.e. is there a functional relationship between the ground state energy (and the Hamiltonian H) for a system and the momentum density $\Pi(\vec{p})$ of the ground state? As we have noted earlier, examination of the Schrödinger equation in momentum space (Eq. 34) reveals that the usual external potential $v(\vec{r})$ which is local in position space is non-local in momentum space. Since the first step in the standard proof of the Hohenberg-Kohn theorem[28] involves establishing that $v(\vec{r})$ is a unique functional of $\rho(\vec{r})$, some have concluded that the extension of the Hohenberg-Kohn ideas to momentum space is not meaningful[29] or even possible. However we observe that the operator for the kinetic energy is a local one in momentum space and the kinetic energy and external potential contributions have in some sense exchanged roles in the discussion. We shall consider several approaches to the problem.[30,31]

Energy as a Functional of the One-particle Density Matrix by Introduction of a Non-Local Potential $v(\vec{r},\vec{r}\,')$

One can establish that the ground state energy is a functional of the one-matrix $\Gamma^{(1)}(\vec{r},\vec{r}\,')$ of the ground state.[32,33] In momentum space, one can work with an external potential term $(v(\vec{p},\vec{p}\,'))$ to establish in an analogous manner to the original Hohenberg-Kohn theorem, the desired unique functional relationship $\Gamma^{(1)}(\vec{p},\vec{p}\,') \leftrightarrow v(\vec{p},\vec{p}\,') \leftrightarrow H$ and the desired variational principle. As in the position space discussion, one must stipulate the requirement that $\Gamma^{(1)}(\vec{p},\vec{p}\,')$ be derivable from an admissible anti-symmetric wavefunction , i.e. be N-representable.[34,35] This requirement must be met both in the proof of the functional relationship and in the choice of admissible trial $\Gamma^{(1)}$'s for application of the variational principle. Furthermore, the trial $\Gamma^{(1)}(\vec{p},\vec{p}\,')$'s must be derivable from some $v(\vec{p},\vec{p}\,')$, i.e. be "v-representable".[36] Since the problem of v-representability has been bypassed in position space by the approach of Levy,[37-39] we shall develop analogous ideas below. First, it is useful to review Levy's approach briefly.

Levy Density Functionals in Position Space

After Hohenberg and Kohn[28] established by their *reductio ad absurdum* argument that there is a one-to-one correspondence (within an additive constant) between the local potential $v(\vec{r})$ and the ground state $\rho(\vec{r})$ (and hence between H and $\rho(\vec{r})$), they developed their second or variational theorem. To do this they introduced the functional

$$E_{HK}[\rho] = \int v(\vec{r})\rho(\vec{r})d\vec{r} + F[\rho] \quad (41)$$

where

$$F[\rho] = \langle\Psi_\rho|T + Vee|\Psi_\rho\rangle \quad (42)$$

and Ψ_ρ is the ground-state eigenfunction of H which has ρ as its electron density. Then the theorem states that $E_{HK}[\rho]$ is greater than or equal to the true ground state energy with equality holding for the true ground-state ρ. However, as mentioned above, the admissible ρ's must be both N- and v-representable (as well as N-normalizable).

Levy[37-39] bypassed the condition of v-representability by introducing a universal functional $Q[\rho] = \inf\langle\Psi_\rho|T + Vee|\Psi_\rho\rangle$. For a given N-normalized ρ, $Q[\rho]$ searches over <u>all</u> antisymmetric N-electron functions Ψ_ρ (which reduce to ρ) and yields the smallest value of the expectation value of T + Vee. We observe that for N- and v-representable ρ, the results of $Q[\rho]$ and $F[\rho]$ agree. Levy's variational functional $E_L[\rho] = \int v(\vec{r})\rho(\vec{r})d\vec{r} + Q[\rho]$ is an upper bound to the true ground state energy for a given $v(\vec{r})$ and is equal to it for the true ground state ρ.

Levy-type Density Functionals in Momentum Space

We introduce a functional $G_v[\Pi] = \inf\langle\Phi_\Pi|v + Vee|\Phi_\Pi\rangle$ which is different for different v and as a result is not universal. $G_v[\Pi]$ searches over all anti-symmetric N-electron wavefunctions Φ_Π in the momentum representation which reduce to Π and yields the smallest value of the expectation value of v + Vee. For a known v, the energy functional $E_v[\Pi] = \int T(\vec{p})\Pi(\vec{p})d\vec{p} + G_v[\Pi]$ is an upper bound to the true ground state energy and is equal to it for the true ground state Π.

Levy-type Density Matrix Functionals

Levy introduced a universal density matrix functional $W[\Gamma^{(1)}(\vec{r};\vec{r}')] = \inf\langle\Psi_\Gamma|Vee|\Psi_\Gamma\rangle$ where $\Gamma^{(1)}$ is an N-representable and N-normalized one-matrix. This functional performs a search over <u>all</u> antisymmetric N-electron wavefunctions Ψ_Γ which reduce to the given $\Gamma^{(1)}$ and provides the smallest value of the expectation value of Vee. This leads to a variational bound to the ground state energy inasmuch as the expectation values of T and U are directly calculable from $\Gamma^{(1)}(\vec{r};\vec{r}')$.

In momentum space, the universal functional $Eee[\Gamma^{(1)}(\vec{p};\vec{p}')=\inf\langle\Phi_\Gamma|Vee|\Phi_\Gamma\rangle$ is defined for N-representable and N-normalized one-matrices. Here the search is performed over all antisymmetric N-electron wavefunctions Φ_Γ in momentum space which reduce to the given $\Gamma^{(1)}(\vec{p};\vec{p}')$. The functional provides the smallest value of the expectation value of Vee. Again we find a variational bound to the ground-state energy because the expectation values of T and U are directly calculable from $\Gamma^{(1)}(\vec{p};\vec{p}')$.

An Effective Mass Approach

By rewriting the kinetic energy in terms of an effective mass operator $m^{-1}(\vec{p})$,

$$T[\Pi] = 1/2 \int m^{-1}(\vec{p})p^2\Pi(\vec{p})d\vec{p} \tag{43}$$

we are able to proceed in a very similar manner to the original reductio ad absurdum approach in order to establish that $E \leftrightarrow H \leftrightarrow m^{-1}(\vec{p}) \leftrightarrow \Pi(\vec{p})$ where $\Pi(p)$ is the N-normalized ground state momentum density which is N-representable and $p^2m^{-1}(\vec{p})$ is unique up to an additive constant. In our discussion we must assume that the external potential $v(\vec{p},\vec{p}')$ obtained from $v(\vec{r})$ is fixed (just as without realizing it in the position space discussion, it was assumed that $m^{-1}(\vec{p})$ was fixed as unity). Furthermore, in order to maintain the concept of the kinetic energy, one should assume that $m^{-1}(\vec{p})$ is of the form $1 + f(\vec{p})$ and that $p^2f(\vec{p})$ is well behaved.

Orbitals and Density Functional Theory

The density functional theory of Hohenberg and Kohn has found its practical implementation through the single-particle equations of Kohn and Sham.[40] Since the total energy expression can be written in the form

$$E[\rho] = T[\rho] + Eext[\rho] + Ec[\rho] + Exc[\rho] \quad , \tag{44}$$

the derivation of the orbital equations of Kohn and Sham requires assumptions about the nature of the relationship of the orbitals with the density. By use of a non-interacting system as reference, (wherein there is a one-to-one correspondence between electrons and spin orbitals) the electron density $\rho(\vec{r})$ can be written in the form

$$\rho(\vec{r}) = \sum_{i=1}^{N} |\phi_i(\vec{r})|^2 \ . \tag{45}$$

With this ansatz for ρ, and the corresponding expression for a kinetic energy,

$$T_s[\rho] = \sum_{i=1}^{N} |\nabla\phi_i|^2 , \tag{46}$$

variation with respect to the $\{\phi_i\}$ leads to a set of equations

$$\{-1/2\nabla^2 + Vc[\rho(\vec{r})] + V(\vec{r}) + Vxc[\rho(\vec{r})]\}\phi_i(\vec{r}) = \varepsilon_i\phi_i(\vec{r}) \tag{47}$$

Here $Vxc[\rho]$ is the exchange-correlation potential obtained as the functional derivative of the exchange correlation energy $Exc[\rho]$,

$$Vxc[\rho(\vec{r})] = \delta\overline{E}xc[\rho]/\delta\rho(\vec{r}) \tag{48}$$

where

$$\overline{E}xc[\rho] = Exc[\rho] + T[\rho] - Ts[\rho] \tag{49}$$

and the ϕ_i corresponding to the N smallest ε_i are used to construct ρ and Ts. The latter expression reflects that the error in using $Ts[\rho]$ has been absorbed into $\overline{E}xc[\rho]$ and hence $Vxc[\rho]$.

We note that the derivation of these equations has required non-interacting wavefunction v-representability[38] as well as N-representability. The latter is easily demonstrated as a number of authors[32,35,41] have shown by construction that any N-normalized, non-negative ρ may be represented by a single Slater determinant which is consistent with a $\rho(\vec{r})$ of the form of Eq. (45). However, the Slater determinants constructed in these existence proofs do not lead to very satisfactory $\Pi(\vec{p})$. In fact, the problem of N-representability given both $\rho(\vec{r})$ and $\Pi(\vec{p})$ has not been solved.

If we use the N-lowest energy solutions of the Kohn-Sham equations to construct $\Pi_s(\vec{p})$ consistent with Ts, i.e.

$$\Pi_s(\vec{p}) = \sum_{i=1}^{N} |\chi(\vec{p})|^2 \tag{50}$$

we must add a correction term, Π_c to Π_s in order to calculate the momentum density $\Pi(\vec{p})$ and hence any momentum space expectation value. The additional term is given by a functional derivative expression,[42-46]

$$\Pi_c(\vec{p}) = \delta Exc[\rho]/\delta\varepsilon_p \tag{51}$$

where $\varepsilon_p = 1/2p^2$. It is interesting to note that the derivation of $\Pi_c(\vec{p})$ was made by addition of a fictitious term $\lambda\Pi(\vec{p})$ to the original Hamiltonian and the application of the Hellman-Feynman theorem to the resulting energy expression.[42] This is similar to the addition of the fictitious term $p^2f(\vec{p})$ in the last section when $f(p) = \lambda p^{-2}$.

LOCAL EXCHANGE POTENTIALS

We are interested in deriving local approximations to the exchange-correlation term in the single-particle equations. We begin by considering the approximation of the non-local exchange contributions to the Hartree-Fock equations. To do this we shall write down the Hartree-Fock equations in momentum space specifically for the closed-shell case of an N-electron atom of nuclear charge Z. This may be done either by using a Slater determinant in the total energy expression (Eq.(36)) and varying for the orbitals or by Dirac-Fourier transformation of the Hartree-Fock equations in position space:

$$[-1/2\nabla^2 + V_{eN} + \sum_j(2J_j-K_j)]\psi_i(\vec{r}) = \varepsilon_i\psi_i(\vec{r}) \tag{52}$$

where the Coulomb and exchange operators are defined respectively by:

$$J_j(\vec{r}) = \int|\psi_j(\vec{r}')|^2 \, |\vec{r}-\vec{r}'|^{-1}d\vec{r}' \tag{53}$$

and

$$K_j(\vec{r})\psi_j(\vec{r}) = [\int\psi_j{*}(\vec{r}')\psi_i(\vec{r}')\,|\vec{r}-\vec{r}'|^{-1}d\vec{r}']\psi_i(\vec{r}) \; . \tag{54}$$

By either approach one obtains the following set of equations[19,24] for the orbitals $\{\chi_i(\vec{p})\}$ and orbital energies $\{\varepsilon_i\}$:

$$\begin{aligned}(1/2p^2-\varepsilon_i)\chi_i(\vec{p}) &= 2Z(2\pi)^{-2}\int d\vec{p}'\, q^{-2}\chi_i(\vec{p}') \\ &- 2(2\pi)^{-2}\int d\vec{p}'\, q^{-2}\sum_j W_{jj}^*(\vec{q})\chi_i(\vec{p}') \\ &+ 2(2\pi)^{-2}\int d\vec{p}'\, q^{-2}\sum_j W_{ij}^*(q)\chi_j(\vec{p}') = 0\end{aligned} \tag{55}$$

where j sums over the N/2 occupied orbitals and

$$\begin{aligned}W_{ij}(\vec{q}) &= \int d\vec{Q}\, \chi_i^*(\vec{Q})\chi_j(\vec{Q}-\vec{q}) \\ &= \underset{3}{F}\{\psi_i^*(\vec{r})\psi_j(\vec{r})\} = \chi_i^*(\vec{q}) \otimes \chi_j(\vec{q}) \; .\end{aligned} \tag{56}$$

We recognize that the first term on the right-hand-side is the electron-nuclear attraction term, the second term is the electronic Coulomb repulsion term (including the self-Coulomb contribution) and the third term is the exchange term (including the self-exchange contribution). It is the latter that we are interested in approximating in such a manner as to develop a consistent and useful set of exchange potentials. It is important to recognize that the kernels of each of the first two terms depend only on the difference between $\vec{p}$ and $\vec{p}'$ while the exchange term is a function of more than just $\vec{q} = \vec{p}-\vec{p}'$.

Lassetre[19] has shown that the integral in the exchange term may be rewritten as follows

$$\int d\vec{p}'\, q^{-2}[\sum_j W_{ij}^*(\vec{q})\chi_j(\vec{p}')] = (4\pi)^{-1}\int d\vec{Q}[\sum_j T_{jj}(\vec{Q},\vec{p})\chi_i(\vec{Q})] \tag{57}$$

where

$$T_{jj}(\vec{Q},\vec{p}) = (4\pi)\int d\vec{q}\, q^{-2}\chi_j^*(\vec{Q}-\vec{q})\chi_j(\vec{p}-\vec{q}) \tag{58}$$

$$= \langle \exp(i\vec{p}\cdot\vec{r}_2)\psi_j(\vec{r}_1)|r_{12}^{-1}|\exp(i\vec{Q}\cdot\vec{r}_1)|\psi_j(\vec{r}_2)\rangle \ . \tag{59}$$

Furthermore, he identified the kernel of the exchange, $T_{jj}(\vec{Q},\vec{p})$ as proportional to the Born-Oppenheimer collision amplitude (A_{BO}) for an elastic collision of an external electron with the system when only the interaction between the two electrons undergoing exchange is included. As a result, methods developed for the approximation of this amplitude in electron scattering calculations may be used to approximate the exchange contribution to the Hartree-Fock equations.

A fruitful approach is to develop A_{BO} in the so-called Bonham series representation.[47] Huo[48] has shown that a reasonably accurate representation of A_{BO} is given by the first three terms of this series. In order that the exchange kernel be Hermitian, the second term is dropped from the expression for $T_{jj}(\vec{Q},\vec{p})$. Then the Bonham-Huo expression[48] takes the form

$$\begin{aligned} T_{jj}(\vec{Q},\vec{p}) &\sim T_{jj}^{BH}(\vec{Q},\vec{p}) \\ &= 4\pi Q^{-2}\langle\psi_j(\vec{r})|\exp[i(\vec{Q}-\vec{p})\cdot\vec{r}]|\psi_j(\vec{r})\rangle \\ &\quad - (4\pi/3)Q^{-4}\langle\nabla^2\psi_j(\vec{r})|\exp[i(\vec{Q}-\vec{p})\cdot\vec{r}]|\psi_j(\vec{r})\rangle \end{aligned} \tag{60}$$

in the case of a bound atomic state of spherical symmetry which we have chosen to simplify the present discussion.

If we define the form factors

$$F_{jj}(\vec{Q}-\vec{p}) = \mathcal{F}_3(|\psi_j(\vec{r})|^2) \tag{61}$$

and

$$G_{jj}(\vec{Q}-\vec{p}) = \mathcal{F}_3(|\nabla\psi_j(\vec{r})|^2) \ , \tag{62}$$

we may rewrite the expression for T_{jj}^{BH} as follows:

$$T_{jj}^{BH}(\vec{Q},\vec{p}) = 4\pi Q^{-2}F_{jj}(\vec{Q}-\vec{p}) - 4\pi Q^{-4}G_{jj}(\vec{Q}-\vec{p}) \ . \tag{63}$$

We note that this approximation to the exchange term now has the same dependence on $\vec{Q}-\vec{p}$ as do the electron-nuclear and Coulomb terms.

Since Q may be identified with the momentum vector of the internal electron in the scattering process, Huo replaces Q^2 by its expectation value $\langle Q^2\rangle$ and Q^4 by $\langle Q^2\rangle^2$. Inverse Fourier transformation of $T_{jj}^{BH}(\vec{Q},\vec{p})$ yields

$$\begin{aligned} V\text{ex}(\vec{r})\psi_i(\vec{r}) &= -(8\pi^3)^{-2}\, d\vec{p}\,\exp(i\vec{p}\cdot\vec{r}) \\ &\quad \times \sum_j \int d\vec{Q}\; T_{jj}^{BH}(\vec{Q},\vec{p})\int dr'\,\exp(-i\vec{Q}\cdot\vec{r}')\psi_i(\vec{r}') \end{aligned} \tag{64}$$

$$= -\sum_j 4\pi\langle Q^2\rangle_j^{-1}\,(|\psi_j|^2 - \langle 3Q^2\rangle_j^{-1}\psi_j\nabla^2\psi_j^*) \tag{65}$$

Before proceeding further we observe that this expression for $Vex(\vec{r})$ is the same for all orbitals and hence will yield orthogonal solutions. However, if we are interested in correcting for the self-interaction error, i.e. imperfect cancellation between the self-Coulomb and self-exchange terms, we may do this in a straight-forward manner because Eq. (65) involves a summation over the orbital index j and we may just subtract the correction for orbital i in $Vex(\vec{r})\psi_i(\vec{r})$. Development and testing of such potentials is in progress in this laboratory. We shall not consider such a correction further in this article.

Huo[48] has suggested two choices for the quantity $\langle Q^2\rangle_j$ appearing in Eq. (65). The first is to use $\langle Q^2\rangle_j = \int Q^2|\chi_j(\vec{Q})|^2 d\vec{Q}$ and the second is to use the same value for each j, namely twice the average kinetic energy in the Thomas-Fermi model. For this choice:

$$\langle Q^2\rangle = T_o[\rho] = 4/9(3\pi^2\rho)^{2/3} = 4/9(6\pi^2\rho_\uparrow)^{2/3} \tag{66}$$

we find

$$Vex(\vec{r}) = -3/2(\frac{6}{\pi}\rho_\uparrow)^{1/3} + \frac{9}{8}\sum_j [\psi_j\nabla^2\psi_j^*][(6\pi\rho_\uparrow^4)^{1/3}] \tag{67}$$

where $\rho_\uparrow$ is the charge density of up spins. We recognize the first term as just the $\alpha = 1$ variant of the $X\alpha$ exchange potential of Slater[49] (Vx1). The second or correction term in Eq. (67) should be compared with the correction term

$$\rho^{-1/3}\ [\frac{4}{3}\rho^{-2}(\nabla\rho)^2 - 2\rho^{-1}\nabla^2\rho] \tag{68}$$

derived by several authors[50-52] with either a coefficient determined empirically or from first principles. Both correction functions suffer from singular behaviour at the origin and increase in an exponential manner for large r. To correct for the latter in Eq. (67), one need only require that the third term in the Bonham-Huo representation possess the correct singularity in the complex Q plane. In this so-called Rudge modification[53] one replaces Q^{-4} by $(Q^2 + \alpha_j^2)^{-2}$ where $\alpha_j^2 = -2\varepsilon_j$. To correct for the former problem for s orbitals, the product of two gradients is used instead of the Laplacian. As a result, the second term in Eq. (65) becomes

$$(\frac{4\pi}{3})\sum_j(\langle Q^2\rangle_j + \alpha_j^2)^{-2} \quad |\nabla\psi_j|^2 \tag{69}$$

We note that one can introduce the Rudge modification in the first term in Eq.(65) as well by replacing Q^{-2} by $(Q^2+\alpha_j^2)^{-1}(1+\alpha_j^2(Q^2+\alpha_j^2)^{-1})$. We shall not do this here because of the desire to maintain a direct identification of the first term with the $X\alpha$ exchange potential.

Use of the modified second term (Eq.(69)), replacement of $\langle Q^2\rangle$ by twice the average kinetic energy in the Thomas-Fermi model, and replacement of α_j^2 by its average value (twice the average ionization potential) leads to the first Huo exchange potential:

$$V_{Huo,1}(\vec{r}) = V_{X1}(\vec{r}) - 4\pi\sum_j |\nabla\psi_j|^2/3[\frac{4}{9}(6\pi^2\rho_\uparrow)+\alpha^2]^2 \qquad (70)$$

Total and individual orbital exchange energies were evaluated from this expression for He, Be and Ne using Hartree-Fock orbitals and compared with the Hartree-Fock values. The results[48] are summarized in Table 1.

Table 1: Comparison[48] of total and individual orbital exchange energies for several approximations (in atomic units).

		H-F	$\langle V_{Huo,1}\rangle$	$\langle VX1\rangle$
He:	$-E_{ex,1s}$	1.0258	0.9892	0.8841
	$-E_{ex}$	1.0258	0.9892	0.8841
Be:	$-E_{ex.1s}$	2.2982	2.3367	1.9745
	$-E_{ex,2s}$	0.3686	0.3469	0.3379
	$-E_{ex}$	2.6668	2.6834	2.3124
Ne:	$-E_{ex,1s}$	6.2214	6.3739	5.3833
	$-E_{ex,2s}$	1.7240	1.5459	1.5007
	$-E_{ex,2p}$	1.3877	1.4254	1.3842
	$-E_{ex}$	12.1085	12.1963	11.0365

We observe that $\langle V_{Huo,1}\rangle$ is a quite accurate approximation and a definite improvement over the $\alpha=1$ potential $\langle V_{X1}\rangle$.

We have solved self-consistently[54,55] the single-particle equations (52) with $V_{Huo,1}(\vec{r})$ as the exchange potential instead of the Hartree-Fock exchange potential. Our results for several atoms are quite encouraging. The orbitals and orbital energies are closer approximations to the Hartree-Fock ones than are the $\alpha=1$ ($X\alpha$) results. Details of these calculations and the results will be published elsewhere.[55]

Since the $V_{Huo,1}$ potential was obtained by approximating the Hartree-Fock exchange operator in the Hartree-Fock equations, it corresponds to the derivation of the V_{X1} potential of Slater[49] and in fact reduces to V_{X1} plus a correction term when the choice of $\langle Q^2\rangle_j = 2$ (kinetic energy) and $\alpha_j^2 = -2IP$ is made for all j. If we had approximated the exchange integral in the Hartree-Fock expression for the energy, and then had varied the resulting total energy with respect to the orbital, we would have obtained a set of single particle equations with an exchange potential, the first term of which corresponds to the Gaspar-Kohn-Sham[40,56] electron gas exchange potential (i.e. $X\alpha$ with $\alpha=2/3$) and a correction term. We have made this derivation[57] and are investigating the self-consistent solutions of the resulting equations.

As we mentioned earlier, there are other choices possible for $\langle Q^2\rangle$ than the Thomas-Fermi expression $T_o[\rho]$ for the kinetic energy. We are currently investigating[58] the use of the gradient expansion of the kinetic energy $T_o[\rho] + T_2[\rho] + T_4[\rho]$, etc. Other choices for $\langle Q^2\rangle_j$ include the use of the orbital kinetic energies[48] $\langle Q^2\rangle_j = \int Q^2 |\chi_j|^2 d\vec{Q}$ or the orbital ionization potentials, $\langle Q^2\rangle_j = -2\varepsilon_j$. A numerical investigation of the former for the exchange energy has been made previously.[48]

Clearly a number of exchange potentials of great promise can be developed starting from the Bonham-Huo expansion. Although the present discussion and the existent work have used these in a position space representation, one could proceed to use them in the momentum space representation. Work along these lines is planned in this laboratory.

ACKNOWLEDGEMENTS

This research was aided by grants from the Natural Sciences and Engineering Research Council of Canada (NSERCC) and the Advisory Research Council of Queen's University. The ideas formulated in this article benefitted from discussions and/or collaboration with a number of individuals through the years, especially I. Absar, R. Benesch, R.E. Brown, P. Kaijser, M. Levy, L. Mendelsohn, J. Mrozek, P. Petelenz, A. Simas, S.R. Singh, A.J. Thakkar, A. Tripathi, W. Westgate, W. Weyrich and M.H. Whangbo.

REFERENCES

1. B.G. Williams, ed., Compton Scattering, (McGraw-Hill, New York, 1977).
2. R.A. Bonham and M. Fink, High Energy Electron Scattering, (Van Nostrand, Princeton, 1974).
3. E. Weigold, ed., Momentum Wave Functions - 1982, (Amer. Inst. Phys., N.Y. 1982).
4. D.W. Devins, ed., Momentum Wave Functions - 1976, (Amer. Inst. Phys., N.Y. 1977).
5. L. Mendelsohn and V.H. Smith, Jr., in: Compton Scattering, ed. B.G. Williams (McGraw-Hill, New York, 1977).
6. I.R. Epstein and A.C. Tanner, in: Compton Scattering, ed. B.G. Williams (McGraw-Hill, New York, 1977).
7. R. Benesch and V.H. Smith, Jr., in Wave Mechanics - the First Fifty Years, ed. W.C. Price S.S. Chissick and T. Ravensdale (Butterworths, London, 1973).
8. R. Benesch, S.R. Singh and V.H. Smith, Jr., Chem. Phys. Lett. 10, 151 (1971).

9. W. Weyrich, P. Pattison and B.G. Williams, Chem. Phys. 41, 271 (1979).
10. A.J. Thakkar, A.M. Simas and V.H. Smith, Jr., Chem. Phys. 63, 175 (1981).
11. V.H. Smith, Jr. in Electron and Magnetization Densities in Molecules and Crystals, ed. P. Becker (Plenum Press, New York 1980).
12. T. Koga and M. Morita, J. Chem. Phys. 77, 6345 (1982).
13. P. Kaijser and V.H. Smith, Jr., Adv. Quantum Chem 10, 37 (1977).
14. V. Fock, Z. Physik. 98, 145 (1935).
15. R. McWeeny and C.A. Coulson, Proc. Phys. Soc. (London) A62, 509 (1949).
16. R. McWeeny, Proc. Phys. Soc. (London) A62, 519 (1949).
17. M.G. Henderson and C.W. Scherr, Phys. Rev. 120, 150 (1960).
18. T. Shibuya and C.E. Wulfman, Proc. Roy. Soc. (London) A286, 376 (1965).
19. E.N. Lassettre, J. Chem. Phys. 58, 1991 (1973).
20. B.K. Novosadov, Opt. Spectrosc. 41, 490 (1976).
21. B.K. Novosadov, J. Molec. Struct. 52, 119 (1979).
22. B.K. Novosadov, J. Molec. Struct. 54, 269 (1979).
23. H.J. Monkhorst and B. Jeziorski., J. Chem. Phys. 71, 5268 (1979).
24. J. Navaza and G. Tsoucaris, Phys. Rev. A 24, 683 (1981).
25. A.M. Simas and V.H. Smith, Jr. (unpublished); A.M. Simas, Diss. Abstr. Int. 43, 438 (1982).
26. P. Kaijser and V.H. Smith, Jr., in Quantum Science: Methods and Structure, ed. J.-L. Calais, O. Goscinski, J. Linderberg and Y. Ohrn (Plenum Press, New York, 1976).
27. V.H. Smith, Jr., in Electron Distributions and the Chemical Bond, ed. P. Coppens and M.B. Hall (Plenum Press, New York, 1982).
28. P. Hohenberg and W. Kohn, Phys. Rev. 136, B864 (1964).
29. J. Callaway and N.H. March, Solid State Physics (in the press).
30. G.A. Henderson, Phys. Rev. A 23, 19 (1981).
31. V.H. Smith, Jr. in Proceedings of the Workshop on Density Functional Theory, Mexico City, October 1980 (unpublished).
32. T.L. Gilbert, Phys. Rev. B 12, 2111 (1975).
33. M. Berrondo and O. Goscinski, Int. J. Quantum Chem. 9S, 67 (1975).
34. A.J. Coleman, Rev. Mod. Phys. 35, 668 (1963).
35. V.H. Smith, Jr. and I. Absar, Isr. J. Chem. 16, 87 (1977).
36. E.G. Larson (unpublished).
37. M. Levy, Proc. Natl. Acad. Sci. USA 76, 6062 (1979).
38. M. Levy, Phys. Rev. A 26, 1200 (1982).
39. K. Freed and M. Levy, J. Chem. Phys. 77, 396 (1982).
40. W. Kohn and L.J. Sham, Phys. Rev. 140, A1133 (1965).
41. J.E. Harriman, Phys. Rev. A 24, 680 (1981).
42. L. Lam and P.U. Platzman, Phys. Rev. B. 9, 5122 (1974).
43. L. Lam and P.U. Platzman, Phys. Rev. B. 9, 5128 (1974).
44. B.Y. Tong and L. Lam, Phys. Rev. A 18, 552 (1978).
45. A. Harmalkar, P.V. Panat and D.G. Kanhere, J. Phys. B: Atom Molec. Phys. 13, 3075 (1980).

46. G.E.W. Bauer, Phys. Rev. B. 27, 5912 (1983).
47. R.A. Bonham, J. Chem. Phys. 36, 3260 (1962).
48. W.M. Huo, J. Chem. Phys. 67, 5133 (1977).
49. J.C. Slater, Quantum Theory of Molecules and Solids, Vol. 4: The Self-Consistent Field for Molecules and Solids (McGraw-Hill, New York, 1974).
50. F. Herman, J.P. Van Dyke and I.B. Ortenburger, Phys. Rev. Letters 22, 807 (1969).
51. N.O. Folland, Phys. Rev. A 3, 1535 (1971).
52. A.D. Becke, Int. J. Quantum Chem. 23, 1915 (1983).
53. M.R.H. Rudge, Proc. Phys. Soc. (London) 85, 607 (1965).
54. J. Mrozek and V.H. Smith, Jr., in: Proceedings of the Workshop on Density Functional Theory, Mexico City, October 1980 (unpublished).
55. J. Mrozek and V.H. Smith, Jr., (to be published).
56. R. Gaspar, Acta Phys. Acad. Sci. Hung. 3, 263 (1954).
57. J. Mrozek, P. Petelenz and V.H. Smith, Jr., (to be published).
58. V.H. Smith, Jr., and J. Mrozek (to be published).

ASPECTS OF DENSITY FUNCTIONAL THEORY

Robert G. Parr

Department of Chemistry
University of North Carolina
Chapel Hill
North Carolina 27514, U.S.A.

1. INTRODUCTION

At the University of North Carolina we began doing research on density functional theory in 1974. The aspects with which we have been involved can be readily identified from the titles of our published papers.[1-52] These have included the discussion of the concepts of electronegativity, atom in molecule, and pressure;[7,15,19,40] establishment of the long-range behavior of the electron density;[1,3] calculations of electronegativity;[4,17] studies of the gradient expansions of the kinetic energy functional and other energy components;[4,11,13,18,20,33,34,42] invention of a new local density functional theory;[12,14] delineation of the exact first-order density matrix functional;[8,10] extensions from pure states to ensembles;[22,23] extensions to momentum space,[26] excited states,[29] time-dependent phenomena,[32,45] and non Born-Oppenheimer systems;[37] relations with information theory;[21] discovery of a new set of orbitals, circulant orbitals, with promise for elucidating connections between wavefunction theory and density functional theory;[31,47] studies of the N and Z dependence of atomic energies and hydrogenic limits;[24,27,36,38,41] derivation of Maxwell reciprocal relations and stability conditions;[43,44,52] proof of a new complementary variational principle;[49] detailed discussion of the problem of non-integral electron number;[30,50] and diverse other problems.[2,5,6,16,25,28,35,39,46,48,51] I have previously written two short reviews of our work.[53,54]

Our research has had two general aims. First, we want to find density functionals of increasing accuracy. To achieve this goal, we use accurate quantum-chemical calculation as appropriate, and we use exact quantum-mechanical analysis as appropriate. If a very accurate density functional procedure ever is found, I suspect that it will involve manipulations with natural spinorbitals, not Kohn-Sham orbitals.

Our second general aim is to elaborate the exact density functional description of molecular properties and molecular behavior; that is, the conceptual foundations of chemistry. The fruits of quantum chemistry must be much more than the successful calculation of molecular properties.

The present text, written in September of 1982, treats the same topics as I discussed in my oral presentation at the Copenhagen Conference, in June, but incorporates several results obtained since the Conference. First I discuss the physical meaning of the chemical potential μ and the hypothesis that the energy E is a continuous function of number of electrons N. Then I make a proposal concerning the standard form for E(N) and describe a model from which follows the geometric mean electronegativity equalization principle. Then I present a statistical mechanical description of an individual atom or molecule and derive a formula for the chemical potential of a air of separated atoms. Finally, I show how the Hartree-Fock problem for the atom Be can be restated as a density functional problem and describe the generalization of this idea to N electrons.

2. THE CHEMICAL POTENTIAL[8,30,50]

In the density functional theory for the ground state of an atom or molecule having N_0 electrons moving in an external potential v, the chemical potential μ is the Lagrange multiplier in the basic variational principle in the theory,

$$\delta(E[\rho]-\mu N[\rho]) = 0 \quad [\text{constant } v] \tag{1}$$

where $E[\rho]$ is the energy as a functional of the density ρ and $N[\rho]$ is the functional $\int\rho d\tau$. There follows both that

$$\mu = [\delta E/\delta\rho]_v \tag{2}$$

and that

$$\mu = [\partial E/\partial N_0]_v \tag{3}$$

For a change from one ground state to another,

$$dE = \mu dN_0 + \langle\rho dv\rangle \tag{4}$$

Without the first term, this is a standard formula from first-order perturbation theory; the first term provides the generalization to include change in particle number.

Physically, μ plays the role for a single atom or molecule analogous to the role of a chemical potential in ordinary macroscopic thermodynamics: it is a constant through the system in equilibrium which measures the escaping tendency of electrons from the systems. Equation (3) shows that it is the slope of the $E(N_0)$ curve at constant v, and there is a whole literature of such curves identifying this slope with the negative of the electronegativity of chemistry.[8] That is all there is to it.

If we compare Eq. (1) with the more mundane principle of conventional quantum chemistry,

$$\delta(\langle\psi|H|\psi\rangle - E\langle\psi|\psi\rangle) = 0 \tag{5}$$

in which the Lagrange multiplier is the energy, we sense the reason why it is important to fully develop the complete consequences of Eq. (1). Physical pictures derived from Eq. (1) are more appealing than those derived from Eq. (5).

But no, that is not all there is to it, because of the subtle and fascinating fact that Eq. (3) requires that the derivative have meaning; it requires that $E(N_0)$ be defined for non-integral as well as integral N_0 values. [This point was featured in the formal presentation at the Copenhagen Conference by Professor Ernest Davidson, and it was also stressed in the discussion remarks at the Conference by Professor Richard Bader and Dr. Yoram Tal.]

In a given case, I see no reason not to assume a continuous $E(N_0)$, with $(\partial E/\partial N_0)_v$ defined for the real (integral N_0) systems of interest. Plots of actual $E(N_0)$ values can be fit with smooth curves, and very many such plots have this property; non-integral atomic populations and non-integral orbital occupations have been in the vocabulary of chemists for decades; workable interpolating models for non-integral N_0 already exist and have proved quantitatively valuable (Thomas-Fermi, $X\alpha$, etc.); an explicit formally exact statistical mechanical (ensemble) description can be written down which has all of the requisite properties. See Section 4 below for this last.

"In a given case", these are very important words in the preceding paragraph. Different actual cases will involve different circumstances; different cases may give different interpolation schemes. The smoothness of the data argues that interpolation is possible and reasonable; it does not of itself imply that interpolation will be unique. Suppose for example that we were concerned with a system made up of real subsystems, say a system of many atoms at great distances from another. Only integral electron numbers could then arise on an individual subsystem; if one wishes to draw full $E(N_0)$ curves one then must draw broken straight lines (as indeed argued at the Conference by Professor Davidson). A detailed treatment of this case has recently been given.[50] But when atoms are actually interacting, as in a molecule or in most other circumstances of interest, smoother curves can result.

Rewriting density functional theory using $\rho(\vec{r}) = N_0\sigma(\vec{r})$, where $\sigma(\vec{r})$ is a shape function normalized to unity, would be helpful for more clearly distinguishing density changes which change N_0 from those which do not.

3. PROPOSED STANDARD FORM FOR E(N), AND GEOMETRIC MEAN ELECTRONEGATIVITY EQUALIZATION PRINCIPLE[40]

Probably mainly because it yields Mulliken's formula,

$$\mu^o = -\frac{I+A}{2} \tag{6}$$

for the neutral species, it has been common to assume a parabolic form for $E(N_0)$, to fit energies of a neutral species S, its monopositive ion S^+, and its mononegative ion S^-; I and A are the ionization potential and electron affinity of S.

However, ionization potentials increase very fast with state of ionization, and successive electron affinities are smaller and smaller, soon becoming zero. We therefore suggest another functional form, a decaying exponential. For an atom of atomic number Z, number of electrons N, write (for N near Z)

$$E(N) = -(\mu^o/\gamma)\exp[-\gamma(N-Z)] + E(\infty) \tag{7}$$

where μ^o, γ and $E(\infty)$ are three constants, μ^o in particular being the chemical potential for the neutral species. Then, in contrast to Eq. (6) one has

$$\mu^o = -\frac{IA}{I-A} \ln \frac{I}{A} \tag{8}$$

Also,

$$\mu^o = -\gamma(e^{\gamma}-1)^{-1}I, \qquad A = e^{-\gamma}I \tag{9}$$

and

$$\mu(N) = \mu^o\exp[-\gamma(N-Z)] \tag{10}$$

The chemical potential itself is exponentially decaying.

From experimental I and A values for 34 neutral atoms, one finds $\gamma = 2.2\pm0.6$. There therefore is a rough rule, that γ is the same for different species. Equations (9) accordingly give, roughly, $\mu^o \sim -0.3I$ and $A \sim 0.1I$. One would like to find a simple physical model giving γ constant, and the value of the constant.

To the extent that Eq. (10) holds for two atoms A and B as they come together to form a molecule AB, with γ the same for A and B, the geometric mean electronegativity principle will hold:

$$\mu_{AB} = \mu_A = \mu_B = \mu_A^o \exp[-\gamma(N_A - Z_A)] = \mu_B^o \exp[-\gamma(N_B - Z_B)] = (\mu_A^o \mu_B^o)^{1/2} \tag{11}$$

Note that this argument in no way constitutes a claim that Eq. (11) is in fact true; it merely establishes sufficient conditions for it to be true. Further detailed studies are clearly called for in this area.

4. STATISTICAL MECHANICAL DESCRIPTION

There is a certain way to define the ground state of an electronic system of interest having a nonintegral number of electrons N. Namely, it is the average state of an individual system in the equilibrium state of an appropriate grand ensemble of equivalent systems, in the limit of zero temperature, the ensemble having an integral number of electrons. This description was first proposed in 1968, in a very interesting paper by Gyftopoulos and Hatsopoulos.[55]

For simplicity let the system of interest be an atom of atomic number Z, number of electrons N, which can exist only in the states $N=N^o=Z$, $E=E^o$; $N=Z-1$, $E=E^o+I$; $N=Z+1$, $E=E^o-A$. Define a grand canonical ensemble with temperature parameter β, chemical potential parameter $\gamma=-\beta\mu$. The grand partition function then is given by the formula

$$\Xi(\beta,\gamma) = e^{-\beta E^o} e^{-\gamma N^o} [1 + e^{\beta A} e^{-\gamma} + e^{-\beta I} e^{+\gamma}] \tag{12}$$

The average number of electrons on the atom is

$$\bar{N} = -\frac{\partial \ln \Xi}{\partial \gamma} = N^o + \frac{e^{\beta A} e^{-\gamma} - e^{-\beta I} e^{+\gamma}}{1 + e^{\beta A} e^{-\gamma} + e^{-\beta I} e^{\gamma}} \tag{13}$$

Clearly this number of electrons can be nonintegral. For the neutral species $\bar{N}=N^o$, and one finds, for any β,

$$2\gamma = \beta(I+A); \qquad \mu^o = -\frac{I+A}{2} \tag{14}$$

which is Mulliken's formula, Eq. (6). For $\bar{N}$ less than but near Z, at $\beta = \infty$ one finds $\mu = -I$; for $\bar{N}$ greater than but near Z at $\beta = \infty$ one finds $\mu = -A$. This is the broken straight-line behavior mentioned in Section 2 above.[50] Note again that this is the case of $\beta = \infty$ and noninteracting atoms, leaving other cases for subsequent discussion.

Suppose now that the system of interest is a pair of atoms at a large distance from another, but in equilibrium. The grand partition function for the A··B system then is a product of partition functions for A and B alone:

$$\Xi_{AB}(\beta,\gamma) = \Xi_A(\beta,\gamma)\Xi_B(\beta,\gamma) \tag{15}$$

Here the common values of β and γ signify the equilibrium condition. Again using $\bar{N} = -\partial \ell n\Xi/\partial\gamma$, we find this time for the neutral species at zero degrees the result

$$\mu^o = -\frac{I_{min}+A_{max}}{2} \tag{16}$$

where I_{min} is the minimum of the ionization potentials I_A, I_B, and A_{max} is the maximum of A_A, A_B.[50]

An alternative derivation of Eq. (16) is instructive, which is patterned after Mulliken's original derivation of Eq. (6). Consider the two separated atom pairs A··B and C··D. Which has the higher electronegativity? This depends on which of

$$A\cdot\cdot B + C\cdot\cdot D \rightarrow (A\cdot\cdot B)^+ + (C\cdot\cdot D)^- \tag{I}$$

and

$$A\cdot\cdot B + C\cdot\cdot D \rightarrow (A\cdot\cdot B)^- + (C\cdot\cdot D)^+ \tag{II}$$

is more favorable. For I, the energy cost is

$$\Delta E_I = I_{min}(AB) - A_{max}(CD) \tag{17}$$

while for II, the cost is

$$\Delta E_{II} = I_{min}(CD) - A_{max}(AB) \tag{18}$$

if $\Delta E_I > \Delta E_{II}$, A··B is more electronegative than C··D. That is, A··B is more electronegative than C··D if

$$I_{min}(AB) + A_{max}(AB) > I_{min}(CD) + A_{max}(CD) \tag{19}$$

Equation (16) then follows, as the natural definition for the negative of the electronegativity of a pair of atoms a large distance from one another.

5. WAVEFUNCTION CALCULATIONS AS DENSITY FUNCTIONAL CALCULATIONS[47]

I now turn to a very different aspect of density functional theory, the question of how to conduct conventional wavefunction calculation in a way to enlighten density functional theory. A recent proposal by Freed and Levy may be mentioned in this connection.[56]

We ourselves have started studies rooted in our observation[31] that the $(1s)^2(2s)^2$ state of Be, in Hartree-Approximation, is equivalent to a description $(\phi_+)^2(\phi_-)^2$, where ϕ_+ and ϕ_- are the circulant transformations of the canonical 1s and 2s orbitals:

$$\phi_+ = (\rho/4)^{1/2}\exp(i\theta), \quad \phi_- = (\rho/4)^{1/2}\exp(-i\theta) \tag{20}$$

Here ρ is the total electron density and the phase factor θ is a real function of r. Orthonormalization conditions on ϕ_+ and ϕ_- give constraints on ρ and θ. Subject to these constraints, the Hartree-Fock energy expectation value can be minimized with respect to ρ and θ, giving two coupled differential equations, one for ρ and one for θ.[47] To write down just the first, it has the following highly perspicuous form.

$$\frac{1}{8}\frac{\nabla\rho\cdot\nabla\rho}{\rho^2} - \frac{1}{4}\frac{\nabla^2\rho}{\rho} - \frac{Z}{r} + \frac{3}{4}\phi(r) + \frac{1}{2}\nabla\theta\cdot\nabla\theta$$

$$- \frac{1}{4}\int\frac{\rho(2)\cos[2\theta(1)-2\theta(2)]}{r_{12}}\,dv_2 = \mu_N + \mu_C \cos 2\theta \tag{21}$$

Here μ_N and μ_C are Lagrange multipliers. ϕ is the classical potential due to ρ; $(3/4)\phi$ is a term to be expected from old ideas of Fermi and Amaldi. $(1/2)\nabla\theta\cdot\nabla\theta$ is a kinetic energy correction due to the orthogonality requirement between shells; when θ is known (from solution of the second equation) this term provides a pseudopotential for the ρ equation. The integral term is the exchange effect. If one neglects the exchange term as well as all terms involving θ, one obtains the differential equation proposed at the Copenhagen Conference by another participant. The deficiencies in such an equation are evident.

We have already discussed these equations in some detail.[47,48] Note that solving them iteratively constitutes a density functional procedure: guess ρ; solve the θ equation; solve the ρ equation for a new ρ; continue until self-consistency is achieved.

The Hartree-Fock method for 2N electrons can be put into analogous form, as a method to find a density ρ and phase factors θ_k for N spinorbitals. In general these phase factors will have complex as well as real parts, but that does not detract from the fact that the resultant method can be regarded as a procedure for starting from a ρ and ending up with the exact Hartree-Fock wavefunction and energy. This idea readily extends to other forms of wavefunctions; we may rephrase any conventional wavefunction determination by representing any orbitals ϕ_k in the form $\phi_k = (\bar{\rho})^{1/2}\exp(i\theta_k)$, and thus put a whole problem in terms of ρ and the θ_k.[47]

Finally, I describe a problem on which we are now actively

working, to find the best single-determinantal description of an electronic system such that every orbital has the same electron density. Harriman has shown that such orbitals exist.[57] Our work with circulant orbitals[31] suggests that they will be very close to the circulant-transformed canonical Hartree-Fock orbitals. We expect little energy loss on imposing this constraint.

ACKNOWLEDGEMENTS

I gratefully acknowledge support from research grants to the University of North Carolina from the National Science Foundation and the National Institutes of Health. My considerable indebtedness to my many collaborators at UNC should be evident from the bibliography.[1-52]

REFERENCES

1. M.M. Morrell, R.G. Parr, and M. Levy, Calculation of ionization potentials from density matrices and natural functions, and the long-range behavior of natural orbitals and electron density, J. Chem. Phys. 62, 549-554 (1975).
2. H. Nakatsuji and R.G. Parr, Variational principles which are functionals of electron density, J. Chem. Phys. 63, 1112-1117 (1975).
3. M. Levy and R.G. Parr, Long-range behavior of natural orbitals and electron density, J. Chem. Phys. 64, 2707-2708 (1976).
4. W.-P. Wang, R.G. Parr, D.R. Murphy, and G. Henderson, Gradient expansion of the atomic kinetic energy functional, Chem. Phys. Lett. 43, 409-412 (1976).
5. W.-P. Wang and R.G. Parr, Statistical atomic models with piecewise exponentially decaying electron densities, Phys. Rev. A 16, 891-902 (1977).
6. P.W. Payne, Density response theory of nonbonded interactions, J. Chem. Phys. 68, 1242-1247 (1978).
7. R.G. Parr, R.A. Donnelly, M. Levy, and W.E. Palke, Electronegativity: the density functional viewpoint, J. Chem. Phys. 68, 3801-3807 (1978).
8. R.A. Donnelly and R.G. Parr, Elementary properties of an energy functional of the first-order reduced density matrix, J. Chem. Phys. 69, 4431-4439 (1978).
9. N.K. Ray, L. Samuels, and R.G. Parr, Studies of electronegativity equalization, J. Chem. Phys. 70, 3680-3684 (1979).
10. R.A. Donnelly, On a fundamental difference between energy functionals based on first- and on second-order density matrices, J. Chem. Phys. 71, 2874-2879 (1979).
11. D.R. Murphy and R.G. Parr, Gradient expansion of the kinetic energy density functional: local behavior of the kinetic energy density, Chem. Phys. Lett. 60, 377-379 (1979).

12. R.G. Parr, S.R. Gadre, and L.J. Bartolotti, Local density functional theory of atoms and molecules, Proc. Natl. Acad. Sci. USA 76, 2522-2526 (1979).
13. D.R. Murphy and W.-P. Wang, Comparative study of the gradient expansion of the atomic kinetic energy functional-neutral atoms, J. Chem. Phys. 72, 429-433 (1980).
14. S.R. Gadre, L.J. Bartolotti, and R.G. Parr, Bounds for Coulomb energies, J. Chem. Phys. 72, 1034-1038 (1980).
15. L.J. Bartolotti and R.G. Parr, The concept of pressure in density functional theory, J. Chem. Phys. 72, 1593-1596 (1980).
16. R.G. Parr and S.R. Gadre, On the basic homogeneity characteristic of atomic and molecular electronic energies, J. Chem. Phys. 72, 3669-3673 (1980).
17. L.J. Bartolotti, S.R. Gadre, and R.G. Parr, Electronegativities of the elements from simple Xα theory, J. Am. Chem. Soc. 102, 2945-2948 (1980).
18. W.-P. Wang, Comparative study of the gradient expansion of the atomic kinetic energy functional-isoelectronic series, J. Chem. Phys. 73, 416-418 (1980).
19. N.K. Ray and R.G. Parr, Diamagnetic shieldings of atoms in molecules and their relation to electronegativity, J. Chem. Phys. 73, 1334-1339 (1980).
20. C.C. Shih, D.R. Murphy, and W.-P. Wang, Gradient expansion of the exchange energy density functional: a complementary expansion of the atomic energy functional, J. Chem. Phys. 73, 1340-1343 (1980).
21. S.B. Sears, R.G. Parr, and U. Dinur, On the quantum-mechanical kinetic energy as a measure of the information in a distribution, Israel J. Chem. 19, 165-173 (1980).
22. S.M. Valone, Consequences of extending 1 matrix energy functionals from pure-state representable to all ensemble representable 1 matrices, J. Chem. Phys. 73, 1334-1349 (1980).
23. S.M. Valone, A one-to-one mapping between one-particle densities and some n-particle ensembles, J. Chem. Phys. 73, 4653-4655 (1980).
24. N.H. March and R.G. Parr, Chemical potential, Teller's theorem, and the scaling of atomic and molecular energies, Proc. Natl. Acad. Sci. USA 77, 6285-6288 (1980).
25. P.K. Acharya, L.J. Bartolotti, S.B. Sears, and R.G. Parr, An atomic kinetic energy functional with full Weizsacker correction, Proc. Natl. Acad. Sci. 77, 6978-6982 (1980).
26. G.A. Henderson, Variational theorems for the single-particle probability density and density matrix in momentum space, Phys. Rev. A 23, 19-20 (1981).
27. J. Katriel and M.R. Nyden, A Comparison between hydrogenic and Thomas-Fermi expectation values, J. Chem. Phys. 74, 1221-1224 (1981).
28. R.G. Parr and A. Berk, The bare-nuclear potential as harbinger for the electron density in a molecule, in Molecular Electrostatic Potentials in Chemistry and Biochemistry, edited by P. Politzer and D.G. Truhlar (Plenum, 1981), 51-62.

29. S.M. Valone and J.F. Capitani, Bound excited states in density functional theory, Phys. Rev. A 23, 2127-2133 (1981).
30. J. Katriel, R.G. Parr, and M.R. Nyden, Concerning the chemical potential of few-electron systems, J. Chem. Phys. 74, 2397-2401 (1981).
31. R.G. Parr and M.-B. Chen, Circulant orbitals for atoms and molecules, Proc. Natl. Acad. Sci. USA 78, 1323-1326 (1981).
32. L.J. Bartolotti, Time-dependent extension of the Hohenberg-Kohn-Levy energy density functional, Phys. Rev. A 24, 1682-1688 (1981).
33. D.R. Murphy, The sixth-order term of the gradient expansion of the kinetic energy density functional, Phys. Rev. A 24, 1682-1688 (1981).
34. L.J. Bartolotti and R.G. Parr, Gradient expansion of the classical Coulomb energy of a charge distribution, J. Chem. Phys. 75, 4553-4555 (1981).
35. M.R. Nyden and P.K. Acharya, Coreless Thomas-Fermi models of atomic structure, J. Chem. Phys. 75, 4567-4571 (1981).
36. Y. Tal, L.J. Bartolotti, and R.F.W. Bader, Universal scaling relations for free and bonded atoms, J. Chem. Phys. 76, 463-467 (1982).
37. J.F. Capitani, R.F. Nalewajski, and R.G. Parr, Non-Born-Oppenheimer density functional theory of molecular systems, J. Chem. Phys. 76, 568-573 (1982).
38. Y. Tal and L.J. Bartolotti, The hydrogenic limit of many-electron atoms, J. Chem. Phys. 76, 2558-2564 (1982).
39. W.-P. Wang, Fixed-shell statistical atomic models with piecewise exponentially decaying electron densities, Phys. Rev. A 25, 2901-2912 (1982).
40. R.G. Parr and L.J. Bartolotti, On the geometric mean principle for electronegativity equalization, J. Am. Chem. Soc. 104, 3801-3803 (1982).
41. Y. Tal and L.J. Bartolotti, On the Z^{-1} and $N^{-1/3}$ expansions of Hartree-Fock atomic energies. J. Chem. Phys. 76, 4056-4062 (1982).
42. L.J. Bartolotti, A new gradient expansion of the exchange energy to be used in density functional calculations on atoms, J. Chem. Phys. 76, 6057-6059 (1982).
43. R.F. Nalewajski and R.G. Parr, Legendre transforms and Maxwell relations in density functional theory, J. Chem. Phys. 77, 399-407 (1982).
44. R.F. Nalewajski and J.F. Capitani, Density functional theory: non Born-Oppenheimer Legendre transforms and Maxwell relations, equilibrium and stability conditions, J. Chem. Phys. 77, 2514-2526 (1982).
45. L.J. Bartolotti, Time-dependent Kohn-Sham density functional theory, Phys. Rev. A, in press.
46. L.J. Bartolotti and P.K. Acharya, On the functional derivative of the kinetic energy density functional, J. Chem. Phys., in press.

47. M.R. Nyden and R.G. Parr, Restatement of conventional wavefunction determination as a density functional procedure, J. Chem. Phys., in press.
48. M.R. Nyden, An orthogonality constrained generalization of the Weizsacker density functional model, J. Chem. Phys. in press.
49. A. Berk, A complementary variational principle for density functional theories, Phys. Rev. A, in press.
50. J.P. Perdew, R.G. Parr, M. Levy, and J.L. Balduz, Jr., Density functional theory for fractional particle number: derivative discontinuities of the energy, Phys. Rev. Lett., in press.
51. P.K. Acharya, Comment on the derivation of atomic kinetic energy functionals with full Weizsacker correction, J. Chem. Phys., in press.
52. R.F. Nalewajski, Reduction of derivatives and simple applications of the Legendre transformed density functional theory, J. Chem. Phys., submitted.
53. R.G. Parr, Density functional theory of atoms and molecules, in Horizons of Quantum Chemistry, edited by K. Fukui and B. Pullman (Reidel, 1980), pp. 5-15.
54. R.G. Parr, Density functional theory, in Electron Distribution and the Chemical Bond, edited by M.B. Hall and P. Coppens (Plenum, 1982), pp. 95-100.
55. E.P. Gyftopoulos and G.N. Hatsopoulos, Quantum-thermodynamic definition of electronegativity, Proc. Natl. Acad. Sci. USA 60, 786-793 (1968).
56. K.F. Freed and M. Levy, Direct first principles algorithm for the universal electron density functional, J. Chem. Phys. 77, 396-398 (1982).
57. J.E. Harriman, Orthonormal orbitals for the representation of an arbitrary density, Phys. Rev. A 24, 680-682 (1981).

ON THE CALCULATION OF POTENTIALS FROM DENSITIES

Scott H. Werden and Ernest R. Davidson

Department of Chemistry BG 10
University of Washington
Seattle, Washington 98195, USA

Below, the linear response of the potential associated with a change in the density is derived for the case of non-interacting Fermions in one dimension. The result is used in iterative calculations to find the potential for a few simple densities.

The derivations of Hohenberg and Kohn[1] and many of the applications of density functional theory assume that any non-negative function can represent the density associated with some potential. Katriel et al.[2] recently explored one example numerically and found that all densities of the type they considered were derivable from some potential. Levy,[3] on the other hand, has constructed some quite reasonable densities which cannot come from any potential, and he has shown how the assumption of the existence of a potential can be avoided in some cases.[4] His construction is tied, however, to the existence of high degeneracy. In his manuscript we examine the situation in one dimension where such high degeneracy is not possible.

Even for non-interacting spin-zero Fermions in one dimension, it is nontrivial to construct the potential related to a given density. If one is given an arbitrary non-negative function $\rho(x)$ on the domain $0 \leq x \leq L$ normalized so that

$$\int_0^L \rho(x)dx = N \tag{1}$$

then this simplified problem is to construct a function $v(x)$ such that

$$\rho(x) = \sum_{i=1}^{N} |\phi_i(x)|^2 \tag{2}$$

and

$$\left\{-\frac{1}{2}\frac{d^2}{dx^2} + v(x)\right\} \phi_i(x) = \varepsilon_i\phi_i(x) \tag{3}$$

$$\varepsilon_i \leq \varepsilon_{i+1} \tag{4}$$

For N = 1 the solution is trivial,

$$v(x) = \varepsilon + \frac{1}{2} \rho^{-1/2} d^2 \rho^{1/2}/dx^2 \tag{5}$$

but for larger N no explicit solution is possible.

In order to compute v by an iterative process, it is necessary to have some estimate of the change δv which would be caused by a change $\delta\rho$ in ρ. The inverse problem has a formal solution. For an arbitrary many-particle system with interaction, the response $\delta\rho(\underset{\sim}{r})$ to a small perturbation in the one-particle potential,

$$\sum_{i=1}^{N} \delta v(\underset{\sim}{r}_i)$$

is just

$$\delta\rho(\underset{\sim}{r}) = \int G(\underset{\sim}{r},\underset{\sim}{r}')\, \delta v(\underset{\sim}{r}')\, d\underset{\sim}{r}' \tag{6}$$

where the Green's function for the density is

$$G(\underset{\sim}{r},\underset{\sim}{r}') = \sum_I{}' \langle\psi|\sum_i \delta(\underset{\sim}{r}_i-\underset{\sim}{r})|\psi_I\rangle\langle\psi_I|\sum_i \delta(\underset{\sim}{r}_i-\underset{\sim}{r}')|\psi\rangle/(E-E_I)+c.c. \tag{7}$$

For non-interacting particles in one dimension this reduces to

$$G(x,x') = \sum_{j=1}^{N} \sum_{k>N} \phi_j^*(x)\phi_k(x)\phi_k^*(x')\phi_j(x')/(\varepsilon_j-\varepsilon_k)+c.c. \tag{8}$$

$$\delta\rho(x) = \int_0^L G(x,x')\, \delta v(x')\, dx' \tag{9}$$

The operator G is hermitian and real. Unfortunately, G does not have an inverse. Clearly, addition of a constant to v does not change ρ and $\int\delta\rho dx$ is zero independent of δv.

That is

$$\int G(x,x')dx' = \int G(x,x')dx = 0 \tag{10}$$

so the constant function

$$f(x) = L^{-1/2} \tag{11}$$

is a normalized eigenfunction of G with eigenvalue zero.

Now, suppose G has the eigenfunction expansion

$$G(x,x') = \Sigma \lambda_i \xi_i(x) \xi_i^*(x') \tag{12}$$

where only non-zero λ_i are included in the sum.

By the Hohenberg-Kohn theorem, G can have only one zero eigenvalue, so we may define

$$P(x',x) = \sum \lambda_i^{-1} \xi_i(x') \xi_i^*(x) + cf(x')f(x) \tag{13}$$

where c is an arbitrary constant and the $\{\xi_i\}$ together with f form a complete set. Then

$$\begin{aligned} \iint dx\ P(x'',x)\ G(x,x') &= \sum \xi_i(x'')\ \xi_i^*(x') \\ &= \delta(x''-x') - f(x'')f(x') \end{aligned} \tag{14}$$

so

$$\iint P(x'',x)\ G(x,x')\ \delta v(x')\ dx'dx = \delta v(x'') - \frac{1}{L} \int_0^L \delta v dx \tag{15}$$

or

$$\int P(x'',x)\ \delta\rho(x) dx = \delta v(x'') - \overline{\delta v} \tag{16}$$

Consequently

$$\delta v(x) = \int P(x,x')\ \delta\rho(x') dx' + \overline{\delta v} \tag{17}$$

where $\overline{\delta v}$ is an arbitrary constant. The integral in (17) is independent of c as long as $\delta\rho$ does not change the normalization of ρ.

With this result it is easy to construct an iterative approximation to δv on a finite grid of x values. If an initial guess v^o to the desired v associated with a given ρ gives the density ρ^o, then the desired change in ρ is

$$\delta\rho = \rho - \rho^o \tag{18}$$

With the guessed v^o one can also easily construct G and then P from

$$P = [G + c^{-1} f(x)f(x')]^{-1} \tag{19}$$

On a finite grid, G is just a matrix so P is a generalized matrix inverse. The improved guess to v is $v^o + \delta v$ where δv is given by (17). In order to get convergence, $\overline{\delta v}$ must be chosen by some consistent criterion. We have usually chosen $\overline{\delta v}$ so that ε_1 became zero at convergence. We also found it necessary to damp the iterative process to prevent wild oscillations.

From (2) it is clear that for a fixed v, $\rho(x)$ becomes nearly constant for large N. Consequently, for a fixed shape $\rho(x)$ rescaled to be correctly normalized, v(x) must become deeper and more highly structured as N increases. This can generally be expected to lead to greater convergence difficulties at larger N.

On the other hand, on a finite, one-dimensional domain, the eigenvalue spectrum will not be degenerate. In all of the examples considered here, the inverse matrix P was well-conditioned. Examples involving a v with multiple minima and nearly degenerate eigenvalues could have been constructed, but they were not explored because of the difficulty with getting accurate numerical integration through a barrier.

In the first sequence of calculations, the density was fixed at the (scaled) density for one particle in a box of width 3 and constant potential. This density is shown in Figure 1. Figure 2 shows the Green's function G(x,x') for this case. Along the diagonal, G is negative corresponding to the expected result that an increase in v at x decreases ρ at x. Far from x = x', the Green's function is positive showing that an increase in v at x increases ρ at x' when $|x-x'|$ is large.

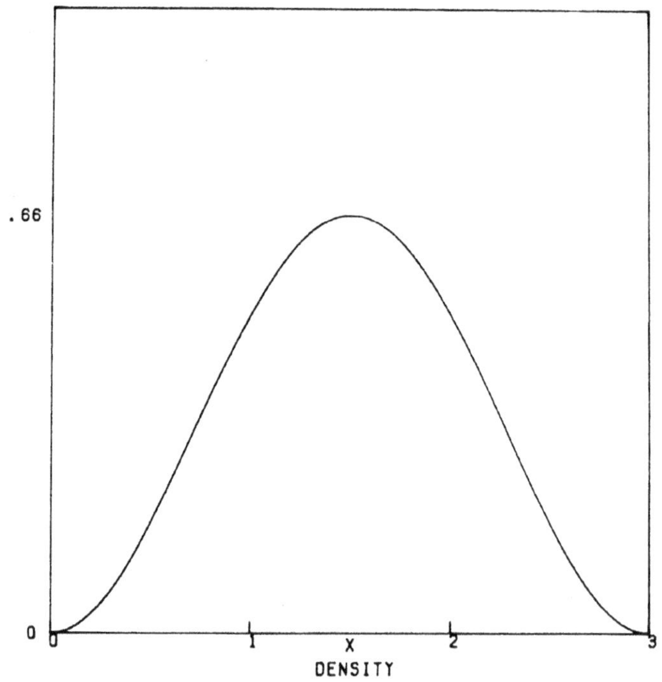

Figure 1. Ground state density for a particle in a box.

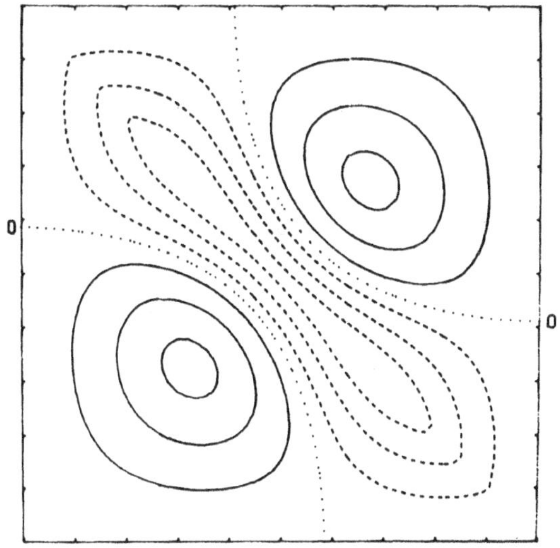

Figure 2. Green's function for the density of one particle in a box.

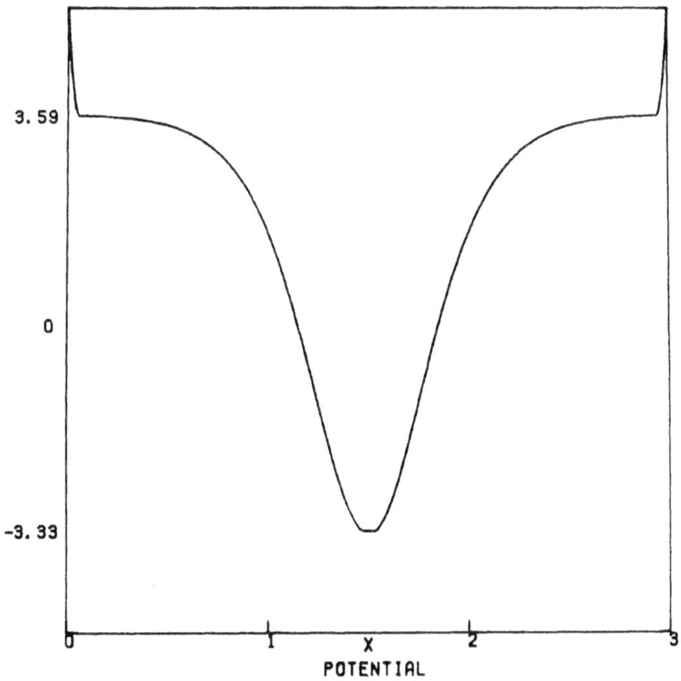

Figure 3. Potential required to give the density shown in Figure 1 for two spin zero fermions in a box.

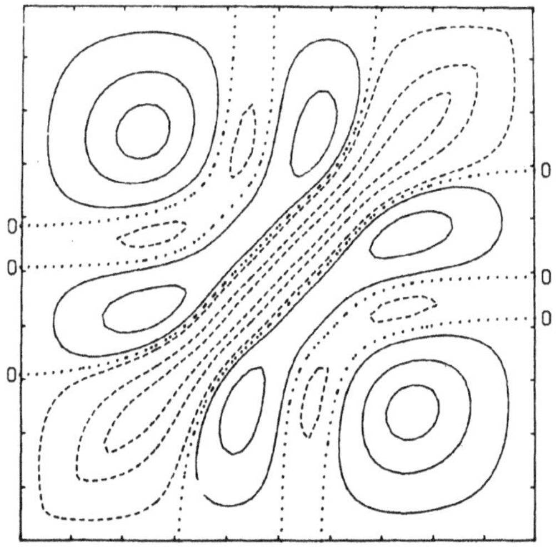

Figure 4. Green's function for two spin zero fermions with the density shown in Figure 1.

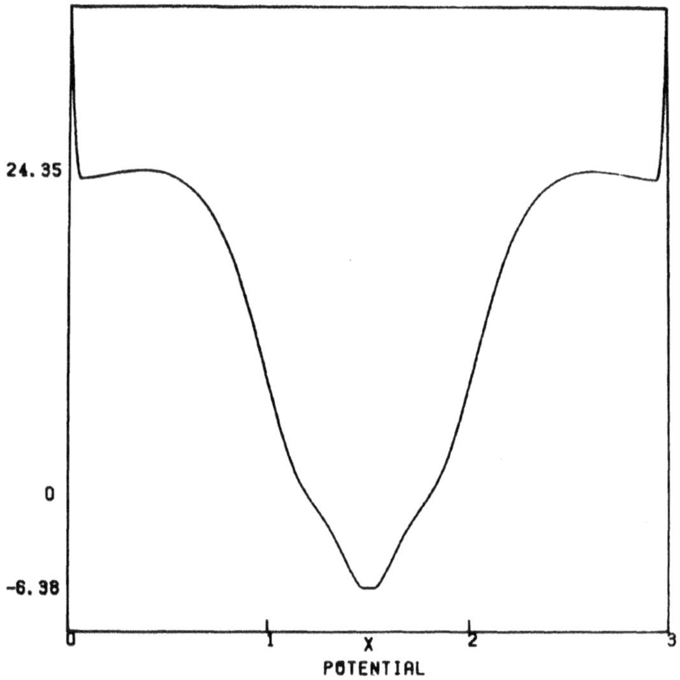

Figure 5. Potential required to give the density shown in Figure 1 for four spin zero fermions in a box.

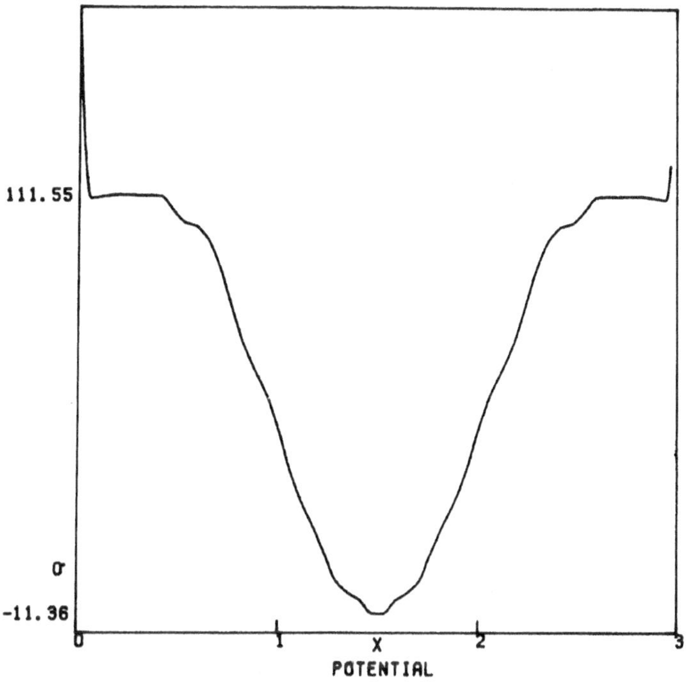

Figure 6. Potential required to give the density shown in Figure 1 for eight spin zero fermions in a box.

Figure 3 shows the potential required to produce the same density for two particles. A constant potential would have given a much flatter density since the second eigenfunction is zero at the center of the box. Consequently, the potential required to get the same density as shown in Figure 1 needed to be most attractive in the center of the box. Figure 4 shows the Green's function for the converged potential. As before, increase of the potential at one point decreases the density at that point. Now, however, increase of the potential produces a ripple effect on the density at other points with the density at the points far away increasing most. The discontinuity in the potential near the boundary is an artifact of the numerical integration scheme. Since the density at the boundary is automatically zero regardless of v, the potential at the boundary is actually undefined.

Figures 5 and 6 show the potentials required to reproduce the density shown in Figure 1 with 4 and 8 particles, respectively. Clearly, as N increases, the potential becomes deeper and more structured. The Green's functions for these potentials are not shown, but they show ripples in the density response with decreasing wavelength

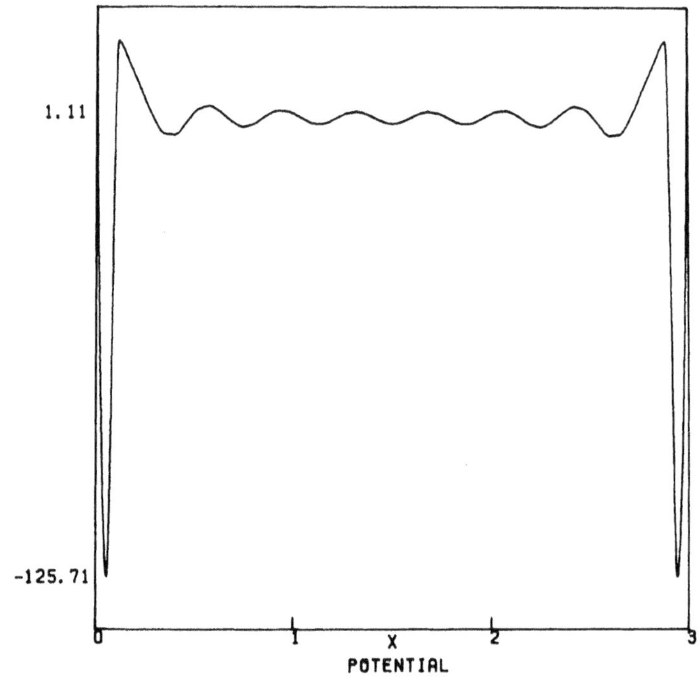

Figure 7. Potential required to produce a constant density for eight spin zero fermions in a box.

as N increases. Only 50 points were used in these numerical integrations and for the eighth eigenfunction there are only 12 points/wavelength; so this last calculation is pushing the limit of the numerical reliability.

A sequence of calculations were done to find a potential which would lead to a constant density except at the end points where the density falls to zero. Figure 7 shows a typical result. This figure for N = 8 shows a wild oscillation near the boundary caused by the abrupt drop in the density. It also shows gentle oscillations elsewhere which are required to produce a constant density.

The last set of calculations also were somewhat artificial. In these calculations, the density was kept fixed at the density shown in Figure 1, but periodic boundary conditions were imposed. With these boundary conditions, the density for a constant potential would be uniform regardless of N. Hence, a very structured potential is required to produce the desired density. Figures 8 and 9 show the required potential for 2 and 8 particles, respectively. The sharp rise near the boundary is required to produce zero density there.

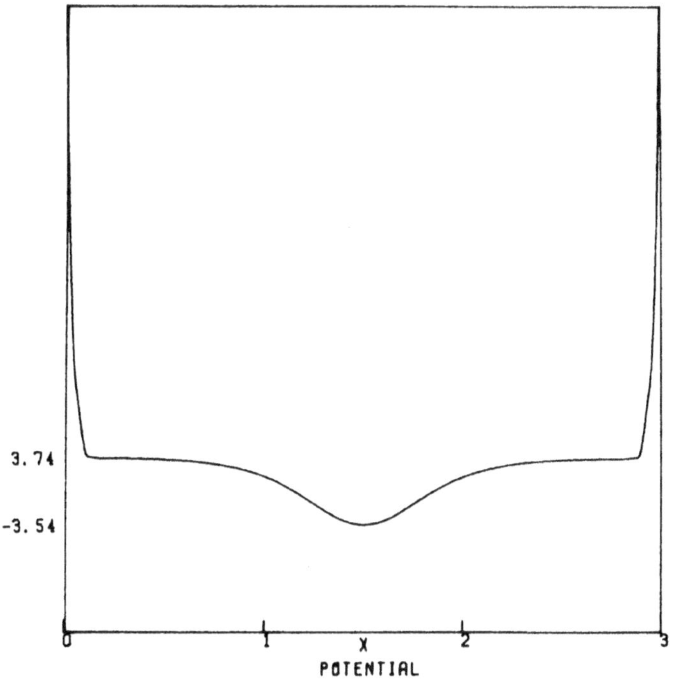

Figure 8. Potential required to produce the density shown in Figure 1 for two spin zero fermions in a box with periodic boundary condition.

With this delta function behavior at the boundary, the rest of the potential is similar to the potential shown in Figures 3 and 6.

The main conclusion from this work is that there is no apparent difficulty in constructing potentials for any density for noninteracting particles in one dimension. Hence, all such densities are probably "V-representable." Interaction between the particles will present a major computational problem but should not change this conclusion. Extension to higher dimension, however, will present the possibility of degeneracy and will cause this construction to fail for certain densities. Certain difficulties associated with two-fold degeneracy were already discussed by Katriel et al.[2] Levy has discussed effects from higher degeneracies.[3]

ACKNOWLEDGMENT

This work was sponsored in part by a grant from the National Science Foundation.

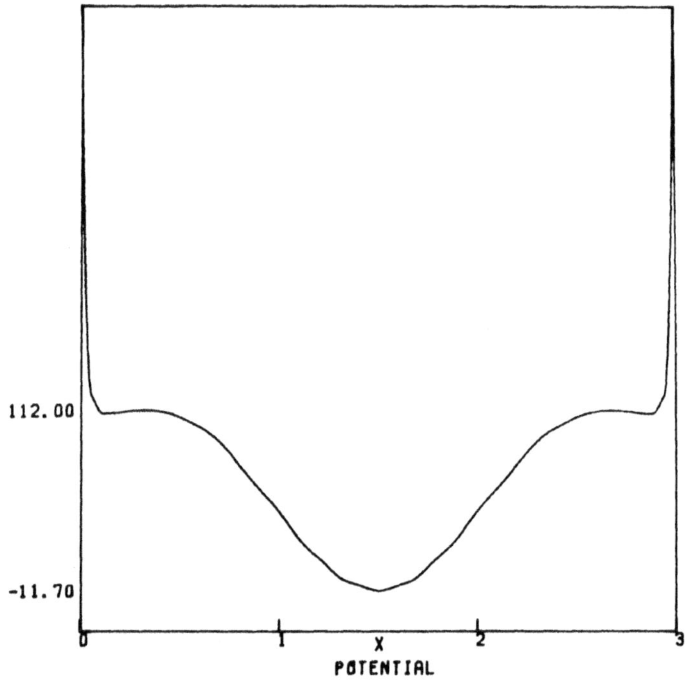

Figure 9. Potential required to produce the density shown in Figure 1 for eight spin zero fermions in a box with periodic boundary conditions.

REFERENCES

1. P. Hohenberg and W. Kohn, Phys.Rev. 136, B864 (1964).
2. J. Katriel, J. Appellof, and E.R. Davidson, Int.J. Quantum Chem. 19, 23 (1981).
3. M. Levy, private communication.
4. M. Levy, Proc.Natl.Acad.Sci. USA 76, 6062 (1979).

THE CHEMICAL POTENTIAL FOR INTERACTING FERMIONS IN A HARMONIC POTENTIAL

Philip Phillips* and Ernest R. Davidson

Department of Chemistry BG-10
University of Washington
Seattle, Washington 98195, U.S.A.

1. INTRODUCTION

One of the key concepts in density functional theory is the chemical potential. For a fixed Hamiltonian in second-quantized form one may define a sequence of ground state energies E(N) for varying integral number, N, of electrons. Although this scheme does not serve to provide a definition of E(N) for non-integral N, the chemical potential is usually defined as

$$\mu = \partial E(N) \partial N \tag{1}$$

and evaluated by a polynomial interpolation among the E(N) for nearby values of N.[1-6]

The difficulties inherent in this approach surface in the simplest model of an atom. For 0, 1 and 2 electrons, E(N) may be approximated by[6]

$$E(0) = 0 \tag{2a}$$

$$E(1) = \varepsilon_1 \tag{2b}$$

$$E(2) = 2\varepsilon_1 + J_{11} \tag{2c}$$

where ε_1 is the 1s energy for charge Z and J_{11} is the Coulomb repulsion integral. A polynomial fit of Eqs. (2a-c) yields

*Danforth-Compton Predoctoral Fellow

$$E(N) = \varepsilon_1 N + \frac{1}{2} N(N-1) J_{11} \tag{3a}$$

and

$$\mu = \varepsilon_1 + (N - \tfrac{1}{2}) J_{11} \tag{3b}$$

for the energy and chemical potential, respectively. Alternatively, E could have been fit with

$$E(N) = \varepsilon_1 N + N{\downarrow}N{\uparrow}J_{11} \tag{4}$$

In Eq. (4) the number of spin-up and spin-down electrons is related by

$$N{\uparrow} + N{\downarrow} = N, \quad 0 \leq N{\downarrow} \leq 1, \quad 0 \leq N{\uparrow} \leq 1 \tag{5}$$

Minimization of E with respect to $N{\uparrow}$ leads to the following equations for E and μ:

$$E(N) = \varepsilon_1 N \qquad 0 \leq N \leq 1 \tag{6a}$$

$$E(N) = \varepsilon_1 N + (N-1) J_{11} \qquad 1 \leq N \leq 2 \tag{6b}$$

$$\mu = \varepsilon_1 \qquad N < 1 \tag{6c}$$

$$\mu = \varepsilon_1 + J_{11} \qquad N > 1 \tag{6d}$$

Clearly, μ is undefined at $N = 1$. Similarly, for $2<N<3$ one expects

$$\mu = \varepsilon_2 + 2J_{12} - K_{12}$$

Thus, the energy may logically be regarded as piece-wise linear in N with discontinuities in the slope precisely at the points where μ is sought.

Quadratic interpolation between $E(N)$ and $E(N \pm 1)$ yields a $\mu(N)$ which is the average of the slopes of a piecewise linear fit on either side of N. This definition has been used by Parr[7] and March.[1-4] March[1] has also discussed the results of fitting a higher degree polynomial through more nearby values $N \pm k$. The results of this approach are limited by the experimental data. March[1] has chosen not to include values of $E(N + k)$ in the range where the extra electrons are unbound (so that $E(N + k - 1) \equiv E(N + k)$).

2. N-BODY HARMONIC OSCILLATOR

Consider the following Hamiltonian for N indistinguishable, spin one-half, fermions in three dimensions

$$H = \sum_i \left(-\frac{1}{2}\nabla_i^2 + \frac{1}{2}\beta^2 N_o r_i^2\right) \pm \frac{1}{2}\beta^2 \sum_{i<j} r_{ij}^2 \qquad (7)$$

As is well known, this equation can be decoupled with the normal coordinates

$$\underset{\sim}{Q}_1 = N^{-1/2} \sum_i \underset{\sim}{r}_i$$

$$\underset{\sim}{Q}_2 = 2^{-1/2} (\underset{\sim}{r}_2 - \underset{\sim}{r}_1)$$

$$\underset{\sim}{Q}_3 = 6^{-1/2} (2\underset{\sim}{r}_3 - \underset{\sim}{r}_2 - \underset{\sim}{r}_1)$$

to produce the Hamiltonian

$$2H = \sum_{i=2}^{N} (P_i^2 + \beta^2(N_o \pm N)Q_i^2) + P_1^2 + N_o\beta^2 Q_1^2 \qquad (8)$$

where

$$P_i = -i\hbar\, \partial/\partial Q_i$$

For spin one-half fermions, the ground state energy for this H is

$$E(N) = \beta N_o^{1/2} \left\{3/2\ (1-\alpha) + \alpha N(j + 3/2) - 2\alpha \binom{j+3}{4}\right\} \qquad (9)$$

with

$$\alpha = (1 \pm N/N_o)^{1/2}$$

and j(N) the smallest integer such that,

$$N \leq 2\binom{j+3}{3}$$

For large N this gives the asymptotic form:[8,9]

$$E(N) \to \frac{1}{4}\beta N_o^{1/2}\alpha(3N)^{4/3} \qquad (10)$$

It is tempting to differentiate Eq. (9) to obtain the 'exact' chemical potential.

$$\mu(N) = \beta \left\{(N_o^{1/2}\alpha)(j+3/2) \pm \frac{1}{2}(N_o^{1/2}\alpha)^{-1} \times \left[-3/2 + N(j + 3/2) - 2\binom{j+3}{4}\right]\right\} \qquad (11)$$

for $N \neq 2\binom{j+3}{3}$.

At large N, Eq. (11) agrees with the derivative of the asymptotic form of E(N):

$$\mu \to \beta \left\{N_o^{1/2}\alpha\,(3N)^{1/3} \pm \frac{1}{8}(N_o^{1/2}\alpha)^{-1}(3N)^{4/3}\right\} \qquad (12)$$

The discontinuities evident in Eq. (11) when $N=2\binom{j+3}{3}$ are reminiscent of those mentioned earlier in Eqs. (6c-d).

It should be remembered, however, that because Eq. (9) is only meaningful for integer N, any function which matches E(N) at all the integers, but produces μ's differing from those defined by Eq. (11), would be equally valid. For example, one could add to Eq. (12) the function $A\sin N\pi$ without changing E; however, μ would change by $A\pi\cos N\pi$. In fact, E could be fit by many smooth curves each of which would give a value of μ for all N.

3. RESULTS

As illustrated by Parr[7] and March,[1] a quadratic fit of E(N) and E(N±1) produces a chemical potential which is equal to $-\chi$ (the Mulliken electronegativity),

$$\mu \simeq -\chi \equiv -\frac{1}{2}(IP + EA) \tag{13}$$

with

$$EA = E(N) - E(N+1)$$

and

$$IP = E(N-1) - E(N)$$

Equation (13) will approximate the 'exact' chemical potential, Eq. (11), unless $j(N+1) \neq j(N-1)$. Further, Eq. (13) will be undefined for a repulsive interparticle potential when $N \geq N_o$ and will differ from Eq. (11) for all N/N_o for which the approximation

$$\partial\alpha/\partial N \simeq \frac{1}{2}[\alpha(N+1) - \alpha(N-1)] \tag{14}$$

breaks down.

Use of a higher degree polynomial fit to more nearby values of $E(N \pm k)$ will increase the likelihood that the range of $N \pm k$ sampled will include points with $j(N+k) \neq j(N-k)$. Of course, if all the $N \pm k$ involve the same j, then the higher degree polynomial should provide an improved approximation to μ.

Cruder estimates of μ could be obtained from linear interpolation between E(N) and E(N ± 1) which approximates μ as either -IP or -EA. Such treatments will differ approximately by

$$IP - EA \simeq \partial\mu/\partial N$$

By analogy with its meaning for atoms and molecules as in Eq. (3b), we will use the symbol J for $\partial\mu/\partial N$. From Eq. (12), the formula for $\partial\mu/\partial N$ for

$$N \neq 2\binom{j+3}{3}$$

is

$$J = \beta\left\{\pm\,(N_o^{1/2}\alpha)^{-1}(j+3/2) - 1/4(N_o^{1/2}\alpha)^{-3} \times\left[-3/2 + N(j+3/2) - 2\binom{j+3}{4}\right]\right\} \quad (15)$$

The asymptotic limit of (15) is

$$J \to \beta\left\{\pm\,(N_o^{1/2}\alpha)^{-1}(3N)^{1/3} - \frac{1}{4}(N_o^{1/2}\alpha)^{-3}\left[-3/2 + \frac{1}{4}\,(3N)^{4/3}\right]\right\} \quad (16)$$

This differs from the derivative of the asymptotic form of μ because j is varied smoothly with N in the latter case but is held constant when $\mu(N)$ is differentiated. Unlike the case of a Coulomb potential, J is positive for an attractive interparticle potential and negative for a repulsive one.

Figure 1 shows graphs of E(N), $\mu(N)$, $-\chi$, and J for $N_o = 1000.5$ and an attractive interparticle potential. Although not illustrated here, it was verified that the relative error between the 'exact' and asymptotic forms of E, μ, and J was small. The step-like behavior of the chemical potential follows from the piece-wise linearity of the energy. The discontinuities between the jth and (j+1)st levels in $\mu(N)$ are analogous to those discussed in Eqs. (6a-d) for atoms with 0, 1, and 2 electrons.

The equations behave differently for a repulsive potential. Figure 2 shows graphs of E, μ, χ, and J for $N_o = 1000.5$ and a repulsive interparticle interaction. In this case, N must be less than N_o as the harmonic well cannot bind more than N_o particles and the nature of the solution changes for $N>N_o$. This change is due to the long-range nature of quadratic repulsions as shown in Eq. (8) which has negative force constants for $N>N_o$. In the intermediate asymptotic range, $N_o>>N>>1$ the formula are identical to those of an attractive interaction except the sign of J is reversed. The maximum μ occurs near $\mu = 0.24\ N_o$, the point at which $\partial\mu_A/\partial N = 0$. It should be noted here that there is no point at which $\partial\mu/\partial N = 0$. Further, E shows a maximum near $8/11\ N_o$ where μ_A is zero. As in the attractive interparticle results the difference between IP and EA is negligible compared to the magnitude of $-\chi$ and μ.

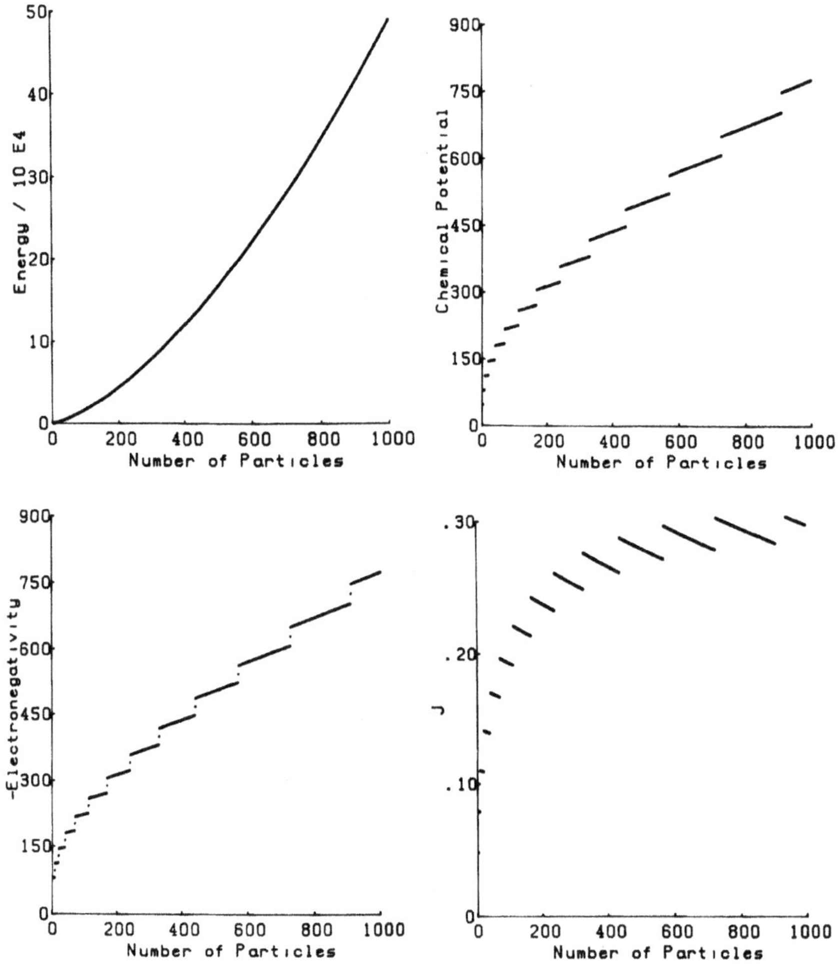

Figure 1. Results for an attractive potential and N_o = 1000.5.

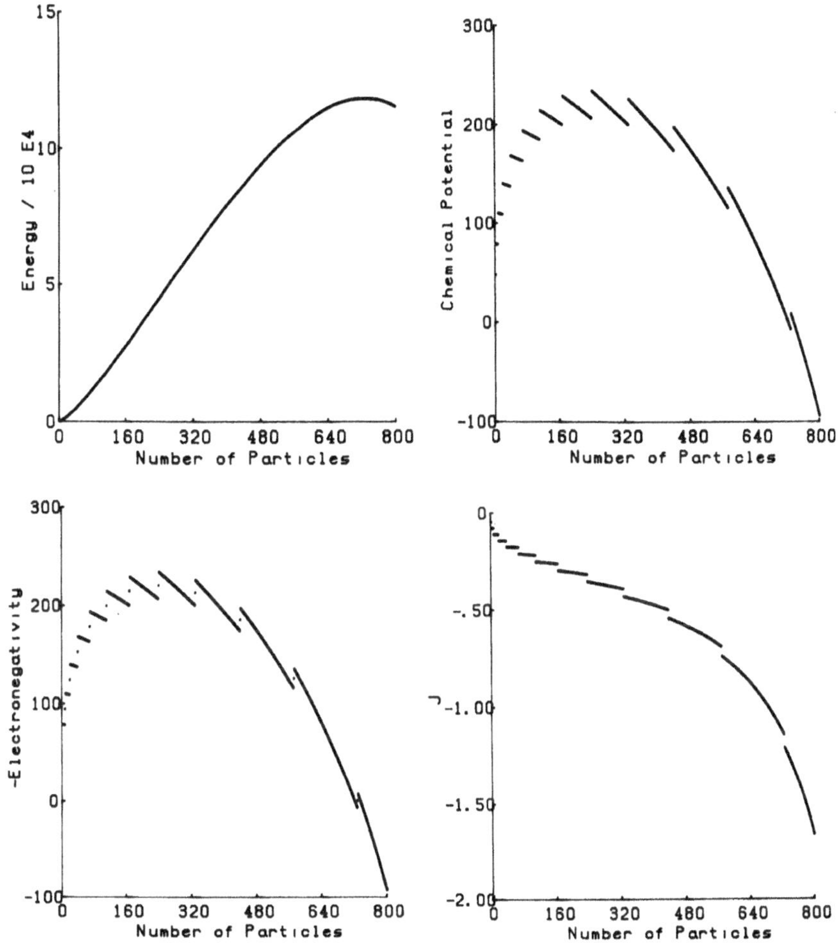

Figure 2. Results for a repulsive potential and N_o = 1000.5.

If this Hamiltonian with a repulsive interparticle interaction is to model a true atomic system, Eq. (7) suggests that N_o must play the role of the nuclear charge. However, the behavior of E and μ for fixed N_o indicates that no choice of N_o will adequately serve to mimic the behavior of neutral atoms. We have, nonetheless, generated a model calculation with N_o varied along with N in the same sense that Z varies with N in a plot of μ for neutral atoms. That is, all derivatives are evaluated for fixed N_o at each N. Figure 3 illustrates the behavior of E, μ, χ and J for $N_o = 11/8$ N, the value for which the asymptotic chemical potential vanishes. In this model $\alpha = (3/11)^{1/2}$, $E \sim N^{11/6}$, and $\mu \sim 0$ because the leading $N^{5/6}$ terms in the asymptotic form cancel each other. In this case, J is large compared to μ and the IP and EA differ by approximately 25%. Since IP and EA express different concepts in real atoms, such a difference between the IP and EA is expected.

4. SPECULATION

Parr et al., have also discussed the chemicel potential defined as the derivative of E with respect to natural orbital occupation and concluded that all such derivatives are equal. This conclusion is fraught with difficulties. One problem may be illustrated as follows. For

$$H = \sum_{i=1}^{N} h(i) + \varepsilon \sum_{i<j}^{N} g(i,j)$$

and a nonseparable g, the ground state wave function has an infinite number of natural orbitals for $\varepsilon \neq 0$. Now consider the limit of small ε (i.e., εg is very small compared with the excitation energy). In this limit

$$\Psi = (N!)^{-1/2} \text{ Det } \{\phi_1(1), \ldots, \phi_N(N)\} + \theta(\varepsilon)$$

where

$$h\phi_k = \lambda_k \phi_k$$

$$\lambda_1 \leq \lambda_2 \cdots \leq \lambda_N < \lambda_{N+1} \leq \cdots$$

and $\lambda_{N+1} - \lambda_N$ is large compared to $\varepsilon g \binom{N}{2}$.

The natural orbitals of Ψ then fall into two sets

$$\chi_k = \sum_{p=1}^{N} c_{pk} \phi_p + \theta(\varepsilon^2) \quad k \leq N$$

$$\chi_k = \sum_{p>N} d_{pk} \phi_p + \theta(\varepsilon^2) \quad k > N$$

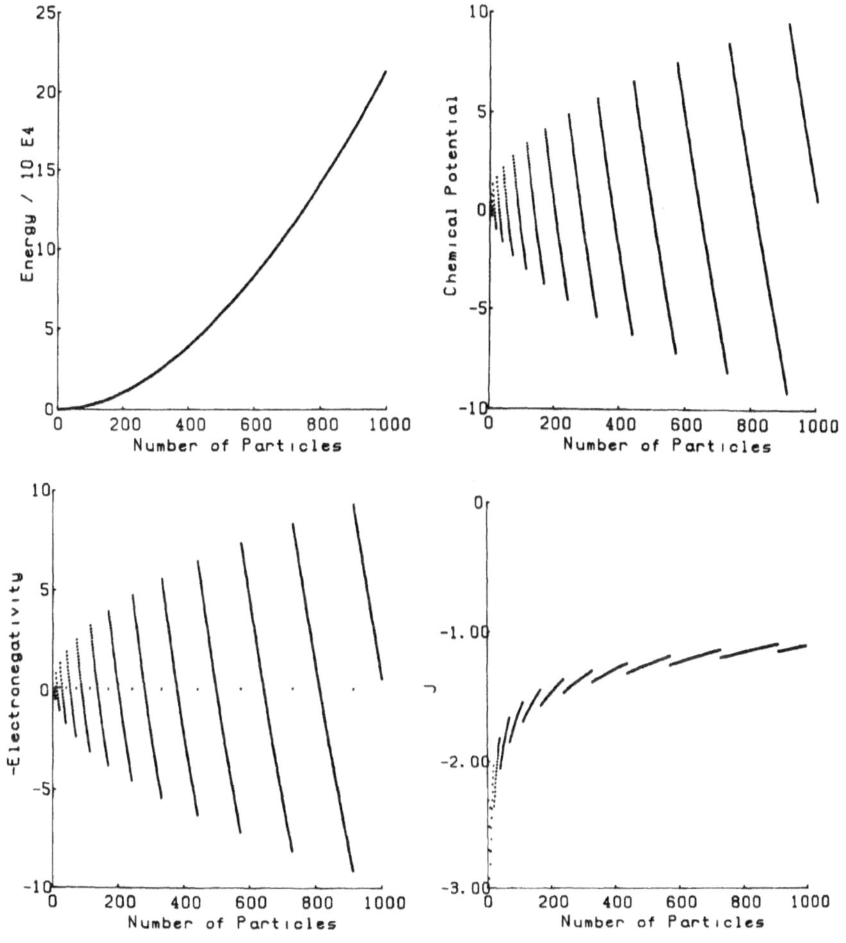

Figure 3. Results for a repulsive potential and $N_o = 11/8$ N.

It seems reasonable that in the limit of small ε,

$$\mu_i = \partial E/\partial n_i \rightarrow \langle \chi_i | h | \chi_i \rangle$$

But clearly

$$\begin{aligned} \langle \chi_i | h | \chi_i \rangle &\leq \lambda_N \quad \text{for } i \leq N \\ &\geq \lambda_{N+1} \quad \text{for } i > N \end{aligned}$$

so $\mu_N \neq \mu_{N+1}$ for small ε in apparant contradiction to Parr's conclusion.

ACKNOWLEDGEMENT

P. Phillips would like to thank Dr. J.C. Ellenbogen for insightful discussions which sparked his initial interest in this problem. We would also like to thank NSF for partially funding this research.

REFERENCES

1. N.H. March, J.Chem.Phys. 76, 1869 (1982).
2. G.P. Lawes, N.H. March, and M.S. Yusaf, Phys.Lett. A 67, 342 (1978).
3. N.H. March and R.F.W. Bader, Phys.Lett. A 78, 242 (1980).
4. N.H. March, Theoretical Chemistry 45 (1981) and references therein.
5. J.A. Alonso and L.A. Girifalco, J.Chem.Phys. 73, 1313 (1980).
6. J. Katriel, R.G. Parr, M.R. Nyden, J.Chem.Phys. 74, 2397 (1981).
7. R.G. Parr, R.A. Donnelly, M. Levy, W.E. Palke, J.Chem.Phys. 68, 3801 (1978).
8. J.M. Lévy-Leblond, Physics Letters 26 A, 540 (1968).
9. E.R. Davidson, Phys.Rev. A 1, 30 (1970).
10. E.R. Davidson, Reduced Density Matrices in Quantum Chemistry, Academic Press, New York, N.Y., 1976.
11. M. Levy, private communication.

DENSITY, TOTAL ENERGY AND CHEMICAL POTENTIAL OF ATOMIC IONS AND SOME MOLECULES

Norman H. March

Theoretical Chemistry Department University of Oxford
1 South Parks Road, Oxford OX1 3TG, England

Renato Pucci

Instituto di Fisica dell'Universitá di Catania
Corso Italia, 57 - 95129 Catania, Italy

1. OUTLINE

In this article we shall review some recent work on the electron density ($\rho(\vec{r})$) description of atomic ions and some molecules with π-electrons. In section 2, we briefly consider first the one-electron treatment of atoms and molecules given by Thomas and Fermi [1-3] in terms of $\rho(\vec{r})$, with its extension to include exchange, first given by Dirac.[4] This treatment is formally generalized to yield the Euler equation of the density functional theory which however contains an as yet unknown functional consisting of kinetic plus exchange and correlation energy. Separation into single-particle kinetic energy allows the Euler equation to be recast into a one-electron framework, following the earlier work of Slater,[5] which was completed by Kohn and Sham.[6] The existence of a one-body potential energy allows approximations of Thomas-Fermi nature in the kinetic energy to be avoided, if desired by direct solution of the Schrödinger equation with the one-body potential (with exchange and correlation suitably approximated).

We turn then to the theory of atomic ions, pointing out that the ground-state energy $E[\rho]$, when evaluated for ions with N electrons and atomic number Z, can be calculated by

a) The $1/Z$ expansion

or

b) the statistical theory in the limit of large N/Z for positive ione (which sums the 1/Z expansion to all orders).

This work enables us to relate the total energy to the chemical potential of the density description (for a precise definition, see Eq. (2.10) below), for isoelectronic sequences.

Following this, we return to $E[\rho]$ and consider how $\rho(\vec{r})$, and hence E, may be usefully represented in terms of a few physical and chemical parameters; in particular the ionization potential and the diamagnetic susceptibility.

The final section on atomic ions treats the chemical potential quite fundamentally and displays its relation to (a) ionization potential and (b) correlation potential energy. While this formally exact relation cannot presently be evaluated to chemical accuracy from many-body theory, it turns out, by considering the asymptotic form of the first-order density matrix, to demonstrate that in Hartree-Fock theory the chemical potential is precisely the orbital energy of the highest occupied one-electron state, i.e. the Koopmans eigenvalue (as shown by Alonso and March[7]).

The subsequent section of this article focus mainly on π-electrons in the linear polyacenes. Again, as for atomic ions, the focal points are electron density, chemical potential and total energy of the π-electrons. Naturally, simple models have to be used to get explicit analytical results: we utilize both two-dimensional electron gas theory in a local approximation and Hückel theory. Contact with experiment is made via the ionization potential as a function of the number of rings in the linear polyacenes. Finally, reference is made to some reference work[8] on simpler molecules relating molecular and atomic ionization potentials and linking these with electron dipole moments.

2. EULER EQUATION OF DENSITY FUNCTIONAL THEORY

The simplest density functional theory, namely the Thomas-Fermi method,[1-3] is based on a variational principle for the total energy E, namely

$$\delta[E - \mu N] = 0 \tag{2.1}$$

where the variation is with respect to the electron density $\rho(\vec{r})$. The Lagrange multiplier μ, which will be identified with the chemical potential, takes care of the normalization condition

$$\int \rho d\vec{r} = N \tag{2.2}$$

the given total number of electrons in the atom or molecule.

The Thomas-Fermi energy is then

$$E = C_K \int \rho^{5/3} d\vec{r} + \int \rho V_N d\vec{r} + \frac{1}{2} \int \rho V_e d\vec{r}: \tag{2.3}$$

$$C_K = \frac{3h^2}{10m} \left(\frac{3}{8\pi}\right)^{2/3}$$

where the first term is the kinetic energy, obtained by employing free electron relations locally, the second is the electron-nuclear potential energy, $V_N(\vec{r})$ being the potential (energy) created by the nuclear framework, while $V_e(\vec{r})$ is the potential energy of the charge cloud $\rho(\vec{r})$, namely

$$V_e(\vec{r}) = e^2 \int \frac{\rho(\vec{r}')}{|\vec{r}-\vec{r}'|} d\vec{r}' \tag{2.4}$$

For molecules, one must add the nuclear-nuclear potential energy which is not however a function of ρ.

Using Eqs. (2.1) and (2.3) the Euler equation is readily obtained as

$$\mu = \frac{5}{3} C_K \rho^{2/3} + V_N + V_e \tag{2.5}$$

This equation already contains one essential feature of the density functional theory: namely that the constant chemical potential μ is made up of the sum of a kinetic term $5/3C_K\rho^{2/3}$ and a Hartree potential energy $V_H=V_N+V_e$, which separately depend on the position $\vec{r}$ in the molecular charge cloud. Eq. (2.5) is the fundamental Thomas-Fermi relation between density and (self-consistent) potential. It is semiclassical, the density ρ being zero in regions which are classically forbidden.

2.1. Introduction of Exchange

An important step forward in the Thomas-Fermi method was taken by Dirac[4] who introduced exchange into the theory. We shall not consider his original argument, but rather we shall generalize the above variational argument, writing the local electron gas exchange energy A as

$$A = -C_e \int \rho^{4/3} d\vec{r} \quad : \quad C_e = \frac{3}{4} e^2 \left(\frac{3}{\pi}\right)^{1/3} \tag{2.6}$$

Using Eq. (2.1) again, but this time adding A to Eq. (2.3). Then one finds

$$\mu = \frac{5}{3} C_K \rho^{2/3} + V_H - \frac{4}{3} C_e \rho^{1/3} \tag{2.7}$$

It is important to note that Eq. (2.7), by analogy with (2.5), can be viewed as a new density-potential relation, with a modified potential

$$V(\vec{r}) = V_H(\vec{r}) - \frac{4}{3} C_e \rho^{1/3} \qquad (2.8)$$

an observation which was made most clearly in the work of Slater.[5] This idea of representing exchange by a potential energy finds a natural generalization in formally exact density functional theory, to which we now turn.

2.2. Formally Exact Euler Equation

The Thomas-Fermi-Dirac theory given above, in which the total energy E was written quite explicitly in terms of the electron density ρ, was formally completed by the theorem of Hohenberg and Kohn,[9] who proved that, as the earlier workers had assumed, the ground-state energy of a many-electron system is indeed a function (al) of the electron density, i.e. $E=E[\rho]$. Thus, one can formally generalize Eqs. (2.3) and (2.6) to read

$$E = T[\rho] + \int \rho V_N \, d\vec{r} + \frac{1}{2} \int \rho V_e \, d\vec{r} + U_{xc}[\rho] \qquad (2.9)$$

where T is now the total kinetic energy, including correlation kinetic energy, as a functional of ρ, while U_{xc} is the potential energy of exchange (x) and correlation (c) interactions. Formally, one can apply the variational principle (2.1) to Eq. (2.9), but, as was especially emphasized by Kohn and Sham,[6] it is often fruitful to introduce an equivalent one-electron problem by writing T in two parts, a single-particle kinetic energy, T_s say, corresponding though to the exact (correlated) electron density $\rho(\vec{r})$, and a correlation kinetic energy which can be lumped with U_{xc} in Eq. (2.9) to give a total exchange and correlation energy $E_{xc}[\rho]$. Then the formally exact (one-electron) Euler equation reads

$$\mu = \frac{\delta T_s}{\delta\rho} + V_H + \frac{\delta E_{xc}}{\delta\rho} \equiv \frac{\delta T}{\delta\rho} + V_H + \frac{\delta U_{xc}}{\delta\rho} \qquad (2.10)$$

We shall utilize both forms (2.10) below. But, for the present, we note with Kohn and Sham[6] that the first line formally defines a generalization of the one-body potential in Eq. (2.8) as

$$V(\vec{r}) = V_H(\vec{r}) + \frac{\delta E_{xc}}{\delta\rho} \qquad (2.11)$$

If we neglect correlation, then $E_{xc} \equiv A$ and using the local density form (2.6) we regain the Dirac-Slater one-body potential energy (2.8).

Naturally, exact knowledge of the 'exchange and correlation' potential energy $\delta E_{xc}/\delta\rho$ in Eq. (2.11) would require exact solution of the many-electron problem posed by the inhomogeneous electron gas in an atom or molecule, which currently is not feasible. Hence progress has come by judicious approximation in this term in Eq. (2.11).

We shall nevertheless be able to exploit merely the existence of such a one-body potential energy $V(\vec{r})$ when we turn to π-electron systems in the latter part of this article.

3. CHEMICAL POTENTIAL AND TOTAL ENERGY OF ATOMIC IONS FROM 1/Z EXPANSION AND STATISTICAL THEORY

As already stressed, although the chemical potential μ in Eq. (2.5) and its generalization (2.10) is built up as the sum of spatially varying components, the sum must be independent of $\vec{r}$ for otherwise electrons could redistribute themselves from one point to another in the atom or molecule to lower the total energy. From this it is clear that μ must be a constant, independent of position and characteristic of the electronic distribution in the atom or molecule.

Next, we note that for atomic ions, the ground-state energy $E[\rho]$ can be expressed finally as $E(N,Z)$, where Z is the atomic number. This can be made explicit by appeal to the 1/Z expansion (see, for example, March[10]):

$$E(N,Z) = Z^2[\varepsilon_o(N)+\frac{1}{Z}\,\varepsilon_1(N) + \frac{1}{Z^2}\,\varepsilon_2(N) + \ldots] \tag{3.1}$$

where ε_o is determined for all N by the Coulomb field problem $\varepsilon_1(N)$ and $\varepsilon_2(N)$ have been determined precisely for small N, and approximately for large N, as discussed further below.

We shall first show that, provided we truncate Eq. (3.1) as

$$E(N,Z) \doteq Z^2\,\varepsilon_o + Z\,\varepsilon_1 + \varepsilon_2 \tag{3.2}$$

then μ can be directly related to E and the number of electrons N, through the coefficients $\varepsilon_n(N)$ (plus their derivatives). The truncation (3.2) can clearly be validated for fixed N by making Z sufficiently large. However, special treatment of the series (3.1) is required when E and N become large together: this will lead us back to the Thomas-Fermi density description (see below).

To obtain μ, we note the thermodynamic result for atomic ions that

$$\mu = \left(\frac{\partial E}{\partial N}\right)_Z \tag{3.3}$$

which is readily understood from the form of Eq. (2.1).

Using Eqs. (3.2) and (3.3) we find immediately

$$\mu = Z^2\,\varepsilon_o' + Z\,\varepsilon_1'\,\varepsilon_2' \tag{3.4}$$

where the primes denote derivatives with respect to N. By eliminating Z between Eqs. (3.2) and (3.3) one obtains

$$\mu = \alpha E + \beta + \gamma \sqrt{\delta + 4\varepsilon_o E} \tag{3.5}$$

The quantities $\alpha(N)$, $\beta(N)$, $\gamma(N)$, and $\delta(N)$ are known from the 1/Z expansion and are explicitly given, in terms of ε_o, ε_1, ε_2, and their derivatives in the work of March and Pucci.[11]

In curve 1, thick line of Figure 1, $|\mu|$ is plotted from Eq. (3.5) for the isoelectronic sequence N=9. The way μ has been approximated (actually in Hartree-Fock theory) will become clearer below when we identify μ_{HF} with the orbital energy of the highest occupied molecular orbital (HOMO). Will return to discuss the broken line in Figure 1 later on in the article.

So far, we have merely asserted that $E=E[\rho]$ in density functional theory can be reduced to the form E(N,Z) in Eq. (3.1). How-

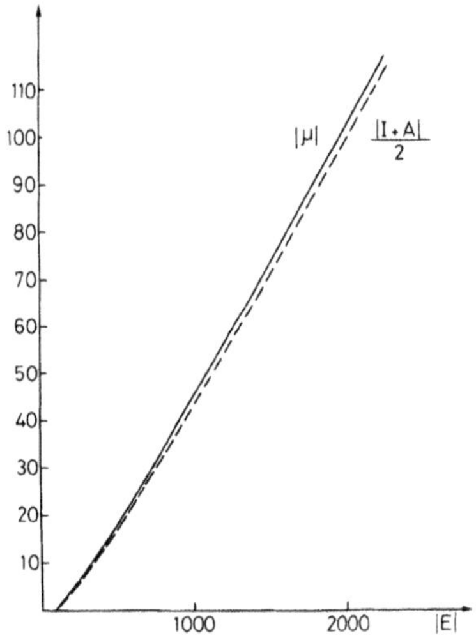

Figure 1. Thick line represents $|\mu|$ versus $|E|$ from Eq. (3.9), with α, β, γ and δ parameters obtained by fitting Hartree-Fock calculations of Clementi[12] for the isoelectronic sequence N=9, with the identification: $|\mu| \equiv |\varepsilon_{HOMO}|$. The broken line represents $|I+A|/2$ (where I and A are defined in Eqs. (3.10) and (3.11)) versus $|E|$ using values of ε_o, ε_1, and ε_2 given in Ref. 11 for the isoelectronic sequence N=9. Atomic units are used.

ever, this statement can be made quite explicit in the simplest density description, namely the Thomas-Fermi method. Then (see, for example, March[10]) it follows that Eq. (2.3) when combined with the self-consistent solution of the Euler Eq. (2.5), yields

$$E(N,Z) = Z^{7/3} f(\frac{N}{Z}) \tag{3.6}$$

where f is known from numerical solutions of the Thomas-Fermi equation for positive atomic ions (see March and White[13]; Tal and Levy[14]).

As stressed by March and White,[13] for heavy positive ions Eq. (3.6) can be regarded as a sum of the 1/Z expansion (3.1) to infinite order, in contrast to the truncated form (3.2) valid for sufficiently large Z for fixed N. Since negative ions are not stable in the Thomas-Fermi theory (essentially because μ for the neutral atom is already identically zero), Eq. (3.6) is valid in the limit N and Z large such that $N/Z \leq 1$.

Combining Eqs. (3.3) and (3.6) it is a straightforward matter to show that $\mu/N^{4/3}$ versus $E/N^{7/3}$ is a universal curve in the Thomas-Fermi limit, the plot being given in Figure 2.

4. DENSITY AND ATOMIC SCATTERING FACTORS FOR CLOSED SHELL ATOMS

Since the ground-state energy $E \equiv E[\rho]$, we next give attention to simple representation of the ground electron density $\rho(\vec{r})$ in atomic ions. It is convenient to use the Fourier transform of ρ, i.e. the atomic scattering factor

$$f(K) = \int e^{i \vec{K} \cdot \vec{r}} \rho(\vec{r}) \, d\vec{r} \tag{4.1}$$

As it is well known[10] the scattering factor f(K) for X-rays from atomic ions with closed shells can be written

$$f(K) = \int_0^\infty 4\pi r^2 \rho(r) \frac{\sin Kr}{Kr} dr \tag{4.2}$$

The small K expansion of Eq. (4.2) is given by[15]

$$f(K) = N + a_2 K^2 \langle r^2 \rangle + a_4 K^4 \langle r^4 \rangle + \ldots + a_{2n} K^{2n} \langle r^{2n} \rangle \tag{4.3}$$

where use has been done of the normalization condition:

$$\int_0^\infty 4\pi r^2 \rho(r) \, dr = N \tag{4.4}$$

and the moments $\langle r^{2n} \rangle$ have been defined by

$$\langle r^{2n} \rangle \equiv \int_0^\infty D(r) r^{2n} dr \tag{4.5}$$

where the radial electron density D(r) is

$$D(r) = 4\pi r^2 \rho(r) \tag{4.6}$$

The term $\langle r^2\rangle$ proportional to K^2 in Eq. (4.3) is known to determine the Langevin-Pauli diamagnetic susceptibility[16] and has, therefore, direct physical significance. From Hartree-Fock calculations on Ar, Kr, Xe and Rn it is found that[17]

$$\langle r^2\rangle \propto Z^{0.75}, \quad \langle r^4\rangle \propto Z^{0.95}, \quad \langle r^6\rangle \propto Z^{1.25} \tag{4.7}$$

This result (4.7) can be interpreted in terms of an exact result for the asymptotic decay of D(r) at large r. The exact asymptotic form of the electron density, known from the work of several groups (see, e.g. Ostenhof et al.[18] and some other recent works [19,20]) to be

$$\rho(r) \sim A r^{\gamma} \exp[-2(2I)^{1/2} r] \tag{4.8}$$

where I is the exact ionization potential.

Noting that in the closed shell atoms the ionization potential I is

$$I \propto Z^{-1/3} \tag{4.9}$$

and by using the asymptotic form (4.8) of the electron density one obtains

$$\frac{\langle r^{2n}\rangle}{\langle r^{2n-2}\rangle} \propto \frac{1}{I} \tag{4.10}$$

at sufficiently large n. The result obtained by combining Eqs. (4.9) and (4.10) is already in fair agreement with the prediction $\langle r^6\rangle/\langle r^4\rangle \propto Z^{0.3}$ from Hartree-Fock results (Eq. 4.6)). Motivated by Eqs. (4.3) and (4.10), we note we can write the identity, for neutral atoms with N=Z,

$$\frac{f(K) - Z - a_2 K^2\langle r^2\rangle}{K^4\langle r^4\rangle} = a_4 \left[1 + \frac{a_6}{a_4}\frac{\langle r^6\rangle}{\langle r^4\rangle} K^2 + \ldots\right] \tag{4.11}$$

If we adopt the form $\langle r^6\rangle/\langle r^4\rangle \propto 1/I$, and note that the next term of $O(K^4)$ involves $\langle r^8\rangle/\langle r^4\rangle \propto 1/I^2$, etc, then we can formally sum the right-hand side of Eq. (4.11) into $H(K^2/I) \equiv G(K/\sqrt{I})$ where, however, H or G cannot presently be expressed analytically. However, in Figure 3, we have plotted the left-hand side of Eq. (4.11) against $K/\sqrt{I}$ for Ar and Kr, using the experimental ionization potentials

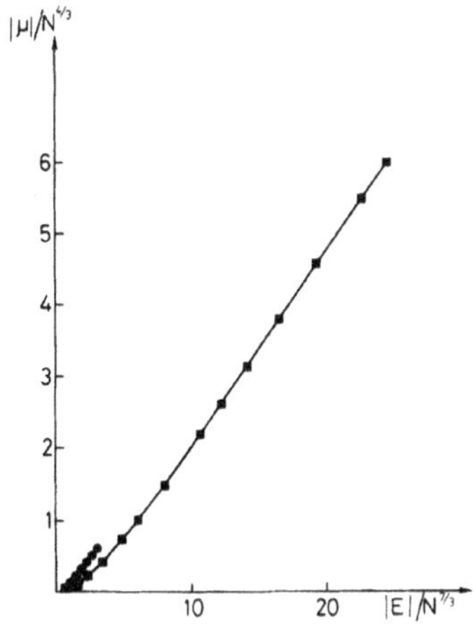

Figure 2. $|\mu|/N^{4/3}$ versus $|E|/N^{7/3}$ (in a.u.) from Eqs. (3.2) and (3.4) using values of f and f' derived from Refs. 13 and 14. The dots show Hartree-Fock values from calculations of Clementi[12] from the isoelectronic sequence N=18 from Ar to Kr.

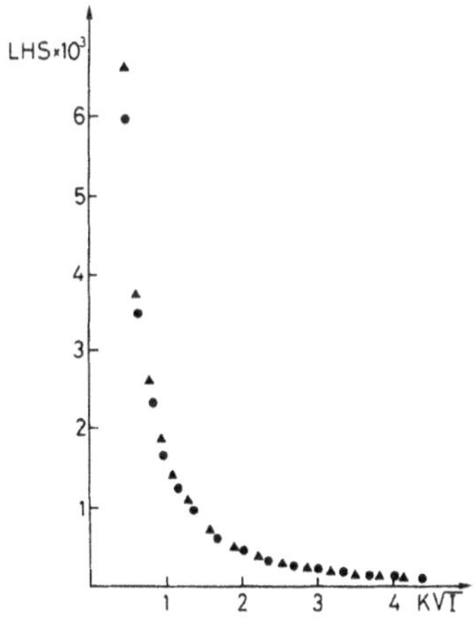

Figure 3. Plot of left-hand side of Eq. (4.11), here called LHS, as a function of $K/\sqrt{I}$ for the neutral atoms Ar (triangles) and Kr (dots). Experimental values for I (in eV), were used together with Hartree-Fock scattering factors.[21]

and the Hartree-Fock values for $f(K)$.[21] It is striking that they lie on the same curve, confirming the scaling predicted above.

Thus, in these atoms, over the entire range of K reported in Ref. 21, the atomic scattering factors can be characterized solely by Z, $<r^2>$, $<r^4>$ and I to graphical accuracy (~1%). The divergent behaviour[17] of $<r^4>$, as a function of (1-N/Z) in the Thomas-Fermi approximation, induce us to believe that $f(K)$ is not crucially dependent of $<r^4>$. This suggest us to analyse the total energy of atomic ions in terms of N, $<r^2>$ and I.

4.1. Total Energy of Atomic Ions Related to Low-order Ionization Potentials and to Diamagnetic Susceptability

From the previous paragraph we know that $f(K)$, which is the Fourier transform of ρ, depends in an important way on the lowest ionization potential I and the mean value $<r^2>$.

Therefore, it follows that since $\rho(\vec{r})$ depends on I and $<r^2>$. we can expect that $E=E[\rho(r)]$ will reduce to the (now approximate) statement that E will be a function of I and $<r^2>$. For one-electron ions, one finds[22] immediately

$$\ell n I = -\ell n <r^2> + \ell n(3/2) \qquad (4.12)$$

We have used[22] the data of Moore[23] and derived the $<r^2>$ values from available Hartree-Fock calculations[24] to effect the generalization of Eq. (4.12) for N>1. In Figure 4, we have plotted $\ell n I$ vs $\ell n <r^2>$ for the isoelectronic sequences, N=6-9 and 11. In each case, except for a few members with N/Z near to unity, we find that an excellent representation is

$$\ell n I = \alpha(N) \ell n <r^2> + \beta(N) \qquad (4.13)$$

Some more examples for different values of N, together with the corresponding $\alpha(N)$ and $\beta(N)$ values, are given in Ref. 22. After some investigation it turns out that (when appropriate) $E(Z,N)$ is most successfully represented by including the first three ionization potentials, and approximating the rest by Eq. (4.13). This yields

$$E(Z,N) = I_N + I_{N-1} + I_{N-2} + \sum_{n=1}^{n-3} <r^2>_{Z,n}^{\alpha(n)} \exp[\beta(n)] \qquad (4.14)$$

In the case of the Ne atom, using Eq. (4.14) together with the values of the parameters $\alpha(n)$ and $\beta(n)$ reported in Ref. 22, we get $E(10,10)=-127.888$ a.u. to compare with the experimental value[25] $E(10,10)=-129.056$ a.u.

Some more equally satisfactory comparisons are given in Ref. 22.

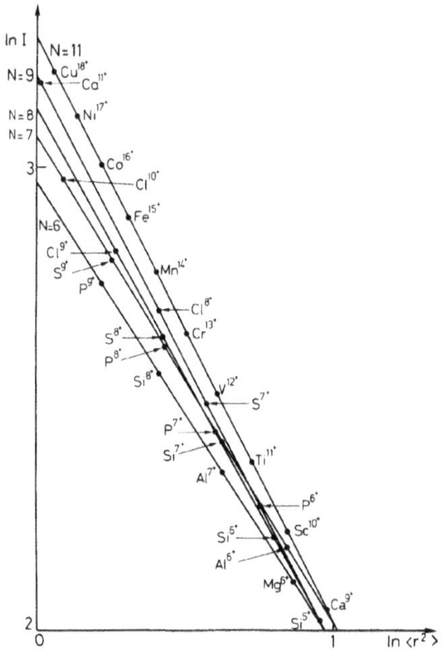

Figure 4. ℓnI vs $\ell n\langle r^2\rangle$ for N=6,7,8,9 and 11. Atomic units are used for both I and $\langle r^2\rangle$. I was taken from Moore's tables.[23] $\langle r^2\rangle$ was obtained from Hartree-Fock calculations.[24]

5. RELATION BETWEEN CHEMICAL POTENTIAL, IONIZATION POTENTIAL AND CORRELATION ENERGY

As we have emphasized, since the chemical potential μ is the same at every point in the electronic cloud, we can choose the most convenient point for its evaluation. Returning to Eq. (2.10), and with the specific case of atoms in mind, we see that $\vec{r}$ taken to infinity removes the Hartree potential $V(\vec{r})$ and hence

$$\mu = \lim_{\vec{r}\to\infty} \left[\frac{\delta T}{\delta\rho} + \frac{\delta U_{xc}}{\delta\rho}\right] \tag{5.1}$$

Evidently the second term on the right-hand side is a full many-body quantity and presently no exact evaluation has proved possible. However, although the kinetic energy T includes correlation kinetic energy, we shall see below that, provided the electron density is known exactly as $\vec{r}$ tends to infinity, then this kinetic contribution to μ can be calculated precisely (as shown by March and Bader;[26] see also Alonso and March[7]).

Essentially, the argument rests on the fact that at large $\vec{r}$ and $\vec{r}'$, the first-order density matrix $\gamma(\vec{r},\vec{r}')$ factorizes to $\rho^{1/2}(\vec{r}) \times \rho^{1/2}(\vec{r}')$ (March and Pucci[27]) and from this the kinetic energy density at large $\vec{r}$ is found to have the von Weizsäcker form $(\hbar^2/8m)(\nabla\rho)^2/\rho$: a result first given by Tal and Bader.[28] We emphasize that this is a many-electron result, provided the exact correlated ρ is used. At large distances, it can be shown that Eq. (4.8) is valid. Evaluating $\delta T/\delta\rho$ as $\vec{r}\to\infty$ then yields

$$\mu = - I + \lim_{\vec{r}\to\infty} \frac{\delta U_{xc}}{\delta\rho} \tag{5.2}$$

At this point, we note that Parr et al.[29] have proposed that the negative of the chemical potential should be equated with electronegativity. If we adopt the Mulliken form (I+A)/2, then it is clear that the many-body term lim $\delta U_{xc}/\delta\rho$, as $\vec{r}\to\infty$, must cancel half of I and have a contribution from the electron affinity A (as stressed by March[30]).

That such a relation between μ and (I+A)/2 is reasonable is shown by referring to Figure 1, where μ and (I+A)/2 are compared by combining Hartree-Fock theory with the 1/Z expansion for atomic ions, using the usual relations

$$I = E(Z,N) - E(Z,N-1) \tag{5.3}$$

and

$$A = E(Z,N+1) - E(Z,N) \tag{5.4}$$

All that remains here is to discuss μ in Hartree-Fock theory therefore.

5.1. Hartree-Fock Equivalence of Chemical Potential and Koopmans Eigenvalue

Two points need to be stressed in applying Eq. (5.1) to the Hartree-Fock theory. The first is that the asymptotic form of the Hartree-Fock electron density is now dominated by the highest occupied one-electron wave function, with eigenvalue $\varepsilon_{Koopmans} \equiv \varepsilon_K$. Therefore, in evaluating $\delta T/\delta\rho$, as $\vec{r}\to\infty$, I in the exact treatment given above becomes $-\varepsilon_K$. It remains to discuss $\delta U_{xc}/\delta\rho$ as $\vec{r}\to\infty$, there being no correlation in Hartree-Fock theory. Using the expression for exchange energy in terms of the first-order (idempotent) density matrix, Alonso and March[7] have shown that this contribution is zero. This is also the conclusion reached in the local exchange approximation given in Eq. (2.6). Thus we have the final result

$$\mu_{HF} = \varepsilon_K \tag{5.5}$$

which has been utilized in constructing Figure 1.

One further important consequence of Eq. (5.5) is to be noted here. The Schrödinger equation for the highest occupied orbital becomes

$$\mu_{HF} = -\frac{\hbar^2}{2m}\frac{\nabla^2\Psi_{HOMO}}{\Psi_{HOMO}} + V(\vec{r}) \tag{5.6}$$

Comparing this with the Euler equation (2.10) in Hartree-Fock theory shows that

$$\frac{\delta T}{\delta\rho}\Big|_{HF} = -\frac{\hbar^2}{2m}\frac{\nabla^2\Psi_{HOMO}}{\Psi_{HOMO}} \tag{5.7}$$

Secondly, the equation of motion of the density matrix $\gamma(\vec{r},\vec{r}\,')$ in Hartree-Fock theory is

$$\nabla^2_{\vec{r}}\,\gamma(\vec{r},\vec{r}\,')-\nabla^2_{\vec{r}\,'}\gamma(\vec{r},\vec{r}\,') = \frac{2m}{\hbar^2}\,[V(\vec{r})-V(\vec{r}\,')]\,\gamma(\vec{r},\vec{r}\,') \tag{5.8}$$

and using Eq. (5.6) for $V(\vec{r})$ it is clear that $\gamma(\vec{r},\vec{r}\,')$ in Hartree-Fock theory is a functional of Ψ_{HOMO}. Of course, the present argument does not display the functional. The only example we can find of such a result lies in the work of Husimi[31] for independent particles moving in 1 dimension in a linear harmonic oscillator potential. Here $\gamma(x,x')$ is known explicitly in terms of γ_{HOMO}, a result already used in the density functional theory of Lawes and March.[32]

6. CHEMICAL POTENTIAL AND ENERGY PER PARTICLE IN π-ELECTRON SYSTEM

6.1. Total Electron Density Decomposed into σ- and π-Contributions

Accepting the existence of a one-body potential $V(\vec{r})$ with the symmetry of the nuclear framework, we can write the electron density $\rho(\vec{r})$ exactly in terms of the one-electron eigenfunctions $\Psi_i(\vec{r})$ of the Hamiltonian

$$H = -\frac{1}{2}\nabla^2 + V(\vec{r})$$

as

$$\rho(\vec{r}) = \sum_{\substack{\text{occupied}\\ \text{states}}} \Psi_i^*(\vec{r})\,\Psi_i(\vec{r}) \tag{6.1}$$

But now we utilize the familiar classification of one-electron wave functions into $\Psi_{i\sigma}$ and $\Psi_{i\pi}$, the former having symmetry about the bonds and the latter being antisymmetric about the plane of the molecule (e.g. benzene). Thus, once the highest occupied one-electron level is fixed in the sum (6.1), we have an operational way of dividing the total density $\rho(\vec{r})$ into a sum of a sigma contribution $\rho_\sigma(\vec{r})$ and a pi contribution ρ_π, i.e.

$$\rho(\vec{r}) = \rho_\sigma(\vec{r}) + \rho_\pi(\vec{r}) \tag{6.2}$$

Of course, this is not to say that experimentally it will be possible to measure ρ_σ and ρ_π separately.

6.2. Kinetic Energy of π Electrons in Terms of π-density

We have argued elsewhere[33] that π-electrons are well treated as a two-dimensional electron gas. If then we take the local approximation

$$T_\pi = \text{const}\int \rho_\pi^2\, d\tau \tag{6.3}$$

we have shown[33] that the Hückel assumption

$$E_{\text{total},\pi} = \sum_{\substack{\text{occupied}\\ \text{states}}} \varepsilon_{i\pi} \tag{6.4}$$

results from the treatment, provided the chemical potential term in the Euler equation is small.*

Naturally, in a final theory, the local density approximation will have to be relaxed, but we shall show below that some interesting consequences, in addition to Eqs. (6.2), (6.3), and (6.4) also flow from such an assumption. We restrict ourselves below to a summary of our recent work on the linear polyacenes.

*If we took $T_{\text{total}} = \text{const}\int \rho^2 d\tau$ then evidently using Eq. (6.2) we could separate into a sum of three terms, σ kinetic energy, π kinetic energy, and σ-π interaction kinetic energy. However, such a treatment would be too crude for σ electrons which are genuinely three-dimensional in character. But it does show that the rigourous separation of ρ into σ and π parts may, in the future, have significant consequences.

6.3. Energy and Chemical Potential of π-electrons in the Linear Polyacenes

Below, we want to make a number of points about the π-electrons, even though in our calculations we shall have to appeal to very simple models.

First of all, empirically, there is an excellent empirical correlation between the observed[34] ionization potential I and the total[35] perimeter p of the linear polyacenes, as shown in Figure 5.

Secondly, we shall argue below that a quite fundamental property of these molecules is the existence of 'saturation' in the energy of the π-electrons as the chain length is allowed to tend to infinity. Precisely, this means that if $E_{\pi}(n)$ is the total π-electron energy as a function of the number of rings n, then

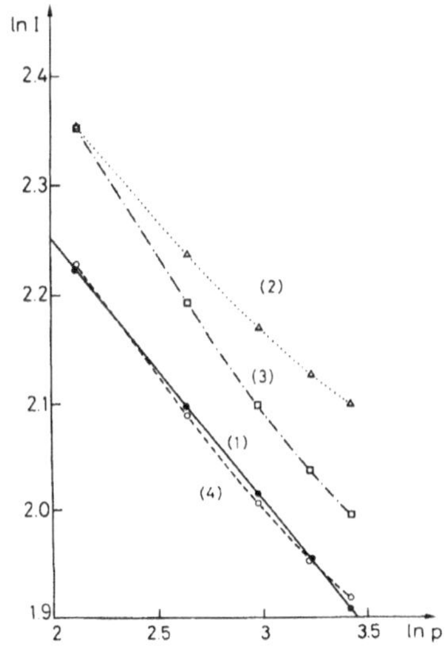

Figure 5. ℓnI vs ℓnp, where I is the ionization potential in eV, and p is the molecular perimeter in Å. Curve (1) shows experimental values; curve (2) Hückel values obtained with α=-7.5 eV and β=-3 eV; curve (3) Hückel values obtained by fitting Buenker et al.[36,37] data; curve (4) Hückel values obtained by fitting the experimental data.

$$\lim_{n\to\infty} \frac{E_\pi(n)}{n} \to \text{constant} \qquad (6.5)$$

We shall argue below that the constant in the relation (6.5) must therefore be one of the fundamental parameters to use in expressing the electronic structure of the linear polyacenes for finite (and often for small) n.

The third point to be made is that the chemical potential of the π-electrons, when we make a specific choice of reference energy from which to measure it, is closely related to the π-electron energy per particle, i.e. again to $E_\pi(n)/n$. The relation, we anticipate to be of the form

$$\lim_{n\to\infty} \mu_\pi(n) = \text{const} \lim_{n\to\infty} \frac{E_\pi(n)}{n} \qquad (6.6)$$

where the two-dimensional electron gas model yields the constant as precisely 2. We have tested this[38] by appeal to Hückel theory and we find a constant from that theory of 2.2, the way the limit is approached being shown in Figure 6.

6.4. Model Calculations for Energy/Particle and Chemical Potential

Returning to the two-dimensional electron gas relation, we make an assumption very close to the perimeter model of Platt.[39] This enables us to relate kinetic energy to the number of π-electrons N_π by

$$T_\pi = \text{constant } N_\pi^2 \qquad (6.7)$$

In a free electron theory therefore, recognizing that we are measuring one-electron energies from the lowest one-electron π-level, we evidently find (cf. Eq. (2.3))

$$\mu = \frac{dT_\pi}{dN_\pi} = 2\frac{T_\pi}{N_\pi} \qquad (6.8)$$

We merely amplify the comment on Figure 6 by saying that we have used Coulson's form[40] of the π-electron levels as a function of the length of the chain to construct this figure. It is gratifying that we find a number 2.2 so close to the prediction of the 2-dimensional electron gas model.

Finally, we return to the empirical connection between ionization potential I and molecular perimeter p. Hückel theory, inevi-

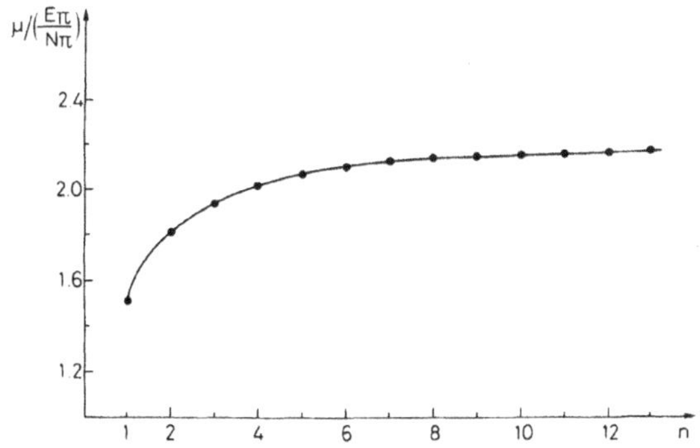

Figure 6. $\mu/(E_{\pi}/N_{\pi})$ vs the number of rings n from Hückel theory. This is calculated from the explicit Hückel formula for the one-electron levels

$$\varepsilon_K = \alpha + \bar{\varepsilon}_K \beta$$

where

$$\bar{\varepsilon}_K = \frac{1}{2}\left[-1+(9+8\cos\frac{n\pi}{n+1})^{1/2}\right]$$

where μ is referred to $\varepsilon_{K=0}$ as the zero of energy and E_{π}/N_{π} is then proportional to β, as is μ.

tably would relate I to the Hückel parameters α and β and the number of rings n.

Therefore, we have enquired whether it is possible, by appropriate choice of Hückel parameters, to fit the experiments. Figure 5 shows that a reasonable fit can be obtained by choosing α=-6.107 eV, β=-3.183 eV. However, it is our clear impression that the simplicity of the experimental plot is not reproduced by any choice of the pair (α,β). We interpret this to mean that electron correlation must be included explicitly in a final theory: an indication that the one-electron models we have appealed to in our work will eventually have to be transcended. This might be done, for example, in the framework of the Pariser-Parr-Pople treatment[41] of electron interactions, and would presumably introduce a further parameter (cf. the Hubbard U of solid state theory) into the treatment. One should not be surprised though if a proper treatment of the further electron localization by electron correlations simplifies the Hückel results, and restores:

(a) The linearity of the experimental $\ell n I$ vs $\ell n p$ plot and

(b) The constant in the Hückel μ-E_π/N relation back towards the two dimensional electron gas.

But these are problems for future work. However, a final point of interest leads us back to the Hartree-Fock atomic discussion in Section 5. There we emphasized that $\mu_{HF}=\varepsilon_{Koopmans}$. We thought it of interest therefore, since in this article we have related μ and the ground-state energy for atomic ions, to show such a relation also holds for the π-electrons in the linear polyacenes in Figure 7. The two curves are characterized by different choices of the Hückel parameters α and β, as given in the caption. In Hückel theory, the significance of the saturation property of the π-electrons, namely $E_\pi/N_\pi\to$ constant as $n\to\infty$, is that it defines $-\beta$, i.e. the Hückel resonance integral. Naturally, as already stressed, the quantity E_π/N_π must eventually be calculated in the presence of electron-electron interactions.

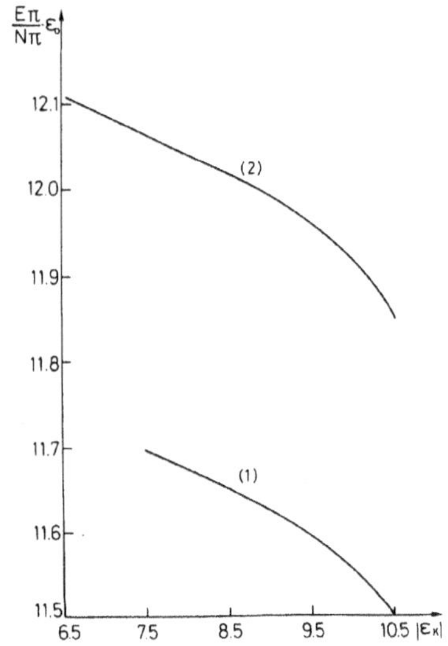

Figure 7. $(E_\pi/N_\pi)-\varepsilon_0$ against the Koopmans ionization potential $|\varepsilon_K|$ from Hückel theory. Curve (1) is obtained with α=-7.5 eV and β=-3 eV; curve (2) corresponds to α and β parameters obtained by fitting Buenker et al.[36,37] data for n=1 and 2.

7. SUMMARY AND CONCLUSIONS

In this paper, we have reviewed some consequences of the density functional theory of atoms and molecules, mainly, but not entirely, within the local density approximation.

We have stressed especially for atomic ions:

(a) The relation between chemical potential and ionization potential, involving though also a many-electron term from the potential energy of electron exchange and correlation.

(b) The simplification of (a) in Hartree-Fock theory, given by Alonso and March[7] that $\mu = \varepsilon_{Koopmans} \equiv \varepsilon_K$ where the latter quantity is evidently the one-electron eigenvalue of the highest occupied molecular orbital (HOMO). This is the precise meaning of ε_K in Hartree-Fock theory; it should not be identified with the (negative of the) ionization potential.

(c) The fact that the total energy can be characterized for atomic ions by the atomic number Z and the chemical potential. This has been demonstrated by direct use of the 1/Z expansion in low-order, and also when this expansion is summed to all orders in the statistical limit, $N \to \infty$, $Z \to \infty$, $N/Z \leq 1$, in a non-relativistic framework.

Some consequences of the density description for molecules have been pointed out, especial attention having been paid to π-electrons in some hydrocarbons. Attention have been drawn to a formally rigorous separation of $\rho(\vec{r})$ into a sum of π and σ contributions, and the potential interest in this for treating π-σ interaction. It has been argued that a fundamental quantity in the linear polyacenes is the energy per π-electron in the limit of an infinite chain. To obtain this energy precisely, of course, will necessitate inclusion of electron correlation.

Finally, we would comment, somewhat more generally, that density functional theory has been employed quite recently by March and Pucci[41] to treat dipole moments of some heteronuclear molecules, which in turn correlate with molecular ionization potentials.

Outstanding problems centre round the treatment of the electron correlation contribution to the chemical potential in both atoms and molecules. No doubt, such a contribution will involve electron affinity. But it may well also involve the ionization potential, and further exploration of these points remain of considerable interest in both density functional theory and quantum chemistry.

REFERENCES

1. L.H. Thomas, The Calculation of Atomic Fields, Proc.Camb.Phil. Soc. 23, 542 (1926).
2. E. Fermi, Un Metodo Statistico per la Determinazione di Alcune Proprietá dell'Atomo, Rend.Accad.Naz.Lincei. 6, 602 (1927).
3. E. Fermi, Eine statistische Methode zur Bestimmung einiger Eigenschaften des Atoms und ihre Anwendung auf die Theorie des periodischen Systems der Elemente, Zeits für Phys. 48, 73 (1928).
4. P.A.M. Dirac, Note on Exchange Phenomena in the Thomas-Fermi Atom, Proc. Cambridge Philos.Soc. 26, 376 (1930).
5. J.C. Slater, A Simplification of the Hartree-Fock Method, Phys. Rev. 81, 385 (1951).
6. W. Kohn and L.J. Sham, Self-Consistent Equations Including Exchange and Correlation Effects, Phys.Rev. 140, A1133 (1965).
7. J.A. Alonso and N.H. March, Relation between Chemical Potential and Ionization Potential in Hartree-Fock Theory, J.Chem. Phys. (in press).
8. J.F. Mucci and N.H. March, Dipole Moments, Molecular Ionization Potentials and Asymptotic Form of Electron Density, J.Chem. Phys. (in press).
9. P. Hohenberg and W. Kohn, Inhomogeneous Electron Gas, Phys.Rev. 136, B864 (1964).
10. N.H. March, "Self-Consistent Fields in Atoms", Pergamon Press, Oxford (1975).
11. N.H. March and R. Pucci, Chemical Potential Related to Total Energy in Isoelectronic Sequences of Positive Ions, J.Chem. Phys. (to be published).
12. E. Clementi, Tables of Atomic Functions, IBM J.Res.Develop. Suppl. 9, 2 (1965).
13. N.H. March and R.J. White, Non-Relativistic Theory of Atomic and Ionic Binding Energies for Large Atomic Number, J.Phys. B. 5, 466 (1972).
14. Y. Tal and M. Levy, Expectation Values of Atoms and Ions: The Thomas-Fermi Limit, Phys.Rev.A. 23, 408 (1981).
15. A.J. Thakkar and V.H. Smith, Form Factors and Total Scattering Intensities for the Helium-like Ions from Explicitly Correlated Wavefunctions, J.Phys.B. 11, 3803 (1978).
16. S. Fraga and G. Malli, "Many-electron Systems: Properties and Interactions," Saunders, Philadelphia (1968).
17. R. Pucci and N.H. March, Some Moments of Radial Electron Density in Closed-shell Atoms and their Atomic Scattering Factors, J.Chem.Phys. 76, 4089 (1982).
18. M. Hoffmann-Ostenhof and T. Hoffmann-Ostenhof, "Schrödinger Inequalities", and Asymptotic behavior of the Electron Density of Atoms and Molecules, Phys.Rev.A. 16, 1782 (1977).
19. J. Katriel and E.R. Davidson, Asymptotic Behavior of Atomic and Molecular Wave-Functions, Proc.Natl.Acad.Sci. USA 77, 4403 (1980).

20. R. Ahlrichs, M. Hoffmann-Ostenhof, T. Hoffmann-Ostenhof and J.D. Morgan III, Bounds on the Decay of Electron Densities with Screening, Phys.Rev.A. 23, 2106 (1981).
21. International Tables for X-ray Crystallography, Kynoch Birmingham (1962).
22. R. Pucci and N.H. March, Total Energy of Atomic Ions Related to Low-order Ionization Potentials and to Diamagnetic Susceptibility, J.Chem.Phys. (in press).
23. C.A. Moore, The Chemical Elements-Ionization Potentials, Natl. Bureau Stand.Circ. 647, 1, 34 (1958).
24. J.B. Mann, Atomic Structure Calculations. II. Hartree-Fock Wavefunctions and Radial Expectation Values: Hydrogen to Lawrencium, Los Alamas Sc.Lab.Rep. LA-3690, 1 (1968).
25. P. Cade, "First and Second Row Hydrides", Tech.Rep.Lab. of Mol.Struc. and Spectra, Chicago (1967).
26. N.H. March and R.F.W. Bader, Relation Between Chemical and Ionization Potentials in Atoms, Phys.Lett.A. 78, 242 (1980).
27. N.H. March and R. Pucci, Asymptotic Form of First-order Density Matrix for Atoms and Molecules, J.Chem.Phys. 75, 496 (1981).
28. Y. Tal and R.F.W. Bader, Studies of the Energy Density Functional Approach, I. Kinetic Energy, Intern.J.Quantum Chem. S12, 153 (1978).
29. R.G. Parr, R.A. Donnelly, M. Levy, and W.E. Palke, Electronegativity: The Density Functional Viewpoint, J.Chem.Phys. 68, 3801 (1978).
30. N.H. March, Electron Density Theory of Atoms and Molecules, J.Phys.Chem. 86, 2262 (1982).
31. K. Husimi, Density Matrix Theory, Proc.Phys.Math.Soc. Japan, 22, 264 (1940).
32. G.P. Lawes and N.H. March, Exact Local Density Method for Linear Harmonic Oscillator, J.Chem.Phys. 71, 1007 (1979).
33. R. Pucci and N.H. March, Kinetic Energy Density Functional of Hückel Theory for Planar Hydrocarbons, J.Chem.Phys. 74, 2936 (1981).
34. R. Boschi, E. Clar, and W. Schmidt, Photoelectron Spectra of Polynuclear Aromatics, J.Chem.Phys. 60, 4406 (1974).
35. J. Trotter, Planar Hydrocarbons, The R.I.C. Lectures Series, 2, 1 (1964).
36. R.J. Buenker, J.L. Whitten, and J.D. Petke, Ab Initio SCF MO and CI Calculations on the Electronic spectrum of Benzene, J.Chem.Phys. 49, 2261 (1968).
37. R.J. Buenker and S.D. Peyerimhoff, Ab Initio SCF Calculations for Azulene and Naphthalene, Chem.Phys.Lett. 3, 37 (1969).
38. R. Pucci and N.H. March, Chemical Potential Related to Energy per π-electron in Linear Polyacenes with Many Rings, (to be published).
39. J.R. Platt, The Chemical Bond, in: "Handbuch der Physik", Vol. 37, S. Flügge, ed., Springer-Verlag, Berlin (1961).
40. C.A. Coulson, Excited Electronic Levels in Conjugated Molecules, Proc.Phys.Soc. 60, 257 (1948).

41. R.G. Parr, "Quantum Theory of Molecular Electronic Structure", Benjamin, Reading (1972).

SOME ASPECTS OF THE ROLE OF SINGLE-PARTICLE DENSITY IN CHEMISTRY

B.M. Deb

Department of Chemistry and
Group of Theoretical Studies
Indian Institute of Technology
Bombay 400 076, India

1. INTRODUCTION

In recent years, several publications (see, e.g., Refs. 1-6) have emphasized the fundamental importance of single-particle (electron) density $\rho(\vec{r})$ in predictive and interpretive theories of chemistry. There are several reasons why this awareness about the electron density has tended to be more and more acute over the past two decades: (1) The sustained belief of a number of workers, strengthened by the Hohenberg-Kohn theorem in 1964, that physicochemical phenomena can perhaps be best understood only through their visualization and interpretation in three-dimensional (3-D) space, in terms of $\rho(\vec{r})$ and associated quantities. (2) The modeling of physicochemical phenomena in terms of quantities in 3-D space enables us to discourse in a "classical" language (see later) within a quantum mechanical framework. This yields new and significant physical insights[1-3,6] which are not readily extractable from the wave function (see also the comment of Bader et al. in para 2, p. 898, Ref. 6). (3) Formalisms which employ $\rho(\vec{r})$ and associated quantities often result in a drastic reduction of computational labour, maintaining similar, if not greater, accuracy as certain ab initio calculations.[7] All these raise the tantalizing possibility of an alternative, hopefully simpler and more practicable, quantum mechanics of atoms, molecules, and solids in terms of $\rho(\vec{r})$ and related quantities, which will bypass the wave function and the Schrödinger equation. Clearly, in order that such an alternative formulation of quantum mechanics of many-electron systems may stand on its own feet, there should be an accurate method for the <u>direct</u> calculation of electron density in many-electron systems, preferably through a single differential or integrodifferential equation.[8] Since this <u>final</u> objective of all density-based studies is obviously the most difficult, we shall defer its discussion until Section 2.7.

The schematic diagram below describes the plan of this article. The three principal aspects of electron density, viz. density functionals, electron fluids and property densities, as well as certain interconnections between them, will be discussed in Section 2. The implications of these aspects in structures, properties, and interactions of atoms, molecules, and solids form the subject matter of Section 3. There will, however, be a slight overlap between Sections 2 and 3.

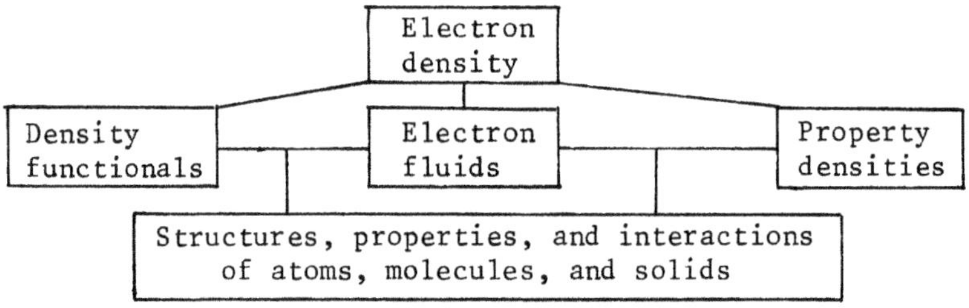

2. THREE ASPECTS OF ELECTRON DENSITY AND THEIR INTERCONNECTIONS

2.1. Internal Stresses in Molecules

Based on a suggestion by Feynman,[9] Deb and Bamzai[10,11] defined an electrostatic stress tensor $S_{\mu\nu}$ (consider a Cartesian framework) in the 3-D space of a molecule in terms of internal fields arising from its nuclear (ρ_N) and electronic (ρ) distributions. Here ν refers to the direction of local force and μ to the direction of the positive outward normal to the area on which the local force acts. Taking the divergence $S_{\mu\nu,\mu}$ as the net electrostatic force density, viz.

$$S_{\mu\nu,\mu} = \rho_N(E_\nu + F_\nu) + \rho(E_\nu + F_\nu) \tag{1}$$

where E_ν and F_ν are ν-th components of nuclear and electronic fields respectively, leads to a solution

$$S_{\mu\nu} = N_{\mu\nu} + L_{\mu\nu} + P_{\mu\nu} = \frac{1}{4\pi}\left(G_\mu G_\nu - \frac{1}{2}\delta_{\mu\nu}\vec{G}\cdot\vec{G}\right) \tag{2}$$

where

$$N_{\mu\nu} = \frac{1}{4\pi}\left(E_\mu E_\nu - \frac{1}{2}\delta_{\mu\nu}\vec{E}\cdot\vec{E}\right) \tag{3}$$

$$L_{\mu\nu} = \frac{1}{4\pi}\left(E_\mu F_\nu + E_\nu F_\mu - \delta_{\mu\nu}\vec{E}\cdot\vec{F}\right) \tag{4}$$

$$P_{\mu\nu} = \frac{1}{4\pi}\left(F_\mu F_\nu - \frac{1}{2}\delta_{\mu\nu}\vec{F}\cdot\vec{F}\right) \tag{5}$$

$$\vec{G} = \vec{E} + \vec{F} = - \nabla V_{mol} \tag{6}$$

V_{mol} being the molecular electrostatic potential.[12] Using the above expressions, one can devise a simple model for interaction between two many-electron systems, I and II, by writing the stress tensor for the whole interacting system[11] as

$$\begin{aligned} S_{\mu\nu}(I+II) &= S_{\mu\nu}(\text{interaction}) + S_{\mu\nu}(I) \\ &\quad + S_{\mu\nu}(II) + S_{\mu\nu}(I,II) \end{aligned} \tag{7}$$

where $S_{\mu\nu}(I,II)$ is a zero-order interaction term which persists even at large distances and

$$\begin{aligned} S_{\mu\nu}(\text{interaction}) &= \frac{1}{4\pi} [G(I)_{\mu} J_{\nu} + G(I)_{\nu} J_{\mu} - \delta_{\mu\nu} \vec{G}(I)\cdot\vec{J}] \\ &+ \frac{1}{4\pi} [G(I)_{\mu} J_{\nu} + G(II)_{\nu} J_{\mu} - \delta_{\mu\nu} \vec{G}(II)\cdot\vec{J}] \\ &+ \frac{1}{4\pi} [J_{\mu} J_{\nu} - \frac{1}{2} \delta_{\mu\nu} \vec{J}\cdot\vec{J}] \end{aligned} \tag{8}$$

Here

$$J_{\mu,\mu} = 4\pi\Delta\rho \tag{9}$$

$\Delta\rho$ being the extent of charge reorganization due to the interaction. Thus, the interaction between systems I and II can be studied by monitoring the local changes in total stress, electrostatic force density, and interaction stress [3 terms, see Eq. (8)] at specified points of interest. For example, in studying chemical binding in H_2 and H_2^+ by this approach,[11,13] a point on the internuclear axis in the antibinding region near a proton is of considerable interest (Fig. 1). In this case, the third term on the right-hand side in Eq. (8) and the total stress $S_{\mu\nu}(I+II)$ pass through extrema* at an internuclear distance (R) close to R_{eq}, whereas the $S_{\mu\nu}$(interaction) vanishes at an $R \simeq R_{eq}$. In this viewpoint, chemical binding in H_2 and H_2^+ occurs due to local variations of electrostatic pressure from point to point so that the total electrostatic force density or the difference force density or both vanish at certain points on the internuclear axis. The third term in $S_{\mu\nu}$(interaction) can clearly depict the consequences of a charge accumulation in the binding region and of a charge contraction toward the proton.

*See Refs. 11 and 13 for details. Due to inaccuracies in the wave functions employed, the locations of these extrema shift when the point of interest is moved.

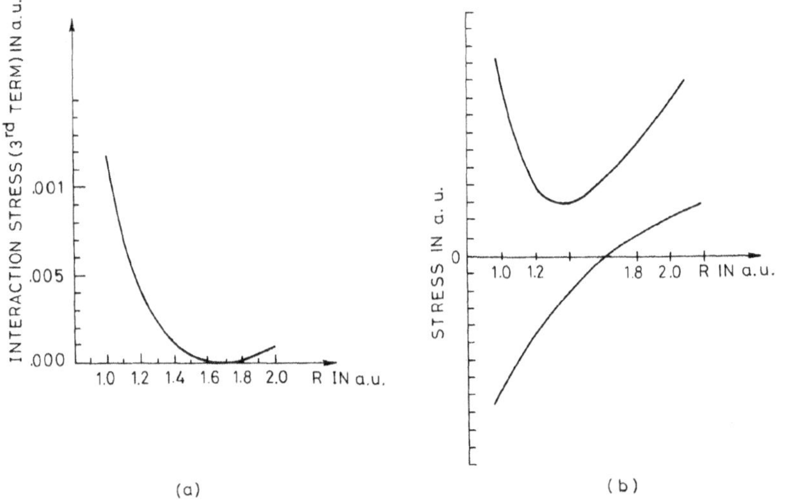

Figure 1. Variations of (a) the third term in the interaction stress, and (b) the total interaction stress (bottom curve) and the total stress (top curve) as functions of the internuclear distance R in the H_2 molecule, for a point 0.30 a.u. distant (on the internuclear axis) from the nearest proton, in the antibinding region. R = 0.8 a.u. at the origin. In (b) the ordinate and its one division are 4.60, 0.005 and 0.0, 0.02 a.u. for the top and bottom curves respectively (Reproduced from Ref. 2. Courtesy, Van Nostrand Reinhold Company, New York).

2.2. The Density Functional Connection and Entry of Quantum Fluid Dynamics

It can be readily argued that for a static stationary state (i.e., a state with null velocity field) of a many-electron system, such as those considered above for H_2 and H_2^+, the net local force density should vanish at all points (other than nuclear sites[14]) in 3-D space and not just at a few points on the internuclear axis. Using the local density functional theory (DFT) of Hohenberg, Kohn, and Sham,[15] one is therefore led to the definition of a comprehensive stress tensor[14] $U_{\mu\nu}$ which is more general than the electrostatic stress tensor. The net local force density is now given by

$$F_\nu = U_{\mu\nu,\mu} = \sum_\mu \frac{\partial}{\partial x^\mu} (U_{\mu\nu}) \tag{10}$$

where

$$U_{\mu\nu} = (T_{\mu\nu} + I_{\mu\nu}) + S_{\mu\nu} \quad (11)$$

$$T_{\mu\nu} = \frac{1}{2}\sum_i \phi_i \nabla_\mu \nabla_\nu \phi_i - \frac{1}{2}\sum_i (\nabla_\mu \phi_i)(\nabla_\nu \phi_i) \quad (12)$$

$$I_{\mu\nu} = p_{xc}\delta_{\mu\nu} \quad (13)$$

and $S_{\mu\nu}$ was given previously. Eq. (11) indicates that the comprehensive stress tensor consists of a quantum part (within brackets) and a classical or electrostatic part. The kinetic tensor $T_{\mu\nu}$ is constructed in terms of the Kohn-Sham (KS) orbitals $\{\phi_i\}$. The p_{xc} term in the exchange-correlation (xc) tensor can be written down explicitly by assuming a local density approximation.[14] It then turns out that, for static stationary states, within the local DFT

$$U_{\mu\nu,\mu} = 0 \quad (14)$$

This may be compared with the Euler or Navier-Stokes equation in fluid dynamics ($\overleftrightarrow{U}$ is a stress tensor), viz.

$$\nabla\cdot\overleftrightarrow{U} = \frac{d\vec{v}}{dt}$$

$$= 0, \quad \text{since } \vec{v} = 0 \quad (15)$$

Further, $U_{\mu\nu}$ can be rewritten in a classically analogous way, i.e.,

$$U_{\mu\nu} = -p\delta_{\mu\nu} + \sigma_{\mu\nu} + S_{\mu\nu} \quad (16)$$

where p can be interpreted as quantum pressure (related to p_{xc}) and $\sigma_{\mu\nu}$ as a viscous stress tensor which is expressible in terms of all the ρ_i's ($\rho_i = |\phi_i|^2$). Pushing this fluid-dynamical analogy further, we may consider[14] the dynamic viscosity (η), the kinematic viscosity (η/ρ) and another viscosity coefficient[16] (ξ), viz.

$$\eta_i = \rho_i/4, \quad \xi_i = \frac{2}{3}\rho_i \quad (17)$$

and the kinematic viscosity (= 1/4) is the same for each KS orbital.

There have been other interesting discussions on the significance of a quantum stress tensor based on its relation with the single-particle density or density matrix. Thus, Bartolotti and Parr [17] propose the relation

$$-\nabla\cdot\overleftrightarrow{\sigma} + \rho\nabla V_{mol} = 0 \quad (18)$$

where $\overleftrightarrow{\sigma}$ is a stress tensor and V_{mol} is the net classical electrostatic potential at a point in space, due to the nuclei and electrons. One also has the density-functional relation

$$\frac{\delta G[\rho]}{\delta\rho} + V_{mol} = \mu \quad (19)$$

reminiscent of a similar equation involving chemical potential in classical thermodynamics. Here the functional $G[\rho]$ consists of the kinetic and xc functionals. Bartolotti and Parr obtained a Gibbs-Duhem relation involving a scalar pressure p, where

$$p = -\frac{1}{3}\, \mathrm{tr}\overleftrightarrow{\sigma} \tag{20}$$

The integral of $\overleftrightarrow{\sigma}$ yields the xc energy plus two times the kinetic energy. This definition of $\overleftrightarrow{\sigma}$ is consistent with the statistical theory of atoms and is exact in the thermodynamic limit of $N = \infty$, N being the number of electrons.

Bader[18] has defined a quantum force as

$$\vec{F}(\vec{x}) = -\nabla\cdot\overleftrightarrow{P}(\vec{x}) \tag{21}$$

where the tensor $\overleftrightarrow{P}$ is a functional of the first-order density matrix. The integral of $\overleftrightarrow{P}$ gives the kinetic energy. Using this definition, Bader obtains a partitioning of the potentials and their associated forces acting within a quantum system, and discusses various aspects of the mechanics of an atom in a molecule.

2.3. Property Densities

The idea of a charge density, energy density or force density can be generalized in terms of a property density (PD) in 3-D space [19,20] Thus, if $p(\vec{r})$ is the PD corresponding to the property P, then

$$P = \int p(\vec{r})d\vec{r} \tag{22}$$

Here $p(\vec{r})$ may be a force density, energy density, charge density, current density, gradient density, dipole moment density (electric and magnetic), polarizability density, magnetizability density, nuclear magnetic shielding density (NMSD), etc. There is a considerable conceptual advantage in analyzing and interpreting a particular property and its physicochemical consequence through appropriate contour maps of its PD in 3-D space. Care has to be taken in this kind of analysis since the PD might have a gauge(origin) dependence, e.g.,

$$\text{Polarizability density:} \quad \overleftrightarrow{\alpha}(\vec{r})\cdot\vec{E} = \rho_1^{E}(\vec{r})\vec{r} \quad , \tag{23}$$

$$\text{Magnetizability density:} \quad \overleftrightarrow{\chi}(\vec{r})\cdot\vec{H} = \frac{1}{2c}\,\vec{r}\times\vec{j}_1^{\,H}(\vec{r}) \tag{24}$$

are both dependent on the choice of origin. In Eq. (23) ρ_1^E is the first-order correction to density in presence of a uniform electric field $\vec{E}$; in Eq. (24) $\vec{j}_1^{\,H}$ is the first-order correction to current density in presence of a uniform magnetic field $\vec{H}$. Jameson and Buckingham have chosen the NMSD which has no explicit origin de-

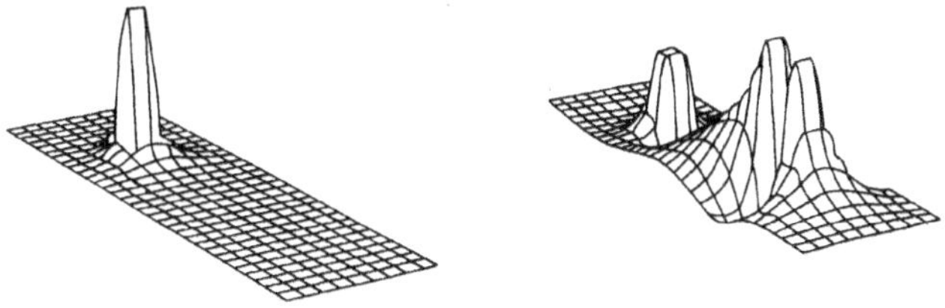

(a) (b)

Figure 2. Maps for isotropic nuclear magnetic shielding density difference for (a) ^{19}F, for HF molecule minus F^- ion, and (b) ^{1}H, for HF molecule minus H atom. The gauge origin is at the F nucleus (Reproduced from Ref. 19. Courtesy, the American Chemical Society).

pendence. The NMSD difference maps may be employed to interpret chemical shifts due to molecule formation, e.g., those in the HF molecule in relation to H atom and F^- ion. One such man is depicted in Fig. 2.

2.4. Many-Electron Quantum Fluid Dynamics in 3-D Space in Terms of Natural Orbitals

The fluid dynamical analogy, drawn through Eq. (15), has been a restricted one because the variable time did not appear there explicitly. Consequently, both the continuity equation and the Euler (or Navier-Stokes) equation of motion were trivially satisfied. However, while attempting to explicitly include time in the fluid dynamical framework, we should note that although the 3-D charge density obeys the continuity equation, it was not possible [21] to derive a fluid dynamical equation of motion in the 3-D space involving the net charge density and the net velocity field. Nevertheless, it is possible to derive the equations of motion through an orbital partitioning of the charge density, in terms of time-dependent Hartree-Fock (TDHF) orbitals,[22] natural orbitals[23,24] (NOs) and KS orbitals.[25] The last two types of orbitals will be discussed in this article.

Consider the NO Hamiltonian defined by Adams.[26] One can show[24] that in presence of external electric and magnetic fields this Hamiltonian becomes (taking Coulomb gauge, $\nabla\cdot\vec{A}=0$ and electronic charge as $-e$)*

*Except Section 2.4, we would employ atomic units throughout this article.

$$\hat{F} = \frac{1}{2m_e}[-\hbar^2\nabla^2 + \frac{2e\hbar}{ic}(\vec{A}\cdot\nabla) + \frac{e^2}{c^2}\vec{A}^2] - e\phi + v + V_{eff} \quad (25)$$

where $\phi(\vec{r},t)$ and $\vec{A}(\vec{r},t)$ are the usual scalar and vector potentials, v is the electron-nuclear potential and V_{eff} is an operator consisting of Coulomb, exchange and correlation parts plus a number operator. According to Adams,[26] if ϕ_k is a time-independent NO, then

$$\hat{F}\phi_k = \varepsilon_k\phi_k; \quad k = 1,\ldots,r; \quad r \geq N \quad (26)$$

and, according to Hirschfelder,[23]

$$\hat{F}\Phi_k = i\hbar\frac{\partial\Phi_k}{\partial t}; \quad k = 1,\ldots,r \quad (27)$$

where Φ_k is a time-dependent NO. This now permits explicit inclusion of time.[24] Writing Φ_k in the polar form

$$\Phi_k(\vec{r},t) = \rho_k(\vec{r},t)^{1/2}\exp[iS_k(\vec{r},t)/\hbar] \quad (28)$$

one obtains the continuity equation

$$\frac{\partial\rho_k}{\partial t} + \nabla\cdot(\rho_k\vec{v}_k) = 0 \quad (29)$$

where

$$\vec{v}_k = \frac{1}{m_e}(\nabla S_k + \frac{e}{c}\vec{A})$$

and an equation of motion,

$$\frac{\partial S_k}{\partial t} + \frac{1}{2m_e}\vec{v}_k^{\,2} - e\phi = -(U_k + V) \quad (30)$$

where the Bohm potential

$$U_k = -\frac{\hbar^2}{2m_e}\frac{\nabla^2\rho_k^{1/2}}{\rho_k^{1/2}} \quad (31)$$

and

$$V = v + V_{eff} \quad (32)$$

Using the relations

$$\vec{E} = -\nabla\phi - \frac{1}{c}\frac{\partial\vec{A}}{\partial t} \quad (33)$$

$$B = \nabla \times \vec{A} \tag{34}$$

$$\nabla(\nabla S_k + \frac{e}{c}\vec{A})^2 = 2[(\nabla S_k + \frac{e}{c}\vec{A})\cdot\nabla](\nabla S_k + \frac{e}{c}\vec{A}) + 2\frac{e}{c}(\nabla S_k + \frac{e}{c}\vec{A}) \times \vec{B} \tag{35}$$

$$\vec{J}_k = \rho_k \vec{v}_k \tag{36}$$

we are led to the Euler-type equation of motion,

$$m_e \frac{\partial \vec{J}_k}{\partial t} + m_e \nabla\cdot(\frac{\vec{J}_k\vec{J}_k}{\rho_k}) + e\rho_k\vec{E} + \frac{e}{c}\vec{J}_k \times \vec{B} = -\rho_k \nabla(U_k + V) \tag{37}$$

Since we have

$$\rho_k \nabla U_k = \nabla\cdot[-\frac{\hbar^2}{4m_e}\nabla(\nabla\rho_k) + \frac{h^2}{4m_e}\frac{(\nabla\rho_k)(\nabla\rho_k)}{\rho_k}] \tag{38}$$

Eq. (37) takes a simpler form

$$\frac{\partial J_k}{\partial t} = -\frac{e}{m_e}(\rho_k\vec{E} + \frac{\vec{J}_k \times \vec{B}}{c}) - \frac{1}{m_e}\rho_k \nabla V + \nabla\cdot\overleftrightarrow{T}_k \tag{39}$$

where the stress tensor $\overleftrightarrow{T}$ is given by

$$\overleftrightarrow{T}_k = (\frac{\hbar}{2m_e})^2\nabla(\nabla\rho_k) + \frac{1}{\rho_k}[\vec{J}_k\vec{J}_k - (\frac{\hbar}{2m_e})^2(\nabla\rho_k)(\nabla\rho_k)] \tag{40}$$

$$\begin{aligned} &= (\frac{\hbar}{2m_e})^2[(\nabla - \frac{e}{\hbar c}i\vec{A})\,\Phi_k^*(\nabla + \frac{e}{\hbar c}i\vec{A})\,\Phi_k \\ &+ (\nabla + \frac{e}{\hbar c}i\vec{A})\,\Phi_k^*(\nabla - \frac{e}{\hbar c}i\vec{A})\,\Phi_k^* \\ &- \Phi_k^*(\nabla + \frac{e}{\hbar c}i\vec{A})(\nabla + \frac{e}{\hbar c}i\vec{A})\,\Phi_k \\ &- (\nabla - \frac{e}{\hbar c}i\vec{A})(\nabla - \frac{e}{\hbar c}i\vec{A})\,\Phi_k^*\,\Phi_k] \end{aligned} \tag{41}$$

Note that since Eq. (29) is valid, one has

$$\frac{\partial\rho}{\partial t} + \nabla\cdot\vec{J} = 0 \tag{42}$$

where

$$\rho(\vec{r},t) = \sum_k \eta_k \rho_k(\vec{r},t); \quad k = 1,\ldots,N \tag{43}$$

$$\vec{J}(\vec{r},t) = \sum_k \eta_k \vec{J}_k(\vec{r},t); \quad k = 1,\ldots,N \tag{44}$$

η_k being the occupation number of the k-th NO.

The three terms on the right-hand side of Eq. (39) represent, respectively, the Lorentz force of the electric and magnetic fields, the body force and the tensorial force. The stress tensor in Eq. (40) can be interpreted as comprising the viscous stress tensor and the turbulence or Reynolds stress tensor. Thus, the N-electron Schrödinger fluid is viewed as consisting of individual fluid components, each corresponding to an NO and having its own charge density and current density. The force governing the motion of each fluid component consists of the Bohm force, the Coulomb force, the xc force, and the Lorentz force due to the external electric and magnetic fields. The local observables, viz. the net charge density and net current density, are obtained by summing over the NOs, with the occupation numbers as weight factors. But, the net velocity field cannot be so obtained. Also, even in the absence of a magnetic field, the net velocity field is <u>not</u> irrotational <u>although each fluid component moves irrotationally</u> in the 3-D space. The motion of the entire fluid or each fluid component can be followed via the Bohm trajectories or quantum-mechanical streamlines which result from integrating the velocity field.[23] As we shall see below, the irrotational character of each velocity component is destroyed by rotation of the nuclear framework of the system.

During rotations, the electrons in a molecule do not rigidly follow the nuclear framework but slip past the nuclei giving rise to a slippage current (see Ref. 27 and earlier references therein). This leads to paramagnetism even in diamagnetic closed-shell molecules because this rotational contribution to the magnetic field (see later) cannot be removed by a gauge transformation.

For a rotating molecule, the Hamiltonian in a molecule-fixed coordinate frame contains an additional Coriolis term

$$- \sum_{j=1}^{N} \vec{\Omega}\cdot\vec{r}_j \times \vec{p}_j \tag{45}$$

where $\vec{\Omega}$ is the angular velocity of the molecule. The Coriolis term can be incorporated into the NO Hamiltonian and the entire analysis can be carried through as before.[24] In a rotating frame (denoted by a prime) we have

$$\vec{v}'_k = \vec{v}_k - \vec{\Omega} \times \vec{r} \tag{46}$$

$$\vec{v}' = \vec{v} - \vec{\Omega} \times \vec{r} \tag{47}$$

$$\vec{j}' = \vec{j} - \rho(\vec{\Omega} \times \vec{r}) \tag{48}$$

The term $\vec{\Omega} \times \vec{r}$ has been called a <u>regular velocity field</u> since it does not have a discontinuity or singularity; it is incompressible but rotational. Takine $\vec{\Omega}$ as a constant vector, one can readily show that the continuity equation is still obeyed, viz.

$$\frac{\partial \rho_k}{\partial t} + \nabla \cdot (\rho_k \vec{v}_k') = 0 \tag{49}$$

$$\frac{\partial \rho}{\partial t} + \nabla \cdot (\rho \vec{v}') = 0 \tag{50}$$

McLachlan and Baker[28] have shown that molecular rotation with angular velocity $\vec{\Omega}$ is equivalent to a magnetic field with vector potential $\vec{A}$ provided

$$\frac{1}{m_e} \frac{e}{c} \vec{A} = \vec{\Omega} \times \vec{r} \tag{51}$$

i.e.,

$$\vec{\Omega} = \frac{1}{2} \frac{e}{m_e c} \vec{B} \tag{52}$$

As shown by Ghosh and Deb,[24] one can also incorporate the effect of electron spin into the above fluid dynamical treatment. In that case (double primed), the expression for each NO current density turns out to be

$$\vec{J}_k'' = \vec{J}_k - \rho_k (\vec{\Omega} \times \vec{r}) + \frac{c}{e} \nabla \times \mu_k \tag{53}$$

The last term in Eq. (53) is known as the spin magnetization moment [29], where

$$(\mu_k)_\ell = \frac{eh}{2m_e c} \Psi_k + \sigma_\ell \Psi_k \; ; \qquad \ell = 1,2,3 \tag{54}$$

σ_ℓ's being the Pauli spin matrices. Writing the spin eigenstate[29] Ψ_k as

$$\Psi_k = \binom{\Phi_k}{0} \tag{55}$$

one obtains the following expressions for the net quantities:

$$\mu_1 = \mu_2 = 0 \quad , \qquad _3 = \frac{eh}{2m_e c} \rho \tag{56}$$

$$\vec{J}'' = \vec{J} - \rho(\vec{\Omega} \times \vec{r}) + \frac{c}{e} \nabla \times \vec{\mu} \tag{57}$$

J" still obeys the continuity equation.

Thus, we see that for a many-electron system the fluid dynamical equations can be brought to the 3-D space in terms of the NO partitioning of the electron density. This interpretation therefore corresponds to an MCSCF wave function. However, for a general many-electron system, the calculation of NOs is a difficult task. Accordingly, we return to the DFT, which is computationally more tractable, and show that the physical content of the above interpretation can be retained in terms of TDKS orbitals. The manner in which this is done establishes a closer link between DFT and quantum fluid dynamics (QFD).

2.5. A Time-Dependent DFT and Its QFD Analog

As a first step, Deb and Ghosh[25] proposed a TDDFT for oscillating external potentials. Let a harmonic electric perturbation $v^*(\vec{r},t)$ be impressed on the static external potential in an N-electron system so that the net external potential is given by

$$v(\vec{r},t) = v_0(\vec{r}) + v^*(\vec{r},t) \tag{58}$$

where

$$v^*(\vec{r},t) = f(\vec{r})[\exp(i\omega t) + \exp(-i\omega t)] \tag{59}$$

$f(\vec{r})$ being a known function. Now, consider only steady or quasi-periodic[30] solutions to the time-dependent Schrödinger equation

$$[H(\vec{R},t) - i\frac{\partial}{\partial t}]\Psi(\vec{R},t) = 0 \tag{60}$$

where H is periodic in time, $\vec{R}$ denotes the set of space and spin coordinates of the system and the operator in square brackets is the steady-state Hamiltonian $\mathcal{H}$ for the system. One can then write

$$\Psi(\vec{R},t) = \Phi(\vec{R},t)\exp(i\varepsilon t) \tag{61}$$

$$\mathcal{H}(\vec{R},t)\Phi(\vec{R},t) = \varepsilon\Phi(R,t) \tag{62}$$

where $\Phi(\vec{R},t)$ is also periodic in time and ε is its quasienergy (real number). The advantage of quasiperiodic wave functions is that the Φ functions obey a time-dependent variation principle corresponding to an energy minimum, where the integration is performed over the whole configuration space and over a time period. One can then prove[25] the Hohenberg-Kohn theorem for the nondegenerate quasiperiodic ground state of the N-electron system. Thus $v(\vec{r},t)$, $H(\vec{R},t)$, $\mathcal{H}(\vec{R},t)$, $\Phi(R,t)$, ε, $E(t) = \langle\phi|H|\phi\rangle$, and any ground-state property are all unique functionals of $\rho(\vec{r},t)$.

With this background, one can derive variationally the TD analog of KS one-particle equation by employing a fluid-dynamical Lagrangian[25]. The Lagrangian density is defined as

$$L(\rho_j,\vec{u}_j) = \frac{1}{2}\sum_{j=1}^{N}\rho_j\vec{u}_j^{\,2} - h[\rho] \tag{63}$$

where ρ_j is the density and $\vec{u}_j$ is the velocity field associated with the j-th orbital. The TD energy of the system may be written as

$$E = \frac{1}{2}\sum_{j=1}^{N}\int\rho_j(\vec{r},t)\vec{u}_j^{\,2}(\vec{r},t)d\vec{r} + E_{int} + Q \tag{64}$$

where the first term on the right-hand side is the macroscopic kinetic energy of the electron fluid, the second term (called intrinsic energy) is the familiar energy density functional and Q is a correction term which may be neglected for small perturbations. We have

$$\rho(\vec{r},t) = \sum_{j=1}^{N}\rho_j(\vec{r},t) \tag{65}$$

$$\rho(\vec{r},t)\vec{u}(\vec{r},t) = \sum_{j=1}^{N}\rho_j(\vec{r},t)\vec{u}_j(\vec{r},t) \tag{66}$$

$$E_{int} = \int h\,[\rho]\,d\vec{r} \tag{67}$$

$$h[\rho] = G[\rho(\vec{r},t)] + v(\vec{r},t)\rho(\vec{r},t)$$

$$+\frac{1}{2}\int\frac{\rho(\vec{r},t)\ (\vec{r}',t)}{|\vec{r}-\vec{r}'|}d\vec{r}' \tag{68}$$

Now, if one (a) minimizes the action integral

$$I = \int dt\int L(\rho_j,\vec{u}_j)d\vec{r} \tag{69}$$

with respect to variations in ρ_j's and $\vec{u}_j$'s, and (b) uses the equation of continuity as a constraint, viz.

$$\frac{\partial\rho_j}{\partial t} + \nabla\cdot(\rho_j\vec{u}_j) = 0\ ;\ j = 1,\ldots,N \tag{70}$$

i.e.

$$\frac{\partial\rho}{\partial t} + \nabla\cdot(\rho\vec{u}) = 0 \tag{71}$$

and (c) replaces $\delta T_s/\delta\rho_j$ by t_j/ρ_j where $t_j = \Psi_j(-\frac{1}{2}\nabla^2)\Psi_j$, then one finally obtains the TDKS equation[25]

$$[-\frac{1}{2}\nabla^2 + V_{eff}]\Psi_j = i\frac{\partial\Psi_j}{\partial t}\ ; \quad j = 1,\ldots N \tag{72}$$

An advantage of this formulation is that minimization is done with respect to quantities which are related separately to the amplitude and phase parts of a wave function.

The QFD interpretation of the TDKA equation may be obtained by writing V_{eff} as

$$V_{eff} = v_{cou} + v_{xc} \tag{73}$$

Then, Eq. (72) leads to the Navier-Stokes equation

$$\frac{\partial}{\partial t}(\rho_j\vec{u}_j) + \nabla\cdot(\rho_j\vec{u}_j\vec{u}_j) = -\rho_j[\nabla W_j + \nabla v_{cou} + \nabla v_{xc}] \tag{74}$$

where the Bohm potential

$$W_j = -\frac{1}{2}\nabla^2|\Psi_j|/|\Psi_j| \tag{75}$$

Using the equation of continuity, Eq. (74) leads to the more compact equation

$$\frac{\partial}{\partial t}(\rho\vec{u}) = \nabla\cdot\overleftrightarrow{U} - \rho\nabla v_{cou} \tag{76}$$

where the stress tensor $\overleftrightarrow{U}$ is given by

$$\overleftrightarrow{U} = -\rho\vec{u}\vec{u} - \sum_{j=1}^{N}\rho_j(\vec{u}_j - \vec{u})(\vec{u}_j - \vec{u}) + \frac{1}{4}\nabla^2\rho\overleftrightarrow{1} - \frac{1}{4}\sum_{j=1}^{N}\frac{(\nabla\rho_j)(\nabla\rho_j)}{\rho_j} + \overleftrightarrow{I} \tag{77}$$

In Eq. (77), the first two terms on the right-hand side arise from the kinetic contribution of the electron fluid and vanish for the static case. The second term, the thermal stress tensor, also vanishes under the assumption of irrotational velocity field. The third and fourth terms represent the Bohm stress tensor while the last $(\overleftrightarrow{I})$ is the xc stress tensor; $\overleftrightarrow{I}$ is the unit dyad. Note that, as in Section 2.4, the mean velocity field $\vec{u}$ is rotational while the component velocity field $\vec{u}_j$ is irrotational.

A similar TDDFT has been proposed by Bartolotti[31]. Using the Dirac-Frenkel variation principle[32] it is possible to derive a QFD variation principle for a TD (periodic) external potential. This last principle is employed to obtain a TD energy density functional which is then minimized with respect to both the amplitude and the phase of the total TD wave function.

As a further extension of their work, Ghosh and Deb[33] have developed a QFD analogy to the relativistic DFT of McDonald and Vosko[34]. The treatment follows the hydrodynamic interpretation of the Dirac equation by Takabayashi[35]. All these considerations are likely to help future practical applications of QFD to many-electron systems (see, e.g., Refs. 36-38).

2.6. A DFT Variation-Perturbation Method for Frequency-Dependent 2^L-Pole Polarizability

The TDDFT and its QFD analog, summarized in Section 2.5, may be used[39] to investigate dynamic (frequency-dependent) polarizabilities of many-electron systems. These quantities are required for describing various optical properties (linear and non-linear) of matter and for long-range interactions between atoms and molecules (see also Section 3.7).

2.6.1. Formalism. The polarizability of a many-electron system is its linear response to an oscillating electric field and can be defined through the induced moment. In the general case of a 2^L-pole moment, the external potential may be written as

$$\begin{aligned} v_{ex}^{L}(\vec{r},t) &= r^L P_L(\cos\theta)\, F_L^o(e^{i\omega t} + e^{-i\omega t}) \\ &= V_L(\vec{r}) F_L^o (e^{i\omega t} + e^{-i\omega t}) \end{aligned} \tag{78}$$

where the term multiplying $V_L(r)$ in Eq. (78) is the oscillating electric field*. Consider now the equations

* For dipole polarizability F_L^o is the field, for quadrupole polarizability it is the field gradient, and so on.

$$\text{Unperturbed}: \quad h_o \Psi_j^o = i\frac{\partial \Psi_j^o}{\partial t} \tag{79}$$

$$\text{Perturbed}: \quad h\,\Psi_j = i\frac{\partial \Psi_j}{\partial t} \tag{80}$$

where

$$h_o = -\frac{1}{2}\nabla^2 + v_N(\vec{r}) + v_{eff}^o(\vec{r}) \tag{81}$$

$$\begin{aligned} h &= h_o + [v_{eff}(\vec{r},t) - v_{eff}^o(\vec{r})] + v_{ex}(\vec{r},t) \\ &= h_o + \nabla v_{eff}(\vec{r},t) + v_{ex}(\vec{r},t) \end{aligned} \tag{82}$$

$v_{ex}(\vec{r},t)$ being the perturbing potential.

The perturbed function may be expanded as

$$\Psi_j(\vec{r},t) = \Psi_j^o(\vec{r},t) + \sum_{n=1}^{\infty} \Psi_j^{(n)}(\vec{r},t)(F_L^o)^n \tag{83}$$

where $\Psi_j^{(n)}$ is the n-th order correction. Substituting Eq. (83) into Eq. (80) and gathering only the first-order terms (following Karplus and Kolker[40]) we have

$$(h_o - i\frac{\partial}{\partial t})\Psi_j^{(1)} + [\Delta v_{eff}^{(1)}(\vec{r},t) + \bar{V}_L(\vec{r},t)]\Psi_j^o = 0 \tag{84}$$

where

$$\bar{V}_L(\vec{r},t) = V_L(\vec{r})[e^{i\omega t} + e^{-i\omega t}] \tag{85}$$

$$\Delta v_{eff}(\vec{r},t) = \Delta v_{eff}^{(1)}(\vec{r},t)F_L^o \tag{86}$$

We also have

$$\rho(\vec{r},t) = \rho^o(\vec{r}) + \delta\rho(\vec{r},t)F_L^o \tag{87}$$

to first order, where

$$\delta\rho(\vec{r},t) = 2\mathrm{Re}\left\{\sum_j \Psi_j^{o*}(\vec{r},t)\Psi_j^{(1)}(\vec{r},t)\right\} \tag{88}$$

and

$$\Delta v_{eff}^{(1)}(\vec{r},t) = \int \frac{\delta\rho(\vec{r}',t)}{|\vec{r} - \vec{r}'|} d\vec{r}' + \delta\rho(\vec{r},t)\frac{\delta v_{xc}}{\delta\rho}\bigg|_{\rho=\rho_0} \tag{89}$$

The last term in Eq. (89) can be derived explicitly[39] within the local density approximation. The time-dependence of $\Psi_j^{(1)}(\vec{r},t)$ and $\Delta v_{eff}^{(1)}(\vec{r},t)$ may be written as

$$\Psi_j^{(1)}(\vec{r},t) = \chi_j^{(1)}(\vec{r},\omega)e^{i(\omega-\omega_j^0)t} + \chi_j^{(1)}(\vec{r},-\omega)e^{-i(\omega+\omega_j^0)t} \tag{90}$$

$$\Delta v_{eff}^{(1)}(\vec{r},t) = \overline{\Delta v}_{eff}^{(1)}(\vec{r},\omega)e^{i\omega t} + \overline{\Delta v}_{eff}^{(1)}(\vec{r},-\omega)e^{-i\omega t} \tag{91}$$

where $\omega_j^0 = \varepsilon_j^0$, in a.u., and

$$\overline{\Delta v}_{eff}^{(1)}(\vec{r},\pm\omega) = \int \frac{\overline{\delta\rho}(\vec{r}',\pm\omega)}{|\vec{r} - \vec{r}'|} d\vec{r}' + \overline{\delta\rho}(\vec{r},\pm\omega)\frac{\delta v_{xc}}{\delta\rho}\bigg|_{\rho=\rho_0} \tag{92}$$

$$\delta\rho(\vec{r},t) = \overline{\delta\rho}(\vec{r},\omega)e^{i\omega t} + \overline{\delta\rho}(\vec{r},-\omega)e^{-i\omega t} \tag{93}$$

Use of Eqs. (90) and (91) in Eq. (84) leads to a pair of coupled time-independent equations

$$(h_0 - \varepsilon_j^0 + \omega)\chi_j^{(1)}(\vec{r},\omega) + [\overline{\Delta v}_{eff}^{(1)}(\vec{r},\omega) + V_L(\vec{r})]\phi_j^0 = 0 \tag{94}$$

$$(h_0 - \varepsilon_j^0 - \omega)\chi_j^{(1)}(\vec{r},-\omega) + [\overline{\Delta v}_{eff}^{(1)}(\vec{r},-\omega) + V_L(\vec{r})]\phi_j^0 = 0 \tag{95}$$

$j = 1,\ldots,N$

Once the functions $\chi_j^{(1)}(\vec{r},\pm\omega) \equiv \chi_{j\pm}^{(1)}$ have been determined by solving Eqs. (94) and (95), the frequency-dependent 2^L-pole polarization is given by

$$\alpha_{2^L}(\omega) = \mathrm{Re}\sum_{j=1}^{N}\left\{\langle\chi_{j+}^{(1)}|V_L|\phi_j^0\rangle + \langle\chi_{j-}^{(1)}|V_L|\phi_j^0\rangle\right\} \tag{96}$$

2.6.2. Variational solution. Eqs. (94) and (95) are identical to the Euler equations

$$\delta L_j(\pm\omega) = 0 \tag{97}$$

where the functionals $L_j(\pm\omega)$ are given by

$$L_j(\pm\omega) = \langle\chi_{j\pm}^{(1)}|h_o-\varepsilon_j^o\pm\omega|\chi_{j\pm}^{(1)}\rangle$$
$$+ 2\mathrm{Re}\langle\chi_{j\pm}^{(1)}|\overline{\Delta v}_{eff}^{(1)}(\vec{r},\pm\omega) + V_L(\vec{r})|\phi_j^o\rangle \qquad (98)$$

Conditions can be defined[39] such that the extremum of L_j will correspond to a minimum. However, instead of minimizing each L_j separately it is better to minimize their sum, viz.

$$L(\pm\omega) = \sum_{j=1}^{N} L_j(\pm\omega)$$
$$= \sum_{j=1}^{N} [\langle\chi_{j\pm}^{(1)}|h_o-\varepsilon_j^o\pm\omega|\chi_{j\pm}^{(1)}\rangle$$
$$+ 2\mathrm{Re}\langle\chi_{j\pm}^{(1)}|\overline{\Delta v}_{eff}^{(1)}(\vec{r},\pm\omega) + V_L(\vec{r})|\phi_j^o\rangle] \qquad (99)$$

This amounts to

$$\delta L[\tilde{\chi}_{j\pm}^{(1)}] = 0 \qquad (100)$$

where the tilde denotes a trial function, subject to the condition

$$\langle\phi_i^o|\chi_{j\pm}^{(1)}\rangle + \langle\chi_{i\pm}^{(1)}|\phi_j^o\rangle = 0 \; ; \; i,j = 1,\ldots,N \qquad (101)$$

Now, let the trial functions be given by*

$$\tilde{\chi}_{j\pm}^{(1)} = f_{j\pm}\phi_j^o \qquad (102)$$

so that the functional L becomes

$$L(\pm\omega) = \sum_{j=1}^{N} [\langle\phi_j^o|\tfrac{1}{2}\nabla f_{j\pm}\cdot\nabla f_{j\pm}\pm\omega(f_{j\pm})^2|\phi_j^o\rangle$$
$$+ 2\mathrm{Re}\langle f_{j\pm}\phi_j^o|\overline{\Delta v}_{eff}^{(1)}(\pm\omega) + V_L(\vec{r})|\phi_j^o\rangle] \qquad (103)$$

*For a discussion of this form, vis-a-vis the presence of nodes, see Ref. 39.

If one writes

$$f_{j\pm} = r^L P_L(\cos\theta) \sum_{k=1}^{M} C_{k\pm}^{(j)} r^{k-1} \qquad (104)$$

where the unknown coefficients $C_{k\pm}^{(j)}$ are frequency-dependent, then Eq. (104) leads to the set of linear equations,

$$\sum_{\ell=1}^{M} A_{k\ell\pm}^{(j)} C_{\ell\pm}^{(j)} = -B_k^{(j)} \quad ; \quad k = 1,\ldots,M \qquad (105)$$

where the A's and B's are linear combinations of certain matrix elements in terms of ϕ_j^o. The 2^L-pole polarizability is now given by

$$\alpha_{2^L}(\omega) = \mathrm{Re} \sum_{j=1}^{N} [\langle\phi_j^o|f_{j+}V_L|\phi_j^o\rangle + \langle\phi_j^o|f_{j-}V_L|\phi_j\rangle] \qquad (106)$$

2.6.3. Calculation of dynamic dipole polarizability of helium atom. For He, one takes

$$f_{\pm} = r\cos\theta \sum_{k=1}^{6} C_{k\pm} r^{k-1} \qquad (107)$$

Using the calculated values of $C_{k\pm}$, $\alpha_2(\omega)$ has been calculated in the frequency range $0.0<\omega<+.65$ a.u. The results, shown in Table 1, are quite satisfactory. Note that our result at zero frequency agrees better with the reference value than other DFT values[41]. This is just a consequence of the fact that our result is a variational one and should therefore be smaller in magnitude than a value obtained by numerical solution of Eqs. (94) and (95) at zero frequency, as done by previous workers. Note, however, that this success with He atom does not necessarily imply that a trial function like that in Eq. (104) would also be successful with heavier noble gas atoms, e.g., Xe, for which a very significant contribution to polarizability comes from the continuum states[42], and these would be poorly described by a function like that in Eq. (104). For such systems, the choice of trial functions is a rather delicate one, requiring a combination of both adequate variational flexibility and requisite continuum character [for calculations of $\alpha_2(\omega)$ for Ne, Ar, Kr, and Xe, see Ref. 43].

2.6.4. A variation-perturbation method to calculate $\alpha_{2^L}(\omega)$ based on a QFD analogy to DFT. More than ten years ago, it was pointed out by Weiner and Askar[44] that in polarizability studies numerical schemes based on QFD appear to be more effective

Table 1. Dynamic Dipole Polarizabilities of He (in a.u.)

Frequency ω	$\alpha_2(\omega)$ DFT-XC[a]	$\alpha_2(\omega)$ DFT-X[a]	$\alpha_2(\omega)$ HF-LXC[a]	$\alpha_2(\omega)$[b]
0.00	1.531 (1.63)[c]	1.597 (1.77)[c]	1.260	1.384
0.10	1.551	1.618	1.273	1.399
0.20	1.615	1.688	1.313	1.448
0.30	1.739	1.824	1.389	1.541
0.40	1.959	2.068	1.516	1.698
0.50	2.377	2.542	1.732	1.971
0.60	3.356	3.718	2.146	2.509

[a] XC : Includes both exchange and correlation (LDA).
X : Includes only exchange.
HF-LXC : A Hartree-Fock-Type calculation using local exchange-correlation terms

[b] Reference values from K.T. Chung, Phys. Rev. 166 : 1 (1968).

[c] Other DFT values from Ref. 41.

than the usual studies. Accordingly, Ghosh and Deb[39] suggested the following approach in terms of QFD in 3-D space (for some earlier works which involved QFD in configuration space, see Refs. 45 and 46).

If one substitutes in Eq. (80) the following polar form for the KS perturbed orbital

$$\Psi_j(\vec{r},t) = R_j(\vec{r},t)\exp[iS_j(\vec{r},t)] \tag{108}$$

then one obtains the equations

$$hR_j + \frac{1}{2}R_j(\nabla S_j)^2 + R_j\frac{\partial S_j}{\partial t} = 0 \tag{109}$$

$$\nabla R_j\cdot\nabla S_j + \frac{1}{2}R_j\nabla^2 S_j = -\frac{\partial R_j}{\partial t}$$

i.e.,

$$\frac{\partial \rho_j}{\partial t} + \nabla\cdot(\rho_j \vec{u}_j) = 0 \tag{70}$$

where $\rho_j = |R_j|^2$, $\vec{v}_j = \nabla S_j$. Using the expressions*

$$R_j = R_j^o + R_j^{(1)} F_L^o + \ldots \tag{110}$$

$$S_j = S_j^o + S_j^{(1)} F_L^o + \ldots \tag{111}$$

in Eqs. (109) and (70) leads to the first-order equations

$$(h_o + \frac{\partial S_j}{\partial t} + \frac{1}{2}\nabla S_j^o\cdot\nabla S_j^o)R_j^{(1)} + (\nabla S_j^o\cdot\nabla S_j^{(1)} + \frac{\partial S_j^{(1)}}{\partial t} + \Delta v_{eff}^{(1)} + \bar{V}_L)R_j^o = 0 \tag{112}$$

$$2\frac{\partial}{\partial t}(R_j^o R_j^{(1)}) + \nabla\cdot[\rho_j^o \nabla S_j^{(1)} + 2R_j^o R_j^{(1)} \nabla S_j^o] = 0 \tag{113}$$

The Lagrangian functional whose Euler-Lagrange equations are Eqs. (112) and (113), corresponding to variations in $R_j^{(1)}$ and $S_j^{(1)}$ respectively, is

$$L_j = \langle\langle R_j^{(1)}|h_o + \frac{\partial S_j^o}{\partial t} + \frac{1}{2}\nabla S_j^o\cdot\nabla S_j^o|R_j^{(1)}\rangle\rangle + 2\langle\langle R_j^{(1)}|\nabla S_j^o\cdot\nabla S_j^{(1)} + \frac{\partial S_j^{(1)}}{\partial t} + \Delta v_{eff}^{(1)} + \bar{V}_L|R_j^o\rangle\rangle + \frac{1}{2}\langle\langle R_j^o|\nabla S_j^{(1)}\cdot\nabla S_j^{(1)}|R_j^o\rangle\rangle \tag{114}$$

---

* R_j^o and S_j^o occur in the expression

$$\Psi_j^o(\vec{r},t) = R_j^o(\vec{r})\exp[iS_j^o(\vec{r},t)]$$

where << >> denotes integration over both space and time. The perturbation, Eq. (77), is written as

$$v_{ex}^{L}(\vec{r},t) = 2V_L(\vec{r})F_L^{o}\cos\omega t$$
$$= \bar{V}_L(\vec{r})F_L^{o} \qquad (115)$$

Now, one can express the time-dependence of $R_j^{(1)}$ and $S_j^{(1)}$ as

$$R_j^{(1)}(\vec{r},t) = r_j^{(1)}(\vec{r},\omega)\cos\omega t \qquad (116)$$

$$S_j^{(1)}(\vec{r},t) = \frac{1}{\omega}s_j^{(1)}(\vec{r},\omega)\sin\omega t \qquad (117)$$

Substitution of Eqs. (116) and (117) into Eq. (114) and integrating over a time period give a functional

$$\begin{aligned} L_j &= \langle r_j^{(1)}|h_o-\varepsilon_j^{o} + \tfrac{1}{2}\nabla s_j^{o}\cdot\nabla s_j^{o}|r_j^{(1)}\rangle \\ &+ 2\langle r_j^{(1)}|s_j^{(1)} + \Delta v'_{eff} + 2V_L|R_j^{o}\rangle \\ &+ \frac{1}{2\omega^2}\langle R_j^{o}|\nabla s_j^{(1)}\nabla\cdot s_j^{(1)}|R_j^{o}\rangle \end{aligned} \qquad (118)$$

where

$$v_{eff}^{(1)}(\vec{r},t) = v'_{eff}(\vec{r},\omega)\cos\omega t \qquad (119)$$

By minimizing the functional L_j or $L(\equiv\Sigma L_j)$, the functions $r_j^{(1)}$ and $s_j^{(1)}$ can be determined. The polarizability is given by

$$\alpha_{2^L}(\omega) = 2\sum_{j=1}^{N}\langle r_j^{(1)}|V_L(\vec{r})|R_j^{o}\rangle \qquad (120)$$

The functions $\chi_{j\pm}^{(1)}$ and $r_j^{(1)}$, $s_j^{(1)}$ are related by[46]

$$\chi_{j\pm}^{(1)} = r_j^{(1)} \pm \frac{1}{\omega}s_j^{(1)}\phi_j^{o} \qquad (121)$$

Note that the QFD treatment deals with two functions, $r_j^{(1)}$ and $s_j^{(1)}$, instead of one function $\chi_j^{(1)}$.

2.7. A Differential Density Equation for Many-Electron Systems

2.7.1. General discussion. At the beginning of this article we had said that the final objective of all density-based studies is an equation for the direct calculation of electron density. The Thomas-Fermi (TF) method represents the earliest attempt in this direction and has been a cornerstone in practically all other attempts to calculate the density, bypassing the wave function. In the method described in this section, the TF terms will again play an important role (for the well-known limitations in the TF density and energy, see Refs. 1, 47, and 48).

In recent years, there have been several attempts to tackle this problem (see Ref. 8 for a review). For example, for closed-shell atoms (1-D case) Lawes and March[49] have proposed the following third order differential equation within a local density approximation:

$$\frac{1}{8}\frac{d^3}{dr^3}(r^2\rho) + \frac{1}{2}r^2\rho\frac{dv}{dr} + (\mu - v)\frac{d}{dr}(r^2\rho)$$

$$= \sum_{\ell} \frac{\ell(\ell+1)(2\ell+1)}{2r^2}\left[r\rho_\ell + \frac{d}{dr}(r^2\rho_\ell)\right] \tag{122}$$

where

$$\rho(r) = \sum_{\ell}(2\ell+1)\rho_\ell(r) \tag{123}$$

and $v(r)$ is the one-body external potential which contains the effects of exchange and correlation through $\rho(r)$. The third-order term is expected to be rather significant for light atoms but its importance might diminish with much heavier atoms. For molecules Lawes and March suggested a second-order linear differential equation whose solution is the first step toward obtaining an approximate density.

So far, attempts to obtain and solve differential equations for $\rho(\vec{r})$ seem to be complicated by non-linearity and/or higher than second-order character of the equations. Consequently, there have been attempts[50] to use trial densities incorporating certain parameters which can be adjusted to make the following energy functional a minimum:

$$E[\rho] = T_0[\rho] + T_2[\rho] + T_4[\rho] + V_{ne}[\rho] + J[\rho] - K[\rho] \tag{124}$$

where T_0 is the kinetic energy of a free electron gas, T_2 is 1/9-th Weizsächer correction and T_4 is a fourth-order correction to the kinetic energy; V_{ne} denotes electron-nuclear attraction; J is electron cloud self-interaction and K is the exchange energy of a free electron gas. The original TF model includes only T_o, V_{ne}, and J whereas the TFD (D for Dirac) model also includes -K. Gázquez and Parr[50] suggested a trial atomic density (without the shell structure)

$$\rho(r) = \frac{C}{(1+\beta r)^n}, \quad \beta = \alpha/n \tag{125}$$

$$\rho(r) \to Ce^{-\alpha r}, \quad \text{as } n \to \infty \tag{126}$$

where C is a normalization constant and α, n are variational parameters. For first-row atoms, this simple approach gives energies within 3% error. But, for Ar, Ni, and Kr the error in energy is large, being 26.7, 105.6 and 235 Hartrees respectively. It seems that the accuracy in energy obtainable with such simple trial densities will not be high and unless one employs a set of piecewise continuous (e.g., exponential) functions[51] or a linear combination of functions[52-54], the shell structure in atomic radial density may not be obtained.

In an interesting work, Hall and Martin[52] suggested the calculation of approximate electron density from a linear combination of expansion functions by minimizing the total energy of the error field (the difference between true and approximate electric fields). This approach is closely related to the Dirichlet minimum-energy theorem in electrostatics and provides a good approximation to the total Coulomb energy of the electrons. Another interesting route towards obtaining the electron density directly is through QDF in 3-D space[24], viz. by solving the continuity equation and the Navier-Stokes equation. However, here also one would encounter a non-linear differential equation and no illustrative density calculation using this approach has been reported so far.

In this section we present a brief account of a recent attempt[54] to develop an approximate second-order non-linear differential equation for the electron density. But, before we do this, it is necessary to remind ourselves that any direct density calculation should satisfy, directly or indirectly, certain conditions (see, e.g., Ref. 55) associated with the antisymmetric nature of the N-electron wave function as well as the latter's short-range and long-range behaviour. These conditions are:

(a) N-representability. (i) $\rho(\vec{r})$ finite, positive semi-definite and differentiable everywhere,

(ii) $\int \rho(\vec{r}) d\vec{r} = N$ (127)

(b) Cusp condition. $\lim_{r_\alpha \to 0}\left(\frac{\partial}{\partial r_\alpha} + 2Z_\alpha\right)\rho_{0\alpha}(r_\alpha) = 0$ (128)

where r_α is the distance from the nucleus of charge Z_α and $\rho_{0\alpha}(r_\alpha)$ is the spherically averaged charge density.

(c) Asymptotic condition. The long-range electron density satisfies the condition

$$\rho(r) \sim \exp[-2(-2\mu_{max})^{1/2} r] \tag{129}$$

where μ_{max} is the least negative eigenvalue of a matrix $\hat{\mu}$ which is a functional of the one-electron and two-electron reduced density matrices[56]; its magnitude is not less than the first ionization energy.

Use of the cusp and asymptotic conditions is likely to reduce the number of adjustable parameters in variational calculations with trial density functions. It should also improve the calculated single-particle (charge and momentum) density as well as energy and other properties. Note that direct density calculations should maintain both high local and global accuracy. Thus, the calculated density should match closely with a good reference density (Hartree-Fock or better than Hartree-Fock) all over 3-D space and the global (integrated) values of energy and other properties should also agree very well with the corresponding reference values. These are obviously stringent conditions.

In a recent approach, which utilizes the concept of a stress tensor and an electrostatic force density, Essen and Bader[57] have derived a fourth-order non-linear differential density equation. For closed-shell atoms and ions the equation may be written as

$$\rho'''' = \frac{3}{\rho}\rho'\rho''' - \frac{7}{\rho^2}(\rho')^2\rho'' + \frac{2}{\rho}(\rho'')^2 + \frac{3}{\rho^3}(\rho')^4 + \frac{1}{r}\left[\frac{10}{\rho}\rho'\rho'' - 4\rho''' - \frac{6}{\rho^2}(\rho')^3\right] - 16\eta\cdot\frac{N-1}{N}\rho^2 \tag{130}$$

where a prime denotes a single differentiation with respect to r, a double prime denotes double differentiation, and so on. Eq. (130) is to be solved subject to the following boundary conditions:

(a) ρ and its derivatives are finite at $r = 0$.

(b) $\rho'(0) = -2Z\rho(0)$: Cusp condition, see Eq. (128).

(c) $\rho'''(0) = 12Z^3\rho(0) - 5Z\rho''(0)$ (131)

$$\text{(d)}\ \rho''''(0) = \frac{1}{5}\left\{\frac{12}{\rho(0)}[\rho''(0)]^2 - 10Z^2\rho''(0) - 72Z^4\rho(0) - 16\eta\cdot\frac{N-1}{N}[\rho(0)]^2\right\} \quad (132)$$

The solution is determined by $Z, N, \rho(0)$ and $\rho''(0)$, and it is exact for the H atom. Efforts are being made[57] to solve Eq. (130).

2.7.2. A new density equation. The approach[54] that we will now describe is an alternative version of DFT. Consider the familiar[15] energy density functional

$$E[\rho] = T[\rho] + E_{xc}[\rho] + \frac{1}{2}\iint \frac{\rho(\vec{r})\rho(\vec{r}')}{|\vec{r} - \vec{r}'|} d\vec{r}d\vec{r}' + \int v(\vec{r})\rho(\vec{r})d\vec{r} \quad (133)$$

where

$$\int \rho(\vec{r})d\vec{r} = N \quad (127)$$

The kinetic energy functional may be written in terms of a kinetic energy density, i.e.,

$$T[\rho] = \int t(\vec{r};\rho)d\vec{r} \quad (134)$$

where

$$t(\vec{r};\rho) = -\frac{1}{2}\nabla^2\rho(\vec{r};\vec{r}')|_{\vec{r}=\vec{r}'} \quad (135)$$

$\rho(\vec{r};\vec{r}')$ being the single-particle (spinless) density matrix. Within the Hartree-Fock approximation, we have[54,58]

$$t(\vec{r};\rho) = -\frac{1}{4}\nabla^2\rho(\vec{r}) + \frac{1}{8}\frac{\nabla\rho(\vec{r})\cdot\nabla\rho(\vec{r})}{\rho(\vec{r})} - \frac{1}{2}\nabla\rho(\vec{r})\cdot\nabla A(\vec{r},\vec{r}')|_{\vec{r}=\vec{r}'} - \frac{1}{2}\rho(\vec{r})\nabla^2 A(\vec{r},\vec{r}')|_{\vec{r}=\vec{r}'} \quad (136)$$

where

$$A(\vec{r},\vec{r}') = [-2C(\vec{r},\vec{r}')]^{1/2} \quad (137)$$

C being the correlation factor[58].

One can now minimize $E[\rho]$ in (133) subject to the normalization condition (127) to obtain the usual Euler equation

$$\frac{\delta E}{\delta\rho} - \mu = 0 \quad (138)$$

This is the deterministic equation for ρ and will be rewritten below subject to a non-local approximation for the kinetic energy density and the Hedin-Lundqvist[59] local approximations for the XC contributions.

One may write the correlation factor as

$$C(\vec{r},\vec{r}\,') = C_{\tilde{\rho}(\vec{r})}(\vec{r},\vec{r}\,') \tag{139}$$

where $\tilde{\rho}$ is a density parameter[58] such that the following sum rule is satisfied:

$$\int \rho(\vec{r}\,')C_{\tilde{\rho}(\vec{r})}(\vec{r},\vec{r}\,')d\vec{r}\,' = -1 \tag{140}$$

Using an expression analogous to that of exchange in a homogeneous electron gas, the correlation factor can be written as[58]

$$C_{\tilde{\rho}(\vec{r})}(\vec{r},\vec{r}\,') = -\frac{9}{2}\frac{j_1^2(\tilde{y})}{\tilde{y}^2} \tag{141}$$

where

$$\tilde{y} = |\vec{r}-\vec{r}\,'|[3\pi^2\tilde{\rho}(\vec{r})]^{1/3} \tag{142}$$

and $j_1(\tilde{y})$ is the first-order spherical Bessel function. The kinetic energy density now becomes

$$t(\vec{r};\rho) = -\frac{1}{4}\nabla^2\rho + \frac{1}{8}\frac{\nabla\rho\cdot\nabla\rho}{\rho} + C_k f(\vec{r})\rho^{5/3} \tag{143}$$

where

$$C_k = \frac{3}{10}(3\pi^2)^{2/3} \tag{144}$$

and

$$f(\vec{r}) = [\tilde{\rho}(\vec{r})/\rho(\vec{r})]^{2/3} \tag{145}$$

is an unknown function (see later) related to the shape of the exchange hole. Let us take

$$\rho(\vec{r}) = |\phi(\vec{r})|^2 \tag{146}$$

so that $\rho(\vec{r})$ is now N-representable. Eq. (138) then becomes a new density equation[54]

$$[-\frac{1}{2}\nabla^2 + v(\vec{r}) + \int \frac{\rho(\vec{r}\,')}{|\vec{r}-\vec{r}\,'|}d\vec{r}\,' + v_{xc}(\vec{r})$$

$$+ \frac{5}{3}C_j g(\vec{r})\rho(\vec{r})^{2/3}]\phi(\vec{r}) = \mu\phi(\vec{r}) \tag{147}$$

where the unknown function

$$g(\vec{r}) = \frac{3}{5}(\vec{r}) + \frac{2}{5}\rho(\vec{r})^{-2/3}\int d\vec{r}'\rho(\vec{r}')\tilde{\rho}(\vec{r}')^{-1/3}\frac{\delta\tilde{\rho}(\vec{r}')}{\delta\rho(\vec{r})} \quad (148)$$

and $\phi(\vec{r})$ may be interpreted as a 3-D 'wave function' of a many-electron system (see also ref. 60). The energy is given by

$$E[\rho] = \frac{1}{8}\int \frac{\nabla\rho\cdot\nabla\rho}{\rho}\,d\vec{r} + C_k\int f(\vec{r})\rho(\vec{r})^{5/3}d\vec{r}$$

$$+ \int v(\vec{r})\rho(\vec{r})d\vec{r} + \frac{1}{2}\iint \frac{\rho(\vec{r})\rho(\vec{r}')}{|\vec{r}-\vec{r}'|}\,d\vec{r}d\vec{r}' + E_{xc}[\rho] \quad (149)$$

$$= \mu N - \frac{1}{2}\iint \frac{\rho(\vec{r})\rho(\vec{r}')}{|\vec{r}-\vec{r}'|}\,d\vec{r}d\vec{r}' + \int\rho(\vec{r})[\varepsilon_{xc} - v_{xc}]\,d\vec{r}$$

$$+ C_k\int[f(\vec{r}) - \frac{5}{3}\,g(\vec{r})]\rho(\vec{r})^{5/3}d\vec{r} \quad (150)$$

2.7.3. Calculation for Ne, Ar, Kr, Xe. As a first step to solving Eq. (147) for the four noble gas atoms, f(r) and g(r) functions are modeled as sums of Gaussians, viz.

$$f(r) = \sum_{i=1}^{n} A_i \exp[-\alpha_i(r-R_i)^2]; \; f(0) = 0, \; f(\infty) = 1 \quad (151)$$

$$g(r) = \sum_{i=1}^{n} A_i \exp[-\beta_i(r-R_i)^2] \quad (152)$$

where $A_i \equiv A_i(Z,n)$ are the peak heights, $R_i \equiv R_i(Z,n)$ are the peak positions, Z is the nuclear charge and n is the number of shells. The exponents $\{\alpha_i\}$ are chosen to reproduce a reasonably good fit (see Fig. 3) with the calculated[54] Hartree-Fock (HF) f(r), while $\{A_i\}$ and $\{R_i\}$ are fitted by functions of Z and n so as to closely reproduce the HF values of these quantities. The exponents in g(r) are adjusted so as to obtain a good fit with the HF density and energy (Fig. 4). With the above input, Eq. (147) can be readily solved numerically (see Ref. 54 for details) by a finite difference method with successive linearization, following an earlier suggestion by Tomishima and Yonei[61]. The resulting density and energy can be made to agree very well with the corresponding HF results (Fig. 5). Note that the maxima in g(r) correspond to minima in the radial density.

The discussion in this section makes it clear that it is possible to obtain a second-order differential equation for the density without directly resorting to a gradient expansion, and it would be

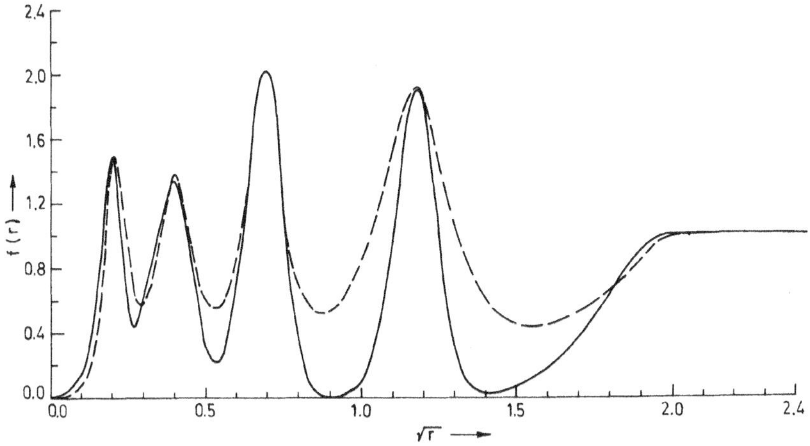

Figure 3. Plots of f(r) for xenon atom, with r in a.u.
_____ : Used in present calculation;
----- : Hartree-Fock.

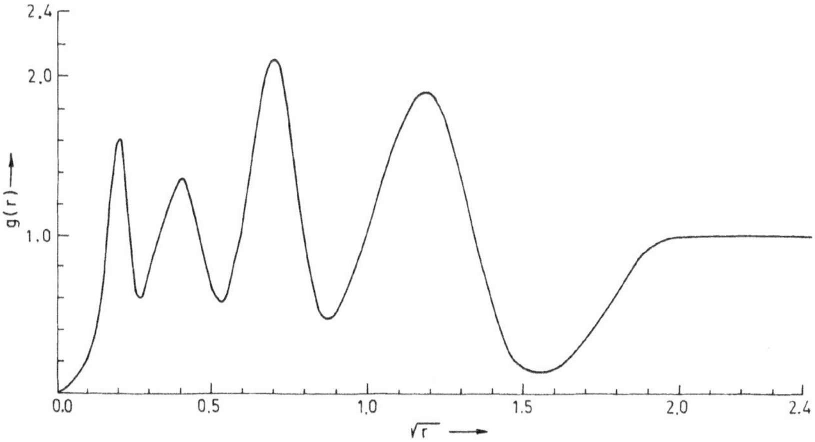

Figure 4. Plot of g(r) for xenon atom, with r in a.u.

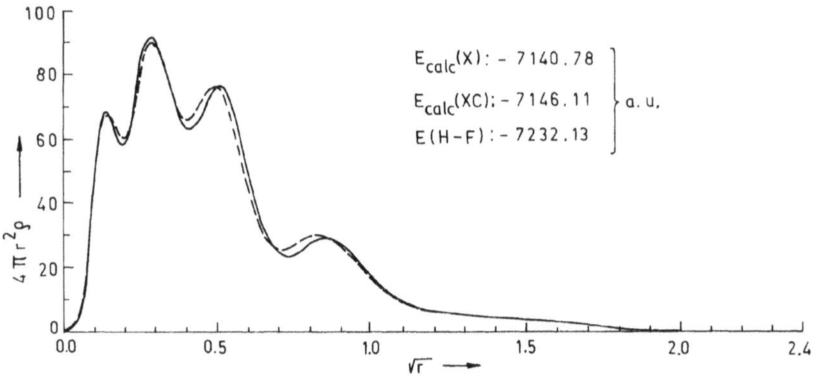

Figure 5. Radial density plots for xenon atom, in a.u.
____ : Present calculation without correlation (merges with the curve with inclusion of correlation);
---- : Hartree-Fock (H-F); X : Includes only exchange;
XC : Includes both exchange and correlation.

much simpler to solve this equation, obtained from a modified statistical model, provided it has the necessary critical input about the unknown function $f(\vec{r})$. An advantage of this equation is that it uses a kinetic energy density functional which shows proper 'local' behaviour. It might also be possible to incorporate symmetry arguments in density studies through $\phi(\vec{r})$, the 3-D 'wave function'. However, note that the inclusion of HF information into f(r) and adjusting the exponents of g(r), to obtain good agreement with HF density and energy, do bring an element of circularity in the computational procedure. Hopefully, this might be removed in future work.

In the next part of the article we will discuss certain other basic and interpretive aspects of single-particle density as well as its further explorations on the electronic structure of atoms, molecules and solids, their various properties, and interactions.

3. OTHER BASIC AND INTERPRETIVE ASPECTS OF ELECTRON DENSITY

3.1. Quantum Topological Theory of Molecular Structure and Change

We begin this part by briefly mentioning the important recent work of Bader et al on the topological properties of single-particle density and a rigorous definition of molecular structure on the basis of these properties (for reviews, see Refs. 6 and 62). This utilizes the catastrophe theory of Rene Thom and, in particular, the idea of critical points[63] in the charge density. We quote from Bader et al.:

"[The] concepts of atoms and bonds may be rigorously defined and given physical expression in terms of the topological properties of the observable distribution of charge for a molecular system. As a consequence of these definitions, one in turn obtains a definition of [molecular] structure and a predictive theory of structural stability. The theory is linked to quantum mechanics ... [because] the atoms so defined represent a class of open quantum subsystems with a unique set of variationally defined properties" (quoted from p. 893, Ref. 6, by permission of the Institute of Physics, London).

"The basis for this approach is provided by the topology of the total charge density in a given molecular system. The essential observation is that the only local maxima of a ground state distribution occur at the positions of the nuclei. The nuclei are therefore identified at point attractors of the gradient vector field of the charge density. The associated basins partition the molecular system into atomic fragments. Each atom is a stable structural unit defined as the union of an attractor and its basin. The common boundary of two neighbouring atomic fragments, the interatomic surface, contains a particular critical point, which generates a pair of gradient paths linking the two neighbouring attractors. The union of this pair of gradient paths and their endpoints is called a bond path. The network of bond paths defines a molecular graph of the system.

"Having defined a unique molecular graph for any molecular geometry, the total configuration space is partitioned into a finite number of regions. Each region is associated with a particular structure defined as an equivalence class of molecular graphs. A chemical reaction in which chemical bonds are broken and/or formed is therefore a trajectory in configuration space which must cross one of the boundaries between two neighbouring structural regions. These boundaries form the catastrophe set of the system which, like a phase diagram in thermodynamics, denotes the points of 'balance' between neighbouring structures .....

"The properties of the topologically defined atoms and their temporal changes .... [can be] identified within a general formulation of subspace quantum mechanics .... [The] quantum mechanical partitioning of a system into subsystems coincides with the topological partitioning: both are defined by the same set of 'zero flux' surfaces. Consequently, the total energy, or any other property, is partitioned into additive atomic contributions.

"[In] general, a definite structure can be assigned to a given molecular system .... [Molecular] structure exists in spite of intermolecular interactions ..." (quoted from p. 8, Ref. 62, by permission of Weizmann Science Press, Jerusalem).

3.2. Relations Between Electrostatic Potential V and Electron Density ρ

Politzer[64,65] has shown that the minima in radial density of ground-state atoms are significant in relations between V and ρ. In particular, the outermost minimum defines a physical boundary between the core and valence regions. In the core region the TF relation, $V = \beta\rho^{2/3}$, is obeyed quite well while for the valence region, $V(Z,r) = \alpha(Z)\rho(Z,r)$ is a better approximation. Using this, the TF energy formula, $E = \frac{3}{7}ZV_0$, where V_0 is the electrostatic potential (ESP) at the nucleus due to the electrons, assumes a more refined expression

$$E = \frac{3}{7}ZV_0 + \frac{3}{28}r_c[V(r_c)]^2 - \frac{1}{7}Kr_c^3[V(r_c)]^{5/2} + \frac{1}{8}Z^{2/3}r_c^3[V(r_c)]^2 \tag{153}$$

where the core radius r_c is the radius of the outermost minimum and $K = 4\pi/\beta^{3/2}$. Substituting the relation

$$V(r_c) = N_v/2r_c \tag{154}$$

in Eq. (153), where N_v is the amount of valence electronic charge, a better expression results:

$$E = \frac{3}{7}ZV + \frac{3N_v^2}{112r_c} - 0.030315r_c^{1/2}N_v^{5/2} + \frac{Z^{2/3}r_cN_v^2}{32} \tag{155}$$

For the atoms Be to Ar, the error in energy (with respect to experimental values) due to Eq. (155) progressively decreases from 3.45 to 0.18%.

The idea of obtaining good estimates of energy (including correlation energy) and electron density starting from an ESP has been explored by several authors. For reviews of these as well as recent applications of ESP to molecular structure and properties (including biomolecules) and its experimental determination from X-ray diffraction data, see Refs. 12 and 66.

3.3. Self-Interaction Correction and Correlation Energy in Density-Functional Theory

In an important paper, Perdew and Zunger[67] have incorporated self-interaction correction (SIC) in local spin density-functional theory (LSD). In the exact DFT, the XC energy of a single fully occupied orbital exactly cancels its self-direct Coulomb energy, i.e.,

$$U[\rho_{\alpha\sigma}] + E_{xc}[\rho_{\alpha\sigma}, 0] = 0 \tag{156}$$

where the orbital density $\rho_{\alpha\sigma} = f_{\alpha\sigma}|\psi_{\alpha\sigma}|^2$ and the occupation number $f_{\alpha\sigma} = 1$; α and σ are quantum numbers, $\sigma(= \pm 1/2)$ denoting the spin. SIC can be introduced by writing

$$E_{xc}^{SIC} = E_{xc}^{approx}[\rho_{\downarrow},\rho_{\uparrow}] - \sum_{\alpha\sigma} \delta_{\alpha\sigma} \tag{157}$$

where $\delta_{\alpha\sigma}$ is the self-interaction of the orbital $\psi_{\alpha\sigma}$, i.e.,

$$\delta_{\alpha\sigma} = U[\rho_{\alpha\sigma}] + E_{xc}^{approx}[\rho_{\alpha\sigma},0] \tag{158}$$

Perdew and Zunger suggest that the most straightforward self-consistent formalism with SIC is the equation

$$[-\frac{1}{2}\nabla^2 + \bar{v}_{\alpha\sigma}^{eff}(\vec{r})]\bar{\psi}_{\alpha\sigma}(\vec{r}) = \varepsilon_{\alpha\sigma}^{SIC}\bar{\psi}_{\alpha\sigma}(\vec{r}) \tag{159}$$

where $\bar{v}_{\alpha\sigma}^{eff}$ is an orbital-dependent potential. While Eq. (159) is valid within the Hohenberg-Kohn theorem, it does not resemble the KS scheme (see Section F of Ref. 67) and can accomodate fractional occupation numbers.

The gains of the SIC-LSD scheme are impressive[67]. For atoms, it gives (i) improved values of exchange energy, correlation energy and total energy*; (ii) very good binding energies of negative ions which are unstable in LSD; (iii) orbital eigenvalues that 'closely approximate physical removal energies, including orbital relaxation effects'[67] (c.f. Koopmans theorem), in contrast to LSD eigenvalues; (iv) correct long-range behaviour of density and the one-body potential; (v) overall improved description of the density and of the shape of the XC hole. For molecules and solids, the SIC scheme can generate accurate a priori pseudopotentials for describing the interaction between valence electron and ions. It can also reduce the LSD underestimate of band gaps in insulators and the LSD overestimate of the cohesive energy[68] of 3d transition metals. In fact, in comparison with atomic HF calculations, SIC-LSD turns out to be simpler and more accurate; it is quite useful in describing[69] the autodetaching states and resonances in systems like H^-, He, etc. However, SIC does not cure all the ills of LSD. Although the SIC-LSD

*For Kr, 44 per cent of the SIC to the total energy comes from 1s orbital. The SIC terms are of similar magnitude as the energy eigenvalues themselves. The LSD generally underestimates E_x by ~10% and overestimates E_c by ~100% due to its inability to treat rapid density variations.

correlation energy is very much improved, it behaves as ln Z, when $Z \to \infty$ for fixed N. The correct description of correlation energy for strongly positive ions remains a difficult problem in DFT[70]. For a recent parametric form of correlation energy, see Ref. 67. For another prescription to improve E_T, E_X and E_C, see Ref. 71.

3.4. Electronic Structure and Properties of Atoms, Molecules, and Solids

A proper account of XC energy, especially in the excited states, is essential in order to calculate multiplet splittings in atoms with incomplete shells. By splitting the total Hilbert space of the system into orthogonal subspaces with different symmetries and by defining a DFT in each subspace, Gunnarsson and Lundqvist[72] have argued that the HKS-DFT can be extended to cover excited states of symmetry (total angular momentum or spin) different from that of the ground state. However, this does not work if the XC functional is taken to be symmetry independent because the LSD misrepresents the XC hole. In view of this, Von Barth[73,74] has generalized DFT to be applicable to states of mixed symmetry, by retaining the LSD approximation. One associates a number of Slater determinants to a given configuration. Then, for each determinant one constructs a linear combination of states of pure symmetry which reduces to this determinant for the non-interacting case. The energy of each linear combination is calculated by using LSD and the multiplet splittings as recovered from these 'mixed-symmetry' energies. For He (1s 2s), $C(p^2)$, $N(p^3)$, and $Si(p^2)$, results are much better than the HF values; compared with experimental multiplet splittings, the largest error is ~23% while the smallest is ~3%. This approach has been further supported by the calculations of Gunnarsson and Jones[75] on the first-row and group IIA atoms and ions. Again, trends in atomic and ionic multiplet structures are reproduced very well, but there are certain difficulties involving configurations with d electrons in particular. Thus, the relative stability of states with greater p- and d-character, and singlet-triplet splittings for sd-configurations tend to be overestimated.

The LSD approximation[72] has also been used by Kitaura et al[76] to calculate energies, geometries, and spectroscopic constants of simple molecules within the LCAO-MO-SCF scheme, by using contracted Gaussian basis sets. For first-row diatomic molecules, the dissociation energies are much better that HF values and agree well with the experimental values, except for F_2 where it is overestimated by 81%. The equilibrium vibration frequencies also agree fairly well with experiment. However, in contrast to the tendency of HF calculations to produce slightly shorter equilibrium bond lengths than the actual ones, the LSD calculations tend to produce slightly longer bond lengths, presumably due to the overestimation of correlation energy. The LSD bond angles for molecules and inversion barrier for NH_3 are somewhat better than the corresponding HF values. Howver, the LSD total energies are still higher than the HF total energies.

In a recent review Schlüter and Sham[77] have summarized several notable successes and limitations of the LSD approximation. For atoms, the calculations provide generally as good as or better than HF values of ionization potential, multiplet energy and hyperfine interaction. Property trends among the elements of periodic table, in particular, Hund's rule on the successive filling of orbitals, are generally reproduced and the correct ground states obtained. Exceptions to this are, e.g. Ti, V, and Co where LSD favours the $d^{n-1}s^1$ configuration over the correct $d^{n-2}s^2$ ground state. For molecules also, LSD scores over HF, e.g., in the reversal of dipole moments in CO and NO, and in the binding of group IIA dimers (Be_2 to Ba_2) where the LSD succeeds but HF does not.

One of the impressive successes of the LSD approximation has been demonstrated with the phase transision and lattice dynamics of ground-state crystalline* silicon[77-79]. Here, the DFT approach is combined with a first-principles pseudopotential approach[78,79] where the 'frozen core' is simulated by a non-empirical, i.e., parameter-free pseudopotential*. These ion-core pseudopotentials do not have sharp singularities and self-consistent numerical solutions of the one-particle Schrödinger equation may be performed with the help of expansion techniques. The silicon calculations lead to the following results:

"Among the six crystal structures that cover about 90% of all existing elemental solids (diamond, hexagonal diamond, β-tin, hcp, bcc and fcc), the calculations show correctly that the diamond structure is the most stable at [zero external pressure and temperature] .... The stable diamond structure is separated by [~20meV] ... per atom from the hexagonal modification. This small energy arises from structural differences beyond the third nearest neighbours, which, for example, separate compound-semiconductors into zinc-blende and wurtzite structures. In the diamond structure, calculated and measured static ground-state properties, such as the equilibrium lattice constant, bulk modulus and cohesive energy, agree to better than a few per cent. The calculations also show correctly that silicon will transform to the β-tin structure under high pressure. The transition volumes, which are extremely sensitive quantities, are reproduced to within 1% of experimental values ... .

---

*No application of DFT to amorphous solids seems to have been attempted so far.

*For the application of first-principles pseudopotentials within the DFT, see Refs. 80-83. Ref. 84 employs non-local exchange and includes core-valence correlation by fitting pseudo-potentials to experimental energies of one-valence-electron ions.

One can calculate the effects of lattice dynamics by assuming adiabatic behaviour and comparing crystal energy differences for various 'frozen-in' phonon distortions. Calculations match measurements to within ... 3% for acoustic and optical phonons on the Brillouin zone surface and at the zone center. In particular, calculations of the very soft transverse acoustic mode, characteristic of shear instabilities in tetrahedral semiconductors, show only a 1% error. Moreover, anharmonic effects are described within 10%, which is remarkable in view of the fact that many theories cannot even reproduce their correct signs.

Overall, we should regard the silicon results as an important confirmation of the validity of the local-density functional approach for studying electronic ground-state properties of systems with s and p valence electrons and with sizable external lattice potential" (quoted from p. 40, Ref. 77, by permission of the American Institute of Physics, New York).

Many other LSD calculations have been made on the bulk and surface properties of solids. LSD can predict ferromagnetism and calculate magnetic moments and susceptibilities. The exchange splitting between 'spin-up' and 'spin-down' bands agrees well with photoemission data for iron and cobalt, but not for nickel[77]; the situation might improve with inclusion of SIC (Section 3.3). The equation of state of metals under high pressure may also be dealt with by the LSD approximation. For semiconductors and insulators, LSD cohesive energies and other bulk ground-state properties are much better than the more cumbersome HF results, e.g., with the rare gas solids where HF give no binding. However, the calculated energy gaps are too small[77], due to the inadequacy of LSD. Also, self-consistent LSD calculations using Green's functions have been made for defects in semiconductors, e.g., silicon, diamond, gallium arsenide, gallium phosphide, etc. In case of silicon vacancy, Jahn-Teller coupling to the lattice produces large energy-lowering distortions which can overcome the Coulomb repulsion of extra electrons in the localized state. This leads to an 'inverted-level' or 'negative U' situation[85,86], which was later verified experimentally. Further, LSD calculations to describe the surfaces of metals and semiconductors have been shown[87] to be capable of determining the small structure-induced charge modulations which generally occur 5-10 Å outside the surface and are only ~10^{-5} times the charge density in the chemical bonds of the crystal. Sander et al[88] have investigated the formation of charge density waves (CDW) in two- and three-dimensional (idealized) jellium, using the local DFT. However, in this context Overhauser[89] points out certain important shortcomings of LSD in case of the ground state of metallic potassium which has spontaneously broken translational symmetry. The (anisotropic) CDW structure for potassium is not reproduced by the LSD approximation to XC, or by replacing the coulomb potentials by screened interactions. Although this indicates a future challenge to the LSD, one should not forget its successes (see also Ref. 90). Thus, it can also explain the occurrence of metal-insulator transitions "as the result of a trade-off between exchange-correlation energy and kinetic energy"[77]. Inclusion of SIC and non-local effects (Section 3.10) may be able to remedy the above drawbacks.

For other pseudopotential DFT calculations of bulk and surface properties of solid and liquid metals, see, e.g., Refs. 91-93. For other applications of DFT to atoms, molecules, and solids, see Refs. 6, 94, 95.

3.5. Quadrupole Polarizabilities and Sternheimer Antishielding Factors

In Section 2.6, we have described a DFT method which can, in principle, calculate the dynamic 2L-pole polarizability of many-electron systems. However, calculations were reported only for L=1*. Recently Stott et al[97] have calculated static (zero-frequency) quadrupole polarizabilities and Sternheimer antishielding factors for a number of closed-shell atoms and ions, using the LSD approximation for treating the linear response of the system to an external perturbation. Taking (-q) as the principal value of the electric field gradient tensor and (q'-q) as the total field gradient acting on a nuclear quadrupole moment, the Sternheimer antishielding factor is given by

$$\gamma_\infty = q'/q \tag{160}$$

while the static quadrupole polarizability is given by

$$\alpha_q = \frac{8\pi}{q}\int_0^\infty r^4\alpha_q(r)dr \tag{161}$$

where

$$q' = 8\pi\int_0^\infty \frac{\alpha_q(r)}{r}dr \tag{162}$$

and the induced density $\alpha_q(r)$ is given by an integral equation[97]. Calculated results indicate that the LSD approximation tends to make a system more polarizable than it actually is, the effect becoming more pronounced with He and Ne than with heavier noble gas atoms. The same is true for the antishielding factors. Furthermore, d-shell ions like Cu^+, Au^+, and Cd^{++} would require the use of SIC

*The L=0 case, corresponding to an external magnetic perturbation, is required in calculating metallic Knight shifts (see, e.g., Ref. 96).

for obtaining good α_q (also perhaps good γ_∞) values and, therefore, provide another sensitive test for the LSD approximation.

3.6. Analysis of Magnetically Induced Current Densities in Aromatic Hydrocarbons

In part 2, we have discussed in some detail the concepts of quantum fluid dynamics in 3-D space, which view a many-electron system through its charge density and current density. Gomes[36-38] has recently examined magnetically induced current densities in benzene and cyclopropenyl cation with a view to place the earlier concept of ring currents on a firmer foundation. The delocalized current in a cyclic molecule is defined as that part of the current density which represents a flow of electronic charge around the ring. In general, the *ab initio* current densities are rather complex and their analysis is not at all straightforward. However, it turns out that although the delocalized current is due largely to the π electrons, one cannot preclude the flow of σ electrons around the ring[36]. The analysis of current density fields is considerably facilitated by the introduction of topological ideas[37,38] (see also Refs. 7, 62, 63). This uses the symmetry properties of the electronic Hamiltonian in presence of magnetic field, the ideas of axial and toroidal vortices[23,98] as well as of stagnation points where current density vanishes but charge density does not. This type of analysis highlights the different contributions from different regions of a molecule to its magnetic properties.

The application of a magnetic field to a homonuclear diatomic molecule (e.g., N_2, F_2) perpendicular to the bond axis reduces the molecular symmetry to a space-time point group[37] isomorphic to D_{2h}. For F_2, the major contribution to the first-order current density seems to involve a $1\pi_u$-$3\sigma_u$ transition while that for N_2 seems to involve both $3\sigma_g$-$1\pi_g$ and $1\pi_u$-$3\sigma_u$ transitions. Thus, there may be an axial vortex in F_2 while in N_2 there may be an superposition of a stagnation point at the origin and an axial vortex.

3.7. Interactions Between Closed-Shell Systems and With a Metal Surface

The electron-gas model for interactions between closed-shell systems, suggested about a decade ago by Gordon and Kim, and others, has undergone several modifications (see Ref. 99 for a review). In the most familiar version of the model, the total interaction is expressed as a sum of Coulomb, kinetic, and XC terms, the last two being given by electron-gas formulas. Earlier calculations indicated that the interaction potentials for most noble gas systems were too deep by 15-20% while those for the helium systems were too deep by nearly 100%. Since the electron-gas formulas give rise to systematic errors, when applied to single atoms, Waldman and Gordon[100] have proposed simple correction factors for the electron-gas kinetic, exchange, and correlation terms of a number of interacting systems. These markedly improve the helium interactions with lighter atoms, but not the interactions involving heavier atoms. Furthermore, the use of an electron-gas Drude model[101] for atom-molecule systems gives good results for spherically averaged potentials, but may not be able to simulate polarization effects arising due to the interaction. A satisfactory density-based theory of intermolecular forces is still awaited, not just for closed-shell systems, but also for long-range interactions involving open-shell systems. For example, recent crossed atomic beam spectroscopic experiments (see, e.g., Ref. 102) have revealed rather interesting structures in the long-range interaction potential between excited He atom (1s 2s, ^1S) and other noble gas atoms in the ground state. Reproduction and interpretation of such features provide a challenge to density-based formalisms.

Gordon et al (see, e.g., Ref. 103) have also applied the electron-gas model to the structures, relative stabilities, and certain properties (interatomic distances, lattice energies, pressure-induced phase transitions) of ionic crystals, e.g., halides and oxides of alkali and alkaline earth metals. The model[103] expresses the binding energy of a crystal as a sum of a Madelung term and a short-range term; the latter consists of a sum of pairwise interactions and many-body corrections which are responsible for the departure from additivity. There are three types of many-body effects in electron-gas theory: (i) non-linearity of the kinetic and XC energies; (ii) size changes in the various atomic charge densities (without changing their spherical symmetry); and (iii) distortion of atomic densities from spherical symmetry, leading to dipolar and/or multipolar interactions. The model, employed by Muhlhausen and Gordon[103] includes (i) and (ii), but not (iii). It calculates the (shrunk) anion density "in a stabilizing potential well whose size is determined self-consistently by the electrostatic potential of the crystal at the anion site"[103]. The corresponding expansion in cation density is negligible for alkali and alkaline earth metal ions. The calculated interatomic distances and lattice energies agree with experimental values within 1-2%. However, the non-linear many-body effects do not appear to be of great significance. This model occasionally fails to predict relative stabilities of various crystal structures and sometimes favours structures of lower coordination number (probably due to the neglect of Pauli repulsions). In case of pressure-induced phase transitions, the model can overestimate the transition pressure by underestimating the stability of the high density form.

Lang[104] has applied the local-density approximation to the van der Waals interaction between noble gas atoms and metal surfaces, by using a jellium model for the metal. Density difference maps[3] are quite helpful here to analyze the nature of the weak bond. Good agreement with experiment was obtained on the values of atomic binding energy, dipole moment, and core-level binding-energy shift. Lang[105] has also employed a jellium model for studying the chemisorption on metals of H, Na, Li, Si, Cl, etc. atoms.

3.8. High-Energy Proton-Atom Collisions Within the TDTF Method

Horbatsch and Dreizler[106] have recently proposed what is probably the first DFT-QFD calculation of atomic collisions, since the early pioneering works of Block[107] and Jensen[108] on the TDTF method. The TDTF equations are:

$$\frac{\partial \rho_0(\vec{r},t)}{\partial t} + \nabla\cdot[\rho_0(\vec{r},t)\nabla\chi(\vec{r},t)] = 0 \tag{163}$$

$$\frac{\partial \chi(\vec{r},t)}{\partial t} + \frac{1}{2}[\nabla\chi(\vec{r},t)]^2 + \frac{5}{3}C_k\rho_0(\vec{r},t)^{2/3} + U(r,t) + \int \frac{\rho_0(\vec{r}',t)}{|\vec{r}-\vec{r}'|}\,d\vec{r}' = 0 \tag{164}$$

Here ρ_0 is the electron density in a time-even picture, χ is a collective velocity potential, and U represents the Coulomb potential of the nuclei which move along a classical trajectory $\vec{R}(t)$. In order to apply the method to high-energy (of the order of 1 MeV) proton-atom scattering*, one proceeds in three steps: (i) calculation of the time evolution of the electron density with the help of TDTF equations; (ii) use of this density to construct a Hartree-Fock-Slater-type effective TD electronic potential; (iii) obtaining the time evolution of electronic orbitals by solving the corresponding effective TD one-particle Schrödinger equation.

Horbatsch and Dreizler have given plots of density contours and velocity fields (c.f. the QFD picture) for proton-argon scattering at 0.4, 0.9, and 2.5 MeV. These pictures reveal that "after the strong disturbance of the electron cloud [caused] by the impinging and receding proton, and oscillating atom [(apparently damped)] is left behind. As the electrons have 'more time' to adapt to the disturbance in the [low-energy] case ...., the disturbance is stronger ..."[106]. Also, in accord with experiment, the ionization probability is much higher at lower energies. However, for a more

*Here one assumes that the relative motion of the nuclei can be represented by a narrow wave packet, leading to an effective decoupling between nuclear motions and electronic excitations. The nuclear motion may then be described by classical equations. Note, however, that this is not a strictly adiabatic picture.

realistic picture one should include the effect of rotational coupling in the density. For a detailed analysis of the density and velocity plots, see Ref. 106

3.9. Further Applications of Virial and Hellmann-Feynman Theorems

In principle, the Hellmann-Feynman theorem[109] is obeyed by local DFT provided[110] that the only implicit (nuclear) coordinate dependence in XC energy is that of $\rho(\vec{r})$. However, in practice, the rather precise core- and valence-polarization which is required is very difficult to attain. Since DFT total energy calculations are often made by freezing core orbitals in their atomic configurations, Harris et al proposed a new force formula by differenting the energy functional with respect to nuclear coordinates "after the frozen core approximation has been made and the frozen core density appears explicitly in the result"[111]. The new formula requires that a modified energy functional be stationary with respect to valence density alone, thereby eliminating the error due to the deviation of the core density from its atomic form. The force on the i-th nucleus (plus core) turns out to be

$$\vec{F}_{vi} = Z_i \nabla\phi_v(\vec{r}_i)|_{\vec{r}_i=0} - Z_{ci} \sum_{j\neq i} Z_{cj} \frac{\vec{r}_i^{\,o} - \vec{r}_j^{\,o}}{|\vec{r}_i^{\,o} - \vec{r}_j^{\,o}|^3}$$
$$- \int d\vec{r}_i \hat{r}_i \frac{\partial\rho_c(r_i)}{\partial r_i} \left\{\phi_v(\vec{r}_i) + V_{xc}[\rho(r_i),\rho_s(\vec{r}_i)]\right\} \qquad (165)$$

where Z_i and $\vec{r}_i^{\,o}$ is the charge and location of nucleus i, ϕ_v is the solution of Poisson's equation for the valence density ρ_v; Z_{ci} is the net charge of core i, $\hat{r}_i = \vec{r}_i/r_i$; ρ_c is the core density; $V_{xc} = \frac{\partial}{\partial\rho}[\rho\varepsilon_{xc}(\rho,\rho_s)]$; and ρ_s is the spin density. The spin-polarization dependence of V_{xc} can be neglected[111]. Calculations on Li_2 and Be_2 indicated that satisfactory correspondence between force and energy can be obtained by minimal variation of a small basis set.

Averill and Painter[110] derive the following DFT virial theorem for an arbitrary cluster of atoms:

$$- \sum_i \vec{r}_i^{\,o} \cdot \nabla_i E = 2T_s + U_{es} - \int \rho(\vec{r})\vec{r}\cdot\nabla v_{xc}(\vec{r})d\vec{r} \qquad (166)$$

where E is the energy, T_s is the part of kinetic energy excluding the XC contribution, U_{es} is the classical electrostatic energy and v_{xc} is the effective xc potential. For H_2, the forces from the virial theorem are in better agreement with the energy gradient than the Hellmann-Feynman forces. Ziesche[112] has also considered the Hellmann-Feynman and virial theorems for an inhomogeneous electron gas, relevant to surface and impurity problems.

In a recent work, Schwarz and Chu[113] have observed that, even by using very simple wave functions such as the extended Hückel-type, it is possible to obtain simple explanations for relativistic bond length changes by using the force concept[2] and density difference maps[3]. This incorporates the phenomenon of relativistic orbital contraction. Orbital contraction contracts bond length except when the relativistic atom is more electronegative[113].

For the definition and analysis of force on a nucleus in a non-stationary state, based on the Ehrenfest theorem, see Ref. 114.

3.10. Non-local Corrections in DFT

The usefulness of a non-local approximation to exchange[58] has been briefly discussed in Section 2.7.2. Alonso and Balbás[115] have also suggested a non-local approximation for Coulomb correlation. The exact XC energy functional may be written as

$$E_{xc}[\rho] = \frac{1}{2} \iint \frac{1}{r_{12}} \rho(\vec{r}_1)\rho(\vec{r}_2)C_{xc}(\vec{r}_1,\vec{r}_2)d\vec{r}_1 d\vec{r}_2 \tag{167}$$

where $C_{xc}(\vec{r}_1,\vec{r}_2)$ is the net pair correlation (Pauli and Coulomb) function, satisfying the limiting conditions

$$\lim_{r_{12}\to\infty} C_{xc}(\vec{r}_1,\vec{r}_2) = 0 \tag{168}$$

$$\lim_{r_{12}\to 0} C_{xc}(\vec{r}_1,\vec{r}_2) = -1 \tag{169}$$

$$\int \rho(\vec{r}_2)C_{xc}(\vec{r}_1,\vec{r}_2)d\vec{r}_2 = -1 \tag{170}$$

One may write

$$C_{xc}(\vec{r}_1,\vec{r}_2) = C_x(\vec{r}_1,\vec{r}_2) + C_c(\vec{r}_1,\vec{r}_2) \tag{171}$$

where C_x and C_c satisfy the limiting conditions

$$\lim_{r_{12}\to\infty} C_x(\vec{r}_1,\vec{r}_2) = 0 \quad ; \quad \lim_{r_{12}\to\infty} C_c(\vec{r}_1,\vec{r}_2) = 0 \tag{172}$$

$$\lim_{r_{12}\to 0} C_x(\vec{r}_1,\vec{r}_2) = -\frac{1}{2} \quad ; \quad \lim_{r_{12}\to 0} C_c(\vec{r}_1,\vec{r}_2) = -\frac{1}{2} \tag{173}$$

$$\int \rho(\vec{r}_2)C_x(\vec{r}_1,\vec{r}_2)d\vec{r}_2 = -1 \quad ; \quad \int \rho(\vec{r}_2)C_c(\vec{r}_1,\vec{r}_2)d\vec{r}_2 = 0 \tag{174}$$

For $C_c(\vec{r}_1,\vec{r}_2)$, the functional form of $C_c^h(\vec{r}_1,\vec{r}_2)$, h denoting a homogeneous electron gas, may be used, but with a density parameter $\tilde{\rho}^*(\vec{r}_1)$ fixed by Eqs. (174). An approximate analytical form of C_c^h is given by

$$C_c^h(\vec{r}_1,\vec{r}_2) = -\frac{1}{2}\exp(-\sqrt{3}\beta r_{12})\cos(\beta r_{12}) \tag{175}$$

where

$$\beta = \{-\pi\rho_h/8\varepsilon_c[\rho_h]\}^{1/2} \tag{176}$$

The splitting of C_{xc} into C_x and C_c, Eq. (171), has certain practical advantages[115].

Wilk and Vosko[116] feel: "One must be careful, when there seems to be agreement between ... [the LSD approximation] and experiment, not to be misled into thinking that non-local corrections to the ... [LSD approximation] are small. Rather, as we have seen for atomic ionisation energies ..., the agreement may be due to a strong cancellation of appreciable non-local effects". For further discussions on improving the LSD approximation, see Ref. 117.

3.11. Miscellanous Applications Involving Electron Density

After Kryachko[118] had questioned the validity of the Hohenberg-Kohn theorem, Levy and Perdew[90] rose to its defence. To the criticism that density does not necessarily contain information on the statistics of particles, Levy and Perdew replied that, for many-electron systems, the Fermi statistics are automatically included because the search for the minimum of $\langle\psi_\rho|\hat{T}+\hat{V}_{ee}|\psi_\rho\rangle$, which is a universal functional of ρ, is carried out only over antisymmetric wave functions.

In this section, we list a number of interesting recent works which involve the electron density. Thus, Pathak and Gadre[119] provided the following two formulas for $\langle p\rangle$ and $\langle p^{-1}\rangle$ from atomic electron densities:

$$\langle p\rangle = \pi\,\frac{3}{4}\left(\frac{3}{\pi}\right)^{1/3}\int\rho^{4/3}(\vec{r})d\vec{r} \tag{177}$$

$$\langle p^{-1}\rangle = \frac{1}{2}\left(\frac{3}{\pi}\right)^{2/3}\int\rho^{2/3}(\vec{r})d\vec{r} \tag{178}$$

Ludena[120] has shown that the exact kinetic energy density (see Section 2.7.2) can be written as a sum of the Weizsäcker term (local) and a non-local term, viz.

$$t(1) = \frac{1}{8}\,\frac{\nabla_1\rho(1)\cdot\nabla_1\rho(1)}{\rho(1)} + \frac{1}{4}\,\frac{\nabla_1\cdot\nabla_{1'}B(1,1')|_{1'\to 1}}{\rho(1)} \tag{179}$$

where B(1,1') depends on the pair correlation function describing the Coulomb and Fermi holes. Freed and Levy[121] have given a formal derivation of the exact universal electron density functional while Riess and Münch[122] extended the Hohenberg-Kohn theorem to subdomains of a bounded (or periodic) many-particle system. The momentum density and Compton profiles for atoms, predicted by TF theory, have been compared[123] with the exact results for the Coulomb potential; for five closed shells there is remarkable agreement between the two treatments. Henderson[124] has demonstrated the existence of Hohenberg-Kohn-type theorems in momentum space, and Koga and Morita[125] have used the integrated Hellmann-Feynman theorem[109] to derive a rigorous relation between the energy and electron density in momentum space. Painter[126] has described a method of solving Poisson's equation, for a localized non-spherical charge distribution expanded in partial waves, as a set of finite-difference equations.

Evans and Sluckin[127] have proposed a DFT for the single-particle densities and thermodynamic properties of classical, many-component, inhomogeneous charged fluids; the theory has been applied to the liquid-vapour interface of a molten alkali halide. Heinrich and Magain[128] have employed DFT to calculate the exact spatial distribution of the pressure at a planar jellium surface, for various densities. Katriel et al[129] have reexamined the concepts of chemical potential and atoms-in-molecules by arguing that atomic and molecular energies may be assumed to be analytic functions of the number (N) of electrons. Alonso and Girifalco[130] related electronegativity to the asymptotic behaviour of the free atom electron density. Finally, Gollisch and Fritsche[131,132] have combined McWeeny's[133] density matrix method with local DFT to make some interesting calculations on the binding characteristics of molecules and clusters.

4. CONCLUSION

This article has continued the theme of our two previous articles[1,5], viz. there is a distinct and desirable possibility of obtaining an alternative quantum mechanics of atoms, molecules, and solids in terms of $\rho(\vec{r})$ and associated quantities. Much still remains to be done within this viewpoint regarding (a) basic quantum mechanical properties of the electron density as well as (b) its various interpretive and predictive aspects. We have discussed here a number of recent works concerning both (a) and (b). The list has not been exhaustive and no doubt other important works would be reported in the near future.

There are quite a few unsolved problems associated with aspects (a) and (b) above, which might seize the minds of quantum theorists for some time. In dealing with the problem of obtaining a single equation for electron density, the search should perhaps concentrate on a second-order differential equation, rather than a higher-order one, not only because the latter may be more difficult to solve but also because the density viewpoint, in principle, should not replace a second-order equation (the Schrödinger equation or its variants) with a higher-order one. Whether a second-order equation with no additional unknown function (see Section 2.7) can be obtained, remains to be seen. A combined attack on this problem through DFT and QFD would be interesting.

Since the density equation is likely to be non-linear, the use of powerful group-theoretical techniques in density-based studies is apparently thwarted and this is clearly a drawback of the density approach in comparison to the wave-function approach. This might introduce the use of non-linear algebraic techniques in a substantial way.

The search for better approximations to the unknown universal functional in DFT requires the inclusion of non-local effects in kinetic and XC terms. The design of proper XC functionals which would also be valid for the limiting case of a one-electron system is important, otherwise spurious two-electron effects would be introduced which might not be wholly corrected by SIC.

On the applications side, a satisfactory DFT treatment for van der Waals forces which would encompass both open- and closed-shell systems is still not available. Density-based formalisms for electron-atom, atom-atom and atom-molecule collisions are eagerly awaited. More trajectory calculations within QFD should yield interesting new insights in chemical dynamics. Excited-state chemistry also remains to be dealt with.

Whatever be the final outcome of the search for a density-based quantum mechanics, this promises much enthusiasm and excitement in the years ahead. Even if the search is not quite successful, it is bound to rejuvenate the whole of quantum chemistry.

ACKNOWLEDGMENT

The author is grateful to the Indian Institute of Technology, Bombay, for financial support to attend this Symposium. It is a pleasure to thank Professor J.P. Dahl and Professor J. Avery for their warm hospitality and additional financial support. This work is supported by a grant from the Department of Atomic Energy, Government of India.

REFERENCES

1. A.S. Bamzai and B.M. Deb, The role of single-particle density in chemistry, Rev. Mod. Phys. 53 , 95, 593 (1981).
2. B.M. Deb (Editor), "The Force Concept in Chemistry", Van Nostrand Reinhold Co., New York (1981).
3. R.F.W. Bader, The nature of chemical binding, in: "The Force Concept in Chemistry", B.M. Deb, ed., Van Nostrand Reinhold Co., New York (1981).
4. P. Becker (Editor), "Electron and Magnetization Densities in Molecules and Crystals", Plenum Press, New York (1980).
5. S.K. Ghosh and B.M. Deb, Densities, density functionals and electron fluids, Phys. Rep. 92, 1 (1982).
6. R.F.W. Bader, T.T. Nguyen-Dang, and Y. Tal, A topological theory of molecular structure, Rep. Prog. Phys. 44, 893 (1981).
7. O. Gunnarsson and R.O. Jones, Density functional calculations for atoms, molecules and clusters, Phys. Scrip. 21, 394 (1980).
8. B.M. Deb, On the direct calculation of electron density in many-electron systems, Curr. Sci. 50, 973 (1981).
9. R.P. Feynman, personal communication to the author.
10. B.M. Deb and A.S. Bamzai, Internal stresses in molecules I. One-electron systems, Mol. Phys. 35, 1349 (1978).
11. B.M. Deb and A.S. Bamzai, Internal stresses in molecules II. A local view of chemical binding in the H_2 molecule, Mol. Phys. 38, 2069 (1979).
12. P. Politzer and D.G. Truhlar (Editors), "Chemical Applications of Atomic and Molecular Electrostatic Potentials", Plenum Press, New York (1981).
13. A.S. Bamzai and B.M. Deb, Internal stresses and chemical binding in the H_2^+ molecule, Int. J. Quantum Chem. 20, 1315 (1981).
14. B.M. Deb and S.K. Ghosh, On some 'local' force densities and stress tensors in molecular quantum mechanics, J. Phys. B 12, 3857 (1979).
15. A.K. Rajagopal, Theory of inhomogeneous electron systems: spin-density-functional formalism, Adv. Chem. Phys. 41, 51 (1980).
16. L.D. Landau and E.M. Lifshitz, "Fluid Mechanics", p. 48, Pergamon Press, New York (1959).
17. L.J. Bartolotti and R.G. Parr, The concept of pressure in density functional theory, J.Chem. Phys. 72, 1593 (1980).
18. R.F.W. Bader, Quantum topology of molecular charge distributions III. The mechanics of an atom in a molecule, J. Chem. Phys. 73, 2871(1980).
19. C.J. Jameson and A.D. Buckingham, Nuclear magnetic shielding density, J. Phys. Chem. 83, 3366 (1979).

20. C.J. Jameson and A.D. Buckingham, Molecular electronic property density functions: the nuclear magnetic shielding density, J. Chem. Phys. 73, 5684 (1980).
21. L. Janossy, Wave mechanics and physical reality III. The many-body problem, Found. Phys. 6, 341 (1976).
22. C.Y. Wong, On the Schrödinger equation in fluid-dynamical form, J. Math. Phys. 17, 1008 (1976).
23. (a) J.O. Hirschfelder, Quantum mechanical equations of change. I, J. Chem. Phys. 68, 5151 (1978).
(b) J.O. Hirschfelder, The angular momentum, creation, and significance of quantized vortices, J. Chem. Phys. 67, 5477 (1977).
24. S.K. Ghosh and B.M. Deb, Quantum fluid dynamics of many-electron systems in three-dimensional space, Int. J. Quantum Chem. 22, 871 (1982).
25. B.M. Deb and S.K. Ghosh, Schrödinger fluid dynamics of many-electron systems in a time-dependent density-functional framework, J. Chem. Phys. 77, 342 (1982).
26. W.H. Adams, Orbital self-consistent-field theory. III. Hamiltonian for the natural orbitals of multiconfiguration wave functions, Phys. Rev. 183, 37 (1969).
27. R.A. Hegstrom and W.N. Lipscomb, Paramagnetism in closed-shell molecules, Rev. Mod. Phys. 40, 354 (1968).
28. A.D. McLachlan and M.R. Baker, A hydrodynamic model for diamagnetic induced currents in molecules, Mol. Phys. 4, 255 (1961).
29. R. Gurtler and D. Hestenes, Consistency in the formulation of the Dirac, Pauli, and Schrödinger theories, J. Math. Phys. 16, 573 (1975)
30. J.M. Okuniewicz, Quasiperiodic pointwise solutions of the periodic, time-dependent Schrödinger equation, J. Math. Phys. 15, 1587 (1974).
31. L.J. Bartolotti, Time-dependent extension of the Hohenberg-Kohn-Levy energy-density functional, Phys. Rev. A 24, 1661 (1981).
32. S.T. Epstein, "The Variation Method in Quantum Chemistry", p. 250, Academic Press, New York (1974).
33. S.K. Ghosh and B.M. Deb, Quantum fluid dynamics within a relativistic density-functional framework, to be published.
34. A.H. McDonald and S.H. Vosko, A relativistic density functional formalism, J. Phys. C 12, 2977 (1979).
35. T. Takabayashi, Relativistic Hydrodynamics of the Dirac Matter Part I. General theory, Progr. Theor. Phys. Suppl. Japan 4, 1 (1957).
36. J.A.N.F. Gomes, Delocalized magnetic currents in benzene, J.Chem. Phys., in press (1983).
37. J.A.N.F. Gomes, Topological elements of the magnetically induced orbital current densities, J. Chem. Phys., in press (1983).

38. J.A.N.F. Gomes, Topological theory of the electronic currents in benzene, to be published.
39. S.K. Ghosh and B.M. Deb, Dynamic polarizability of many-electron systems within a time-dependent density-functional theory, Chem. Phys. 71, 295 (1982).
40. M. Karplus and H.J. Kolker, A variation-perturbation approach to the interaction of radiation with atoms and molecules, J.Chem. Phys. 39, 1493 (1963).
41. M.J. Stott and E. Zaremba, Linear response theory within the density-functional formalism: Application to atomic polarizabilities, Phys. Rev. A 21, 12 (1980), erratum in Phys. Rev. A 22, 2293 (1980).
42. P.W. Langhoff and C.T. Corcoran, Stieltjes imaging of photoabsorption and dispersion profiles, J. Chem. Phys. 61, 146 (1974).
43. S.K. Ghosh and B.M. Deb, A density-functional calculation of dynamic dipole polarizabilities of noble gas atoms, Theoret. Chim. Acta 62, 209 (1983).
44. J.H. Weiner and A. Askar, Particle method for the numerical solution of the time-dependent Schrödinger equation, J. Chem. Phys. 54, 3534 (1971).
45. L.J. Bartolotti and J. Tyrrell, An alternative functional to determine dynamic multipole polarizabilities, Mol. Phys. 36, 79 (1978).
46. L.J. Bortolotti and S.T. Epstein, Comments on variational approximations within the hydrodynamic model, Mol. Phys. 38, 1311 (1979).
47. N.H. March, "Self-consistent Fields in Atoms", Pergamon Press, New York (1975).
48. E.H. Lieb, Thomas-Fermi and related theories of atoms and molecules, Rev. Mod. Phys. 53, 603 (1981).
49. G.P. Lawes and N.H. March, An approximate differential equation for calculating the electron density in closed shell atoms and in molecules, Phys. Scrip. 21, 402 (1980).
50. J.L. Gázquez and R.G. Parr, Two-parameter statistical model for atoms, J. Chem. Phys. 68, 2323 (1978).
51. W.P. Wang and R.G. Parr, Statistical atomic models with piecewise exponentially decaying electron densities, Phys. Rev. A 16, 891 (1977).
52 G.G. Hall and D. Martin, Approximate electron densities for atoms and molecules, Israel J. Chem. 19, 255 (1980).
53. P. Csavinszky, Introduction of the shell structure into the the Thomas-Fermi-Dirac and Thomas-Fermi-Dirac-Weizsäcker energy-density functionals for neutral atoms, Phys. Rev. A 24, 1215 (1981).
54. B.M. Deb and S.K. Ghosh, A new method for the direct calculation of electron density in many-electron systems. I. Application to closed-shell atoms, Proceedings of the Fourth International Congress in Quantum Chemistry, Int. J. Quantum Chem. 23, 1 (1983).

55. V.H. Smith and I. Absar, Basic concepts of quantum chemistry for electron density studies, Israel J. Chem. 16, 87 (1977).
56. E.R. Davidson, "Reduced Density Matrices in Quantum Chemistry", Academic Press, New York (1976).
57. H. Essen and R.F.W. Bader, paper presented at the Fourth International Congress in Quantum Chemistry, 1982.
58. J.A. Alonso and L.A. Girifalco, Non-local approximation to the exchange potential and kinetic energy of an inhomogeneous electron gas, Phys. Rev. A 17, 3735 (1978).
59. L. Hedin and B.I. Lundqvist, Explicit local exchange-correlation potentials, J. Phys. C 4, 2064 (1971).
60. P. Gombas, "Die Statistische Theorie des Atoms", Springer Verlag, Vienna (1949).
61. Y. Tomishima and K. Yonei, Solution of the Thomas-Fermi-Dirac equation with a modified Weizsäcker correction, J. Phys. Soc. Japan 21, 142 (1966).
62. R.F.W. Bader, Y. Tal, S.G. Anderson, and T.T. Nguyen-Dang, Quantum topology: theory of molecular structure and its change, Israel J. Chem. 19, 8 (1980).
63. K. Collard and G.G. Hall, Orthogonal trajectories of the electron density, Int. J. Quantum Chem. 12, 623 (1977).
64. P. Politzer, Electrostatic potential-electronic density relationships in atoms, J. Chem. Phys. 72, 3027 (1980).
65. P. Politzer, Electrostatic potential-electronic density relationships in atoms. II, J. Chem. Phys. 73, 3264 (1980).
66. G. Náray-Szabó, A. Grofcsik, K. Kósa, M. Kubinyi, and A. Martin, Simple calculation of electrostatic isopotential maps from bond fragments, J. Comp. Chem. 2, 58 (1981).
67. J.P. Perdew and A. Zunger, Self-interaction correction to density-functional approximations for many-electron systems, Phys. Rev. B 23, 5048 (1981).
68. V.L. Moruzzi, J.F. Janak, and A.R. Williams, "Calculated Electronic Properties of Metals", Pergamon Press, New York (1978).
69. J.P. Perdew, J.H. Rose, and H.B. Shore, Density functional theory of autodetaching states, J. Phys. B 14, L233 (1981).
70. J.P. Perdew, E.R. McMullen, and A. Zunger, Density-functional theory of the correlation energy in atoms and ions: A simple analytic model and a challenge, Phys. Rev. A 23, 2785 (1981).
71. J.L. Gázquez, E. Ortiz, and J. Keller, Electron gas model for inhomogeneous systems, Int. J. Quantum Chem. Symp. 13, 377 (1979).
72. O. Gunnarsson and B.I. Lundqvist, Exchange and correlation in atoms, molecules, and solids by the spin-density-functional formalism, Phys. Rev. B 13, 4274 (1976).
73. U. von Barth, Local-density theory of multiplet structure, Phys. Rev. A 20, 1693 (1979).
74. U. von Barth, Local-density theory of exchange and correlation energies for valence electrons, Phys. Scrip. 21, 585 (1980).

75. O. Gunnarsson and R.O. Jones, Extensions of the LSD approximation in density functional calculations, J. Chem. Phys. 72, 5357 (1980).
76. K. Kitaura, C. Satoko, and K. Morokuma, Total energies of molecules with the local density functional approximation and Gaussian basis sets, Chem. Phys. Lett. 65, 206 (1979).
77. M. Schlüter and L.J. Sham, Density functional theory, Phys. Today, p. 36, February (1982).
78. M.T. Yin and M.L. Cohen, Microscopic theory of the phase transformation and lattice dynamics of Si, Phys. Rev. Lett. 45, 1004 (1980).
79. A. Zunger, Ground-state properties of crystalline silicon in a density-functional pseudopotential approach, Phys. Rev. B 21, 4785 (1980).
80. A. Zunger and M.L. Cohen, First-principles nonlocal-pseudopotential approach in the density-functional formalism: Development and application to atoms, Phys. Rev. B 18, 5449 (1980).
81. A. Zunger and M.A. Ratner, On the first principles Hartree-Fock and local density pseudopotentials, Chem. Phys. 30, 423 (1978).
82. A. Zunger, Spin-dependent correlated atomic pseudopotentials, Phys. Rev. B 22, 649 (1980).
83. P.K. Lam, M.L. Cohen, and A. Zunger, Analytic representation for first-principles pseudopotentials, Phys. Rev. B 22, 1698 (1980).
84. H. Preuss, H. Stoll, U. Wedig, and Th. Krüger, Combination of pseudopotentials and density functionals, Int. J. Quantum Chem. 19, 113 (1981).
85. G.B. Bachelet, G.A. Baraff, and M. Schlüter, Defects in diamond: The unrelaxed vacancy and substitutional nitrogen, Phys. Rev. B 24, 4736 (1981).
86. G.B. Bachelet, H. Greenside, G.A. Baraff, and M. Schlüter, Structural-energy calculations based on norm-conserving pseudopotentials and localized Gaussian orbitals, Phys. Rev. B 24, 4745 (1981).
87. D.R. Hamann, Surface charge densities and atom diffraction, Phys. Rev. Lett. 46, 1227 (1981).
88. L.M. Sander, J.H. Rose, and H.B. Shore, Charge-density waves in two- and three-dimensional jellium, Phys. Rev. B 21, 2739 (1980).
89. A.W. Overhauser, Insights in many-electron theory from the charge density wave structure of potassium, plenary lecture, Fourth International Congress in Quantum Chemistry, D. Reidel, Dordrecht, in press.
90. M. Levy and J.P. Perdew, In defense of the Hohenberg-Kohn theorem and density functional theory, Int. J. Quantum Chem. 21, 511 (1982).
91. J.A. Alonso, D.J. González, and M.P. Iñiguez, Electron density in simple metals. Relation to bulk and surface properties, J. Phys. F 10, 1995 (1980).

92. J.R. Chelikowsky, Thomas-Fermi-pseudopotential approach for calculating the static properties of simple metals, Phys. Rev. B 21, 3074 (1980).
93. R. Evans and M. Hasegawa, A self-consistent theory of inhomogeneous liquid metals: Calculations of the electron and ion density profiles and the liquid-vapour surface tension of the alkali metals, J. Phys. C 14, 5225 (1981).
94. D.D. Koelling, Self-consistent energy band calculations, Rep. Prog. Phys. 44, 139 (1981).
95. U. Gupta and A.K. Rajagopal, Density functional formalism at finite temperatures with some applications, Phys. Rep. 87, 259 (1982).
96. E. Zaremba and D. Zobin, Knight shifts in simple metals: A single-ion approach, Phys. Rev. Lett. 44, 175 (1980).
97. (a) M.J. Stott, E. Zaremba, and D. Zobin, Quadrupole polarizabilities and Sternheimer antishielding factors in the density functional theory, Can. J. Phys. 60, 210 (1982)
(b) G.D. Mahan, Modified Sternheimer equation for polarizability, Phys. Rev. A 22, 1780 (1980).
98. C.T. Corcoran and J.O. Hirschfelder, The magnetic susceptibility of BH, J. Chem. Phys. 72, 1524 (1980).
99. M.J. Clugston, The calculation of intermolecular forces. A critical examination of the Gordon-Kim model, Adv. Phys. 27, 893 (1978).
100. M. Waldman and R.G. Gordon, Scaled electron gas approximation for intermolecular forces, J. Chem. Phys. 71, 1325 (1979).
101. M. Waldman and R.G. Gordon, Generalized electron gas-Drude model theory of intermolecular forces, J. Chem. Phys. 71, 1340 (1979).
102. P.E. Siska, One-electron model potential calculations of van der Waals forces. I. He*($2^1S,2^3S$)+Ne,Ar,Kr,Xe, J. Chem. Phys. 71, 3942 (1979).
103. C. Muhlhausen and R.G. Gordon, Electron-gas theory of ionic crystals, including many-body effects, Phys. Rev. B 23, 900 (1981).
104. N.D. Lang, Interaction between closed-shell systems and metal surfaces, Phys. Rev. Lett. 46, 842 (1981).
105. N.D. Lang, Density-functional studies of chemisorption on simple metals, AIP Conf. Proc. 61, 137 (1980)
106. M. Horbatsch and R.M. Dreizler, Time dependent Thomas-Fermi approach to atomic collisions I, Z. Phys. A 300, 119 (1981).
107. F. Bloch, Bremsvermögen von Atomen mit mehreren Elektronen, Z. Phys. 81, 363 (1933).
108. H. Jensen, Eigenschwingungen eines Fermi-Gases und Anwendung auf die Blochische Bremsformel für schnelle Teilchen. Z. Phys. 106, 620 (1937).
109. S.T. Epstein, The Hellmann-Feynman theorem, in: "The Force Concept in Chemistry", B.M. Deb, ed., Van Nostrand Reinhold Co., New York (1981).
110. F.W. Averill and G.S. Painter, Virial theorem in the density-functional formalism: Forces in H_2, Phys. Rev. B 24, 6795 (1981).

111. J. Harris, R.O. Jones, and J.E. Müller, Force calculations in the density functional formalism, J. Chem. Phys. 75, 3904 (1981).
112. P. Ziesche, The virial and Hellmann-Feynman theorems of an inhomogeneous electron gas, J. Phys. C 13, 3625 (1980).
113. W.H.E. Schwarz and S.Y. Chu, Relativistic contributions to ionization energies and bond lenghts, to be published.
114. S. Srebrenik and R. Pauncz, The notion of the force: analysis of the non-stationary case, Israel J. Chem. 19, 272 (1980).
115. J.A. Alonso and L.C. Balbás, A nonlocal approximation to the correlation energy of inhomogeneous electron systems, Phys. Lett. 81A, 467 (1981).
116. L. Wilk and S.H. Vosko, Estimates of non-local corrections to total, ionisation and single-particle energies, J. Phys. C 15, 2139 (1982).
117. J.L. Gázquez and E. Ortiz, Self-interaction and interelectronic exchange in the electron gas model for atoms, Chem. Phys. Lett. 77, 186 (1981).
118. E.S. Kryachko, Hohenberg-Kohn theorem, Int. J. Quantum Chem. 18, 1029 (1980).
119. R.K. Pathak and S.R. Gadre, Estimation of $\langle p \rangle$ and $\langle p^{-1} \rangle$ from atomic electron densities, J. Chem. Phys. 74, 5925 (1981).
120. E.V. Ludena, On the nature of the correction to the Weizsäcker term, J. Chem. Phys. 76, 3157 (1982).
121. K.F. Freed and M. Levy, Direct first principles algorithm for the universal electron density functional, J. Chem. Phys. 77, 396 (1982).
122. J. Riess and W. Münch, The theorem of Hohenberg and Kohn for subdomains of a quantum system, Theoret. Chim. Acta 58, 295 (1981).
123. M.S. Yusaf, G.P. Lawes, and N.H. March, Asymptotic forms of atomic scattering factors and momentum densities, Int. J. Quantum Chem. 17, 833 (1980)
124. G.A. Henderson, Variational theorems for the single-particle probability density and density matrix in momentum space, Phys. Rev. A 23, 19 (1981).
125. T. Koga and M. Morita, Energy-density relations in momentum space, Theoret. Chim. Acta 59, 639 (1981).
126. G.S. Painter, Calculation of the multipole potential from an arbitrary localized charge distribution, Phys. Rev. B 23, 1624 (1981).
127. R. Evans and T.J. Sluckin, A density functional theory for inhomogeneous charged fluids: Application to the surfaces of molten salts, Mol. Phys. 40, 413 (1980).
128. J. Heinrichs and P. Magain, Density-functional theory of the pressure distribution at metal surfaces, Phys. Stat. Sol. (b) 110, 495 (1982).
129. J. Katriel, R. G. Parr, and M.R. Nyden, Concerning the chemical potential of few-electron systems, J. Chem. Phys. 74, 2397 (1981).

130. J.A. Alonso and L.A. Girifalco, Atomic electronegativity from density functional theory, J. Chem. Phys. 73, 1313 (1980).
131. H. Gollisch and L. Fritsche, Density matrix calculations for molecules and clusters I. Theoretical foundations, J. Phys. B 14, 4441 (1981).
132. H. Gollisch, Density matrix calculations for molecules and clusters II. First-row diatomic molecules, J. Phys. B 14, 4451 (1981).
133. R. McWeeny, Some recent advances in density matrix theory, Rev. Mod. Phys. 32, 335 (1960).

THE MECHANICS OF AND AN EQUATION FOR THE ELECTRONIC CHARGE DENSITY

R.F.W. Bader and Hanno Essén

Department of Chemistry
McMaster University
Hamilton, Ontario, Canada L8S 4M1

Through a study of the topological theory of atoms in molecules, one is led to a description of the mechanics of the charge density in terms of a stress tensor. The stress tensor $\overleftrightarrow{\sigma}$ is totally determined by the first-order density matrix. Quantum mechanics yields a differential force law which equates the divergence of σ to the force density derived from the gradient of the total potential energy operator. We partition each side of the differential force law into a contribution determined by the charge density $\rho(\underset{\sim}{r})$ and another determined by correlation. By identifying the $\rho(\underset{\sim}{r})$ parts we obtain a vector differential equation for the charge density. Taking the divergence of both sides of this equation yields a fourth-order partial differential equation determining ρ in terms of the external potential.

1. INTRODUCTION

The topological properties of a molecular charge distribution yield a single unified theory of molecular structure, one which defines atoms, bonds structure and the mechanisms of structural change. The same theory defines all average properties of an atom in a molecule.

While an atom is defined by a property of the charge density and while this definition allows for a unique specification of all of its properties, these properties are determined by the state function. It has been possible to show, however, that a knowledge of just the first-order density matrix is sufficient to determine all the atomic properties. Thus one has a situation in which the atom itself is defined by a property of the charge density, while its properties require knowledge of the first-order density matrix. Thus there is a gap in this theory corresponding to finding the relationship between the properties of the atoms and the distribution of charge. This is the problem of density functional theory. It is the purpose of this paper to explore what we have learned about the relationships existing between the charge density and the properties of the system.

If our perspective and our approach to this problem are somewhat different it is because we have approached it as chemists and carried with us the belief that the properties of some total system are understandable in terms of the properties of its constituent atoms. By first determining how atoms and the other chemical concepts which make up the molecular structure theory are defined in terms of the charge density, we are able to bring to bear all of the qualitative knowledge that has been accumulated through the study of chemistry. What is to be stressed is that the present approach is based entirely on the properties we have observed for molecular charge distributions and upon our attempts to understand these properties using quantum mechanics.

Both the topological and the quantum mechanical facets of this theory have been recently reviewed.[1,2] We give here only a brief riview of the essential ideas which lead to a definition of the average properties of an atom in a molecule. The topological definition of an atom as the union of an attractor and its basin as defined in terms of the gradient vector field of the charge density ρ, imparts to the atom a particular mechanical property. This property is that the atom is bounded by a surface through which there is no flux in the gradient vector of ρ. As a consequence, the action integral for an atom equals zero, as it does for an isolated system and their action integrals exhibit the same variational properties. The result is that a single principle of quantum mechanics, the principle of stationary action,[3] defines the average properties of both the total system and of an atom in this total system. Hence all of the theorems of quantum mechanics which govern the relationships between the average values of a system such as Heisenberg's equation of motion, the virial and hypervirial theorems, apply to the atom in a molecule as well as to the total system. The same variation principle yields, of course, Schrödinger's equation. Within the Lagrangian formulation of the mechanics of Schrödinger's field one may define an energy momentum tensor whose components summarize the principal properties of this field. Through these properties one may obtain expressions for the forces which act locally on an element of charge density. By taking the virial of this force one obtains an expression for the local energy density. These equations provide the fundamental description of the mechanics governing the charge density. This description is given in terms of the stress tensor which is totally determined by the first-order density matrix, $\Gamma^{(1)}(\tilde{r},\tilde{r}')$. The integration of these local quantities over the basin of an atom yields expressions for their average atomic values, expressions which are identical to those obtained from the atomic statement of the principle of stationary action.

This paper is primarily concerned with the mechanics of ρ as determined by the first-order density matrix and with the consequences of the relationships between ρ and $\Gamma^{(1)}$ for density functional theory.

2. THE MECHANICS OF THE CHARGE DENSITY AND OF AN ATOM IN A MOLECULE

Any atomic property F is the atomic average of an effective singleparticle density $\rho_F(\underset{\sim}{r})$

$$F(\Omega) = \int_\Omega d\underset{\sim}{r}\ \rho_F(\underset{\sim}{r}) \tag{1}$$

where

$$\rho_F(\underset{\sim}{r}) = N \int d\tau'\ \tfrac{1}{2}\{\Psi^*\hat{F}(\underset{\sim}{r})\Psi + c.c.\} \tag{2}$$

In Eq. (2), the integration over $d\tau'$ implies the summation over all spin coordinates and an integration over the coordinates of all electrons but one, whose coordinates are denoted by the vector $\underset{\sim}{r}$. Eq. (1) is a fundamental statement obtained from the theory. The mode of integration in Eq. (2) is not arbitrary but is instead imposed by the atomic statement of the principle of stationary action. It is the result one obtains when one asks the question, 'are there subsystems of a total system which possess a physically realizable definition.' One notes that if the operator $\hat{F}$ is set equal to the unit operator, the expression in Eq. (2) yields the charge density. Properties such as the average force acting on an element of charge that are defined in terms of the Heisenberg equation of motion in terms of a commutator average, are obtained in a corresponding manner in this theory from Eq. (3)

$$\rho_F(\underset{\sim}{r}) = N \int d\tau'\ \tfrac{1}{2}\{\Psi^*[\hat{H},\hat{F}'(\underset{\sim}{r})]\Psi + c.c.\} \tag{3}$$

where

$$\hat{F} = (i/\hbar)[\hat{H},\hat{F}'(\underset{\sim}{r})] \tag{4}$$

For example, the average force exerted on the electrons is determined by the operator $-\nabla\hat{V}$, where $\hat{V}$ is the total potential energy operator. This operator is, from Heisenberg's equation of motion, given by

$$\frac{d\hat{\underset{\sim}{p}}}{dt} = (i/\hbar)\ [\hat{H},\hat{\underset{\sim}{p}}] = -\nabla\hat{V} \tag{5}$$

If $[\hat{H},\hat{F}'(\underset{\sim}{r})]$ generates two-electron interactions, then $\rho_F(\underset{\sim}{r})$ is the density of the corresponding property F at $\underset{\sim}{r}$ as determined by the average interaction of the electron at $\underset{\sim}{r}$ with the other electrons in the system.

Atomic properties have the necessary property of yielding the value of F for the total system when summed over all the atoms in the system,

$$\bar{F} = \langle\hat{F}\rangle = \sum_{\Omega} F(\Omega) \tag{6}$$

The vector current density associated with the property F is

$$\underset{\sim}{J}_F(\underset{\sim}{r},t) = (N\hbar/4mi) \int d\tau' \, \{\Psi^*\nabla(\hat{F}\Psi) - (\nabla\Psi^*)(\hat{F}\Psi)\} \tag{7}$$

The total time rate of change of a subsystem average value $F(\Omega,t)$ is

$$\frac{dF(\Omega,t)}{dt} = \frac{N}{2}\left\{\frac{i}{\hbar}\langle[\hat{H},\hat{F}]\rangle_{\Omega} + c.c.\right\} + \int dS(\underset{\sim}{r}) \left\{\frac{\partial S(\underset{\sim}{r})}{\partial \tau}\rho_F - (\underset{\sim}{J}_F \cdot \underset{\sim}{n}(\underset{\sim}{r}) + c.c.)\right\} \tag{8}$$

We have assumed for simplicity of expression that $\hat{F}$ has no implicit time dependence. The commutator average in Eq. (8) is the subsystem projection of the time rate of change of F for the total system. Thus corresponding to the Heisenberg equation

$$d\hat{F}/dt = (i/\hbar)[\hat{H},\hat{F}] \tag{9}$$

one has

$$\langle d\hat{F}/dt\rangle_{\Omega} + c.c. = (i/\hbar)\{\langle[\hat{H},\hat{F}]\rangle_{\Omega} + c.c.\} \tag{10}$$

The principle of stationary action generalised to apply to any region of space bounded by a surface of zero flux in $\nabla\rho$ may be written as

$$\delta W_{12}[\Psi,\Omega] = \varepsilon\int_{t_1}^{t_2} dt\{\langle d\hat{F}/dt\rangle_{\Omega} + c.c.\}/2 \tag{11}$$

that is, as the time integral of the temporal rate of change in the generator evaluated over just the space of the atom. The variation in the action integral $W_{12}[\Psi,\Omega]$ is brought about through the action of the operator $\hat{F}$, the generator of an infinitesmal unitary transformation on the state function. The change in atomic action for an infinitesmal time integral yields

$$\delta L[\Psi,\Omega] = (\varepsilon/2)\{(i/\hbar)\langle[\hat{H},\hat{F}]\rangle_{\Omega} + c.c.\} \tag{12}$$

for the variation of the atomic Lagrangian integral. For a stationary state the corresponding variation of the atomic energy functional is given by Eq. (13).

$$\delta E[\psi,\Omega] = (-\varepsilon/2)\{(i/\hbar)\langle[\hat{H},\hat{F}]\rangle_{\Omega} + c.c.\}/\langle\psi,\psi\rangle_{\Omega} \tag{13}$$

One notes that since both these expressions are derived from the variation of the atomic action integral they do not contain any contributions from the atomic surface, the unique variational property of the atomic class of subsystems. Thus the variational expressions for the atomic functionals $L[\Psi,\Omega]$ and $E[\psi,\Omega]$ are identical in form and content to the corresponding expressions for the total system.

3. VARIATIONAL DERIVATION OF THE ATOMIC FORCE LAW

If the generator $\hat{F}$ in Eq. (12) is set equal to $\hat{p}$, the momentum operator associated with the electronic coordinate $\underset{\sim}{r}$, the variation principle yields the integrated force law

$$\underset{\sim}{F}(\Omega,t) = m\int_{\Omega} d\underset{\sim}{r}(\partial \underset{\sim}{J}(\underset{\sim}{r})/\partial t) - \int dS(\underset{\sim}{r})\overleftrightarrow{\sigma}(\underset{\sim}{r})\cdot\underset{\sim}{n}(\underset{\sim}{r}) \tag{14}$$

where

$$\overleftrightarrow{\sigma}(\underset{\sim}{r}) = \frac{N\hbar^2}{4m}\int d\tau'\{\nabla(\nabla\Psi^*)\Psi + \Psi^*\nabla(\nabla\Psi) - \nabla\Psi^*\nabla\Psi - \nabla\Psi\nabla\Psi^*\} \tag{15}$$

The quantity $\overleftrightarrow{\sigma}$ is a symmetric dyadic, referred to as the stress tensor. Its dot product with $\underset{\sim}{n}(\underset{\sim}{r})$ gives the density of the force exerted on the element of the atomic surface which is specified by its outwardly directed normal $\underset{\sim}{n}(\underset{\sim}{r})$. The stress tensor $\overleftrightarrow{\sigma}(\underset{\sim}{r})$, which has the dimension of an energy density plays a dominant role in the description of the mechanical properties of an atom. It may be alternatively expressed as a functional of the one-density matrix as

$$\overleftrightarrow{\sigma}(\underset{\sim}{r}) = (\hbar^2/4m)\{(\nabla\nabla + \nabla'\nabla') - (\nabla\nabla' + \nabla'\nabla)\}\Gamma^{(1)}(\underset{\sim}{r},\underset{\sim}{r}')\Big|_{\underset{\sim}{r}=\underset{\sim}{r}'} \tag{16}$$

This variation of the Lagrangian yields the atomic average of the total force acting on the electronic density at $\underset{\sim}{r}$. Eq. 14 is the generalization of Ehrenfest's theorem to an atom. It constitutes the quantum analog of Newton's equation of motion in classical mechanics expressed in terms of a vector current density and a stress tensor, both defined in real space.

The differential or local statement of the atomic force law is obtained from the divergence equations satisfied by the energy momentum stress tensor of Schrödinger's field. The expression for the force density obtained in this manner is given in Eq. (17)

$$\underset{\sim}{F}(\underset{\sim}{r},t) = m\,\partial \underset{\sim}{J}(\underset{\sim}{r})/\partial t - \nabla\cdot\overleftrightarrow{\sigma}(\underset{\sim}{r}) \tag{17}$$

where

$$\underset{\sim}{F}(\underset{\sim}{r},t) = N\int d\tau'\ \Psi^*(-\nabla\hat{V})\Psi \tag{18}$$

Eq. (17) is the differential form of the integrated atomic force law given in Eq. (14).

$$\mathbf{F}(\Omega,t) = \int_\Omega d\mathbf{r}\ \mathbf{F}(\mathbf{r},t) \tag{19}$$

The force density $F(r,t)$ is seen to be the total force exerted on the electron at position $\mathbf{r}_i$

$$-\nabla_i \hat{V} = \sum_\alpha Z_\alpha \nabla_i (|r_i - X_\alpha|)^{-1} - \sum_{j\neq i} \nabla_i (|\mathbf{r}_i - \mathbf{r}_j|)^{-1} \tag{20}$$

averaged over the motion of the remaining electrons in the system to yield

$$\mathbf{F}(\mathbf{r},t) = \left[\sum_\alpha Z_\alpha \nabla(|\mathbf{r} - \mathbf{X}_\alpha|)^{-1}\right]\rho(\mathbf{r}) - 2\int d\mathbf{r}'\ \nabla(|\mathbf{r} - \mathbf{r}'|)^{-1}\ \Gamma^{(2)}(\mathbf{r},\mathbf{r}') \tag{21}$$

[where $\Gamma^{(2)}(r,r')$ is the diagonal element of the two-density matrix normalized to $N(N-1)/2$]. While $\mathbf{F}(r,t)$ clearly involves two-body forces it is totally determined by $\mathbf{J}(\mathbf{r})$ and $\overleftrightarrow{\sigma}(r)$, which are functionals of the one-density matrix $\Gamma^{(1)}(\mathbf{r},\mathbf{r}')$ evaluated at the point $\mathbf{r}=\mathbf{r}'$ in real space. The force density $\mathbf{F}(\mathbf{r},t)$ may be interpreted as the total force exerted on the element of charge density located at $\mathbf{r}$ and obeying an equation of motion in real space, Eq. (17). This equation is identical in form and physical content to Cauchy's first equation of motion of classical continuum mechanics.

For a stationary state the expression for the average force acting on the electronic charge of an atom is

$$\mathbf{F}(\Omega) = -\int dS(\mathbf{r})\ \overleftrightarrow{\sigma}(\mathbf{r})\cdot\mathbf{n}(\mathbf{r}) \tag{22}$$

In general physical terms, the average force exerted on some enclosed volume is equal to the negative of the integral of the pressure acting on each element of its surface - a statement of the physical content of Eq. (22). This same equation demonstrates that the effect of the environment on an open quantum system Ω is determined by just the flux in the forces across the boundary of Ω.

4. ATOMIC VIRIAL THEOREM

By setting the generator $\hat{F} = -\mathbf{r}\cdot\hat{p}$ in the variation of $L[\Psi,\Omega]$, one obtains the atomic statement of the virial theorem,

$$-2T_K(\Omega) = V_b(\Omega) + V_s(\Omega) + V_p(\Omega) + \frac{\hbar^2}{4m}\int_\Omega d\mathbf{r}\ \nabla^2\rho(\mathbf{r}) \tag{23}$$

In Eq. (23), $V_b(\Omega)$ is the virial of the forces exerted on an electron over the basin of the atom

$$V_b(\Omega) = \int_\Omega d\mathbf{r} \int d\tau'\{-\mathbf{r}\cdot\nabla\hat{V}\}\Psi^*\Psi \tag{24}$$

$V_s(\Omega)$ is the virial of the forces exerted on the surface of the atom

$$V_s(\Omega) = \int dS(\underset{\sim}{r})\underset{\sim}{r}\cdot\overset{\leftrightarrow}{\sigma}(\underset{\sim}{r})\cdot\underset{\sim}{n}(\underset{\sim}{r}) \tag{25}$$

and $V_{\dot{p}}(\Omega)$, specific to a time-dependent system, is the virial of the forces arising from the time rate of change of the momentum density $m\underset{\sim}{J}$ throughout the atom

$$V_{\dot{p}}(\Omega) = -m \int_\Omega d\underset{\sim}{r}\ \underset{\sim}{r}\cdot\partial\underset{\sim}{J}/\partial t \tag{26}$$

The average kinetic energy of an atom in a molecule $T(\Omega)$ may be defined in terms of either of the single-particle kinetic energy densities given in Eq. (27)

$$\begin{aligned}(-\hbar^2/4m)\{\nabla^2 + \nabla'^2)\Gamma^{(1)}(\underset{\sim}{r},\underset{\sim}{r}')\}_{\underset{\sim}{r}=\underset{\sim}{r}'}\\ = (\hbar^2/2m)\{\nabla\cdot\nabla'\Gamma^{(1)}(\underset{\sim}{r},\underset{\sim}{r}')\}_{\underset{\sim}{r}=\underset{\sim}{r}'} - (\hbar^2/4m)\ \nabla^2\rho(\underset{\sim}{r})\end{aligned} \tag{27}$$

an equation which may be stated in abbreviated notation as

$$K(\underset{\sim}{r}) = G(\underset{\sim}{r}) + L(\underset{\sim}{r}) \tag{28}$$

The quantity $L(r)$ is the Lagrangian density of a quantum mechanical system. Since its integration over the volume of an atom vanishes, one finds that the average kinetic energy of an atom $T(\Omega)$ is equal to the average of either kinetic energy density

$$T(\Omega) = \int_\Omega d\underset{\sim}{r}\ K(\underset{\sim}{r}) = \int_\Omega d\underset{\sim}{r}\ G(\underset{\sim}{r}) \tag{29}$$

While the values of the individual contributions to the virial of an atom are dependent upon the choice of origin for the vector $\underset{\sim}{r}$, their sum, which determines the total atomic virial

$$V(\Omega) = V_b(\Omega) + V_s(\Omega) + V_p(\Omega) \tag{30}$$

is independent of the choice of origin.

By taking the virial of $F(r,t)$ in Eq. (17) one obtains an expression for the local virial relationship

$$\underset{\sim}{r}\cdot\underset{\sim}{F}(\underset{\sim}{r},t) = m\underset{\sim}{r}\cdot\partial\underset{\sim}{J}/\partial t - \underset{\sim}{r}\cdot\nabla\cdot\overset{\leftrightarrow}{\sigma} \tag{31}$$

using the identity

$$\nabla\cdot(\underset{\sim}{r}\cdot\overset{\leftrightarrow}{\sigma}) = \mathrm{Tr}\ \overset{\leftrightarrow}{\sigma} + \underset{\sim}{r}\cdot\nabla\cdot\overset{\leftrightarrow}{\sigma} \tag{32}$$

Where $\mathrm{Tr}\ \overset{\leftrightarrow}{\sigma}$ denotes the trace or spur of the stress tensor, Eq. (31) gives

$$\underset{\sim}{r}\cdot\underset{\sim}{F}(\underset{\sim}{r},t) = m\underset{\sim}{r}\cdot\partial\underset{\sim}{J}/\partial t + \mathrm{Tr}\,\overleftrightarrow{\sigma} - \nabla\cdot(\underset{\sim}{r}\cdot\overleftrightarrow{\sigma}) \tag{33}$$

The trace

$$\begin{aligned}\mathrm{Tr}\,\overleftrightarrow{\sigma}(\underset{\sim}{r}) &= -K(\underset{\sim}{r}) - G(\underset{\sim}{r}) = -2K(\underset{\sim}{r}) - (\hbar^2/4m)\,\nabla^2\rho(\underset{\sim}{r})\\ &= -2G(\underset{\sim}{r}) + (\hbar^2/4m)\,\nabla^2\rho(\underset{\sim}{r})\end{aligned} \tag{34}$$

Its integral over an atomic volume yields the unique value

$$\int_\Omega d\underset{\sim}{r}\;\mathrm{Tr}\,\overleftrightarrow{\sigma}(\underset{\sim}{r}) = -2T(\Omega) \tag{35}$$

Using the final identity given in Eq. (34), Eq. (33) may be expressed as

$$-2G(\underset{\sim}{r}) = \underset{\sim}{r}\cdot F(\underset{\sim}{r},t) + \nabla\cdot(\underset{\sim}{r}\cdot\overleftrightarrow{\sigma}(\underset{\sim}{r})) - m\underset{\sim}{r}\cdot\partial\underset{\sim}{J}(\underset{\sim}{r})/\partial t - (\hbar^2/4m)\,\nabla^2\rho(\underset{\sim}{r}) \tag{36}$$

Integration of Eq. (36) over the atomic volume yields, term for term, the atomic virial theorem for a time dependent system, Eq. (23). Thus Eq. (36) is, in terms of its derivation and in terms of its integrated form, a local expression of the virial theorem.[4] The atomic virial theorem provides the basis for the definition of the average energy of an atom.

5. THE ENERGY OF AN ATOM IN A MOLECULE

The potential energy of a molecular system may be expressed in terms of the virials of the corresponding forces exerted on the electrons and nuclei as

$$\sum_i (-\underset{\sim}{r}_i\cdot\nabla_i\hat{V}) + \sum_\alpha (-\underset{\sim}{X}_\alpha\cdot\nabla_\alpha\hat{V}) = \hat{V} \tag{37}$$

or as

$$\sum_i (\underset{\sim}{r}_i\cdot\hat{\underset{\sim}{F}}_i) + \sum_\alpha (\underset{\sim}{X}_\alpha\cdot\hat{\underset{\sim}{F}}_\alpha) = \hat{V} \tag{38}$$

The force operator $\hat{F}_\kappa$ for particle κ is the classical force exerted on this particle by all the particles in the system. Correspondingly, one may consider the virial of this force, $\underset{\sim}{r}_\kappa\cdot\underset{\sim}{F}_\kappa$, to be the potential energy operator for particle κ since, according to Eq. (38) the sum of such operators for all the particles in the system yields the total potential energy operator $\hat{V}$. The first summation in Eq. (38) - the sum of the potential energy operators for the electrons - should, together with $\hat{T}$, define the electronic Hamiltonian of the system and this in turn its electronic energy E_e. Thus

$$E_e = \langle\hat{H}_e\rangle = \langle(\hat{T} + \sum_i \underset{\sim}{r}_i\cdot\hat{\underset{\sim}{F}}_i)\rangle \tag{39}$$

The importance of expressing the electronic energy in the form of Eq. (39) is that it enables one to express $\hat{H}_e$ and hence E_e as sums over single particle contributions:

$$\hat{H}_e = \sum_i \hat{H}_i = \sum_i (-\tfrac{1}{2}\nabla_i^2 - \underset{\sim}{r}_i \cdot \nabla_i \hat{V}) \tag{40}$$

and

$$E_e = N\langle\hat{H}_i\rangle = N \int d\underset{\sim}{r} \int d\tau' \ \{\psi^*(-\tfrac{1}{2}\nabla_i^2 - \underset{\sim}{r}_i \cdot \nabla_i \hat{V})\psi\} \tag{41}$$

Thus the electronic energy is N times the average energy of a single electron, and E_e may be expressed as an integral over an effective single-particle energy density

$$E_e = \int d\underset{\sim}{r} \ (K(\underset{\sim}{r}) + V(\underset{\sim}{r})) \tag{42}$$

In Eq. (46), $G(\underset{\sim}{r})$ is the kinetic energy density and $V(\underset{\sim}{r})$ is the virial field:

$$V(\underset{\sim}{r}) = N \int d\tau' \ \{\psi^*(-\underset{\sim}{r} \cdot \nabla\hat{V})\psi\} \tag{43}$$

the field which determines the atomic virial $V(\Omega)$. The virial field $V(\underset{\sim}{r})$ is N times the potential energy density of one electron at $\underset{\sim}{r}$ as determined by its average energy of interaction with all the other particles in the system. $V(\underset{\sim}{r})$ is an exact prescription of the "average field" experienced by a single electron in a many-particle system.

Thus the electronic energy since it may be expressed as an integral over effective single-particle kinetic and potential densities may be defined for an atom by restricting the integration in Eq. (42) to the volume Ω

$$E_e(\Omega) = T(\Omega) + V_b(\Omega) \tag{44}$$

Since the virial theorem applies to the atom in a molecule one has

$$E_e(\Omega) = -\int_\Omega d\underset{\sim}{r} \ K(\underset{\sim}{r}) = -\int_\Omega d\underset{\sim}{r} \ G(\underset{\sim}{r}) = -T(\Omega) = \tfrac{1}{2}V_b(\Omega) \tag{45}$$

In summary, through the concept of the average force exerted on an electron at $\underset{\sim}{r}$ - the force density $\underset{\sim}{F}(\underset{\sim}{r})$ of the differential form of the equation of motion, Eq. (17) - one may define a corresponding single-particle potential energy density as the virial of this force - the virial field. The atomic average of the virial field is related to an average kinetic energy through the atomic virial theorem. The kinetic and potential energies so obtained define the electronic energy of the atom, $E_e(\Omega)$. The energy $E_e(\Omega)$ satisfies the same theorems as does E_e for the total system since both are defined in terms of the same variational principle with the same generator.

The atomic energies $E_e(\Omega)$, $T(\Omega)$, and $V(\Omega)$ possess a most important property: The extent of constancy of the atom as it is manifested in the real space of various systems is reflected in the corresponding constancy in its contribution to the total energy of a system, the limit of this property being a constant contribution to the total energy when the atom is perfectly transferable between systems. The observation that regions of space bounded by a zero flux surface satisfied this requirement with respect to an energy as defined in Eq. (44) formed the original basis for the identification of such spatial regions with the chemical atoms of a system[5]. This observation may now be predicted from theory as a consequence of the properties of the stress tensor.

In a stationary state the stress tensor determines the force density $\underset{\sim}{F}(\underset{\sim}{r})$

$$\underset{\sim}{F}(\underset{\sim}{r}) = -\nabla\cdot\overleftrightarrow{\sigma}(\underset{\sim}{r}) \tag{46}$$

and, since the virial field is the virial of this force, it determines $V(\underset{\sim}{r})$ as well;

$$V(\underset{\sim}{r}) = \underset{\sim}{r}\cdot\underset{\sim}{F}(\underset{\sim}{r}) = -\underset{\sim}{r}\cdot\nabla\cdot\overleftrightarrow{\sigma}(\underset{\sim}{r}) \tag{47}$$

The integral of $V(\underset{\sim}{r})$ yields the virial of the forces over the basin of the atom $V_b(\Omega)$. The local expression for the virial theorem may be used to obtain a virial density $V'(\underset{\sim}{r})$ that integrates to the total virial of the atom $V_b(\Omega) + V_s(\Omega)$. This is given by

$$V'(\underset{\sim}{r}) = \underset{\sim}{r}\cdot\underset{\sim}{F}(\underset{\sim}{r}) + \nabla\cdot(\underset{\sim}{r}\cdot\overleftrightarrow{\sigma}(\underset{\sim}{r})) = \mathrm{Tr}\ \overleftrightarrow{\sigma}(\underset{\sim}{r}) \tag{48}$$

From the virial theorem for an atom or for the total system Eq. (48) is seen to yield an expression for the energy density $E_e(\underset{\sim}{r})$ in terms of the trace of the stress tensor:

$$E_e(\underset{\sim}{r}) = \tfrac{1}{2}\mathrm{Tr}\ \overleftrightarrow{\sigma}(\underset{\sim}{r}) \tag{49}$$

and

$$E_e(\Omega) = \tfrac{1}{2}\int_\Omega d\underset{\sim}{r}\ \mathrm{Tr}\ \overleftrightarrow{\sigma}(\underset{\sim}{r}) = \int_\Omega d\underset{\sim}{r}\ (G(\underset{\sim}{r}) + V'(\underset{\sim}{r})) \tag{50}$$

where Ω may refer to an atomic volume or to the volume of the total system.

The energy density $E_e(\underset{\sim}{r})$, the virial fields $V(\underset{\sim}{r})$ or $V'(\underset{\sim}{r})$ and the force density $\underset{\sim}{F}(\underset{\sim}{r})$ all involve two-electron interactions and require $\Gamma^{(2)}(\underset{\sim}{r},\underset{\sim}{r}')$ for their direct evaluation. Yet from the above relationships, they may all be expressed in terms of the stress tensor that is a functional of the first-order density matrix. The kinetic energy density is determined by the trace of the stress tensor.

The diagonal element of $\Gamma^{(1)}(\underset{\sim}{r},\underset{\sim}{r}')$ determines the charge density at the point $\underset{\sim}{r}=\underset{\sim}{r}'$. If the properties of the first-order density matrix are identical over the basin of a given atom in two different molecular systems, then the atom's distribution of charge will be identical in the two systems as will its spatial distribution of energy, Eq. (49), and its potential and kinetic components, Eqs. (48) and (34). In such a case, the atom will appear identical in the two different systems and will contribute identical amounts to the total energies of both systems. It is an observation which we have now made in many systems that when the distribution of charge over an atom in two different systems remains sensibly unchanged then the energy and other properties of the atom remain unchanged to a corresponding degree. Thus while it is necessary at the moment to use $\Gamma^{(1)}$ as a bridge to illustrate the constancy of properties that coincides with the constancy of the distribution of charge, our observations tell us that this bridge may some day be abandoned, and replaced with a direct functional dependence between the energy and the charge density ρ over the basin of an atom.

A density functional theory must possess the properties exhibited by the expression for $E_e(\underset{\sim}{r})$ in Eq. (49): when the distribution $\Gamma^{(1)}(\underset{\sim}{r},\underset{\sim}{r}')$ in the neighbourhood of $\underset{\sim}{r}=\underset{\sim}{r}'$ and hence $\rho(\underset{\sim}{r})$ are identical over well-defined regions of space in two different systems, these regions must contribute identical amounts to the energies of both systems. This observation, which has been shown to be a quantum mechanical property of a molecular system, is the physical basis and the raison d'être for the belief that a density functional theory for the energy can be found.

Thus there is a relation between the virial field and the distribution of charge in a molecular system as was originally proposed on the basis of the observed properties of the topological atoms.[5] They are related through $\Gamma^{(1)}(\underset{\sim}{r},\underset{\sim}{r}')$ by the quantum mechanical stress tensor $\overleftrightarrow{\sigma}$.

Absolute transferability of an atom and its properties between systems is probably never obtained. This is unimportant. The limit of perfect transferability or the approach to this limit serves to establish the principle that an atom and its properties are defined by the distribution of charge in real space, and the extent to which its properties change is directly determined by the extent to which its distribution of charge changes.

The stress tensor $\overleftrightarrow{\sigma}$ determines the mechanics of the electronic charge distribution. The force density and energy density are expressible in terms of this tensor. Of particular importance is the differential statement of the force law which for a stationary state is

$$N \int d\tau' \, \psi^*(-\nabla\hat{V})\psi = -\nabla\cdot\overleftrightarrow{\sigma}(\underset{\sim}{r})$$
$$= \rho(\underset{\sim}{r})\sum_\alpha Z_\alpha \nabla(|\underset{\sim}{r} - \underset{\sim}{X}_\alpha|)^{-1}$$
$$- 2\int d\underset{\sim}{r}'\Gamma^{(2)}(\underset{\sim}{r},\underset{\sim}{r}')\nabla(|\underset{\sim}{r} - \underset{\sim}{r}'|)^{-1} \quad (51)$$

The divergence of $\overleftrightarrow{\sigma}(\underset{\sim}{r})$ equals ρ times the gradient of the so-called electrostatic potential $\Phi(\underset{\sim}{r})$

$$\Phi(\underset{\sim}{r}) = -\sum_\alpha Z_\alpha (|\underset{\sim}{r} - \underset{\sim}{X}_\alpha|)^{-1} + \int d\underset{\sim}{r}' \rho(\underset{\sim}{r}')(|\underset{\sim}{r} - \underset{\sim}{r}'|)^{-1} \quad (52)$$

plus a contribution arising from the correlated motions of the electrons. Thus the force density is a sum of two forces; one is conservative and the other, determined by the correlated motions of the electrons, is not. Since $\overleftrightarrow{\sigma}$ is expressible in terms of the first-order density matrix, it is clear that $-\nabla\cdot\overleftrightarrow{\sigma}$ extracts from the second-order density matrix only that information regarding electron correlation which is essential for determining the physics. In what follows we argue that $-\nabla\cdot\overleftrightarrow{\sigma}$ may itself be expressed as a sum of two contributions, corresponding to the conservative and non-conservative contributions to the force derived from the operator $-\nabla\hat{V}$. This partitioning of the divergence of $\overleftrightarrow{\sigma}$ yields separate equations for the conservative and non-conservative components of the total force density. The former is free of correlation and is expressible as a functional of the charge density. From it one may obtain a Poisson equation for $\rho(\underset{\sim}{r})$, whose solution is determined by specifying the external potential.

Deb and Ghosh,[6] and Bartolotti and Parr[7] have previously defined and examined the properties of stress tensors in many-electron systems, but from the point of view of density functional theory.

6. SEPARATION OF CORRELATION FROM $\Gamma^{(1)}$ AND $\Gamma^{(2)}$

In this section we will show how both $\Gamma^{(1)}$ and the diagonal elements of $\Gamma^{(2)}$ can be written as sums of two parts; the first, local parts, being determined by $\rho(\underset{\sim}{r})$ and the second, "non-local", requiring knowledge of six-dimensional functions. (We will be working with the spin-averaged, i.e. spin-independent versions of these quantities since our operators have this property.) In the past McWeeny has written the diagonal elements of $\Gamma^{(2)}$, here denoted

$$\rho_2(\underset{\sim}{r}_1,\underset{\sim}{r}_2) \equiv \Gamma^{(2)}(\underset{\sim}{r}_1,\underset{\sim}{r}_2;\underset{\sim}{r}_1,\underset{\sim}{r}_2) \quad (53)$$

as

$$\rho_2(\underset{\sim}{r}_1,\underset{\sim}{r}_2) = \rho(\underset{\sim}{r}_1)\rho(\underset{\sim}{r}_2)[1 + f(\underset{\sim}{r}_1,\underset{\sim}{r}_2)] \quad (54)$$

More recently Ludena[8] has given $\Gamma^{(1)}$ a similar form

$$\Gamma^{(1)}(\underset{\sim}{r},\underset{\sim}{r}') = [\rho(\underset{\sim}{r})\rho(\underset{\sim}{r}')]^{1/2} [1 + B(\underset{\sim}{r},\underset{\sim}{r}')/\rho(\underset{\sim}{r})\rho(\underset{\sim}{r}')]^{1/2} \qquad (55)$$

For our purposes it is more convenient to introduce a new function β defined through the following expression

$$\Gamma^{(1)}(\underset{\sim}{r},\underset{\sim}{r}') = [\rho(\underset{\sim}{r})\rho(\underset{\sim}{r}')]^{1/2} + \beta(\underset{\sim}{r},\underset{\sim}{r}') \qquad (56)$$

where thus

$$\beta(\underset{\sim}{r},\underset{\sim}{r}) \equiv 0 \qquad (57)$$

This β is then related to Ludena's B through

$$\beta = \frac{1}{2} B(\rho\rho')^{-1/2} [1 - \frac{1}{4}(B/\rho\rho') + \frac{1}{8}(B/\rho\rho')^2 - \ldots] \qquad (58)$$

One could use a similar trick to rewrite ρ_2 in (54) but in our opinion this does not separate out correlation in an optimal way as will be discussed below. Instead we proceed as follows: define the function $\Phi^{(2)}$ through

$$\Gamma^{(2)}(\underset{\sim}{r}_1,\underset{\sim}{r}_2;\underset{\sim}{r}_1',\underset{\sim}{r}_2') = \frac{1}{2}\frac{N-1}{N}\Gamma^{(1)}(\underset{\sim}{r}_1,\underset{\sim}{r}_1')\Gamma^{(1)}(\underset{\sim}{r}_2,\underset{\sim}{r}_2') + \Phi^{(2)}(\underset{\sim}{r}_1,\underset{\sim}{r}_2;\underset{\sim}{r}_1',\underset{\sim}{r}_2') \qquad (59)$$

It then follows, from our definitions, that

$$\int\Phi^{(2)}(\underset{\sim}{r}_1,\underset{\sim}{r}_2;\underset{\sim}{r}_1',\underset{\sim}{r}_2)dv_2 \equiv 0 \qquad (60)$$

For the diagonal elements we introduce the notation

$$\phi(\underset{\sim}{r}_1,\underset{\sim}{r}_2) = \Phi^{(2)}(\underset{\sim}{r}_1,\underset{\sim}{r}_2;\underset{\sim}{r}_1,\underset{\sim}{r}_2) \qquad (61)$$

and thus we finally get

$$\rho_2(\underset{\sim}{r}_1,\underset{\sim}{r}_2) = \frac{1}{2}\frac{N-1}{N}\rho(\underset{\sim}{r}_1)\rho(\underset{\sim}{r}_2) + \phi(\underset{\sim}{r}_1,\underset{\sim}{r}_2) \qquad (62)$$

where now

$$\int\phi(\underset{\sim}{r}_1,\underset{\sim}{r}_2)dv_1 = \int\phi(\underset{\sim}{r}_1,\underset{\sim}{r}_2)dv_2 \equiv 0 \qquad (63)$$

This way of partitioning ρ_2 into a ρ and a correlation (including, of course, exchange, i.e. Fermi-correlation) dependent part has many desirable properties. For a one-electron system where, by definition there is no correlation, ρ_2 must vanish identically. Eq. (54) thus requires $f \equiv -1$ in the hydrogen atom while eq. (62) achieves this with $\phi \equiv 0$. The factor $(N-1)/N$, long ago introduced by Fermi and Amaldi in another similar context, corresponds in traditional terminology to a smeared out "correlation hole". This terminology is somewhat misleading, however, since the only purpose of this factor is to account for the fact that an electron does not interact with itself, a fact which has nothing to do with real correlation defined as the tendency of two different electrons to avoid being in the same place.

Using Eq. (56) and (62) the energy of a many-electron system can be written as a functional

$$E_o[\rho] + E_{corr}[\beta,\phi] \tag{64}$$

If one neglects the correlation part E_{corr}, one obtains an explicit density functional. The Euler equation resulting from this functional can be shown to give the Hartree-Fock result for a two-electron ground state (Tal and Bader[9]). Since the (restricted) Hartree-Fock ground state of a two-electron system by definition has no correlation this fact strongly favours our identification of ϕ with the correlation part of ρ_2.

7. AN EQUATION FOR THE CHARGE DENSITY?

In this section we will derive a differential equation for the electronic charge density $\rho(\underset{\sim}{r})$ based on the differential force law (51) and the assumption that the functions β and ϕ, introduced in the previous section, are equivalent in a sense made more precise below.

If we introduce (56) into the expression

$$\overleftrightarrow{\sigma}(\underset{\sim}{r}) = -\nabla\nabla'\Gamma^{(1)}(\underset{\sim}{r},\underset{\sim}{r}')\Big|_{\underset{\sim}{r}=\underset{\sim}{r}'} + \frac{1}{4}\nabla\nabla\rho(\underset{\sim}{r}) \tag{65}$$

for the stress tensor we get the following partitioning of the force density

$$\underset{\sim}{F}(\underset{\sim}{r}) = -\nabla\cdot\overleftrightarrow{\sigma} = \underset{\sim}{F}^{d}[\rho] + \underset{\sim}{F}^{nd}[\beta] \tag{66}$$

In Eq. (66) (we write the Hessian of ρ as $\nabla\nabla\rho$)

$$\underset{\sim}{F}^{d} = -\frac{1}{4}\left[\nabla(\nabla^2\rho) + \frac{\nabla\rho}{\rho}\left(\frac{\nabla\rho\cdot\nabla\rho}{\rho} - \nabla^2\rho\right) - \frac{\nabla\rho\cdot(\nabla\nabla\rho)}{\rho}\right] \tag{67}$$

is determined by ρ, the diagonal of $\Gamma^{(1)}$ and $\underset{\sim}{F}^{nd}$ by β, its non-diagonal part. On the other side of the differential force law we introduce Eq. (62) for ρ_2 and thus get another partitioning of the force density

$$\underset{\sim}{F}(\underset{\sim}{r}) = \underset{\sim}{F}^{e}[\rho] + \underset{\sim}{F}^{xc}[\phi] \tag{68}$$

into an electrostatic, conservative part

$$\underset{\sim}{F}^{e} = \rho(\underset{\sim}{r})\nabla\left[\sum_{\alpha}\frac{Z_\alpha}{|\underset{\sim}{r} - \underset{\sim}{X}_\alpha|} + \frac{N-1}{N}\int\frac{\rho(\underset{\sim}{r}')}{|\underset{\sim}{r} - \underset{\sim}{r}'|}\,dv'\right] \tag{69}$$

and a part determined by correlation (including exchange), $\underset{\sim}{F}^{xc}$. Note that the factor (N-1)/N is present because the field is not that acting on an external test charge, but one acting on an element of itself, i.e., on one of the N electrons in the system. The field is thus an average field from N-1 electrons and the nucleus. In this way one achieves the correct asymptotic behaviour for large $\underset{\sim}{r}$ (see e.g. Essén[10]) and one also automatically gets the exact result for a one-electron system.

The basic assumption that we now make, in order to find an equation for ρ, is that we can separately identify the ρ-parts and the correlation parts of the force densities (66) and (68), i.e. that

$$\underset{\sim}{F}^{d}[\rho] = \underset{\sim}{F}^{e}[\rho] \tag{70}$$

This may be an approximation but there is reason to believe that it is not crude, since it does not involve neglect of correlation, but rather identification of two different manifestations of correlation. Our experience with kinetic energy densities indicates that β contributes most in regions where many orbitals contribute to the charge density (Tal and Bader[9]). This is also where the (Fermi) correlation will be most important and thus the contribution of ϕ to the potential energy. These are our intuitive arguments in favour of Eq. (70).

Taking the divergence of both sides of Eq. (70) yields the following partial differential equation for ρ

$$\begin{aligned}\frac{1}{\rho}\nabla^2\nabla^2\rho - \frac{3}{\rho^2}\nabla\rho\cdot\nabla(\nabla^2\rho) + \frac{4}{\rho^3}\nabla\rho\cdot(\nabla\nabla\rho)\cdot\nabla\rho + \frac{3}{\rho^3}\nabla^2\rho(\nabla\rho\cdot\nabla\rho)\\ - \frac{1}{\rho^2}(\nabla^2\rho)^2 - \frac{1}{\rho^2}\,\mathrm{Tr}[(\nabla\nabla\rho)^2] - \frac{3}{\rho^4}(\nabla\rho\cdot\nabla\rho)^2\\ = 16\pi\Big(\sum_\alpha Z_\alpha\delta(\underset{\sim}{r} - \underset{\sim}{R}_\alpha) - \frac{N-1}{N}\rho\Big)\end{aligned} \tag{71}$$

This equation is a bit complicated but, for a spherically symmetric atom the problem simplifies greatly. Eq. (70) can then be written as the equality of the radial components

$$F_r^d = F_r^e \tag{72}$$

where

$$F_r^d = -\frac{1}{4}\Big[\frac{\partial^3\rho}{\partial r^3} - \frac{2}{\rho}\frac{\partial\rho}{\partial r}\frac{\partial^2\rho}{\partial r^2} + \frac{1}{\rho^2}\Big(\frac{\partial\rho}{\partial r}\Big)^3 + \frac{2}{r}\Big(\frac{\partial^2\rho}{\partial r^2} - \frac{1}{\rho}\Big(\frac{\partial\rho}{\partial r}\Big)^2\Big) - \frac{2}{r^2}\frac{\partial\rho}{\partial r}$$

and (73)

$$F_r^e = -\frac{\rho(r)}{r^2}\Big[Z - 4\pi\,\frac{N-1}{N}\int_0^r \rho(r')r'^2\,dr'\Big] \tag{74}$$

Taking the divergence of both sides gives (here prime denotes differentiation with respect to r)

$$\rho'''' = \frac{3}{\rho}\rho'\rho''' - \frac{7}{\rho^2}(\rho')^2\rho'' + \frac{2}{\rho}(\rho'')^2 + \frac{3}{\rho^3}(\rho')^4 + \frac{1}{r}\left(\frac{10}{\rho}\rho'\rho'' - \frac{6}{\rho^2}(\rho')^3 - 4\rho'''\right) - 16\pi\frac{N-1}{N}\rho^2 \qquad (75)$$

Eqs. (72) - (75) together with the assumption that ρ and its derivatives remain finite when $r \to 0$ give the following boundary conditions[11]

$$\rho'(0) = -2Z\rho(0) \qquad (76)$$

$$\rho'''(0) = 12Z^3\rho(0) - 5Z\rho''(0) \qquad (77)$$

$$\rho''''(0) = \frac{1}{5}[12(\rho''(0))^2/\rho(0) - 10Z^2\rho''(0) - 72Z^4\rho(0) - 16\pi\frac{N-1}{N}\rho^2(0)] \qquad (78)$$

A solution of the fourth order differential Eq. (75) is thus determined by the four parameters:

$$Z,\ N,\ \rho(0),\ \rho''(0) \qquad (79)$$

Naturally one must also demand that ρ normalizes to N and it seems likely that this can only be achieved for some (possibly discrete) values of $\rho(0)$ and $\rho''(0)$. Since the Pauli principle has not been built into these equations it can be expected that unphysical Bosonic solutions will exist. At the time of writing we are programming these equations and eagerly awaiting the results.

REFERENCES

1. R.F.W. Bader, T.T. Nguyen-Dang, and Y. Tal, Rep. Prog. Phys. 44, 831 (1981).
2. R.F.W. Bader and T.T. Nguyen-Dang, Adv. Quantum Chem. 14, 63 (1981).
3. J. Schwinger, Phys. Rev. 82, 914 (1951).
4. The local and integrated relations for $\overleftrightarrow{\sigma}$ and the force and virial were first obtained by W. Pauli in Encyclopedia of Physics (editor, S. Flügge) vol. V (Springer, Berlin, 1958).
5. R.F.W. Bader and P.M. Beddall, J. Chem. Phys. 56, 3320 (1972).
6. B.M. Deb and S.K. Ghosh, J. Phys. B 12, 3857 (1979).
7. L.J. Bartolotti and R.G. Parr, J. Chem. Phys. 72, 1593 (1980).
8. E.V. Ludeña, J. Chem. Phys. 76, 3157 (1982).
9. Y. Tal and R.F.W. Bader, Int. J. of Quantum Chem. Symp. 12, 153 (1980).
10. H. Essén, Int. J. of Quantum Chem. 21, 717 (1982)
11. The first of these conditions, Eq. (76), is recognized as the Bingel-Steiner cusp condition on $\rho(\underline{r})$. W.A. Bingel, Z. Naturforsch. 18a, 1249 (1963); E. Steiner, J. Chem. Phys. 39, 2365 (1963).

THE QUALITATIVE THEORY OF THE CURRENT DENSITY IN MOLECULES*

J.A.N.F. Gomes

Departamento de Quimica
Faculdade de Ciencias
4000 Porto
Portugal

This paper introduces a global qualitative theory of the electronic current induced in molecules by an external, constant, homogeneous magnetic field. The theory rests on the topological analysis of the vector field which is divergenceless and on certain symmetry constraints imposed by the magnetic field.

The fundamental topological elements for the analysis are the singularities of the vector field, for which a classification is presented. A major role is played by the set of all singular lines which form the so called stagnation graph.

Two examples of the current density field in diatomic molecules are discussed.

1. INTRODUCTION

Maps of the probability current density are currently used to give an intuitive picture of the way in which the electronic charge in a molecule reacts to an external magnetic field. (Throughout this paper, the field is assumed to be constant and homogeneous.) These maps are in general rather complex and little use is made of them. One of the reasons for the complexity and the difficulty of interpretation of the maps comes from the fact that they attempt to give a two dimensional representation of a three dimensional vector field, $R^3 \to R^3$. Frequently, the maps shown refer to planes of symmetry of

*The research reported here was supported by INIC(Lisbon).

the molecule with the magnetic field perpendicular to it; in this very particular condition the current vector is parallel to the plane and is reduced to a $R^2 \to R^2$ map usually displayed as two planar contour plots, one for the direction and the other for the intensity of the current. A difficulty of a different sort arises when orbital current density components are plotted as these need not be divergenceless.

The qualitative theory presented here gives a global description of the three dimensional vector field in terms of some simple elements. The backbone of the description is the stagnation graph formed by the set of singular lines of the vector field. Associated with each vortical singular line exists a vortex which is encased into a separatrix. Separatrices are closed surfaces associated either with saddle singular lines or with a pair of isolated singular points.

Hirschfelder[1-3] discussed the axial and toroidal vortices which are associated with the nodal lines of one-electron statefunctions. For a many-electron system, the analysis in higher dimensional space was made very rigourously by Riess.[4-5] Alternatively, each natural orbital component may be studied in ordinary space.[2] In this case the exchange current[7] should be considered to form a complete orbital current which is divergenceless and can therefore be submitted to the same general topological analysis that is discussed below. The present author[8] has discussed the properties of the complete orbital current; in particular, the orbital contribution of important molecular properties is the same when computed with the standard or the complete orbital current density.

The complete orbital current density, $\underline{j}_k^c(\underline{r})$, associated with the natural orbital $\psi_k(\underline{r})$ is defined as the sum

$$\underline{j}_k^c(\underline{r}) = \underline{j}_k(\underline{r}) + \underline{j}_k^{exch}(\underline{r}) \tag{1}$$

of the standard orbital current density,

$$\underline{j}_k(\underline{r}) = (ie\hbar/2m_e)(\psi_k^* \nabla \psi_k - \psi_k \nabla \psi_k^*) - (e^2/m_e)\,\underline{A}(\underline{r})\,\psi_k^*\psi_k \tag{2}$$

and the exchange orbital current, $\underline{j}_k^{exch}(\underline{r})$. This is best defined in terms of the potential associated with orbital source function $S_k(\underline{r})$,

$$\underline{j}_k^{exch}(\underline{r}) = \nabla V_k(\underline{r}) \tag{3}$$

$$V_k(\underline{r}) = \frac{1}{4\pi} \int d^3\underline{r}' \frac{S_k(\underline{r}')}{|\underline{r}'-\underline{r}|} \tag{4}$$

The source function

$$S_k(\underline{r}) = \nabla\cdot\underline{j}_k(\underline{r}) \tag{5}$$

may be expressed in terms of the non-local part, $\hat{N}$, of the hamiltonian associated with orbital $\psi_k(\underline{r})$,

$$S_k(\underline{r}) = (-2\ e/\hbar)\ \mathrm{Im}(\psi_k^*\ \hat{N}\ \psi_k) \tag{6}$$

The topological analysis of the vector field depends on the study of the field near its singular points, the points where it vanishes,

$$\underline{j}(\underline{r}_s) = 0 \tag{7}$$

(Henceforth the notation $\underline{j}$ is used indiscriminately for the total or orbital current; in this last case, the complete orbital current is in general used to assure the simplifications deriving from its divergencelessness.)

Near the singular point $\underline{r}_s$, the field may be described by the gradient tensor,

$$\underline{\underline{D}} = (\nabla\underline{j})_{\underline{r}=\underline{r}_s} \tag{8}$$

A classification of the singularities is made in terms of the eigenvalues of the $\underline{\underline{D}}$-tensor which add to zero for divergenceless fields. (For a more detailed account see references 9 and 10).

(a) Isolated singularities when the $\underline{\underline{D}}$-tensor has three non-zero eigenvalues. Two of the eigenvalues must have the same sign (for its real parts, in the complex case) and the associated eigenvalues define locally a separatrix. The separatrices are surfaces filled by asymptotic lines, i.e. by lines of current which originate and terminate at singular points. The third eigenvector defines an asymptotic line which crosses the separatrix at the critical point but has no particular interest.

(b) Stagnation lines are formed by singular points when the $\underline{\underline{D}}$-tensor has only two non-zero eigenvalues. The eigenvector associated with the zero eigenvalue defines the local direction of the singular line. Two types of flow near the stagnation line may exist depending on the nature of the eigenvalues.

(b.i) Saddle lines are the snagnation lines formed by singular points for which the eigenvalues of the $\underline{\underline{D}}$-tensor are real.

(b.ii) Vortical lines are the stagnation lines formed by singular points with complex eigenvalues.

The set of all stagnation lines forms the so called stagnation graph of the molecule which may be disconnected.

(c) Critical points are singular points with three zero eigenvalues and correspond to the points of the stagnation lines where a change of regime occurs which may involve a branching of the line. These points give the physical location of the vertices of the stagnation graph.

The major regimes of flow are sketched in Fig. 1.

In Section 2, a global topological analysis of the current density field is introduced, which depends on the combination of the topological elements just introduced. Finally, in Section 3, two examples are discussed showing the power of the theory to extract global information about the current density field from simple maps.

2. THE TOPOLOGICAL ANALYSIS

The topological analysis proposed here leads to a partition of the 3-dimensional physical space into sub-domains, each one filled by a vortex of electronic currents flowing around a vortical line. Vortices are encased into separatrices which are closed surfaces filled by asymptotic lines of current. The global qualitative theory

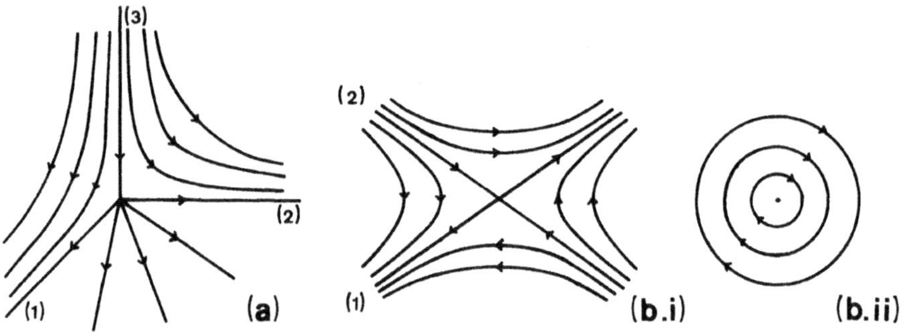

Fig. 1. Major regimes of flow near (a) an isolated singularity, (b.i) a saddle line and (b.ii) a vortical line. In the last two cases, a section perpendicular to the stagnation line. The plane defined by (1)-(2) in (a) and the two planes defined by (1), (2) and the stagnation line in (b.i) are tangent to separatrices.

depends on the properties of the singularities studied in Section 1; furthermore, the properties of the branchings of the stagnation lines constituting the stagnation graph and the topology of the separatrices must be considered.

For planar vector fields a topological index[11] may be associated with each singularity. It is basically the number of turns that the tip of the vector makes around the origin when a small circle around the singular point is traced. It may easily be checked that the cases displayed in Fig. 1.b (i) and (ii) have indices -1 and +1, respectively. In this obvious way an index may be associated with a stagnation line, ±1 being the most common values.

Consider a vortical line as it branches out at a critical point into a number of saddle lines and new vortical lines. A topological sphere may be constructed around the critical point such that the vector field is everywhere tangent to the surface. The Poincaré-Hopf theorem[11] requires the sum of the indices of all singularities on the surface to equal the Euler characteristic, $\chi = 2$. Since the vortical line has an index +1, the sum of the indices of the stagnation lines into which it branches at the critical point must equal +1. This implies that the number of vortical lines must exceed the number of saddle lines by one. This argument may be generalized[10] to all classes of branchings with the general result that the index total is always conserved.

Consideration of a sphere of a radius large relative to the molecular dimensions suggests that the field of currents in this region may be described as consisting of a single vortex, the vortical line crossing the sphere at the poles. The index theorem above allows this vortical line to branch out generating a more or less complicated stagnation graph in the region closer to the nuclei. In this region other connected parts of the stagnation graph may form, each with a closed vortical line and its allowed branchings. Associated with each of this connected parts, there is a toroidal vortex which is encased into a (topologically) spherical separatrix. This separatrix is defined by two isolated singularities which play the role of asymptotic source and sink.

Associated with an infinite vortical line or with a finite portion of a vortical line, there is an axial vortex encased into a non-spherical separatrix. The outermost circulations form an unbounded axial vortex associated with the vortical line which extends to infinity north- and southwards. The new type of separatrix is defined by one or more saddle lines.

To clarify the concepts discussed here, an example is created in Fig. 2 where the snagnation graph has a single connected part.

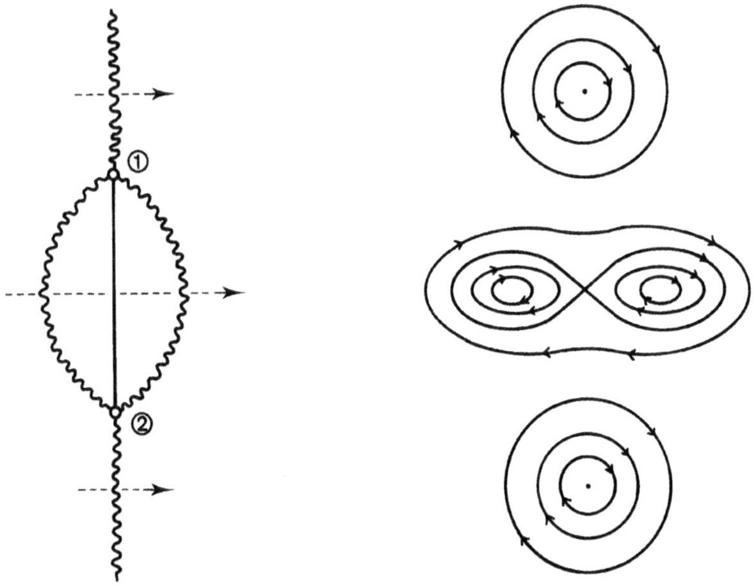

Fig. 2. The stagnation graph and the associated circulations in three representative sections. The curly lines in the graph are vortical lines and the straight line is a saddle stagnation line. The patterns of flow displayed on the r.h.s. show that the unbounded axial vortex surrounds another two axial vortices. These bounded vortices are encased by the separatrices associated the saddle line that exists between critical points (1) and (2). The index theorem may be easily checked at critical points (1) or (2). The index +1 associated with the outer vortical line equals the sum (+1-1+1) of the indices associated with the three lines into which it branches at the critical points.

3. DISCUSSION OF TWO EXAMPLES

Simple examples for the application of the ideas discussed above are provided by the maps of the currents induced in diatomic molecules by external magnetic fields, which were calculated by Lipscomb and coworkers.[12] Two particular molecules are considered here, LiH and BH, with the external magnetic field perpendicular to the internuclear axis.

The data available refers to the current density in a single

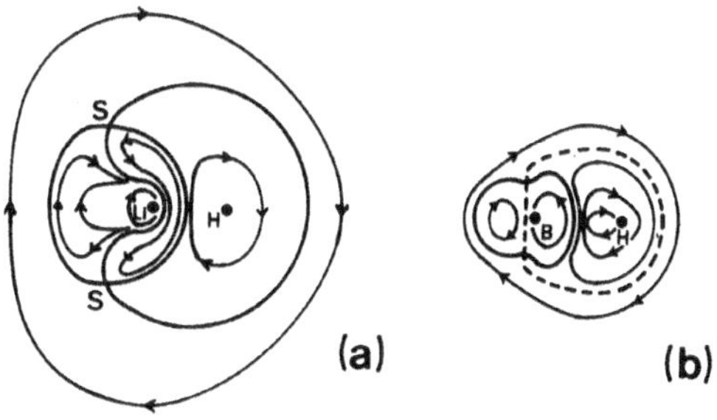

Fig. 3. Maps of the current density showing the separatrices between vortices. (a) Total current in LiH induced (1st order) by a perpendicular magnetic field. Three axial vortices are encased by the separatrices generated by the two saddle lines (S). The external lines of current are associated with the unbounded vortex. (b) Standard 2σ orbital current of BH. A toroidal vortex is encased into a (topologically) spherical separatrix. The isolated asymptotic line (dashed) crosses the separatrix at two isolated singularities. (Based on results obtained by Lipscomb et al.[12] Clockwise circulation corresponds to a diamagnetic current.)

plane perpendicular to the external field. Due to symmetry, the currents are planar but this is guaranteed only in this plane. The representation of the current density at points outside this plane is more complicated due to the extra degree of freedom. One of the aims of the qualitative theory summarized here is to give a global description of the current density field which will be helpful in reading such simple maps.

In the first example of LiH, Fig. 3.a, the planar map shows the existence of three bounded axial vortices encased into separatrices associated with saddle lines that cross the plane at the two points (S). These saddle lines must be finite. Somehow, they will merge into the single vortical line that extends to infinity. Two hypothesis are suggested in Fig. 4. Both are compatible with the data in Fig. 3.a but (b) is the most likely as no symmetry condition does appear to exist constraining the merger of the five stagnation lines to occur at a single critical point. Of course, there is no information about the location of the critical points.

In the second example, Fig. 3.b, the lines of current refer

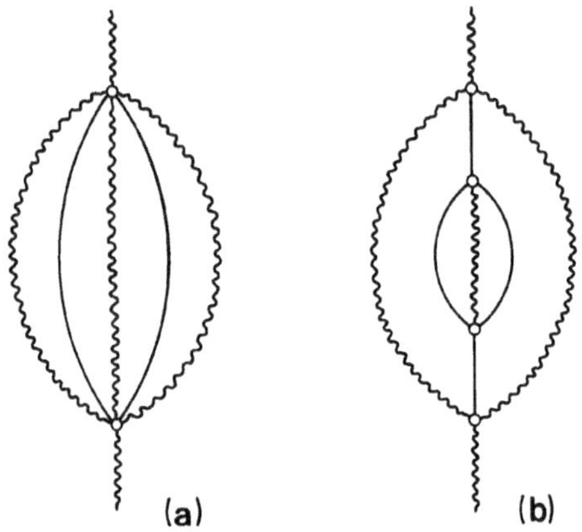

Fig. 4. Hypothetical stagnation graphs for LiH (Fig. 3.a). The two graphs are compatible with the map of currents in Fig. 3.a as they do not differ in the central region.

to the standard current of the 2σ orbital in BH. This current field does not conform to the assumptions made above; in fact, to guarantee the divergencelessness, the complete orbital current should be formed after the calculation of the exchange currents. In this particular case, a major change of the qualitative features of the current density field seems unlikely. The planar map in the figure suggests that there is a single bounded vortex, a toroidal vortex flowing around a vortex line which, for a standard orbital current, coincides with an orbital nodal line. The isolated asymptotic line (dashed) crosses the spherical separatrix at the two isolated singularities of the system.

The two examples discussed here have very simple current density patterns. They should however show the power of the qualitative theory presented earlier to interpret and extend the information actually contained in the maps. The definition of subdomains containing the different vortices is helpful in the break down of molecular properties into local contributions. An application of this sort to the definition of the delocalized current in a molecule has been reported.[8] The separation thus obtained between localized and delocalized currents fits well with the discussion based on model wavefunctions.[13] Kutzelnigg and coworkers[14,15] developed a theory of molecular magnetic properties using localized molecular orbitals. It is a coupled Hartree-Fock theory using an individual gauge for each localized orbital in a way that makes it computationally very effective. An interesting outcome of the application of this theory to a number of small molecules[15] is that the new terms (as compared with the traditional theories) named "exchange corrections" and "resonance corrections" give small contributions to the magnetic susceptibility. The magnetic properties are thus approximately additive over the set of localized orbitals. However, the concept of orbital localization should not be confused with the use made here of the word localization (or localized) as applied to the current density. Orbital current densities may show very large local non conservation ($\nabla \cdot \underline{j}_k \neq 0$); the use of the orbital exchange currents was proposed[7] to compensate for this and it has been shown[8] that the orbital contribution to most properties is unchanged by this procedure. The orbital current densities are additive to form the total, many-electron current density when natural orbitals (or, at least, the pseudo-natural orbitals associated with the particular many-electron function) are used.

REFERENCES

1. D.F. Heller and J.O. Hirschfelder, J. Chem. Phys. 66, 1929 (1977).
2. J.O. Hirschfelder, J. Chem. Phys. 67, 5477 (1977).
3. J.O. Hirschfelder, J. Chem. Phys. 68, 5151 (1978).
4. J. Riess and H. Primas, Chem. Phys. Lett. 1, 545 (1968).
5. J. Riess, Ann. Phys. (N.Y.) 57, 301 (1970); ib., Phys. Rev. D 2, 647 (1970).
6. J. Riess, Ann. Phys. (N.Y.) 67, 347 (1971); ib., Helv. Phys. Acta 45, 1067 (1972); ib., Phys. Rev. B 13, 3862 (1976).
7. P.W. Atkins and J.A.N.F. Gomes, Molec. Phys. 32, 1063 (1976).
8. J.A.N.F. Gomes, J. Chem. Phys. 1983.
9. J.A.N.F. Gomes, J. Chem. Phys. 1983.
10. J.A.N.F. Gomes, to be published.
11. J.W. Milnor, Topology from a Differential Point of View, U.P. Virginia, Charlottesville, Va., 1965.
12. W.N. Lipscomb in Theoretical Chemistry, Ed. W.B. Brown, Physical Chemistry, series one, v.1, Butterworths, London, 1972, p. 167.
13. J.A.N.F. Gomes, Molec. Phys. 40, 765 (1980).
14. W. Kutzelnigg, Isr. J. Chem. 19, 193 (1980).
15. M. Schindler and W. Kutzelnigg, J. Chem. Phys. 76, 1919 (1982).

THE CONSTRAINED SEARCH APPROACH, MAPPINGS TO EXTERNAL POTENTIALS, AND VIRIAL-LIKE THEOREMS FOR ELECTRON-DENSITY AND ONE-MATRIX ENERGY-FUNCTIONAL THEORIES

Mel Levy

Department of Chemistry and
Quantum Theory Group
Tulane University
New Orleans, Louisiana 70118
U.S.A.

1. INTRODUCTION

Quantitative predictions by means of electronic wavefunctions, within the framework of the Schroedinger equation, continues to be quite cumbersome for systems large enough to be of interest because the dimensions of wavefunctions grow spacially as three times the number or electrons. Density functional theory provides an attractive alternative to wavefunctional theory because the electron density possesses only three dimensions no matter how large the system. Similarly, the reduced spacial one-matrix possesses only six dimensions regardless the size of the system. Furthermore, a formal justification for density functional theory arises from the fact, which is by now well-known, that a ground-state electron density contains implicitly all the information embedded within its ground-state wavefunction. Specifically, as proved by Hohenberg and Kohn, a ground-state electron density contains sufficient information to determine the more complicated ground-state wavefunctions.[1]

In this chapter, we shall analyze the relationship between the ground-state density and the ground-state wavefunction, and the relationship between the ground-state density and its corresponding spin-free local external potential. We shall also discuss the universal variational functionals in density and one-matrix theories for ground-state energy calculations, and we shall present rigorous virial-like equalities and bounds. Finally, we shall compare wavefunction, two-matrix, one-matrix,[2-5] and Kohn-Sham formulations, and we shall analyze the validity of noninteger occupation numbers within the Kohn-Sham theory.[6]

2. THE RELATIONSHIP BETWEEN A GROUND-STATE ELECTRON DENSITY AND ITS CORRESPONDING GROUND-STATE WAVEFUNCTION AND ENERGY

First of all, it is important to point out immediately that very many wavefunctions generally yield a given electron density. For instance, a single Slater determinant of spin-orbitals can always be found to yield the same density as the given ground-state wavefunction which is usually composed of an infinite number of Slater determinants. If only one wavefunction were to always yield a given ground-state density, then there would be no need for a Hohenberg-Kohn theorem because the existence of a ground-state density to ground-state wavefunction mapping would be trivially established. Namely, just find that wavefunction which yields the density. That wavefunction which fits this prescription would then automatically be identified as the ground-state wavefunction associated with the given ground-state density.

The realization that there is, in general, a many-to-one relationship between wavefunctions and a given ground-state density makes the problem more difficult, but the variational principle solves it for us by the following "constrained search" approach[5] to density functional theory:

Notice that all those antisymmetric wavefunctions that yield a given density possess the same expectation value with respect to any local external potential (the electron-nuclear attraction operator is an example of a local external potential). Consequently, of all those wavefunctions which yield the ground-state density, the ground-state wavefunction distinguishes itself as the one which minimizes the sum of the expectation values of the kinetic ($\hat{T}$) plus electron-electron repulsion ($\hat{V}ee$) operators. Hence, for the formal determination of the ground-state wavefunction from a ground-state electron density ρ, just compute the integral $\langle\Psi_\rho|\hat{T} + \hat{V}ee|\Psi_\rho\rangle$ for each and every antisymmetric wavefunction Ψ_ρ which possesses the given ground-state density ρ. That Ψ_ρ which yields the minimum, Ψ_ρ^{min}, is the ground-state wavefunction associated with the given ρ. If degeneracies exist such that more than one Ψ_ρ gives this minimum, then all of the ground-state wavefunctions may be obtained, one at a time, by the above procedure. Furthermore, all of the wavefunctions which give the minimum must obviously be ground-states of the same local external potential.

Once each Ψ_ρ^{min} has been determined, then by means of $\hat{H}\,\Psi_\rho^{min} = E\,\Psi_\rho^{min}$, it follows directly that the total multiplicative potential operator, $\hat{V}$, is obtained, within an additive constant, from Ψ_ρ^{min} by performing

$$V-E = -\,\hat{T}\,\Psi_\rho^{min}/\Psi_\rho^{min} \qquad (1)$$

Moreover, since $\hat{V}$ usually vanishes at infinity, E is given by

$$E = \underset{\text{all } r_i \to \infty}{\text{Lim}}\ \hat{T}(\vec{r}_1 \ldots \vec{r}_N)\ \Psi_\rho^{min}(\vec{r}_1 \ldots \vec{r}_N)\ /\Psi_\rho^{min}(\vec{r}_1 \ldots \vec{r}_N) \quad (2)$$

so that

$$\hat{V} = -\hat{T}\ \Psi_\rho^{min}/\Psi_\rho^{min} + \text{Lim}\ \hat{T}(\vec{r}_1 \ldots \vec{r}_N)\ \Psi_\rho^{min}(\vec{r}_1 \ldots \vec{r}_N)\ / \quad (3)$$

$$\Psi_\rho^{min}(\vec{r}_1 \ldots \vec{r}_N),\ \text{all } \vec{r}_i \to \infty$$

Finally, once $\hat{V}$ has been determined, the local external potential $\hat{v}$ may be obtained by

$$\sum_{j=1}^{N} \hat{v}(\vec{r}_j) = \hat{V} - \hat{V}ee \quad (4)$$

(Note that the spin in the N-electron wavefunction has been surpressed to simplify notation.)

So, by formal construction,[5] we have just witnessed how the ground-state density contains sufficient information to determine the ground-state wavefunction, the external potential, and the ground-state energy. In this section we have emphasized the "constrained search" approach[5] which is a reformulation of the Hohenberg-Kohn orientation. We shall return to the "constrained search" approach when we discuss the variational methods.

3. THE MAPPING OF GROUND-STATE DENSITIES TO HAMILTONIANS[8]

Consider N interacting electrons in a local spin-independent external potential $\hat{v}$. (The external potential for a molecule consists of the electron-nuclear attraction operator, which is a coulomb potential, but it is important to note that we shall not necessarily restrict $\hat{v}$ to be a coulomb potential). The corresponding Hamiltonian is

$$\hat{H} = \hat{T} + \hat{V}ee + \sum_{j=1}^{N} v(\vec{r}_j) \quad (5)$$

The following assertion of the Hohenberg-Kohn theorem vividly illustrates the central role of the electron density in quantum chemistry and solid state physics: The ground-state density must change when the local external potential, v(r), changes by more than an additive constant.

We see that the marriage between density and external potential is an intimate one. In fact, in the last section it was shown quite explicitly how the external potential may be obtained, in a formal way, from its ground-state density. But, can one map a ground-state density to its external potential explicitly and exactly in a more practical manner? Well, yes and no. The best one can do (which is indeed aesthetically pleasing) is to map a set of M ground-state densities to a corresponding set of M Hamiltonians, where each density is associated with a different Hamiltonian in the form of Eq. (5). At the start, it is assumed that we do not know which density belongs with which $\hat{H}$. (Incidentally, it is important to note that for each coulomb $\hat{v}$ there are an infinite number of wrong densities with the right cusp conditions. In particular, a density might obey the electron-nuclear cusp conditions for a given coulomb $\hat{v}$ and yet be the ground-state density for some noncoulomb $\hat{v}$.) Following is the introduction to the theorem which achieves the mappings:

Label the M Hamiltonians $\hat{H}_1, \hat{H}_2, \ldots \hat{H}_M$ and the corresponding ground-state densities $\rho_1, \rho_2, \ldots \rho_M$, where ρ_i is the ground-state density of H_i. The functional relationship between the H's and the ρ's shall be given by the study of

$$G^{\alpha,\beta,\ldots\omega}_{1,2,\ldots M} = \int d\vec{r}\, [\hat{v}_1(\vec{r})\rho_\alpha(\vec{r})+\hat{v}_2(\vec{r})\rho_\beta(\vec{r})+\ldots+ \hat{v}_M(\vec{r})\rho_\omega(\vec{r})] \quad (6)$$

where the superscripts in G are formed by permuting its subscripts. There are thus M! possible G's.

THEOREM: G achieves its unique minimum when $\alpha=1$, $\beta=2, \ldots \omega=M$.

Namely

$$\text{Min}\; G^{\alpha,\beta,\ldots\omega}_{1,2,\ldots M} = G^{1,2,\ldots N}_{1,2,\ldots N} = \int d\vec{r}[\sum_{i=1}^{M} \hat{v}_i(\vec{r})\rho_i(\vec{r})] \quad (7)$$

PROOF: We begin along the lines of Eqs. (1-4) and Eq. (8) in reference 5c or along the lines of Eqs. (2.3-2.4) in reference 9. As with the original Hohenberg-Kohn theorem, the existence of an inequality in Eq. (8) of reference 5c or in Eq. (2.4) of reference 9 dictates that $\hat{v}$ is a unique functional of ρ. But, Eq. (8) and Eq. (2.4) imply something much more as well. The direction of the inequality in these equations shall allow us to achieve our objective: By the variational theorem

$$\langle\Psi_\alpha|\hat{H}_1|\Psi_\alpha\rangle + \langle\Psi_\beta|\hat{H}_2|\Psi_\beta\rangle + \ldots + \langle\Psi_\omega|\hat{H}_M|\Psi_\omega\rangle \quad (8)$$

$$\geq \langle\Psi_1|\hat{H}_1|\Psi_1\rangle + \langle\Psi_2|\hat{H}_2|\Psi_2\rangle + \ldots + \langle\Psi_M|\hat{H}_M|\Psi_M\rangle$$

where the Ψ's on the left-hand-side of Eq. (8) are permutations of the Ψ's on the right-hand-side. The equality clearly holds iff $\alpha=1$, $\beta=2,\ldots,\omega=M$. In any case, the sum of the $<\hat{T} + \hat{V}ee>$'s on the left-hand-side must equal the sum of the $<\hat{T} + \hat{V}ee>$'s on the right-hand-side. Hence it follows that

$$\int d\vec{r}\ [v_1(\vec{r})\rho_\alpha(\vec{r}) + \hat{v}_2(\vec{r})\rho_\beta(\vec{r}) + \ldots + \hat{v}_M(\vec{r})\rho_\omega(\vec{r})] \tag{9}$$

$$\geq \int d\vec{r}\ [\hat{v}_1(\vec{r})\rho_1(\vec{r}) + \hat{v}_2(\vec{r})\rho_2(\vec{r}) + \hat{v}_M(\vec{r})\rho_M(\vec{r})]$$

which is the desired result. That is,

$$G^{\alpha,\beta,\ldots\omega}_{1,2,\ldots M} \leq G^{1,2,\ldots M}_{1,2,\ldots M} \tag{10}$$

The equality obviously applies iff $\alpha=1$, $\beta=2,\ldots$, $\omega=M$. The theorem is thus proved. Therefore, the ground-state ρ's are matched to their $\hat{H}$'s through the minimization of the functional $G^{\alpha,\beta,\ldots\omega}_{1,2,\ldots M}$ by the optimum of the superscripts. The density denoted by the superscript is thereby the ground-state density for the external potential denoted by the corresponding subscript directly below the superscript in the minimum G.

4. FORMAL UNIVERSAL VARIATIONAL FUNCTIONALS FOR $< \hat{T} + \hat{V}ee >$

A. The Density

A variational principle involving the electron density was established indirectly by the original Hohenberg-Kohn theorem.[1] The existence of a proper universal functional for $< T + Vee >$ of a trial ρ may be established quite directly by following in the spirit of the development in Section 2. Without further delay, then, let us simply define the universal functional $Q[\rho]$ as follows:[5]

$$Q[\rho] = \text{Min} <\Psi_\rho \mid \hat{T} + \hat{V}ee \mid \Psi_\rho>. \tag{11}$$

$Q[\rho]$ evaluates $<\hat{T} + \hat{V}ee>$ for each and every wavefunction Ψ_ρ that yields the fixed trial ρ. The minimum in the expectation value search is the number assigned to $Q[\rho]$. (Even though wavefunctions are employed in the formal definition of $Q[\rho]$, $Q[\rho]$ is definitely a functional of ρ; input a ρ and $Q[\rho]$ will output a number.) The variational principle, with ρ as a trial density for $\hat{v}$ (r), is established immediately. Just call Ψ_ρ^{min} that wavefunction which satisfies the search in Eq. (11). Then[5]

$$\int \hat{v}(\vec{r})\rho(\vec{r})\, d\vec{r} + Q[\rho] = \langle \Psi_\rho^{min} | \hat{H} | \Psi_\rho^{min} \rangle \geq E_v \tag{12}$$

where E_v is the ground-state energy for an $\hat{H}$ of form Eq. (2) with external potential $\hat{v}$. Furthermore, it is evident that the equality in Eq. (12) is achieved when the trial ρ is equal to the true ground-state density. Moreover, Parr, Donnelly, Levy, and Palke[10] have identified electronegativity with the Lagrange multiplier associated with the normalization constraint in Eq. (12).

Observe that Q[ρ] does not require that the trial ρ be a ground-state of some local external potential. In other words, unlike the indirect F[ρ] of Hohenberg and Kohn, v-representability[11] is not required in the definition of Q[ρ]; F[ρ], on the other hand, is defined only for v-representable ρ. Specifically, F[ρ] is the $\langle \hat{T} + \hat{V}ee \rangle$ of that ground-state wavefunction which yields ρ. In any case, F[ρ] = Q[ρ] when ρ is v-representable so that Q[ρ] is really an explicit formal display of the Hohenberg-Kohn functional F[ρ] which is only implicitly defined. Moreover, the following explicit formal display of a universal exchange-correlation functional, $E_{xc}[\rho]$, is consistent with Q[ρ], compatible with Kohn-Sham theory, and does not require ρ to be either interacting or non-interacting v-representable:[5b]

$$E_{xc}[\rho] = Q[\rho] - 1/2 \iint \rho(\vec{r}_1)\rho(\vec{r}_2) r_{12}^{-1} d\vec{r}_1 d\vec{r}_2 - \mathrm{Min} \langle \Phi_\rho | \hat{T} | \Phi_\rho \rangle \tag{13}$$

where the Φ_ρ's are those antisymmetric wavefunctions which integrate to the trial ρ. In Section V we shall argue that fractional occupation numbers should result in the optimum equations of Kohn-Sham theory when the Φ_ρ^{min}, associated with the last term in Eq. (13), does not turn out to be a single determination.

In closing Part A, the reader should be made aware of the fact that Bartolotti[12] has recently utilized Eq. (11) in his time-dependent density functional theory which is based upon a hydrodynamic analogy to the Schroedinger equation, and Henderson[13] has put forth a counterpart of Q[ρ] for his density functional theory in momentum space. Finally, we note that excited-state density functional formulations have been reported by Theophilou,[14] Katriel,[15] Valone and Capitani,[16] and Levy.[17] It can be shown that Theophilou's formulation can be looked upon as a constrained ensemble search along the lines of Valone's ensemble extension[18] of the definition of Q[ρ] in Eq. (11).

B. The One-Matrix

The one-matrix (or first-order reduced dentity matrix)[19] contains more information than the density so that it is not surprising (using hindsight) that a variational principle exists,[2-4] for local external potentials, which involves the one-matrix, γ, and several groups have spearheaded work on universal variational functionals of γ for $\langle\hat{V}ee\rangle$. Specifically, Gilbert[2] established the existence of a functional of γ for a nonlocal external potential by extending the original Hohenberg-Kohn theorem to the nonlocal case (note that a local external potential may be looked upon as a special case of a non-local external potential). Berrondo and Goscinski,[3] with an orientation towards the implementation of Green's functions, added a non-local external potential to the N-body Hamiltonian in Eq. (5), and then derived a variational principle involving the one-matrix for a local external potential by eliminating the non-local external source. Donnelly and Parr[4] perceived that the original Hohenberg-Kohn theorem immediately implies the existence of a universal variational functional of γ. In particular, Donnelly and Parr utilized the fact that the Hohenberg-Kohn theorem implies that no two γ's may be identical if they belong to local external potentials which differ by more than an additive constant.

Given a pure-state N-representable[20,21] trial γ, the proper universal functional of γ for $\langle\hat{V}ee\rangle$ is[5]

$$W[\gamma] = \mathrm{Min}\ \langle\Psi_\gamma|\hat{V}ee|\Psi_\gamma\rangle \qquad (14)$$

The search in Eq. (14) is conducted over all those antisymmetric wavefunctions which yield the fixed trial γ. With $W[\gamma]$ thus defined, the variational principle follows at once:

$$E_v \leq \int dx[-1/2\nabla^2 + \hat{v}]\ \gamma(x'|x) + W[\gamma(x'|x)] \qquad (15)$$

where x signifies space-spin coordinate. As with $Q[\rho]$, it goes without saying that $W[\gamma]$ has to be approximated in actual calculations.

For γ, the requisite of pure-state N-representability[21] is more severe than the requisite of ensemble N-representability.[20] If a γ is pure-state representable then it is automatically ensemble representable, but the converse is not always true. With this in mind, Valone, by taking into consideration Gilbert's analysis, has extended the search in Eq. (14) to include all those ensemble density matrices which give the trial γ. Valone[18] has also expressed a reformulation with the reducing basis of Harriman.[22] For γ to be ensemble representable, it is necessary and sufficient that all the occupation numbers of γ be on the closed interval [0,1] when γ is assumed to be normalized to the number of electrons N.[20]

Significantly, a meaningful approximation to $W[\gamma]$ has already been indirectly made. Lieb has very recently proven that when an ensemble representable γ is inserted into the Fock energy functional (which can be expressed entirely in terms of γ), the energy is rigorously bounded below by the Hartree-Fock energy, even when γ contains fractional occupation numbers.[23] Moreover, given a γ constructed from a certain one-electron basis set, it is entirely possible that the resultant energy could actually be lower than that of the best single determinant wavefunction constructed of orbitals expanded in this particular one-electron basis.[23]

5. COMPARISON OF ONE-MATRIX, KOHN-SHAM, AND SLATER'S X-α FUNCTIONAL THEORIES

By one-matrix functional theory, we shall mean a formulation which yields, upon energy minimization, that one matrix which arises from the exact interacting ground-state wavefunction.[2-5] Specifically, the ground-state γ is related to the ground-state Ψ by

$$\gamma(x'|x) = N \int \Psi^*(x', x_2, \ldots x_N)\, \Psi(x, x_2, \ldots x_N)\, dx_2 \ldots dx_N \quad (16)$$

Minimization of the right-hand-side of Eq. (15) yields the ground-state γ.

The γ in Eq. (16) can be expanded in terms of its natural spin-orbitals:[24]

$$\gamma(x'|x) = \sum_i d_i\, \lambda_i^*(x')\, \lambda_i(x) \quad (17)$$

The natural spin-orbitals, λ_i, are the eigenfunctions of γ and d_i are the corresponding eigenvalues.

We shall now discuss the formally exact Kohn-Sham theory in terms of the "constrained search" approach[5] to density functional theory, as exemplified by Eq. (11). Assume that a given trial ρ for an interacting system is simultaneously a ground-state density of some auxiliary noninteracting system. Then, the ground-state kinetic energy, which we shall call $T_q[\rho]$, has to be given by[5,25a,26]

$$T_q[\rho] = \text{Min} \langle \Phi_\rho | \hat{T} | \Phi_\rho \rangle \quad (18)$$

Kohn-Sham only considers ρ's which are noninteracting ground-states, but this is not always the case. We shall, consequently, define $T_q[\rho]$ to be the kinetic energy of ρ even when ρ is not a noninteracting ground-state.[5,25a] Next, in the spirit of Kohn-Sham, we choose to put forth the following variational expression for an interacting system, with an even number of electrons, by taking into consideration Eq. (12):

$$E_v \leq \int dx[-1/2\nabla^2 + \hat{v}]\ \tilde{\gamma}(x'|x) + \omega[\tilde{\gamma}] \quad (19)$$

where

$$\tilde{\gamma}(x'|x) = \sum_k n_k \phi_k^*(x')\phi_k(x); \quad 0 \leq n_k \leq 1 \quad (20)$$

where[5b,25a,17]

$$\omega[\tilde{\gamma}] = Q[\rho] - T_q[\rho] \quad (21)$$

and where

$$\rho(\vec{r}) = \int \tilde{\gamma}(x|x)ds \quad (22)$$

with s denoting spin.

The "pseudo" one-matrix $\tilde{\gamma}$ is assumed to be expanded in terms of the orthonormal spin-orbitals ϕ_k. To meet pure-state N-representability requirements for an even number of electrons, it is sufficient that the n_k in Eq. (20) be degenerate in pairs.[21]

The label "pseudo" has been attached to $\tilde{\gamma}$ because the optimum $\tilde{\gamma}$ should almost never turn out to be the same as the corresponding optimum γ. In fact, the optimum $\tilde{\gamma}$ would have to turn out to be idempotent (N occupied ϕ_k, each with $n_k = 1$) when the optimum ρ is a nondegenerate noninteracting ground-state density (noninteracting v-representable) because a noninteracting ground-state wavefunction is a single determinant. As implied above, noninteracting v-representability was assumed in the original Kohn-Sham paper[6] even though the authors did not concern themselves with the possibility of fractional occupation numbers. In any case, Eq. (19) certainly allows for the possibility of noninteger occupation numbers in the Kohn-Sham scheme, even though only integers should result in most instances. Exceptions include atomic multiplet problems when holes exist below the Fermi level. (For a related discussion concerning fractional occupation numbers consult Appendices A and B in the study by Perdew and Zunger.)[25a]

Even though the Kohn-Sham formulation does not yield the optimum γ, the formulation is nevertheless exact, in principle, in that the true ground-state energy and density are obtained. The X-α formulation of Slater[7] is actually a predecessor to the exact Kohn-Sham formulation as summarized in its general form by Eq. (19). In a way, the X-α scheme can be looked upon as either an approximation to Eq. (19) or to Eq. (15). I like to classify X-α as an approximation to Eq. (19) because the X-α occupation numbers often turn out to be unity in ground-state calculations, but strong arguments could be made for either classification.

Eq. (19) can be expressed as

$$E_v \leq \int dx[-1/2\nabla^2 + \hat{v}]\, \tilde{\gamma}(x'|x) \tag{23}$$

$$+ 1/2 \iint \rho(\vec{r}_1)\rho(\vec{r}_2)|\vec{r}_1 - \vec{r}_2|^{-1} d\vec{r}_1 d\vec{r}_2 + E_{xc}[\rho]$$

where[5b,25a,17]

$$E_{xc}[\rho] = \omega[\tilde{\gamma}] - 1/2 \iint \rho(\vec{r}_1)\rho(\vec{r}_2)|\vec{r}_1 - \vec{r}_2|^{-1} d\vec{r}_1 d\vec{r}_2 \tag{24}$$

In X-α theory, $E_{xc}[\rho]$ is approximated by the simple local functional*

$$E_{xc}[\rho] = \alpha\, C \int \rho(\vec{r})^{4/3}\, d\vec{r} \tag{25}$$

where C is a constant and α is an adjustable parameter. Much more elaborate approximations to $E_{xc}[\rho]$ are now, of course, in common use as discussed in various review articles and in other chapters in this book. But, whatever approximations are used for $E_{xc}[\rho]$, they generally must mimic the constrained searches featured within this chapter.

Denote Ψ_v, γ_v, $\tilde{\gamma}_v$, and ρ_v as the ground-state functions of $\hat{H}$ with external potential $\hat{v}$. According to the "constrained search" approach, the exact $E_{xc}[\rho_v]$ is not purely potential energy. Indeed, $E_{xc}[\rho]$ must contain positive kinetic energy because the search dictates that[5,17]

$$T_q[\rho_v] < \langle\Psi_v|\,\hat{T}\,|\Psi\rangle \tag{26}$$

The magnitude of this positive kinetic energy, which equals $\langle\Psi_v|\,\hat{T}\,|\Psi_v\rangle$ $T_q[\rho_v]$, is expected to be of the order of magnitude of the correlation energy, as rationalized in the following development which exploits the method of integration over a coupling constant, a technique devised by Langreth and Perdew[29] and by Gunnarsson and Lundqvist[30] to elucidate the exchange-correlation hole.

Let us form a Hamiltonian analogous to Eq. (5), but now multiply $\hat{V}ee$ by a coupling constant λ:

$$\hat{H}(\lambda) = \hat{T} + \lambda\hat{V}ee + \sum_{j=1}^{N} \hat{v}(\lambda, \vec{r}_j) \tag{27}$$

*For a completely local density functional theory, see references 27 and 28.

For each λ ($0 \leq \lambda \leq 1$), $\hat{v}(\lambda, \vec{r})$ is adjusted so that the ground-state density corresponding to $\lambda=1$, ρ_v, is always obtained, even though the ground-state wavefunction, $\Psi(\lambda)$, depends on λ (note that according to the notation $\Psi(\lambda=1)$ and Ψ_v are identical). Utilization of the Hellmann-Feynman theorem,

$$[\partial E(\lambda)/\partial\lambda] = \langle\Psi(\lambda)|\partial\hat{H}(\lambda)/\partial\lambda|\Psi(\lambda)\rangle \tag{28}$$

followed by integration over λ, produces the appropriate formal expression for $E_{xc}[\rho_v]$

$$E_{xc}[\rho_v] = \int_0^1 \langle\Psi(\lambda)|\hat{V}ee|\Psi(\lambda)\rangle - 1/2 \iint \rho_v(\vec{r}_1)\rho_v(\vec{r}_2)|\vec{r}_1 - \vec{r}_2|^{-1} d\vec{r}_1 d\vec{r}_2 \tag{29}$$

Implementation of the trapezoid rule as an approximation for the integration over λ on the right-hand-side of Eq. (29) gives[17]

$$E_{xc}[\rho_v] = 1/2[\langle\Psi(\lambda=1)|\hat{V}ee|\Psi(\lambda=1)\rangle - \langle\Psi(\lambda=0)|\hat{V}ee|\Psi(\lambda=0)\rangle - \iint \rho_v(\vec{r}_1)\rho_v(\vec{r}_2)|\vec{r}_1 - \vec{r}_2|^{-1} d\vec{r}_1 d\vec{r}_2] \tag{30}$$

which in turn yields, after a few algebraic manipulations, the desired relationship:[17]

$$\langle\Psi_v|\hat{T}|\Psi_v\rangle - T_q[\rho_v] = \langle\Psi(\lambda=0)|\hat{H}(\lambda=1)|\Psi(\lambda=0)\rangle - \langle\Psi(\lambda=1)|\hat{H}(\lambda=1)|\Psi(\lambda=1)\rangle = |E_{corr}| \tag{31}$$

where E_{corr} signifies correlation energy. (By the way, notice that in density functional theory the Kohn-Sham single determinant often replaces the Hartree-Fock determinant as the basis for the definition of correlation energy.)

Perdew and Zunger[25a] have shown that $Q[\rho_v]$ as defined by Eq. (11), $T_q[\rho_v]$ as defined by Eq. (18), and $E_{xc}[\rho_v]$ as defined by Eq. (24) and expressed in Eq. (29), directly imply that the exchange-correlation energy of a single fully-occupied orbital in $\tilde{\gamma}_v$ must cancel exactly its self-direct coulomb energy. Partly with this in mind, Perdew and Zunger have created methods for incorporating self-interaction corrections (SIC) into any density functional for the energy.[25] In particular, many of the anomalies of the local spin-density approximation have been removed by their SIC methods.

6. VIRIAL-LIKE ENERGY-DENSITY AND ENERGY-ONE MATRIX RELATIONS

In the previous sections we discussed the formal variational functionals for computing ground-state energies and for the generation of the corresponding optimum densities and one-matrices. It should be evident that the external potential $\hat{v}(\vec{r})$ plays a very crucial role.[31] Once the desired optimum function (either density or one-matrix) is obtained, are there known exact simple relationships involving the optimum function, E_v, and $\hat{v}(\vec{r})$ which might serve as checks on the calculations? Well, there is no known operational energy-density equality.[32] We shall, however, present a fairly tight rigorous energy-density bound in this section. But first it is important to reveal, because it is not very well-known, that the exact ground-state electronic energy, for an interacting $\hat{H}$ of form Eq. (5), can be expressed in terms of γ_v for any local external $\hat{v}(\vec{r})$, not just for a coulomb $\hat{v}(\vec{r})$, by means of the following exact and simple virial-like formula:[8]

$$E_v = 1/2 \int \nabla^2(\vec{r})\gamma_v(x'|x)dx + \int d\vec{r}\hat{v}(\vec{r})\rho_v(\vec{r}) - \int d\vec{r}\rho_v(\vec{r})[\partial\hat{v}(\lambda^{-1}\vec{r}/\partial\lambda]_{\text{at } \lambda=1} d\vec{r} \tag{32}$$

where λ is a uniform scale factor of the electronic coordinates. The derivation of Eq. (32) follows along the usual lines of approach. Specifically, consideration is made of the fact that the ground-state wavefunction must be optimum with respect to a uniform scaling of its coordinates and consideration is made of the fact that $\hat{T}$ is homogeneous of degree -2 and $\hat{V}ee$ is homogenous of degree -1.

Let us now leave γ_v and focus entirely upon ρ_v to obtain our energy-density bound.[8] With this in mind, assume that the interacting ρ_v is simultaneously the ground-state density of the following auxiliary Kohn-Sham noninteracting Hamiltonian with the one-body potential $\hat{v}'$:

$$\hat{H}' = \hat{T} + \sum_{j=1}^{N} \hat{v}'(j) \tag{33}$$

Let the antisymmetric ground-state wavefunction of $\hat{H}'$ be called Ψ'. Then the homogeneity of $\hat{T}$ implies

$$\langle\Psi'|\ \hat{T}\ |\Psi'\rangle = -\ 1/2 \int d\vec{r}\ \rho(\vec{r})[\partial v'(\lambda^{-1}\vec{r})/\partial\ \lambda]_{\text{at } \lambda=1} \tag{34}$$

But

$$\langle\Psi'|\ \hat{T}\ |\Psi'\rangle \leq -1/2 \int \nabla^2(\vec{r})\gamma\ (x'|x)dx \tag{35}$$

because Ψ' has to distinguish itself, from all the other antisymmetric wavefunctions which yield ρ_v, by minimizing the expectation value of $\hat{T}$ (it should be clear that

$$\langle\Psi'|\ \hat{T}\ |\Psi'\rangle = T_q[\rho_v].)$$

Hence, from Eqs. (34) and (35)

$$1/2 \int \nabla^2(\vec{r})\gamma_v(x'|x)dx \leq 1/2 \int d\vec{r}\rho(\vec{r})[\partial\hat{v}(\lambda^{-1}\vec{r})/\partial\lambda]_{\text{at } \lambda=1} \tag{36}$$

Finally, when Eq. (36) is substituted into Eq. (32), the anticipated relation follows:[8]

$$E_v \leq \int d\vec{r}\hat{v}(\vec{r})\rho_v(\vec{r}) + 1/2 \int d\vec{r}\rho_v(\vec{r})[\partial\hat{v}'(\lambda^{-1}\vec{r})/\partial\lambda]_{\text{at } \lambda=1} - \int d\vec{r}\rho_v(\vec{r})[\partial v(\lambda^{-1}\vec{r})|\partial\lambda]_{\text{at } \lambda=1} \tag{37}$$

The bound in Eq. (37) is expected to be reasonably tight. Indeed, the arguments in Section 5 dictate that the right-hand-side exactly exceeds E_v by the magnitude of the correlation energy when $\langle\Psi(\lambda)|\hat{V}ee|\Psi(\lambda)\rangle$ is linear in λ, from $\lambda=0$ to $\lambda=1$.

We end this section by reporting that semirigorous bounds are obtained if an N-electron atom is visualized as having been formed by gradual increases in the nuclear charge with simultaneous additions of electrons. This visualization presented by Levy and Tal leads to recursion relations involving E and $\langle r^{-1}\rangle$. Specifically, given the $\langle r^{-1}\rangle$'s for the first N atoms, E's for the first N atoms are obtained. One such recursion relation is[32g,33]

$$E_N \leq -\sum_{k=1}^{N} (d\vec{r}r^{-1}\rho(K,H\text{-}F) + \varepsilon_o^K \leq E_N^{H-F} \tag{38}$$

where E_K denotes the ground-state energy of a neutral atom with K electrons, E_K^{H-F} is the corresponding Hartree-Fock energy, $\rho(K,H\text{-}F)$ the corresponding Hartree-Fock density, and ε_o^K is the zero-order perturbation coefficient for K electrons (ε_o^K is known exactly from the one-electron atomic problem). The validity of the semirigorous bounds in Eq. (38) is supported through Ar, in Table 1. The left-side inequality arises, in part, from the assumption that electron affinities are mostly positive and the right-side inequality arises, in part, from the assumption that the N-electron correlation energy is somewhat larger in magnitude than the corresponding sum, from K=1 to K=N, of electron affinities. In any case, the numbers from Eq. (38) always lie between the Hartree-Fock energies and the exact non-relativistic energies, and they generally fall closer to the latter.[32g,33]

Table 1. Correlated Energies from Hartree-Fock Densities. (Energies Are Expressed in Hartrees.)

Atom	Exact	Eq. (38)[c] (Hartree-Fock Densities)	Hartree-Fock Energy[d]
H	-0.500	-0.500	-0.500
He	-2.904	-2.875	-2.862
Li	-7.477[a]	-7.465	-7.433
Be	-14.666[a]	-14.624	-14.573
B	-24.652[a]	-24.628	-24.529
C	-37.842[a]	-37.818	-37.689
N	-54.585[a]	-54.529	-54.401
O	-75.061[a]	-75.038	-74.809
F	-99.722[a]	-99.682	-99.409
Ne	-128.93[b]	-128.79	-128.55
Na	-162.24[b]	-162.17	-161.86
Mg	-200.04[b]	-199.98	-199.61
Al	-242.34[b]	-242.31	-241.88
Si	-289.35[b]	-289.33	-288.85
P	-341.24[b]	-341.20	-340.72
S	-398.10[b]	-398.05	-397.50
Cl	-460.15[b]	-460.04	-459.48
Ar	-527.55[b]	-527.32	-526.82

[a] Obtained from reference 39.
[b] See page 88 in H.S. Schaefer III, "The Electron Structure of Atoms and Molecules", Addison-Wesley, Reading (1972). The uncertain Lamb correction has been ignored.
[c] Integrals are obtained from reference 40.
[d] Obtained from reference 40.

In concluding this section, it is important to note that quantum topology and subspace formulations have been developed in a promising manner to express the total electronic "energy density" as a functional of the one-matrix.[34]

7. SOME CLOSING REMARKS

The full N-electron wavefunction increases in complexity quite rapidly as N grows. Alternatives to wavefunctional theory, therefore, have been sought for some time. The two-matrix, whose dimensionality does not grow with N, appeared for a while to be a marvellous possible alternative, because the variational functional of the two-matrix is simple and known explicitly in terms of the reduced Hamiltonian. Fulfilment of the sufficient requisites for N-representability of a trial two-matrix is so difficult,[19,35] however, that there is now some pessimism concerning the future of direct variational calculations with the two-matrix. This brings us to one-matrix[2-5] and density[1] functional theories. For the one-matrix, ensemble N-representability is more easily satisfied than pure-state N-representability, but both are really little problems,[20,21] and the density creates no problem at all.[2] In fact, a very specific prescription has recently been presented for the construction of a single Slater determinant of orthonormal orbitals from any density that is positive-definite and integrable.[36]

The universal variational functionals in one-matrix[2-4] and density[1] functional theories may be defined explicitly by means of the "constrained searches" as given by Eqs. (11) and (14).[5] Those functionals, of course, have to be approximated in actual calculations. What calculations have been carried out thus far? Well, while the number of calculations with $Q[\rho]$ implicitly in mind have been growing by leaps and bounds, through approximations to $E_{xc}[\rho]$, almost nothing has been done to approximate $W[\gamma]$ for the purpose of introducing electron-electron correlation, and one should study $W[\gamma]$ as well as $E_{xc}[\rho]$.

One advantage of $E_{xc}[\rho]$ over $W[\gamma]$ results from the fact that γ_{γ} has to be more complex than ρ_{γ}. On the other hand, $E_{xc}[\rho]$ is complicated by the fact that it must contain a positive kinetic energy contribution (of the order of the total correlation energy) as well as a large negative potential energy contribution. $W[\rho]$, on the other hand, has to contain only pure potential energy. But then again, one can say that the necessary existence of a positive kinetic contribution to $E_{xc}[\rho]$ just means that approximations to $E_{xc}[\rho]$ have to provide that much less negative energy. Specifically, the correlation part of $W[\gamma]$ must be made to provide about twice as much negative energy as the correlation part of $E_{xc}[\rho]$.

Often when the expression "density functional theory" is employed, it is the formal Kohn-Sham theory and its variations which are automatically assumed. But perhaps Kohn-Sham theory (of which $E_{xc}[\rho]$ is a part) is really a synthesis of a "pure density functional theory" and a "pure one-matrix functional theory" because Kohn-Sham generates, upon optimization, $\tilde{\gamma}_v$ and not γ_v; $\tilde{\gamma}_v$ is more complicated than ρ_v and less complicated than γ_v. In this connection, I have found in my discussions that many of us who have "grown up" with γ_v and natural spin-orbitals[24] are a bit skeptical of exact Kohn-Sham theory. We ask, how can correlation energy be obtained with an idempotent $\tilde{\gamma}_v$? Well, as reviewed in Section V in the present chapter $\tilde{\gamma}_v$ does yield, in principle, the true ground-state energy, E_v, even though $\tilde{\gamma}_v$ is not the same as the true ground-state one-matrix, γ_v. On the other side of the coin, I feel that many of us who have "grown up" with Kohn-Sham theory, and are active in the use of $\tilde{\gamma}$, may not yet have full appreciation of the possibilities of γ, natural spin-orbitals, and $W[\gamma]$.[10,2-5]

Finally, it gives me great pleasure to thank the organizers for a most exciting symposium and to thank my collaborators on the works discussed within: Professors Robert G. Parr, Yoram Tal, John P. Perdew, William E. Palke, Robert Donnelly, and Stephen C. Clement.

REFERENCES

1. P. Hohenberg and W. Kohn, Phys.Rev. 136, B864 (1964).
2. T.L. Gilbert, Phys.Rev. B12, 2111 (1975).
3. M. Berrondo and O. Goscinski, Int.J.Quantum Chem. S9, 67 (1975).
4. R.A. Donnelly and R.G. Parr, J.Chem.Phys. 69, 4431 (1978).
   R.A. Donnelly, J.Chem.Phys. 71, 2874 (1979).
5a. M. Levy, Proc.Natl.Acad.Sci. USA 76, 6062 (1979).
5b. M. Levy, Bull.Am.Phys.Soc. 24, 626 (1979).
5c. M. Levy, J.Chem.Phys. 70, 1573 (1979).
6. W. Kohn and L.J. Sham, Phys.Rev. 140, A 1133 (1965).
7. J.C. Slater, "The Self-Consistent Field for Molecules and Solids", McGraw-Hill, New York (1974).
   J.W.D. Connolly, "Semi-Empirical Methods of Electronic Structure Calculation", Part A, Plenum, N.Y. (1977).
8. M. Levy, Phys.Rev.A, submitted for publication.
9. P.W. Payne, J.Chem.Phys. 71, 490 (1979).
10a. R.G. Parr, R.A. Donnelly, M. Levy and W.E. Palke, J.Chem.Phys. 68, 3801 (1978).
10b. R.G. Parr, "Density Functional Theory of Atoms and Molecules", in "Horizons of Quantum Chemistry", eds. K. Fukui and B. Pullman, D. Reidel (1980).
11. Term coined by E.G. Larson.
12. L.J. Bartolotti, Phys.Rev.A. (in press).
13. G. Henderson, Phys.Rev.A 23, 19 (1981).
14. A.K. Theophilou, J.Phys. C 12, 5419 (1979).

15. J. Katriel, J.Phys. C 13, L375 (1980).
16. S.M. Valone and J.F. Capitani, Phys.Rev. A 23, 2127 (1981).
17. M. Levy, Phys. Rev. B (to be published).
18. S.M. Valone, J.Chem.Phys. 73, 1344, 4653 (1980).
19. For a comprehensive monograph, see E.R. Davidson, "Reduced Density Matrices in Quantum Chemistry", Academic, New York (1976).
20. A.J. Coleman, Rev.Mod.Phys. 35, 668 (1963).
21. D.W. Smith, Phys.Rev. 147, 896 (1966).
22. J. Harriman, Phys.Rev. A 17, 1249, 1257 (1978).
23. E.H. Lieb, Phys.Rev.Lett. 46, 457 (1981).
24a. P.O. Löwdin, Phys.Rev. 97, 1474 (1955).
24b. P.O. Löwdin and H. Shull, Phys.Rev. 101, 1730 (1956).
25a. J.P. Perdew and A. Zunger, Phys.Rev. B 23, 5048 (1981).
25b. A. Zunger, J.P. Perdew and G.L. Oliver, Solid State Commun. 34, 933 (1980).
25c. J.P. Perdew, Chem.Phys.Lett. 64, 127 (1979).
25d. A. Zunger and M.L. Cohen, Phys.Rev. B 18, 5449 (1978).
26. J.K. Percus, Int.J.Quant.Chem. 13, 89 (1978).
27. R.G. Parr, S.R. Gadre and L.J. Bartolotti, Proc.Natl.Acad.Sci. USA 76, 2522 (1979).
28. S.R. Gadre, L.J. Bartolotti and N.C. Handy, J.Chem.Phys. 72, 1034 (1980).
29. D.C. Langreth and J.P. Perdew, Phys.Rev. B 15, 2884 (1977).
30. O. Gunnarsson and B.I. Lundqvist, Phys.Rev. B 13, 4274 (1976).
31. In this connection, see the chapters dealing with energies and electrostatic potentials at the nuclei in "Chemical Applications of Atomic and Molecular Electrostatic Potentials", eds. P. Politzer and D.G. Truhlar, Plenum, New York (1980).
32. There are, however, approximate relations. In particular, see
32a. S. Fraga, Theoret.Chim.Acta. 2, 406 (1964).
32b. R. Gaspar, Int.J.Quantum Chem. 1, 139 (1967).
32c. P. Politzer and R.G. Parr, J.Chem.Phys. 61, 4258 (1974).
32d. P. Politzer, J.Chem.Phys. 64, 4239 (1976).
32e. Reference 8.
32f. Reference 10a.
32g. M. Levy and Y. Tal, J.Chem.Phys. 72, 3416 (1980).
32h. M. Levy and Y. Tal, J.Chem.Phys. 73, 5168 (1980).
32i. Y. Tal and M. Levy, Phys.Rev. A 23, 408 (1981).
32j. Reference 5c.
33. M. Levy, S. Clement and Y. Tal, in "Chemical Applications of Atomic and Molecular Electrostatic Potentials", eds. P. Politzer and D.G. Truhlar, Plenum, New York (1980).
34a. R.F.W. Bader, J.Chem.Phys. 73, 2871 (1980) and references within.
34b. R.F.W. Bader, Y. Tal, S.E. Anderson, T.T. Nguyen-Dang, Israel J.Chem. 19, 8 (1980).
34c. S. Srebrenik, Int.J.Quant.Chem. S9, 375 (1975).
35. W.B. Mcrae and E.R. Davidson, J.Math.Phys. 13, 1527 (1972).

36. J. Harriman, "Orthonormal Orbitals for the Representation of an Arbitrary Density", WIS-TCI-606, 1980. (Theoretical Chemistry Institute, University of Wisconsin, Madison, Wisconsin).
37. P.O. Löwdin, Phys.Rev. 97, 1474 (1955).
38. P.O. Löwdin and H. Shull, Phys.Rev. 101, 1730 (1956).
39. G.C. Lie and E. Clementi, J.Chem.Phys. 60, 1275 (1974).
40. C.F. Fischer, "The Hartree-Fock Method for Atoms", John Wiley and Sons, New York (1972).

SELF-INTERACTION CORRECTION

John P. Perdew

Department of Physics and Quantum Theory Group
Tulane University
New Orleans, LA 70118, U.S.A.

1. INTRODUCTION

An electron can interact coulombically with other electrons but not with itself. Approximations for many-electron systems which violate this simple principle are candidates for self-interaction correction.

The earliest density functional approach, the Thomas-Fermi approximation, was purged of self-interaction by Fermi and Amaldi.[1] Later self-consistent-field theories constructed the electron density $n(\underline{r})$ and energy E from orbitals $\psi_{\alpha\sigma}(\underline{r})$ and occupation numbers $f_{\alpha\sigma}$, where $\sigma = \uparrow$ or $\downarrow$:

$$n(\underline{r}) = \sum_{\alpha\sigma} n_{\alpha\sigma}(\underline{r}) \tag{1}$$

$$n_{\alpha\sigma}(\underline{r}) = f_{\alpha\sigma}|\psi_{\alpha\sigma}(\underline{r})|^2 \tag{2}$$

$$\int d^3r|\psi_{\alpha\sigma}(\underline{r})|^2 = 1 \tag{3}$$

$$0 \leq f_{\alpha\sigma} \leq 1 \tag{4}$$

$$\sum_{\alpha\sigma} f_{\alpha\sigma} = N \tag{5}$$

where N is the total number of electrons. In the Hartree-Fock approximation, the electron-electron repulsion energy is made up of direct electrostatic

$$\frac{1}{2}\int d^3r \int d^3r' \frac{n(\underline{r})n(\underline{r}')}{|\underline{r}-\underline{r}'|} \qquad (6)$$

and exchange

$$E_x = -\frac{1}{2}\sum_{\sigma}\sum_{\alpha,\alpha'} f_{\alpha\sigma}f_{\alpha'\sigma}\int d^3r\int d^3r' \frac{\psi^*_{\alpha\sigma}(\underline{r})\psi^*_{\alpha'\sigma}(\underline{r}')\psi_{\alpha'\sigma}(\underline{r})\psi_{\alpha\sigma}(\underline{r}')}{|\underline{r}-\underline{r}'|} \qquad (7)$$

terms. The $\alpha' = \alpha$ terms in (7) constitute a self-exchange energy

$$-\frac{1}{2}\sum_{\alpha\sigma}\int d^3r\int d^3r' \frac{n_{\alpha\sigma}(\underline{r})n_{\alpha\sigma}(\underline{r}')}{|\underline{r}-\underline{r}'|} \qquad (8)$$

which dominates the total exchange energy for atoms[2] (Table 1). Thus the Hartree approximation, which retains only these terms, involves a self-interaction correction to (6). On the other hand, the Hartree-Fock approximation is naturally self-interaction free and requires no afterthought correction.

There are two distinct reasons why the Hartree-Fock approximation may be undesirable: (1) The exchange integrals like (7) are difficult. They can be approximated by functionals of the electron density, as in the $X\alpha$[3] or local density approximations,[4] at the cost of introducing a self-interaction error. There are several schemes for correcting this error.[5-9] (2) The Hartree-Fock approximation neglects dynamical correlation which may be as important as exchange for the low-density valence electrons which determine many of the interesting chemical and physical properties of atoms, molecules and solids. The theorems of modern density functional theory[4,10,11] sanction the hope of including these effects within a self-consistent field calculation. Some approximate density functionals for exchange and correlation are naturally self-interaction free but rather difficult to implement,[12] while others like the local spin density (LSD) approximation,[4,11] or its generalization as a gradient expansion,[13,14] are easily-implemented but contain a self-interaction error. Again there are several competing brands of self-interaction correction.[15-20]

Table 1. Hartree-Fock total, exchange, and self-exchange energies for lighter inert gas atoms. The correlation energy is shown for comparison. (After Refs. 2 and 35.) Energies in eV.

Atom	Total	Exchange	Self-exchange	Correlation
He	-77.9	-27.9	-27.9	-1.1
Ne	-3,497.8	-329.5	-269.2	-10.4
Ar	-14,334.7	-821.3	-617.8	-19.9
Kr	-74,883.4	-2,561.9	-1,653.6	

Here we will review the theory and applications of a particular method of self-interaction correction known as SIC.[17-20] As far as we know, it is the only method which satisfies all four of the following properties: (1) It can be applied to correlation as well as exchange; the two are treated on an equal footing. (2) It could be used in order to correct nonlocal functionals as well as local ones, and would yield no correction to the exact functional. (3) It is fully self-consistent: the one-electron potentials are derived variationally from the total energy. (4) It is exact for isolated electrons, and gives the exact energy per electron in systems where the density varies slowly over space. For electronic states which are either localized or localizable by unitary transformation, SIC yields a number of improvements over LSD in the density, the total energy and its separate exchange and correlation pieces, and especially the orbital energy eigenvalues which more closely approximate physical removal energies. However, there are some practical difficulties in applying SIC to extended orbitals, which will also be discussed.

2. THEORY

A. A Property of the Exact Exchange-Correlation Functional

We will work within the spin-density version[11] of density functional theory. The total density is the sum of up- and down-spin densities

$$n(\underline{r}) = n_{\uparrow}(\underline{r}) + n_{\downarrow}(\underline{r}) \tag{9}$$

and the energy functional is

$$E_v[n_{\uparrow},n_{\downarrow}] = \int d^3r\ v(\underline{r})n(\underline{r}) + T[n_{\uparrow},n_{\downarrow}] + \frac{1}{2}\int d^3r \int d^3r'\ \frac{n(\underline{r})n(\underline{r}')}{|\underline{r}-\underline{r}'|} + E_{xc}[n_{\uparrow},n_{\downarrow}] \tag{10}$$

where T and E_{xc} are the non-interacting kinetic and the exchange-correlation energies respectively. The ground-state energy and spin-densities are found by minimizing (10) subject to the constraint $\int d^3r\ n(\underline{r}) = N$.

We intend to prove that, if $n_{\alpha\sigma}(\underline{r})$ of (2) is a one-electron density ($f_{\alpha\sigma} = N = 1$), then

$$\frac{1}{2}\int d^3r \int d^3r'\ \frac{n_{\alpha\sigma}(\underline{r})n_{\alpha\sigma}(\underline{r}')}{|\underline{r}-\underline{r}'|} + E_{xc}[n_{\alpha\sigma},0] = 0 \tag{11}$$

i.e. the direct electrostatic self-interaction of an orbital is precisely cancelled by its exchange-correlation self-interaction.

The proof is immediate if $n_{\alpha\sigma}(\underline{r})$ is a possible ground-state density for some local external potential v(r), since in this case the left-hand side of (10) identically equals the sum of the first two terms on the right.

The exchange energy of (7) satisfies

$$\frac{1}{2}\int d^3r\int d^3r' \frac{n_{\alpha\sigma}(\underline{r})n_{\alpha\sigma}(\underline{r}')}{|\underline{r}-\underline{r}'|} + E_x[n_{\alpha\sigma},0] = 0 \tag{12}$$

so (11) amounts to a statement that the self-correlation of an orbital is zero:

$$E_c[n_{\alpha\sigma},0] = 0 \tag{13}$$

A general proof of the obvious results (11) and (13) will require an appeal to Levy's "constrained search" approach[21] to density functional theory:

Define the universal functional

$$Q[n_\uparrow,n_\downarrow] = \min \langle\hat{T} + \hat{V}_{ee}\rangle \tag{14}$$

which searches the set of all antisymmetric wavefunctions yielding the given spin densities and delivers the minimum expectation value for the sum of the kinetic and electron-electron repulsion operators. Clearly

$$E_v[n_\uparrow,n_\downarrow] = Q[n_\uparrow,n_\downarrow] + \int d^3r\ v(\underline{r})n(\underline{r}) \tag{15}$$

Similarly define

$$T[n_\uparrow,n_\downarrow] = \min \langle\hat{T}\rangle \tag{16}$$

and

$$E_{xc}[n_\uparrow,n_\downarrow] = Q[n_\uparrow,n_\downarrow] - T[n_\uparrow,n_\downarrow] - \frac{1}{2}\int d^3r\int d^3r' \frac{n(\underline{r})n(\underline{r}')}{|\underline{r}-\underline{r}'|} \tag{17}$$

For a one-electron system, $\hat{V}_{ee} = 0$ so Q = T and (11) follows from (17). The one-electron density $n_{\alpha\sigma}(\underline{r})$ does not have to be a possible ground-state density.

B. The SIC Method

Many simple approximations for the exchange-correlation energy fail to satisfy the exact result (11). Consider for example the local spin density approximation

$$E_{xc}^{LSD}[n_\uparrow,n_\downarrow] = \int d^3r\ n(\underline{r})\ \varepsilon_{xc}(n_\uparrow(\underline{r}),n_\downarrow(\underline{r})) \tag{18}$$

where $\varepsilon_{xc}(n_\uparrow,n_\downarrow)$ is the exchange-correlation energy per particle of an electron gas with uniform spin densities $n_\uparrow$ and $n_\downarrow$. This approximation becomes exact for densities that vary slowly over space, but in general a local functional cannot precisely cancel a nonlocal one to satisfy (11).

The most accurate density functional calculations are performed in the Kohn-Sham scheme,[4] in which the kinetic energy functional T is treated exactly by the introduction of auxiliary orbitals and occupation numbers:

$$T = \sum_{\alpha\sigma} f_{\alpha\sigma} \int d^3r \psi^*_{\alpha\sigma}(\underline{r})(-\tfrac{1}{2}\nabla^2)\psi_{\alpha\sigma}(\underline{r}) \tag{19}$$

The orbitals which minimize the total energy, including some approximation $E_{xc}^{approx}[n_\uparrow,n_\downarrow]$ for exchange and correlation, satisfy the self-consistent Schrödinger equation

$$\left\{-\tfrac{1}{2}\nabla^2 + v(\underline{r}) + \int d^3r' \frac{n(\underline{r}')}{|\underline{r}-\underline{r}'|} + v_{xc}^{\sigma,approx}(\underline{r};[n_\uparrow,n_\downarrow])\right\}\psi_{\alpha\sigma}(\underline{r}) = \varepsilon_{\alpha\sigma}\psi_{\alpha\sigma}(\underline{r}) \tag{20}$$

where the exchange-correlation potential is the functional derivative

$$v_{xc}^{\sigma,approx}(\underline{r};[n_\uparrow,n_\downarrow]) = \frac{\delta}{\delta n_\sigma(\underline{r})} E_{xc}^{approx}[n_\uparrow,n_\downarrow] \tag{21}$$

For example, in the LSD of Eq. (18)

$$v_{xc}^{\sigma,LSD}(\underline{r};[n_\uparrow,n_\downarrow]) = \frac{\partial}{\partial n_\sigma}[n\varepsilon_{xc}(n_\uparrow,n_\downarrow)] \tag{22}$$

According to (11), each orbital should carry zero self-interaction. If $E_{xc}^{approx}[n_\uparrow,n_\downarrow]$ fails to satisfy this condition, define the self-interaction corrected exchange-correlation energy[18]

$$E_{xc}^{SIC} = E_{xc}^{aprox}[n_\uparrow,n_\downarrow] - \sum_{\alpha\sigma}\left\{\frac{1}{2}\int d^3r \int d^3r' \frac{n_{\alpha\sigma}(\underline{r})n_{\alpha\sigma}(\underline{r}')}{|\underline{r}-\underline{r}'|} + E_{xc}^{approx}[n_{\alpha\sigma},0]\right\} \tag{23}$$

which does satisfy (11) for each occupied orbital.

Now seek the orbitals which minimize

$$\tilde{E} = \int d^3r\, v(\underline{r})n(\underline{r}) + T + \frac{1}{2}\int d^3r \int d^3r' \frac{n(\underline{r})n(\underline{r}')}{|\underline{r}-\underline{r}'|} + E_{xc}^{SIC} \tag{24}$$

with T given by (19), subject to the constraint (3) of normalization.

The Euler equation is

$$\frac{\delta}{\delta\psi^*_{\alpha\sigma}(\underline{r})}\left\{\tilde{E} - \sum_{\alpha'\sigma'} f_{\alpha'\sigma'}\varepsilon_{\alpha'\sigma'}\int d^3r'\,|\psi_{\alpha'\sigma'}(\underline{r}')|^2\right\} = 0 \tag{25}$$

or

$$\left\{-\frac{1}{2}\nabla^2 + v(\underline{r}) + \int d^3r'\,\frac{n(\underline{r}')}{|\underline{r}-\underline{r}'|} + v_{xc}^{\alpha,\sigma SIC}(\underline{r})\right\}\psi_{\alpha\sigma}(\underline{r})$$

$$= \varepsilon_{\alpha\sigma}\psi_{\alpha\sigma}(\underline{r}) \tag{26}$$

where the exchange-correlation potential contains self-interaction corrections:

$$v_{xc}^{\alpha\sigma,SIC}(\underline{r}) = v_{xc}^{\sigma,approx}(\underline{r};[n_\uparrow,n_\downarrow]) - \left\{\int d^3r'\,\frac{n_{\alpha\sigma}(\underline{r}')}{|\underline{r}-\underline{r}'|}\right.$$

$$\left. + v_{xc}^{\uparrow,approx}(\underline{r};[n_{\alpha\sigma},0])\right\} \tag{27}$$

The orbital energy eigenvalues of (26) have a well-defined meaning. Differentiation of (24) with respect to the occupation number $f_{\alpha\sigma}$, holding the orbitals and the other occupation numbers fixed gives

$$\frac{\partial\tilde{E}}{\partial f_{\alpha\sigma}} = \int d^3r\,\psi^*_{\alpha\sigma}(\underline{r})\left\{-\frac{1}{2}\nabla^2 + v(\underline{r}) + \int d^3r'\,\frac{n(\underline{r}')}{|\underline{r}-\underline{r}'|}\right.$$

$$\left. + v_{xc}^{\alpha\sigma,SIC}(\underline{r})\right\}\psi_{\alpha\sigma}(\underline{r}) = \varepsilon_{\alpha\sigma} \tag{28}$$

Starting from the orbitals that satisfy (3) and (26), a small variation of the $\{f_{\alpha\sigma}\}$ and the $\{\psi_{\alpha\sigma}\}$ (with the normalization of the latter preserved) results in a first variation of the energy

$$\delta\tilde{E} = \sum_{\alpha\sigma}\varepsilon_{\alpha\sigma}\delta f_{\alpha\sigma} \tag{29}$$

Now seek the occupation numbers which minimize (24), subject to the constraints (4) and (5). Minimization of E subject to the number-conserving constraint of (5) is equivalent to minimization without this constraint of $\tilde{E} - \mu N$ where μ is the chemical potential:

$$\sum_{\alpha\sigma}(\varepsilon_{\alpha\sigma} - \mu)\delta f_{\alpha\sigma} \geq 0 \tag{30}$$

which is an inequality because of the remaining constraint (4). In light of (4), equation (30) implies an "aufbau principle" for the ground-state: All orbitals with $\varepsilon_{\alpha\sigma} < \mu$ have $f_{\alpha\sigma} = 1$, and those with $\varepsilon_{\alpha\sigma} > \mu$ have $f_{\alpha\sigma} = 0$. Fractional occupation ($0 < f_{\alpha\sigma} < 1$) is possible only for orbitals with $\varepsilon_{\alpha\sigma} = \mu$, if there are any. (This principle holds if each $\varepsilon_{\alpha\sigma}$ has a continuous dependence on the $\{f_{\alpha\sigma}\}$. It is valid in the LSD, SIC-LSD, Hartree and Hartree-Fock approximations.)

C. Exact Properties of LSD Preserved by SIC

While LSD is not self-interaction free, it is a good approximation for many purposes because its self-interaction is fairly small: The direct electrostatic energy can be estimated from the close inequality[22]

$$\frac{1}{2}\int d^3r \int d^3r' \frac{n(\underline{r})n(\underline{r}')}{|\underline{r}-\underline{r}'|} \leq 1.092\, N^{2/3} \int d^3r\, n^{4/3}(\underline{r}) \tag{31}$$

while the LSD exchange energy is[11]

$$E_x^{LSD}[n_\uparrow, n_\downarrow] = -0.9305 \int d^3r [n_\uparrow^{4/3}(\underline{r}) + n_\downarrow^{4/3}(\underline{r})] \tag{32}$$

For a single occupied orbital ($N = 1$, $n = n_\uparrow = n_{\alpha\sigma}$, $n_\downarrow = 0$), the near cancellation between (31) and (32) is evident. Correlation tends to assist this cancellation. For the more tightly-bound orbitals, correlation is much smaller than exchange and the residue of the cancellation between (31) and (32) is small but positive. Thus the self-interaction correction of (23), when applied to LSD, lowers the calculated total energy.

LSD is exact for slowly-varying densities. Consider the limit of jellium, an electron gas of uniform density. The solutions of the LSD single-particle equation (20) are plane waves, orbitals extended over a large volume Ω. From (31) and (32), the self-interaction corrections to the energy and one-electron potential from an orbital extended over volume Ω are of order $\Omega^{-1/3}$ and vanish as $\Omega \to \infty$. Thus plane waves are also self-consistent solutions of the SIC single-particle equation (26), and SIC gives the exact energy per electron in the limit of slowly-varying density (unless the SIC equations also admit other, localized solutions of lower energy).

LSD has been remarkably useful outside its domain of formal validity. It gives reasonable results for densities that do not vary slowly on the scale of the local Fermi wavelength or screening length, i.e. for nearly all densities of physical interest. This unexpected success has been attributed[23] to the fact that LSD satisfies an exact conservation law: the exchange-correlation hole around an electron represents a deficit of one electron, i.e. the electron and its hole together constitute a neutral object.

If the ground-state of the interacting system can be connected adiabatically to that of a noninteracting system with the same density, the exchange correlation energy can be written as the electrostatic interaction of each electron with its positively-charged hole:[3,24]

$$E_{xc} = \frac{1}{2}\int d^3r \int d^3r' \frac{n(\underline{r})\rho(\underline{r},\underline{r}')}{|\underline{r}-\underline{r}'|} \tag{33}$$

where $\rho(\underline{r},\underline{r}')$ is the density at $\underline{r}'$ of the hole around an electron at $\underline{r}$, and

$$\int d^3r'\rho(\underline{r},r') = -1 \tag{34}$$

The LSD and SIC holes can be deduced from (18) and (23):

$$\rho^{LSD}(\underline{r},\underline{r}') = \rho_u(n_\uparrow(\underline{r}),n_\downarrow(\underline{r});|\underline{r}'-\underline{r}|) \tag{35}$$

$$\rho^{SIC}(\underline{r},\underline{r}') = \rho_u(n_\uparrow(\underline{r}),n_\downarrow(\underline{r});|\underline{r}'-\underline{r}|) - \sum_{\alpha\sigma} \frac{n_{\alpha\sigma}(\underline{r})}{n(\underline{r})}[n_{\alpha\sigma}(\underline{r}') + \rho_u(n_{\alpha\sigma}(\underline{r}),0;|\underline{r}'-\underline{r}|)] \tag{36}$$

where $\rho_u(n_\uparrow,n_\downarrow;|\underline{r}'-\underline{r}|)$ is the hole in an electron gas with uniform spin densities $n_\uparrow$ and $n_\downarrow$. Since ρ_u satisfies the sum rule (34), so obviously does ρ^{LSD}. From (36) we find

$$\int d^3r'\rho^{SIC}(\underline{r},\underline{r}') = -[\sum_{\alpha\sigma} f_{\alpha\sigma}n_{\alpha\sigma}(\underline{r})]/n(\underline{r}) \tag{37}$$

Under the condition of adiabatic connection, all the occupation numbers are zero or one and the right-hand side of (37) reduces to -1. Thus, like LSD, SIC satisfies the exact sum rule (34).

For tightly-bound orbitals, SIC gives a much more realistic description of the shape of the hole than LSD does.[20] Fig. 1 shows the exchange hole (no correlation) about an electron located at distance r from the nucleus in the neon atom, for two different values of r. The full curves are exact (from Eq. (7)), while the dashed and dotted curves represent the SIC and LSD approximations respectively. Note that the observation point $\underline{r}' = \underline{r} + \underline{R}$ lies on a line with the electron and the nucleus.

D. SIC Orbital Energies and Long-Range Behavior

Consider the physical removal energy, defined as the energy needed to remove an electron from orbital $\alpha\sigma$, allowing for relaxation of the other orbitals. According to (29), this is the integral of the orbital energy over the occupation number

$$\Delta E_{\alpha\sigma}|_{rel} = -\int_0^1 df_{\alpha\sigma}\, \varepsilon_{\alpha\sigma}(f_{\alpha\sigma}) \tag{38}$$

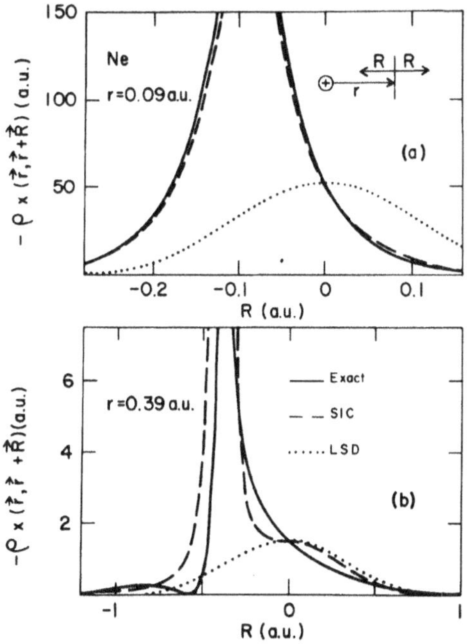

Fig. 1. Exchange hole in the neon atom. (From Ref. 20.)

which might be approximated as $-\varepsilon_{\alpha\sigma}(1/2)$ using a transition state of half occupancy.[3] Equation (38) is a general self-consistent field result, and applies to the LSD,[3,25] Hartree and Hartree-Fock approximations as surely as to SIC, once a continuation of the total energy to the interval $0 \le f_{\alpha\sigma} \le 1$ has been chosen.

A "self-interaction free" theory is one which satisfies (11), with $n_{\alpha\sigma}(\underline{r}) = f_{\alpha\sigma}|\psi_{\alpha\sigma}(\underline{r})|^2$ and $0 \le f_{\alpha\sigma} \le 1$. This category includes not only SIC but also the Hartree and Hartree-Fock approximations in their most natural continuation (equations (6) and (7)) to fractional occupation numbers. In a self-interaction free theory, the orbital energy eigenvalue $\varepsilon_{\alpha\sigma}$ in (38) depends on $f_{\alpha\sigma}$ largely through physical relaxation effects, i.e. through the effect on orbital $\alpha\sigma$ of the relaxation of the other orbitals. This effect is easily described: When an electron is removed from an atomic orbital $\alpha\sigma$, i.e. when $f_{\alpha\sigma} \to 0$, the remaining orbitals are drawn closer to the nucleus, which becomes more effectively screened. This makes $\varepsilon_{\alpha\sigma}(1/2) > \varepsilon_{\alpha\sigma}(1)$, i.e. in a self-interaction free theory one expects that

$$\Delta E_{\alpha\sigma}\Big|_{rel} \le -\varepsilon_{\alpha\sigma}(1) \tag{39}$$

Since the screening effect from physical relaxation is often small, the inequality (39) is often close to an equality. In a self-interaction free theory, a single ground-state calculation yields an estimate of all the physical removal energies.

In contrast the spurious self-interaction in the LSD one-electron potential gives $\varepsilon_{\alpha\sigma}$ a strong but spurious dependence on $f_{\alpha\sigma}$. Thus in LSD accurate removal energies can only be found by the more elaborate means of transition-state or change-in-selfconsistent-field calculations. In particular, LSD band-structure calculations are in error due to self-interaction.

It is not hard to show, using formulas like (31) and (32), that the LSD self-interaction error of the orbital energy eigenvalue is about six times the self-interaction error from that orbital in the total energy.[20] While the left-hand side of (31) tends to cancel the right-hand side of (32), their functional derivatives cannot cancel so effectively due to the different powers of n which appear (n^2 vs. $n^{4/3}$). Thus self-interaction corrections to the LSD one-electron potential (see equation (27)) can lower the orbital energy eigenvalue significantly.

Note that (38) is an equation for the <u>relaxed</u> energy difference. It does <u>not</u> imply that the unrelaxed energy difference is

$$\Delta E_{\alpha\sigma}\big|_{unrel} = -\varepsilon_{\alpha\sigma}(1) \tag{40}$$

Koopmans' theorem (40) is true in the Hartree-Fock approximation, but in SIC there is a well-defined positive correction to the right-hand side of (40). At least for atomic orbitals, this positive correction tends to cancel the negative orbital relaxation energy, making the SIC eigenvalue at full occupancy a better approximation to the physical removal energy than to the unrelaxed one.[20]

We turn now to the long-range behavior of the one-electron potential and density in an N-electron atom of nuclear charge Z. Far from the nucleus, the one-electron potential tends to

$$-\frac{Z^*}{r} \tag{41}$$

where $Z^* = Z - N + 1$ in a self-interaction free theory and $Z - N$ in LSD. The calculated electron density falls off as

$$[n(r)]^{1/2} = k\, r^{\beta}\, e^{-\alpha r} \tag{42}$$

where $\beta = Z^*/\alpha - 1$, $\alpha = (-2\varepsilon_{max})^{1/2}$, and ε_{max} is the greatest energy eigenvalue of the occupied orbitals.[18] From (39)

$$I \le -\varepsilon_{max} \tag{43}$$

where I is the first ionization potential. Thus

$$[n(r)]^{1/2} \leq k\, r^{\beta'}\, e^{-\alpha' r} \tag{44}$$

where $\beta' = Z^*/\alpha' - 1$, and $\alpha' = (2I)^{1/2}$. Eq. (44) with $Z^* = Z - N + 1$ agrees with rigorous bounds on the long-range behavior of the density.[26,27] Self-interaction free theories are correct far from the nucleus, where LSD is not.

The worst self-interaction errors of LSD appear in the orbital energy eigenvalue, and not in the total energy or the electron density. Consider for example a one-electron system, in which the exact one-electron hamiltonian is given by the first two terms of (20). The remaining two terms

$$\int d^3r' \frac{n_{\alpha\sigma}(\underline{r}')}{|\underline{r}-\underline{r}'|} + v_{xc}^{\sigma,\mathrm{approx}}(\underline{r};[n_{\alpha\sigma},0]) \tag{45}$$

constitute a spurious self-interacting potential, which is displayed for the LSD hydrogen atom in Fig. 2(a). Over most of the volume of the atom, this potential is large (about 6eV out of a total binding energy of 13.6eV) but approximately constant, so it contaminates the LSD orbital energy eigenvalue far more than it contaminates the density profile (Fig. 2(b)). However at very large distances from the nucleus the spurious self-interacting potential falls slowly to zero (like 1/r), and so contaminates the long-range behavior of the density.

Note well that all conclusions of this section, except Eq. (38) assume that the effective potential of the one-electron Schrödinger equation tends to zero at large r, at it does in the LSD, SIC-LSD, Hartree and Hartree-Fock approximations.

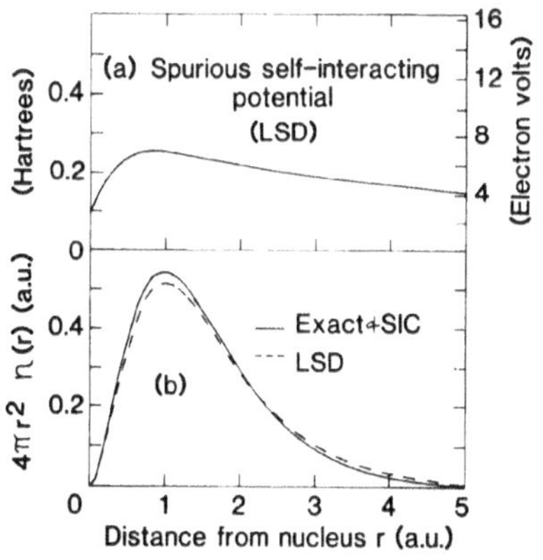

Fig. 2. The hydrogen atom. (a) Spurious self-interacting potential. (b) Radial distribution of the electron density.

E. Is SIC a Density Functional Theory?

So far we have thought of SIC as a self-interaction correction to a density functional approximation for exchange and correlation, such as LSD. We could equally well think of SIC as the Hartree method corrected by an approximation for exchange and correlation between different orbitals. Like the Hartree method, SIC has an orbital-dependent potential (27) which can generate non-orthogonal orbitals. In the atomic calculations of section 3, the orbital overlaps are always small and Schmidt orthogonalization has only negligible effect on the total energy, but there could be situations in which this non-orthogonality causes trouble.

Clearly SIC does not fit into the Kohn-Sham formalism of (20) in which the effective one-electron potential is independent of the orbital quantum numbers α. We will show[20] that, nevertheless, the self-consistent SIC method is a density functional approximation and is wholly sanctioned by the Hohenberg-Kohn theorem. (These conclusions and the arguments which support them apply with equal force to the Hartree and Hartree-Fock methods.)

As in section 2.A, we appeal to the "constrained search" idea.[21] Note that the SIC total energy (24) is manifestly a functional of the orbitals and occupation numbers, and define the SIC total energy as a functional of the spin densities by

$$E^{SIC}[n_\uparrow,n_\downarrow] = \min \widetilde{E}[\{f_{\alpha\sigma}\},\{\psi_{\alpha\sigma}\}] \qquad (46)$$

which searches all orbitals and occupation numbers yielding the given spin densities and delivers the minimum. Then for any choice of spin densities integrating to N electrons,

$$E^{SIC}[n_\uparrow,n_\downarrow] \geq \widetilde{E}(\text{self-consistent}) \qquad (47)$$

where the right-hand side is (24) evaluated using the self-consistent orbitals and occupation numbers. For the self-consistent $n_\uparrow$ and $n_\downarrow$, (47) becomes an equality.

F. Problems and Promise of SIC in Extended Systems

The SIC method can be implemented easily in LSD calculations for atoms and perhaps small molecules. What problems may arise for more extended systems?

First consider the tightly-bound core orbitals in a crystal. These can be represented equivalently (under unitary transformation) by Bloch orbitals extended over the whole crystal or Wannier orbitals localized on individual atoms. The Hartree-Fock exchange energy (7) is invariant under unitary transformation of the occupied orbitals, but the SIC exchange-correlation energy (23) is not. The extended orbitals have no self-interaction and so can be self-consistent solutions of both the SIC equation (26) and the LSD equation (20). But the SIC equation (26) will also admit localized orbitals as self-consistent solutions, and the self-interaction correction will give these localized solutions a lower total energy and lower orbital eigenvalues.[18] These corrections, which are easily implemented, will improve the calculated total energy and band-structure.

The valence-electron problem can be separated from the core-electron problem via electron-ion pseudopotentials. *A priori* pseudopotentials for valence electrons can be based on either the LSD or SIC methods. The resulting bare pseudopotentials are usually very similar; the most significant differences between LSD and SIC pseudopotentials arise from differences in screening by the valence electrons.[28]

There are practical problems in the application of SIC to more extended orbitals, especially those in the partially-filled bands of metals. How in practice can (26) be solved without either the spherical symmetry of tightly-bound core orbitals or the Bloch symmetry usually attributed to valence orbitals? Development of an easily-implemented, self-interaction free method of band-structure calculation is one of the greatest problems now facing density functional theory. (For further discussion, see section 4.)

3. APPLICATIONS

A. Preface

Self-consistent self-interaction corrections to LSD have been applied[19,20,28-30] with good success to a number of problems in atomic physics and to the band-structure of insulating crystals. These calculations have been performed in the central field approximation, in which all the orbital densities $n_{\alpha\sigma}(\underline{r})$ are replaced by their spherical averages. The electron gas correlation energy input to (18) is a parametrization[20] which matches Ceperley's[31] accurate Monte Carlo numerical values for low and metallic densities onto the exact high density limit.[32] Thus discrepancies between theory and experiment should accurately reflect the errors of the LSD and SIC approximations.

Before describing the impressive improvements of SIC over LSD, it is worth noting that SIC is not a cure-all. Some of the errors of LSD have a true many-body origin which has no connection with self-interaction. Altered boundary conditions can produce split-off modes such as surface plasmons[14,24] which are not going to be successfully described by the correlation energy of a uniform electron gas. There are also strong correlations between degenerate or nearly degenerate orbitals that are not properly described by either LSD or SIC. Thus SIC does not correct the errors of the LSD description of s-d and s-p interconfigurational energies.[20,33,34] And for strongly positive ions, where all the orbitals in a shell are nearly degenerate with each other and energetically well-separated from the other shells, both LSD and SIC predict a qualitatively wrong behavior (ℓnZ) for the correlation energy as the nuclear charge $Z \to \infty$, instead of the correct behavior (Z^0 or Z^1 depending on N).[30]

We do not know of any problems for which SIC results are significantly worse than LSD. We review here some problems for which they are significantly better. We start with those objects of traditional veneration in density functional theory, the total energy and electron density. Next come the orbital energy eigenvalues and bandstructure, which assume a new physical significance in SIC. Finally we consider negative ions and autodetaching states, which become accessible to realistic self-consistent calculation only within a self-interaction free, correlated theory.

B. Total Energies of Atoms

The "experimental" total energy of an N-electron atom is the sum of the ionization potentials

$$-\sum_{i=1}^{N} I_i \tag{48}$$

corrected for relativistic and reduced-mass effects. In Table 2 we compare "experimental" total energies[35] with Hartree-Fock,[36] LSD and SIC values.[20] The Hartree-Fock values are of course higher than experiment, and the LSD values are higher still, with an error about twice that of Hartree-Fock. The SIC values, on the other hand, are lower than experiment, with an error magnitude far less than that of Hartree-Fock for the lightest atoms but comparable to that of Hartree-Fock for Ne and Ar.

Exchange energies are compared in Table 3. Taking the Hartree-Fock values[2] as a standard, we see that LSD consistently underestimates the magnitude of the exchange energy by 10-15%. SIC greatly improves the exchange energy, but tends to overestimate its magnitude slightly. This accounts for most of the error in the SIC total energy.

Correlation energies are compared with "experimental" values[35] in Table 4. LSD overestimates the magnitude of the correlation energy by as much as 100-200%. The SIC correlation energies,

Table 2. Calculated ground-state total energies of atoms compared with experiment, in eV. (After Ref. 20). The Hartree-Fock (HF) calculations are spin-restricted.

Atom	HF	LSD	SIC	Expt.
H	-13.6	-13.0	-13.6	-13.6
He	-77.9	-77.1	-79.4	-79.0
Li	-202.2	-199.8	-204.2	-203.5
Be	-396.5	-393.0	-399.8	-399.1
B	-667.4	-662.5	-672.0	-670.8
N	-1,480.2	-1,472.7	-1,488.9	-1,485.3
F	-2,704.9	-2,696.6	-2,720.7	-2,713.5
Ne	-3,497.8	-3,488.9	-3,517.6	-3,508.1
Na	-4,404.2	-4,392.6	-4,426.1	-4,414.7
Mg	-5,431.5	-5,418.3	-5,456.4	-5,443.2
Al	-6,581.5	-6,566.0	-6,608.8	-6,594.0
P	-9,271.0	-9,251.1	-9,303.7	-9,285.1
Ar	-14,334.7	-14,310.5	-14,378.3	-14,354.6

$$E_c^{SIC} = \int d^3r\, n(\underline{r})\ \varepsilon_c(n_\uparrow(\underline{r}),n_\downarrow(\underline{r})) - \sum_{\alpha\sigma} \int d^3r\, n_{\alpha\sigma}(\underline{r})\ \varepsilon_c(n_{\alpha\sigma}(\underline{r}),0) \tag{49}$$

overestimate the magnitude by only about 20%. Contrary to widespread opinion, the uniform electron gas correlation is relevant to small systems like atoms, once it has been purged of self-interaction.

Table 3. Exchange energies of atoms, in eV. (after Ref. 20)

Atom	LSD	SIC	HF
H	-6.9	-8.5	-8.5
He	-23.2	-27.9	-27.9
Ne	-297.6	-337.8	-329.5
Ar	-755.8	-842.4	-821.3
Kr	-2,407.5	-2,632.0	-2,561.9

Table 4. Correlation energies of atoms, in eV. (after Ref. 20)

Atom	LSD	SIC	Expt.
H	-0.6	-0.0	-0.0
He	-3.0	-1.5	-1.1
Be	-6.0	-3.1	-2.6
Ne	-19.9	-11.4	-10.4
Mg	-23.9	-13.6	-11.6
Ar	-38.4	-22.3	-19.9

Fig. 3 shows the LSD and SIC correlation energies for neutral atoms as a function of atomic number Z. Note the change of slope that occurs in the first transition series at chromium, as electrons of anti-parallel spin start to be added to the d shell. Except for such fine details, the LSD and SIC correlation energies for all atoms and ions may be described by a simple analytic model.[30]

C. Cohesive Energy of Metals

The cohesive energy is the small (1 - 6eV) difference of two large numbers: the energy of the free atom and the energy per atom of the solid. Although LSD-calculated total energies of atoms and

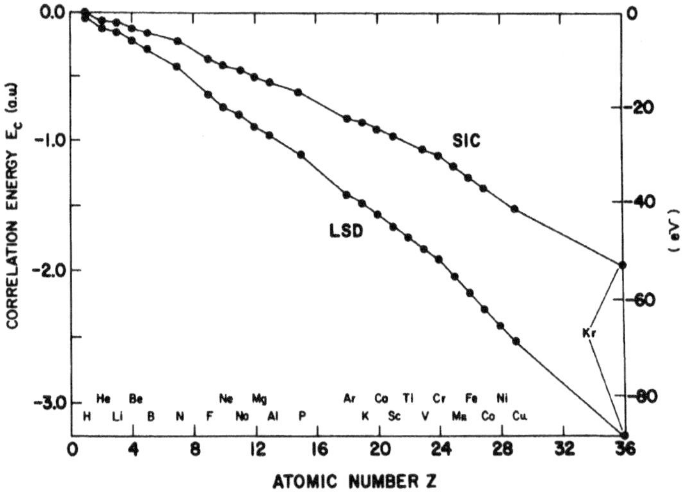

Fig. 3. Correlation energies of neutral atoms. (From Ref. 20.)

solids are in error due to self-interaction, their difference yields a reasonable estimate of the cohesive energy in most metals.[37] In potassium for example, where the LSD self-interaction error of the atomic energy is 73eV, the LSD cohesive energy is only 1eV, in good agreement with experiment. The explanation is of course that nearly all the self-interaction error is carried by the ionic core, which is essentially the same for the solid as for the atom, while the self-interaction error of the K 4s orbital is much smaller than the cohesive energy.

The 3d transition metals are a notable exception to this rule. LSD gives accurate cohesive energies at both ends of the 3d series, but overestimates them by as much as 70% near the middle.[37]

As we cross the transition series from calcium to copper, the 3d band is progressively filled, and the orbitals shrink in response to the increasing nuclear charge. As a result the LSD self-interaction of each 3d orbital grows progressively, from about 0.2eV to about 1.2eV. These orbitals are not inert core orbitals until we reach copper; in fact they hybridize strongly with the 4s orbitals. It seems likely then that the s-d valence shell carries a greater LSD self-interaction in the transition atom than it does in the solid, leading to the LSD overestimation of the cohesive energy.[19,20]

D. Electron Density in Atoms

As is evident in Fig. 2(b), self-consistent LSD electron densities are too diffuse. Self-interaction corrections to the one-electron potential counter this error by drawing the density in closer to the nucleus.

The polarizability of an atom describes its density response to a weak, uniform external electric field. In comparison with LSD, SIC polarizabilities tend to be slightly closer to experiment for the rare gas atoms, but slightly further away for the group IIA divalents.[38]

In section 3.B we saw how SIC reduces the LSD correlation energies of atoms by about a factor of two, bringing them into much better agreement with experiment. If this is not a fluke, we should expect a similar improvement in the density response to the correlation potential, i.e. the difference between the electron density calculated self-consistently with and without correlation.

Fig. 4 shows the density response to the correlation potential for the helium atom in LSD and SIC calculations.[20] This figure also shows the difference between configuration interaction[39] (CI) and Hartree-Fock densities, which can be regarded as the exact response to the correlation potential. The SIC results are clearly more real-

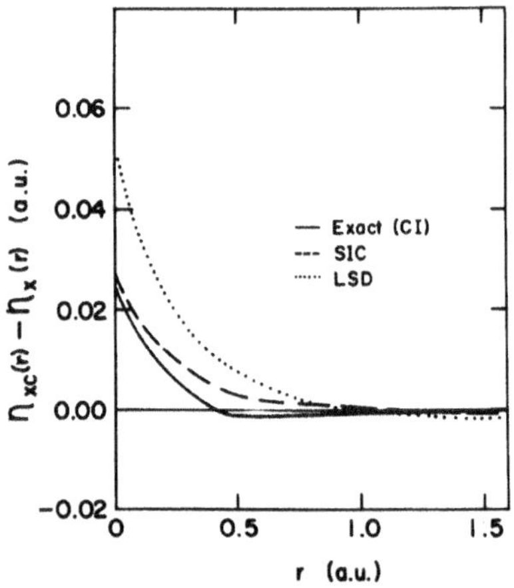

Fig. 4. Density response to the correlation potential in the helium atom. (From Ref. 20.)

istic than LSD, although both approximations miss the CI tendency to deplete the density in the intermediate region 0.5 a.u. $\leq$ r $\leq$ 1.0 a.u.

E. Orbital Energy Eigenvalues and Removal Energies

As discussed in section 2.D, the orbital energies (multiplied by −1) should approximate the relaxed energy changes due to electron removal in SIC, but not in LSD. This conclusion is supported by Table 5, which compares the eigenvalue of the least-bound occupied orbital to the measured first-ionization potential[40] in a series of atoms. On the average, the SIC eigenvalues agree with experiment even better than the Hartree-Fock eigenvalues do. They seem to satisfy the inequality (39) almost as an equality. In contrast, the LSD eigenvalues only capture about 60% of the ionization potential.

Qualitatively similar conclusions apply when the ionization process leaves the atom in an excited state. Table 6 displays a detailed comparison between ground-state orbital eigenvalues and relaxed energy differences due to hole creation in Ar. First consider the relaxed energy differences $\Delta E_{\alpha\sigma}|_{rel}$ in comparison with measured removal energies:[41] The SIC values show the best overall agreement with experiment, while the Hartree-Fock[41] values are somewhat in error for the outer orbitals (due to neglect of correlation) and the LSD values are slightly in error for the inner orbitals (due to self-interaction).

Table 5. Comparison of ε_{max}, the least negative eigenvalue of the occupied orbitals, with the measured first ionization potential I. Energies in eV. (After Ref. 20)

Atom		$-\varepsilon_{max}$		I
	HF	LSD	SIC	Expt.
H	13.6	7.3	13.6	13.6
Li	5.3	3.2	5.4	5.4
Na	5.0	3.1	5.1	5.1
K	4.0	2.6	4.3	4.3
N	15.4	8.3	14.9	14.5
P	10.7	6.3	10.0	10.5
Cr	6.5	4.0	6.7	6.8
Mn	6.7	4.6	7.1	7.4
He	25.0	15.5	25.8	24.6
Ne	23.1	13.5	22.9	21.6
Ar	16.1	10.4	15.8	15.8
Kr	14.3	9.4	14.0	14.0

Table 6. Comparison of ground-state orbital eigenvalues $\varepsilon_{n\ell}$, relaxed energy differences $\Delta E_{n\ell}|_{rel}$, and measured electron removal energies in atomic argon. Energies in eV. (After Ref. 20)

Hole state		$-\varepsilon_{n\ell}$			$\Delta E_{n\ell}\|_{rel}$		
$n\ell$	HF	LSD	SIC	HF	LSD	SIC	Expt.
1s	3,227.0	3,097.0	3,220.0	3,195.0	3,179.0	3,195.0	3,206.0
2s	335.0	294.0	316.0	325.0	310.0	311.0	
2p	260.0	230.0	257.0	249.0	248.0	251.0	249.0
3s	34.8	24.0	30.2	33.2	29.9	30.1	29.9
3p	16.1	10.4	15.8	14.8	15.7	15.8	15.8

Next compare the orbital eigenvalues to the measured removal energies in Table 6: The Hartree-Fock[41] eigenvalues overestimate the removal energies for inner orbitals (due to neglect of relaxation and correlation) while the LSD eigenvalues seriously underestimate the removal energies from all orbitals. In contrast, the SIC eigenvalues yield a good estimate of the removal energies. Finally, note that the self-interaction free theories (Hartree-Fock and SIC) satisfy the inequality (39), and SIC satisfies it more nearly as an equality.

F. Band Gaps in Insulators

Neither the Hartree-Fock nor the LSD approximations are really satisfactory for band-structure calculations. The former predicts band gaps for insulators that are too large,[42] often by as much as a factor of 2-5 relative to experiment, while the latter yields band gaps that are too small[43] by 20-40%.

The LSD band gaps are too small because the states at the bottom of the gap, being typically more localized than those at the top of the gap, carry a greater positive self-interaction error in their orbital energy eigenvalues. Exactly the same error occurs in LSD calculations for the free atom.

We show in Table 7, along with the LSD band-structure error $\Delta E_{gap} = E_{gap}^{LSD} - E_{gap}^{expt}$ for rare-gas solids, the difference Δ_{SIC}

Table 7. Band gaps in rare gas solids, in eV. The first two columns show the experimental and LSD values for the energy gap between the occupied and unoccupied single-particle bands. The third column shows ΔE_{gap}, the error of the LSD gap, and the last column shows Δ_{SIC}, the self-interaction error of the eigenvalue for the highest occupied state in the atom. (After Refs. 19 and 20.)

Solid	E_{gap} Expt.	E_{gap} LSD	ΔE_{gap}	Δ_{SIC}
Ne	21.4	11.2	-10.2	9.9
Ar	14.2	8.3	-5.9	5.8
Kr	11.6	6.8	-4.9	4.9
Xe	9.3			4.2

between the LSD and SIC eigenvalues of the outermost atomic orbital forming the valence-band maximum (e.g. 2p, 3p, 4p for Ne, Ar and Kr respectively). The correction Δ_{SIC} is expected to be accurate since the p-like valence bands in the solid are atomic-like, narrow, and well-separated from the other bands. Δ_{SIC} was calculated using the same local exchange-correlation functionals used in the corresponding band-structure calculations.[43] Table 7 clearly shows that the self-interaction of the p orbital accounts for almost 100% of the LSD error in the insulating gap of the solid.[19,20] This result is also confirmed by an actual SIC band-structure calculation for solid argon.[20,44]

G. Negative Ions

The negative ions are a sensitive test of any many-electron theory, since a large fraction of their binding energy is due to correlation. In LSD the one-electron equations (20) do not admit self-consistent solutions for negative ions:[45] The eigenvalue of the extra electron iterates to a positive value because of the incorrect long-range behavior of the one-electron potential discussed in section 2.D. (If however the extra electron is artificially confined to a region near the atom, the LSD total energy is lower than for the neutral atom, and the difference is a rough estimate of the true binding energy.[46]) This kind of misbehavior cannot occur in a self-interaction free theory like Hartree-Fock or SIC: Because of the inequality (39), the eigenvalue of the extra electron will always be negative if binding is allowed by the total energy.

Table 8 shows Hartree-Fock and SIC binding energies of negative ions, calculated from total energy differences,[19,20] in comparison with experiment.[47] The Hartree-Fock values[48] grossly underestimate the binding, and in some cases predict an instability for the experimentally stable negative ions, while the SIC values are remarkably close to experiment, giving us additional confidence in the SIC description of correlation.

Table 8. Binding energies of negative ions in eV (or electron affinities of the corresponding neutral atoms). (After Refs. 19 and 20.)

Ion	Hartree-Fock	SIC	Expt.
H^-	-0.3	0.7	0.75
O^-	-0.5	1.6	1.46
F^-	1.4	3.6	3.40
Cl^-	2.6	3.8	3.62

H. Relativistic SIC: Electron Affinities of the Elements

Correlation effects are always important in the binding energy of negative ions, but relativistic effects turn out to be equally important for the heavier ions.

The LSD and SIC approximations require that the z-component of an electron's spin be a good orbital quantum number, so these approximations do not blend well with the full Dirac equation. Koelling and Harmon[49] have proposed a modified Dirac equation, with the spin-orbit interaction averaged out, for use with LSD. This equation has been used to perform relativistic SIC calculations for all the ground-state neutral atoms and negative ions with $Z < 86$.[50] The electron affinity of each atom has been calculated as the difference between SIC total energies of the neutral atom and its negative ion.

The calculated affinities[50] are compared with experimental values[47] in Fig. 5 (a) and (b). The discrepancies are usually small and exhibit a pattern of "interconfigurational energy error"[51] which is familiar from LSD and SIC calculations of atomic first ionization potentials:[20,33,34] The calculated values are nearly exact for re-

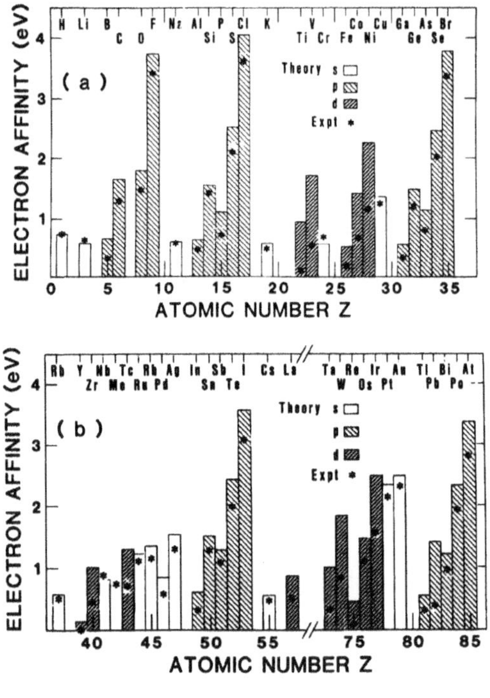

Fig. 5. Theoretical (histogram bars) and experimental (asterisks) electron affinities. (From Ref. 50.)

moval of an s electron, a few tenths of an eV too large for removal of a p electron, and as much as leV too large (but with considerable scatter) for removal of a d electron.

The rare earth atoms are not included in Fig. 5, since "almost nothing is known"[47] about their negative ions - not even whether they exist or not. Relativistic SIC calculations were attempted[50] for the negative ions of Ce, Pr, Nd and Gd, assuming a configuration the same as the neutral atom plus one extra 5d electron. Self-consistent solutions were found for all four, although the calculated affinities were too small to be trusted for all except possibly Ce (calculated affinity = 0.8 eV). The neutral atom configurations are $4f^{k-3}\ 5d^1\ 6s^2$ for Ce and Gd, and $4f^{k-2}\ 5d^0\ 6s^2$ for Pr and Nd. The calculated affinities are significantly smaller for the latter configuration, probably because 4f electrons shield the nuclear potential more efficiently than 5d electrons.

The existence of stable doubly-negative ions in the free state, such as O^{2-} and Te^{2-}, is also experimentally uncertain.[47] No self-consistent SIC solution was found for either species. For the ten-electron O^{2-} closed shell configuration, the nuclear charge Z was varied continuously:[50] self-consistent solutions were found only for $Z > 8.5$, and positive second affinities only for $Z > 8.7$.

Nonrelativistic calculations were also performed for a few species to monitor the relativistic contribution to the electron affinity.[50] It appears that relativity <u>enhances</u> the affinity for s electrons (as in the alkali and noble metals), and <u>suppresses</u> the affinity for p (as in the halogens) and d electrons (as in transition metals). The relativistic effects are small in the alkalis and halogens, but dramatic in the transition and noble metals.

Both Cs and Au can bind an extra 6s electron, but the relativistic effects are far greater in Au where the 6s orbital is considerably more contracted and feels a much stronger nuclear potential. The non-relativistic calculated affinity of Au is only 1.5 eV, about half of which comes from correlation energy. The relativistic affinity is 2.5 eV, in good agreement with the experimental value of 2.31 eV.[47] Thus the extra stabilization of s orbitals due to the relativistic mass increase applies even to the most loosely-bound valence orbitals.

The spin-orbit splitting was not included in the calculations of Fig. 5. While it could be important for the non-s affinities of the heavier atoms, its importance is not discernible from Fig. 5. If included at all, it should be included self-consistently.

I. Autodetaching States

The LSD and SIC methods admit self-consistent solutions which look like excited electronic states as well as ground states. The excited-state solutions are not blessed by as many theorems as the ground-state ones, but they can still be quite realistic, as we saw for neutral atoms in Table 6 of section 3.E. Among the excited states of atoms and ions are some which are energetically permitted to decay by emission of an electron--the autodetaching states. For example, the autodetaching states of a negative ion A^- show up as resonances in the scattering of electrons from the neutral atom or molecule A.

Table 9 compares Hartree-Fock and SIC total energies for the lowest-lying $(ns)^2\,{}^1S$ resonances in H^-[29] with "exact" values calculated from close-coupling equations.[52] No self-consistent LSD solutions were found. The Hartree-Fock solutions place each $(ns)^2$ resonance of H^- above the threshold for excitation of the $(ns)^1$ state in H, while the SIC solutions place each resonance below threshold, in good agreement with the "exact" results. SIC also predicts the existence of other known resonances (e.g. $(2p)^2$, $(3p)^2$), and admits no self-consistent solution for those non-resonant configurations (e.g. $(1s)^1\,(2s)^1$) which were tried. SIC calculations of autodetaching states in a few other atoms have also been performed.[50]

Autodetaching states can be regarded as ghostly remnants of bound states which have moved up into the continuum due to a weakening of the nuclear attraction. Consider for example an ion with $N \leq 2$ electrons and variable nuclear charge Z. Fig. 6 shows the total energy $\tilde{E}\,(Z,N)$ calculated by the SIC method. By (28), the slope of

Table 9. Total energies in eV of the $(1s)^2\,{}^1S$ ground-state and some $(ns)^2\,{}^1S$ autodetaching states in H^-. The "exact" $(2s)^2$ and $(3s)^2$ energies are the results of close-coupling calculations. (After Ref. 29)

Parent state (H)	Exact	Daughter state (H^-)	HF	SIC	Exact
1s	-13.61	$(1s)^2$	-13.28	-14.32	-14.36
2s	-3.40	$(2s)^2$	-3.29	-3.80	-4.03
3s	-1.51	$(3s)^2$	-1.45	-1.75	-1.88
4s	-0.85	$(4s)^2$	-0.81	-1.01	

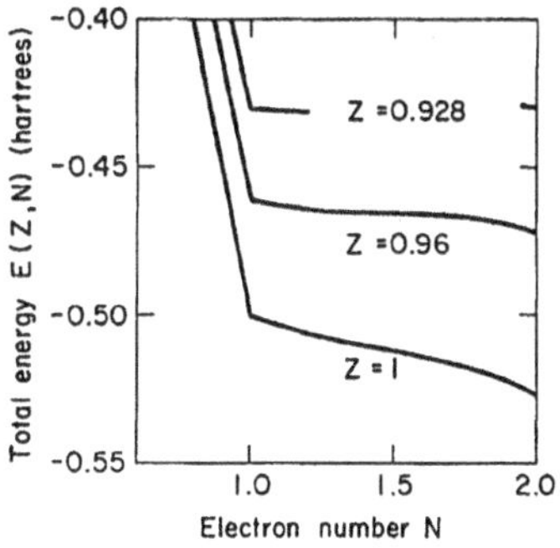

Fig. 6. Total SIC energy for $N \leq 2$ and $Z \leq 1$. (From Ref. 29.)

$\tilde{E}$ vs. N gives the eigenvalue of the orbital being populated (1s↑ for $N \leq 1$, and 1s↓ for $1 < N \leq 2$). At $Z = 1$, the $(1s)^2$ configuration lies below the $(1s)^1$ configuration, with an energy difference which correctly approximates the binding energy of H^-. As Z is reduced, the binding energy of the second electron decreases until it vanishes at $Z = 0.93$, where the $(1s)^2$ configuration becomes an autodetaching state.[29]

J. Other Applications

Self-interaction corrections have been applied to calculate quantum defects in Rydberg atoms.[53] Here SIC is clearly important, since a highly excited valence electron samples the long-range behavior of the one-electron potential. For divalent Rydberg atoms, SIC was used only to generate a core potential; the two valence electrons were then coupled by standard many-body techniques.

SIC has also been applied to the core electrons on a Li impurity in metallic Aℓ.[54] The resulting core orbitals are more contracted than the LSD core orbitals, but not enough to significantly affect the scattering of Fermi surface electrons from the impurity.

The moving electrons in an atom or solid interact not only through their instantaneous Coulomb repulsion but also through magnetic and retarded Coulomb interactions, which together make up the transverse-photon electron or Breit interaction. The local density approximation for the transverse-photon electron energy in an atom displays a large relative error,[55] varying from 150% for He to 50% for Fm. Upon self-interaction correction,[56] these errors are dramatically reduced* to values ranging from 25% for He to 1% for Fm.

*For further discussion, see the article "Local Density Aproximation for Relativistic Exchange Energies" by A.H. MacDonald (this volume).

Nearly all of the correction can be associated with the 1s orbitals.

4. RECENT DEVELOPMENTS

A. Simplified Self-Interaction Correction of the LSD Orbital Energies

For many practical purposes, and especially for band structure calculations, the worst problem with the LSD approximation is the discrepancy between orbital energy eigenvalues and electron removal energies. As we have seen, this discrepancy is greatly narrowed by SIC, at least in atoms. Here we will reduce the SIC method to its simplest viable form, and then generalize the result to solids and other extended systems.[57]

We begin by noting that the self-exchange correction to the potential in Eq. (27) is

$$-v_x^{\uparrow,LSD}(\underline{r};[n_{\alpha\sigma},0]) = 1.241\, n_{\alpha\sigma}^{1/3}(\underline{r}) \tag{50}$$

The self-direct correction in Eq. (27) is a nonlocal functional of the orbital density which, according to Eq. (31), also has a local density approximation:

$$\begin{aligned} -\int d^3r'\, \frac{n_{\alpha\sigma}(\underline{r}')}{|\underline{r}-\underline{r}'|} &= -\frac{\delta}{\delta n_{\alpha\sigma}(\underline{r})}\left\{\frac{1}{2}\int d^3r'\int d^3r''\, \frac{n_{\alpha\sigma}(\underline{r}')n_{\alpha\sigma}(\underline{r}'')}{|\underline{r}'-\underline{r}''|}\right\} \\ &\approx -\frac{\delta}{\delta n_{\alpha\sigma}(\underline{r})}\left\{1.085N^{2/3}\int d^3r'\, n_{\alpha\sigma}^{4/3}(\underline{r}')\right\} \\ &= -\frac{4}{3}(1.085)n_{\alpha\sigma}^{1/3}(\underline{r}) - \frac{2}{3}(1.085)\int d^3r'\, n_{\alpha\sigma}^{4/3}(\underline{r}') \end{aligned} \tag{51}$$

The numerical factor 1.092 has been replaced by 1.085 in order to make the inequality (31) an equality for exponential densities $n(r) = ae^{-br}$, and N has been set equal to 1 for an occupied orbital. The last term on the far right of Eq. (51) is just a constant - it represents the attempt of the local approximation to mock up the long-range tail $-1/r$.

Thus the complete self-interaction correction to the potential seen by orbital $\alpha\sigma$ is approximately

$$-0.206n_{\alpha\sigma}^{1/3}(\underline{r}) - \mu_c^{\uparrow}(n_{\alpha\sigma}(\underline{r}),0) - \frac{2}{3}(1.085)\int d^3r'\, n_{\alpha\sigma}^{4/3}(\underline{r}') \tag{52}$$

The first term here is the small residue of the cancellation of two larger terms. It tends to cancel further against $-\mu_c^{\uparrow}(n_{\alpha\sigma}(\underline{r}),0)$, the negative of the correlation potential which is also small.

Neglecting these terms, we are left with a self-interaction correction to the effective potential which is constant in space but orbital-dependent. This correction shifts each unphysical LSD orbital energy into better agreement with the physical removal energy,

$$\Delta E_{\alpha\sigma}\Big|_{rel} \approx -\varepsilon_{\alpha\sigma}^{SIC} \approx -\varepsilon_{\alpha\sigma}^{LSD} + 0.723\int d^3r\, n_{\alpha\sigma}^{4/3}(\underline{r}) \qquad (53)$$

while it leaves each LSD orbital $\psi_{\alpha\sigma}(\underline{r})$ unchanged.

Eq. (53) has been derived for a single atom, and is not size-consistent. However, it suggests[57] a size-consistent add-on self-interaction correction to the LSD orbital energy:

$$\Delta E_{\alpha\sigma}\Big|_{rel} \approx -\varepsilon_{\alpha\sigma}^{LSD} + 0.104\int d^3r\,|\psi_{\alpha\sigma}(\underline{r})|^2\, \frac{\hat{u}\cdot\nabla\rho(\underline{r},\varepsilon)}{\rho(\underline{r},\varepsilon)}\Big|_{\varepsilon=\varepsilon_{\alpha\sigma}^{LSD}} \qquad (54)$$

where $\hat{u}$ is unit vector in the direction of $\nabla n(\underline{r})$, and

$$\rho(\underline{r},\varepsilon) = \sum_{\alpha\sigma} |\psi_{\alpha\sigma}(\underline{r})|^2 \delta(\varepsilon - \varepsilon_{\alpha\sigma}) \qquad (55)$$

is the local density of states. All inputs to Eq. (54) are to be calculated self-consistently in the LSD approximation.

Eqs. (53) and (54) have been tested in numerical calculations for 16 atoms ranging from hydrogen to mercury; in comparison with experimental values[58] for the electron removal energies $\Delta E_{\alpha\sigma}|_{rel}$ from all occupied orbitals, they agree as well as the full, original, self-consistent SIC orbital energies $-\varepsilon_{\alpha\sigma}^{SIC}$. Typical results for the nickel atom are shown in Table 10.

Consider the application of Eq. (54) to the deep core orbitals in a crystal. The orbital density may be described either in the Wannier representation ($|\psi_{n\ell}|^2$ on one atom and zero elsewhere) or in the Bloch representation ($|\psi_{n\ell}|^2/N_a$ on each of the N_a atoms). In either case, Eq. (54) gives the same removal energy. The physical analog of this invariance is the fact that a long-lived core hole remains localized on one atom for a long time on the microscopic scale, and yet becomes delocalized over longer times as it hops to neighboring atoms.

The new self-interaction correction of Eq. (54) can also be regarded[57] as an approximation of the first-order effect of the difference between Dyson's quasiparticle operator[59] and the LSD effective potential. It remains to be tested in actual band structure calculations.

B. On the Exact Kohn-Sham Orbital Energies

The LSD orbital energy eigenvalues do not closely approximate electron removal energies. Is this a defect only of LSD, or also

Table 10. Experimental electron removal energies $\Delta E_{n\ell}|_{rel}$ (from Ref. 58) in atomic Ni ($3d^8 4s^2$), compared with LSD and original SIC (Ref. 20) orbital energies $\varepsilon_{n\ell}$ and with the results of Eqs. (53) and (54). (All energies in eV. To facilitate comparison with experiment, the calculations were done relativistically using the expressions summarized in Ref. 50. All energies were averaged over the magnetic quantum numbers.) (From Ref. 57.)

Orbital	expt. $\Delta E_{n\ell}\|_{rel}$	$-\varepsilon_{n\ell}^{LSD}$	$-\varepsilon_{n\ell}^{SIC}$	Eq. (53)	Eq. (54)
1s	8348.	8168.	8375.	8326.	8329.
2s	1024.	977.	1013.	1012.	1012.
2p	876.	843.	888.	878.	876.
3s	125.	110.	121.	127.	121.
3p	81.9	71.0	81.4	84.6	81.1
3d	16.9	9.3	18.5	17.7	17.1
4s	9.0	5.9	8.7	9.8	8.5

of the exact Kohn-Sham orbital formalism to which LSD is an approximation? Will improved approximations[12,60] to the exact Kohn-Sham potential significantly improve the calculated band structures of crystals, or not?

The answer[57] appears to be that the unphysical LSD orbital energies closely approximate the exact Kohn-Sham values, apart from an overall constant shift of all the occupied levels (roughly equal to the LSD self-interaction of the least-bound electron). In other words, a local effective potential $v_{eff}(\underline{r})$ of the Kohn-Sham type can yield the correct density but not also the correct removal energies; only an orbital-dependent (or nonlocal) potential can do both. This conclusion is based on the "exchange only" approximation, which is justified for tightly-bound orbitals where correlation is negligible compared to exchange.

Talman and co-workers[61] have constructed an exact realization of the Kohn-Sham potential for "exchange only" in atoms, i.e. they have found the local potential $v_{eff}(\underline{r})$ whose orbitals yield an absolute minimum of the Hartree-Fock energy functional. (The Hartree-Fock orbitals, which arise from a nonlocal effective potential $v_{eff}(\underline{r},\underline{r}')$, yield a lower minimum, but only by less than fifty parts per million.[61] Note also that the Talman potential tends to zero as $r \to \infty$, like the LSD, SIC and Hartree-Fock potentials; the exact Kohn-Sham potential whose orbital energies satisfy Eq. (38) could conceivably tend to a different constant in this limit.)

Table 11 compares the exact Kohn-Sham orbital energies[61] $\varepsilon_{\alpha\sigma}$ with LSD approximate values and with removal energies, all calculated for "exchange only" in the Ar atom. (Similar results were found for other open - and closed-shell atoms.) The removal energies $\Delta E_{\alpha\sigma}|_{rel}$ were taken from Hartree-Fock total energy differences, calculated with relaxed Hartree-Fock orbitals.[41] The ground-state Hartree-Fock orbital energies,[41] which by Koopmans' theorem equal the unrelaxed total energy differences $\Delta E_{\alpha\sigma}|_{unrel}$, are also displayed. Note that, when the LSD orbital energies are all shifted by the same constant, the results are close to the exact Kohn-Sham orbital energies, but the latter do not accurately predict the removal energies. From these results we can conclude in particular that LSD will give an accurate approximation to the occupied Kohn-Sham energy bands in a solid (apart from a constant shift of all the bands), but this band structure will not predict the correct removal energies.

C. Fractional Electron Number and the Separated Atom Limit

Consider a system composed of one H atom and one Li atom, infinitely separated, in an otherwise empty universe. Fig. 7 displays the energy of this system as a function of $\Delta N_{Li} = (N_{Li} - 3)$, the number of electrons transferred from the H to the Li in a trial density variation. A separate self-consistent calculation employing fractional[20,25] occupation numbers $f_{\alpha\sigma}$ has been performed for each atom.

Table 11. Orbital energies $\varepsilon_{n\ell}$ and electron removal energies $\Delta E_{n\ell}|_{rel}$ for the Ar atom. HF: Hartree-Fock (Ref. 41). KS-X: Exact Kohn-Sham for "exchange-only" (Ref. 61). LSD-X: Local spin density approximation for "exchange-only". (All energies in eV.) (From Ref. 57.)

Orbital	HF		KS-X	LSD-X	
$n\ell$	$-\varepsilon_{n\ell}$	$\Delta E_{n\ell}\|_{rel}$	$-\varepsilon_{n\ell}$	$-\varepsilon_{n\ell}$	$-\varepsilon_{n\ell}+7.1$
1s	3227.	3195.	3114.	3094.	3101.
2s	335.	325.	303.	292.	299.
2p	260.	249.	238.	228.	235.
3s	34.8	33.2	29.7	22.6	29.7
3p	16.1	14.8	16.1	9.0	16.1

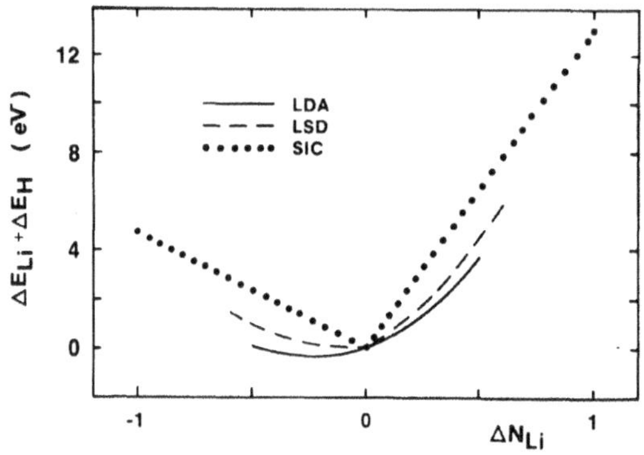

Fig. 7. Change in total energy of two atoms, Li and H, separated by an infinite distance, when $\Delta N_{Li} = (N_{Li} - 3)$ electrons are transferred from H to Li.

In the local density approximation,[4] the energy is differentiable at $\Delta N_{Li} = 0$ and as a result the energy minimizes incorrectly at $\Delta N_{Li} \cong -0.25$, i.e. at the configuration $Li^{+0.25}\ H^{-0.25}$. In LSD, a derivative discontinuity appears which minimizes the energy correctly at $\Delta N_{Li} = 0$, the neutral-atoms configuration. In SIC, the derivative discontinuity is magnified and the curve of energy vs. ΔN_{Li} becomes essentially two straight line segments.

The exact result is in fact two straight line segments: When the Hohenberg-Kohn[10] theorem is extended[62,*] to an isolated open system described by a statistical mixture, the curve of average energy E vs. average electron number N is a series of straight line segments with slope discontinuities at integer N. From this theorem, a number of interesting conclusions may be drawn[62,*] about the neutrality of well-separated atoms, the maximum separation between two atoms which can support an ionic ground state, the chemical potential $\mu = \partial E/\partial N$ for an atom, the highest occupied Kohn-Sham orbital energy, the limiting constant value of the Kohn-Sham potential far from the center of an atom, and violations of the "aufbau principle".

Evidently (from Fig. 7) the self-interaction error of the LSD total energy is much worse for fractional occupation $0 < f_{\alpha\sigma} < 1$ than it is for integer occupation $f_{\alpha\sigma} = 1$. The reason is evident from a comparison of Eqs. (31) and (32), evaluated for $n_\uparrow(\underline{r}) = f_{\alpha\sigma}|\psi_{\alpha\sigma}(\underline{r})|^2$ and $n_\downarrow(\underline{r}) = 0$.

---

*See also the article by R.G. Parr in this volume.

D. Other and Future Work

Recent experiments[63] have not confirmed the observation of doubly-charged atomic negative ions like O^{2-} and Te^{2-}. These new experimental results agree with the SIC prediction[50] discussed in section 3.H.

A recent study[64] confirms and illustrates the remarks made in section 2.D about the variation of the orbital energy $\varepsilon_{\alpha\sigma}$ with occupation number $f_{\alpha\sigma}$ in LSD and SIC as well as in the Hartree-Fock approximation.

An interesting method has been presented[65] which incorporates SIC into self-consistent band structure calculations for insulating crystals. The essence of the idea is to use localized Wannier orbitals to make the self-interaction correction to the total energy. The Wannier functions are constructed from delocalized Bloch orbitals which belong to a periodic orbital-dependent potential. The band structure calculated in this way for LiCℓ shows a remarkable improvement in the band gap and core levels. The method may also be applicable to semiconductors.

Since the Hartree-Fock method is naturally self-interaction free and independent of the choice of orbital representation, it has been suggested[66] that the Hartree-Fock expression for the energy as a functional of the orbitals could be generalized to include correlation by replacing the kernel $-1/|\underline{r}-\underline{r}'|$ by a spin- and density-dependent effective interaction. While it remains untested, this is an interesting and suggestive idea, especially if a similar result should emerge for the effective one-electron potential. It may be possible to prove a Kohn-Sham-like theorem for a nonlocal effective potential, and such a potential is just what is needed to improve the agreement between orbital energy eigenvalues and physical electron removal energies, as discussed in sections 4.A and B.

ACKNOWLEDGMENTS

Sections 1-3 in preliminary form were presented as an invited talk at the International Workshop on Electronic Density Functionals (U.N.A.M., Mexico City, November 1980). Special thanks are due to my co-authors Alex Zunger, Jim Rose, Herb Shore, Lee Cole, Ed McMullen, Mike Norman, Bob Parr, Mel Levy and Jose Balduz, who all made major contributions to this work. The Support of the National Science Foundation, Grant No. DMR 80-16117, is gratefully acknowledged.

REFERENCES

1. E. Fermi and E. Amaldi, Mem. Accad. Ital. 6, 117 (1934).
2. J.C. Slater and J.H. Wood, Int. J. Quantum Chem. 4, 3 (1971).
3. J.C. Slater, "The Self-Consistent Field for Molecules and Solids," McGraw-Hill, New York (1974).
4. W. Kohn and L.J. Sham, Phys. Rev. 140, A1133 (1965).
5. R.D. Cowan, Phys. Rev. 163, 54 (1967).
6. D.A. Liberman, Phys. Rev. 171, 1 (1968).
7. I. Lindgren, Int. J. Quantum Chem. 5, 411 (1971).
8. B.H. Brandow, Adv. Phys. 26, 651 (1977).
9. M. Gopinathan, Phys. Rev. A 15, 2135 (1977); T.J. Tseng, S.H. Hong and M.A. Whitehead, J. Phys. B 13, 4101 (1980).
10. P. Hohenberg and W. Kohn, Phys. Rev. 136, B864 (1964).
11. U. von Barth and L. Hedin, J. Phys. C 5, 1629 (1972).
12 O. Gunnarsson, M. Jonson and B.I. Lundqvist, Phys. Rev. B 20, 3136 (1979); J.A. Alonso and L.A. Girifalco, Phys. Rev. B 17, 3735 (1978).
13. D.J.W. Geldart and M. Rasolt, Phys. Rev. B 13, 1477 (1976).
14. D.C. Langreth and J.P. Perdew, Phys. Rev. B 21, 5469 (1980).
15. G.W. Bryant and G.D. Mahan, Phys. Rev. B 17, 1744 (1978); G.D. Mahan, Phys. Rev. B 22 , 3102 (1980) and Phys. Rev. A 22, 1780 (1980).
16. H. Stoll, C.M.E. Pavlidou and H. Preuss, Theoret. Chim. Acta 149, 143 (1978); H. Stoll, E. Golka and H. Preuss, ibid. 55, 29 (1980).
17. A. Zunger and M.L. Cohen, Phys. Rev. B 18, 5449 (1978).
18. J.P. Perdew, Chem. Phys. Lett. 64, 127 (1979).
19. A. Zunger, J.P. Perdew and G.L. Oliver, Solid State Commun. 34, 933 (1980).
20. J.P. Perdew and A. Zunger, Phys. Rev. B 23, 5048 (1981).
21. M. Levy, Proc. Natl. Acad. Sci. USA 76, 6062 (1979).
22. R.S. Gadre, L.J. Bortolotti and N.C. Handy, J. Chem. Phys. 72, 1034 (1980).
23. O. Gunnarsson and B.I. Lundqvist, Phys. Rev. B 13, 4274 (1976).
24. D.C. Langreth and J.P. Perdew, Phys. Rev. B 15, 2884 (1977).
25. J.F. Janak, Phys. Rev. B 18, 7165 (1978).
26. M.M. Morrell, R.G. Parr and M. Levy, J. Chem. Phys. 62, 549 (1975).
27. Y. Tal, Phys. Rev. A 18, 1781 (1978); 21, 2186 (E) (1980).
28. A. Zunger, Phys. Rev. B 22, 649 (1980).
29. J.P. Perdew, J.H. Rose and H.B. Shore, J. Phys. B 14, L233 (1981).
30. J.P. Perdew, E.R. McMullen and A. Zunger, Phys. Rev. A 23, 2785 (1981).
31. D.M. Ceperley, Phys. Rev. B 18, 3126 (1978); D.M. Ceperley and B.J. Alder, Phys. Rev. Lett 45, 566 (1980).
32. M. Gell-Mann and K.A. Brueckner, Phys. Rev. 106, 364 (1957); S. Misawa, Phys. Rev. 140, A1645 (1965).
33. O. Gunnarsson and R.O. Jones, Solid State Commun. 37, 249 (1981).

34. L.A. Cole and B.N. Harmon, J. Appl. Phys. 52, 2149 (1981).
35. A. Veillard and E. Clementi, J. Chem. Phys. 49, 2415 (1968).
36. C. Froese Fischer, "The Hartree-Fock Method for Atoms," Wiley, New York (1977).
37. V.L. Moruzzi, J.F. Janak and A.R. Williams, "Calculated Electronic Properties of Metals," Pergamon, New York (1978).
38. E. Zaremba (unpublished).
39. S.Jagannathan, Ph.D. thesis, University of Georgia, Athens, Ga., 1979 (unpublished); D.W. Smith, S. Jagannathan and G.S. Handler, Int. J. Quantum Chem. S13, 103 (1979).
40. C.E. Moore, Nat. Bur. Stand. (U.S.) Ref. Data Ser. 34, 1 (1970).
41. P.S. Bagus, Phys. Rev. 139, A619 (1965).
42. A.B. Kunz, Phys. Rev. B 12, 5890 (1975).
43. See e.g. Refs. (27)-(31) of Ref. 20.
44. A. Zunger (unpublished).
45. K. Schwarz, Chem. Phys. Lett. 57, 605 (1978).
46. H.B. Shore, J.H. Rose and E. Zaremba, Phys. Rev. B 15, 2858 (1978).
47. H. Hotop and W.C. Lineberger, J. Phys. Chem. Ref. Data 4, 539 (1975).
48. E. Clementi and C. Roetti, A. Data Nucl. Data Tables 14, 177 (1974); L.C. Green, M.M. Mulder, M.N. Lewis and J.W. Woll, Phys. Rev. 93, 757 (1954).
49. D.D. Koelling and B.N. Harmon, J. Phys. C 10, 3107 (1977).
50. L.A. Cole and J.P. Perdew, Phys. Rev. A 25, 1265 (1982).
51. J. Harris and R.O. Jones, J. Chem. Phys. 68, 3316 (1978).
52. G. Seiler, R. Oberoi and J. Callaway, Phys. Rev. A 3, 2006 (1971); J. Callaway, Phys. Rep. 45, 89 (1978).
53. J.A. Armstrong, S.S. Jha and K.C. Pandey, Phys. Rev. A 23, 2761 (1981).
54. L.A. Cole, Ph.D. thesis, Dartmouth College, Hanover, New Hampshire (1979).
55. A.H. MacDonald and S.H. Vosko, J. Phys. C 12, 2977 (1978).
56. J.P. Perdew and L.A. Cole, J. Phys. C 15, L905 (1982).
57. J.P. Perdew and M.R. Norman, Phys. Rev. B 26, 5445 (1982).
58. K.D. Sevier, At. Data and Nucl. Data Tables 24, 323 (1979).
59. L.J. Sham and W. Kohn, Phys. Rev. 145, 561 (1966).
60. D.C. Langreth and M.J. Mehl, Phys. Rev. Lett. 47, 446 (1981).
61 J.D. Talman and W.F. Shadwick, Phys. Rev. A 14, 36 (1976); K. Aashamar, T.M. Luke and J.D. Talman, At. Data and Nucl. Data Tables 22, 443 (1978).
62. J.P. Perdew, R.G. Parr, M. Levy and J.L. Balduz, Phys. Rev. Lett. 49, 1691 (1982).
63. D. Spence, W.A. Chupka and C.M. Stevens, Phys. Rev. A 26, 654 (1982).
64. L. Wilk and S.H. Vosko, J. Phys. C 15, 2139 (1982).
65. R.A. Heaton, J.G. Harrison and C.C. Lin, Solid State Commun. 41, 827 (1982).
66. J.F. Dobson and J.H. Rose, J. Phys. C 15, L1183 (1982).

SOME APPROACHES TO DENSITY FUNCTIONAL THEORY

Eugene S. Kryachko

Institute for Theoretical Physics
Kiev-130, USSR 252130

Ad discendum, non ad docendum.

During the last few years, much attention has been focused on density functional theory, a theory which originated with the early work of Fermi,[2] Dirac,[3] Slater[4-6] and Gombas.[7]* This is well illustrated by the present Symposium.

A large volume of the literature has been devoted to density functional theory.[9-123] This theory has some advantages in the quantum theory of atoms, molecules and the solid state, mainly because of its one-particle picture of the description of many-body systems which leads to greater essential simplicity in computational problems. Density functional theory is a powerful approach whose possibilities are not yet exhausted. Like other theories it has both achievements and shortcomings, the achievements being on a level with other well-known approaches such as the Hartree-Fock approximation, the Multiconfiguration Self-Consistent Field Theory etc. The author feels that the increasing interest in density functional theory - this is confirmed in Fig. 1 - is connected to a certain extent with unsuccessful attempts to solve the N-representability problem. There is certainly some disappointment among quantum chemists related to the noncomputability of necessary N-representability conditions, a problem which has been extensively studied by Coleman, Davidson, Garrod, Percus, Erdahl and others. Here the author expresses his firm belief that the N-representability will bear the cross, but it needs a fresh new spirit. Ceterum censeo praefationem non esse seribendam.

*Also see Rajagopal.[8]

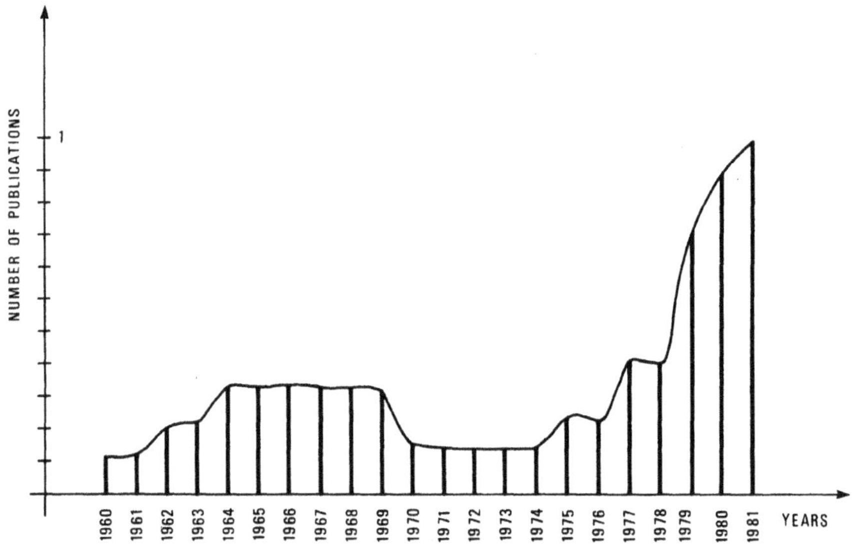

Fig. 1. The number of papers published, in general, in the Journal of Chemical Physics, the International Journal of Quantum Chemistry and the Physical Review per year. Number denoted by 1 corresponds to the level of 1981.

Ab haedis segregare oves.

1. INTRODUCTION

Like any theory, density functional theory is based on some propositions, axioms and theorems. In the literature there exists a viewpoint that density functional theory has a correct mathematical foundation based on the Hohenberg-Kohn theorem,[15] and that this is the formal justification of density functional theory. However, from the historical chronology we can notice that the first results of density functional theory appeared about fifty years before the birth of the Hohenberg-Kohn theorem, and we think that nothing would have happened to density functional theory if this delivery had been slightly delayed.

What does the Hohenberg-Kohn theorem represent? This theorem states the existence of a universal energy density functional $E_{HK}[\rho]$ for a fixed but arbitrary N-particle system, where $\rho(\vec{r})$ is the density of the system. $E_{HK}[\rho]$ is decomposed as follows:

$$E_{HK}[\rho] = \int d^3\vec{r} V_{ext}(\vec{r})\rho(\vec{r}) + F[\rho] \quad (1)$$

where $V_{ext}(\vec{r})$ is the external local potential, and $F[\rho]$ is a functional only of $\rho(\vec{r})$ and not of $V_{ext}(\vec{r})$. So, $F[\rho]$ is also universal in the sense that it need be computed only once for all time and for all many-body systems. Then an energy of the ground state and the ground-state density $\rho(\vec{r})$ are determined from the variational principle:

$$\frac{\delta F[\rho]}{\delta\rho(\vec{r})} + V_{ext}(\vec{r}) - \varepsilon = 0 \qquad (2)$$

The Hohenberg-Kohn theorem is only an existence theorem and valid for variationally determined ground-state densities. There exist some disagreements about the content[13] of the Hohenberg-Kohn theorem. Often the Hohenberg-Kohn theorem is interpreted as addressing itself to the mapping between ground-state densities and local external potentials, that is to the question of whether by arbitrary density ρ is V-representable. One asks whether a local external potential $V(\vec{r})$ exists, such that a given $\rho(\vec{r})$ is its ground-state density. Extensive studies of this question have been made via the discrete matrix analysis[16,17] and topological methods.[18]

The Hohenberg-Kohn theorem is a very general and very strong theorem, so there exists the point of view that its proof is based on fallacious arguments and that the theorem itself is not true.[19] In fact, in what manner does a density $\rho(\vec{r})$ (and the universal functional $F[\rho]$) reflect all the properties of the many-electron system under consideration?[20] Can it reflect the antisymmetry of its wavefunction, its point symmetry, i.e. all the information lying in the 2-matrices (second order density matrices) and connected with the N-representability problem?[21,22] Notice that if a universal functional exists then there exists a functional for both bosons and fermions, i.e. a density determines the quantization rule for a given many-body system. Also in their proof, Hohenberg and Kohn omitted $H_0 = T + U$ and concentrated the attention on $V_{ext}(\vec{r})$ only, which, as they found, is a unique functional of a density. Based on this they erroneously concluded that the matrix element $\langle\Psi|H_0|\Psi\rangle$ is also a unique functional of a density. In this connection there arose many reformulated proofs of the Hohenberg-Kohn theorem,[23-27] and it has been claimed[23] that a rigorous proof of the theorem requires the use of more powerful tools of functional analysis. Also some generalizations of density functional theory and, in particular, the Hohenberg-Kohn theorem, on the complete first order density matrix and a spin-density were proposed in the literature.[12,13,34-45]

In contrast to the existence theorem of Hohenberg and Kohn, Levy[26] proposed a constructive approach. But in Levy's reformulation of the universal functional $E_L[\rho]$, the 1-1 mapping between densities and wavefunctions is lost. Valone[27] rediscovered such a mapping. The mapping $\rho \to \Psi_\rho$, as in algebra is "the mapping in" but not "the mapping on". This is confirmed by Gilbert[28] and Harriman[29] (see also Kryachko,[20] p. 1032). Also the continuity and the differentiability of $E_{LV}[\rho]$ is not established. The Levy-Valone approach is very constructive and we would like to present their ideas.

The N-particle density matrix $D_N(x_1,\ldots,x_N;\ x_1',\ldots,x_N')$ may be represented in an expansion form based on a complete set of orthonormal, antisymmetric (in the case of fermions), N-particle functions $\{\Psi_K(x_1,\ldots,x_N)\}$:

$$D_N(x_1,\ldots,x_N;\ x_1',\ldots,x_N') = \sum_K W_K \Psi_K(x_1,\ldots,x_N)\bar{\Psi}_K(x_1',\ldots,x_N') \quad (3)$$

where $W_K \geq 0$ for all K, and $\sum_K W_K = 1$.

Now we denote a set of such N-matrices by E_N and define subsets there as follows:

$$L_\rho(N) \equiv \left\{ D_N \in E_N \middle| \int dx_2,\ldots,dx_N\ D_N(x,x_2,\ldots,x_N;\ x,x_2,\ldots,x_N) = \rho(x) \right\}$$

$$L_\rho(N) \subseteq E_N \quad (4)$$

for a given fixed density $\rho(x)$. Notice that for a fixed ρ the set $L_\rho(N)$ is not empty.[28,29] Levy and Valone then define the energy density functional $E_{LV}[\rho]$ as follows:

$$E_{LV}[\rho] \equiv \min \left\{ Tr(HD_N) \middle| D_N \in L_\rho(N) \right\} \quad (5)$$

Evidently, this approach does not resolve completely the problem of the direct construction of the energy density functional and the problem of its universality. These problems are opened but they require more extensive investigation.

Another constructive approach to the energy density functional without any reference to a wavefunction set has been proposed.[46,47] As a matter of fact, the approach proposed resembles the harmonic composition of the idea of Kohn-Sham quasiparticles[30] and Levy-Valone approach.

Let us formulate briefly the basic concepts of the approach mentioned above. For definiteness we consider an N-electron system in a given external local field $V_{ext}(\vec{r})$, provided, for example, by atomic nuclei charges or charges in a molecule under consideration. Let $\rho_N(\vec{r})$ be a one-particle density of this system of electrons, exact or approximate, which satisfies a normalization condition:

$$\int d^3\vec{r}\ \rho_N(\vec{r}) = N \quad (6)$$

If, in a given external field, there is a single electron (by this we provide a so-called V-representability if ρ_N is V-representable[26,27]), then the normalization condition has the following form

$$\int d^3\vec{r}\ \rho_1(\vec{r}) = 1 \tag{7}$$

Eqs. (6) and (7) provide the possibility of describing $\rho_N(\vec{r})$ in terms of $\rho_1(\vec{r})$, namely, we suppose that there exist N quasiparticles (their statistics are immaterial) by analogy with the Kohn-Sham quasiparticles,[30] each being described by the corresponding density $\rho_1^v(\vec{r})$ ($v = 1,2,\ldots,N$) with the normalization, Eq. (7):

$$\rho_N(\vec{r}) = \sum_{v=1}^{N} \rho_1^v(\vec{r}) \tag{8}$$

Thus, there follows a mapping between $\rho_N(\vec{r})$ and a set of $\{\rho_1^1,\rho_1^2,\ldots,\rho_1^N\}$. Notice, that the expansion given by Eq. (8) is slightly different from the well-known Löwdin's expansion based on natural orbitals.[48] Therefore, if $M(N,\rho_1)$ is the one-particle density ρ_1 -set for a fixed V_{ext}, for each ρ_N there exists a subset $M(V,\rho_N) \subset \otimes^N M(V,\rho_1)$ such that

$$M(V,\rho_N) = \begin{cases} \left\{\rho_1^1,\rho_1^2,\ldots,\rho_1^N\right\} \mid \rho_1^v \in M(V,\rho_1),\ v = 1,2,\ldots,N \\ \\ \rho_N = \sum_{v=1}^{N} \rho_1^v \end{cases} \tag{9}$$

Obviously, the set $M(V,\rho_N)$ is not empty since, in particular, for a given ρ_N one can usually choose a division of the type $\rho_1^v = \rho_N/N$ for all $v = 1,2,\ldots,N$. The space $M(V,\rho_N)$ is convex. So there exist the ordinary methods of the complete characterization of this set in the literature, for example, with the help of extremal points.

The expansions proposed in Eq. (8), we construct as follows

$$\begin{aligned} \rho_N(\vec{r}) &= \sum_{v=1}^{N} \rho_1^v(\vec{r}) \\ \rho_1^v(\vec{r}) &= \rho_v(\vec{r}) - \rho_{v-1}(\vec{r}) \end{aligned} \tag{10}$$

Further we may define the density functional $E_V[\rho_N;\{\rho_1^v\}_v]$ approximating from above the universal density functionals $E_V^N[\rho_N]$. In particular, the approach proposed above has the following advantages:

(i) It reduces to a system of N coupled functional Euler's equations which is, in fact, complicated but within the framework of the approximation chosen they are solved from the bottom to the top.

(ii) This construction may be implemented recurrently beginning from a one-electron problem.

(iii) This approach provides the first term of the kinetic energy part of the density functional which is the Weizsacker term.[49-51]

(iv) The proposed construction of the approximating quasiparticle energy density functional may be applied within the framework of some approximation for the exchange-correlation pseudopotential part to obtain the "universal density functional in a given approximation" as a minimum over all divisions I of ρ_N. In particular, then we may also solve the problem of whether a density functional is, in fact universal.

2. FIRST APPROACH: ALGEBRAIC-GEOMETRICAL ANSATZ

In this section we shall use the pioneering ideas proposed by Rowe, Rosensteel and collaborators[52-57] to investigate, from the algebraic-geometrical point of view, the properties of densities and to represent in this spirit the essence of the Hohenberg-Kohn theorem. This idea nontrivially intersects with the recent paper by Harriman on the geometry of density matrices.[58]

It is well known that the Hartree-Fock theory is fundamental among all many-electron theories for many reasons but, perhaps, the most significant is the fact that it describes the dynamical behaviour of the first order density. In this light we may say that the Hohenberg-Kohn theorem provides the generalization of the Hartree-Fock theory. As the latter the Hohenberg-Kohn theorem lies beyond Fermi and Bose statistics, and this is a stumbling stone, a reef, for many quantum chemists and solid state physicists, in particular, for the present author.

In general, most physical observables correspond to one-electron operators, i.e. they are linear functions of the density. And as a corrollary of this, these physical observables are reasonably well described within the framework of the Hartree-Fock approach, which is based on the following constraints on densities expressing the law of electron number conservation and the idempotency of a density:

$$\mathrm{Tr}\, D^1 = N, \quad (D^1)^2 = D^1 \tag{11}$$

where a density is defined from N-electron wavefunction $\Psi(x_1, x_2, \ldots, x_N)$ in the following manner $(x_i = (\vec{r}_i, s_i))$:

$$D^1(x;x') = \int dx_2 \ldots dx_N \Psi(x, x_2, \ldots, x_N) \bar{\Psi}(x', x_2, \ldots, x_N) \tag{12}$$

Then a density $\rho(x)$ is defined by the diagonal of $D^1(x;x')$:

$$\rho(x) = D^1(x;x) \qquad (13)$$

The idempotency of D^1 restricts the class of N-electron wavefunctions to Slater determinants. And this is, in many cases, an undesirable constraint.

Following Rosensteel and Rowe[52], let us consider the unitary group U(n) where n is the dimension of a chosen one-electron basis set L. Here, for simplicity, we limit ourselves to a finite dimensional basis set which, in fact, is what we use in practice. Let us consider the set of unitary transformations, acting on this basis set L. This action is transitive on a subset of L which consists of N-electron Slater determinants, i.e. if Φ is any given normalized determinant of the form

$$\Phi = \varphi_1 \wedge \varphi_2 \wedge \ldots \wedge \varphi_N \qquad (14)$$

then every other normalized determinant is of the following form

$$g\Phi = g\varphi_1 \wedge g\varphi_2 \wedge \ldots \; g\varphi_N \qquad (15)$$

for some $g \varepsilon U(n)$.

Therefore the Hartree-Fock N-electron states lie on a single orbit of U(n) in the Grassman manifold. That is, by the Konstant-Souriau classification theorem for symplectic manifolds (below we define the symplectic structure) with a transitive canonical Lie group action, the Slater determinants are in one-to-one mapping with som coadjoint orbit of U(n).[59,60]

Now let us consider an arbitrary 1-density (first order density) D^1 corresponding to a Slater determinant. Then the coadjoint orbit is given by the following expression

$$P_{D^1} = \left\{ gD^1g^{-1} \,|\, g\varepsilon U(n) \right\} \qquad (16)$$

Every 1-matrix lying on this orbit corresponds to some Slater determinant, i.e. is determinantal, and conversely, every 1-matrix associated with a Slater determinant is contained on the orbit P_{D^1}. Indeed, the coadjoint action, i.e. the so called moment map

$$D^1 \rightarrow gD^1g^{-1} \qquad (17)$$

refers to the transformation in the density induced by a change in L associated with $g\varepsilon U(n)$. In turn, every determinant may be obtained from a given determinant with the help of a unitary tranformation acting on the basis set L.

Now let us reformulate this geometrically. First we consider P_{D^1} the orbit of determinantal densities, as the surface contained within the space of all 1-densities, the Hermitian nxn matrices with traces equal to N. Then the generalized dynamic Hartree-Fock solution (Here GDHF, and TDHF[52]) is defined as a curve constrained to the orbit surface and tangent everywhere to the Hartree-Fock Hamiltonian (Fig. 2). Physically, the constraint to a U(n)-orbit surface means, by definition, the restriction to one-electron dynamical behaviour. This is a corrollary of the fact that the generators of the unitary group are one-electron operators. Moreover, the picture proposed is applied to every U(n)-orbit surface. So we are, together with Rosensteel, Rowe et.al., led naturally to consider the continuum of all possible orbits P_{D^1}, each of which determines one-electron constrained dynamical behaviour, Thus we may propose a framework for the desirable generalizations of conventional GDHF.

As proved in references 52, 56 and 57 an arbitrary orbit surface P_{D^1} is in a 1-1 mapping with the coset space of U(n) modulo to the isotropy subgroup for D^1

$$H_{D^1} = \left\{h\varepsilon U(n)\,|\,hD^1h^{-1} = D^1\right\} \tag{18}$$

Therefore, this mapping is given by

$$\begin{aligned} U(n)/H_{D^1} &\to P_{D^1} \\ gH_{D^1} &\to gD^1g^{-1} \end{aligned} \tag{19}$$

For the orbit of determinants, the isotropy subgroup is given by the unitary transformations which separately leave invariant the

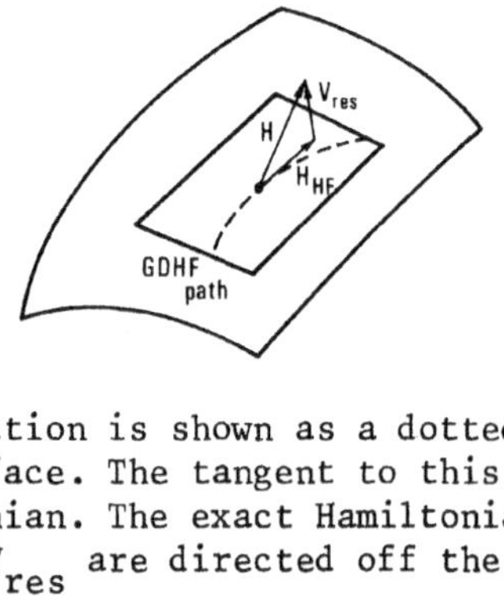

Fig. 2. The GDHF solution is shown as a dotted curve in the determinantal surface. The tangent to this path is the Hartree-Fock Hamiltonian. The exact Hamiltonian H and the residual interaction V_{res} are directed off the tangent plane.[54,55]

subspaces of occupied (particles) and nonoccupied (holes) orbital states. Thus, in this case $H_{D^1} = U(N)\times U(n-N)$. The dimension of the surface $P_{D^1} \equiv U(n)/H_{D^1}$ is given by $\dim U(n) - \dim H_{D^1} = n^2 - \dim H_{D^1}$. So, for the determinantal orbit $\dim P_{D^1} = 2N(n-N)$, which is twice the number of particle-hole pairs. Obviously, the algebraic structure defined earlier,[61-64] is directly connected with the term of the isotropy group.

Geometry on each orbit is determined by the Lie algebra u(n) of the unitary group, the skew Hermitian n×n matrices. A basis for this n^2 dimensional real Lie algebra is given by the matrices:[52]

$$
\begin{aligned}
iQ_{\alpha\beta} &\equiv 2^{-1/2} i(E_{\alpha\beta} + E_{\beta\alpha}), \quad \alpha \geq \beta \\
iP_{\alpha\beta} &\equiv 2^{-1/2}(E_{\alpha\beta} - E_{\beta\alpha}), \quad \alpha > \beta
\end{aligned}
\tag{20}
$$

where $E_{\alpha\beta}$ denotes the n×n matrix whose sole nonzero entry is at the interaction of row α with column β. An arbitrary element X of u(n) is a linear combination of $iQ_{\alpha\beta}$ and $iP_{\alpha\beta}$ with real coefficients. Each such element $X\in u(n)$ defines an element of the group U(n) by exponentiation $\exp X \in U(n)$. Each element $X\in u(n)$ defines a curve $\gamma_X(t)$ through the point D^1 and lying completely on the surface P_{D^1}, $\gamma_X(t) \equiv \exp(tX) D^1 \exp(-tX)$. The tangent vector to the curve γ_X at D^1 may be identified with X itself.

In order to complete the geometrical picture, it only remains to define the symplectic structure, in terms of which the Hamiltonian dynamics is defined. The symplectic form ω at the point gD^1g^{-1} in the orbit P_{D^1} is an antisymmetric form defined on pairs of tangent vectors at gD^1g^{-1},[52]

$$
\omega_{gD^1g^{-1}}(X,Y) \equiv -i\mathrm{Tr}(gD^1g^{-1}[X,Y])
\tag{21}
$$

where X and $Y\in u(n)$ are regarded as tangent vectors at gD^1g^{-1}. Because of the symplectic geometry on the coadjoint orbits of densities,[52] one-body dynamics on these orbits is defined, which is the generalization and abstract definition of the Hartree-Fock approximation.

Now in this spirit let us consider the formulation of the Hohenberg-Kohn theorem. The theorem states that the energy of the ground state and the ground-state density $\rho(\vec{r})$ are determined from the variational principle, Eq. (2). For an arbitrary element $g\in U(n)$ the moment map is given by

$$
\rho \to g\rho g^{-1}
\tag{22}
$$

in accordance with the moment map for D^1 associated with a given ρ. Obviously, the moment map induces the following transformation in our variational principle

$$\frac{\delta F[g\rho g^{-1}]}{\delta(g\rho g^{-1})(\vec{r})} + V_{ext}(\vec{r}) - \varepsilon = 0 \qquad (23)$$

And because $V_{ext}(\vec{r})$ is independent of $\rho(\vec{r})$, we may conclude that the ground-state energy $E_{g.s.}$ derived from this variational principle is a function of the coadjoint orbit of ρ (read D^1), and $E_{g.s.}$ is degenerate with a degeneracy which is equal to the dimension of this coadjoint orbit. But from general considerations[52] it is known that in all cases the orbit surface is even dimensional. Therefore, $E_{g.s.}$ is, at least, doubly degenerate. But this statement contradicts the general principles of quantum mechanics.

In the framework of the Levy-Valone approach to the density functional, we may consider the set $L_\rho(N)$ and the set $L_{g\rho g^{-1}}(N)$ for a given $g \in U(n)$. Then the Levy-Valone definition of the density functional can be rewritten in the following form

$$E_{LV}[\rho] = \min_{\text{orbit } \rho} \; \min_{g \in U(n)/H_\rho} \; \min_{h \in H_\rho} \left\{ \langle H, D^N \rangle \,|\, ghD^N(gh)^{-1} \in L_{g\rho g^{-1}}(N) \right\} \qquad (24)$$

Let us consider D_0^1 and ρ_0 of the ground state for a given many-electron system. We may define the corresponding orbit surface P_0. There arise the following unsolved problems which, evidently, are of great importance to an understanding of the basic concepts of the density functional theory:

I. What is the structure of P_0?
II. The orbit surface P_0 is an even dimensional and, at least, two different 1-matrices D_0^1 and $\tilde{D}_0^1$ are contained on this surface. What is a state corresponding $\tilde{D}_0^1$? And, in general, what mapping exists between the element of P_0 and the real N-electron states?
III. What is the method for obtaining the minimum minimorum on P_0 to determine the energy of the ground state?

The author feels that the solution of these problems would permit us to understand more profoundly the essence of the Hohenberg-Kohn theorem and the universality of the energy density functional.

3. SECOND APPROACH: APPROXIMATIONS TO THE ENERGY DENSITY FUNCTIONAL

In this section we shall discuss briefly approximations to the kinetic exchange and correlation term in the energy density functional, speculations which were and are developed actively long before the birth of the Hohenberg-Kohn theorem.

Firstly we consider the kinetic energy term. Following Alonso and Girifalco[65] we have for the kinetic energy

$$E_K = -\frac{1}{2}\int d^3\vec{r}_1 \cdot \nabla_1^2 D^1(\vec{r}_1;\vec{r}_1')\Big|_{\vec{r}_1=\vec{r}_1'} \tag{25}$$

Let us define $D^1(\vec{r}_1;\vec{r}_1')$ in the following manner

$$D^1(\vec{r}_1;\vec{r}_1') = A(\vec{r}_1;\vec{r}_1')[\rho(\vec{r}_1)\rho(\vec{r}_1')]^{1/2} \tag{26}$$

Now if we introduce a kinetic energy density $e_K(\vec{r}_1)$ such that

$$E_K \equiv \int d^3\vec{r}_1 e_K(\vec{r}_1) \tag{27}$$

then we obtain[65]

$$\nabla_1^2 D^1(\vec{r}_1;\vec{r}_2) = [\rho(\vec{r}_1)\rho(\vec{r}_2)]^{1/2}\nabla_1^2 A(\vec{r}_1;\vec{r}_2) -$$

$$\frac{1}{4} A(\vec{r}_1;\vec{r}_2)[\rho(\vec{r}_1)]^{-3/2}[\rho(\vec{r}_2)]^{1/2}|\nabla\rho(\vec{r}_1)|^2 + \frac{1}{2} A(\vec{r}_1;\vec{r}_2) \times$$

$$[\rho(\vec{r}_1)]^{-1/2}[\rho(\vec{r}_2)]^{-1/2}\nabla^2\rho(\vec{r}_1) +$$

$$[\rho(\vec{r}_1)]^{-1/2}[\rho(\vec{r}_2)]^{1/2}\nabla_1 A(\vec{r}_1;\vec{r}_2)\cdot\nabla\rho(\vec{r}_1) \tag{28}$$

In particular

$$e_K(\vec{r}_1) = -\frac{1}{2}\rho(\vec{r}_1)\nabla_1^2 A(\vec{r}_1;\vec{r}_2)\Big|_{\vec{r}_1=\vec{r}_2} + \tag{29}$$

$$\frac{1}{8} A(\vec{r}_1;\vec{r}_1)\frac{|\nabla\rho(\vec{r}_1)|^2}{\rho(\vec{r}_1)} - \frac{1}{4} A(\vec{r}_1;\vec{r}_2)\ {}^2\rho(\vec{r}_1) -$$

$$\frac{1}{2}\nabla_1 A(\vec{r}_1;\vec{r}_2)\Big|_{\vec{r}_1=\vec{r}_2}\cdot\nabla\rho(\vec{r}_1)$$

The main problem which arises here is the determination of the physically relevant approximation of $A(\vec{r}_1;\vec{r}_2)$. Notice, that above the formula is written correctly without any approximations.

It is necessary to point out that this field of investigating the study of properties of the so called correlation factor $A(\vec{r}_1;\vec{r}_2)$, is almost undeveloped. We feel sure that this approach will help to unify the approximate treatments of the kinetic energy term examined in the literature. In particular, in this light we may investigate all known gradient expansions[49,66-78] and discuss their physical relevance and their connection with the Hartree-Fock approach, namely in what way they go beyond the Hartree-Fock theory. From Eq. (29) it follows that we may write the exact expression for the kinetic energy density $e_K(\vec{r})$ if we know the following functions as functionals of $\rho(\vec{r})$:

$$A(\vec{r}_1;\vec{r}_1) \tag{30a}$$

$$\nabla_1 A(\vec{r}_1;\vec{r}_2)|_{\vec{r}_1=\vec{r}_2} \tag{30b}$$

$$\nabla_1^2 A(\vec{r}_1;\vec{r}_2)|_{\vec{r}_1=\vec{r}_2} \tag{30c}$$

It is evident that a basis of these investigations is provided by the extension of the Kato cusp condition,[79,80-83] and by perturbation theory.[84-86]

Let us consider the Hartree-Fock approach. From Eq.(26) and the idempotency of $D^1(\vec{r}_1;\vec{r}_1')$ it follows that the following constraint must be imposed on $A(\vec{r}_1;\vec{r}_1')$:

$$A(\vec{r}_1;\vec{r}_1) = 1 \tag{31a}$$

and

$$A(\vec{r}_1;\vec{r}_2) = \int d^3\vec{r}_3 A(\vec{r}_1;\vec{r}_3)\rho(\vec{r}_3)A(\vec{r}_3,\vec{r}_2) \tag{31b}$$

Eq. (31a) requires that the diagonal elements of the A-matrix are equal to one, and Eq. (31b) gives an integral equation for the A-kernel. Obviously, the trivial solution of this Eq. (31b) is a δ-function, namely:

$$A(\vec{r}_1;\vec{r}_2) = \frac{\delta(\vec{r}_1-\vec{r}_2)}{[\rho(\vec{r}_1)\rho(\vec{r}_2)]^{1/2}} \tag{32}$$

and this is the trivial approximation for Coulomb hole. The author feels that although this solution is the most trivial, it emphasizes at the same time the real singular behaviour of $A(\vec{r}_1;\vec{r}_2)$. In the given approximation the kinetic energy density has the following form

$$e_K(r_1) = \frac{1}{8}\frac{|\nabla\rho(\vec{r}_1)|^2}{\rho(\vec{r})} - \frac{1}{4}\nabla^2\rho(\vec{r}_1) -$$

$$\frac{1}{2}\nabla_1 A(\vec{r}_1;\vec{r}_2)|_{\vec{r}_1=\vec{r}_2}\cdot\nabla\rho(\vec{r}_1) - \frac{1}{2}\nabla_1^2 A(\vec{r}_1;\vec{r}_2)|_{\vec{r}_1=\vec{r}_2}\cdot\rho(\vec{r}_1) \tag{33}$$

This allows us to give at once the first two terms in kinetic energy expansion:

$$\text{first} - \quad \frac{1}{8}\frac{|\nabla\rho|^2}{\rho}$$
$$\text{second} - \quad -\frac{1}{4}\nabla^2\rho \qquad (34)$$

The first term lies in the best agreement with the recent investigations of the kinetic energy gradient expansion.[49,75-77] But we give this term in the Hartree-Fock approach, and, in general, it is incorrect to state that the coefficient in front of the term $\frac{|\nabla\rho|^2}{\rho}$ is equal to $\frac{1}{8}$. Strictly speaking, this coefficient is equal to $A(\vec{r}_1;\vec{r}_1)$, and it is interesting to investigate a range of values of the A-diagonal.

The first two terms given in Eq. (34) are formally the exact expression for the kinetic energy in the Hartree-Fock approximation. However it must be stressed that these terms are not a gradient expansion.[65] Notice that the approach presented above, where we follow Alonso and Girifalco, is similar to that of March and Young.[88]

Let us now consider briefly the exchange part of the density functional. All that is needed here, is the introduction of the correlation factors $c_\alpha(\vec{r}_1;\vec{r}_2)$ and $c_\beta(\vec{r}_1;\vec{r}_2)$ which connect the spin components of 2-matrix with that of 1-matrix:[88]

$$c_\gamma(\vec{r}_1;\vec{r}_2) = \frac{D^2_{\gamma\gamma}(\vec{r}_1,\vec{r}_2)}{D^1_\gamma(\vec{r}_1)D^1_\gamma(\vec{r}_2)} - 1 \qquad (35)$$

$$\gamma = \alpha,\beta$$

It is evident that the functions $A(\vec{r}_1;\vec{r}_2)$ and $c_\alpha(\vec{r}_1;\vec{r}_2)$ and $c_\beta(\vec{r}_1;\vec{r}_2)$ are connected by the relation

$$A(\vec{r}_1;\vec{r}_2) = [-\,2c(\vec{r}_1;\vec{r}_2)]^{1/2} \qquad (36)$$

In particular, in the approximation of a homogeneous electron gas given by Seitz[89] and McWeeny,[90] we have

$$A(\vec{r}_1;\vec{r}_2) = \frac{3}{Y}\left[\frac{\pi}{2Y}\right]^{1/2} J_{3/2}(Y) \qquad (37)$$

where

$$Y = r_{12}[3\pi^2\rho]^{1/3} \qquad (38)$$

Then, in the first approximation, we have[91]

$$e_K(\vec{r}_1) = \frac{1}{8}\frac{|\nabla\rho(\vec{r}_1)|^2}{\rho(\vec{r}_1)} - \frac{1}{4}\nabla^2\rho(\vec{r}_1) + \frac{1}{4}(3\pi^2)^{2/3}[\rho(\vec{r}_1)]^{5/3} \quad (39)$$

Thus, we have the well known term which is proportional to $\rho^{5/3}$! The reason for this lies in the electron gas approximation which was chosen. The author feels that a considerable improvement of the kinetic density functional results when we choose the correlation factor in the following form[65]

$$c_{\tilde{\rho}}(\vec{r}_1;\vec{r}_2) = -\frac{9}{2}\left[\frac{\sin\tilde{Y} - \tilde{Y}\cos\tilde{Y}}{\tilde{Y}^3}\right]^2$$

$$\tilde{Y} = r_{12}[3\pi^2\tilde{\rho}]^{1/3} \quad (40)$$

and $\tilde{\rho}$ is determined by the integral constraint:

$$\int d^3\vec{r}_2\ \rho(\vec{r}_2)c_{\tilde{\rho}}(\vec{r}_1;\vec{r}_2) = -1 \quad (41)$$

for the advanced description of the Coulomb hole.

The exchange-correlation term is of great importance in quantum chemistry and solid state physics and this is the reason why a large volume of literature is devoted to various speculations in this field. However, the exchange-correlation part of the energy density functional cannot be treated exactly within the framework of presently known approaches, including the Hartree-Fock method, the Slater local-density approximation-Xα-method, the Kohn-Sham local-exchange-correlation method, the local-spin-density approximation, or model exchange-correlation potential approaches.[39-43,92-102] The methods mentioned above provide practicable computational schemes and are very successful, notwithstanding restrictions of applicability for, in particular, the local-density approach, and the gradient corrections demonstrated in the literature.[92,93]

4. THIRD APPROACH: FOUR-DIMENSIONAL DENSITY

In this section we shall consider a one-particle density which is dependent on not only the space coordinates, but also on some scaling parameter Z corresponding to the nuclear charges

$$Z_\alpha e = Zz_\alpha e$$

By means of the Hellmann-Feynman theorem[103-108] we obtain an expression for the ground-state energy of a many-electron system as a functional of $\rho(x,y,z,Z)$:[106,109]

$$E = \sum_{\alpha<\beta} \frac{Z_\alpha Z_\beta e^2}{r_{\alpha\beta}} - \sum_\alpha Z_\alpha e^2 \int_{Z=0}^{Z=1} dZ \int d^3\vec{r}\, \frac{\rho(\vec{r},Z)}{r_\alpha} \tag{42}$$

Let us consider a diatomic molecule with nuclear charges Z_A and Z_B. We define the new variables[110]

$$\zeta = (Z_A Z_B)^{1/2}, \quad y = \left(\frac{Z_A}{Z_B}\right)^{1/2}$$

In these variables we let

$$Z_A = \zeta y, \text{ and } Z_B = \zeta y^{-1}$$

Let us suppose that a given diatomic molecular system has a Hamiltonian $H = H(\Lambda)$, where Λ is a parameter (or parameters). Then from the integrated Hellmann-Feynman formula[108] the energy difference between two molecular states $\Psi(\lambda_1)$ and $\Psi(\lambda_2)$ is determined by the following formula

$$\Delta E \equiv E(\lambda_2) - E(\lambda_1) = \int_{\lambda_1}^{\lambda_2} d\Lambda \langle\Psi(\Lambda)|\frac{\partial H(\Lambda)}{\partial\Lambda}|\Psi(\Lambda)\rangle \tag{43}$$

where λ_1, λ_2 are two values of the parameter Λ.

ζ is a parameter which enters the Hamiltonian H only in its electron-nuclear part

$$V = \frac{\zeta^2}{|\vec{R}_A - \vec{R}_B|} - \zeta \sum_{i=1}^{N} \left(\frac{y}{r_{iA}} + \frac{1}{y r_{iB}}\right) \tag{44}$$

where

$$r_{iA} = |\vec{r}_i - \vec{R}_A|$$

$$r_{iB} = |\vec{r}_i - \vec{R}_B|$$

$\vec{r}_i$ is the position vector of the ith electron, and N is a number of electrons. Then

$$\frac{\partial V}{\partial\zeta} = \frac{2\zeta}{|\vec{R}_A - \vec{R}_B|} - \sum_{i=1}^{N} \left(\frac{y}{r_{iA}} + \frac{1}{y r_{iB}}\right) \tag{45}$$

Following Wilson[106] we have

$$\frac{\partial W}{\partial\zeta} = \frac{2\zeta}{|\vec{R}_A - \vec{R}_B|} - \int d^3\vec{r} \left[\frac{y\rho(\vec{r},\zeta)}{|\vec{r} - \vec{R}_A|} + \frac{\rho(\vec{r},\zeta)}{y|\vec{r} - \vec{R}_B|}\right] \tag{46}$$

Using the Hellmann-Feynman theorem we obtain

$$E = \frac{\zeta^2}{|\vec{R}_A-\vec{R}_B|} - \int d^3\vec{r}\int d\zeta\left[\frac{y}{|\vec{r}-\vec{R}_A|} + \frac{1}{y|\vec{r}-\vec{R}_B|}\right]\rho(\vec{r},\zeta) \tag{47}$$

Now within the framework of the atoms-in-molecules-concept we introduce the atomic-like potentials

$$V_A(\zeta) = \frac{\zeta}{|\vec{R}_A-\vec{R}_B|} + \tilde{V}_A(\zeta) \tag{48a}$$

$$V_B(\zeta) = \frac{\zeta}{|\vec{R}_A-\vec{R}_B|} + \tilde{V}_B(\zeta) \tag{48b}$$

where

$$\tilde{V}_A(\zeta) = -y\int d^3\vec{r}\,\frac{\rho(\vec{r},\zeta)}{|\vec{r}-\vec{R}_A|} \tag{48c}$$

$$\tilde{V}_B(\zeta) = -\frac{1}{y}\int d^3\vec{r}\,\frac{\rho(\vec{r},\zeta)}{|\vec{r}-\vec{R}_B|} \tag{48d}$$

Thus finally we have

$$E = \int_0^{\zeta} d\zeta[V_A(\zeta) + V_B(\zeta)] = \frac{\zeta^2}{|\vec{R}_A-\vec{R}_B|} + \int_0^{\zeta} d\zeta[\tilde{V}_A(\zeta) + \tilde{V}_B(\zeta)] \tag{49}$$

so that the energy of a diatomic molecule is determined by a sum of contributions of atomic-like parts.

We now restrict ourselves to homonuclear diatomic molecules ($Z_A = Z_B = Z = \zeta$, $y = 1$). Then

$$E = \int_0^{Z} dZ'[V_A(Z') + V_B(Z')] =$$

$$\frac{1}{2} Z[V_A(Z) + V_B(Z)] -$$

$$\frac{1}{2}\int_0^{Z} dZ'\left\{\left[Z'\frac{\partial V_A}{\partial Z'} - V_A\right] + \left[Z'\frac{\partial V_B}{\partial Z'} - V_B\right]\right\} \tag{50}$$

i.e. we obtain the well known Politzer-Parr formula.[111]

In the case of a heteronuclear diatomic ($Z_A > Z_B$) we obtain

$$\int_0^{\zeta} d\zeta V_A(\zeta) = \int_0^{Z_A} d\zeta V_A(\zeta) - \int_{Z_A}^{\zeta} d\zeta V_A(\zeta)$$

$$\int_0^{\zeta} d\zeta V_B(\zeta) = \int_0^{Z_B} d\zeta V_B(\zeta) + \int_{Z_B}^{\zeta} d\zeta V_B(\zeta)$$

Then

$$E = \left[\int_0^{Z_A} d\zeta V_A(\zeta) + \int_0^{Z_B} d\zeta V_B(\zeta)\right] + \left[\int_{Z_B}^{\zeta} d\zeta V_B(\zeta) - \int_{Z_A}^{\zeta} d\zeta V_A(\zeta)\right] \quad (51)$$

or

$$E = \frac{\zeta^2}{|\vec{R}_A-\vec{R}_B|} + \left[\int_0^{Z_A} d\zeta \widetilde{V}_A(\zeta) + \int_0^{Z_B} d\zeta \widetilde{V}_B(\zeta)\right] +$$

$$\left[\int_{Z_B}^{\zeta} d\zeta \widetilde{V}_B(\zeta) - \int_{Z_A}^{\zeta} d\zeta \widetilde{V}_A(\zeta)\right] \quad (51a)$$

or

$$E = \frac{Z_A Z_B}{|\vec{R}_A-\vec{R}_B|} + K_A \widetilde{E}_A(Z_A) + K_B \widetilde{E}_B(Z_B) \quad (51b)$$

where

$$\widetilde{E}_A(Z_A) = \int_0^{Z_A} d\zeta \widetilde{V}_A(\zeta), \; \widetilde{E}_B(Z_B) = \int_0^{Z_B} d\zeta \widetilde{V}_B(\zeta)$$

$$K_A = \int_0^{\zeta} d\zeta \widetilde{V}_A(\zeta) \Big/ \int_0^{Z_A} d\zeta \widetilde{V}_A(\zeta) \quad (51c)$$

$$K_B = \int_0^{\zeta} d\zeta \widetilde{V}_B(\zeta) \Big/ \int_0^{Z_B} d\zeta \widetilde{V}_B(\zeta) \quad (51d)$$

As we assumed, $Z_A > Z_B$. It is evident that $Z_A > \zeta > Z_B$, and

$$K_A = 1 - \int_{\zeta}^{Z_A} d\zeta \widetilde{V}_A(\zeta) \Big/ \int_0^{Z_A} d\zeta \widetilde{V}_A(\zeta)$$

$$K_B = 1 + \int_{Z_B}^{\zeta} d\zeta \widetilde{V}_B(\zeta) \Big/ \int_0^{Z_B} d\zeta \widetilde{V}_B(\zeta)$$

Thus finally we obtain

$$E = \frac{Z_A Z_B}{|\vec{R}_A-\vec{R}_B|} + \widetilde{E}_A(Z_A) + \widetilde{E}_B(Z_B) + \Delta\widetilde{E}_{AB}(Z_A,Z_B) \quad (52)$$

where

$$\Delta\widetilde{E}_{AB}(Z_A,Z_B) = \int_{Z_B}^{\zeta} d\zeta \widetilde{V}_B(\zeta) + \int_{\zeta}^{Z_A} d\zeta \widetilde{V}_A(\zeta) \quad (52a)$$

or

$$\Delta\tilde{E}_{AB}(Z_A,Z_b) = \frac{1}{2}\left\{\int_{Z_B}^{Z_A} d\zeta V_{AB}^{+}(\zeta) + \left[\int_{\zeta}^{Z_A} d\zeta - \int_{Z_B}^{\zeta} d\zeta\right] V_{AB}^{-}(\zeta)\right\} \quad (52b)$$

where

$$\tilde{V}_{AB}^{\pm}(\zeta) = \tilde{V}_A(\zeta) \pm \tilde{V}_B(\zeta)$$

The expression obtained above can without any difficulties be generalized to polyatomic molecules. The result is that

$$E = \sum_A K_A Z_A V_{0,A}^{mol} \quad (53)$$

Eq. (53) is completely identical with the Politzer formula,[112] but our approach permits us to determine the constants K_A which, obviously, differ from their values for a free atom, and this is confirmed by a number of calculations.[113-116] Thus we see that K_A depends on the surroundings of atom A.

Notice finally that the approach proposed allows us to treat some isoelectronic processes such as $N_2 \to CO$.[108] Finally we would like to point out that Eq. (47) represents E as a density functional. Therefore, it offers another way for examining the ideas of Hohenberg and Kohn.

5. EPILOGUE

Get at the root.
Koz'ma Prutkov.

The author specially chooses the Kozma Prutkov's apophthegm as the beginnings of the epilogue for the review presented. Naturally the scope of this paper does not allow the author to address the larger question of what the density functional theory represents today. But the author does not purpose this object!

Much of literature on the density functional theory dissembles its essence, its root. Now the moment has come when the density-functional-theory workers must concentrate their efforts on the fundamental principles of the theory if they are to achieve successes in the near future.

Once Thor Heyerdahl said that any philosopher digs a hole in the neighbourhood of himself and does not always scramble out of his hole to see what has happened in the rest of the holes. This is the reason, perhaps, why density-functional-theory workers must look at the theory developed from other sides. And there are many sides! For example, there is the momentum space approach to the density functional theory, here the papers by V. Smith,[117] Lassettre[118-123] and others are very useful. In fact, in the momentum space approach the kinetic energy functional is trivial, and, at least, we do not experience obstacles here and we need only to write correctly, within the framework of any approximation, the Coulomb-exchange-correlation part of the density functional. In general, we need to scramble out of our holes in density functional theory before they become transformed to "black holes".

ACKNOWLEDGMENTS

The author is indebted to Professors P.-O. Löwdin, O. Goscinki, J.-L. Calais, J. Avery and J.P. Dahl for their kind invitations to take part in the Fourth International Congress in Quantum Chemistry as Symposium Leader and to attend the Symposium "Local Density Approximations in Quantum Chemistry and Solid State Physics". The author is very much obliged to Professors Jean-Louis Calais, Robert G. Parr, and Dr. Mel Levy for useful discussions of his previous paper on the Hohenberg-Kohn theorem. Finally, the author's thanks to Professors E.R. Davidson, J.E. Harriman, R.G. Parr, R.A. Donnely, R.F.W. Bader, O. Gunnarsson, G. Rosensteel, W. Bingel, M. Hoffmann-Ostenhof, S.T. Epstein, W. Münch, M. Levy, S. Valone, S. Henderson, Z.J. Horak for providing their works, and he thanks Dr. Serge Beznosjuk for useful collaboration.

REFERENCES

1. L.H. Thomas, Proc. Cambridge Phil. Soc. 23, 542 (1927).
2. E. Fermi, Rend. Acad. Naz. Lincei 6, 602 (1927).
3. P.A.M. Dirac, Proc. Cambridge Phil. Soc. 26, 376 (1930).
4. J.C. Slater, Phys. Rev. 35, 210 (1930).
5. J.C. Slater, Phys. Rev. 81, 385 (1951).
6. J.C. Slater, "The Self-Consistent Field for Molecules and Solids", McGraw-Hill Book Co., New York (1974).
7. P. Gombas, "Die Statistische Theorie des Atoms und ihre Anwendungen", Springer-Verlag, Vienna (1949).
8. N.H. March, "Self-Consistent Fields in Atoms", Pergamon Press, New York (1975).
9. N.H. March, Adv. Phys. 6, 1 (1957).
10. N.H. March, W.H. Young, S. Sampanthar, "The Many-Body Problem in Quantum Mechanics", Cambridge University Press (1967).
11. R.G. Parr, p. 5 in: "Horizons of Quantum Chemistry", K. Fukui, B. Pullman, eds., D. Reidel Publ. Co., Amsterdam (1981).
12. A.S. Bamzai, B.M. Deb, Rev. Mod. Phys. 53, 95 (1981).
13. A.K. Rajagopal, Adv. Chem. Phys. 41, 59 (1980).
14. D.A. Kirzhnits, Yu.E. Lozovik, G.V. Shpatakovskaya, Uspekhi Fiz. Nauk 117, 3 (1975).
15. P. Hohenberg, W. Kohn, Phys. Rev. B136, 864 (1964).

16. S.T. Epstein, C.M. Rosenthal, Theoretical Chemistry Institute Reports WIS-TSI-483, University of Wisconsin (1975).
S.T. Epstein, C.M. Rosenthal, J.Chem. Phys. 64, 247 (1976).
17. J. Katriel, C.J. Appellof, E.R. Davidson, Int. J. Quantum Chem. 19, 293 (1981).
18. J. Tal, R.F.W. Bader, J. Erkku, Phys. Rev. A21, 1 (1980).
19. M. Berrondo, O. Goscinski, Int. J. Quantum Chem., Symp. 9, 67 (1975).
20. E.S. Kryachko, Int. J. Quantum Chem. 18, 1029 (1980).
21. A.J. Coleman, Rev. Mod. Phys. 35, 668 (1963).
22. E.R. Davidson, "Reduced Density Matrices in Quantum Chemistry", Academic Press, New York (1976).
23. W. Münch, J. Riess, H. Premas, Preprint, Laboratory of Physical Chemistry, Swiss Federal Institute of Technology, Zurich (1967).
24. J. Riess, W. Münch, Theor. Chim. Acta 58, 295 (1981).
25. V. Mikolas, A. Tomasek, Phys. Lett. 64A, 109 (1977).
26. M. Levy, Proc. Natl. Acad. Sci. USA 76, 6062 (1979).
27. S.M. Valone, J. Chem. Phys. 73, 4653 (1980).
28. T.L. Gilbert, Phys. Rev. B12, 2111 (1975).
29. J.E. Harriman, Theoretical Chemistry Institute Report WIC-TCI-642, University of Wisconsin (1980).
J.E. Harriman, Phys. Rev. A24, 680 (1981).
30. W. Kohn, L.J. Sham, Phys. Rev. A140, 1133 (1965).
31. M. Berrondo, O. Goscinski, Phys. Rev. 184, 10 (1969).
32. A.W. Overhauser, Phys. Rev. B2, 874 (1970).
33. A.W. Overhauser, Phys. Rev. B3, 1188 (1971).
34. R.G. Parr, R.A. Donnelly, M. Levy, W.E. Palke, J. Chem. Phys. 68, 3801(1978).
35. R.A. Donnelly, R.G. Parr, J. Chem. Phys. 69, 4431 (1978).
36. R.A. Donnelly, J. Chem. Phys. 71, 2874 (1979).
37. S. Srebrenik, J. Chem. Phys. 63, 4841 (1975).
38. J. Tal, R.F.W. Bader, Int. J. Quantum Chem. Symp. 12, 153 (1978).
39. S. Larsson, J.-L. Calais, Chem. Phys. Lett. 14, 12 (1972).
40. O. Gunnarsson, M. Jonson, B.I. Lundqvist, Phys. Rev. B20, 3136 (1979).
41. O. Gunnarsson, R.O. Jones, J. Chem. Phys. 72, 5357 (1980).
42. O. Gunnarsson, R.O. Jones, Physica Scripta 21, 394 (1980).
43. P.W. Payne, J. Chem. Phys. 71, 490 (1979).
44. G.A. Henderson, Phys. Rev. A23, 19 (1981).
45. M. Levy, Phys. Rev. (in press).
46. S.A. Beznosjuk, Ph.D. Thesis, Tomsk State University (1978).
47. S.A. Beznosjuk, E.S. Kryachko (to be published).
48. P.-O. Löwdin, Phys. Rev. 97, 1509 (1955).
49. P.K. Acharya, L.J. Bartolotti, S.B. Sears, R.G. Parr, Proc. Natl. Acad. Sci. USA 77, 6978 (1980).
50. S.B. Sears, R.G. Parr, U. Dinur, Isr. J. Chem. 19, 165 (1980).
51. J.L. Gazquez, E.V. Ludena, Chem. Phys. Lett. 83, 145 (1981).

52. D.J. Rowe, A. Ryman, G. Rosensteel, Phys. Rev. A22, 2362 (1980).
53. G. Rosensteel, E. Ihrig, J. Math. Phys. 21, 2297 (1980).
54. G. Rosensteel, D.J. Rowe, Phys. Rev. A24, 673 (1981).
55. G. Rosensteel, Phys. Rev. A23, 2794 (1981).
56. G. Rosensteel, unpublished.
57. G. Rosensteel, unpublished.
58. J.E. Harriman, Phys. Rev. A17, 1249, 1257 (1978).
59. D.J. Simms, N.M.J. Woddhouse, "Lectures on Geometric Quantization, Lecture Notes in Physics", vol. 53, Springer, Berlin (1976).
60. B. Konstant, "Quantization and Unitary Representations, Lecture Notes in Mathematics", vol. 170, Springer, Berlin (1970).
61. E.S. Kryachko, Int. J. Quantum Chem. 13, 719 (1978).
62. G.G. Dyadyusha, E.S. Kryachko, Int. J. Quantum Chem. 19, 251 (1981).
63. G.G. Dyadyusha, E.S. Kryachko, Int. J. Quantum Chem. 19, 505 (1981).
64. G.G. Dyadyusha, E.S. Kryachko, Fizika Molecules 7, 3 (1979).
65. J.A. Alonso, L.A. Girifalco, Phys. Rev. B17, 3735 (1978).
66. J.C. Stoddart, A.M. Beattice, N.H. March, Int. J. Quantum Chem. 4S, 35 (1971).
67. C.F. von Weizsacker, Z. Physik, 96, 431 (1935).
68. A.K. Kompaneets, E.S. Pavlovskii, Soviet Phys. JETP 4, 328 (1957).
69. D.A. Kirzhnitz, Soviet Phys. JETP 5, 64 (1957).
70. W.-P. Wang, R.G. Parr, D.R. Murphy, G.A. Henderson, Chem. Phys. Lett. 43, 409 (1976).
71. J. Tal, R.F.W. Bader Int. J. Quantum Chem. 12S, 153 (1978).
72. H.B. Huntington, Phys. Rev. B20, 3165 (1979).
73. G.L. Oliver, J.P. Perdew, Phys. Rev. A20, 397 (1979).
74. D.R. Murphy, R.G. Parr, Chem. Phys. Lett. 60, 377 (1979).
75. W.-P. Wang, J. Chem. Phys. 73, 416 (1980).
76. D.R. Murphy, W.-P. Wang, J. Chem. Phys. 72, 429 (1980).
77. S.B. Sears, R.G. Parr, U. Dinur, Isr. J. Chem. 19, 165 (1980).
78. I.M. Reznik, V.M. Shatalov, Ukranian Phys. J. 26, 1187 (1981).
79. T. Kato, Commun. Pure Appl. Math 10, 151 (1957).
80. E. Steiner, J. Chem. Phys. 39, 2365 (1963).
81. W. Bingel, Z. Naturforsch. 18a, 1249 (1963).
82. M. Hoffmann-Ostenhof, T. Hoffmann-Ostenhof, W. Thirring, J. Phys. B11, L571 (1978).
83. M. Hoffmann-Ostenhof, R. Seiler, Phys. Rev. A23, 21 (1981).
84. V. Fock, Det K. Nor. Vidensk. Selsk. Forh. 31, 145 (1958).
85. R.J. White, F.H. Stillinger, J. Chem. Phys. 52, 5800 (1970).
86. Z.J. Horak, J. Phys. B14, 937 (1981).
87. N. March, W. Young, Proc. Phys. Soc. London 72, 182 (1958).
88. E.S. Kryachko, Adv. Quantum Chem. 14, 1 (1981).
89. F. Seitz, "Modern Theory of Solids", McGraw-Hill, New York (1940).
90. R. McWeeny, p. 3 in: "The New World of Quantum Chemistry", B. Pullman and R.G. Parr, eds., D. Reidel, Boston (1976).

91. E.S. Kryachko, Int. J. Quantum Chem. 22, 661 (1982).
92. H. Shore, J. Rose, E. Zaremba, Phys. Rev. B15, 2858 (1977).
93. N. Elyashar, D. Koelling, Phys. Rev. B15, 3620 (1977).
94. F. Herman, J.P. Van Dyke, I.B. Ortenburger, Phys. Rev. Lett. 22, 807 (1969).
95. L. Kllinman, Phys. Rev. B10, 2221 (1974).
96. M. Basolt, D.G. Geldart, Phys. Rev. Lett. 35, 1234 (1975).
97. J. Migdalek, W.E. Baylis, Phys. Rev. A22, 22 (1980).
98. R. Latter, Phys. Rev. 99, 510 (1955).
99. M. Berrondo, O. Goscinski, Chem. Phys. Lett. 81, 75 (1981).
100. A.W. Overhauser, Phys. Rev. 132, 874 (1970).
101. A.W. Overhauser, Phys. Rev. 133, 1188 (1971).
102. J.L. Gazquez, E. Ortiz, Chem. Phys. Lett. 77, 186 (1981).
103. P.-O. Löwdin, Adv. Chem. Phys. 2, 207 (1959).
104. A.C. Hurley, Proc. Roy. Soc. A226, 170, 179, 193 (1954).
105. A.C. Hurley, Proc. Roy. Soc. A235, 224 (1956).
106. E.B. Wilson, Jr., J. Chem. Phys. 36, 2232 (1962).
107. P.-O. Löwdin, J. Mol. Spectr. 3, 46 (1959).
108. S.T. Epstein, A.C. Hurley, R.E. Wyatt, R.G. Parr, J. Chem. Phys. 47, 1275 (1967).
109. A.A. Frost, J. Chem. Phys. 37, 1147 (1962).
110. E.S. Kryachko, Intern. J. Quantum Chem. 21, 1123 (1982).
111. P. Politzer, R.G. Parr, J. Chem. Phys. 61, 4258 (1974).
112. P. Politzer, J. Chem. Phys. 70, 1067 (1979).
113. H. Teruya, T. Anno, J. Chem. Phys. 72, 6044 (1980).
114. T. Anno, J. Chem. Phys. 72, 782 (1980).
115. J.C. Slater, Phys. Rev. 36, 57 (1930); J.C. Slater, "Quantum Theory of Atomic Structure", McGraw-Hill, New York (1960). Vol. 1, pp. 368-372.
116. S. Fraga, TCA 2, 406 (1964); E.B. Priestley, S. Fraga, Canad. J. Phys. 44, 1933 (1966).
117. P. Kaijser, V.H. Smith, Jr., Adv. Quantum Chem. 10, 37 (1977).
118. E.N. Lassettre, J. Chem. Phys. 43, 4479 (1965).
119. E.N. Lassettre, J. Chem. Phys. 57, 4357 (1972).
120. E.N. Lassettre, J. Chem. Phys. 58, 1991 (1973).
121. E.N. Lassettre, J. Chem. Phys. 64, 4375 (1976).
122. E.N. Lassettre, J. Chem. Phys. 70, 3468 (1979).
123. M. Huo, E.N. Lassettre, J. Chem. Phys. 72, 2374 (1980).

EXCHANGE-CORRELATION ENERGY FUNCTIONALS

IN THE DENSITY FUNCTIONAL FORMALISM

O. Gunnarsson

Max-Planck-Institut für Festkörperforschung
D-7000 Stuttgart, Federal Republic of Germany

R. O. Jones

Institut für Festkörperforschung
der Kernforschungsanlage Jülich
D-5170 Jülich, Federal Republic of Germany

The accuracy of the local spin density (LSD) approximation is discussed, with particular reference to ionization energies, transfer energies and the multiplet structure of atoms. Although trends are generally reproduced very well, some less satisfactory results are pointed out. An exact relationship between the exchange-correlation energy functional and the pair correlation function is used to make the relative success of the LSD approximation more plausible. This approach suggests ways of constructing improved functionals and results obtained with functionals with a non-local density dependence are reviewed.

1. INTRODUCTION

The work surveyed in the present volume shows clearly that the description of exchange and correlation effects using a local energy functional leads to very satisfactory results in a variety of contexts. In some cases, the results are so impressive that there is a tendency to overlook the fact that the local spin density (LSD) approximation is, in fact, an approximation. However, other calculations have indicated deficiencies in the approximation and the rapid improvement in calculational techniques in recent years has made it a major remaining source of error and uncertainty.

In the present chapter, we examine some of the errors which result from the use of the LSD approximation, with particular ref-

Table 1. Total energies of first-row atoms (Ry). Experimental energies E^{exp} and the estimated relativistic contributions ΔE^{rel} are from Ref. 1, Hartree-Fock energies with relativistic corrections from Ref. 2. Non-relativistic LSD calculations with HLG functional (Ref. 3).

	E^{exp}	ΔE^{rel}	E_{HF}^{corr}	E_{LSD}^{nr}	$E^{exp} - E_{HF}^{corr}$
Li	-14.957	-0.001	-14.866	-14.706	0.091
Be	-29.339	-0.004	-29.150	-28.909	0.188
B	-49.318	-0.011	-49.070	-48.653	0.248
C	-75.715	-0.025	-75.404	-74.958	0.310
N	-109.228	-0.050	-108.856	-108.298	0.373
O	-150.225	-0.090	-149.716	-149.064	0.508
F	-199.618	-0.150	-198.982	-198.336	0.635

erence to atoms and ions, for which there is an extensive literature. A knowledge of the ionization energies, for example, permits the total energies to be determined[1] and accurate Hartree-Fock (HF) energies are also available.[2] We show experimental total energies for some first-row atoms in Table 1. Also shown are the estimated relativistic contributions to the total energy and the total energies obtained using HF and the HLG spin density functional.[3] The last column, the difference between experimental and corrected HF energies, is the correlation energy focused on by many authors. There are several points of interest. Firstly, both HF and LSD methods lead to departures from experiment which are much greater than would be acceptable for any binding energies of interest. It is clear that both methods would require a substantial cancellation of errors in any determination of energy differences. Secondly, the HF total energies are somewhat closer to experimental values than are the LSD values. This is also true for the LSD parametrizations due to von Barth and Hedin,[4] Gunnarsson and Lundqvist,[3] and Vosko, Wilk and Nusair.[5]

In this chapter, we should like to address the following questions: (a) How good is the error cancellation in HF and LSD calculations of energy differences in atoms? (b) What can be done to improve the absolute values of the total energies? In the case of HF, configuration interaction (CI) provides a systematic means for doing this. A comparable scheme is not available for the LSD approximation, but we shall describe a way of viewing the exchange-correlation energy functional which makes the success of the local approximation easier to understand and indicates the direction possible improvements may take.

2. ENERGY DIFFERENCES IN ATOMS

a. Ionization Energies

The energy required to remove a single electron is accurately known for most atoms.[6] In Fig. 1 we compare the values for some light atoms with HF energies and with values calculated using local density (LD) and LSD functionals.[7] The calculated values are found from differences between calculations for the ground states of atom and ion, respectively, $I_p = E_0^{N-1} - E_0^N$. Agreement with experimental trends is significantly better using the LSD approximation than with either the LD functional or the HF method. In particular, cases involving half-filled shells are more accurately given, an example being the break between N and O in an otherwise smooth curve. In N, the p-electron removed has its spin parallel to the others ($p\uparrow\uparrow\uparrow \rightarrow p\uparrow\uparrow$), whereas in O it is anti-parallel ($p\uparrow\uparrow\uparrow\downarrow \rightarrow p\uparrow\uparrow\uparrow$) and more weakly bound. The HF approximation, which neglects correlations between anti-parallel spins, does not describe such energy differences adequately.

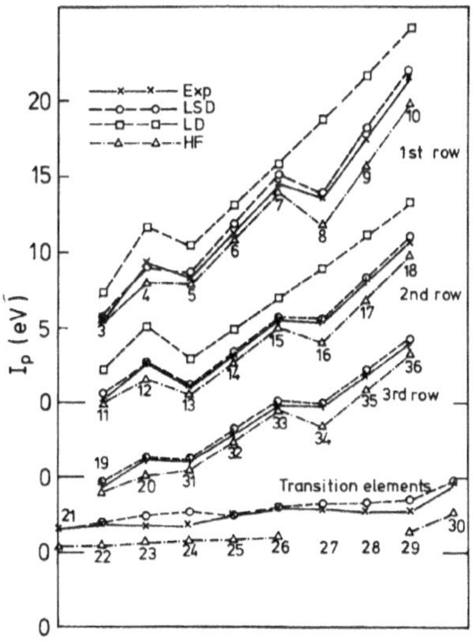

Fig. 1. The first ionization potential, I_p(eV), of atoms in the HF, LD and LSD approximations, compared with experiment (see Ref. 6). The zero of energy is shifted 5eV, 10eV and 15eV for the second row, the third row and the transition elements, respectively. The LD results for the first and second rows are increased by an additional 2eV.

b. Transfer Energies

When atoms bind to form a molecule or solid, there will in general be a change in the relative s, p, d ... contributions to the wave function. Such energy transfers can be studied in atoms by calculating the energy required to change an s- to a p-electron, for example the $2s^2 2p^3$ ground state of N into the $2s^1 2p^4$ excited state. Since these states are the lowest-lying states of symmetries ^4S and ^4P, respectively, density functional calculations are justified by the extension to the Hohenberg - Kohn principle discussed by Gunnarsson and Lundqvist.[3]

In Fig. 2, we compare experimental s-p transfer energies for first-row atoms and ions with results of LSD calculations.[8] It is clear that overall trends are given well, particularly for the energy required to remove an s-electron, but the absolute deviations from experiment are significant for s-p transfer energies in O and F. Corresponding HF results are not generally available, but Sasaki and Yoshimine[9] have performed non-relativistic calculations for C. The result is an sp-transfer energy Δ_{sp} of 0.179 Ry, compared with the experimental (non-relativistic) value of 0.307 Ry. In this case as well, the neglect of correlations between anti-parallel spins leads to a substantial deviation from experiment.

In Fig. 3, we show corresponding values for sd-transfer energies in iron-series atoms and ions.[8] In these systems, the energy required to remove an s-electron is also reproduced well, but the LSD approximation overestimates the relative stability of configurations with more d-electrons. For example, the stability of the

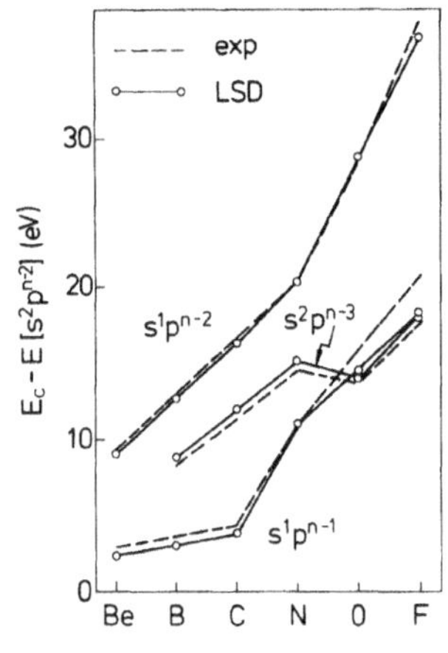

Fig. 2. Calculated and experimental energies, relative to s^2p^{n-2} atomic ground state, of low-lying states in first-row atoms and ions.

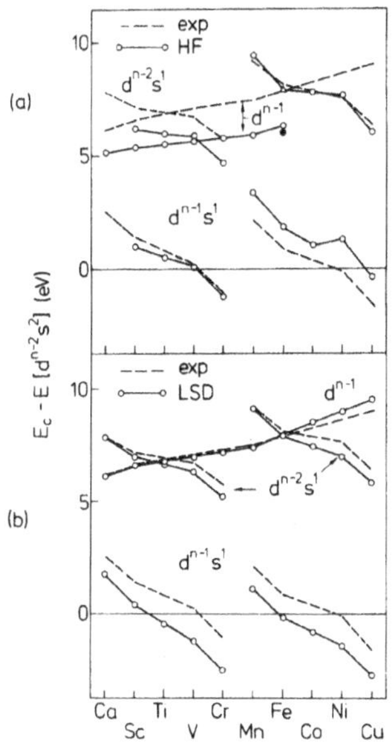

Fig. 3. Calculated and experimental energies, relative to $d^{n-2}s^2$ atomic state, of low-lying states in iron series atoms and ions (a) HF calculations (Ref. 2) (b) LSD values. Experimental values are indicated by the broken curves. (Ref. 6). See Ref. 8 for more details.

$d^{n-1}s^1$ states, relative to the $d^{n-2}s^2$ states in the atoms, is overestimated by approximately 1eV and there is a corresponding, though smaller, effect in the ions.

Fig. 3 also shows that the deviations from experiment, although considerable, are much more systematic than those of Hartree-Fock calculations. In particular, the break in the centre of the series, which results from the spin-flip required when the d-shell is more than half-filled, is poorly reproduced by the latter.

The standard method of improving the HF approximation is to increase the number of determinants used to describe the wave function and it is interesting to examine the scale of the calculations required for accurate energy determinations. Sasaki and Yoshimine[9] used an extensive basis set (35 orbitals, up to $\ell = 6$, expanded in turn in terms of 47 Slater-type functions) and single, double, triple and quadruple excitations. For first-row atoms, where up to 2650 configurations are involved, the calculated correlation energies are only 3 - 5% different from the experimental values and the ^5S - ^3P splitting is in error by slightly less than 0.1eV.

Bauschlicher and Shavitt[10] studied the basis set dependence of both the total energy and the quintet-triplet splittings in carbon. To compute the splitting to within 0.05eV of the experimental value requires at least a triple-zeta basis with two sets of polarization functions and single and double excitations. It can be seen that, even for the lightest atoms with few electrons, the incorporation of correlation effects using configuration interaction is far from trivial. Calculations of the sd-transfer energy in iron-series atoms have been performed recently by Bauschlicher et al.[11] There is a substantial improvement over HF results, but the agreement with experimental s-d splittings is noticeably better in the first half of the series, Sc - Cr, where the two states in question have different numbers of paired electrons, than for Mn - Cu.

Density functional calculations have also been performed for other series of atoms.[12-14] Although the LSD approximation gives a good description of trends in all cases, there is a systematic tendency to overestimate the stability of p- or d-occupancy relative to s and of f-occupancy relative to d, in the last case by approximately 2eV. Essential to the reproduction of the experimental trends in all these series is the use of the <u>spin</u> density functional. If the spin-free (LD) approximation is used, qualitatively inferior results are obtained.

c. Multiplet Structure

The calculation of multiplet structures using the density functional formalism requires care. In our previous discussion of the LSD approximation, the only input required to calculate the energy of the lowest-lying state of a specific symmetry is the total density and the spin-density. It is not difficult, however, to find examples where this leads to serious inconsistencies for the multiplet structure.[15] In the spirit of the HF approximation, Ziegler et al[16] suggested that local density calculations be performed only for single determinant functions. In general, such a function is not an eigenstate of $\hat{L}^2$ and $\hat{S}^2$ and a multiplet description requires a linear combination of determinants. The total energy calculated from a single determinant can be represented as a weighted sum of multiplet energies. Provided that the number of determinants with different energies equals the number of multiplets, the energies of the latter can be extracted. The justification of this technique requires a generalization of the density functional theory to states of mixed symmetry.[15]

The use of this scheme to describe multiplet structures has been very encouraging. In atoms, ions and small molecules, it leads in general to qualitative improvements in the description of multiplet structures.[15,16,8] The method can only be applied, however, if the number of determinants is at least as great as the number of multiplets. A counter-example is provided by the first excited configuration of benzene,[16] which gives rise to six states but only four determinants with different energies. Although some splittings can still be found, a complete picture is not obtainable and this is likely to be the case in many larger molecules with relatively low symmetry.

d. Dependence on Functional Form

In discussing the results of LSD calculations, it must be borne in mind that a variety of functional forms has been used.[3-5] Energies calculated using some of these have been compared recently.[17,18] Atomic total energies for different electron gas parametrizations are very similar, the errors for the most accurate form[5] being only 1-3% greater than those for the HLG functional.[3] Ionization energies and sp-transfer energies show generally improved agreement with experiment.[17,18] Markedly different ionization and transfer energies result, however, if the Xα (α = 0.7) form[19] is used. The Xα method shares the Hartree-Fock tendency to overestimate the relative stability of states with large spin polarizations and this is particularly evident in the sp-transfer in C and in the ionization energies of Be and O.[17]

It is also interesting to compare multiplet structures calculated with density functional methods with HF values and with experiment. As noted by Slater,[20] the HF method leads to energy spreads of multiplets which are substantially greater than experiment. In a CI calculation, the perturbative interaction of higher-lying configurations will reduce this range. Wood[21] has carried out multiplet structure calculations using the Xα (with different values of α) and von Barth - Hedin[4] LSD parametrizations and compared the results with HF and experimental values for eight light atoms and their ions. The results show that the Xα functional reproduces the HF splittings well and in a fashion only weakly dependent on the choice of α. This is a clear indication that correlation effects are not included in the Xα approach. The LSD calculations show, however, much improved agreement with experimental multiplet splittings. Wood also noted that the total energies obtained with both Xα and LSD functionals were unsatisfactory.

3. EXCHANGE-CORRELATION ENERGY AND THE PAIR-CORRELATION-FUNCTION

The results surveyed in the previous section suggest that the LSD approximation gives a remarkably good picture of trends in ionization and transfer energies in atoms. In view of the simplicity of the approximation and the strong density variations in an atom, this is rather surprising. The earliest derivations[22,4] of the LD approximation assumed that the variations of electron density are small over a distance of the order $k_F(n(r))^{-1} \equiv (3\pi n(r))^{-1/3}$. Each electron then "sees" an essentially homogeneous system, which leads to the LD approximation

$$E^{xc}_{LD} \equiv \int d\underline{r}\; n(\underline{r})\; \varepsilon^{xc}(n(\underline{r})) \qquad (1)$$

where $\varepsilon^{xc}(n)$ is the exchange-correlation (xc-) energy per particle of a homogeneous system with density n. For most systems, of interest, the density does not satisfy this assumption and an alternative justification is necessary.

Using an integration-over-the-coupling-constant technique, the xc-functional can be written[23]

$$E^{xc}[n] = \frac{e^2}{2} \int d\underline{r}\; n(\underline{r}) \int d\underline{r}' \frac{1}{|\underline{r}-\underline{r}'|}\; n^{xc}(\underline{r},\underline{r}'-\underline{r}) \qquad (2)$$

with

$$n^{xc}(\underline{r},\underline{r}-\underline{r}') = n(\underline{r}') \int_o^1 d\lambda [g(\underline{r},\underline{r}',\lambda) - 1] \qquad (3)$$

where $g(\underline{r},\underline{r}',\lambda)$ is the exact pair-correlation function for a system with density $n(\underline{r})$ and the coulomb interaction $\lambda e^2/|\underline{r}-\underline{r}'|$. The density in Eq. (3) is that corresponding to the physical value of λ (unity). The so-called xc-hole n^{xc} describes how the electrons are kept apart by exchange and correlation effects. Given an electron at $\underline{r}$, the probability of finding another at a short distance $|\underline{r}'-\underline{r}|$ is reduced. This leads to the xc-energy, which in Eq. (2) is described as the interaction between an electron and its xc-hole. The LD approximation can now be obtained by assuming

$$n^{xc}_{LD}\; (\underline{r},\underline{r}'-\underline{r}) = n(\underline{r}) \int_o^1 d\lambda [g_h(\underline{r},\underline{r}',\lambda,n(\underline{r})) - 1] \qquad (4)$$

where $g_h(\underline{r},\underline{r}',\lambda,n(\underline{r}))$ is the pair-correlation function of a homogeneous system with density $n(\underline{r})$ and coupling constant λ. Inserting Eq. (4) into Eq. (2) leads to the expression (1) for the LD approximation.

We now discuss the LD approximation using Eq. (2-4).[3,24] As an illustration, we consider a free atom and include exchange effects only, which makes it possible to compare the LD approximation with an exact solution (see Fig. 4). Since an atom has a rapidly varying density and is a far from homogeneous system, is is not surprising that the two curves in Fig. 4 are very different. In particular, the approximate hole is spherically symmetric, while the exact hole is very asymmetric. A variable substitution $\underline{R} = \underline{r}'-\underline{r}$ in (2) yields

$$E^{xc}[n] = \frac{e^2}{2} \int d\underline{r}\; n(\underline{r}) \int_o^\infty dR\; R^2 \frac{1}{R} \int d\Omega\; n^{xc}(\underline{r},\underline{R}) \qquad (5)$$

This equation shows that the xc-energy depends only on the spherical

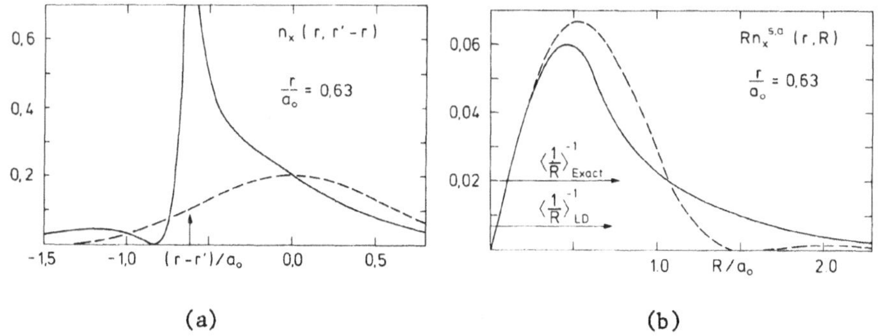

Fig. 4. The exact (Hartree-Fock, full line) and LSD (broken line) exchange hole for a spin-up electron in a nitrogen atom at $r = 0.63\ a_0$. The LSD hole has been calculated from a spin-generalized version of Eq. 4 using the HF pair-correlation function of a homogeneous medium. In Fig. a, $\underline{r}' \parallel \underline{r}$ and $\underline{r}'-\underline{r} = 0$ gives the electron position. The arrow shows the nuclear position. Fig. b shows the spherical average (around the electron) of the hole. The area under the curve is proportional to the exchange energy and <1/R> is defined in Eq. (7).

average of $n^{xc}(\underline{r},\underline{R})$. This average is also shown in Fig. 4. It illustrates that the differences between the exact and approximate holes are largely due to the non-spherical components of the hole, which according to (5) give no contribution to E^{xc}. In addition, there is a sum-rule stating that the xc-hole should contain unit charge

$$\int d\underline{r}'\ n^{xc}(\underline{r},\underline{r}'-\underline{r}) = -1 \tag{6}$$

This sum-rule is satisfied in the LD approximation. As shown in Fig. 4, the sum-rule leads to a partial cancellation of positive and negative errors in $E^{xc}[n]$. Because of this sum-rule, we can consider $-n^{xc}(\underline{r},\underline{r}'-\underline{r})$ as a normalized weight factor and define the size of the xc-hole as

$$\langle \frac{1}{R} \rangle_r = -\int d\underline{R}\ \frac{n^{xc}(r,\underline{R})}{|\underline{R}|} \tag{7}$$

which leads to

$$E^{xc}[n] = -\frac{e^2}{2}\int d\underline{r}\ n(\underline{r}) \langle \frac{1}{R} \rangle_r \tag{8}$$

This formula shows that E^{xc} is insensitive to the details of n^{xc} and depends only on the size of the xc-hole, provided that the sum-rule (6) is satisfied.

To illustrate the importance of the sum-rule, we consider the approximation[22]

$$E^{xc}[n] = E^{xc}_{LSD}[n] - \frac{1}{4}\int d\underline{r} \int d\underline{r}' \, K^{xc}(\underline{r}-\underline{r}', n(\tfrac{\underline{r}+\underline{r}'}{2}))(n(\underline{r}) - n(\underline{r}'))^2 \tag{9}$$

where the kernel K^{xc} is related to the dielectric function of a homogeneous medium. This approximation is exact for a density

$$n(\underline{r}) = n_o + \Delta n(\underline{r}) \tag{10}$$

in the limit $|\Delta n(\underline{r})| \ll n_o$, while the LSD approximation would require in addition $|\nabla n(\underline{r})| \ll [n(\underline{r})]^{4/3}$. However, for a system with strongly varying density, this approximation may be worse than the LSD approximation and for a free atom it gives an infinite energy.[24] Such a result could not be obtained unless there is a dramatic violation of either the sum-rule or the non-negativity of the probability $n(\underline{r}) - n^{xc}(\underline{r},\underline{r}'-\underline{r})$ of finding an electron at $\underline{r}'$ given that there is one at $\underline{r}$.

4. NON-LOCAL FUNCTIONALS

a. Functionals based on Pair-Correlation Functions

In recent years a variety of approximate functionals with a non-local density dependence has been proposed and tested, and we consider first two which are based on Eqs. (2) and (3) and which fulfil the sum-rule (6). In the LD approximation, we assume that the xc-hole $n^{xc}(\underline{r},\underline{r}'-\underline{r})$ depends only on the charge density at the electron. It would seem more appropriate to assume[24] that n^{xc} depends on a suitable average $\overline{n(\underline{r})}$,

$$n^{xc}(\underline{r},\underline{r}'-\underline{r}) = \overline{n(\underline{r})} \int d\lambda [g_h(\underline{r}-\underline{r}',\lambda,\overline{n(\underline{r})}) - 1] \tag{11}$$

It is possible to choose the weight factor which determines $\overline{n(\underline{r})}$ in such a way[24] that the functional reduces to Eq. (9) in the limit of almost constant density, where this equation is exact. However, (11) satisfies the sum-rule automatically and leads to results which are very different from Eq. (9) for strong density variations. In particular, it yields improvements over the LSD approximation for the total energy of atoms. Applications of this functional are discussed in Ref. 24.

An alternative approximation is obtained if we keep the proper prefactor $n(\underline{r}')$ in Eq. (3), leading to the so-called weighted density (WD) approximation

$$n^{xc}(\underline{r},\underline{r}'-\underline{r}) = n(\underline{r}') \, G(|\underline{r}-\underline{r}'|, \tilde{n}(\underline{r})) \tag{12}$$

where $\tilde{n}(\underline{r})$ is chosen to satisfy the sum-rule (6). Different forms have been proposed for $G(\underline{r},n)$,[24,25,7] and we discuss now an analytic form[7] which has advantages over a form based on the pair-correlation function of the homogeneous electron gas.[24,25] The analytic form is computationally simpler and can be chosen to yield exact results in certain limits. We assume[7]

$$G(\underline{r},n) = C(n)\ (1 - \exp(-\lambda(n)/|r|^5)) \tag{13}$$

where C and λ are parameters to be determined. The function G behaves as $|r|^{-5}$ for large distances, which is needed to obtain the proper image potential in this type of approximation. For a homogeneous system with the density n, we require that the functional should fulfil the sum-rule for $\tilde{n}(\underline{r}) = n$ and should give the exact xc-energy. This leads to two equations

$$n \int d\underline{r}\ G(|\underline{r}|,n) = -1 \tag{14}$$

$$\frac{e^2}{2} \int d\underline{r}\ \frac{1}{|\underline{r}|}\ G(|\underline{r}|,n) = \varepsilon^{xc}(n) \tag{15}$$

which are sufficient to determine the two parameters $C(n)$ and $\lambda(n)$.

This functional is exact in several limiting cases: (i) for a homogeneous system; (ii) for a hydrogen atom it gives an exact cancellation of the electron self-interaction; (iii) for an atom it gives the correct behaviour of the exchange-correlation energy density far from the nucleus, $\varepsilon^{xc}(r) = -e^2/(2r)$; (iv) far outside a surface it gives the image potential $\varepsilon^{xc}(z) = -e^2/(4z)$. The LSD approximation gives qualitatively wrong answers for (iii) and (iv), and the cancellation in (ii) is only approximate. Since (ii) is fulfilled, the WD approximation provides a self-interaction correction in the sense of Ref. 26 (see Sec. 4.b). In Fig. 5, we show an application of this approximation to the total energy of atoms.[7] The errors in the WD approximation are about one order of magnitude smaller than those in the LSD approximation. However, the sd-transfer energies in iron series atoms are only marginally better than those given by LSD and atomic ionization energies are somewhat worse.[7]

A serious defect of the LSD approximation is that it underestimates systematically the band gap in semiconductors. Although the one-electron eigenvalues ε_i are Lagrange parameters and have no physical significance in general, they satisfy the equation

$$\partial E/\partial n_i = \varepsilon_i(n_i) \tag{16}$$

where n_i is an occupation number.[27] Excitation energies can then be written

$$\Delta E = \int_{n_o}^{n_1} dn_i\ \varepsilon_i(n_i) \tag{17}$$

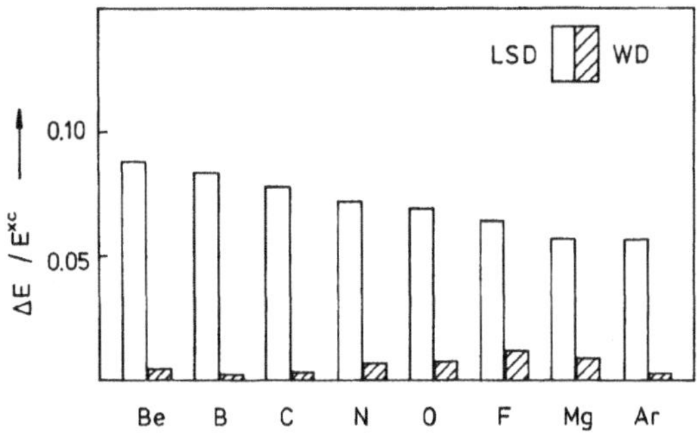

Fig. 5. The error in the total energy of a free atom as a fraction of the xc-energy in the LSD and WD approximations. For details of the calculation, see Ref. 7.

For an extended system, the excitation energy and the eigenvalue are equivalent[27] and the band gap is a quantity which should be obtainable from a density functional calculation. Although the absolute errors in LSD calculations for semiconductors are similar to those in other systems, the effect on many properties is drastic, because the band gap is small and the relative error large. The necessity for a non-local potential to describe the silicon band structure, for example, is evident from the work of Kane,[28] who showed that local potentials gave errors of 0.5 - 0.7eV in band gaps and 10 - 20% in effective masses. Recently, Kerker[29] applied a simplified version of the functional described above to the case of silicon. The experimental band gap (1.1 eV) is obtained, while the LSD value is in error by a factor of two, and other improvements were also found. The WD approximation has also been applied recently to Rh[30] and to Ni.[31]

b. Self-interaction Corrections

In seeking improvements to the LSD approximation, it is natural to focus attention on the imperfect cancellation of the unphysical electrostatic self-interaction by the xc-term. It might be expected that better results would follow if the self-interaction could be cancelled explicitly and a reasonable approximation made for the remainder. Evidence for this is provided by the results of Lindgren,[32] who showed that such a scheme reproduces the HF exchange potential better than the LSD approximation and yields a good approximation to the HF eigenvalues and electron binding energies. Assuming that correlation can be treated in the same way,[26] one obtains the functional

$$E_{SIC} = E_{LSD} - \sum_{\alpha\sigma} \delta_{\alpha\sigma} \tag{18}$$

where E_{LSD} is the LSD energy functional and

$$\delta_{\alpha\sigma} = \frac{e^2}{2} \int d\underline{r} \int dr' \frac{\rho_{\alpha\sigma}(\underline{r})\rho_{\alpha\sigma}(\underline{r}')}{|\underline{r}-\underline{r}'|} + E^{xc}_{LSD}[\rho_{\alpha\sigma},1] \tag{19}$$

is the self-interaction correction (SIC) for the orbital α with spin α and charge density $\rho_{\alpha\sigma}(\underline{r})$. The first term in (19) is the self-interaction energy and the second the LSD approximation to the xc-energy of a fully spin-polarized system with density $\rho_{\alpha\sigma}(\underline{r})$. This functional is exact for a one-electron system and it leads to substantial improvements over LSD results for atomic total energies.[26]

Although the WD and SIC functionals lead to much improved total energies, transfer and ionization energies are generally no better than the corresponding LSD values. This is rather disappointing, since it is energy differences which are experimentally accessible. The improved total energies in SIC can, in fact, be traced[33] to the much better treatment of the core electrons, which play a minor role in most physical and chemical processes of interest.

The correction for the spurious self-interaction has an interesting consequence for the one-electron eigenvalues ε_i. Although the eigenvalue does depend on the occupation number due to the effects of self-consistency, the effect is much smaller than if the LSD is used. As a result, the correspondence between excitation energies and eigenvalues (see Eq. 17) is much improved.[26] Band structure calculations performed with an SIC potential for LiCl[34] give a band gap of 9.9 eV, much closer to the experimental value (9.4 eV) than the LSD value of 6.3 eV.

c. Wave-vector Analysis

In an early calculation of the correlation energy of an infinite electron gas, Noziéres and Pines[35] demonstrated the value of separating contributions from regions with different momentum transfers. It is then possible to make approximations which are appropriate for the different regions. In the context of the exchange-correlation energy functional, a similar approach was taken by Langreth and Mehl,[36] who carried out the wave-vector decomposition[37]

$$E^{xc} = \int d\underline{k}\; \varepsilon^{xc}(\underline{k}) \tag{20}$$

The LSD approximation is valid for large k and the corresponding decomposition of the lowest order gradient expansion shows good agreement with the exact expression for k-values which are not too small. For values of k close to zero, however, the gradient expansion leads to a substantial overestimate. Langreth and Mehl[36] therefore proposed a functional of the form

$$E^{xc} = E^{xc}_{LSD} + \int d\underline{k}\ \varepsilon^{xc}_{grad}(\underline{k})\ \Theta(k - \xi^{-1}) \tag{21}$$

where $\varepsilon^{xc}_{grad}(\underline{k})$ is the wave-vector decomposition of the lowest order gradient correction. The length ξ is the range over which the density varies. Its inverse ξ^{-1}, the wave-vector cutoff, was defined in terms of a single parameter, adjusted so that the LD approximation was improved for the surface energy of a jellium surface and for certain atomic energies.[36] Further tests of this functional would be most interesting.

5. CONCLUDING REMARKS

The success of local density approximations in describing a wide variety of physical phenomena is amply documented in this volume. Although the original formulation of the LD approximation assumed small departures from homogeneity, arguments based on the pair-correlation function show that it can also be expected to give reasonable results for systems which are far from homogeneous. Energy differences are generally better than those given by Hartree-Fock calculations, particularly when spin-flips are involved. The experience with non-local functionals has, however, been rather mixed. Dramatic improvements in total energies have resulted, but energy differences such as transfer and ionization energies have remained a stubborn problem. Perhaps it is being overoptimistic to expect that such subtle effects can be described by a relatively simple approximation, but the challenge remains.

REFERENCES

1. C.W. Scherr, J.N. Silverman, and F.A. Matsen, Phys. Rev. 127, 830 (1962).
2. See, for example, S. Fraga, K.M.S. Saxena, and B.W.N. Lo, Atomic Data 3, 323 (1971); E. Clementi and C. Roetti, At. Data Nucl. Data Tables 14, 177 (1974).
3. This functional is based on the paramagnetic term calculated by L. Hedin and B.I. Lundqvist, J. Phys. C 4, 2064 (1971) and the spin-dependent term of O. Gunnarsson and B.I. Lundqvist, Phys. Rev. B 13, 4273 (1976).
4. U. van Barth and L. Hedin, J. Phys. C 5, 1629 (1972).
5. S.H. Vosko, L. Wilk, and M. Nusair, Can. J. Phys. 58, 1200 (1980).
6. C.E. Moore, "Atomic Energy Levels", Natl. Bur. Stand. (U.S.), Circ. 467, Vol. I (1949); Vol. II (1952); Vol. III (1958).
7. O. Gunnarsson and R.O. Jones, Phys. Scr. 21, 394 (1980).
8. O. Gunnarsson and R.O. Jones, J. Chem. Phys. 72, 5357 (1980).
9. F. Sasaki and M. Yoshimine, Phys. Rev. A 9, 17 (1974).
10. C.W. Bauschlicher, Jr. and I. Shavitt, J. Am. Chem. Soc. 100, 739 (1978).

11. C.W. Bauschlicher, Jr., S.P. Walch, and H. Partridge, J. Chem. Phys. 76, 1033 (1982).
12. J. Harris and R.O. Jones, J. Chem. Phys. 70, 830 (1979).
13. R.O. Jones, J. Chem. Phys. 71, 1300 (1979).
14. D. Glötzel (unpublished) has calculated d-f transfer energies in rare earth and actinide atoms.
15. U. von Barth, Phys. Rev. A 20, 1693 (1979).
16. T. Ziegler, A. Rauk, and E.J. Baerends, Theor. Chim. Acta 43, 261 (1977).
17. R.O. Jones, J. Chem. Phys. 76, 3098 (1982).
18. L. Wilk and S.H. Vosko, J. Phys. C 15, 2139 (1982). See also G. Painter, Phys. Rev. B 24, 4264 (1981).
19. J.C. Slater, "Quantum Theory of Molecules and Solids", McGraw-Hill, New York (1974) Vol. IV.
20. J.H. Wood, J. Phys. B 13, 1 (1980); J.C. Slater, "Quantum Theory of Atomic Structure", McGraw-Hill, New York (1960), Vol. I, Sec. 15.
21. J.H. Wood, J. Phys. B 13, 1 (1980).
22. P. Hohenberg and W. Kohn, Phys. Rev. 136, B 864 (1964); W. Kohn and L.J. Sham, Phys. Rev. 140, A 1133 (1965).
23. J.P. Perdew and D.C. Langreth, Solid State Commun. 17, 1425 (1975); Gunnarsson and Lundqvist (Ref. 3). A similar expression was derived by J. Harris and R.O. Jones, J. Phys. F 4, 1170 (1974).
24. O. Gunnarsson, M. Jonson and B.I. Lundqvist, Solid State Commun. 24, 765 (1977); Phys. Rev. B 20, 3136 (1979).
25. J.A. Alonso and L.A. Girifalco, Solid State Commun. 24, 135 (1977); Phys. Rev. B 17, 3735 (1978).
26. A. Zunger, J.P. Perdew, and G.L. Oliver, Solid State Commun. 34, 933 (1980); J.P. Perdew and A. Zunger, Phys. Rev. 23, 5048 (1981).
27. See, for example, J.F. Janak, Phys. Rev. B 18, 7165 (1978).
28. E.O. Kane, Phys. Rev. B 4, 1910 (1971).
29. G.P. Kerker, Phys. Rev. B 24, 3468 (1981).
30. G. Borstel, M. Neumann, and W. Braun, Phys. Rev. B 23, 3113 (1981).
31. C.S. Wang, Bull. Am. Phys. Soc. 27, 407 (1982).
32. I. Lindgren, Int. J. Quantum Chem. 5, 411 (1971).
33. O. Gunnarsson and R.O. Jones, Solid State Commun. 37, 249 (1981).
34. R.A. Heaton, J.G. Harrison, and C.C. Lin, Solid State Commun. 41, 827 (1982).
35. P. Nozières and D. Pines, Phys. Rev. 111, 442 (1958).
36. D.C. Langreth and M.J. Mehl, Phys. Rev. Lett. 47, 446 (1981).
37. D.C. Langreth and J.P. Perdew, Phys. Rev. B 15, 2884 (1977).

AN APPROACH TO ELECTRON CORRELATION

BEYOND THE LOCAL SPIN DENSITY APPROXIMATION

L. Fritsche and H. Gollisch

Institut für Theoretische Physik
der Technischen Universität Clausthal
and Sonderforschungsbereich 126 Göttingen/Clausthal
D-3392 Clausthal-Zellerfeld
Federal Republic of Germany

We analyse the general expression for the electron-electron interaction in terms of spin-dependent correlation factors which have already been discussed by other authors in earlier quantum chemical studies. By using this expression for a system of N interacting electrons moving in some external potential and by invoking the Hohenberg-Kohn theorem, we obtain on minimizing the total energy of the system a set of one-particle equations which are formally identical to those derived by Kohn and Sham. However, the exchange-correlation potential emerging from our derivation does not suffer from any of the discrepancies typical of the local spin density (LSD-)potentials which are at present used within the Kohn-Sham scheme. Most important, our potential yields the correct 1/r long range dependence for free atoms and correctly cancels the classical Coulomb potential in the limiting case of a hydrogen molecule as the internuclear distance of the two atoms tends to infinity. Application of an approximate form of this potential to first and second row atoms gives very satisfactory agreement with experimental data on total energies, first and second ionization potentials and on electron affinities. The latter quantities may be viewed as a particularly sensitive test of the improvement that has been achieved beyond potentials based on electron gas theory. Moreover, we have calculated (3d, 4s)-charge exchange energies for atoms of 3d series and also found a remarkable gain in accuracy compared to LSD calculations.

1. INTRODUCTION

The concept of configuration interaction (CI) put forward by Hartree and others in the thirties (see e.g. ref. 1) has given rise to a wealth of different techniques used in treating electron correlation. A common feature of these methods, either variational or perturbational/diagrammatic, is that they are based on suitably chosen sets of orbitals in terms of which the interaction matrix elements can be expressed. A specific technique is characterized then mainly by its method of computing the total electron-electron interaction energy from these matrix elements.

An alternative to this approach consists in expressing the electron correlation energy as a functional of the electronic charge density $\rho(\underline{r})$. This idea has its conceptual roots in early papers by Wigner[2] and others. According to the theorem of Hohenberg and Kohn[3] for the ground state of an N-electron system, it is always possible to find an external potential in which the charge density of N non-interacting electrons is identical with that of the original system of N interacting electrons moving under the influence of a different external potential. The charge density of the non-interacting system can be expressed as a sum of one-particle contributions $|\psi_n(\underline{r})|^2$. The kinetic energy E_{kin} can be written accordingly, and hence the total energy E becomes a sum of E_{kin} and some functional of $\rho(r)$. Again invoking the Hohenberg-Kohn theorem, Kohn and Sham[4] conclude in a subsequent paper that E must be identical with the total energy of the original system if the charge densities agree. Since E attains a minimum in the ground state, E must be stationary against variation of the orbitals ψ_n. This requirement leads to one-particle Schrödinger equations to be obeyed by the ψ_n's. Exchange and correlation are described in these equations by a spin-dependent potential $U_{xc}^{(s)}(\underline{r})$ where s refers to the spin direction.

Of course, these equations have to be solved self-consistently. Apart from certain technicalities to be observed in performing this task, the Kohn-Sham approach offers the advantage that one is only dealing with N different orbitals in contrast to CI-based methods which always involve a large set of orbitals.

In turn, the obvious disadvantage of the Kohn-Sham scheme consists in the fact that the functional $E_{xc}[\rho]$ describing the exchange and correlation energy is not well known. Most practitioners of the scheme use expressions for $E_{xc}[\rho]$ and local spin density (LSD) potentials $U_{xc}^{(s)}(\underline{r})$ which have been derived from electron gas theory where $\rho(\underline{r})$ is required to be slowly varying. (See e.g. Gunnarsson and Lundqvist[5] and references therein). Clearly, no realistic system consisting of atoms meets such a requirement. Notwithstanding this rather disturbing inconsistency, the Kohn-Sham approach in conjunction with improved LSD-potentials has in the recent past led to a surprisingly quantitative description of atomic, molecular, and solid state properties. (See e.g. Harris and Jones,[6] Yin and Cohen,[8] and Schlüter and Sham[8]). However, there have remained three obvious discrepancies of principal nature which are typical of the LSD-potentials used at present within the Kohn-Sham scheme:

1. For free atoms these potentials do not exhibit a $1/r$ long range behaviour as they should.

2. At solid surfaces they do not change into the $1/z$ long range behaviour of an image potential.

3. They do not correctly cancel the classical Coulomb potential in the limiting case when the atoms of a hydrogen molecule are successively pulled apart.

It is the objective of the present paper to derive an expression for $E_{xc}[\rho]$ which is entirely general and applies to any particle density and in particular to any degree of inhomogeneity. This will be the subject of Sect. 2. Our notation follows closely that of McWeeny,[9] specifically in the use of correlation factors $f_{s's}(\underline{r}',\underline{r})$. These factors are subject to constraints associated with the N-representability of the true N-electron wavefunction. In Sect. 3 we derive one-particle equations much along the lines of the paper by Kohn and Sham.[4] It will be shown that the potential $U_{xc}^{(s)}(\underline{r})$ obtained in the process of this derivation does not suffer from the deficiencies mentioned above in points 1 and 3. In Sect. 4 we assume an approximate form for $f_{s's}(\underline{r}',\underline{r})$ which conforms to the N-representability constraints and guarantees the correct image potential dependence of $U_{xc}^{(s)}(\underline{r})$ at solid surfaces. Sect. 5 is devoted to practical applications of this approximation. In self-consistently solving the associated one-particle equations and computing the total energies, ionization potentials and affinities for 13 first and second row atoms, we can show that the new approach to exchange and correlation leads to an overall improvement over the present state theory, most significantly for the electron affinities. Moreover, we have calculated the energy differences $\Delta E = E[3d^{n-1}\,4s^1] - E[3d^{n-2}\,4s^2]$ for atoms of 3d-series. There is a remarkable gain in accuracy compared to LSD-results. In trying to go beyond the LSD approximation Gunnarsson and Jones[10] have recently applied an idea similar to ours, but for unidentified reasons obtained only little improvement.

2. GENERAL FORM OF THE EXCHANGE-CORRELATION ENERGY

In the following Hartree units will be used throughout. The electron-electron interaction energy of a not spin-polarized N-electron system can generally be written in the form

$$\langle U_{e-e}\rangle = \frac{1}{2}\int \rho(\underline{r}) \int \frac{\rho_{N-1}(\underline{r}',\underline{r})}{|r'-r|}\, d^3r'd^3r$$

where $\rho_{N-1}(\underline{r}',\underline{r})$ is the particle density of N-1 electrons if one electron is known to be at $\underline{r}$. The above expression is identical with the interaction energy of a statistical ensemble of N classical point charges.

If we explicitely introduce the spin index s, where s = ± 1 refers to "spin-up" and "spin-down", respectively, the above expression takes the form

$$\langle U_{e-e}\rangle = \frac{1}{2}\sum_{s}\int \rho_s(\underline{r})\int \frac{\rho_{N-1}^{(s)}(\underline{r}',\underline{r})}{|\underline{r}'-\underline{r}|}\, d^3r' d^3r \qquad (1)$$

here $\rho_{N-1}^{(s)}(\underline{r}',\underline{r})$ denotes the particle density of N-1 electrons of any spin orientation if one electron is known to have the position $\underline{r}$ and spin orientation s. Likewise, $\rho_s(r)$ is the particle density of all electrons with spin orientation s. Hence the total particle density is given by

$$\rho(\underline{r}) = \sum_{s} \rho_s(\underline{r})$$

which integrates to

$$N = \sum_{s} N_s$$

According to its definition $\rho_{N-1}^{(s)}$ integrates to N-1 particles irrespective of the spin orientation of the particle at $\underline{r}$:

$$\int \rho_{N-1}^{(s)}(\underline{r}',\underline{r})\, d^3r' = N-1 \qquad \text{for all } \underline{r} \qquad (2)$$

We partition $\rho_{N-1}^{(s)}$ into the contributions $\rho_{s's}(\underline{r}',\underline{r})$ associated with the spin orientations s' of those N-1 electrons:

$$\rho_{N-1}^{(s)}(\underline{r}',\underline{r}) = \sum_{s'} \rho_{s's}(\underline{r}',\underline{r}) \qquad (3)$$

Hence (1) may be rewritten:

$$\langle U_{e-e}\rangle = \frac{1}{2}\sum_{s',s}\int \frac{\rho_s(\underline{r})\rho_{s's}(\underline{r}',\underline{r})}{|\underline{r}'-\underline{r}|}\, d^3r' d^3r \qquad (4)$$

We further introduce the two-particle density $\rho_{s's}^{(2)}(\underline{r}',\underline{r})$ defined by

$$\rho_{s_1s_2}^{(2)}(\underline{r}_1,\underline{r}_2) = N(N-1)\int |\Psi(\underline{q}_1,\underline{q}_2,\underline{q}_3,\ldots,\underline{q}_N)|^2\, d^4q_3\ldots d^4q_N \qquad (5)$$

where

$$\underline{q} := (\underline{r},s)\ ; \qquad \int\ldots d^4q := \sum_{s}\int\ldots d^3r$$

and $\Psi(\underline{q}_1,\underline{q}_2,\ldots,\underline{q}_N)$ is the N-electron wavefunction.

By using (5) the electron-electron interaction energy can alternatively be written:

$$\langle U_{e-e}\rangle = \frac{1}{2}\sum_{s',s}\int \frac{\rho^{(2)}_{s's}(\underline{r}',\underline{r})}{|\underline{r}'-\underline{r}|}\, d^3r'd^3r \tag{6}$$

Since Ψ is antisymmetric, it changes sign on exchanging $\underline{q}$ and $\underline{q}'$, and hence - according to (5) - $\rho^{(2)}_{s's}$ must have the property:

$$\rho^{(2)}_{s's}(\underline{r}',\underline{r}) = \rho^{(2)}_{ss'}(\underline{r},\underline{r}') \tag{7a}$$

and in particular

$$\rho^{(2)}_{s's}(\underline{r}',\underline{r}) \equiv 0 \qquad \text{if } s' = s \text{ and } \underline{r}' = r \tag{7b}$$

It is convenient to introduce pair correlation functions $G_{s's}(\underline{r}',\underline{r})$ defined by

$$\rho_{s's}(\underline{r}',\underline{r}) = \rho_{s'}(\underline{r}')G_{s's}(\underline{r}',\underline{r}) \tag{8}$$

Inserting (8) into (4) we obtain

$$\langle U_{e-e}\rangle = \frac{1}{2}\sum_{s',s}\int \frac{\rho_{s'}(\underline{r}')\rho_s(\underline{r})G_{s's}(\underline{r}',\underline{r})}{|\underline{r}'-\underline{r}|}\, d^3r'd^3r \tag{9}$$

A comparison with (6) shows that

$$G_{s's}(\underline{r}',\underline{r}) = \frac{\rho^{(2)}_{s's}(\underline{r}',\underline{r})}{\rho_{s'}(\underline{r}')\rho_s(\underline{r})} \tag{10}$$

Because of (7a,b) $G_{s's}(\underline{r}',\underline{r})$ has the property

$$G_{s's}(\underline{r}',\underline{r}) = G_{ss'}(\underline{r},\underline{r}') \tag{11a}$$

and in particular

$$G_{ss}(\underline{r},\underline{r}) \equiv 0 \qquad \text{for } \underline{r}' = r \tag{11b}$$

Because of (8) the latter property leads to

$$\rho_{ss}(\underline{r}',\underline{r}) \equiv 0 \qquad \text{if } \underline{r}' = r$$

which reflects the existence of a Fermi-hole around $\underline{r}$: If there is one electron at $\underline{r}$ with spin orientation s, then the particle density of all other electrons of the same spin orientation is zero at $\underline{r}$.

Following the notation of McWeeny[9] we introduce correlation factors $f_{s's}(\underline{r}',\underline{r})$ defined by

$$G_{s's}(\underline{r}',\underline{r}) = 1-f_{s's}(\underline{r}',\underline{r}) \tag{12}$$

(Note that distinct from McWeeny's notation we have chosen the opposite sign in front of $f_{s's}$.)

Combining (2), (3), (8) and (12) we arrive at sum rules to be obeyed by these factors:

$$\sum_{s'} \int \rho_{s'}(\underline{r}') f_{s's}(\underline{r}',\underline{r})\, d^3r' = 1 \tag{13a}$$

Moreover, as follows from (11a)

$$f_{s's}(\underline{r}',\underline{r}) = f_{ss'}(\underline{r},\underline{r}') \tag{14}$$

Because of (11b) and (12) the existence of a Fermi-hole is expressed by the property

$$f_{ss}(\underline{r}',\underline{r}) \equiv 1 \qquad \text{for } \underline{r}' = r \tag{13b}$$

Inserting (12) into (9) we may rewrite the electron-electron interaction energy in the form:

$$\langle U_{e-e} \rangle = E_C + E_{xc} \tag{15a}$$

where

$$E_C = \frac{1}{2} \int \frac{\rho(\underline{r}')\rho(\underline{r})}{|\underline{r}'-\underline{r}|}\, d^3r' d^3r \tag{15b}$$

and

$$E_{xc} = \sum_s \int \rho_s(\underline{r}) \varepsilon_{xc}^{(s)}(\underline{r}) d^3r \tag{15c}$$

with

$$\varepsilon_{xc}^{(s)}(\underline{r}) = -\frac{1}{2} \sum_{s'} \int \frac{\rho_{s'}(\underline{r}') f_{s's}(\underline{r}',\underline{r})}{|\underline{r}'-\underline{r}|}\, d^3r' \tag{15d}$$

The quantity E_C may be referred to as "classical Coulomb energy" and E_{xc} as "exchange-correlation energy".* In the spirit of the latter definition $\varepsilon_{xc}^{(s)}$ has to be viewed as exchange-correlation energy per particle. We want to emphasize that the detailed form of E_{xc} and $\varepsilon_{xc}^{(s)}$ which we have arrived at is entirely general and does not rest on any kind of approximation.

---

*We do not follow here the terminology of density functional theory where the total exchange-correlation energy is defined as containing an additional kinetic energy portion.

In deriving the equations (13a,b) and (14) we had to refer to the existence of an antisymmetric N-electron wavefunction. Hence these equations for $f_{s's}(\underline{r}',\underline{r})$ have to be regarded as N-representability constraints.

3. ONE-PARTICLE EQUATIONS

According to the theorem of Hohenberg and Kohn[3] the kinetic energy of an interacting N-electron system in the ground state can be calculated by resorting to an auxiliary system of N non-interacting fermions which has the same particle density $\rho(\underline{r})$. The kinetic energy of that system can be written

$$\langle T\rangle = \sum_{k,s} \int \psi^*_{ks}(\underline{r})(-\tfrac{1}{2}\underline{\nabla}^2)\psi_{ks}(\underline{r})d^3r \tag{16}$$

where the functions $\psi_{ks}(\underline{r})$ are solutions to a one-particle Schrödinger equation

$$H_s\ \psi_{ks}(\underline{r}) = \varepsilon_{ks}\ \psi_{ks}(\underline{r}) \tag{17a}$$

with

$$H_s = -\tfrac{1}{2}\underline{\nabla}^2 + U_s(\underline{r}) \tag{17b}$$

The potential $U_s(\underline{r})$ has to be chosen such that

$$\rho(\underline{r}) = \sum_{k,s} |\psi_{ks}(\underline{r})|^2 \tag{18}$$

is identical with the particle density of the interacting N-electron system under consideration. The sum in (18) runs over the N lowest lying states.

If the electrons move under the influence of an external potential $U_n(\underline{r})$, set up by atomic nuclei, for example, the contribution of this potential to the total energy is

$$\langle U_n\rangle = \int \rho(\underline{r})\ U_n(\underline{r})d^3r \tag{19}$$

Combining (15a-c), (16), (18) and (19) we can put the total energy of the interacting N-electron system in the form

$$E = \sum_{k,s} \int \psi^*_{ks}(\underline{r})[-\tfrac{1}{2}\underline{\nabla}^2 + U_n(\underline{r}) + \tfrac{1}{2}U_c(\underline{r}) + \varepsilon^{(s)}_{xc}(\underline{r})]\psi_{ks}(\underline{r})d^3r$$

where (20)

$$U_c(\underline{r}) = \int \frac{\rho(\underline{r}')}{|\underline{r}'-\underline{r}|}d^3r'$$

In the ground state E must be stationary against variation of the functions $\psi_{ks}(\underline{r})$. This is the case, as has been shown by Kohn and

Sham,[4] if $\psi_{ks}(\underline{r})$ satisfies the one-particle equation (17a). Applying their derivation to our expression (20) we obtain for $U_s(\underline{r})$

$$U_s(\underline{r}) = U_n(\underline{r}) + U_c(\underline{r}) + U_{xc}^{(s)}(\underline{r}) \tag{21a}$$

The exchange-correlation potential results in the form

$$U_{xc}^{(s)}(\underline{r}) = -\sum_{s'} \int \frac{\rho_{s'}(\underline{r}')}{|\underline{r}'-\underline{r}|} \Big(f_{s's}(\underline{r}',\underline{r}) + \frac{1}{2}\sum_{s''} \rho_{s''}(\underline{r})\rho_{s'}(\underline{r}') \frac{\delta f_{s''s'}(\underline{r}',\underline{r})}{\delta \rho_s(\underline{r})}\Big)\, d^3r' \tag{21b}$$

where $\delta f_{s''s'}(\underline{r}',\underline{r})/\delta\rho_s$ denotes the functional derivative of $f_{s''s'}(\underline{r}',\underline{r})$.

The above potential can easily be seen to have a very important property. If the N-electron system consists of only one electron with spin orientation s, we have

$$\int \rho_{s'}(\underline{r}')\, d^3r' = \delta_{s's}$$

and the sum rule (13a) reduces to

$$\int \rho_s(\underline{r}')\, f_{ss}(\underline{r}',\underline{r})\, d^3r' = 1$$

This equation can only be satisfied together with the preceding one if

$$f_{ss}(\underline{r}',\underline{r}) \equiv 1$$

Hence (21a) attains the form

$$U_{xc}^{(s)}(\underline{r}) = -\int \frac{\rho_s(\underline{r}')}{|\underline{r}'-\underline{r}|}\, d^3r'$$

which is identical with $U_c(\underline{r})$ except for the sign. The resulting potential (21a) for a one-electron system is therefore identical with the external potential, as it should.

As has been pointed out by McWeeny,[9] the particle densities $\rho_{s's}(\underline{r}',\underline{r})$ must have the property

$$\rho_{s's}(\underline{r}',\underline{r}) \to (1-\delta_{s's}\frac{1}{N_{s'}})\ \rho_{s'}(\underline{r}')$$

if the N-electron system does not extend to infinity in all directions and if $\underline{r}$ is taken along a direction crossing a periphery of the system. It follows then from (8) and (12) that

$$f_{s's}(\underline{r}',\underline{r}) \to \delta_{s's}\frac{1}{N_{s'}} \tag{22}$$

and therefore

$$\frac{\delta f_{s's}(\underline{r}',\underline{r})}{\delta \rho_s} \to 0$$

Because of these properties we have for systems with a closed periphery, i.e. for atoms, molecules, and clusters when $|\underline{r}'-\underline{r}| \to |\underline{r}|$:

$$U_{xc}^{(s)}(\underline{r}) \to -\frac{1}{|\underline{r}|}\left[\sum_{s'} \delta_{s's} \frac{1}{N_{s'}} \int \rho_{s'}(\underline{r}')\, d^3r'\right]$$

that is

$$U_{xc}^{(s)} \to -\frac{1}{|\underline{r}|}$$

As already mentioned in point 3 of Sect. 1, this is another important property which is closely connected with the existence of negative ions. The approximate form of $f_{s's}(\underline{r}',\underline{r})$ which will be discussed in the ensuing section, must therefore guarantee the asymptotic behaviour (22) if one is interested in calculating relatively accurate electron affinities along with other atomic properties.

If the system under consideration has the form of a half-space the correlation factors (if finite) must be required to have the additional long range behaviour

$$f_{s's}(\underline{r}',\underline{r}) \to |\underline{r}|^{-5} \tag{23}$$

for a direction perpendicular to the planar boundary of that system. According to Gunnarsson and Jones[10] this property of $f_{s's}(\underline{r}',\underline{r})$ guarantees that the image charge associated with the electron at $\underline{r}$ (being outside the half-space in that limit) moves towards the surface of the system as $|\underline{r}|$ increases. In this case $U_{xc}^{(s)}(\underline{r})$ shows the correct long range behaviour of an image charge potential.

4. CHOICE OF AN APPROXIMATE FORM OF THE CORRELATION FACTORS

To facilitate practical calculations on atoms we have chosen the following approximate form of the correlation factors

$$f_{s's}(\underline{r}',\underline{r}) = \left[1 + \left(\frac{|\underline{r}'-\underline{r}|}{r_{s'so}(\underline{r})}\right)^2\right]^{-5/2} \qquad \text{for } s' = s \tag{24a}$$

and

$$f_{s's}(\underline{r}',\underline{r}) \equiv 0 \qquad \text{for } s' \neq s \tag{24b}$$

Obviously this choice conforms to the general requirements (13b) and (23). As follows from (24b) we do not allow for the formation of a hole around $\underline{r}$ in the density of electrons which have the opposite spin orientation as the one at $\underline{r}$. This does, however, not mean that we are going to include only exchange in our calculations. Since (24a) may be expected to be a better approximation to the true Fermi hole than can be simulated by only N orbitals used in the Hartree-Fock theory, our calculation of E_{xc} includes correlation effects between electrons of like spin which are not contained in the analogous Hartree-Fock expression $E_x^{(HF)}$. Hence, according to the generally accepted terminology, the difference between the total Hartree-Fock energy and our value of E must be interpreted as correlation energy.

In view of (24b) we may in the following drop the second index s'.

The length $r_{so}(\underline{r})$ in (24a) must be determined numerically by employing the sum rule (13a) which because of (24a,b) reduces to

$$\int \rho_s(\underline{r}') \left[1 + \left(\frac{|\underline{r}'-\underline{r}|}{r_{so}(\underline{r})}\right)^2\right]^{-5/2} d^3r' = 1 \tag{25}$$

If $\rho_s(\underline{r})$ is substantially different from zero only within a finite region and if $\underline{r}$ is large compared to the average radius of that region, we have from (24a)

$$\int \rho_s(\underline{r}')\, f_s(\underline{r}',\underline{r})\, d^3r' \to \left(\frac{r_{so}(\underline{r})}{|\underline{r}|}\right)^5 \int \rho_s(\underline{r}')\, d^3r'$$

According to (25) the left-hand side is equal to one, hence

$$\frac{r_{so}(\underline{r})}{|\underline{r}|} \to N_s^{-1/5}$$

and therefore

$$f_s(\underline{r}',\underline{r}) \to \frac{1}{N_s}$$

as required in (22).

The correlation factor $f_s(\underline{r}',\underline{r})$ attains a particularly simple form in cases where $\rho_s(\underline{r})$ varies relatively slowly around $\underline{r}$ within a range of radius $r_{so}(\underline{r})$. It is legitimate then to approximate $\rho_s(\underline{r}')$ by its first Taylor polynomial

$$\rho_s(\underline{r}') = \rho_s(\underline{r}) + (\underline{r}'-\underline{r})\cdot\underline{\nabla}\rho_s(\underline{r}) \tag{26}$$

which on insertion into the sum rule (25) yields

$$\rho_s(\underline{r}) \int \left[1 + \left(\frac{|\underline{r}'-\underline{r}|}{r_{so}(\underline{r})}\right)^2\right]^{-5/2} d^3r' = 1 \tag{27a}$$

The integral gives $\frac{4\pi}{3}r_{so}^3$, and hence $r_{so}(\underline{r})$ can in this case simply be calculated from

$$\frac{4\pi}{3}\rho_s(\underline{r})\, r_{so}^3(\underline{r}) = 1 \tag{27b}$$

which happens to be the definition of the Wigner radius. Because of our assumption (24b) the potential $U_{xc}^{(s)}(\underline{r})$ reduces to

$$U_{xc}^{(s)}(\underline{r}) = -\int \frac{\rho_s(\underline{r}')}{|\underline{r}'-\underline{r}|}\left(f_s(\underline{r}',\underline{r}) + \frac{1}{2}\rho_s(\underline{r})\frac{\delta f_s(\underline{r}',\underline{r})}{\delta\rho_s}\right) d^3r' \tag{28}$$

By using the approximation (26) we obtain

$$\int \frac{\rho_s(\underline{r}')}{|\underline{r}'-\underline{r}|} f_s(\underline{r}',\underline{r})\, d^3r' = \frac{4\pi}{3}\rho_s(\underline{r})\, r_{so}^2(\underline{r}) \tag{29a}$$

Because of (27b) only $r_{so}(\underline{r})$ depends directly on $\rho_s(\underline{r})$. we may therefore write

$$\frac{\delta f_s}{\delta\rho_s} = \frac{\partial f_s}{\partial r_{so}}\,\frac{\delta r_{so}(\rho_s)}{\delta\rho_s} \tag{29b}$$

where

$$\frac{\delta r_{so}(\rho_s)}{\delta\rho_s} = -\frac{1}{3}\frac{r_{so}(\underline{r})}{\rho_s(\underline{r})} \tag{29c}$$

and

$$\frac{\partial f_s}{\partial r_{so}} = \frac{5|\underline{r}'-\underline{r}|^2}{r_{so}^3(\underline{r})}\left[1 + \left(\frac{|\underline{r}'-\underline{r}|}{r_{so}(\underline{r})}\right)^2\right]^{-7/2} \tag{29d}$$

Inserting these results (29a-d) into (28) $U_{xc}^{(s)}(\underline{r})$ attains the form

$$U_{xc}^{(s)}(\underline{r}) = -\frac{3}{2}\alpha\left(\frac{3}{\pi}\right)^{1/3}[2\rho_s(\underline{r})]^{1/3} \tag{30}$$

where

$$\alpha = \frac{4}{9} \left(\frac{2\pi^2}{9}\right)^{1/3} = 0.58$$

This expression agrees - apart from a minor difference in the prefactor α - with the leading term of all versions of local spin density potentials used at present.

The numerical calculations to be discussed in Sect. 5 are based on the form (28) for $U_{xc}^{(s)}(\underline{r})$ where $f_s(\underline{r}',\underline{r})$ is given by (24a) and $r_{so}(\underline{r})$ is determined via the sum rule (27a). Only in deriving the functional derivative of $f_s(\underline{r}',\underline{r})$ we have made use of (29c) which is a consequence of the Taylor approximation (26) via eq. (27b). (It should be noted that (29b) and (29d) do not depend on this approximation.) However, it appears that the inaccuracy involved in calculating the functional derivative $\delta f_s/\delta\rho_s$ this way does not seriously affect the total expression for $U_{xc}^{(s)}(\underline{r})$.
We quote two reasons for this:

1. The contribution of the second term in (28) which contains $\delta f_s/\delta\rho_s$, is approximately only one third of that of the first term.

2. If one calculates $U_{xc}^{(s)}(\underline{r})$ according to (28) and rearranges this function such that it attains a form analogous to (30):

$$U_{xc}^{(s)}(\underline{r}) = -\frac{3}{2}\,\alpha_{eff}\,(\rho_s)\,\left(\frac{3}{\pi}\right)^{1/3}\,[2\rho_s(\underline{r})]^{1/3}$$

the resulting function $\alpha_{eff}(\rho_s)$ shows only a relatively weak variation within a range of ρ_s which is of physical importance. This is illustrated in Fig. 1 for the example of single atoms of Na and Cu. Since the numerical value of α_{eff} does not enter in the derivation of (29c), it is this weak variation of $\alpha_{eff}(\rho_s)$ which a posteriori justifies the calculation of $\delta f_s/\delta\rho_s$ on the basis of approximation (26).

Notwithstanding the above justification, $\delta f_s/\delta\rho_s$ has to be modified as $\underline{r}$ moves across the periphery of an atom, since - according to (22) - $\delta f_s/\delta\rho_s$ must asymptotically go to zero. This behaviour has been simulated in our calculations by multiplying the second term in (28) by a switch function

$$S(r) = 1 - [Q_s(r)-N_s+1]\;\theta(Q_s(\underline{r})-N_s+1)$$

where Q_s is defined

$$Q_s(r) = 4\pi \int_0^r \rho_s(r')r'^2 dr'$$

and $\theta(Q_s(r)-N_s+1)$ is the Heaviside step function. The results discussed in the ensuing section are therefore based on the following form of $U_{xc}^{(s)}$:

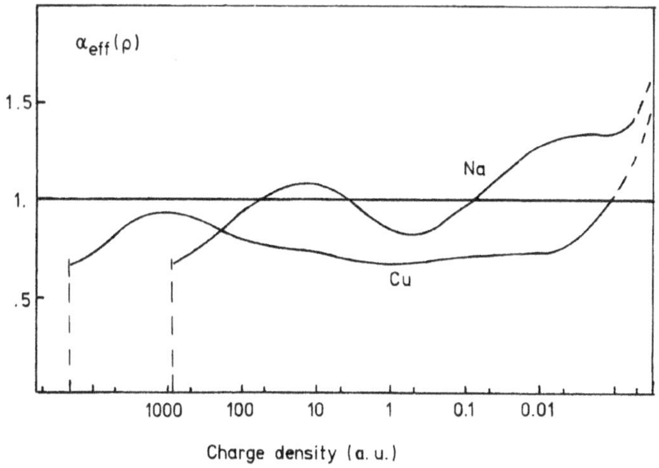

Fig. 1. Effective exchange factor $\alpha_{eff} = U_x^{(s)}/(\frac{3}{2}(\frac{6\rho_s}{\pi})^{1/3})$ (for majority spin) as a function of the charge density $2\rho_s$ for Na and Cu

$$U_{xc}^{(s)}(\underline{r}) = -\int \frac{\rho_s(\underline{r}')}{|\underline{r}'-\underline{r}|}[1 + \chi_s(\underline{r}',\underline{r})]^{-5/2} \times \left(1 - \frac{5}{6} S(Q_s) \frac{\chi_s(\underline{r}',\underline{r})}{1+\chi_s(\underline{r}',\underline{r})}\right) d^3r'$$

where

$$\chi_s(\underline{r}',\underline{r}) = \left(\frac{|\underline{r}'-\underline{r}|}{r_{so}(\underline{r})}\right)^2$$

It should be noted that $U_{xc}^{(s)}(\underline{r})$ is a local potential notwithstanding the fact that $r_{so}(\underline{r})$ has to be determined from an integral equation. E_{xc} has been calculated according to (15c,d) and (24a).

5. RESULTS

We have carried out self-consistent calculations for 13 first and second row atoms. Where it appeared to be useful for an assessment of the relative merits, we have listed our results together with those of the Hartree-Fock (HF-) theory and those based on the same one-particle equation and total energy expression as ours but $U_{xc}^{(s)}(\underline{r})$ and E_{xc} replaced by the analogous LSD-expressions derived from electron gas theory by v. Barth and Hedin.[11]

Table 1 shows the total energies of the atoms, indicating a small gain in accuracy of our results versus the HF-theory.

Table 1. Comparison of calculated total energies (in Ryd) with experimental[1)] and Hartree-Fock[2)] (HF) data

	this work	HF	exp.
B	-48.99	-49.06	-49.31
C	-75.33	-75.38	-75.71
N	-108.76	-108.80	-109.22
O	-149.63	-149.62	-150.22
F	-198.91	-198.82	-199.61
Ne	-257.22	-257.09	-258.11
Na	-323.84	-323.72	-324.88
Mg	-399.43	-399.23	-400.67
Al	-483.98	-483.75	-485.50
Si	-577.95	-577.71	-579.85
P	-681.65	-681.44	-684.05
S	-795.28	-795.01	-798.29
Cl	-919.25	-918.96	-923.03

1) Veillard and Clementi [14]
2) Clementi and Roetti [15]

Only the first three atoms, B, C and N are exceptions, reflecting a general deficiency of the approximate form (24a,b) of our correlation factors for few-electron systems. This might mainly be due to the complete neglect of a correlation hole for electrons of like spin. As has to be expected, the total energies are in general slightly above the experimental values. The kinetic energy is obviously somewhat too large, as a result of which the virial theorem is only fulfilled within an accuracy of 1 %. First ionization potentials are shown in Table 2. Except for Si and P, the HF-values are generally smaller than ours, whereas those obtained with LSD potentials are larger and show less satisfactory agreement with the experiments. Essentially the same can be said with reference to second ionization potentials listed in Table 3.

As already mentioned in Sect. 3 electron affinities are expected to demonstrate more clearly the progress achieved by our approach. Fig. 2 shows a comparison between our results, HF data[12] and recommended experimental values according to Hotop and Lineberger.[13] LSD potentials generally fail to describe stable negative ions.

We have finally calculated energy differences which occur on exchanging a 3d-electron with a 4s-electron in an atom of the 3d-series. The results are shown in Fig. 3 together with those obtained by Gunnarsson[10] and with experimental data. The comparison is clearly in favour of our results.

Table 2. Comparison of experimental[1] ionization potentials (eV) with results of Hartree-Fock[2] (HF) and LSD-theory and the present calculation

	this work	HF	LSD	exp.
B	7.92	7.93	9.02	8.30
C	10.97	10.8	12.20	11.26
N	14.03	13.9	15.43	14.54
O	12.87	11.9	14.46	13.61
F	17.06	15.7	18.60	17.42
Ne	20.91	19.9	22.69	21.56
Na	5.68	4.96	5.77	5.14
Mg	7.88	6.60	8.12	7.64
Al	5.70	5.50	6.42	5.98
Si	7.59	7.64	8.71	8.15
P	9.59	9.99	11.00	10.55
S	9.88	9.03	11.19	10.36
Cl	12.15	11.8	13.74	13.01

1) Moore [16]
2) Clementi and Roetti [15]

In conclusion we would like to emphasize once again that the one-particle equations (17a) together with (17b) and (21a,b) provide a rigorous scheme for the calculation of ground state properties of N-electron systems. The relative success of our simple approximation to the correlation factors $f_{s's}(\underline{r}',\underline{r})$, seems to suggest that any kind of greater sophistication within that concept might provide more progress than has been gained from electron gas theory during the past decade.

Table 3. Some second ionization potentials[1] (eV)

	this work	HF	LSD	exp.
B	23.02	23.4	25.00	25.15
C	24.05	24.0	25.15	24.38
N	29.25	29.2	30.61	29.61
O	34.38	34.6	36.00	35.15
F	34.07	33.3	35.76	34.98
Ne	40.27	39.3	42.14	41.07
Na	46.24	45.7	48.34	47.29

1) Same references as in table 2

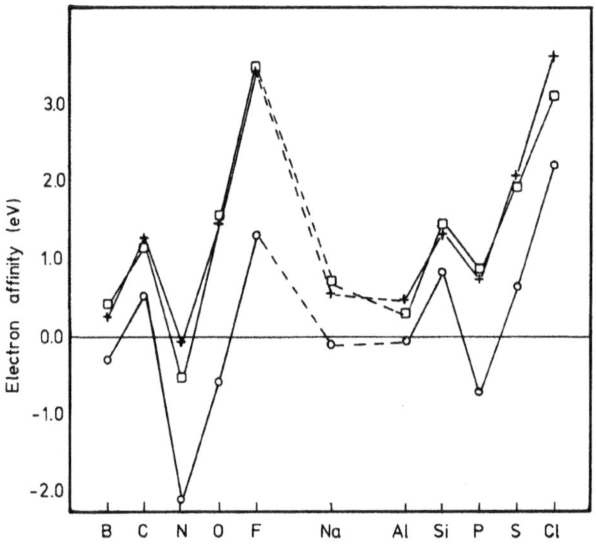

Fig. 2. Electron affinities of first and second row elements. Our results (□) are compared with Hartree-Fock data (○, Fraga e.a. [12]) and recommended experimental data (+, Hotop and Lineberger [13])

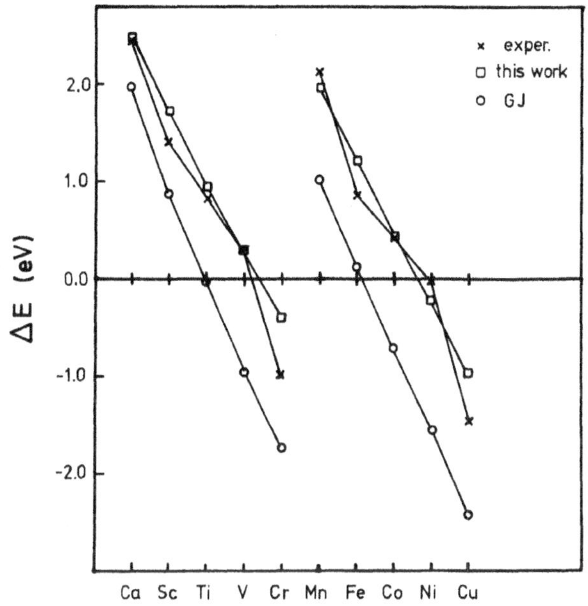

Fig. 3. Energy differences $\Delta E = E(3d^{n-1}4s^1) - E(3d^{n-2}4s^2)$ for the atoms of the 3d series. Our results (□) are compared with those of Gunnarsson and Jones (○ , [10]). The experimental values are weighted multiplet averages of the respective lowest lying states (x)

REFERENCES

1. D.R. Hartree, W. Hartree, and B. Swirles, Phil. Trans. Roy. Soc. (London) A 238, 229 (1939).
2. E. Wigner, Phys. Rev. 46, 1002 (1934).
3. P. Hohenberg and W. Kohn, Phys. Rev. 136, B 864 (1964).
4. W. Kohn and L.J. Sham, Phys. Rev. 140, A 1133 (1965).
5. O. Gunnarsson and B.I. Lundqvist, Phys. Rev. B 13, 4274 (1976).
6. J. Harris and R.O. Jones, J. Chem. Phys. 70, 830 (1979).
7. M.T. Yin and M.L. Cohen, Phys. Rev. Lett. 45, 1004 (1980).
8. M. Schlüter and L.J. Sham, Physics Today 35, 36 (1982).
9. R. McWeeny, Rev. Mod. Phys. 32, 335 (1960).
10. O. Gunnarsson and R.O. Jones, Physica Scripta 21, 394 (1980).
11. U. von Barth and L. Hedin, J. Phys. C 5, 1629 (1972).
12. S. Fraga, J. Karwowski, and K.M.S. Saxena, Handbook of Atomic Data, Elsevier Sci. Publ. Cy., Amsterdam (1976).
13. H. Hotop and W.C. Lineberger, J. Phys. Chem. Ref. Data 4, 539 (1975).
14. A. Veillard and E. Clementi, J. Chem. Phys. 49, 2415 (1968).
15. E. Clementi and C. Roetti, Atomic Data and Nuclear Data Tables 14, 177 (1974).
16. C.E. Moore, National Bureau of Standards, Circular 467 (1948).

PAIR CORRELATION ENERGIES AND LOCAL SPIN-DENSITY FUNCTIONALS

A. Savin, H. Stoll, and H. Preuss

Institut für Theoretische Chemie, Universität Stuttgart
D-7000 Stuttgart 80, West Germany

Some ways of defining pair correlation energies by means of local spin-density functionals are shown. Trends are deduced from numerical examples on atoms and molecules.

1. INTRODUCTION

It is well-known that the configuration interaction (CI) method and the local spin-density (LSD) approximation have different merits in the computation of the correlation energy. While the former shows the possibility of systematic improvement, the latter has the advantage of economy. A compromise between these two ways of obtaining the correlation energy is to do the calculation partly by CI, partly by LSD methods.

It is not clear, however, how to avoid a double counting of the correlation energy, when the CI and LSD values are added. A possible approach to this problem is the assumption that a limited multiconfiguration self-consistent field (MCSCF) calculation gives correlation energy contributions that are not included in certain LSD functionals. Such methods have the advantage that the MCSCF calculation can be kept to a very small number of configurations. Calculations on small molecules[1,2] have shown good agreement with experimental results. However, choosing the most important configurations in a way which does not interfere with a given LSD functional is a problem which has not been solved yet.

Another possible approach would be the partitioning of pair correlation energies between the CI and the LSD calculation; thus, it would be possible to gain accuracy for chemically interesting pairs, without fully neglecting the contribution of the other pairs. While the concept of pair energies in CI methods has been the object of thorough research (see, e.g. Refs. 3 and 4), there exists no similar definition in LSD theory.

This paper tries to show that reasonable definitions of pair energies can be made within the LSD formalism, and these definitions give pair energies with an accuracy which is comparable with that for total energies in the LSD approximation.

2. METHOD AND DEFINITIONS

The following notation will be used:

$$E_c[\rho] = \int \rho \varepsilon_c(\rho_+,\rho_-)\, dv$$

for the LSD correlation energy; $\rho(=\rho_+ + \rho_-)$ for the density, ρ_+,ρ_- for the partial densities of spin + and spin −; ε_c is the electron-liquid correlation energy per particle in the parametrization of Vosko et al.[5]

The contribution of a spinorbital, ψ_i, to the density is $\rho_i=\psi_i^2$; $i=1,\ldots,N$. The self-interaction correction[6] for the spinorbital ψ_i is e_i. The correlation energy of the pair of spinorbitals i and j is denoted by e_{ij}.

The following differences are used for the calculations of the e_{ij}:

$$D_i[\rho] = E_c[\rho] - E_c[\rho-\rho_i]$$

$$D_{ij}[\rho] = E_c[\rho] - E_c[\rho-\rho_i-\rho_j] \quad (i\neq j)$$

They show the effect of ρ_i (or $\rho_i+\rho_j$) when added to the "background" density $\rho-\rho_i$ (or $\rho-\rho_i-\rho_j$).

We assume that, in terms of e_{ij} and e_i, the correlation energy, E_c, and the differences D_i, D_{ij} can be written as:

$$E_c = 1/2 \sum_{k,l}{}' e_{kl} + \sum_k e_k \tag{1}$$

$$D_i = \sum_k{}' e_{ki} + e_i \tag{2}$$

$$D_{ij} = \sum_k{}' e_{ki} + \sum_k{}' e_{kj} - e_{ij} + e_i + e_j \tag{3}$$

where the summations are over occupied spinorbitals and the primes mark the omission of terms with equal indices.

If the definition of Perdew and Zunger[6] is used for the self-energies, $e_i = E_c[\rho_i]$, then the following pair energies and sums over pair energies can be obtained:

- the total correlation energy:[6]

$$1/2 \sum_{k,l}' e_{kl} = E_c[\rho] - \sum_k e_k \tag{4}$$

from Eq. (1); this definition will be denoted by "N";

- the "N-1" definition:

$$\sum_k' e_{ki} = D_i[\rho] - e_i \tag{5}$$

from Eq. (2);

- the "N-2" definition:

$$e_{ij} = -D_{ij}[\rho] + S_i + S_j + e_i + e_j \tag{6}$$

where

$$S_i = \sum_k' e_{ki} = (N-3)^{-1} \{\sum_k' D_{ki}[\rho] - S - \sum_k e_k - (N-2)\, e_i\}$$

$$S = \sum_{k,l}' e_{kl} = (2N-3)^{-1} \{ \sum_{k,l}' D_{kl}[\rho] - (2N-2) \sum_k e_k\}$$

from Eq. (3).

It is possible to avoid any assumption about the explicit form of the e_i with the following definition of the pair energies, which can be obtained by using Eqs. (2) and (3):

$$e_{ij} = D_i[\rho] + D_j[\rho] - D_{ij}[\rho] \tag{7}$$

this will be called the "N-1&N-2" definition.

Of course, it is possible to combine different definitions. For example, by subtracting the intra-pair energy calculated in the N-1&N-2 definition from the N-1 sum it is possible (and indeed more economical than by using the N-2 definition) to obtain the sum of inter-pair energies ("N-1 - N-1&N-2"). The valence energy can be summed up from pair energies, but there are also other schemes which require less computational effort, e.g. that of subtracting from the total energy (N definition) all the pair energies of the core orbitals (N-1 definition); as the core intra-pair energy is subtracted twice, this must be corrected by adding a term which can be deduced from the N-1&N-2 definition ("N - (N-1) + N-1&N-2").

As it can be expected that only a few LSD pair energies have to be replaced by CI pair energies, it is possible to subtract only those pair energies from the total LSD correlation energy ("N", Eq. (4)).

In a system of isolated pairs the following property holds, for all the definitions presented up to now: the total correlation energy of the whole system equals the sum of the e_{ij} for the individual pairs. Actually one may look at the definitions given above as extensions to the case of overlapping densities. Such extensions are already known from the definitions of self-interaction corrections.[6-8] The straightforward extension of the self-interaction correction of Perdew and Zunger[6] to the pair energies:

$$e_{ij} = E_c[\rho_i + \rho_j] - e_i - e_j$$

would give, however, a wrong number-of-particles dependence, in the case of the homogeneous electron liquid: the correlation energy per particle would increase as $N^{2/3}$. This is not the case for the definitions (4)-(7).

The total correlation energy per particle, calculated by summing up the pair energies of the definitions (5)-(7) remains finite, but does not give the exact $\varepsilon_c(\rho)$ in the case of the homogeneous electron liquid. This inconvenience might be removed by defining a new density functional with correct limiting behavior. One might ask on the other hand, whether it would not be better to abandon trying to correctly describe the homogeneous electron liquid, and introduce instead an empirical functional, adjusted for atoms and molecules. As the scope of this paper is merely to study the effect of Eqs. (5)-(7), none of these possibilities was taken into account. Based on the analysis of the electron-liquid behaviour of Eqs. (5)-(7) it is expected that the N-1 and the N-2 definitions give similar results, smaller than the N definition, but larger than the N-1&N-2 definition. This trend has been confirmed by the calculations which will be presented in this paper.

3. COMPUTATIONAL DETAILS

Experience in our group has shown that the modification of the Hartree-Fock orbitals by the correlation potential has little effect on the correlation energy computed in the LSD approximation. Thus, we have used the restricted Hartree-Fock (RHF) densities ρ and ρ_i, in order to determine the E_c, the D_i's, and the D_{ij}'s. The following basis sets were used in the RHF calculation:

- Huzinaga,[9] for Li-Ne,
- Roos et al.,[10] for Mg and Ar,
- Huzinaga,[9] with the contraction coefficients of Dunning and Hay,[11] and the polarization functions of the "small" basis of Ahlrichs et al.,[12] for the first-row hydrides.

The density functionals E_c, D_i, D_{ij} were evaluated by three-dimensional numerical integration.

The geometries of the hydrides were taken from Ref. (12).

For the purpose of discussing correlation energies, positive quantities will be considered ($-E_c, -e_{ij}$, etc.). All values will be given in 10^{-3} Hartree.

4. NUMERICAL TESTS

The dependence of the correlation energy on the atomic number (Z) is not well described by LSD methods in four-electron atoms.[13] We find, however, that the LSD pair energies show a certain tendency to reproduce CI characteristics:

- the 1s intra-pair energy (roughly constant in CI) increases less pronouncedly with Z, than the 2s-energy (nearly linear with A in CI[14]);

- the N-2 inter-pair energies almost coincide with the CI values.[15]

We now turn to the discussion of some closed-shell atoms (Be, Ne, Mg, Ar). The following trends of the second-order "basis limit" calculations[16,17] for intra-pair energies (Fig. 1) are well reproduced:

- the ordering of the pair energies for each atom;

- for the outer s orbitals (2s in Be, 3s in Mg) larger energies than for the inner ones (2s in Ne, Mg, and Ar, 3s in Ar);

- nearly constant energies for inner shell orbitals.

However, the LSD pair energies show some errors:

- the 1s values are systematically too large;

- the LSD np energies are too small; this is connected to the fact that the ns and np densities are rather similar, but the CI ns and np energies differ by a factor of 2.

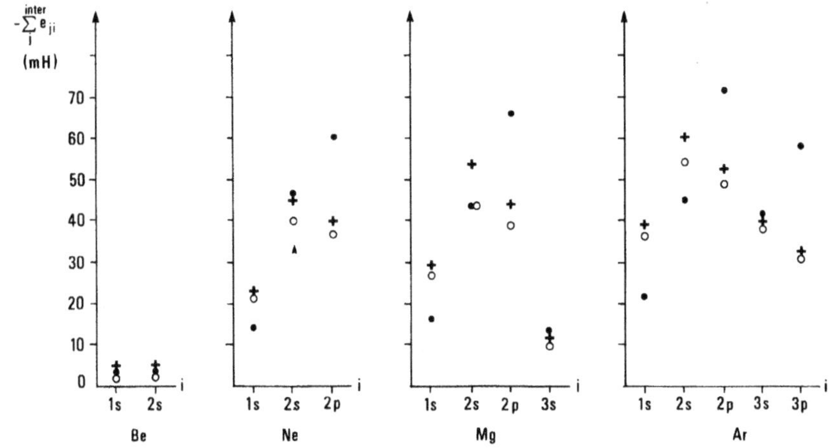

Figure 1. Intra-pair correlation energies in neutral atoms.
•: Second-order pair energies (from Ref. 16 for Be, extrapolated values for Ne, Mg, Ar[17]).
o: N-2.
x: N-1&N-2.

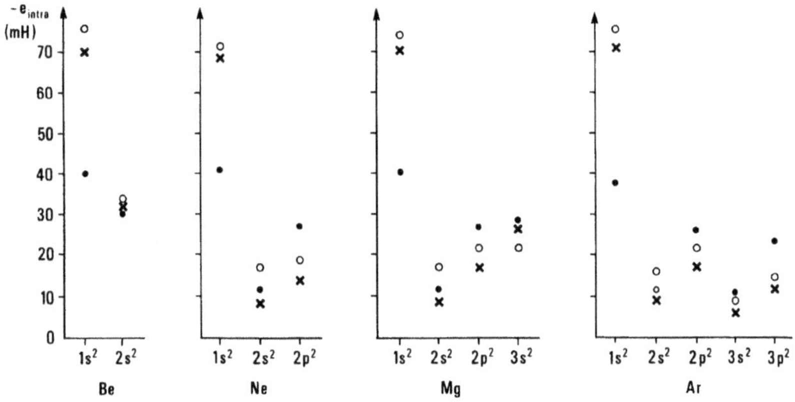

Figure 2. Sum of inter-pair correlation energies for spinorbital i.
•: Second-order ([16] for Be, [17] for Ne, Mg, Ar).
o: N-2
x: (N-1)-(N-1&N-2).

The case of Be is of special interest. It is known that the second-order values (even in the "basis-set limit", used in Fig. 1) cannot entirely reproduce the 2s correlation energy of Be, due to a degeneracy effect (see, e.g. [18]). Thus, it has to be stressed that the LSD functional has not reproduced the entire 2s intra-pair energy (46 mH[15]).

In Fig. 2 the sum of all inter-pair energies for orbital i (i=1s,2s,2p,3s,3p) is shown. The only larger discrepancy appears again in the case of the p orbitals, where the LSD results are too small.

We now present results for pair energies of some first row hydrides. The orbitals were localized using the method of Foster and Boys.[19]

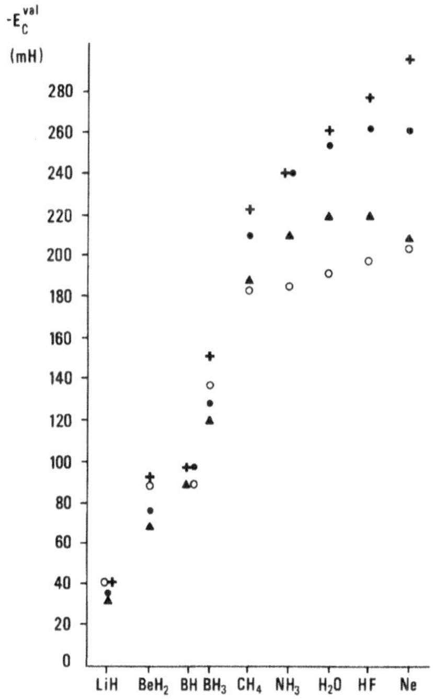

Figure 3. Valence shell correlation energies in localized representation.
.: CEPA ([21] for LiH, otherwise "standard" basis[12]).
Δ: CEPA ("small" basis[12] with exception of LiH: IEPA[20]).
O: N-2.
+: N-(N-1)+(N-1&N-2).

In Fig. 3 valence energies are plotted. The Ne atom is included, also in the localized representation. A comparison is made with coupled electron pair approximation (CEPA) results, obtained in two different calculations,[12] one described in [12] as being performed with the "small" basis set (of triple zeta plus polarization quality) and a larger one, with the "standard" basis set. For LiH values are taken from an independent electron pair approximation (IEPA)[20] and CEPA[21] calculation. The latter uses a larger basis set than the "standard" one.

It is perhaps interesting to note that in the ten-electron systems (CH_4, NH_3, H_2O, HF, Ne) the LSD method predicts an increase of the valence correlation energy with increasing Z, as does the better of the two CEPA calculations.

An analysis of the individual contributions are shown in the following figures. In Figs. 4 and 5 the intra-pair energies are shown for the bonding and non-bonding orbitals, respectively. The N-1&N-2 values are not shown because they parallel the N-2 values by a few mH. The LSD and CI trends are similar, as can be seen, e.g., by comparing the value of the B-H bonding orbital intra-pair energy in BH and BH_3. A discrepancy (of less than 10 mH) appears, however, in the compounds where the bonding orbital is localized on the H atom. The LSD definition yields values which are close to the H^- value (42 mH[22]), while CEPA predicts lower values.

The sum of the singlet and triplet CEPA inter-pair energies are compared in Fig. 6 to the corresponding LSD values. The LSD values are too small, but the correct order is predicted; in particular, it is correctly described that the correlation energy is smaller when bonding orbitals are involved than in the case of non-bonding orbitals.

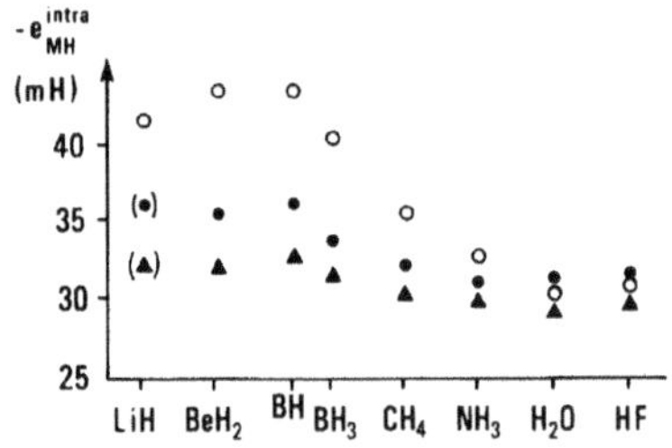

Figure 4. Intra-pair correlation energies in MH_n molecules; the M-H localized molecular orbital.
.: CEPA ([21] for LiH, otherwise "standard" basis[11]).
Δ: IEPA ([20] for LiH, otherwise "small" basis[11]).
o: N-2.

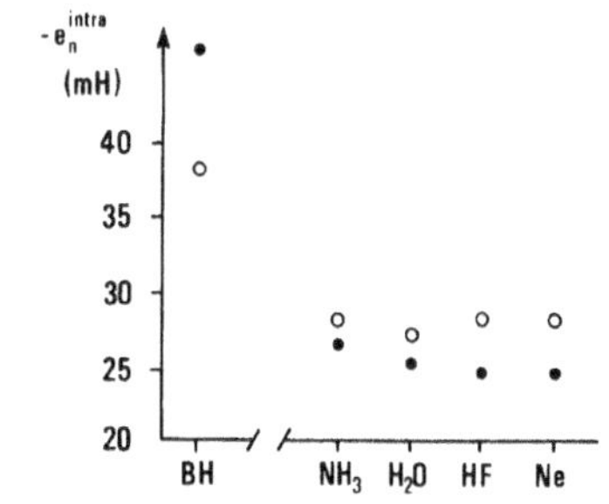

Figure 5. Intra-pair correlation energies for non-bonding localized molecular orbitals.
.: CEPA ("standard" basis[12]).
o: N-2.

5. CONCLUSION

In all cases studied so far, it has been observed that trends are correctly predicted by the LSD pair correlation energies defined in this paper. While the possibility of refining the density functional leaves these definitions open to further improvement, it may be stated that a combination of CI and LSD pair energies seems to be feasible.

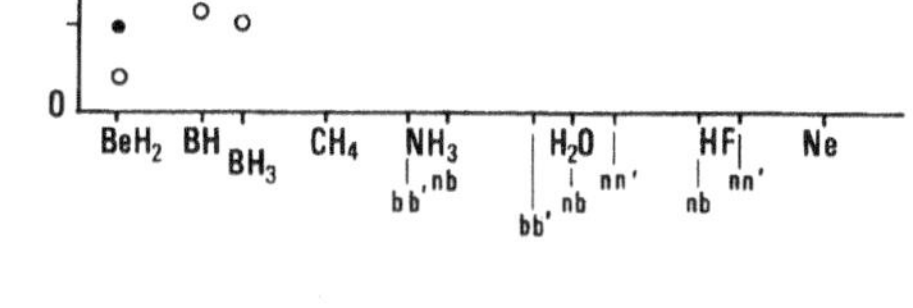

Figure 6. Sum of singlet and triplet inter-pair correlation energies, between localized pairs (b: bonding, n: non-bonding).
.: CEPA ("standard" basis[12]).
o: N-2.

ACKNOWLEDGEMENT

One of us (A.S.) is grateful to the Fonds der Chemischen Industrie for financial support.

REFERENCES

1. G.C. Lie and E. Clementi, J.Chem.Phys. 60, 1275, 1288 (1974).
2. R. Montagnani, P. Riani, and O. Salvetti, Theor.Chim.Acta 60, 399 (1982).
3. W. Kutzelnigg, in "Methods of Electronic Structure Theory", H.F. Schaefer III, Ed. (Plenum, New York, 1977), p. 129.
4. W. Meyer, in "Methods of Electronic Structure Theory", H.F. Schaefer III, Ed. (Plenum, New York, 1977), p. 413.
5. S.H. Vosko, L. Wilk, and M. Nusair, Can.J.Phys. 58, 1200 (1980).
6. J.P. Perdew and A. Zunger, Phys.Rev. B23, 5048 (1981).
7. H. Stoll, C.M.E. Pavlidou, and H. Preuss, Theor.Chim.Acta 49, 143 (1978).
8. H. Stoll, E. Golka, and H. Preuss, Theor.Chim.Acta 55, 29 (1980).
9. S. Huzinaga, J.Chem.Phys. 42, 1293 (1965).
10. B. Roos, A. Veillard, and G. Vinot, Theor.Chim.Acta 20, 1 (1971).
11. T.H. Dunning and P.J. Hay, in "Methods of Electronic Structure Theory", H.F. Schaefer III, Ed. (Plenum, New York, 1977), p. 1.
12. R. Ahlrichs, F. Driessler, H. Lischka, V. Staemmler, and W. Kutzelnigg, J.Chem.Phys. 62, 1235 (1975).
13. J.P. Perdew, E.R. McMullen, and A. Zunger, Phys.Rev. A23, 2785 (1981).
14. J. Linderberg and H. Shull, J.Mol.Spectr. 5, 1 (1960).
15. G.A. Petersson and S.L. Licht, J.Chem.Phys. 75, 4556 (1981).
16. T.P. Eggarter and E. Eggarter, J.Phys. B: Atom.Molec.Phys. 11, 1157 (1978).
17. K. Jankowski, P. Malinowski, and M. Polasik, J.Phys. B: Atom. Molec.Phys. 12, 3157 (1979).
18. T.P. Eggarter and E. Eggarter, J.Phys. B: Atom.Molec.Phys. 11, 3635 (1978).
19. J.M. Foster and S.F. Boys, Rev.Mod.Phys. 32, 300 (1960).
20. M. Jungen and R. Ahlrichs, Theor.Chim.Acta 17, 339 (1970).
21. W. Meyer and P. Rosmus, J.Chem.Phys. 63, 2356 (1975).
22. E. Clementi, J.Chem.Phys. 38, 2252 (1962).

ON THE EXTENSION OF THE STATISTICAL ENERGY EXPRESSION TO MULTI-DETERMINANTAL WAVE FUNCTIONS

Tom Ziegler

Department of Chemistry
University of Calgary
Calgary, Alberta, T2N 1N4
Canada

1. INTRODUCTION

It is shown in section 2 of this chapter, based on previous work by Ziegler, Rauk and Baerends,[1] that the statistical energy expression due to Slater[2] only can be justified by the statistical exchange approximation for the special case, where the n-electron wave function is represented by a single Slater determinant.

This is indeed a severe limitation, since a variety of phenomena in molecular physics as well as solid state physics require a description of the n-electron wave function in terms of multi-determinantal wave functions, even at the lowest level of approximation.

An extension[3] of the statistical energy to multi-determinantal wave function is however possible, if the energy expression of the multi-determinantal wave function with respect to the exact Hamiltonian can be brought on a form where it appears as a sum of energies due to Slater determinants. The extension is discussed in section 3 for electronic multiplets and in section 4 for correlated wave functions of use in a description of bond dissociation.

2. THE STATISTICAL ENERGY EXPRESSION AND THE STATISTICAL EXCHANGE APPROXIMATION

The non-relativistic n-electron Hamiltonian is given in atomic units by:

$$H^{o} = \sum_{i} h(\vec{r}_i) + \frac{1}{2}\sum_{i}\sum_{j}{}' g(r_{ij}) + \frac{1}{2}\sum_{a}\sum_{b}{}' Z_a Z_b / |\vec{R}_a - \vec{R}_b| \qquad (2.1)$$

Here

$$h(r_i) = -\frac{1}{2}\sum_i \nabla_i^2 - \sum_i\sum_a Z_a/|\vec{r}_i-\vec{R}_a| \tag{2.2}$$

is the sum of the operators representing the electronic-kinetic energy and the electron-nucleus attraction. Further:

$$\frac{1}{2}\sum_i\sum_j{}' g(r_{ij}) = \frac{1}{2}\sum_i\sum_j{}' 1/|\vec{r}_i-\vec{r}_j| \tag{2.3}$$

accounts for the electron-electron repulsion. The last term in Eq. (2.1) is due to the nucleus-nucleus interaction. The total energy of an n-electron wave function $\Psi(\vec{\tau}_1, \vec{\tau}_2,\ldots,\vec{\tau}_n)$ with respect to the Hamiltonian H^o can be expressed in a compact way in terms of the first order density matrix[4]

$$\rho(\vec{\tau}_1,\vec{\tau}_1') = n\int\Psi(\vec{\tau}_1',\vec{\tau}_2,\vec{\tau}_3,\ldots,\vec{\tau}_n)\ \Psi^*(\vec{\tau}_1,\vec{\tau}_2,\vec{\tau}_3,\ldots,\vec{\tau}_n)d\vec{\tau}_2 d\vec{\tau}_3\ldots d\vec{\tau}_n \tag{2.4}$$

and the second order density matrix[4]

$$\rho_2(\vec{\tau}_1,\vec{\tau}_2|\vec{\tau}_1',\vec{\tau}_2') = n(n-1)\int\Psi(\vec{\tau}_1',\vec{\tau}_2',\vec{\tau}_3,\ldots,\vec{\tau}_n) \times\Psi^*(\vec{\tau}_1,\vec{\tau}_2,\vec{\tau}_3,\ldots,\vec{\tau}_n)d\vec{\tau}_3 d\vec{\tau}_n \tag{2.5}$$

as

$$E^o = \int_{\tau_1'\to\tau_1} \rho_1(\vec{\tau}_1',\vec{\tau}_1)h(\vec{r}_1)d\vec{\tau}_1 + \frac{1}{2}\int\rho_2(\vec{\tau}_1,\vec{\tau}_2)g(r_{12})d\vec{\tau}_1 d\vec{\tau}_2 + V_{nn} \tag{2.6}$$

Here $\rho_2(\vec{\tau}_1,\vec{\tau}_2)$ is the diagonal part of $\rho_2(\vec{\tau}_1',\vec{\tau}_2'|\vec{\tau}_1,\vec{\tau}_2)$. The nucleus-nucleus interaction of Eq. (2.1) is written in Eq. (2.6) as V_{nn}. Both $\rho_1(\vec{\tau}_1',\vec{\tau}_1)$ and $\rho_2(\vec{\tau}_1,\vec{\tau}_2)$ are given in terms of the generalized coordinate $\vec{\tau}_1$ for an electron at position $\vec{r}_i$ with the spin coordinate $\vec{s}_i$.

It will be of use for the subsequent discussion to write both matrices in terms of their spin part and space part. We have[4]:

$$\rho_1(\vec{\tau}_1',\vec{\tau}_1) = \rho_1^{\alpha}(\vec{r}_1',\vec{r}_1)\alpha(\vec{s}_1')\alpha(\vec{s}_1) + \rho_1^{\beta}(\vec{r}_1',\vec{r}_1)\beta(\vec{s}_1')\beta(\vec{s}_1) \tag{2.7}$$

and

$$\rho_2(\vec{\tau}_1,\vec{\tau}_2) = \rho_2^{\alpha\alpha}(\vec{r}_1,\vec{r}_2)\alpha^2(\vec{s}_1)\alpha^2(\vec{s}_2) + \rho_2^{\alpha\beta}(\vec{r}_1,\vec{r}_2)\alpha^2(\vec{s}_1)\beta^2(\vec{s}_2) + \rho_2^{\beta\alpha}(\vec{r}_1,\vec{r}_2)\beta^2(\vec{s}_1)\alpha^2(\vec{s}_2) + \rho_2^{\beta\beta}(\vec{r}_1,\vec{r}_2)\beta^2(\vec{s}_1)\beta^2(\vec{s}_2) \tag{2.8}$$

Here $\rho_1^{\alpha}(\vec{r}_1)$ and $\rho_1^{\beta}(\vec{r}_1)$ are the probabilities of finding an electron at $\vec{r}_1$ with α-spin and β-spin respectively. Further, $\rho_2^{\alpha\alpha}$, $\rho_2^{\beta\beta}$ give the probability of finding two electrons of the same spin (α and β, respectively) at $(\vec{r}_1,\vec{r}_2)$, and $\rho_2^{\alpha\beta}$, $\rho_2^{\beta\alpha}$ the probability of finding two electrons of different spin at $(\vec{r}_1,\vec{r}_2)$. We have in Eq. (2.7) and Eq. (2.8) omitted terms containing $\alpha(\vec{s}_1)\beta(\vec{s}_1)$ and $\alpha(\vec{s}_2)\beta(\vec{s}_2)$, since their contribution to the total energy with respect to the spin free Hamiltonian H^o is zero.

Slater[2] has suggested an approximation to Eq. (2.6) in which the energy only depends on ρ_1. This energy expression is given by:

$$E[\rho_1] = \int_{\tau_1'\to\tau_1} \rho(\vec{\tau}_1',\vec{\tau}_1)h(\vec{r}_1)d\vec{\tau}_1 + \frac{1}{2}\int\rho_2(\vec{\tau}_1)\rho_1(\vec{\tau}_2)g(r_{12})d\vec{\tau}_1 d\vec{\tau}_2 + V_{nn}$$

$$- \frac{3}{4}k\int[\rho_1^{\alpha}(\vec{r}_1)^{4/3}]d\vec{r}_1 - \frac{3}{4}k\int[\rho_1^{\beta}(\vec{r}_1)^{4/3}]d\vec{r}_1 \qquad (2.9)$$

Here $k = 3\alpha_{ex}(3/4\pi)^{1/3}$. The factor α_{ex} is an adjustable parameter (the exchange scale factor) with a value between 2/3 and 1.0.

The reduction of E^o to $E[\rho_1]$ was justified by Slater using the statistical exchange approximation (see later), and Slater referred to $E[\rho_1]$ as the statistical energy expression. The approximate energy $E[\rho_1]$ has been the basis for the X_α - scheme or the Hartree-Fock-Slater (HFS) method implemented during the last decade[5,9] on digital computers for calculations on molecules. It is however not in general realized that the use of the statistical exchange approximation only will reduce E^o to $E[\rho_1]$ in the rather limited case where the n-electron wave function Ψ is approximated by a single Slater determinant.

Let us, in order to see this, substitute Eq. (2.8) into Eq. (2.9). A little manipulation provides the following expression suitable for a discussion of the statistical energy and the statistical exchange approximation:

$$E^o = \int_{\tau_1'\to\tau_1} \rho_1(\vec{\tau}_1',\vec{\tau}_1)h(\vec{r}_1)d\vec{\tau}_1 + \frac{1}{2}\int\rho_1(\vec{\tau}_1)\rho_1(\vec{\tau}_2)g(r_{12})d\vec{\tau}_1 d\vec{\tau}_2 + V_{nn}$$

$$+ \frac{1}{2}\int\rho_1^{\alpha}(\vec{r}_1)\rho_x^{\alpha\alpha}(\vec{r}_1,\vec{r}_2)\alpha^2(\vec{s}_1)\alpha^2(\vec{s}_2)g(r_{12})d\vec{\tau}_1 d\vec{\tau}_2$$

$$+ \frac{1}{2}\int\rho_1^{\beta}(\vec{r}_1)\rho_x^{\beta\beta}(\vec{r}_1,\vec{r}_2)\beta^2(\vec{s}_1)\beta^2(\vec{s}_2)g(r_{12})d\vec{\tau}_1 d\vec{\tau}_2$$

$$+ \frac{1}{2}\int\rho_1^{\alpha}(\vec{r}_1)\rho_x^{\alpha\beta}(\vec{r}_1,\vec{r}_2)\alpha^2(\vec{s}_1)\beta^2(\vec{s}_2)g(r_{12})d\vec{\tau}_1 d\vec{\tau}_2$$

$$+ \frac{1}{2}\int\rho_1^{\beta}(\vec{r}_1)\rho_x^{\beta\alpha}(\vec{r}_1,\vec{r}_2)\beta^2(\vec{s}_1)\alpha^2(\vec{s}_2)g(r_{12})d\vec{\tau}_1 d\vec{\tau}_2 \qquad (2.10)$$

Here

$$\rho_x^{\alpha\alpha}(\vec{r}_1,\vec{r}_2) = \frac{\rho_2^{\alpha\alpha}(\vec{r}_1,\vec{r}_2)}{\rho_1^{\alpha}(\vec{r}_1)} - \rho_1^{\alpha}(\vec{r}_2) \tag{2.11}$$

and

$$\rho_x^{\alpha\beta}(\vec{r}_2,\vec{r}_2) = \frac{\rho_2^{\alpha\beta}(\vec{r}_1,\vec{r}_2)}{\rho_1^{\alpha}(\vec{r}_1)} - \rho_1^{\beta}(\vec{r}_2) \tag{2.12}$$

Analogous expressions are obtained for the function $\rho_x^{\beta\beta}$ by replacing α with β in Eq. (2.11), and for $\rho_x^{\beta\alpha}$ by permuting α and β in Eq. (2.12). The statistical exchange expression by Slater involves the terms in Eq. (2.10) containing $\rho_x^{\alpha\alpha}$ and $\rho_x^{\beta\beta}$. The specific expression for the exchange approximation is given by:[1,2]

$$\frac{1}{2}\int\rho_1^{\alpha}(\vec{r}_1)\rho_x^{\alpha\alpha}(\vec{r}_1,\vec{r}_2)g(r_{12})d\vec{r}_1 d\vec{r}_2 = -\frac{3}{4}k\int\rho_1^{\alpha}(\vec{r}_1)[\rho_1^{\alpha}(\vec{r}_1)^{1/3}]d\vec{r}_1 \tag{2.13}$$

with a similar expression involving $\rho_x^{\beta\beta}$.

The substitution of Eq. (2.10) now provides:

$$E \cong E[\rho_1] + E^{\alpha\beta} \tag{2.14}$$

The first term in Eq. (2.14) is simply the statistical energy expression of Eq. (2.9) with the second term given as:

$$\begin{aligned} E^{\alpha\beta} &= \frac{1}{2}\int\rho_1^{\alpha}(\vec{r}_1)\rho_1^{\alpha\beta}(\vec{r}_1,\vec{r}_2)g(r_{12})\alpha^2(\vec{s}_1)\beta^2(\vec{s}_2)d\vec{\tau}_1 d\vec{\tau}_2 \\ &+ \frac{1}{2}\int\rho_1^{\beta}(\vec{r}_1)\rho_x^{\beta\alpha}(\vec{r}_1,\vec{r}_2)g(r_{12})\beta^2(\vec{s}_1)\alpha^2(\vec{s}_2)d\vec{\tau}_1 d\vec{\tau}_2 \end{aligned} \tag{2.15}$$

The energy of Eq. (2.14) is not identical to the statistical energy $E[\rho_1]$ unless $E^{\alpha\beta}$ is zero. This is the case for a single determinantal wave function, where one has:

$$\rho_2^{\alpha\beta}(\vec{r}_1,\vec{r}_2) = \rho_1^{\alpha}(\vec{r}_1)\rho_1^{\beta}(\vec{r}_2), \rho_2^{\beta\alpha}(\vec{r}_1,\vec{r}_2) = \rho_1^{\beta}(\vec{r}_1)\rho_1^{\alpha}(\vec{r}_2) \tag{2.16}$$

and correspondingly $\rho_x^{\alpha\beta} = \rho_x^{\beta\alpha} = 0$.

The term $E^{\alpha\beta}$ is in general not zero for multi-determinantal wave functions representing multiplets or correlated wave functions. It is in those cases as already mentioned not consistent with the statistical exchange approximation to calculate the statistical energy from a simple substitution of ρ_1 into $E[\rho_1]$ of Eq. (2.9).

We shall however in the next two sections show, that it is possible to evaluate the statistical energy for correlated wave functions as well as multiplets by other means.

3. THE STATISTICAL ENERGY EXPRESSION FOR ELECTRONIC MULTIPLETS

Electronic multiplets are in general represented by a sum of determinantal wave functions, thus $\rho_x^{\alpha\beta}$ and $\rho_x^{\beta\alpha}$ of Eq. (2.12) will be non-zero. The statistical energy for the multiplet can as a consequence in those cases not be evaluated by substituting the related first order density matrix ρ_1 into Eq. (2.9).

To see this let us consider the simple electronic configuration $(a)^1(b)^1$ where one electron is placed in each of the two non-degenerate levels a,b. We can construct four determinantal wave functions corresponding to this configuration as:

$$D_1 = |\overset{+}{a}\,\overset{+}{b}|,\quad D_2 = |\bar{a}\,\bar{b}|,\quad D_3 = |\overset{+}{a}\,\bar{b}|,\quad D_4 = |\bar{a}\,\overset{+}{b}|$$

where (+,-) represent α- and β-spin respectively. The configuration give rise to one triplet and one singlet with the spin eigenfunctions given by:

$${}^3_1\Psi = D_1,\quad {}^3_0\Psi = \sqrt{\tfrac{1}{2}}\,\{D_3+D_4\},\quad {}^3_{-1}\Psi = D_2,\quad {}^1_0\Psi = \sqrt{\tfrac{1}{2}}\,\{D_3-D_4\}$$

The energy expression for the singlet ${}^1_0\Psi$ and the triplet ${}^3_0\Psi$ according to Eq. (2.10) will have contributions from $\rho_x^{\alpha\beta}$, $\rho_x^{\beta\alpha}$. Thus $E^{\alpha\beta}$ of Eq. (2.14) will be different from zero. We shall not attempt a direct evaluation of $E^{\alpha\beta}$ in order to resolve the triplet-singlet separation. Rather, the energy expressions for the singlet and the triplet will be reformulated in a way more suitable for a calculation of the statistical energy of the two states as discussed next.

The energies of the four determinantal wave functions with respect to the Hamiltonian H^o of Eq. (2.1), $E^o(D_1)$, $E^o(D_2)$, $E^o(D_3)$ and $E^o(D_4)$, and the corresponding triplet- and singlet energy, E^o_T,E^o_S, are related in the following way:

$$E^o(D_1) = E^o(D_2) = E^o_T\ ;\quad E^o(D_3) = E^o(D_4) = \frac{1}{2}E^o_S + \frac{1}{2}E^o_T \tag{3.1}$$

We can in turn express both the singlet energy E^o_S and the triplet Energy E^o_T as a sum of energies due to single determinantal wave functions. Thus by solving the system of equations (3.1), the following relations result:

$$E^o_S = 2E^o(D_3) - E^o(D_1);\quad E^o_T = E^o(D_1) \tag{3.2}$$

The corresponding statistical energy expression for the triplet as well as the statistical energy expression of the singlet is now well defined, since both the triplet energy expression of Eq. (3.2) and the singlet energy expression of Eq. (3.2) is in terms of energies due to single determinantal wave functions, thus:

$$E_S = 2E(D_3) - E(D_1) \tag{3.3}$$

$$E_T = E(D_1) \tag{3.4}$$

In Eq. (3.3) and Eq. (3.4) E_S, E_T represent the statistical energies of the singlet and the triplet respectively and $E(D_i)$ the statistical energy of the single determinantal wave function D_i.

Calculations[1] according to the above procedure on triplet energies and singlet energies (relative to the ground states) for the first excited states of H_2CO, C_2H_2, C_4H_6, CO and N_2 are presented in Table 1. The comparison with experiment is in general good. It should be mentioned that separate pairs of orbitals (a,b) can be obtained for the singlet and the triplet by separate optimizations of the two energy expressions given in Eq. (3.3) Eq. (3.4) respectively.

Consider next the general case of a multiplet, where the energies of all possible determinantal wave functions D_i with respect to the Hamiltonian H^o are given by $E^o(D_i)$ and the multiplet energies by H_j^o. The two sets of energy expressions are related by:

$$E^o(D_i) = \sum_j C_{ij} H_j^o \tag{3.5}$$

where the coefficients C_{ij} in most cases are given by group theory. The procedure for calculating multiplet energies by the statistical energy expression is now to invert the set of equations (3.5) to:

$$H_j^o = \sum_i A_{ji} E^o(D_i) \tag{3.6}$$

so that the expression for each multiplet energy appears as a sum of energies due to single determinantal wave functions. By evaluating the well defined statistical energy of each of the determinants, $E(D_i)$, the statistical energy expression for the multiplet takes the form:

$$H_j = \sum_i A_{ji} E(D_i) \tag{3.7}$$

A more detailed account of the method, and a discussion of the condition under which Eq. (3.5) can be inverted to Eq. (3.6), is given in Ref. 1. The procedure has also been used by von Barth[10] and Gunnarsson et. al.[11] to calculate the multiplet energies of several atomic systems.

Table 1. Calculation[a] of triplet and singlet transition energies (eV)

Compound	Transition	E_T	E_T^{exp}	E_S	E_S^{exp}
H_2CO	$2b_2 \rightarrow 2b_1$	3.18	3.21[b]	3.70	3.50[b]
C_2H_2	$\pi \rightarrow \pi^*$	4.57	4.60	7.45	7.6
C_4H_6	$1b_g \rightarrow 2a_u$	3.31	3.22	5.05	5.9
CO	$5\sigma \rightarrow 2\pi$	5.81	6.3	7.99	8.4
N_2	$3\sigma_g \rightarrow 1\pi_g$	7.83	8.1	9.13	9.3
N_2	$2\sigma_u \rightarrow 1\pi_g$	10.68	11.1	12.34	12.8

[a]References to experimental values are given in Ref. 1
[b]From Ref. 1.

We conclude that the statistical energy of a multi-determinantal wave function corresponding to an electronic multiplet can be evaluated by first bringing the energy expression for the multiplet on a footing where it appears as a sum of energy expressions due to single determinantal wave functions. We shall show in the next section that the same procedure can be used to calculate the energies for some types of correlated wave functions.

4. THE STATISTICAL ENERGY EXPRESSION FOR PAIRED EXCITATION MULTICONFIGURATION WAVE FUNCTIONS

It is well known that a single determinantal wave function, although adequate for molecules in their equilibrium geometry is insufficient for a discussion of bond formation and bond dissociation. Goddard et al.[12] have in connection with their Generalized Valence Bond method (GVB) shown, that one needs wave functions of the paired excitation multiconfiguration type (PEMC) in order to give a proper description of bond formation and bond dissociation.

A general PEMC wave function can be written as:

$$\Psi = \sum_L C_L \Psi_L \tag{4.1}$$

where each determinant Ψ_L consists of one or more pairs of orbitals $(\phi_j, \bar{\phi}_j)$, as well as a number of unpaired orbitals with spin up, $\overset{+}{\phi}_i$, and spin down, $\bar{\phi}_l$, respectively. Thus the determinant Ψ_L can be expressed as:

$$\Psi_L = |\overset{+}{\phi}_1\bar{\phi}_1 \ldots \overset{+}{\phi}_j\bar{\phi}_j \ldots \overset{+}{\phi}_i \ldots \bar{\phi}_l \ldots| \tag{4.2}$$

The additional determinants in Eq. (4.2) are generated from Ψ_L by a substitution of one or more of the pairs $(\overset{+}{\phi}_i, \bar{\phi}_i)$ by different pairs of couples $(\overset{+}{\phi}_\nu, \bar{\phi}_\nu)$, whereas the unpaired orbitals are common to all determinants in Eq. (4.2). We shall now show that it is possible to derive a statistical energy expression for a PEMC wave function. The energy for the PEMC wave function Ψ of Eq. (4.1), with respect to the exact Hamiltonian H^o of Eq. (2.1), can be written as:

$$E^o = \sum_L C_L^2 <\Psi_L|H|\Psi_L> + \sum_{KL}\sum C_K' C_L <\Psi_K|H|\Psi_L> + V_{nn} \tag{4.3}$$

The statistical energy expression corresponding to the first sum in Eq. (4.3) is already well defined, since each term refers to the energy of a single Slater determinant. Thus, $<\Psi_L|H|\Psi_L> = E[\rho_L]$, where ρ_L is the first order density matrix derived from Ψ_L. We shall now as the next step show, that the second term in Eq. (4.3) too can be written as a sum of energies due to Slater determinants.

Each individual term $<\Psi_L|H|\Psi_K>$ in Eq. (4.3) is zero for a PEMC wave function unless Ψ_L and Ψ_K differ in only one pair of orbitals. In that case, $\Psi_L = |\{\alpha\}\{\beta\}\overset{+}{\phi}_\nu\bar{\phi}_\nu|$, and $\Psi_K = |\{\alpha\}\{\beta\}\overset{+}{\phi}_\mu\bar{\phi}_\mu|$, where $\{\alpha\}$ is the set of orbitals with α-spin common to both Ψ_L and Ψ_K, and $\{\beta\}$ the corresponding set with β-spin. It is now easily shown after introducing the four determinants:

$$\Phi_1 = |\{\alpha\}\{\beta\}\overset{+}{\phi}_\nu\overset{+}{\phi}_\mu|; \quad \Phi_2 = |\{\alpha\}\{\beta\}\bar{\phi}_\nu\bar{\phi}_\mu|$$

$$\Phi_3 = |\{\alpha\}\{\beta\}\overset{+}{\phi}_\nu\bar{\phi}_\mu|; \quad \Phi_4 = |\{\alpha\}\{\beta\}\bar{\phi}_\nu\overset{+}{\phi}_\mu|$$

that:

$$<\Psi_L|H|\Psi_K> = \frac{1}{2}<\Phi_3|H|\Phi_3> + \frac{1}{2}<\Phi_4|H|\Phi_4> - \frac{1}{2}<\Phi_2|H|\Phi_2> - \frac{1}{2}<\Phi_1|H|\Phi_1> \tag{4.4}$$

Equation (4.4) now has the desired form. Thus in the statistical exchange approximation indicated by the subscript (ex), one has:

$$<\Psi_L|H|\Psi_K>_{ex} = \frac{1}{2}E[\rho_3] + \frac{1}{2}E[\rho_4] - \frac{1}{2}E[\rho_1] - \frac{1}{2}E[\rho_2] \tag{4.5}$$

with ρ_1, ρ_2, ρ_3 and ρ_4 as the first order density matrices due to Φ_1, Φ_2, Φ_3 and Φ_4 respectively. By using the definition of $E[\rho]$ given in Eq. (2.9), a collection of terms in Eq. (4.5) provides the following explicit expression for $<\Psi_L|H|\Psi_K>_{ex}$:

$$\langle\Psi_L|H|\Psi_K\rangle_{ex} =$$

$$\sum_{\sigma} \frac{3}{8} k\{\int|\rho_T^{\sigma}(\vec{r}_1) + \frac{1}{2}[\phi_{\mu}(\vec{r}_1)\phi_{\mu}(\vec{r}_1) + \phi_{\nu}(\vec{r}_1)\phi_{\nu}(\vec{r}_1)]^{4/3}|d\vec{\tau}_1$$

$$+ \int|\rho_T^{\sigma}(\vec{r}_1) - \frac{1}{2}[\phi_{\mu}(\vec{r}_1)\phi_{\mu}(\vec{r}_1) + \phi_{\nu}(\vec{r}_1)\phi_{\nu}(\vec{r}_1)]^{4/3}|d\vec{\tau}_1$$

$$- \int|\rho_T^{\sigma}(\vec{r}_1) + \frac{1}{2}[\phi_{\mu}(\vec{r}_1)\phi_{\mu}(\vec{r}_1) - \phi_{\nu}(\vec{r}_1)\phi_{\nu}(\vec{r}_1)]^{4/3}|d\vec{\tau}_1$$

$$- \int|\rho_T^{\sigma}(\vec{r}_1) - \frac{1}{2}[\phi_{\mu}(\vec{r}_1)\phi_{\mu}(\vec{r}_1) - \phi_{\nu}(\vec{r}_1)\phi_{\nu}(\vec{r}_1)]^{4/3}|d\vec{\tau}_1 \qquad (4.6)$$

In Eq. (4.6) the index σ runs over α-spin as well as β-spin. Further,

$$\rho_T^{\alpha}(\vec{r}_1) = \frac{1}{2}(\rho_L^{\alpha}(\vec{r}_1) + \rho_K^{\alpha}(\vec{r}_1)), \qquad \rho_T^{\beta}(\vec{r}_1) = \frac{1}{2}(\rho_L^{\beta}(\vec{r}_1) + \rho_K^{\beta}(\vec{r}_1)) \qquad (4.7)$$

where $\rho_L^{\alpha},\rho_K^{\beta}$ are the densities of electrons with α-spin due to Ψ_L and Ψ_K respectively, and $\rho_L^{\beta},\rho_K^{\beta}$ the corresponding densities for electrons of β-spin. With the statistical energy expression for a PEMC wave function:

$$E = \sum_L C_L^2 E[\rho_L] + \sum_{LK}\sum{}' C_{KL}\langle\Psi_K|H|\Psi_L\rangle_{ex} + V_{nn} \qquad (4.8)$$

where $\langle\Psi_L|H|\Psi_K\rangle_{ex}$ is given by Eq. (4.6), one can now discuss bond dissociation in a way equivalent to the GVB-method due to Goddard et al.[12] We shall in the next section as a practical example discuss the dissociation of H_2.

4.1 THE DISSOCIATION OF THE H_2 MOLECULE

The groundstate configuration for the H_2 molecule is given by $(\sigma_g)^2$. Here

$$\sigma_g(\vec{r}_1) \xrightarrow[R_{ab}\to\infty]{} \sqrt{\tfrac{1}{2}}\{\chi_a(\vec{r}_1) + \chi_b(\vec{r}_1)\} \qquad (4.9)$$

Thus at infinite internuclear separation the orbital $\sigma_g(\vec{r}_1)$ will approach the simple linear combination of the two 1s orbitals $\chi_a(\vec{r}_1),\chi_b(\vec{r}_1)$, centered on hydrogen atom a and hydrogen atom b respectively. The statistical energy for $\Psi_g = |\overset{+}{\sigma}_g\overset{-}{\sigma}_g|$ is given by:

$$E|\rho_g| = \int_{\tau_1'\to\tau_1} \rho_g(\vec{\tau}_1',\vec{\tau}_1)h(\vec{r}_1)d\vec{\tau}_1 + \frac{1}{2}\int\rho_g(\vec{\tau}_1)\rho_g(\vec{\tau}_2)g(r_{12})d\vec{\tau}_1 d\vec{\tau}_2$$

$$- \frac{3}{4}k\int 2|\frac{1}{2}\rho_g(\vec{r}_1)^{4/3}|d\vec{r}_1 + V_{nn} \qquad (4.10)$$

Here ρ_g is the first order density matrix related to Ψ_g. Further, use has been made of the fact that $\rho_g^{\alpha}=\rho_g^{\beta}=\frac{1}{2}\rho_g$ in the exchange part of Eq. (4.10).

At the dissociation limit:

$$\rho_g(\vec{r}_1) \xrightarrow[R_{ab}\to\infty]{} \rho_a(\vec{r}_1) + \rho_b(\vec{r}_1) \qquad (4.11)$$

where $\rho_a(\vec{r}_1)=\chi_a(\vec{r}_1)\chi_a(\vec{r}_1)$ and $\rho_b(\vec{r}_1)=\chi_b(\vec{r}_1)\chi_b(\vec{r}_1)$. It is clear from Eq. (4.10) and Eq. (4.11) that the energy $E[\rho_g]$ at infinite nuclear separation will have the following asymptotic behaviour:

$$E[\rho_g] \xrightarrow[R_{ab}\to\infty]{} E_\infty[\rho_g] = \sum_{\gamma}^{a,b}\{\int_{\tau_1'\to\tau_1} \rho(\vec{\tau}_1',\vec{\tau}_1)h(\vec{r}_1)d\vec{\tau}_1 + V_{nn}$$

$$+ \frac{1}{2}\int\rho_\gamma(\vec{\tau}_1)\rho_\gamma(\vec{\tau}_2)g(r_{12})d\vec{\tau}_1 d\vec{\tau}_2 - \frac{3}{4}k\int 2|\frac{1}{2}\rho_\gamma(\vec{r}_1)^{4/3}|d\vec{r}_1\} \qquad (4.12)$$

where the index γ runs over the centers a,b. The energy $E_\infty[\rho_g]$ of Eq. (4.12) does not correspond to the statistical energy of two separate hydrogen atoms given by:

$$\sum_{\gamma}^{a,b} E[\rho_\gamma] = \sum_{\gamma}^{a,b}\{\int_{\tau_1'\to\tau_1} \rho_\gamma(\vec{\tau}_1',\vec{\tau}_1)h(\vec{r}_1)d\vec{\tau}_1$$

$$+ \frac{1}{2}\int\rho_\gamma(\vec{\tau}_1)\rho_\gamma(\vec{\tau}_2)g(r_{12})d\vec{\tau}_1 d\vec{\tau}_2 + V_{nn}$$

$$- \frac{3}{4}k\int|\rho_\gamma(\vec{r}_1)^{4/3}|d\vec{r}_1\} \qquad (4.13)$$

The two energy expressions in Eq. (4.12) and Eq. (4.13) differ in their exchange part. It is further clear that:

$$E_\infty[\rho_g] > \sum_{\gamma}^{a,b} E[\rho_\gamma] \qquad (4.14)$$

To assure a proper dissociation one must consider the PEMC wave function

$$\Psi = C_1|\overset{+}{\sigma}_g\overset{-}{\sigma}_g| + C_2|\overset{+}{\sigma}_u\overset{-}{\sigma}_u| \tag{4.15}$$

The $\sigma_u(\vec{r}_1)$ orbital is the "anti-bonding" counterpart to $\sigma_g(\vec{r}_1)$ with the property:

$$\sigma_u(\vec{r}_1) \xrightarrow[R_{ab}\to\infty]{} \sqrt{\tfrac{1}{2}}\{\chi_a(\vec{r}_1) - \chi_b(\vec{r}_1)\} \tag{4.16}$$

The statistical energy due to Ψ of Eq. (4.15) is given according to Eq. (4.8) as:

$$E = C_1^2\,E[\rho_g] + C_2^2\,E[\rho_u] + 2C_1C_2\langle\Psi_g|H|\Psi_u\rangle_{ex} \tag{4.17}$$

where ρ_u is the first order density matrix related to $\Psi_u = |\overset{+}{\sigma}_u\overset{-}{\sigma}_u|$. At infinite separation of the two hydrogen atoms

$$E[\rho_g] = E[\rho_u] = E_\infty[\rho_g] \tag{4.18}$$

since from Eq. (4.9) and Eq. (4.16) one has that $\rho_g = \rho_u$ at $R_{ab}\to\infty$. Thus, if the two coefficients (C_1, C_2) in Eq. (4.17) are to be determined from a variational calculation, one must have at $R_{ab}\to\infty$ that $|C_1| = |C_2| = \sqrt{\tfrac{1}{2}}$. Since in addition, as we shall see next, $\langle\Psi_g|H|\Psi_u\rangle_{ex}$ is positive, one will have $C_1 = -C_2 = \sqrt{\tfrac{1}{2}}$ at $R_{ab}\to\infty$. The matrix element $\langle\Psi_g|H|\Psi_u\rangle_{ex}$ is according to Eq. (4.6) given by:

$$\langle\Psi_g|H|\Psi_u\rangle_{ex} = \frac{3}{4}\int\{|\tfrac{1}{2}\rho_u(\vec{r}_1) + \tfrac{1}{2}\rho_g(\vec{r}_1)^{4/3}| - |\tfrac{1}{2}\rho_g(\vec{r}_1)^{4/3}|$$

$$- |\tfrac{1}{2}\rho_u(\vec{r}_1)^{4/3}|\}d\vec{r}_1 \tag{4.19}$$

or by making use of Eq. (4.9) and Eq. (4.16)

$$\langle\Psi_g|H|\Psi_u\rangle_{ex} \xrightarrow[R_{ab}\to\infty]{} \frac{3}{4}k\sum_{\gamma}^{a,b}\{\int|\rho_\gamma(\vec{r}_1)^{4/3}|d\vec{r}_1 - 2\int|\tfrac{1}{2}\rho_\gamma\ (\vec{r}_1)^{4/3}|d\vec{r}_1\} \tag{4.20}$$

It is clear from Eq. (4.19) that $\langle\Psi_g|H|\Psi_u\rangle_{ex}$ is positive for any interatomic distance R_{ab}.

A substitution of Eq. (4.18) and Eq. (4.20) into Eq. (4.17) along with the condition that $C_1 = -C_2 = \sqrt{\tfrac{1}{2}}$ at $R_{ab}\to\infty$, finally provides:

$$E \xrightarrow[R_{ab}\to\infty]{} \sum_{\gamma}^{a,b} \{ \int_{\tau_1'\to\tau_1} \rho_\gamma(\vec{\tau}_1',\vec{\tau}_1)h(\vec{r}_1)d\vec{\tau}_1$$

$$+ \frac{1}{2}\int\rho_\gamma(\vec{\tau}_1)\rho_\gamma(\vec{\tau}_2)g(r_{12})d\vec{\tau}_1 d\vec{\tau}_2 + V_{nn}$$

$$- \frac{3}{4}\,k\int 2|\tfrac{1}{2}\rho_\gamma(\vec{r}_1)^{4/3}|d\vec{r}_1\}$$

$$- \frac{3}{4}\,k \sum_{\gamma}^{a,b} \{\int|\rho_\gamma(\vec{r}_1)^{4/3}|d\vec{r}_1 - 2\int|\tfrac{1}{2}\rho_\gamma(\vec{r}_1)^{4/3}|d\vec{r}_1\} \tag{4.21}$$

or after a collection of terms:

$$E \xrightarrow[R_{ab}\to\infty]{} \sum_{\gamma}^{a,b} E|\rho_\gamma| \tag{4.22}$$

We have thus shown, that the general expression for the statistical energy of a PEMC wave function given in Eq. (4.8) provides the correct dissociation limit in the case of H_2 for the simple PEMC wave function, defined by Eq. (4.15).

5. CONCLUSION

It has been shown that the statistical energy expression due to Slater in general only can be justified by the statistical exchange approximation for a single determinantal wave function.

The statistical energy for a multi-determinantal wave function can nevertheless be evaluated in those cases, where the energy of the multi-determinantal wave functions with respect to the exact Hamiltonian can be expressed as a sum of energies due to Slater determinants. The extension has been applied to electronic multiplets and correlated wave functions of the paired excitation multiconfiguration type. It was shown that the theory provided good estimates for triplet-singlet separations and assured a proper dissociation of the H_2 molecule.

ACKNOWLEDGEMENTS

This research has been financially supported by the Natural Sciences and Engineering Research Council of Canada (NSERC).

REFERENCES

1. T. Ziegler, A. Rauk, and E.J. Baerends, Theor. Chim. Acta 43, 261 (1977).
2. J.C. Slater, Quantum Theory of Molecules and Solids, Vol. 4, McGraw-Hill, New York, 1974.
3. Other possible extensions have been discussed by: L. Noodleman, J. Chem. Phys. 74, 5737 (1981).
4. R. McWeeny and B.T. Sutcliffe, Methods of Molecular Quantum Mechanics, Academic Press, 1969.
5. K.H. Johnson, J. Chem. Phys. 45, 3085 (1966).
6. E.J. Baerends, D.E. Ellis, and P. Ros, Chem. Phys. 2, 41 (1973).
7. H. Sambe and R.H. Felton, J. Chem. Phys. 62, 1122 (1975).
8. B.I. Dunlap, J.W.D. Connolly, and J.R. Sabin, Int. J. Quant. Chem. S 11, 81 (1977).
9. J. Tylicki, R.J. Hood, G.A. Gallup, and C.J. Eckhardt, J. Chem. Phys. 66, 3745 (1977).
10. U. von Barth, Phys. Rev. A 20, 1693 (1979).
11. O. Gunnarsson and R.O. Jones, Physica Scripta 21, 394 (1980).
12. F.W. Bobrowicz and W.A. Goddard III, Modern Theoretical Chemistry, Vol. 3, H.F. Schaefer III, Ed., Plenum Press, New York, 1977.

THE WEIZSACKER TERM, ITS CORRECTIONS AND VIRIAL PARTITIONING IN DENSITY FUNCTIONAL THEORY

Eduardo V. Ludeña

Centro de Quimica
Instituto Venezolano de Investigaciones Cientificas
I.V.I.C., Apartado 1827, Caracas 1010-A
Venezuela

1. INTRODUCTION

The Thomas-Fermi term[1]

$$t_{TF}(1) = \frac{3}{10}(3\pi^2)^{2/3} \rho(1)^{5/3} \tag{1}$$

has been traditionally regarded as the most important one in the kinetic energy density expansion. As this term contains the correct expression for an infinite homogeneous electron gas, it has been argued[2-5] that those corrections which account for the finite size of a system and for the non-uniformity of the charge density ρ, should involve terms of the form $\nabla\rho$ and of higher derivatives of ρ.

In this perspective, the Weizsacker term[2] given by

$$t_W(1) = \frac{1}{8}\frac{\nabla_1\rho(1)\nabla_1\rho(1)}{\rho(1)} \tag{2}$$

is a first order correction to the Thomas-Fermi term. Let us notice that for constant ρ, this term is zero. Hence, it becomes important precisely for those systems which are finite and inhomogeneous.

The essence of a development such as the gradient expansion lies in the contention that a finite system may be treated, in a zeroth order approximation, as an infinite homogeneous electron gas and that its description may be gradually refined by the addition of terms that include the missing physical effects. Although this approach has lead to interesting results,[6] it has also presented a number of difficulties.[7,8]

In a recent work, Sears et al[9] have advanced an alternative view which stresses the locality and the non-locality of the kinetic energy terms. They have shown from basic quantum mechanics coupled to some information theory concepts, that the Weizsacker term is the "local" term in the kinetic energy density expansion. This conclusion has also been reached by Gázquez and Ludeña.[10]

The importance of an analysis based on local and non-local terms has been recently stressed by us in a discussion of the nature of the correction to the Weizsacker term.[11] In particular, the non-local part of the kinetic energy has been shown to be related to the statistical correlation factor.

In the present work we review some of the basic ideas of this approach. In Section 2 we discuss the representation of the reduced first order density operator $\gamma^1(1,2)$ in terms of the density ρ and a pair correlation function $G(1,2)$. In Section 3, we discuss the representation of the exact kinetic energy density as a functional of ρ and $G(1,2)$. We show that the form of the Thomas-Fermi term is easily obtained, from the correction to the Weizsacker term, using the plane-wave homogeneous electron gas correlation function.[12] In Section 4 we review the possibility of introducing an approximate functional, defined in terms of ρ and $G(1,2)$ for the total energy[12] and show that the Dirac-Slater exchange term corresponds to the plane-wave correlation function correction in this approximate functional. Finally, in Section 5, we show that the virial fragments or quasi atoms which have been shown by Bader et al[13] to be the basic building blocks of molecular systems, are also intrinsic elements of density functional theory.[14]

2. A REPRESENTATION OF $\gamma^1(1,2)$ IN TERMS OF ρ AND $G(1,2)$

The reduced first order density operator $\gamma^1(1,2)$ is related to the charge density $\rho(1)$ through the following equation:

$$\rho(1) = \gamma^1(1,1) \tag{3}$$

For this reason, it is of interest to search for an expression of the form

$$\gamma^1(1,2) = [\rho(1)]^{1/2} [\rho(2)]^{1/2} G(1,2) \tag{4}$$

Obviously, $G(1,2)$ is a function which must satisfy the condition

$$G(1,1) = 1 \tag{5}$$

It has been shown[11] that $G(1,2)$ is given by

$$G(1,2) = [1 + B(1,2)/\rho(1)\rho(2)]^{1/2} \tag{6}$$

where

$$B(1,2) = \phi^{*T}(1)\gamma^1 d(1,2)\gamma^1\phi^*(2) \tag{7}$$

In this expression, $\phi(2)$ is a column vector formed by the orthonormal spin-orbital set $\{\phi_i(2),\ i=1,\ldots,m\}$

$$\phi^*(2) = \begin{bmatrix} \phi_1^*(2) \\ \vdots \\ \phi_m^*(2) \end{bmatrix} \tag{8}$$

and $\phi^{*T}(1)$ is its transpose vector. In general, $m > n$ where n is the number of electrons. γ^1 stands for the reduced first order density matrix and d(1,2) is a matrix whose elements are

$$d_{ij}(1,2) = \det \begin{bmatrix} \phi_i(2) & \phi_i(1) \\ \vdots & \\ \phi_j(2) & \phi_j(1) \end{bmatrix} \tag{9}$$

Because of Eq. (9)

$$d_{ij}(1,1) = 0 \tag{10}$$

and consequently

$$B(1,1) = 0 \tag{11}$$

Therefore, using Eqs. (10) and (11) one has that Eq. (5) is satisfied.

In order to ascertain the nature of G(1,2) it is important to give a statistical interpretation[15] to $\rho(1)$ and $B(1,2)/\rho(1)\rho(2)$.

The charge density operator $\rho(1)$ is the mean value for finding $1,2,\ldots,n$ particles in the volume Ω_1 when the remaining particles are in the volume complement $R^3-\Omega_1$. The dependence of ρ upon the coordinate [1] is obtained when the volume Ω_1 is contracted to the point [1]. Hence, the charge density operator $\rho(1)$ represents an average number of particles at point [1] when no condition is imposed on the remaining particles. On the other hand, when one of the particles is at a fixed position [2], then the conditional average number of particles at position [1] is given by

$$\rho_c(1;2) = 2\,\gamma^2(1,2;1,2)/\rho(2) \tag{12}$$

where γ^2 is the reduced second order density operator. The quotient

$$\bar{f}(1,2) = \rho_c(1;2)/\rho(1) \tag{13}$$

has been defined[15] as the statistical correlation factor. In view of Eq. (12) it may be written as

$$\bar{f}(1,2) = 2\,\gamma^2(1,2;1,2)/\rho(1)\rho(2) \tag{14}$$

In principle, γ^2 and ρ are exact expressions obtained from an exact wavefunction. However, if the same definition is used for approximate wavefunctions, then $\bar{f}(1,2)$ does not represent the full statistical correlation among particles. In particular for a doubly occupied single Slater determinant, $f(1,2)$ is merely the correlation factor for particles with the same spin since in this approximation the motion of electrons with different spins is not correlated.[16]

For a single Slater determinant, γ^2 is defined by

$$\gamma^2(1,2;1,2) = \frac{1}{2}\,[\rho(1)\rho(2) - \gamma^1(1,2)\gamma^1(2,1)] \tag{15}$$

From Eqs. (14) and (15) we get the following expression for the statistical correlation factor:

$$\bar{f}(1,2) = [1 - \gamma^1(1,2)\gamma^1(2,1)/\rho(1)\rho(2)] \tag{16}$$

Using Eq. (16) and the fact that $\gamma^1(1,2)$ is symmetrical, we obtain

$$\gamma^1(1,2) = [\rho(1)]^{1/2}\,[\rho(2)]^{1/2}\,[1 - \bar{f}(1,2)]^{1/2} \tag{17}$$

Comparing Eqs. (17) and (6) we arrive at the conclusion that for a doubly occupied single Slater determinant

$$\bar{f}(1,2) = -\,B(1,2)/\rho(1)\rho(2) \tag{18}$$

Hence, in this case, the function $G(1,2)$ is defined purely in terms of the statistical correlation factor:

$$G(1,2) = [1 - \bar{f}(1,2)]^{1/2} \tag{19}$$

In general, $\bar{f}(1,2)$ may be separated into two components which describe the correlation among electrons with different and equal spins: $\bar{f}_{\neq}(1,2)$ and $f_{=}(1,2)$, respectively. For the present case,

$$f_{\neq}(1,2) = 0 \tag{20}$$

Thus, $G(1,2)$ depends upon the statistical correlation among particles with the same spin so that it only describes the Fermi hole.[16]

For a more general case, such as a linear combination of Slater determinants, $G(1,2)$ describes the part of the Fermi and Coulomb correlation which can be given in terms of γ^1. The dependence of G(1,2) upon the first order density matrix may be readily seen expressing G(1,2) in its orbital representation.[12] Using Eqs. (6) and (7) we obtain

$$G(1,2) = \left\{ \frac{\sum_{i=1}^{m}\sum_{j=1}^{m}\sum_{k=1}^{m}\sum_{\ell=1}^{m} \gamma^1_{ik}\gamma^1_{\ell j}\ \phi_i^*(1)\phi_j^*(2)\phi_k(2)\phi_\ell(1)}{\sum_{i=1}^{m}\sum_{k=1}^{m} \gamma^1_{ik}\ \phi_i^*(1)\phi_k(1) \times \sum_{j=1}^{m}\sum_{\ell=1}^{m} \gamma^1_{j\ell}\ \phi_j^*(2)\phi_\ell(2)} \right\}^{1/2} \tag{21}$$

Let us notice that G(1,2) is not a local functional of the density matrix as the summations in the numerator run over coupled indixes of the matrix elements $\gamma^1_{ik}\gamma^1_{\ell j}$.

Equation (21) may be used in two different contexts. It may be employed in order to analyze the nature of electron exchange and Coulomb correlation for highly accurate atomic and molecular configuration interaction wavefunctions. In this respect, an analysis of Bunge's highly accurate wavefunction for atomic Be is currently under way. But it may also be employed in order to obtain approximate functional expressions for G(1,2). In what follows we explore this second possibility.

Let us assume that m = n, or in other words, that we are dealing with a single Slater determinant. Furthermore, let us assume that

$$\gamma^1_{ij} = \delta_{ij} \quad , \ \forall\, i,j \tag{22}$$

For an arbitrary orthonormal set $\{\phi_i\}$, this condition does not generally satisfy the variational principle. Strictly speaking, only when $\{\phi_i\}$ is formed by the canonical H-F orbitals, γ^1 is the identity matrix. Hence, if we require this condition to hold, say, for planewaves, it is clear that the Slater determinant formed by these functions does not correspond to an energy minimum, since it does not satisfy the variational principle.

Using Eq. (22) and the fact that the spin orbitals are products of spatial and spin parts,

$$\phi_i(1) = u_i(1)v_{m_{s_i}}(1) \tag{23}$$

we obtain for G(1,2),

$$G(1,2) = \left\{ \frac{\sum_{i=1}^{n}\sum_{j=1}^{n} \delta_{m_{s_i},m_{s_j}}\ u_i^*(1)u_j^*(2)u_i(2)u_j(1)}{\sum_{i=1}^{n} u_i^*(1)u_i(1) \times \sum_{j=1}^{n} u_j^*(1)u_j(1)} \right\}^{1/2} \tag{24}$$

This expression is amenable to a number of simplifications depending upon the types of orbitals that we use for the set $\{u_i\}$. For example, when the plane-wave approximation is introduced,

$$u_i(1) = \frac{1}{\sqrt{V}} e^{i\vec{k}_i \cdot \vec{r}_1} \tag{25}$$

we obtain[17] from Eq. (24)

$$G(1,2) = \frac{1}{n} \left\{ \sum_{i=1}^{n} \sum_{j=1}^{n} \delta_{m_{s_i}, m_{s_j}} e^{i(\vec{k}_i - \vec{k}_j) \cdot (\vec{r}_2 - \vec{r}_1)} \right\}^{1/2} \tag{26}$$

which, under the usual replacement of summation over orbital indexes by integration over momentum, yields the homogeneous electron gas pair correlation function:[17]

$$G(r) = 3 \frac{(\sin k_m r - k_m r \cos k_m r)}{(k_m r)^3} \tag{27}$$

In deriving this equation, the coordinate transformation

$$\vec{r} = \vec{r}_2 - \vec{r}_1 \ , \quad \vec{R} = \frac{1}{2} (\vec{r}_1 + \vec{r}_2) \tag{28}$$

has been used; k_m denotes the momentum at the Fermi level.

Using Eq. (27) we show in the following section that the Thomas-Fermi term is the plane-wave correction to the Weizsacker term.

3. THE KINETIC ENERGY DENSITY AS A FUNCTIONAL OF $\rho(1)$ AND $G(1,2)$

The exact expression for the kinetic energy density functional given in terms of $\gamma^1(1,2)$ is

$$t(1) = \frac{1}{2} \nabla_1 \nabla_2 \gamma^1(1,2)|_{2 \to 1} \tag{29}$$

In view of Eq. (4), we can rewrite t(1) as follows:

$$t(1) = \frac{1}{2} (\{\nabla_1 [\rho(1)]^{1/2}\}^2 G(1,1)$$

$$+ [\rho(1)]^{1/2} \nabla_1 [\rho(1)]^{1/2} [\nabla_1 G(1,2) + \nabla_2 G(1,2)]_{2 \to 1}$$

$$+ \rho(1) \nabla_1 \nabla_2 G(1,2)|_{2 \to 1}) \tag{30}$$

Noticing that

$$\nabla_1 G(1,2)\big|_{2\to 1} = \nabla_2 G(1,2)\big|_{2\to 1} = 0 \tag{31}$$

and using Eq. (5) we can simplify this equation to

$$t(1) = \frac{1}{8}\frac{\nabla_1\rho(1)\nabla_1\rho(1)}{\rho(1)} + \frac{1}{2}\rho(1)\nabla_1\nabla_2 G(1,2)\big|_{2\to 1} \tag{32}$$

The Weizsacker term is the first one in this expression. It is a local term as it only depends upon the density $\rho(1)$. The second term, however, contains G(1,2) and, hence, it is non-local. From the discussion in the previous section concerning the nature of G(1,2), we may conclude that the exact non-local correction to the kinetic energy depends upon that part of the Fermi and Coulomb holes that can be described in terms of γ^1.

Using Eq. (32) it is now easy to obtain the Thomas-Fermi-term as a correction to the Weizsacker term. Because of the coordinate transformation given by Eq. (28), the gradient operator product $\nabla_1\nabla_2$ becomes

$$\nabla_1\nabla_2 = -\nabla_r^2 + \frac{1}{4}\nabla_R^2 \tag{33}$$

Substituting Eq. (33) into Eq. (32) we have

$$t(1) = t_W(1) + \frac{1}{2}\rho(1)\left(-\nabla_r^2 + \frac{1}{4}\nabla_R^2\right)G(R,r)\Big|_{\substack{r\to 0\\ R\to r_1}} \tag{34}$$

In general, G(1,2) goes into G(R,r) upon this coordinate transformation. In the case of a plane-wave expansion of a single Slater determinant we observe, however, that the pair correlation function given by Eq. (27) is a function of r alone. Thus, in this particular approximation we have

$$t(1) = t_W(1) - \frac{1}{2}\rho(1)\nabla_r^2 G(r)\big|_{r\to 0} \tag{35}$$

Introducing explicitly Eq. (27) into Eq. (35) and taking the limit, we obtain

$$t(1) = t_W(1) + \frac{3}{10}\rho(1)\,k_m^2 \tag{36}$$

The Thomas-Fermi term is obtained through the usual identification[17]

$$k_m = (3\pi^2\rho)^{1/3} \tag{37}$$

which yields for Eq. (36)

$$t(1) = t_W(1) + \frac{3}{10}(3\pi^2)[\rho(1)]^{5/3} \tag{38}$$

Caution should be employed concerning this "derivation" of the Thomas-Fermi term. The density ρ in Eq. (37) corresponds to an infinite homogeneous electron gas; it has a constant value which is related to the constant values of the momentum at the Fermi level. In Eq. (38), on the other hand, ρ is assumed to be a non-constant function of coordinate [1]. This is a necessary condition for the non-vanishing of the Weizsacker term. However, if $\rho = \rho(1)$, then $k_m = k_m(1)$. It follows, therefore, that the simple form of G(r) given by Eq. (27) should be changed to G(R,r). Hence, Eq. (34) should be used instead of Eq. (35) for the evaluation of the kinetic energy. In conclusion, these basic difficulties make it unlikely for Eq. (38) to be taken as an adequate starting point for variational calculations.

The situation is different for Eq. (34) where it is clearly seen that a variational approach, based on modeling of the pair function G(R,r), is feasible.

4. AN APPROXIMATE TOTAL ENERGY DENSITY FUNCTIONAL

The exact total energy is given by

$$E = \int [t(1) + v(1)]\, d1 \tag{39}$$

where t(1) is the kinetic energy functional defined by Eq. (29), or equivalently by Eq. (32), and v(1) is the potential energy functional which is written as

$$v(1) = -\sum_{\alpha} \frac{Z_\alpha \rho(1)}{r_{1\alpha}} + \int d2\, \frac{\gamma^2(1,2;1,2)}{r_{12}} \tag{40}$$

It is seen that although the electron-nuclear attraction only depends upon $\rho(1)$, the electron-electron repulsion involves the reduced second order density operator $\gamma^2(1,2;1,2)$.

In Section 2 it was shown that t(1) can be written in terms of $\rho(1)$ and G(1,2). The same reduction is not possible for v(1). The reason for this is that the exact expression for the statistical correlation factor involves $\gamma^2(1,2;1,2)$ so that using Eq. (14) and Eq. (40) we have for the electron-electron repulsion energy

$$\frac{1}{2} \int d1 \int d2\; \rho(1)\rho(2)\, \frac{\bar{f}(1,2)}{r_{12}} \tag{41}$$

The physical interpretation is now simple: this term represents the electrostatic energy between charge distribution $\rho(1)$ and $\rho(2)$ when this interaction is weighed by the statistical correlation factor

$\bar{f}(1,2)$. But in general, $G(1,2)$ can only describe part of this statistical correlation[11,12] and therefore it becomes impossible to describe the exact electron-electron repulsion energy in terms of ρ and $G(1,2)$ alone. Nevertheless, an approximation of this term through $G(1,2)$ is not altogether inadequate.

Let us remember that for the Hartree-Fock case we have from Eq. (19)

$$\bar{f}_{HF}(1,2) = 1 - [G_{HF}(1,2)]^2 \tag{42}$$

and, hence, Eq. (41) may be rewritten as

$$\frac{1}{2}\int d1\int d2\,\frac{\rho(1)\rho(2)}{r_{12}} - \frac{1}{2}\int d1\int d2\,\frac{\rho(1)\rho(2)[G_{HF}(1,2)]^2}{r_{12}} \tag{43}$$

The first term of this expression is, of course, the Coulomb energy; the second, the exchange energy.

These considerations have prompted us[12] to make a separation of the total potential energy functional into two parts: one that can be described in terms of ρ and $G(1,2)$ and another which must necessarily include $\gamma^2(1,2;1,2)$. This is done by adding and subtracting the same integral to $v(1)$:

$$v(1) = v_o(1) + v_1(1) \tag{44}$$

where

$$v_o(1) = -\sum_\alpha \frac{Z_\alpha \rho(1)}{r_{1\alpha}} - \frac{1}{2}\int d2\,\frac{B(1,2)}{r_{12}} \tag{45}$$

and

$$v_1(1) = \int d2\,\frac{[\gamma^2(1,2;1,2) + \frac{1}{2}B(1,2)]}{r_{12}} \tag{46}$$

In this expression, $B(1,2)$ is the function defined by Eq. (7). Using Eq. (6) and (45) we obtain an explicit expression for $v_o(1)$ as a functional of ρ and $G(1,2)$:

$$v_o(1) = -\sum_\alpha \frac{Z_\alpha \rho(1)}{r_{1\alpha}} + \frac{1}{2}\int d2\,\frac{\rho(1)\rho(2)}{r_{12}} - \frac{1}{2}\int d2\,\frac{\rho(1)\rho(2)[G(1,2)]^2}{r_{12}} \tag{47}$$

In view of Eqs. (47) and (32) we have advanced[12] the following approximate functional for the total energy:

$$E[\rho,G] \equiv \int d1[t(1) + v_o(1)] \tag{48}$$

This functional is exact for the Hartree-Fock case. Also, it has been shown[12] that the function

$$I(1) \equiv \int d2[\gamma^2(1,2;1,2) + \tfrac{1}{2} B(1,2)] \tag{49}$$

is small and locally negative. The correction term $v_1(1)$ represents, therefore, the electrostatic energy produced by a small and negative charge distribution.

An approximate evaluation of the exchange potential given by

$$V_x(1) \equiv -\frac{1}{2} \int d2 \frac{\rho(2)[G(1,2)]^2}{r_{12}} \tag{50}$$

can be performed by substituting in Eq. (50) the homogeneous electron gas correlation function given by Eq. (27). This leads to an exchange potential equal to

$$V_x(1) = \left(\frac{3}{2\pi}\right) k_m \tag{51}$$

Making the identification between k_m and $\rho(1)$ discussed in Section 3, we finally obtain the Dirac-Slater term[17,18]

$$V_x(1) = -3\left(\frac{3}{8\pi}\right)^{1/3} [\rho(1)]^{4/3} \tag{52}$$

5. VIRIAL PARTITIONING IN DENSITY FUNCTIONAL THEORY

Starting from the exact energy functional expressed in terms of the total wavefunction, Bader et al.[13] have been able to show that there occurs a partitioning of R^3 into fragments bound by surfaces defined by the condition

$$\nabla\rho(r)\cdot\vec{n} = 0 \qquad r \in S(\Omega) \tag{53}$$

These fragments are particularly interesting in view of the fact that the virial theorem is satisfied by the fragment energies. In molecular systems each contains an atomic nucleus; this gives rise to a description of a molecule as a collection of quasi-atoms.

The additivity rules for the properties of chemical functional groups have found a solid basis on the theory of virial fragments. In addition, interesting developments in the description of the chemical bond have been made possible through the study of the topological properties of the density as related to these fragments.[19]

For all these reasons, it becomes important to establish connections between density functional theory and Bader's virial fragment theory. In what follows we review some of these connections which are given in detail elsewhere.[14]

Let us assume, in the spirit of the Hohenberg-Kohn theorem,[20] that the total energy E may be expressed as a functional of ρ. Using $E[\rho]$ we define the functional

$$H[\rho] = E[\rho] - \mu N[\rho] \tag{54}$$

where $N[\rho]$ is the number of particles and μ is a Lagrange multiplier that incorporates the normalization condition. It is convenient to express this equation in terms of the energy density $\xi(r)$ and the number of particles density $n(r)$:

$$H[\rho] = \int[\xi(r) - \mu n(r)]dr \tag{55}$$

The exact functional dependence of $\xi(r)$ on ρ and its higher derivatives is not known. We treat in what follows two cases. The first, when the energy density is given by

$$\xi(r) = \xi(r;\rho,\nabla\rho) \tag{56}$$

and the second, when the following dependence holds

$$\xi(r) = \xi(r;\rho,\nabla\rho,\nabla^2\rho) \tag{57}$$

The first approximation is rather limited and does not correspond to the actual quantum mechanical problem.[13] It is in this case, however, that the surface condition of Eq. (53) arises as a natural boundary condition of the problem. In the second case, on the other hand, although condition (53) must be satisfied, the variation leads to a more complicated expression.

A. <u>The Energy Density as a Functional of ρ and $\nabla\rho$</u>

Let us consider the functional

$$H[\rho,\Omega] \equiv \int_\Omega [\xi(r;\rho,\nabla\rho) - \mu\,\rho(r)]dr \tag{58}$$

where Ω denotes a subvolume of R^3. Letting

$$\bar{\rho}(r) = \rho(r) + \varepsilon\,\eta(r) \tag{59}$$

where $\bar{\rho}(r)$ is a trial density, $\rho(r)$ the exact density, ε a parameter and $\eta(r)$ an arbitrary function which need not vanish at the boundary $S(\Omega)$, we obtain for the variation of this functional

$$\delta H[\rho,\Omega] \equiv \frac{dH}{d\varepsilon}\bigg|_{\varepsilon=0} = \oint_{S(\Omega)} \eta(r)(\nabla_\rho \xi\cdot\vec{n})dS \tag{60}$$

if the Euler-Lagrange equation

$$\frac{\partial\xi}{\partial\rho} - \nabla\cdot\nabla_\rho \xi - \mu = 0 \tag{61}$$

is satisfied in Ω. In the above equations,

$$\nabla_\rho = \hat{i}\,\frac{\partial}{\partial\rho_x} + \hat{j}\,\frac{\partial}{\partial\rho_y} + \hat{k}\,\frac{\partial}{\partial\rho_z} \tag{62}$$

and

$$\rho_x = \frac{\partial\rho}{\partial x}\text{, etc.} \tag{63}$$

The vanishing of the variation in Eq. (60) implies that

$$\nabla_\rho \xi(r)\cdot\vec{n} = 0\ ,\ r \in S(\Omega) \tag{64}$$

It has been shown that when ρ and $\nabla\rho$ appear in ξ as a sum of products of the type $\rho^k(\nabla\rho\cdot\nabla\rho)^\ell$, then Eq. (64) is identical to Eq. (53). Hence, the virial fragment condition is a natural boundary condition of this problem.

The vanishing of the variation in Eq. (60) merely shows that the energy density functional cannot be simply expressed in terms of ρ and $\nabla\rho$. The reason is, as Bader has pointed out,[13] that the variation of the exact energy functional over Ω does not vanish when condition (53) is impossed.

B. The Energy Density as a Functional of ρ, $\nabla\rho$ and $\nabla^2\rho$

In this case, the trial functional becomes

$$H[\rho,\Omega] = \int_\Omega [\xi(r;\rho,\nabla\rho,\nabla^2\rho) - \mu\,\rho(r)]dr \tag{65}$$

Carrying out the variation of this functional in the subvolume Ω, the following surface integrals are obtained:[14]

$$\delta H[\rho,\Omega] = \oint_{S(\Omega)} \eta(r)(\nabla_{\rho}\xi\cdot\vec{n})dS$$

$$+ \oint_{S(\Omega)} dS \sum_{x} \left[\frac{\partial\xi}{\partial\rho_{xx}} \eta_x - \eta \frac{\partial}{\partial x}\left(\frac{\partial\xi}{\partial\rho_{xx}}\right)\right] n_x \tag{66}$$

Obviously, the Euler Lagrange equation

$$\frac{\partial\xi}{\partial\rho} - \nabla\nabla_{\rho}\xi + \nabla\cdot B\xi - \mu = 0 \tag{67}$$

where

$$B\xi = \frac{\partial}{\partial x}\left(\frac{\partial\xi}{\partial\rho_{xx}}\right)\hat{i} + \frac{\partial}{\partial y}\left(\frac{\partial\xi}{\partial\rho_{yy}}\right)\hat{j} + \frac{\partial}{\partial z}\left(\frac{\partial\xi}{\partial\rho_{zz}}\right)\hat{k} \tag{68}$$

must also be satisfied within Ω. When the full gradient expansion for ξ is introduced it has been shown[14] that the vanishing of the first surface integral in Eq. (66) is equivalent to imposing condition (53). The second integral does not vanish, however, and leads to the expression

$$\delta H[\rho,\Omega] = \oint_{S(\Omega)} (F\nabla\eta - \eta\nabla F)\cdot\vec{n}dS \tag{69}$$

where

$$F \equiv \frac{\partial\xi}{\partial\rho_{xx}} = \frac{\partial\xi}{\partial\rho_{yy}} = \frac{\partial\xi}{\partial\rho_{zz}} \tag{70}$$

It is clear, therefore, that although Eq. (53) is a necessary condition, it is not sufficient to guarantee the vanishing of the variation. Eq. (69) is more in accord with the exact quantum mechanical result which leads in fact to a non-vanishing commutator.[13]

The interesting conclusion which may be drawn from the present treatment is that virial partitioning also appears as a basic ingredient in density functional theory. However, in order to further develop this approach it will be necessary to investigate the relationship between F given in Eq. (70) and the pair correlation function G(1,2) discussed in the previous sections.

6. DISCUSSION

The appeal of density functional theory lies in that, in principle, it provides a more convenient way for solving the many-body problem of atomic and molecular quantum mechanics. However, the functional for the total energy density expressed in terms of ρ alone, remains as yet elusive. Furthermore, in a variational treatment of this problem, it will probably be necessary to incorporate additional conditions in order to guarantee the n-representability of the density.[21]

In the present work, we have developed an approach which is based on a description of the reduced first order density operator $\gamma^1(1,2)$ in terms of the density ρ and of a "pair correlation function" G(1,2). In particular we have shown that the exact kinetic energy functional acquires a rather simple form: it consists of a local (Weizsacker) term plus a non-local contribution which involves G(1,2). We have derived from this non-local expression the Thomas-Fermi kinetic energy term using the homogeneous electron gas approximation for G(1,2).

Because the pair correlation function G(1,2) represents only that part of the total exchange and Coulomb correlation which may be given in terms of γ^1, it is not possible to write the total energy functional solely in terms of ρ and G(1,2). For this reason we have advanced an approximate functional for the total energy, where this restriction is satisfied. For a single Slater determinant it corresponds to the H-F functional although, in general, it goes beyond the H-F case in the sense that it incorporates additional correlation effects.

An advantage of the present formulation comes about from the possibility of modeling G(1,2). As a simple example of this modeling, we have obtained the Dirac-Slater exchange term when G(1,2) is the pair correlation function for a homogeneous electron gas. In this respect, some recent developments such as the model Coulomb hole introduced in $X\alpha$ theory,[22] or the expansion of the pair function in a set of hydrogenic wavefunctions,[23] may lead to substantial simplifications. Finally, we have shown that the virial partitioning obtained for the exact energy, when it is expressed as a functional of the n-electron wavefunction,[13] also follows from the variation of energy functionals described in terms of the density ρ and of its derivatives. The formulation of a density functional theory for virial fragments may be computationally and conceptually convenient and would be an important step toward the direct description of chemical functional groups.

REFERENCES

1. L.H. Thomas, Proc. Cambridge Phil. Soc. 23, 542 (1927); E. Fermi, Z. Phys. 48, 73 (1928).
2. C.F.V. Weizsacker, Z. Phys. 96, 431 (1935).
3. D.A. Kirzhnitz, Zhur. Eksp. Teor. Fiz. 32, 115 (1957) [Sov. Phys. JETP 5, 64 (1957)].
4. C.H. Hodges, Can. J. Phys. 51, 1428 (1973).
5. W. Kohn and L. Sham, Phys. Rev. A 140, 1133 (1965).
6. D.M. Murphy and W-P. Wang, J. Chem. Phys. 72, 429 (1980); W-P. Wang, R.G. Parr, D.R. Murphy, and G.A. Henderson, Chem. Phys. Lett. 43, 409 (1976); see also: L.J. Bartolotti and R.G. Parr, J. Chem. Phys. 75, 4553 (1981).
7. Y. Tal and R.F.W. Bader, Int. J. Quant. Chem. Symp. 12, 153 (1978)
8. D.R. Murphy, Phys. Rev. A 24, 1682 (1981).
9. S.B. Sears, R.G. Parr, and U. Dinur, Israel J. Chem. 19, 165 (1980); see also: K.P. Acharya, L.J. Bartolotti, S.B. Sears, and R.G. Parr, Proc. Natl. Acad. Sci., U.S.A. 77, 6978 (1980); J.L. Gázquez and J. Robles, J. Chem. Phys. 76, 1467 (1982).
10. J.L. Gázquez and E.V. Ludeña, Chem. Phys. Lett. 83, 145 (1981).
11. E.V. Ludeña, J. Chem. Phys. 76, 3157 (1982).
12. E.V. Ludeña, "The Thomas-Fermi Term as the Simplest Correction to the Weizsacker Term". Submitted for publication (1982).
13. S. Srebrenik and R.F.W. Bader, J. Chem. Phys. 63, 3945 (1975); S. Srebrenik, R.F.W. Bader, and T.T. Nguyen-Dang, J. Chem. Phys. 68, 3667 (1978); R.F.W. Bader, S. Srebrenik, and T.T. Nguyen-Dang, J. Chem. Phys. 68, 3680 (1978).
14. E.V. Ludeña and V. Mujica, Intern. J. Quant. Chem. 21, 927 (1982).
15. G. Sperber, Intern. J. Quant. Chem. 5, 177 (1971).
16. R. McWeeny, Rev. Mod. Phys. 32, 335 (1960).
17. J.C. Slater, Quantum Theory of Atomic Structure, (McGraw-Hill, New York, 1960), Vol. II, p. 323-332.
18. P.A.M. Dirac, Proc. Cambridge Philos. Soc. 26, 376 (1930).
19. R.F.W. Bader, S.E. Anderson, and A.J. Duke, J. Am. Chem. Soc. 101, 1389 (1979); R.F.W. Bader, T.T. Nguyen-Dang, and Y. Tal, J. Chem. Phys. 70, 4316 (1979); R.F.W. Bader, Y. Tal, S.E. Anderson, and T.T. Nguyen-Dang, Israel J. Chem. 19, 8 (1980).
20. P. Hohenberg and W. Kohn, Phys. Rev. B 136, 864 (1964).
21. T. Tung Nguyen-Dang, E.V. Ludeña, and Y. Tal, "Variation of the Energy Functional of the First-order Density Matrix", Submitted for publication in the J. Chem. Phys.
22. S. Maloni and M.A. Whitehead, Phys. Rev. A 23, 2150 (1981); see also R.O. Jones, J. Chem. Phys. 76, 3098 (1982).
23. P.W. Payne, Proc. Natl. Acad. Sci. U.S.A., 77, 6293 (1980).

THE SECOND ORDER REDUCED DENSITY MATRIX AND THE CHEMICAL BOND

William L. Luken and John C. Culberson

Paul M. Gross Chemical Laboratory
Duke University
Durham
North Carolina 27706, U.S.A.

The Fermi hole is described as a function of one electron which is parametrically dependent on the position of a 'probe electron.' The sensitivity of the Fermi hole to the position of the probe electron is described by the Fermi hole mobility function. Localized molecular orbitals analogous to Wannier functions may be determined from the Fermi hole of a self-consistent field (SCF) electronic wavefunction. Comparison of the Fermi hole mobility function for an SCF wavefunction to that of a free-electron gas may be used to partition a molecule into electronic domains. The Fermi hole and Coulomb hole of a correlated wavefunction may be specified by natural representations which depend on the position of the probe electron.

1. DENSITY MATRICES

The first and second order reduced density matrices for an N-electron wavefunction $\Psi(1,2,\ldots N)$ are defined as

$$\gamma(1,1') = N \iint \Psi(1,2,\ldots N)\ \Psi^*(1',2,\ldots N)\ d(2)\ldots d(N) \tag{1}$$

and

$$\Gamma(1,2;1',2') = \frac{N(N-1)}{2} \iint \Psi(1,2,3,\ldots N) \times\Psi^*(1',2',3,\ldots N)\ d(3)\ldots d(N) \tag{2}$$

respectively.[1-3] These may be expressed in terms of spin functions and functions of spatial co-ordinates. Thus, the first order reduced density matrix is given by

$$\gamma(1;1') = \rho_\alpha(\vec{r}_1;\vec{r}_{1'})\ \alpha_1\ \alpha_{1'}^* + \rho_\beta(\vec{r}_1;\vec{r}_{1'})\ \beta_1\ \beta_{1'}^* \tag{3}$$

where α and β are the m_s=+1/2 and m_s=-1/2 spin functions, $\rho_\alpha(\vec{r};\vec{r}')$ is the first order spatial density matrix for electrons with m_s=+1/2, and $\rho_\beta(\vec{r};\vec{r}')$ is the corresponding matrix for electrons with m_s=-1/2.

By integrating over the spins in $\gamma(1;1)$, we obtain the one-electron density function

$$\rho_1(\vec{r}) = \rho_\alpha(\vec{r};\vec{r}) + \rho_\beta(\vec{r};\vec{r}) \tag{4}$$

Likewise, the two-electron density function may be defined as

$$\rho_2(\vec{r}_1,\vec{r}_2) = \rho_{\alpha\alpha}(\vec{r}_1,\vec{r}_2;\vec{r}_1,\vec{r}_2) + \rho_{\alpha\beta}(\vec{r}_1,\vec{r}_2;\vec{r}_1,\vec{r}_2) + \rho_{\beta\alpha}(\vec{r}_1,\vec{r}_2;\vec{r}_1,\vec{r}_2) + \rho_{\beta\beta}(\vec{r}_1,\vec{r}_2;\vec{r}_1,\vec{r}_2) \tag{5}$$

where

$$\begin{aligned}\Gamma(1,2;1',2') = {} & \rho_{\alpha\alpha}(\vec{r}_1,\vec{r}_2;\vec{r}_{1'},\vec{r}_{2'})\ \alpha_1\ \alpha_{1'}^*\ \alpha_2\ \alpha_{2'}^* \\ & + \rho_{\alpha\beta}(\vec{r}_1,\vec{r}_2;\vec{r}_{1'},\vec{r}_{2'})\ \alpha_1\ \alpha_{1'}^*\ \beta_2\ \beta_{2'}^* \\ & + \rho_{\beta\alpha}(\vec{r}_1,\vec{r}_2;\vec{r}_{1'},\vec{r}_{2'})\ \beta_1\ \beta_{1'}^*\ \alpha_2\ \alpha_{2'}^* \\ & + \rho_{\beta\beta}(\vec{r}_1,\vec{r}_2;\vec{r}_{1'},\vec{r}_{2'})\ \beta_1\ \beta_{1'}^*\ \beta_2\ \beta_{2'}^* \end{aligned} \tag{6}$$

The first order density matrix and the one-electron density function have been extremely useful in the calculation and interpretation of molecular structure and properties. Applications of these quantities include the evaluation of the expectation values of one-electron operators,[4] calculation of the Fock operators for self-consistent field (SCF) wavefunctions,[5] determination of natural orbitals,[2,6-8] virial partitioning,[9,10] and local density functional theory.[11,12] Most attempts to give a physical interpretation to the second order density matrix, however, have been characterized as having failed to provide much insight.[13]

Recent investigations[14-17] have examined the properties of the second order reduced density matrix and the two-electron density function for molecules. Much of this work has been based on the Fermi hole,[18,19]

$$\Delta_{\alpha\alpha}(\vec{r}_1;\vec{r}_2) = \rho_\alpha(\vec{r}_1) - 2\ \rho_{\alpha\alpha}(\vec{r}_1,\vec{r}_2)/\rho_\alpha(\vec{r}_2) \tag{7}$$

where $\rho_\alpha(\vec{r}) = \rho_\alpha(\vec{r};\vec{r})$, and $\rho_{\alpha\alpha}(\vec{r}_1,\vec{r}_2) = \rho_{\alpha\alpha}(\vec{r}_1,\vec{r}_2;\vec{r}_1,\vec{r}_2)$

Additional work has involved the conditional density function,

$$\rho_c(\vec{r}_1|\vec{r}_2) = 2\ \rho_2(\vec{r}_1,\vec{r}_2)/\rho_1(\vec{r}_2) \tag{8}$$

and the Coulomb hole

$$\Delta_{\alpha\beta}(\vec{r}_1;\vec{r}_2) = \rho_{\alpha}(\vec{r}_1) - 2\,\rho_{\alpha\beta}(\vec{r}_1,\vec{r}_2)/\rho_{\beta}(\vec{r}_2) \qquad (9)$$

where $\rho_{\beta}(\vec{r}) = \rho_{\beta}(\vec{r};\vec{r})$ and $\rho_{\alpha\beta}(\vec{r}_1,\vec{r}_2) = \rho_{\alpha\beta}(\vec{r}_1,\vec{r}_2;\vec{r}_1,\vec{r}_2)$

As shown in the remainder of this article, these quantities, which are derived from the second order reduced density matrix, can provide useful information about electronic localization, electronic mobility, and chemical bonding in molecules.

2. PROPERTIES OF THE FERMI HOLE

The Fermi hole for a closed shell single determinant wavefunction $\Psi(1,2,\ldots N)$ is given by

$$\Delta_{\alpha\alpha}(\vec{r}_1;\vec{r}_2) = |\ f(\vec{r}_1;\vec{r}_2)\ |^2 \qquad (10)$$

where $f(\vec{r}_1;\vec{r}_2)$ is the Fermi orbital defined as

$$f(\vec{r}_1;\vec{r}_2) = \sum_{i=1}^{N/2} \phi_i(\vec{r}_1)\ g_i(\vec{r}_2)^* \qquad (11)$$

The functions $\phi_i(\vec{r})$ are the spatial parts of the spin-orbitals occupied in Ψ, and the coefficients $g_i(\vec{r})$ are defined as

$$g_i(\vec{r}) = \phi_i(\vec{r})\ [2/\rho_1(\vec{r})]^{1/2} \qquad (12)$$

The Fermi hole and the Fermi orbital may be regarded as functions of $\vec{r}_1$ which are parametrically dependent on the position $\vec{r}_2$ of a 'probe electron'. The shape of the Fermi hole and the Fermi orbital changes as the position of the probe electron changes. These changes, however, do not occur uniformly. The space of the probe electron may be divided into three types of regions:

1. regions where the Fermi hole is insensitive to the position of the probe electron;
2. regions where the Fermi hole follows the probe electron; and
3. regions where the Fermi hole changes rapidly in response to small changes in the position of the probe electron.

The first and third types of behavior are demonstrated in a 16mm movie based on an SCF wavefunction for the water molecule.[14] This type of behavior is expected for isolated pairs of electrons separated by repulsive barriers.[15,16] The third type of behavior is characteristic of a free electron gas.[16]

A quantitative measure of the dependence of the shape of the Fermi hole on the position of the probe electron is provided by the Fermi hole mobility function,[16]

$$F(\vec{r}) = F_x(\vec{r}) + F_y(\vec{r}) + F_z(\vec{r}) \tag{13}$$

where

$$F_s(\vec{r}) = \lim_{\varepsilon \to 0} [1 - \langle f(\vec{t};\vec{r}) | f(\vec{t};\vec{r}+\varepsilon\vec{s}) \rangle_{\vec{t}}]/\varepsilon^2 \tag{14}$$

and $\vec{s}$ is a unit vector along the s axis. Substitution of equation 11 into equation 14 leads to[16]

$$F_s(r) = [2/\rho_1(\vec{r})^2] \sum_{i>j} [\phi_i(\partial\phi_j/\partial s) - \phi_j(\partial\phi_i/\partial s)]^2 \tag{15}$$

where s = x, y, or z. For simplicity, we have assumed that the spatial orbitals $\phi_i(\vec{r})$ in equation 15 are all real.

Figure 1 shows the Fermi hole mobility function $F(\vec{r})$ for a slice through the formaldehyde molecule. This figure is based on a double zeta SCF wavefunction[20] for the ground electronic state of the formaldehyde molecule. The same function based on an STO-3G SCF wavefunction[21] is shown in Figure 2. Comparison of Figures 1 and 2 shows that the atomic orbital basis set has a large effect on the second order reduced density matrix. In spite of this, there are certain essential features common to both of these figures:

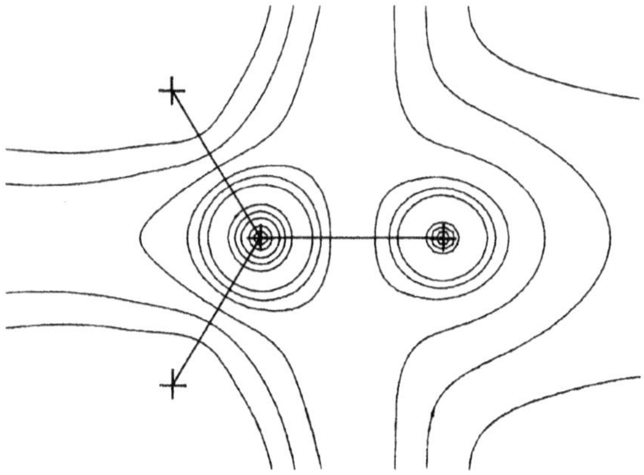

Fig. 1. The Fermi hole mobility function for formaldehyde, based on the double zeta SCF wavefunction of ref. 20. The contour lines indicate values of 0.1, 0.2, 0.4, 1.0, 2.0, and 4.0 $bohr^{-2}$, starting with the outermost contour, rising to maxima enclosing the carbon and oxygen nuclei, and falling to minima at the carbon and oxygen nuclei. There is a small local minimum near the midpoint of the C-O axis.

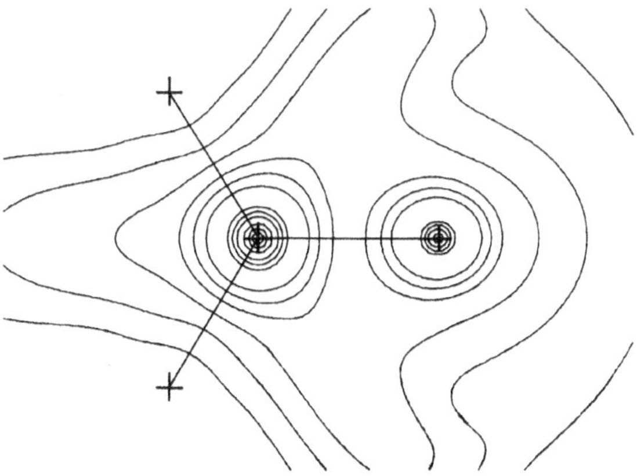

Fig. 2. The Fermi hole mobility function for formaldehyde, based on the STO-3G minimal basis SCF wavefunction. The contour levels are the same as in Fig. 1. As in Fig. 1, there is a local minimum between the carbon and oxygen nuclei.

1. The mobility function approaches a minimum at or near each of the nuclei.
2. The carbon and oxygen nuclei are each enclosed by a ridge which separates the valence region from the inner shell regions.
3. Additional ridges separate the protons from each other, as well as from the valence region around the oxygen atom.

The Fermi hole mobility function shown in Figures 1 and 2 may be compared to that of a uniform density free-electron gas. If the Fermi hole is approximated with a spherical gaussian shape, then the Fermi hole mobility (equation 14) for a free-electron gas with density ρ is given by

$$F_0(\rho) = (3\pi/4)(\rho/2)^{2/3} \qquad (16)$$

In contrast to the SCF Fermi hole mobilities shown in Figures 1 and 2, $F_0(\rho_1)$ has a maximum at each of the nuclei. Consequently, the difference $F(\vec{r}) - F_0(\rho_1)$, as shown in Figures 3 and 4, is negative near each of the nuclei. In addition, this difference is negative in a region between the carbon and oxygen nuclei, and in the region associated with the lone pair orbitals of the oxygen atom. In each of these negative regions, the Fermi hole of the SCF wavefunction is less sensitive to the position of the probe electron than it would be in a free-electron gas of the same density. The

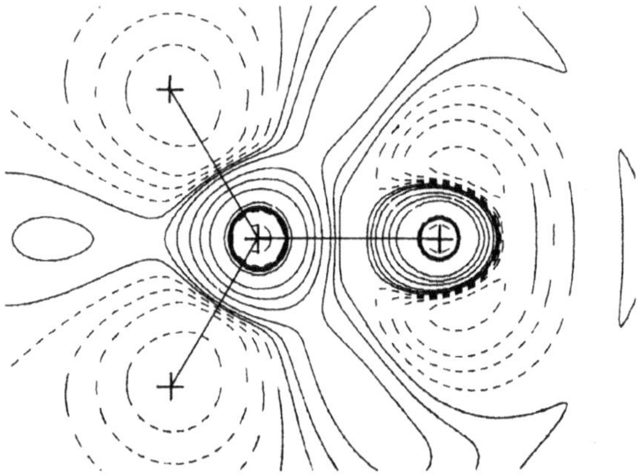

Fig. 3. The difference $F(\vec{r}) - F_0(\rho_1)$ between the SCF Fermi hole mobility function and the free-electron gas approximation for formaldehyde. This figure is based on the STO-3G minimal basis SCF wavefunction. The solid contour lines indicate the same values specified in Fig. 1. The broken contour lines represent values of 0, -0.1, -0.2, -0.4, -1.0, and -10.0 bohr^{-2}. The carbon and oxygen nuclei are located at very deep minima, and there is a shallow minimum between the carbon and oxygen nuclei.

shape of the Fermi hole is relatively stable when the probe electron is in one of these negative regions. The positive ridges which separate the negative regions indicate places where the Fermi hole changes rapidly from the stable shape characteristic of one negative region to that of another.

3. LOCALIZED MOLECULAR ORBITALS

When the probe electron is located in one of the regions where $F(\vec{r}) < F_0(\rho_1)$, the relatively stable shape characteristic of that region tends to resemble a conventional localized molecular orbital.[23-27] This is demonstrated in Figures 5 and 6, which compare an oxygen atom lone pair localized molecular orbital to the corresponding Fermi hole. The Fermi orbital represented in Figure 5 has an overlap of 0.994707 with the localized molecular orbital shown in Figure 6.

The similarity between the Fermi orbitals and localized

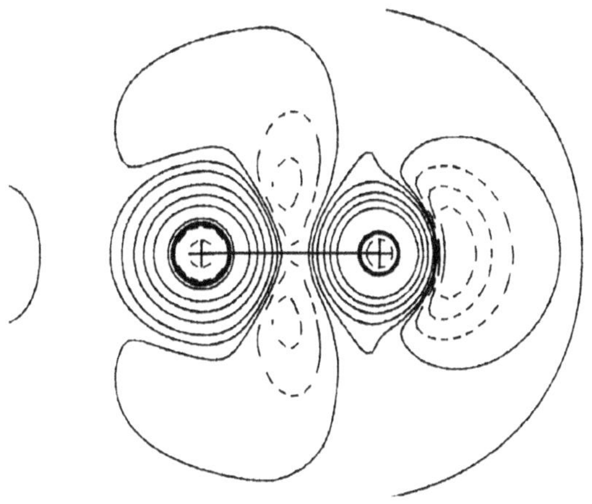

Fig. 4. The same as Figure 3, for a plane containing the C-O axis, perpendicular to the plane shown in Figure 3.

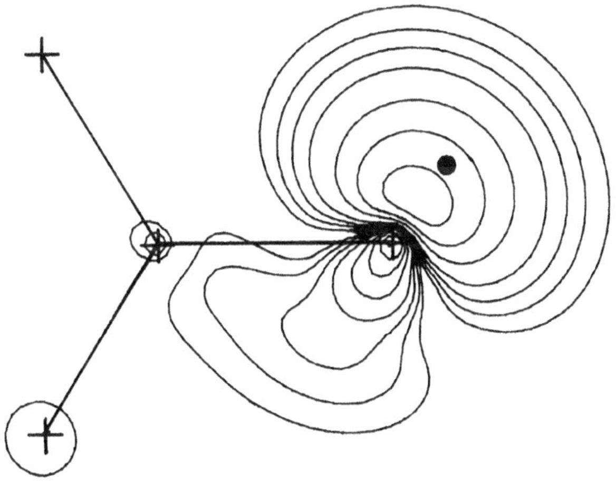

Fig. 5. The Fermi hole for the STO-3G SCF wavefunction for the formaldehyde molecule. The location of the probe electron is indicated by a solid circle. The contour lines indicate densities of 0.004, 0.01, 0.02, 0.04, 0.1, 0.2, 0.4, and 1.0 electrons per cubic bohr.

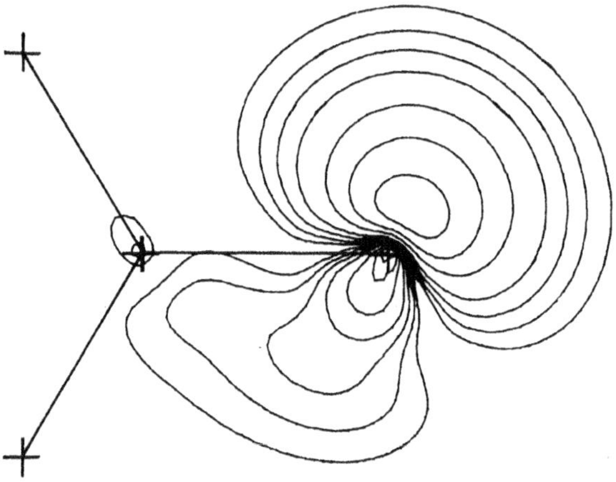

Fig. 6. A localized molecular orbital for formaldehyde. This is one of the oxygen lone pair orbitals determined by application of Boys' criterion to the STO-3G SCF wavefunction for formaldehyde. The contour levels are the same as in Fig. 5.

molecular orbitals can be exploited to calculate localized molecular orbitals based directly on the Fermi hole. Thus equation 11 may be interpreted as a direct transformation from canonical molecular orbitals to localized molecular orbitals. It is necessary to specify N/2 positions of the probe electron to determine a set of N/2 localized molecular orbitals based on equation 11. This set of positions may be expected to include the positions of each of the nuclei. Additional positions may be determined from inspection of a figure like Figure 3. For example, the minima in $F(\vec{r}) - F_0(\rho_1)$ may be chosen for these points. The precise details of how these points are chosen are not critical because the Fermi hole is relatively insensitive to the position of the probe electron near the minima in $F(\vec{r}) - F_0(\rho_1)$.

A set of N/2 Fermi orbitals will generally not be mutually orthogonal. The symmetric orthogonalization procedure[26,27] may be used to determine an orthonormal set of localized molecular orbitals based on the Fermi orbitals. The resulting set of localized molecular orbitals form the molecular analogy to the Wannier functions of solid state physics.[28-30]

The relationship between Fermi orbitals and localized molecular orbitals may be simplified further if the molecular orbitals are defined by linear combinations of suitable basis functions (e.g. atomic orbitals),

$$\phi_i(\vec{r}) = \sum_u C_{iu}\, X_u(\vec{r}), \quad i = 1 \text{ to } N/2 \tag{17}$$

If one of the basis functions $X_u(\vec{r})$ dominates a region of space Ω_u, that is, $|X_u(\vec{r})| \gg |X_v(\vec{r})|$ for all $v \neq u$ and all $\vec{r}$ in Ω_u, then the Fermi orbital may be approximated by

$$f(\vec{r};\vec{r}_u) \cong f_u(\vec{r}) = \sum_{i=1}^{N/2} C_{iu}\, \phi_i(\vec{r})/C_u \tag{18}$$

for any point $\vec{r}_u$ in Ω_u, with C_u determined by

$$C_u^{\,2} = \sum_{i=1}^{N/2} |C_{iu}|^2 \tag{19}$$

The localized orbital $f_u(\vec{r})$ may also be written as

$$f_u(\vec{r}) = \sum_v d_{uv}\, X_v(\vec{r}) \tag{20}$$

where

$$d_{uv} = \sum_{i=1}^{N/2} C_{iu}\, C_{iv}/C_u \tag{21}$$

This approximation is demonstrated in Table 1, which compares the C-H bond localized molecular orbital of formaldehyde as determined by Boys' criterion, the energy criterion (ref. 24), equation 11, and equation 18.

Application of equations 18 or 20 requires the identification of suitable atomic orbitals $X_u(r)$. This is accomplished more easily with a minimal basis set than it is with an extended atomic orbital basis set. In general, the 1s atomic orbitals may be used to determine localized orbitals for inner shells and bonds which involve a hydrogen atom. In order to determine localized orbitals for lone pairs and bonds not involving a hydrogen atom, it may be necessary to express the atomic orbital basis set in terms of appropriate atomic hybrid orbitals (e.g. sp, sp^2, sp^3) instead of the usual cartesian atomic orbitals (s, p_x, p_y, and p_z). In such cases, it may be easier to use equation 11 than equations 18 or 20.

If there is excessive ambiguity in the selection of atomic orbitals required to apply equation 18, or in selection of the probe electron positions required to apply equation 11, it may be preferable to use a conventional unitary transformation method[24-26] to determine a set of localized molecular orbitals. In such cases, equations 11 or 18 may be used as a preliminary step to reduce the number of transformations required by such methods to reach convergence.

Table 1. Comparison of Fermi orbitals and localized molecular orbitals for the formaldehyde molecule. The following coefficients specify the composition of a C-H bond localized molecular orbital as determined by Boys' criterion to the corresponding orbital determined by the energy criterion of localization (ref. 24), the Fermi orbital determined by equation 11, and the approximate Fermi orbital determined by equation 18

Atom	orbital	Boys	Energy	Fermi	Equation 18
Hyd-1	1s	-0.316928	-0.314568	-0.322143	-0.322375
	1s'	-0.236648	-0.230457	-0.261079	-0.265507
Hyd-2	1s	0.035858	0.034761	0.071915	0.076629
	1s'	0.100335	0.096456	0.137049	0.144429
Carbon	1s	0.003703	0.002126	0.004446	0.004403
	1s'	0.084699	0.055305	0.096618	0.095710
	2s	-0.266482	-0.272481	-0.226817	-0.225636
	2s'	-0.183512	-0.184118	-0.161461	-0.159095
	$2p_x$	-0.382002	-0.385259	-0.404347	-0.402413
	$2p_x'$	-0.044686	-0.046227	-0.041872	-0.040015
	$2p_y$	0.0	0.0	0.0	0.0
	$2p_y'$	0.0	0.0	0.0	0.0
	$2p_z$	0.244099	0.234529	0.229381	0.221925
	$2p_z'$	0.026402	0.027216	0.022138	0.022120
Oxygen	1s	-0.000505	-0.000221	-0.000580	-0.000559
	1s'	-0.011706	-0.005869	-0.013087	-0.012609
	2s	0.032921	0.025798	0.031761	0.030492
	2s'	0.027234	0.024482	0.020760	0.021335
	$2p_x$	0.054412	0.029499	0.138829	0.165050
	$2p_x'$	0.079138	0.067326	0.123740	0.136377
	$2p_y$	0.0	0.0	0.0	0.0
	$2p_y'$	0.0	0.0	0.0	0.0
	$2p_z$	-0.053241	-0.042901	-0.064928	-0.058898
	$2p_z'$	-0.017322	-0.015548	-0.019532	-0.018039

4. ELECTRONIC MOBILITY

In addition to describing the localization of electrons within a molecule, the Fermi hole mobility function F(r), and the mobility difference $F(\vec{r}) - F_0(\rho_1)$, also describe electronic delocalization within a molecule. This may be used to develop a molecular model of electrical conductivity. This model is based on the use of $F(\vec{r}) - F_0(\rho_1)$ to partition a molecule into a number of domains. Each domain possesses one or more interior regions where $F(\vec{r}) - F_0(\rho_1)$ is negative. These interior regions are enclosed by an exterior region where $F(\vec{r}) - F_0(\rho_1)$ is positive. The boundaries of the domains are determined by the points where $F(\vec{r}) - F_0(\rho_1)$ passes through a maximum along a path between two minima.

Ideally, the positive region should completely enclose a negative interior region with a unique point where $F(\vec{r}) - F_0(\rho_1)$ reaches a local minimum. In some cases, however, it may be necessary to divide a negative region among two or more domains, or to permit a single domain to possess two or more local minima. For example, the formaldehyde molecule, shown in Figures 3 and 4, may be partitioned into six domains: two inner shell domains, two C-H bond domains, a C-O double bond domain, and an oxygen atom lone pair domain. The oxygen lone pair domain could be divided into two smaller domains, each with a unique minimum and a single pair of electrons, but the resulting pair of domains would be connected by a large negative window. Likewise, the domain associated with the C-O double bond may be divided into two domains separated by the plane of the nuclei. These two domains would be separated by very small values of $F(\vec{r}) - F_0(\rho_1)$. Consequently, it seems more appropriate to describe this region as a single four-electron domain than a pair of two-electron domains.

A more complex example is provided in Figures 7 and 8, which show the values of $F(\vec{r}) - F_0(\rho_1)$ for two parallel slices through a cyclic α,β enone, $C_4H_4O_2$. These figures are based on an STO-3G SCF wavefunction. The geometrical parameters used in these calculations are summarized in Table 2. All nuclei except the methylene protons are assumed to lie in the plane shown in Figure 7. This figure shows six inner shell domains, two C-H bond domains, three C-C bond domains, and three C-O bond domains. In addition, there is a lone pair domain associated with each of the oxygen atoms. The lone pair domain of the carbonyl oxygen atom shows two minima in this plane.

Figure 8 represents the planes 1.30 bohr above and below that shown in Figure 7. One of the methylene C-H domains is prominent in this figure, as well as the lone pair domain for the oxygen atom in the ring. The inner shell domains do not appear in this plane, and there are no negative contours associated with any of the C-C or C-O single bonds. The double bonds, however, are more prominent in Figure 7 than in Figure 8.

The values of $F(\vec{r}) - F_0(\rho_1)$ which separate the C-C and C-O domains associated with the acrolein-like fragment are not very large. Consequently, it is possible to consider this region as a large conjugated domain occupied by five electrons of each spin. This domain may be extended to include seven electrons of each spin if the lone pair domain of the carbonyl oxygen is also included.

The positive regions associated with the boundaries of these domains may be interpreted as implying the presence of real or

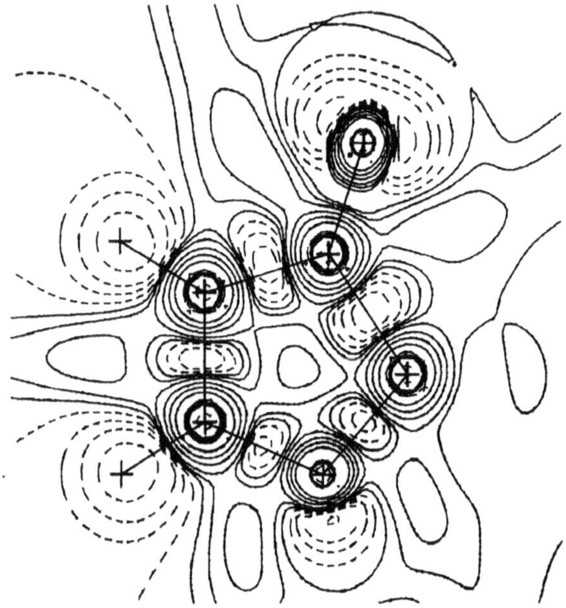

Fig. 7. The Fermi hole mobility difference $F(\vec{r}) - F_0(\rho_1)$ for a cyclic α,β enone $C_4H_4O_2$. This figure is based on an STO-3G SCF wavefunction, assuming a planar geometry. The plane shown here contains all of the nuclei except the methylene protons. The contour levels are the same as in Fig. 3.

effective repulsive potentials. Such repulsive potentials may retard or inhibit the exchange of electrons across the boundaries of the domains. These potentials may include effective local potentials related to the anti-symmetric form of the electronic wavefunction. Such potentials have been proposed in the local density functional interpretation of atomic shell structure.[31] This type of potential would be associated with the large values of $F(\vec{r})$ enclosing each of the inner shell domains in Figures 3, 4, and 7.

Each domain may be associated with one or more localized molecular orbitals. A domain which is associated with a single localized molecular orbital represents a strongly localized pair of electrons. This type of domain is represented by the inner shell domains and C-H bond domains of the formaldehyde molecule.

A domain associated with two or more localized molecular orbitals, or equivalently, two or more domains connected by large

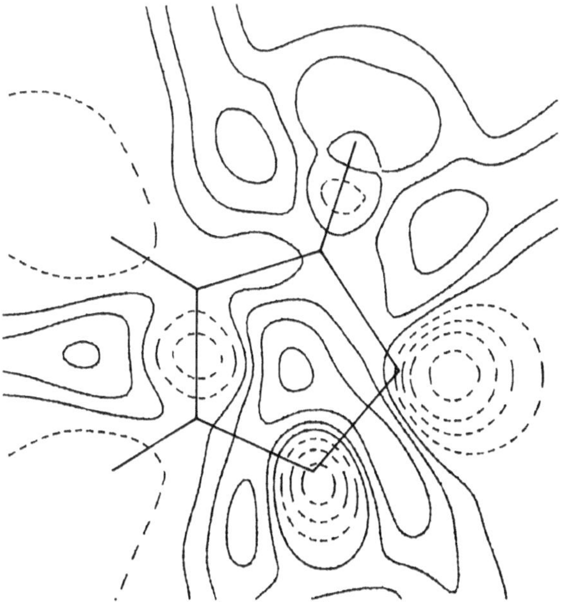

Fig. 8. The same as Figure 7, except that the plane shown here is parallel to, and 1.3 bohr above or below the plane shown in Figure 7.

negative windows, provides evidence of electronic delocalization. Examples of this include the C-O bond domain and the oxygen atom lone pair domain(s) of the formaldehyde molecule, and the conjugated pi domain of the α,β enone.

5. CORRELATED WAVEFUNCTIONS

Most of the preceding material has been limited to density matrices based on closed shell SCF wavefunctions. Such wavefunctions may provide satisfactory descriptions of the ground electronic states of many molecules in their equilibrium nuclear geometries. A closed shell SCF wavefunction, however, cannot describe a covalent bond forming from or breaking into a pair of radical fragments. In order to obtain a complete and accurate description of molecular structure, it is necessary to include the effects of electron correlation.[2,32]

The Fermi hole for a correlated wavefunction cannot be described

Table 2. Geometrical parameters for the cyclic α,β enone, $C_4H_4O_2$ shown in Figures 7 and 8. The molecule is assumed to have C_s symmetry. The plane of the methylene atoms (C_4, H_3, and H_4) bisects the C_3 - C_4 - O_2 angle. The bond lengths are given in atomic units (bohr).

Atom	Atom	Distance	Atom	Atom	Atom	Angle
C_1	H_1	2.079	H_1	C_1	C_2	120°
C_2	H_2	2.079	C_1	C_2	H_2	120°
C_1	C_2	2.589	C_1	C_2	C_3	106°
C_2	C_3	2.779	C_2	C_3	O_1	125.5°
C_3	O_1	2.287	C_2	C_3	C_4	109°
C_3	C_4	2.949	C_3	C_4	O_2	102°
C_4	O_2	2.665	C_4	O_2	C_1	102°
O_2	C_1	2.737	H_3	C_4	H_4	109.5°
C_4	H_3	2.079				
C_4	H_4	2.079				

by equation 10. Instead, the natural form of the Fermi hole for a correlated wavefunction is given by[17]

$$\Delta_{\alpha\alpha}(\vec{r}_1;\vec{r}_2) = \sum_i n_{\alpha\alpha i}(\vec{r}_2) \mid f_{\alpha\alpha i}(\vec{r}_1;\vec{r}_2) \mid^2 \tag{22}$$

where the Fermi hole natural orbital occupation numbers $n_{\alpha\alpha i}(\vec{r})$ sum to unity for all r. Likewise, the Coulomb hole defined by equation 9 may be expressed in the natural representation

$$\Delta_{\alpha\beta}(\vec{r}_1;\vec{r}_2) = \sum_i n_{\alpha\beta i}(\vec{r}_2) \mid f_{\alpha\beta i}(\vec{r}_1;\vec{r}_2) \mid^2 \tag{23}$$

where the Coulomb hole natural orbital occupation numbers $n_{\alpha\beta i}(\vec{r})$ sum to zero for all r.

If a correlated wavefunction is dominated by a closed shell Slater determinant, then the Fermi hole natural orbitals should include one with an occupation number near unity. The occupation numbers for the remaining Fermi hole natural orbitals, as well as those of the Coulomb hole natural orbitals, will be clustered about zero, including both positive and negative values. For example, a set of Fermi hole natural orbitals for a configuration interaction (CI) wavefunction for the water molecule[17] was found to have one orbital with an occupation number of 0.98861. The remaining Fermi hole natural orbital occupation numbers ranged from 0.01482 to -0.01461. The Coulomb hole natural orbital occupation numbers ranged from

0.12302 to -0.12309. These results were obtained for a probe electron located on one of the protons. This CI wavefunction was based on a double zeta set of atomic orbitals, and had a total energy of -76.12697 hartrees. This energy includes 84% of the correlation energy recoverable with this basis set.

The process of rupturing a covalent bond to create a pair of radical fragments cannot be described by a closed shell Slater determinant. The simplest example of this process is the dissociation of a hydrogen molecule into two hydrogen atoms. In this case, equation 2 reduces to

$$\Gamma(1,2;1',2') = \Psi(1,2)\ \Psi^*(1',2') \tag{24}$$

In addition, there is no Fermi hole because there is only one electron of each spin. Consequently, the Coulomb hole and the conditional density function (equation 8) form complementary descriptions of the electronic structure of the hydrogen molecule, each representing a single electron. These may be described as the probe electron, and the responding electron. Each of these functions depends on the position of the probe electron, but their sum is independent of this position.

The dependence of the conditional density function $\rho_c(\vec{r}_1|\vec{r}_2)$ for the hydrogen molecule on the position of the probe electron and inter-nuclear separation has been presented in a 16mm movie based on CI wavefunctions.[15] One of the contour maps used to make this movie is shown in Figure 9. The CI wavefunctions used to make this movie are based on a 4s/2p set of primitive gaussian orbitals. Orbital exponential factors of 19.2406, 2.89915, 0.65341, and 0.17758 were used for the 1s gaussian functions,[20] and exponential factors of 3.0 and 0.5 were used for each set of 2p gaussian functions.

The CI wavefunctions used to produce this movie and Figure 9 were each calculated in the following way:

1. A closed shell SCF wavefunction was calculated. The occupied σ_g orbital and the lowest vacant σ_u orbital were determined by this calculation.
2. A CI wavefunction was calculated based on the two orbitals determined in step 1 and 16 additional virtual orbitals determined by making symmetry-adapted linear combinations of atomic orbitals orthogonal to these orbitals and each other by the Schmidt procedure. This set of additional orbitals included 4 orbitals of σ_g symmetry, four orbitals of σ_u symmetry, and two sets each of π_g and π_u symmetries. These orbitals were used to form 46 Slater determinants, including all symmetry allowed orbital occupancies except those with two σ orbitals (other than those determined in step 1) singly occupied.

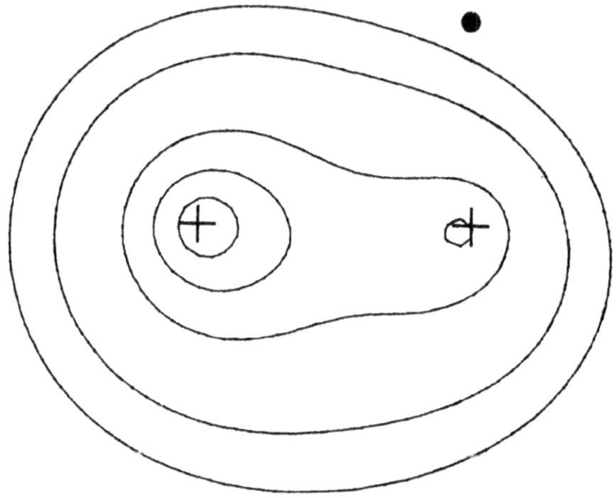

Fig. 9. The conditional density function $\rho_c(\vec{r}_1|\vec{r}_2)$ for the hydrogen molecule. The internuclear separation is 1.41 bohr. The probe electron $(\vec{r}_2)$, indicated by the solid circle, is located 1.0 bohr from one proton, and 1.7286 bohr from the other proton. The contour lines indicate densities of 0.025, 0.0431, 0.1, 0.15, and 0.2 electrons per cubic bohr.

3. The 16 highest occupancy natural orbitals determined by the CI wavefunction of step 2 were used to calculate a second CI wavefunction composed of 16 Slater determinants, each having a single doubly occupied orbital.

Steps 1 to 3 were repeated at each of 15 internuclear separations from 1.41 bohr (equilibrium) to 9.87 bohr. The energies of the wavefunctions obtained in these calculations are reported in Table 3.

At internuclear separations greater than 4 bohr, the conditional density function represents essentially two hydrogen atoms with small instantaneous polarization effects. If the probe electron is near one of the protons, then the conditional density function is strongly restricted to the vicinity of the other proton. As the internuclear separation decreases from 3 bohr to 2 bohr, the conditional density develops a secondary maximum on the proton near the probe electron. At the equilibrium internuclear separation, the conditional density is shared by both protons as it systematically tries to avoid the probe electron.

Table 3. Energies of hydrogen molecule wavefunctions. See the text for a description of basis sets, orbitals, and configurations. All energies are in hartrees except E_{bind}, which is in kcal/mole. The internuclear separations are given in bohr

R(H-H)	E_{SCF}	E_{CI-1}	E_{NO-CI}	E_{bind}
1.41	-1.13147	-1.163953	-1.166515	109.3
1.60	-1.12390	-1.158358	-1.160195	105.3
1.80	-1.10808	-1.144508	-1.145270	96.0
2.00	-1.08820	-1.126765	-1.126944	84.5
2.20	-1.06653	-1.107970	-1.107984	72.6
2.40	-1.04437	-1.089752	-1.089765	61.1
2.60	-1.02249	-1.072922	-1.073032	50.6
2.82	-0.99929	-1.056484	-1.056662	40.4
3.00	-0.98118	-1.044800	-1.045008	33.1
3.50	-0.93557	-1.020514	-1.020729	17.8
4.23	-0.88124	-1.002230	-1.002306	6.3
5.00	-0.83744	-0.995173	-0.995196	1.8
6.00	-0.79712	-0.992745	-0.992754	0.3
8.00	-0.75650	-0.992309	-0.992311	0.0
9.87	-0.74160	-0.992318	-0.992319	0.0

6. SUMMARY AND CONCLUSION

The second-order reduced density matrix contains much information related to localization and delocalization in electronic wavefunctions. The Fermi hole for an SCF wavefunction provides a direct and simple method for transforming canonical SCF molecular orbitals to localized SCF molecular orbitals. The resulting Fermi hole localized molecular orbitals are the molecular equivalent of the Wannier functions of solid state physics. These orbitals may be calculated as a preliminary step in the calculation of conventional localized molecular orbitals.

The sensitivity of the Fermi hole to the position of the probe electron is highly variable. Comparison of the Fermi hole mobility function $F(\vec{r})$ to the corresponding free-electron gas approximation $F_0(\rho)$ can be used to partition a molecule into a number of electronic domains, each associated with one or more localized molecular orbitals. A strongly localized pair of electrons is characterized by the association of a single localized molecular orbital with a specific domain. Delocalized or mobile electrons, such as those of a conjugated pi system, correspond to a domain which is associated with two or more localized molecular orbitals.

The Fermi hole and Coulomb hole for a correlated wavefunction may be expressed in terms of natural expansions which are parametrically dependent upon the co-ordinates of the probe electron. The corresponding occupation numbers sum to unity and to zero, respectively.

ACKNOWLEDGMENTS

Acknowledgment is made to the donors of the Petroleum Research Fund, administered by the American Chemical Society, for partial support of this research. Acknowledgment is also made to the Research Corporation for partial support of this research.

The calculations reported here were performed using the electronic structure program MOLECI developed in this laboratory. The gaussian orbital integral routines used in this program were adapted from GAUSS70.[33] Most of the calculations reported here were performed on a PDP-11/44 in the Chemistry Department at Duke University. The authors thank Mr. David Beratan and Mr. Joseph Leonard for assistance with these calculations.

REFERENCES

1. K. Husimi, Proc. Phys. Math. Soc. Japan 22, 264 (1940).
2. P.O. Löwdin, Phys. Rev. 97, 1474, 1490, 1509 (1955); Adv. Chem. Phys. 2, 207 (1959).
3. R. McWeeny, Proc. Roy. Soc. (London) A 232, 114 (1955); Rev. Mod. Phys. 32, 335 (1960).
4. R. McWeeny and B.T. Sutcliffe, Methods of Molecular Quantum Mechanics, (Academic, New York, 1969).
5. J.A. Pople, Trans. Faraday Soc. 49, 1375 (1953).
6. P.O. Löwdin and H. Shull, Phys. Rev. 101, 1730 (1956).
7. W. Kutzelnigg, J. Chem. Phys. 40, 3640 (1964).
8. W. Meyer, Int. J. Quant. Chem. Symp. 5, 341 (1971).
9. R.F.W. Bader and P.M. Beddall, J. Chem. Phys. 56, 3320 (1972).
10. R.F.W. Bader, P.M. Beddall, and J. Peslak, J. Chem. Phys. 58, 557 (1973).
11. A.K. Rajagopal, Adv. Chem. Phys. 41, 59 (1980).
12. R.G. Parr, in Horizons of Quantum Chemistry, edited by K. Fukui and B. Pullman (Reidel, Dordrecht, 1980).
13. E.R. Davidson, Reduced Density Matrices in Quantum Chemistry, (Academic, New York, 1976).
14. W.L. Luken and D.N. Beratan, Electron Correlation and the Chemical Bond, (Freewater Productions, Duke University, Durham, NC, 1980).
15. W.L. Luken and D.N. Beratan, Theor. Chim. Acta (Berlin), in press.
16. W.L. Luken and J.C. Culberson, Int. J. Quant. Chem. Symp. 16, to be published.

17. W.L. Luken, Int. J. Quant. Chem., to be published.
18. E.P. Wigner and F. Seitz, Phys. Rev. 43, 804 (1933); 46, 509 (1934).
19. J.C. Slater, Phys. Rev. 81, 385 (1951).
20. L.C. Snyder and H. Basch, Molecular Wavefunctions and Properties, (Wiley-Interscience, New York, 1972).
21. W.J. Hehre, R.F. Stewart, and J.A. Pople, J. Chem. Phys. 51, 2657 (1969).
22. J.E. Lennard-Jones and J.A. Pople, Proc. Roy. Soc. (London) A 202, 166 (1950).
23. S.F. Boys, Rev. Mod. Phys. 32, 296 (1960); S.F. Boys and J.M. Foster, Rev. Mod. Phys. 32, 300 (1960).
24. C. Edmiston and K. Ruedenberg, Rev. Mod. Phys. 35, 457 (1963).
25. C. Trindle and O. Sinanoglu, J. Chem. Phys. 49, 65 (1968).
26. J.M. Leonard and W.L. Luken, submitted to Theor. Chim. Acta (Berlin).
27. P.O. Löwdin, J. Chem. Phys. 18, 356 (1950).
28. G. Wannier, Phys. Rev. 52, 191 (1937).
29. J.C. Slater, J. Chem. Phys. 19, 220 (1951).
30. J.C. Slater, Quantum Theory of Molecules and Solids, Vol. I, (McGraw-Hill, New York, 1963).
31. M. Nyden and R.G. Parr, submitted to J. Chem. Phys; M. Nyden, submitted to J. Chem. Phys.
32. O. Sinanoglu, J. Chem. Phys. 36, 706, 3198 (1962).
33. W.J. Hehre, W.A. Lathan, R. Ditchfield, M.D. Newton, and J.A. Pople, GAUSSIAN 70, QCPE 236, Indiana University, Bloomington.

CHARACTERISTIC FUNCTION APPROACH TO DENSITY MATRIX CALCULATIONS

Metin Demiralp

Applied Mathematics Department, Marmara
Scientific and Industrial Research Institute
P.O. Box 21, Gebze-Kocaeli, Turkey

In this paper the system of ordinary differential equations which determines the density matrix elements is converted to a system of algebraic equations with the aid of a Laplace transform. Reorganization of the resulting equations makes it possible to deal with an inversion problem. The relation between the characteristic function and the solutions of these algebraic equations is demonstrated.

The evaluation of characteristic function by Padé approximants and the discussions for the convergence of Padé approximant sequences by using some properties of Stieltjes series and simple applications for illustrative purposes complete the paper.

1. INTRODUCTION

Using matrix notation, we can write the following equation for the density matrix $\hat{\rho}(t)$

$$i\hbar \frac{\partial \hat{\rho}}{\partial t} = \hat{H}\,\hat{\rho} - \hat{\rho}\hat{H} \; ; \quad \hat{\rho}(0) = A \tag{1}$$

where $\hbar$ and $\hat{H}$ denote respectively Planck's constant and the matrix representation of the Hamiltonian of the quantum mechanical system which is under consideration.[1]

If the matrix $\hat{H}$ does not vary with time, the formal solution of the Eq. (1) can be given as follows[2]

$$\hat{\rho} = e^{-i\tau\hat{H}}\, A\, e^{i\tau\hat{H}} \tag{2}$$

where the reduced time τ is defined as below

$$\tau = t/\hbar \tag{3}$$

Let us now define the following entities

$$x_j(\tau) = e^{-i\tau\hat{H}} u_j \tag{4}$$

$$\hat{\rho}_{jk} = x_j x_k^* \tag{5}$$

where u_j denotes the jth Cartesian unit vector in an infinite dimensional Euclidean space and the symbol * stands for the Hermitian conjugate. We can then write

$$\hat{\rho} = \sum_{j,k=1}^{\infty} A_{jk} \hat{\rho}_{jk} \tag{6}$$

However, as a short look at the structure of Eq. (4) shows, the following equation holds for x_j

$$\frac{dx_j}{d\tau} = -i\hat{H}x_j \ ; \quad x_j(0) = u_j \tag{7}$$

In fact, this equation is the matrix representation of Schrödinger's equation and is superior to the Eq. (1) due to the fact that a linear form is more suitable than the bilinear structure in Eq.(1) for the use of the characteristic function approach.[3] However, the infinite sum in Eq. (6) may be troublesome because of the convergence problem in the series, and this situation may generally necessitate the use of rational approximants of matrix series, which therefore may cause the loss of the attractiveness of Eq. (7).

On the other hand, without making an infinite expansion one can obtain an equation similar to (7) directly from Eq. (1), by suitable replacements among the elements of $\hat{\rho}$ and $\hat{H}$. Indeed, if we reorder the elements of $\hat{\rho}$ in an infinite vector and represent the resulting coefficient matrix which appears after the readjustment of the equation for this new ordering by M we can write

$$\frac{dr}{d\tau} = iMr \ ; \ r(0) = a \tag{8}$$

where a stands for the readjusted vector form of the matrix A. The infinite dimensional r can be denoted "density vector" since it contains all the information of the density matrix, in other words, all elements of the density matrix are included in "density vector".

2. SOLUTION FOR DENSITY VECTOR VIA CHARACTERISTIC FUNCTION

Let us, now, take the Laplace transform of Eq. (8)

$$(isI + M)\bar{r} = ia \tag{9}$$

where s and I represent the Laplace transform parameter and infinite dimensional unit matrix. The formal solution of this equation is

$$\bar{r}(s) = i(isI + M)^{-1}a \tag{10}$$

$\bar{r}_j$, the jth element of density vector can be expressed as

$$\bar{r}_j(s) = i\, u_j^T(isI + M)^{-1}a \tag{11}$$

or identically as

$$\bar{r}_j(s) = \frac{i}{2}\,[V_j^*(isI + M)^{-1}V_j - u_j^*(isI + M)^{-1}u_j - a^*(isI + M)^{-1}a] \tag{12}$$

where we assume that M is Hermitian and the symbol * denotes the Hermitian conjugate. The definition of V_j is as follows

$$V_j = u_j + a \tag{13}$$

The self-adjointness of M is more than an assumption. As a matter of fact, after a careful investigation of the structure of M, one can easily show that it is a reality.

The characteristic function, the zeros of which are the desired eigenvalues of the eigenvalue problem under consideration, can be defined as follows

$$\Delta(\phi,\Lambda|\lambda) = (\phi,[I-\lambda\Lambda]^{-1}\phi)^{-1} \tag{14}$$

where the scalar λ and the operator Λ appear in the following eigenvalue problem

$$(I-\lambda\Lambda)f = 0 \tag{15}$$

The functions f and ϕ represent respectively the eigenfunction of Eq. (15) and a normalized basis function which may be arbitrarily chosen except that it must satisfy the accompanying boundary conditions of Eq. (15).[3]

We can now rewrite the equation (12) in terms of characteristic function

$$\bar{r}_j(s) = \frac{-1}{2s}\,[\Delta(u_j,M|\tfrac{i}{s}) + a^*a\Delta(\alpha,M|\tfrac{i}{s}) - V^*V\Delta(\nu_j,M|\tfrac{i}{s})] \tag{16}$$

where the vector ν_j and α are defined at below

$$\nu_j = (V_j^* V_j)^{-1/2} V_j \tag{17}$$

$$\alpha = (a^* a)^{-1/2} a \tag{18}$$

3. PADE APPROXIMANTS FOR THE CHARACTERISTIC FUNCTION

The explicit expression for the characteristic function given by (14) is a special case of a characteristic function definition given in Ref. 3. In other words one can obtain the definition (14) by replacing Λ and W in Ref. 3 with I and Λ respectively. Therefore theorems given in the last two works of the author,[3,4] one of which is a slight generalization of the other, are valid for all Δs appearing in Eq. (16), in the case of a positive definite M. Hence, one can begin by expanding the reciprocal of the following characteristic function into the inverse powers of s as follows

$$\Delta^{-1}(\xi, M|\frac{i}{s}) = \sum_{j=0}^{\infty} \Delta_j(\xi, M)(\frac{i}{s})^j \tag{19}$$

where ξ stands for any one of the unit vectors ν_j, u_j, α and the dependence of Δ_j on ξ and M can be given as follows[3]

$$\Delta_j(\xi, M) = \xi^* M^j \xi \tag{20}$$

Then, one can easily prove that the serial expansion given by Eq. (19) forms a Stieltjes series if the coefficients Δ_js exist for all positive j values and M is a positive definite matrix. Therefore the Padé approximant sequences as [L+M/L]; M>-1, L=0,1,2,.. are convergent and are efficient tools for the rapid calculation of the density vector.[3,4]

The Padé approximant [L/M] is defined as the ratio of two polynomials as follows[5]

$$[L/M] = \frac{P_L(i/s)}{Q_M(i/s)} \; ; \; Q_L(0) = 1 \tag{21}$$

where the coefficients in these polynomials depend on Δ_js. Therefore one can get the following approximation

$$\Delta(\xi, M\,\frac{i}{s}) \approx \frac{Q_L(i/s)}{P_{L+M}(i/s)} \quad ; \quad M>-1 \tag{22}$$

However, the use of this approximation in Eq. (16) implies that

$$\bar{r}_j(s) \approx \frac{1}{2s}\,\frac{q_L(i/s)}{P_{L+M}(i/s)} \quad ; \quad M>-1 \tag{23}$$

where q_M and P_{L+M} represent respectively Mth and (L+M)th order polynomials of (i/s). Separating into partial fractions we can write

$$\frac{q_L(i/s)}{P_{L+M}(i/s)} = \sum_{k=1}^{L+M} \frac{\beta_k^{(M)}}{\gamma_k(i/s)-1} + \sigma_o^{(M)}(\delta_{M,0} + \delta_{M,-1}) + \frac{i}{s}\sigma_1^{(M)}\delta_{M,-1} \tag{24}$$

This however implies that

$$\bar{r}_j(s) \approx \frac{1}{2}\sum_{k=1}^{L+M} \frac{\beta_k^{(M)}}{i\gamma_k - s} + \frac{1}{2s}(\delta_{M,0} + \delta_{M,-1})\sigma_o^{(M)} + \frac{i}{2s^2}\sigma_1^{(M)}\delta_{M,-1} \tag{25}$$

By utilizing an inverse Laplace transform we have

$$r_j(\tau) \approx \frac{-1}{2}\sum_{k=1}^{L+M} \beta_k^{(M)} e^{i\gamma_k\tau} + \frac{1}{2}(\delta_{M,0} + \delta_{M,-1})\sigma_o^{(M)} + \frac{i\tau}{2}\delta_{M,-1}\sigma_1^{(M)} \tag{26}$$

Since the density vector elements must not increase unboundedly, either the last term of Eq. (26) must be zero or the value -1 of M must be discarded. The latter is preferable because it may always be realized, hence

$$r_j(\tau) \approx \frac{-1}{2}\sum_{k=1}^{L+M} \beta_k^{(M)} e^{-i\gamma_k\tau} + \frac{1}{2}\delta_{M,0}\sigma_o^{(M)} \quad ; \quad M>0 \tag{27}$$

As can be seen easily the approximate evaluation of r_j needs only $\sigma_o^{(M)}$; γ_ks, the zeros of certain polynomials; and $\beta_k^{(M)}$s; all of which are positive as can easily be proved.[5]

Let us now investigate the structure of M. For this purpose, we can use a diagonalization transformation in Eq. (1). If λ_j denotes the jth eigenvalue of $\hat{H}$, one can easily show that the jth row and kth column element of the right hand side of Eq. (1) is multiplied by $(\lambda_j-\lambda_k)$ after this transformation. By changing the transformed density matrix $\tilde{\rho}$ into an infinite vector form as follows

$$\tilde{\rho}^T = [\rho_{11}, \rho_{12}, \rho_{21}, \rho_{13}, \rho_{22}, \rho_{31}, \ldots] \tag{28}$$

and using a diagonal matrix $\tilde{M}$ whose elements are 0, $\lambda_1-\lambda_2, \lambda_2-\lambda_1, \ldots$ respectively one can easily arrive at the result

$$\frac{d\tilde{\rho}}{d\tau} = i\tilde{M}\tilde{\rho} \quad ; \quad \tilde{\rho}(0) = \tilde{a} \tag{29}$$

This, however, implies that the matrix M is not positive definite. On the contrary, it is singular, and has the eigenvalue pairs with the same magnitude but opposite signs. Therefore all the above discussions may not hold for the characteristic functions which appear in Eq. (16) because of the fact that the positive definiteness of M in the theorems mentioned before[3,4] is not a necessary but only a sufficient condition. In other words, by using other aspects of

Padé approximant theory[5,6] one may, perhaps, prove that some Padé approximant sequences for the characteristic function converge without necessitating the positive definiteness of M. However, the following decomposition of the characteristic function makes it possible to use the theorems mentioned before to construct a convergent scheme to evaluate the characteristic function of any self-adjoint matrix (or operator) Λ

$$\Delta^{-1}(\phi,\Lambda|\lambda) = (1+\lambda^2)(1-\frac{\lambda}{2})\Delta^{-1}(\phi,I+\Lambda^2|\frac{\lambda^2}{1+\lambda^2})$$
$$-(1+\lambda^2)\frac{\lambda}{2}(\phi,[I+\Lambda]^2\phi)\Delta^{-1}(\chi_1,I+\Lambda^2|\frac{\lambda^2}{1+\lambda^2})$$
$$-(1+\lambda^2)\frac{\lambda}{2}(\phi,\Lambda^2\phi)\Delta^{-1}(\chi_2,I+\Lambda^2|\frac{\lambda^2}{1+\lambda^2}) \tag{30}$$

In this equation the basis functions χ_1 and χ_2 are defined as below

$$\chi_1 = (\phi,[I+\Lambda]^2\phi)^{-1/2}\,[I+\Lambda]\phi \tag{31}$$

$$\chi_2 = (\phi,\Lambda^2\phi)^{-1/2}\Lambda\,\phi \tag{32}$$

One thing may be problematic in these definitions, which is that ϕ may be coincidentally chosen as the eigenfunction of Λ for the eigenvalues 0 or, if it appears, -1. However, such a choice implies that the eigenvalue problem can be reduced to another eigenvalue problem without such eigenvalues.

Now, our previous considerations apply the right hand side characteristic functions of Eq. (30) without worries about the convergence.

4. EVALUATION OF SPECTRAL COEFFICIENTS

Δ_is can be denoted "Spectral Coefficients" due to the fact that they have all information about the spectral properties of the problem under investigation.

Let us now try to find the relation between the matrices M and $\hat{H}$ by defining the following entities

$$\tilde{r}_j^T\ [\rho_{1j},\rho_{2j-1},\ldots,\rho_{j1}] \quad ; \quad j = 1,2,\ldots \tag{33}$$

$$r^T = [\tilde{r}_1^T,\tilde{r}_2^T,\ldots] \tag{34}$$

$$\tilde{a}_j^T = [a_{1j},a_{2j-1},\ldots,a_{j1}] \quad ; \quad j = 1,2,\ldots \tag{35}$$

$$a^T = [\tilde{a}_1^T,\tilde{a}_2^T,\ldots] \tag{36}$$

$$M = \begin{bmatrix} M_{11} & M_{12} & \cdots\cdots \\ M_{21} & M_{22} & \cdots\cdots \\ \cdots\cdots & & \end{bmatrix} \tag{37}$$

where M_{jk} denotes a $(j \times k)$ matrix.

If we compare the coefficient of $a_{\ell m}$ in the serial expansion of ρ_{jk} in terms of as in Eq. (2) with the same thing obtained from the formal solution of Eq. (8)

$$r = e^{i\tau M} a \tag{38}$$

we can write

$$u_j^T e^{-i\tau H} u_\ell u_m^T e^{i\tau H} u_k = u_{q(j,k)}^T e^{i\tau M} u_{q(\ell,m)} \tag{39}$$

where u_j is the jth Cartesian unit vector in infinite dimensional Euclidean space and the definition of $q(j,k)$ is as follows

$$q(j,k) = \frac{(j+k-1)(j+k-2)}{2} + j \quad ; \quad j,k = 1,2,\ldots \tag{40}$$

If we define the following vectors and matrix

$$\xi_j^T = [\xi_{1j}, \xi_{2j-1}, \ldots, \xi_{j1}] \quad ; \qquad j = 1,2,\ldots \tag{41}$$

$$\xi^T = [\xi_1^T, \xi_2^T, \ldots.] \tag{42}$$

$$X_{jk} = \xi_{jk} \; ; \quad j,k = 1,2,\ldots \tag{43}$$

we arrive at the following equation after the multiplication of each sides of Eq. (39) by $\xi_{kj}\xi_{\ell m}$

$$\mathrm{tr}(X\, e^{-i\tau H} X e^{i\tau H}) = \xi^* e^{i\tau M} \xi \tag{44}$$

The comparison of the coefficients of τ^j in the serial expansion of both sides of Eq. (44) implies

$$\Delta_j(\xi, M) = \sum_{k=0}^{j} (-1)^{j-k} \mathrm{Tr}(XH^{j-k} XH^k) \tag{45}$$

This form is more convenient for calculations than the previous one and it is attractive due to the fact that it does not necessarily need the matrix representations for investigation of operators.

5. AN ILLUSTRATIVE EXAMPLE

Let us consider the one dimensional anharmonic oscillator whose Hamiltonian is as follows

$$\hat{H} = -\frac{d^2}{dx^2} + x^2 + \varepsilon x^4 \tag{46}$$

If we want to find the elements of the density matrix for the system which is initially in the pure ground state of harmonic oscillator, we can write the initial value of the density matrix acting on a function F in Hilbert space spanned by harmonic oscillator eigenfunctions as follows[7]

$$\hat{\rho}(0)F(x) = \frac{1}{\sqrt{\pi}} e^{-\frac{x^2}{2}} \int_{-\infty}^{\infty} e^{-\frac{y^2}{2}} F(y)dy \tag{47}$$

This, however, shows that all we have to do is to evaluate the following integrals

$$J_{jk}^{(\ell)} = \int_{-\infty}^{\infty} e^{-\frac{x^2}{2}} H_j(x)\, \hat{H}^{\ell}\, e^{-\frac{x^2}{2}} H_k(x)dx \tag{48}$$

for the construction of the characteristic function, hence, for the evaluation of density matrix elements.

For illustrative purposes, we shall be contented with only the evaluation of the following characteristic function and not make further calculations toward density matrix elements due to the fact that all we want to show is the convergence of Padé approximants

$$\Delta(\xi, I + M^2 | \frac{\lambda^2}{1+\lambda^2}) = \sum_{j=0}^{\infty} \Delta_j(\xi, I + M^2) \frac{\lambda^{2j}}{(1+\lambda^2)^j} \tag{49}$$

where M is as previously mentioned and ξ is defined in terms of Kronecker's delta as follows

$$\xi_{jk} = \delta_{j1}\delta_{k1} \tag{50}$$

The definition of Δ_j given by Eq. (20) makes it possible to write

$$\Delta_j(\xi, I + M^2) = \sum_{k=0}^{2j} \binom{j}{k} \Delta_{2k}(\xi, M) \tag{51}$$

However, this result, Eq. (45) and the following integrals are sufficient to construct the Padé table for characteristic function

$$J_k = \frac{1}{\sqrt{\pi}} \int_{-\infty}^{\infty} e^{-\frac{x^2}{2}} \hat{H}^k e^{-\frac{x^2}{2}} dx \tag{52}$$

The comparison of some Padé approximants as a function of λ is given for $\varepsilon = 0.01$ in Fig. 1.

6. CONCLUSION AND REMARKS

The relation between the Laplace transform of the density matrix and the characteristic function can be used also for the evaluation of the expectation values of certain operators. Several convergence acceleration techniques are being investigated to improve the effectiveness of the approach presented. To this end certain transformations among several characteristic functions of different arguments are investigated.

Another point is to improve effective and rapidly convergent techniques for the calculations of spectral coefficients which may be very complicated depending on the nature of the system under investigation.

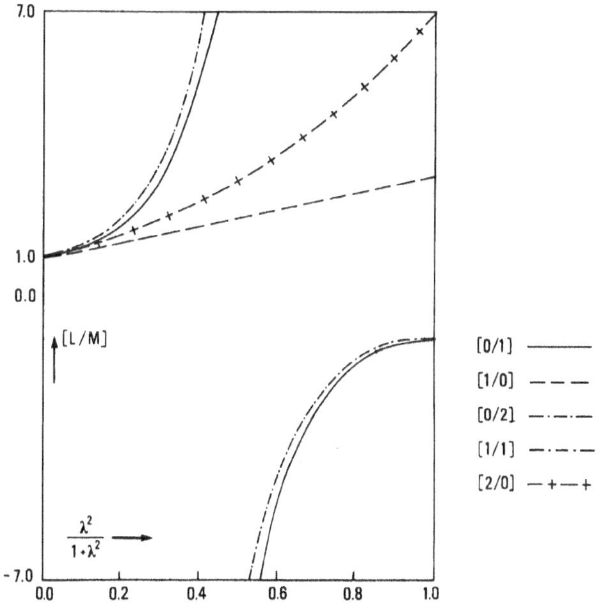

Fig. 1. Comparison of several Padé approximants for the reciprocal of the characteristic function $\Delta(\xi, I + M^2 \frac{\lambda^2}{\lambda^2+1})$ in the case where $\varepsilon = 0.01$.

The critical values of $(\lambda^2/1+\lambda^2)$ are 0.4959, 0.4888 and 0.4889 for [0/1], [1/1] and [0/2] respectively.

REFERENCES

1. A.S. Davydov, "Quantum Mechanics", Pergamon Press, New York (1965).
2. P. Lancaster, "Theory of Matrices", Academic Press, New York (1969).
3. M. Demiralp, A characteristic function approach to the discrete spectrum of electrically charged particles, J. Math. Phys. (in print).
4. M. Demiralp, Exponential factor optimization in characteristic function method for electrically charged particles, presented at IV Int. Congress of Quant. Chem., Uppsala (June 1982).
5. G.A. Baker, Jr., "Essentials of Padé Approximants", Academic Press, New York (1975).
6. G.A. Baker, Jr., and J.L. Gammel, "The Padé Approximants in Theoretical Physics", Academic Press, New York (1970).
7. L.I. Schiff, "Quantum Mechanics", McGraw-Hill, New York (1968).

ON THE SYSTEMATIC ASSESSMENT OF CORRELATION EFFECTS IN LOCAL DENSITY MODELS*

J.R. Sabin†

Kemisk Institut, Odense Universitet
Campusvej 55, DK-5230 Odense M
Denmark

S.B. Trickey

Quantum Theory Project
Departments of Physics and Chemistry
University of Florida
Gainesville, Florida 32611
U.S.A.

Since the early discussions and approximate applications of the Hohenberg-Kohn theorem, the question of the extent and nature of correlation contributions in various local density models has been studied, discussed, and disputed. We review a broad and representative sample of calculations, and show by systematic assessment that much of the confusion arises from inadequate attention to the distinction between single-particle or orbital excitation energies and total energies. On the basis of the normative definition of correlation in terms of total energies

$$E_{corr.} = E_{exact,nonrel} - E_{HF}$$

we suggest guidelines for efficient exploration of the problem and apply them to the extant data.

*Supported in part by the U.S. National Science Foundation under DMR 79-09721

†Permanent address: Quantum Theory Project, Departments of Physics and Chemistry, University of Florida, Gainesville, Florida 32611, U.S.A.

1. INTRODUCTION AND REVIEW

A recent paper[1] entitled "Is There Correlation in Xα?" is perhaps the latest in a long series of studies of a tantalizing question by a remarkably diverse and illustrious group of authors. Such studies date at least to Tong and Sham's 1966 numerical work on atoms,[2] which was based on the Kohn-Sham formulation[3] in 1965 of an approximation to the content of the Hohenberg-Kohn theorem,[4] and implicitly even in Slater's original 1951 paper[5] in which he identified his suggestion of a $\rho^{1/3}$ contribution to a one electron SCF equation as an exchange contribution. The history of these studies is one of divergent results, apparent contradictions, controversy, and dispute.

A reasonably representative sample of the literature must be digested to illustrate the point. We ask our readers to bear with us. We shall quote from various workers at some length in order to attempt to avoid misrepresentation.

Tong and Sham[2] studied atomic He, Ne, Ar, Kr, Li, Na, K, Rb, and O in terms of total energy, ionization potential (from total energy differences), and what they identified as exchange energies and correlation energies. The latter arise from inclusion in their model of an exchange-like contribution and also a correlation-like contribution of uniform electron gas form. (We use the phrases "exchange-like" and "correlation-like" to identify the origins of terms in model theories without prejudice as to whether the antecedent content of those terms carries over to the model in which they are invoked. We return to this distinction below.) Tong and Sham remark that "with some correlation effect taken into account" the calculated total energies are "in fair agreement with experiment but not as good as HF (Hartree-Fock)" (Ref. 2, p. 3). For an estimate of correlation energy they use

$$E_{corr} = E_{TS} - E_{X\alpha}(2/3) \tag{1}$$

where $E_{TS} = E_{Tong-Sham}$ is the total energy from their model and $E_{X\alpha}(2/3)$, in contemporary terms, is the Xα total energy with $\alpha = 2/3$. (Their model reduces to Xα with $\alpha = 2/3$ when their correlation-like contributions are removed.) On the basis of the various comparisons they conclude that in their model the "exchange-approximation is satisfactory" but that their "approximation to the correlation energy is poor".

A numerical study of ionization energies of SF_6 by Connolly and Johnson[7] is reasonably representative of the view that the Xα orbital eigenvalues (calculated in the so-called transition state, See Ref. 6) in particular correspond to a very reliable approximation to Hartree-Fock orbital eigenvalues. That such a view should arise is not at all surprising given that Slater originally introduced the $\rho^{1/3}$ ansatz as an approximation to the HF exchange potential.[5] Indeed Connolly and Johnson state early on that "the solutions of the Hartree-Fock equations can be approximated to a high degree of accuracy by the introduction of a local exchange potential". They describe the Xα transition state procedure for calculating one-electron excitation energies by pointing out the role of Koopmans' theorem in HF calculations, then go on to say "In the Xα approximation to the Hartree-Fock theory Koopmans' theorem no longer holds. However, an analogous theorem holds ..." The article concludes that the excitation energies and orbital symmetry ordering show that the Xα method provides a prescription whose accuracy" ... is close to that of an exact Hartree-Fock calculation." The values of α used by Connolly and Johnson were determined by Schwarz and Connolly[8] from matching the atomic total energy from Xα with the HF value. The title "Approximate Numerical Hartree-Fock Method for Molecular Calculations" of the Schwarz and Connolly paper[8] is of course a direct claim that Xα contains no correlation. The emphasis in these and many similar works is clearly on orbital (one-electron) properties with HF as the standard of merit.

However, both fairness and historical accuracy require one to note that not long after these papers appeared[7,8] Johnson, Norman and Connolly[9] published a review of Xα multiple-scattering techniques and applications in which the $\rho^{1/3}$ term is called the "Xα statistical approximation to exchange-correlation" and it is remarked that the Xα model"..., in conjunction with the multiple-scattered-wave procedure for solving the one-electron equations, has several advantages over the Hartree-Fock method and is therefore not just an approximation to the latter approach." To what the Xα scheme is an approximation those authors do not volunteer any speculation. A similar side-stepping is found in Mintmire and Sabin's comparison of LCAO-HF and LCAO-Xα on the CO molecule.[10] They interpret their results as demonstrating that LCAO-Xα is as reliable for calculating molecular properties as LCAO-HF, "... at least in terms of predicting experimental results." They venture no conclusions as to the origin of this putative reliability, though they do devote some circuitous discussion to the distinction between approximate HF and approximate HK theory.

Recently Wood[11] has presented a numerical study of multiplet energies in selected light atoms ($Z < 41$). Wood refers to Xα with $\alpha = 2/3$ as the "normal statistical exchange" and observes that Xα with α somewhat larger than 2/3 "evolved from observations that adjustments in α frequently led to better agreement between SCF one-electron energies and those determined by experiment ... Because of the improved agreement, it was natural - although probably incorrect - to believe that the use of α corresponded to the inclusion of correlation within the SCF problem ... If the correlation is present then we should obtain better agreement between calculation and experiment than is obtained in the HF comparison ... (Xα) reproduces the HF results quite well and in a fashion more or less independent of the choice of α." Wood then states that these results provide a convincing argument that Xα does not contain correlation beyond the normal statistical exchange and concludes the paragraph with "Of course, there was never any intent that the statistical approximation include correlation." Next he argues that his Local Spin Density (LSD)[12] results "... are in far better agreement with experiment then those obtained from the HF and the statistical exchange approximation which means that the LSD includes the correlation in an accurate fashion." In his conclusion there is the unqualified statement that Xα does not include correlation. We defer an analysis of Wood's methodology and conclusions except to note here his strong emphasis on single determinants, orbitals, and single-particle properties.

By contrast, Salahub et al.[1] have considered the ground state symmetry and the ordering of low-lying ionization potentials of O_3 as computed in HF and Xα (LCAO). They conclude from numerical data that the success of Xα (relative to experiment) vs. HF for this system indicates the presence of a "smoothing" of the relative sizes of the orbital-by-orbital contributions to the exchange-like part of the total energy. A similar smoothing in rigorous theory would arise from configuration mixing, so they speculate that perhaps Xα does include correlation in that sense. En route to their tentative conclusion they argue that because at least two determinants are required to get an acceptable ozone ground state and therefore a single determinant of symmetry-adapted functions is inadequate, it is "... clear that the accurate values for the IP's in Xα theory do not arise from taking account of correlation in the detailed sense just outlined." They also calculate so-called "Koopmans' Theorem," single-determinant, or eigenvalue IP's from both the Xα and the HF orbitals. From the roughly 85% agreement between the two sets of IP's and the consequently identical misordering they conclude that "the Xα MO's must be quite similar to those from HF ..." Again, we defer analysis except to note the emphasis here on single determinants and one-particle energies.

In a comparison of HF and local-density energy bands for rare gas crystals, Kunz and Mickish[13] speculated that it might be reasonable to argue that statistical exchange bands actually include some correlation effects and should not therefore be expected to agree with their HF results (bands gaps and band widths). However, they found the whole matter puzzling since, after they incorporated correlation and polarization effects (via the electron polaron model) the comparison between the HF one-electron properties they calculated and those from local density theory was not improved.

The perspective one gains from a survey of the literature on local-density calculations of cohesive energies, equilibrium lattice constants, and crystalline PV relations (or, analogously, dissociation energies, equilibrium bond lengths, and force constants in molecules) is almost equally muddled. For example, Janak, Moruzzi, and Williams[14] studied such ground state bulk properties of eight cubic crystals using the Gunnarson, Lundqvist, and Wilkins[15] or Hedin and Lundqvist[16] LSD functional. They found (from calculated static lattice quantities plus lattice corrections) cohesive energies within 20% of experimental values, lattice constants within 6%, and bulk moduli within 10%. They explicitly identified their work as being an exploration of an approximate and rather satisfactory realization of Hohenberg-Kohn theory, hence one which included some measure of correlation. They also noted that a major error in the atomic total energies they found was due to spatially rapid density variations in the atomic cores so that to the degree such core contributions are unaltered in crystal formation, a careful cohesive energy calculation can achieve a cancellation of this error. They also noted that the PV relation and equilibrium lattice constant are more sensitive tests of exchange correlation modeling than the cohesive energy since the former quantities do not depend on the atomic total energy at all.

Koelling[17] has argued somewhat to the contrary, that cohesive properties of solids "... cannot be a very sensitive test as the exchange correlation contribution to the total energy is a small one and its basic variation is the dominating $\rho^{1/3}$ component. It is possible to see that inclusion of a correlation term is required ... (It) is probably not possible to distinguish between the various homogeneous electron gas results utilised ..." (Note that Koelling uses "correlation terms" for what we prefer to identify as a "correlation-like term.") It should be noted that earlier in his review, Koelling introduces the $X\alpha$ method as an approximation to Hartree-Fock separate and distinct from local density methods and states that Wood[11] has shown (via his multiplet studies) that $X\alpha$ does not include correlation effects.

Quite the contrary interpretation has been put forth by Trickey et al. on the basis of a long series of systematic $X\alpha$ calculations on the fcc rare gas crystals (Ne, Ar, Kr, Xe).[18-20] In Hartree-Fock there can be no binding at all in these systems. In contrast the $X\alpha$ calculations showed binding of better than merely the right order of magnitude except for Neon - which was overbound ! As with the Moruzzi et al. metals work,[14b] they found very good comparison with experimental lattice constants (off by less than 5%) and PV relations (except for Ne for which the lattice constant differed from experiment by 13%). The calculations were done with the α parameter fixed at α_{VT},[21] except for Ne where $\alpha = 2/3$ was also used. Neither value, it should be noted, depends on the HF problem for the atom. In contrast to the success with bulk properties, the calculated one-electron excitation properties are very poor, even if one abandons the ordinary energy-band interpretation of the eigenvalues in favor of a ΔE_{SCF} scheme.[22] Because $X\alpha$ did what HF could not do for the ground state yet treated single particle excitations poorly, Trickey et al. concluded that Xα and similar local density prescriptions behaved "... in rough fashion as the rigorous HK density functional would." They interpreted such models as being related to rigorous HF theory as "... heuristic, intuitive prescriptions ... developed out of computational experience ... which simulate in a rough way much of the behaviour of the exact but unknown HK functional. The more realistic the simulation, the more nearly will the model be appropriate only for the ground state properties and not the single-particle excitations of the system."[20]

The binding in the fcc rare gas crystals was thought by some to be spurious, an artifact of the muffin tin methods used by Trickey et al. However, two different non-muffin-tin Xα calculations on Ne_2 also gave overbinding, thus laying to rest that particular misgiving.[23] An example of a more general study of Xα predictions of ground state properties in diatomic molecules is the work of Dunlap, Connolly, and Sabin.[24] In analogy with the work on crystals just discussed, they calculated the dissociation energy, equilibrium bond length, and vibrational frequency. Gaussian orbitals were used in an LCAO-Xα procedure and α was fixed at 0.7 for the seven molecules studied: H_2, B_2, C_2, N_2, CO, O_2, and F_2. Compared to experiment they found dissociation energies in error by as little as 4% (C_2) and as much as 93% (over binding in F_2). They noted, however, that in every case where comparative results were available, HF errors were of greater magnitude. Also the Xα dissociation energies appeared to be superior to the LSD results,[24] a point also made by Baerends and Ros.[25] The calculated equilibrium bond lengths were in remarkable agreement with experiment (4.3% too long for H_2, 2.6% too short for F_2, and for the other molecules, exact agreement with experiment !). Two aspects of the interpretation they placed on their results merit attention in the present context. First, both the Xα and the LSD calculations predicted the ground state of C_2 to be $^3\Pi_u$, a state which is found experimentally to lie about 0.1 eV above the actual ground state ($^1\Sigma_g^+$). The problem is one left-right correlation. Dunlap et al. comment that local density models "... can attempt to treat correlation (including left-right such as this) in an average way by adjusting the exchange-correlation functional ... However, for C_2, the triple-bond electrons and (the) localized singlet-coupled electrons have very different left-right correlation, perhaps untreatable by any local density functional." Second, they conclude that their ground state properties calculations support the view" ... that, compared to experiment, the local density models are excellent one-electron models."

The question of correlation also rises in the calculation of so-called other-than-energy (OTE) quantities (those which require some form of matrix element after completion of the SCF problem). The most oft-studied of these are probably the X-ray form factor and the Compton profile. Euwema and Suratt[26] studied both quantities in atomic Ne, Ar, and Kr. With the Xα parameter set to α_{VT}[21]

the HF and Xα form factors differed by less than 1% (at $\alpha = 2/3$ the largest deviation was less than 2%). On the other hand, the Xα Compton profile, especially in Ne, was high (with respect to the HF value, which itself should be above experiment because of correlation effects) at zero momentum transfer, a deviation in the wrong direction if one expected somehow the inclusion of correlation in Xα. Sabin and Trickey[27] studied Compton profiles from four local density schemes in atomic He, Li, Be, B and Ne. Their results at zero momentum transfer (J(0)) did not confirm Euwema and Suratt's analysis, since the best correlated wavefunction and best HF calculations both gave J(0) below the experimental value by almost exactly the same amount, about 1.3%. The Xα (α_{VT}) value for J(0), by comparison, was less than 0.3% high. This apparently gratifying finding was vitiated completely by their assessment that no one of the local density models could be shown to be preferable as compared with HF or correlated wavefunction results and a subsequent non-parametric statistical ranking test[20] which confirmed their evaluation. They concluded with the speculation that the X-ray form factors from Xα or similar schemes were of substantially better quality than Compton profiles from the same methods because "... the present local density models are suitable only for the calculation of the electron density and of 'bulk' properties (e.g. total energies) and not one-electron properties." They based this argument on the fact that the form factor depends only on the charge density (diagonal part of the one-particle reduced density matrix) while the Compton profile depends on the entire one-matrix to generate the momentum-space electronic density.

There is not even agreement in the literature on the proper derivation of the momentum-space electronic density in these local density models. Euwema and Suratt,[26] Sabin and Trickey,[27] Smith and Sabin,[28] Kaijser and Sabin,[29] Rath et al.[30], and a host of other authors (see Refs. 28 and 29 for other examples) made the connection by simple insertion of the local-density spin-orbitals into the natural spin-orbital form of the one-matrix.[31] A rather different procedure was employed by Lam and Platzman[32] to identify the momentum density directly from the (unknown) Hohenberg-Kohn density functional (or more explicitly, its exchange-correlation part). The resulting expression has a term like the HF momentum density (but with the Fourier transform of the local density spin-orbitals of course) plus a correction term which none of the authors cited above[26-30] included. Tong and Lam[33] used the Lam-Platzman development to calculate Compton profiles for atomic Ne, Ar, and Kr. In all three cases J(0) was slightly higher than that found from HF and somewhat below the experimental value. In the one case (Ne) for which a CI wavefunction was available, the Tong-Lam, HF, and CI results were very nearly indistinguishable, differing at most by ± 0.015, while the deviation from experiment was as much as twice as large. Since the Tong-Lam work is supposed to include correlation (by inclusion of correlation-like terms in the model energy functional) at the level of an approximate realization of Hohenberg-

Kohn theory, the inconclusive nature of their results illustrates the murky status of the question of inclusion of correlation.

2. CORRELATION: DEFINITIONS, MANIFESTATIONS, STANDARDS OF ASSESSMENT

The customary and accepted definition of correlation energy is[34]

$$E_{corr} = E_{exact} - E_{RHF} \tag{2}$$

where E_{exact} is the exact expectation value of the non-relativistic, clamped-nucleus Hamiltonian and E_{RHF} is the restricted Hartree-Fock expectation value for that same Hamiltonian. There are several grounds (beyond those of historical and customary usage in almost all of the many fermion literature) for insisting that any attempt to delineate correlation effects in a local-density model must begin with this definition.

First, of course, there is the fundamental result of the Hohenberg-Kohn theorem itself: if an exact LD model were to be constructed, the HK theorem guarantees that the model would deliver $E_{corr.}$ exactly for the system ground state. The HK theorem, in contrast, does not address the quasi-particle excitation spectrum. Second, basing the examination of correlation inclusion on the total energy difference (2) avoids the ambiguity contained in any quasi-particle treatment of a many-particle system as to the proper choice of bare quasi-particles. Thus, for example, Kurtz and Öhrn[35] have calculated ionization energies and electron affinities by a second-order perturbative treatment of the self-energy operator based on an unperturbed propagator that includes all relaxation contributions to the ionization energy. Clearly the unperturbed or bare quasi-particle spectrum in this case does not refer to the uncorrelated reference state in (2).

Third, use of the historical definition (2) as the benchmark for the meaning of "correlation" frees the discussion from the kind of confusion with which we have already had to cope in our distinguishing between correlation-like terms in a model expression for the total energy and the actual inclusion of correlation effects. To include correlation in a model it is not sufficient that the model include some term or terms which arose from effects beyond RHF for some specimen systems, e.g. the homogeneous electron gas. The historical definition (2) enables us to see what such terms in LD models do if we recall that one may always separate[34] the total second-order reduced density matrix (rdm) into HF and correlation parts:

$$\Gamma^{(2)}(x_1x_2|x_1'x_2') = \tfrac{1}{2}\begin{vmatrix} \rho_{HF}(x_1|x_1') & \rho_{HF}(x_1|x_2') \\ \rho_{HF}(x_2|x_1') & _{HF}(x_2|x_2') \end{vmatrix} + \Gamma^{(2)}_{corr}(x_1x_2|x_1'x_2') \quad (3)$$

Here ρ_{HF} is the usual HF invariant or first order rdm. In Rydberg au, with $g_{\mu\nu} = 2Z_\mu Z_\nu/r_{\mu\nu}$, $g_{1\nu} = -2Z_\nu/r_{1\nu}$, and $g_{12} = -2/r_{12}$, one has the direct and exchange contributions

$$E_{HF} = E_D + E_X \quad (4)$$

with

$$E_D = \tfrac{1}{2}\sum_{\mu\nu} g_{\mu\nu} + \int dx_1[p_1^2 + \sum_\nu g_{1\nu}]\,\rho_{HF}(x_1|x_1') \quad (5a)$$

$$+ \tfrac{1}{2}\int dx_1dx_2\,\rho_{HF}(x_1|x_1)\,\rho_{HF}(x_2|x_2)\,g_{12}$$

$$E_X = -\tfrac{1}{2}\int dx_1dx_2\,\rho_{HF}(x_1|x_2)\,\rho_{HF}(x_2|x_1)\,g_{12} \quad (5b)$$

Then $E = E_{HF} + E_{corr}$ with

$$E_{corr} = \int dx_1dx_2\,[\frac{2}{N-1}(p_1^2 + \sum_\nu g_{1\nu}) + g_{12}]\Gamma^{(2)}_{corr}(x_1x_2|x_1'x_2') \quad (6)$$

LD models typically do a different separation, motivated in part by the HK theorem and in part by the lack of any reasonably utilitarian (not to say convincing!) density functional for the kinetic energy. Thus the typical LD separation is presented as

$$E = E_H + E_{XC} \quad (7)$$

Here E_H is of Hartree form, namely the <u>same form</u> as the HF E_D (5a) but <u>not</u> with the same first order rdm <u>in general</u> while E_{XC} is in principle of the form of E_X plus the two-body coulombic part of E_{corr}. That is

$$E_{XC} = \int dx_1dx_2\,[\Gamma^{(2)}_{corr}(x_1x_2|x_1'x_2') - \tfrac{1}{2}\rho_{HF}(x_1|x_2)\,\rho_{HF}(x_2|x_1)]g_{12} \quad (8)$$

which in typical LD theory is presented as approximated via the density ρ_{LD} as

$$E_{XC} \approx \tfrac{1}{2}\int dx_1\,\rho_{LD}(x_1|x_1)\,U_{LD(x_1)} \quad (9)$$

and the particular model is defined by specification of U_{LD}. (Obviously LSD theories simply keep track of the spin density where we have used the density itself for simplicity.) By exclusion one deduces that LD theories also insist on the approximation

$$E_H \approx E_D + \int dx_1 dx_2 \left[\frac{2}{N-1}\left(p_1^2 + \sum_\upsilon g_{1\upsilon}\right)\right] \times \Gamma^{(2)}_{corr}(x_1 x_2 | x_1' x_2') \quad (10)$$

All extant LD schemes deviate from depending on the density only to write their kinetic energy term as a single particle rdm comprised of the lowest N spin-orbitals for an N-body system

$$\rho_{LD}(x_1|x_1') = \sum n_i \, \phi_i^*(x_1) \, \phi_i(x_1') \quad (11)$$

Thus, current LD schemes may be analyzed as making the approximation (9) as well as an approximation in (10) which is

$$\begin{aligned} &\int dx_1 \left[p_1^2 + \sum_\nu g_{1\nu}\right] \rho_{LD}(x_1|x_1') \\ &\quad + \tfrac{1}{2}\int dx_1 dx_2 \, \rho_{LD}(x_1|x_1) \, \rho_{LD}(x_2|x_2) \, g_{12} \\ &\approx \int dx_1 \left[p_1^2 + \sum_\nu g_{1\nu}\right] \\ &\quad \times \left[\rho_{HF}(x_1|x_1') + \frac{2}{N-1}\int dx_2 \, \Gamma^{(2)}_{corr}(x_1 x_2|x_1' x_2')\right] \\ &\quad + \tfrac{1}{2}\int dx_1 dx_2 \, \rho_{HF}(x_1|x_1) \, \rho_{HF}(x_2|x_2) \, g_{12} \end{aligned} \quad (12)$$

We may say therefore that to the extent, if any, that any current LD model is including correlation it is doing so <u>both</u> via the choice of U_{LD} in (9) and via the reshaping (with respect to HF) of the orbitals which occur in (11) in order to make (12) work. Even then there is no <u>a priori</u> way to ascertain the extent to which an error in E_{XC} (9) is compensated via an error in E_H ((10) and (12)), so that the correlation effects if present are entangled in subtle and interdependent ways in any LD total energy expression. It is for this reason that we argued above that it is highly misleading and very possibly incorrect to take an expression for the exchange or correlation energy in the uniform electron gas, include that expression in the construction of an LD model and claim to have included exchange or correlation. In fact what is contrived is an approximation to both exchange and correlation effects of a quality which is unknown <u>a priori</u>. Even at the level of Xα, the origins of a contribution do not tell unequivocally what that contribution is doing in the model.

We suggest therefore that there are strong arguments for requiring that any careful assessment by a numerical calculation of correlation inclusion in a LD model must meet at least these guidelines:

A) minimization of the model total energy expression and comparison of the result with the best available value of E_{exact};

B) elimination or, at least, control of basis set and similar methodological effects;

C) utilization of E_{RHF} from a RHF calculation which itself is as free of methodological effects as possible;

D) avoidance of inappropriate or unjustified imposition of single-determinant methodology in LD scheme (e.g. LD eigenvalues do not enjoy a Koopmans' theorem).

Since one rarely if ever has E_{exact} at hand, application of these guidelines frequently requires some careful use of experimental data. Immediately one point is evident: any comparison of calculated LD one-electron levels with experimental values is out of bounds unless it is based on total energy optimization (including geometry optimization for molecules and solids). This proscription extends to Slater transition state and similar molecular calculations which do not proceed from an optimized geometry since such transition-state calculations are supposed to yield orbital eigenvalues which approximate self-consistent total energy differences (ΔE_{SCF} values).

Judicious comparison of total energies with experiment may be furthered by cautious use of cohesive or binding energies, provided that the calculation treats the bound and separated species on the same footing. If such an equitable procedure is followed, one may have some hope that basis set effects, for example, might be subtracted out. Such a procedure may also be helpful in escaping the worst effects of an inadequate treatment of an isolated atom; see Sec. 3. Caution is the watchword however. In this connection, it may be noted that the total energy (in either LD or HF) is frequently less sensitive to subtle deficiencies in a basis than are the orbital energies, so that our insistence on the total energy as the normative quantity has practical value as well as the merit of principle. Even so, achievement of so-called convergence with respect to basis set size and parameters is an essential ingredient (cf. guideline B).

3. CONSEQUENCES

Perhaps the first and most personally honest thing to note is that virtually all LD calculations of energy bands in solids and many LD calculations of orbital energies in molecules which are said to be "in good agreement with experiment" (or some variant of this familiar litany) stand on very shaky ground when matters of correlation arise. Calculations which adjust the α in $X\alpha$ to achieve sensible energy bands, for example the recent work on Ne by Khan and Callaway,[36] can be both ingenious and very useful, but they say nothing about the extent to which some LD model copes with correlation since the adjustments needed in such a calculation typically yield very unphysical equilibrium total energies and geometries.[21]

We turn to the list of example references which we surveyed at the outset. The Tong-Sham atomic calculation seems to illustrate three points, one major and two of lesser importance. The major point is that atomic studies with LD models show just how hard it is to get such a model to reproduce E_{corr} (equation (2) above) at all: Tong and Sham's most elaborate model does not produce, in any instance, as low a total energy as HF ! No discussion of correlation makes much sense in this context, a problem of acute embarrassment and urgency we believe, to aficionados of LD methods. One of the minor points is that Tong and Sham used rather different definitions of exchange and correlation energies from those we have invoked, so that even if their models had behaved well (i.e. produced a markedly lower total energy than HF) their reported correlation energies might well not have been directly useful. The other minor point is simply that the total density $\rho(x|x)$ is not a good measure of the correlation inclusion in an LD model, even though $\rho(x|x)$ is a ground state property within the purview of the HK theorem. Tong and Sham compared the ρ from their model with that from HF. The differences were imperceptible on the scale of the plot for Ne for example, in spite of the total energy failure just discussed.

The SF_6 ionization energy study by Connolly and Johnson[7] provides a convenient example of what has become a common analysis. The local density transition state orbital energies were compared with HF eigenvalues and with experimental ionization energies. The ordering predicted by Xα was closer to the best available HF calculation (which itself was rather far from the authentic HF limit) than either of two less extensive calculations or two CNDO calculations. The Xα results were also in better agreement overall with experimental ionization energies than were the best HF eigenvalues. These comparisons led the authors to conclude that Xα accuracy was close to that of the exact HF treatment. This deduction rests on an implicit judgement that the experimental ionization energies in SF_6 are well described by the exact HF eigenvalues. That implicit claim, if true, would lead inexorably to the statement that the ionization energies in SF_6 are insensitive to correlation effects, in which case the Xα calculations reported shed no light whatsoever on the question of the inclusion of correlation in an LD scheme. If that implicit claim is not true, one might think that at least some mimicking of correlation by Xα had been found, since transition state excitation energies correspond to the lowest order (in occupation numbers) contribution to ΔE_{scf} in Xα. Such a hope is unjustified, however, in view of our guidelines A or D (Sec. 2) since neither the geometry of the neutral nor the ion was optimised.

The purpose of this critique is most emphatically not <u>ex post</u>

facto refereeing, but the presentation of a particularly accessible example of the futility of using orbital energy studies of this sort to address such an elusive question as the inclusion or reproduction of correlation effects. The extraordinary degree of that elusiveness is well illustrated by the tentative, noncommital discussion of Mintmire and Sabin[10] (c.f. the first two paragraphs of Sec. 3, Ref. 10). That discussion has its origins in the very realistic CO binding energy curve found in LCAO-Xα versus the much too shallow HF curve. It is argued there variously that a) the HF curve is too shallow because a single determinant cannot describe the binding of CO well, but that b) it would be misleading to associate the improved binding in Xα with correlational energy, since c) Xα should be viewed as approximate HK theory and not d) as an ansatz which contains systematic corrections beyond HF. (Reference 11 is incorrect, by the way, in stating that the HF energy functional "... depends on the determinantal wavefunction and not on the first order density matrix as in the Xα model." Obviously Xα depends on $\Gamma^{(1)}(1|1)$ and not on $\Gamma^{(1)}(1|1')$; Löwdin showed long ago[30] that HF is uniquely determined by $\Gamma^{(1)}(1|1')$.) It appears that Mintmire and Sabin intended to illustrate the utility of Xα vis a vis interpretation and prediction of chemical properties irrespective of the underlying theoretical structure. With the agnostic view that Xα is better than HF, is not semiempirical, but is not correlated either, these authors seem to suggest a *reductio ad absurdum* in which LD theory is disjoint from the rest of quantum theory. Their calculated data suggest somewhat otherwise when evaluated against our guidelines. Guideline A and B are satisfied while C is met at most in a restricted sense by use of the same basis for both the Xα and HF calculations. Compared with experimental data for the ground state binding energy, bond length, and fundamental frequency, the best available CI results are not too good; Xα results are better in each instance where a comparison can be made. This remark also holds for the calculated values of the four lowest ionization energies in CO. The Xα ionization energies were calculated at the energy-optimized ground state geometry by use of the Slater transition state. The possibility of a change in bond length was not explored, so that guideline A was not really shown to have been met. Nevertheless, the substantial success of the calculation might lead one to expect that, if such performance were realized on many systems, then the LD model could be said to be capable of reproducing correlation contributions.

Wood's work on multiplets[11] was put forward as, among other things, a clear counter example to that proposition for Xα and a confirmation for LSD. In terms of the guidelines we have suggested, however, his calculations must be interpreted much more cautiously with regard to inclusion of correlation because of some subtle conflicts with guidelines A and D. The deviation from A is that Wood used the same orbitals for all multiplet levels rather than optimising them for each level in turn. At least for low-lying excitations we surmise that this is not a severe limitation.

More significantly, however, Wood performed all his calculations by means of the diagonal sum rule so that he explicitly assumed that HF, Xα, and LSD are all single determinantal theories. Since no form of exact wavefunction has ever been associated rigorously with any of the currently popular LD or LSD schemes, either Wood's calculations are internally inconsistent in their interpretation or else they require an interpretation which goes substantially beyond LD theory. The assertion of internal inconsistency follows from the fact noted in connection with Mintmire and Sabin's work[10] that an energy optimized single determinant leads necessarilly to HF theory and hence excludes correlation. Wood's finding (recall Sec. 1) that LSD multiplets are in better agreement with experiment than HF because "... LSD includes correlation in an accurate fashion" is then a contradiction.

The only way out of this contradiction is to introduce the speculation that the LSD model generates an effective Hamiltonian. That is, if the model LSD functional is

$$E_{LSD} = E_{LSD}[\rho_\uparrow, \rho_\downarrow] \tag{13}$$

and

$$\delta E_{LSD}/\delta\rho_\uparrow = 0 \Rightarrow H_{eff,\uparrow}(\underline{r}_1)\{\phi_\uparrow(\underline{r}_1)\} = \{\varepsilon\}\{\phi_\uparrow(\underline{r}_1)\} \tag{14}$$

with similar for spin down, then the interpretation of Wood's LSD multiplet calculation as including correlation in single determinants constructed from the $\{\phi\}$ requires the conjecture

$$\langle\{D\}|H'|\{D\}\rangle/\langle\{D\}|\{D\}\rangle \simeq E_{LSD,CG} + \{\Delta_{LSD}\} \tag{15}$$

$E_{LSD,CG}$ = multiplet center of gravity evaluated via (13) (see Ref. 11, Sec. 4) (16a)

$\{\Delta_{LSD}\}$ = exchange-like and correlation-like contributions to particular multiplets (16b)

$\{D\}$ = manifold of single determinants (16c)

Note that one must conjecture the existence of an effective many body Hamiltonian H' which combines with a single determinant to yield a reliable estimate of the exact multiplet energies. (The single determinants are necessary because the diagonal sum rule was used.) We find such speculation implausible. At the very least, if it were correct it would be an inelegant way to rescue density functional theory by invocation of H', and {D} as computational artifacts (beyond the usual ones, namely, {ε} and {ρ}). Such a construct would intermingle density functional theory and wavefunction theory in a way which to us loses entirely the clarity and simplicity of the HK theorem.

We are forced to the reluctant conclusion that Wood's calculations achieve good multiplet energies from LSD by employment of the LSD orbitals in a manner which is not susceptible to straightforward interpretation. Assessment of the Xα results is even more problematical to us, since there is strong evidence (see discussion of Refs. 18-20 below) that the Xα total energies do not behave at all like the HF total energies, while the Xα orbital eigenvalues (the Xα{ε} analogous to (14)) do not of course behave at all like HF eigenvalues. We suspect, but cannot demonstrate, that the single determinant diagonal sum rule calculation used by Wood somehow biases his Xα results toward an agreement with HF which other, more straightforeward calculational evidence would not support.

The Salahub, Lamson and Messmer[1] paper which provoked this study tends to the contrary argument to that of Wood, yet uses single determinant constructs and conceptualizations throughout. Now, the generalized valence bond (GVB) treatment shows that O_3 cannot have a single determinant of symmetry adapted orbitals as its ground state. Therefore, Salahub et.al. argue that the good agreement of Xα O_3 ionization potentials with experiment cannot have resulted from Xα having taken "... into account correlation in the detailed sense just outlined" (i.e. GVB). The argument amounts to a declaration that Xα is a single determinant theory, an unexamined assumption which our guideline D attempts to proscribe.

Salahub et.al. go on to ascribe, tentatively, the success of Xα in O_3 (and perhaps other systems) to what they call the orbital-by-orbital smoothing of the exchange-like term in Xα relative to the exchange term in HF. (They identify explicitly the exchange-like term in Xα as pure exchange (c.f. their discussion on p. 431 of Ref. 1) even though in their Introduction they use the phrase exchange-correlation potential.) They argue that this smoothing accomplishes a sort of orbital stabilization relative to HF which would also, roughly speaking, be accomplished by configurational mixing. We suggest that the whole notion of examination of individual orbitals is suspect (c.f. guideline D) because it assumes that the occupied orbital and orbital energy manifolds {ϕ} and {ε} are somehow physically significant in an LD theory.

In fact, the {ϕ} and {ε} are simply artifacts for the construction of $\rho(\underline{r})$, with the ordering of occupance of the {ϕ} being given by Fermi statistics being applied to {ε}. It is commonplace to state, to the contrary, to wit that the original Kohn-Sham paper[3] "... showed that in principle it is possible to determine the exact one-particle properties of a many electron system by using a set of Schrödinger-type equations."[37] Such statements are incorrect, for Kohn and Sham never discussed the interpretation of either the {ε} or the $\{\phi\}$, nor did they propose anything like a Koopmans' theorem or any single determinantal concepts. Indeed, their Introduction emphasizes total energies, cohesive energies, elastic constants and the like; $\{\phi\}$ and $\{\varepsilon\}$ occur only as devices to build the central quantity, namely $\rho(\underline{r})$.

The disagreement between LD and HF energy bands in rare gas crystals discussed by Kunz and Mickish[13] also illustrates the difficulty in interpretation of the manifold of an LD calculation as being energy bands in the textbook sense. By now it is well known (see Ref. 20 for example) that the Slater transition state in $X\alpha$ does not differ from the ground state for an extended system, so that if one must use orbital eigenvalues to estimate one-electron properties such as a band gap one is stuck with using ground state orbital eigenvalues. Any discussion of correlation in a crystal from an LD model along these lines can nowadays be seen to be in trouble from the outset and it was that difficulty which motivated our guideline D in part.

Studies of cohesive energies and related quantities such as lattice constants and PV curves certainly avoid difficulty with guideline D. Both the work of Janak et.al.[14] and of Trickey et.al.[18-20] on such quantities seem to meet guidelines A and D and to some extent, B. The difficulty with B is that both bodies of work used the muffin-tin approximation. While Worth et.al.[23] and Moruzzi et.al.[14b] have argued that non-muffin-tin effects cannot be critical in closed packed systems, we know of no numerical tests of that contention. Within the limitations of the muffin-tin approximation both groups of workers did meet guideline B with regard to Brillouin zone scans, partial wave decomposition truncation, etc.

Guideline C has not been met at all and in fact cannot be met at this time, for reliable crystalline RHF calculations of cohesive energies and related quantities are simply unavailable. The argument advanced by Trickey et.al.[20] and quoted in Sec. 1 (to the effect that $X\alpha$ and its ilk are heuristic simulations of HK theory) rests on the observation that one knows without calculation the qualitative behavior of the RHF cohesive energy curve for a rare gas crystal, to wit: that curve is purely repulsive. Since the rare gas crystals are bound (except for He which does not bind under its own vapor pressure; its study is a specialty in its own right) and $X\alpha$ so predicts, the argument runs that $X\alpha$ must be doing what RHF cannot do, include the correlation. Thus guideline C is avoided, not literally met. The IBM workers[14] could not employ this circumvention since their work focussed almost entirely on metals.

Within the caveats just discussed, both groups of workers have characterized the calculated cohesive and pressure-volume properties as giving strong evidence that local-density models do in fact reproduce fairly well certain correlational effects. It is important to understand what such a characterization does not include, to wit: no claim is made that, referred to the same zero of energy, the local-density total energy vs. volume curve will lie below the corresponding RHF result. The total energy of atomic Ne (i.e. the infinite volume limit of the crystal) is some three Rydbergs higher in a local-density calculation than in RHF (see Ref. 2 for an early encounter with this fact; the result remains valid for models currently in vogue) when the LD parameters are set (i.e. $\alpha = 2/3$ in $X\alpha$) to yield at least a fairly decent description of the crystal. The clear deduction is that the entire LD total energy vs. volume curve lies above the RHF curve, but that the shape of the former is substantially more realistic than the latter.

LD calculations therefore simulate or mimic correlation contributions to crystalline cohesion by removing from consideration the biggest source of difficulty, namely intra-atomic correlation contributions. We have already remarked in our discussion of Tong and Sham's early work on atoms,[2] on the pervasiveness of this difficulty. If and when one succeeds in constructing an LD theory which achieves a substantial improvement in atomic total energies while leaving the crystalline properties largely unaltered, then calculations analogous to those by Trickey et.al. [18-20] and Moruzzi et.al.[14] might well be argued convincingly as taking correlation into account. The needed improvement will almost surely have to come from realistic treatment of core electrons since it is they alone which can influence the atomic energy noticeably yet be almost unnoticed in the crystalline or molecular cohesive energy.

Difficulties with the behavior of molecular orbitals and their effects on correlation manifested themselves in the Dunlap, Connolly, and Sabin[24] study of binding in diatomic molecules. Those authors discussed the prediction, by both $X\alpha$ and LSD, of the ground state symmetry of C_2 as $^3\Pi_u$ rather than the experimentally found $^1\Sigma_g^+$ in terms of a failure to treat left-right correlation suitably. A CI calculation[38] achieves the $^1\Sigma_g^+$ ground state by fractional promotion out of the (otherwise closed) $2\sigma_u$ orbital, a form of occupance which is energetically unfavorable in LD or LSD models (since the orbitals which generate the ground state energy fill exactly in order of their eigenvalues.)

There is, however, an even more acute problem implicit in the work of Dunlap et al.,[24] namely how does one identify the ground state symmetry legitimately in an LD or LSD theory ? What is usually done, of course, is to label the symmetry according to the orbital occupancy, just as if one were using a determinantal theory. Strict adherence to guideline D, however, would require that one use the density and, if need be, various calculations of excitation energies via total energy differences to identify allowed transitions. Even the calculation of such total energy differences is a bit suspect, however, even for exact HK theory since the latter applies only to non-degenerate ground states. A question thus arises. Suppose one had an approximate HK theory of known, arbitrarily high quality and suppose that one eschewed molecular orbital concepts as one should in HK theory. Could one identify the ground state symmetry ?

The Dunlap, Connolly, and Sabin diatomic study[24] does appear to support the interpretation advanced above that LD theories do emulate or mimic correlation contributions. Our guidelines are more or less well met by these calculations (except for the ground state symmetry question discussed above) and they do show better binding energies, force constants, and bond lengths than HF. However, the severe overbinding found in F_2 and O_2 illustrates very pointedly the uncontrolled and unpredictable nature of the Xα approximation (and, by inference, other, more complicated LD and LSD models).

The entire area of OTE quantities seems to us to be of almost no utility in assessment of correlation contributions to LD models at present. Even for atoms, the best LD total energies do not surpass the HF values as we have noted, while the HF and LD form factors are hardly distinguishable.[26] As noted in Sec. 1, the Compton Profile problem is in an even poorer state since construction of the proper momentum density is an unresolved question which we suspect has gone largely unnoticed. One potentially interesting aspect of the Lam and Platzman[32] analysis of the proper momentum density is that it might be possible to use their result to unravel the structure of the effective one particle rdm in a particular LD model. Since the momentum density is (c is a constant unimportant to the discussion)

$$\rho(\underline{p}) = c \int d\underline{r}_1 d\underline{r}_2 \exp[i\,\underline{p} \cdot (\underline{r}_1 - \underline{r}_2)]\, \Gamma^{(1)}(\underline{r}_1|\underline{r}_2) \qquad (17)$$

(with $\Gamma^{(1)}$ the true one particle rdm for the problem) and since Lam and Platzman have provided a functional derivative procedure for constructing $\rho(\underline{p})$ in LD theory independent of executing the Fourier transform (17), it might be possible to learn something about the form of $\Gamma^{(1)}$ from that $\rho(\underline{p})$ and (17).

Most of the other LD calculations of Compton profiles [26-30] ignored the Lam and Platzman analysis and assumed the one particle rdm to be the simplest possible generalization of the ordinary density. This assumption amounts to invoking a single-determinantal construction in violation of our guideline D. This violation, plus the criticism of such a proceedure which Lam and Platzman's analysis constitutes, leaves us with the conclusion that irrespective of their other interest, the extant Compton profile calculations can shed little light on the question of correlation inclusion.

4. CONCLUSIONS

Our assessment has necessitated a lengthy and arduous inspection tour of a great deal of the literature. A number of interesting papers have been omitted (for example, the decade old note by Collins et.al.[39] or the recent work by Jones[40]) even then, simply because illustration of the various aspects of the subject did not require contemplation of every paper no matter how noteworthy otherwise.

We have attempted to make several points convincingly. First, it is very difficult to perform numerical calculations which can serve as effective tests of the questions of the extent and nature of correlation inclusion. Second, much of the published work, while interesting on its own merits, is not instructive as to the correlation question in spite of explicit or implicit assertions to the contrary. Third, there is some evidence for the simulation of certain correlational effects, especially in metals and molecular crystals, but, fourth, there are significant (and embarrassing) problems with such simple things as atomic total energy values (for LD or LSD vs. HF) and the proper way to obtain the momentum density in LD theory.

ACKNOWLEDGEMENTS

We are grateful to the Editors of this volume, Jens Peder Dahl and John Avery, for this opportunity to indulge in a pre-emptive first strike. Helpful conversations with Murray Daw, Barry Kunz, and Yngve Öhrn are acknowledged with thanks. Part of this work was done while SBT was a guest at the Quantum Chemistry Group, University of Uppsala, for which opportunity he thanks Per-Olov Löwdin.

REFERENCES

1. D.R. Salahub, S.H. Lamson, and R.P. Messmer, Chem.Phys.Lett. 85, 430-433 (1982).
2. B.Y. Tong and L.J. Sham, Phys.Rev. 144, 1-4 (1966).
3. W. Kohn and L.J. Sham, Phys.Rev. 140, A1133-A1138 (1965).
4. P. Hohenberg and W. Kohn, Phys.Rev. 136, B864-B871 (1964).
5. J.C. Slater, Phys.Rev. 81, 385-390 (1951).
6. J.C. Slater, Adv. in Quantum Chem. 6, 1-92 (1972).
7. J.W.D. Connolly and K.H. Johnson, Chem.Phys.Lett. 10, 616-622 (1971).
8. K. Schwarz and J.W.D. Connolly, J.Chem.Phys. 55, 4710-4714 (1971).
9. K.H. Johnson, J.G. Norman, and J.W.D. Connolly in "Computational Methods for Large Molecules and Localized States in Solids," F. Herman, A.D. McLean, and R.K. Nesbet, editors (Plenum, New York, 1973), pp. 161-201, see p. 163 especially.
10. J.W. Mintmire and J.R. Sabin, Chem.Phys. 50, 91-103 (1980).

11. J.H. Wood, J.Chem.Phys. B Atom.Molec.Phys. 13, 1-14 (1980).
12. O. Gunnarsson and B.I. Lundqvist, Phys.Rev. B 13, 4274-4298 (1976).
13. A.B. Kunz and D.J. Mickish, Phys.Rev. B 8, 770-794 (1973).
14. a) J.F. Janak, V.L. Moruzzi, and A.R. Williams, Phys.Rev. B 12, 1257-1261 (1975). b) V.L. Moruzzi, A.R. Williams, and J.F. Janak, Phys.Rev. B 15, 2854-2857 (1977). c) see also J.F. Janak and A.R. Williams "Calculated Models of Metals" (Pergamon, Oxford, 1978).
15. O. Gunnarsson, B.I. Lundqvist, and J.W. Wilkins, Phys.Rev. B 10, 1319-1327 (1974).
16. L. Hedin and B.I. Lundqvist, J.Phys. C Solid State Phys. 4, 2064-2083 (1971).
17. D.D. Koelling, Rept.Prog.Phys. A 4, 139-212 (1981).
18. S.B. Trickey, F.R. Green, and F.W. Averill, Phys.Rev. B 8, 4822-4832 (1973).
19. J.R. Sabin, J.P. Worth, and S.B. Trickey, Phys.Rev. B 11, 3658-3661 (1975).
20. S.B. Trickey and J.P. Worth, Inter.J. Quantum Chem. S11, 529-538 (1977).
21. K. Schwarz, Phys.Rev. B 5, 2466-2468 (1972).
22. S.B. Trickey, A.K. Ray, and J.P. Worth, Physica Status Solidi B 106, 613-620 (1981).
23. J.P. Worth, B.I. Dunlap, and S.B. Trickey, Chem.Phys. Letters 55, 168-170 (1978).
24. B.I. Dunlap, J.W.D. Connolly, and J.R. Sabin, J.Chem.Phys. 71, 4993-4999 (1979).
25. E.J. Baerends and P. Ros, Inter.J. Quantum Chem. S12, 169-190 (1978).
26. R.N. Euwema and G.T. Surratt, J.Phys. C Solid State Physics 7, 3655-3660 (1974).
27. J.R. Sabin and S.B. Trickey, J.Phys. B Atom.Molec.Phys. 8, 2593-2600 (1975).
28. V.H. Smith, Jr., and J.R. Sabin, J.Phys. B Atom.Molec.Phys. 11, 385-389 (1978).
29. P. Kaijser and J.R. Sabin, J.Chem.Phys. 74, 559-563 (1981).
30. J. Rath, C.S. Wang, R.A. Tawil, and J. Callaway, Phys.Rev. B 8, 5139-5142 (1973).
31. P.-O. Löwdin, Phys.Rev. B 97, 1474-1489 (1955).
32. L. Lam and P.M. Platzman, Phys.Rev. B 9, 5122-5127 (1974).
33. B.Y. Tong and L. Lam, Phys.Rev. A 18, 552-558 (1978).
34. P.-O. Löwdin, Adv. Physics 5, 1-172 (1956), see especially p. 31.
35. H. Kurtz and N.Y. Öhrn, J.Chem.Phys. 69, 1162-1167 (1978).
36. M.A. Kahn and J. Callaway, Phys. Letters A 76, 441-444 (1980).
37. A.K. Theophilou, J.Phys. C Solid State Physics 12, 5419-5430 (1979).
38. P.F. Fougere and R.K. Nesbit, J.Chem.Phys. 44, 285-298 (1966).
39. T.C. Collins, R.N. Euwema, and A.B. Kunz, Inter.J. Quantum Chem. S6, 459-461 (1972).
40. R.O. Jones, J.Chem.Phys. 76, 3098-3101 (1982).

RELATIVISTIC DENSITY FUNCTIONAL THEORY

E.K.U. Gross and R.M. Dreizler

Institut für Theoretische Physik der Universität
Frankfurt, Robert-Mayer-Str. 8
D-6000 Frankfurt-Main, Fed. Rep. Germany

1. INTRODUCTION

Density functional methods have proven to be a successful tool in the treatment of a variety of atomic, molecular and solid state physics problems. The basic idea is to represent the total energy, as well as other quantities of interest, by functionals of the one particle density. According to the theorem of Hohenberg and Kohn,[1] this should in principle be possible even for the exact ground state energy. For relativistic systems, a similar statement has been demonstrated by Rajagopal.[2] The exact groundstate energy functional however is not known. Therefore, we have to rely on approximate forms.

In terms of the one- and two-particle density matrices the exact total energy of an interacting Coulomb-system can be written as

$$E = E_{kin} + E_{en} + E_{ee} + E_{ex} + E_{cor}$$

with (natural units $\hbar = c = 1$ are used throughout this paper)

$$E_{kin} = \int d^3r \int d^3r' \delta(\underline{r}-\underline{r}') \left[- \frac{\Delta_r}{2m} \rho(\underline{r},\underline{r}') \right] \equiv \int d^3r \tau(\underline{r}) \qquad (1)$$

$$E_{en} = \int d^3r \rho(\underline{r}) v_n(\underline{r}) \qquad (2)$$

$v_n(\underline{r})$ is the nuclear potential corresponding to either atomic, molecular or solid state situations.

$$E_{ee} = \frac{e^2}{2} \int d^3r \int d^3r' \frac{\rho(\underline{r})\rho(\underline{r}')}{|\underline{r}-\underline{r}'|} \qquad (3)$$

$$E_{ex} = -\frac{e^2}{2}\int d^3r\int d^3r' \frac{\rho(\underline{r},\underline{r}')\rho(\underline{r}',\underline{r})}{|\underline{r}-\underline{r}'|} \equiv \int d^3r\, e_x(\underline{r}) \tag{4}$$

$$E_{cor} = \frac{e^2}{2}\int d^3r\int d^3r' \frac{\rho^{(2)}(\underline{r}\underline{r}',\underline{r}\underline{r}')}{|\underline{r}-\underline{r}'|} - E_{ee} - E_{ex} \tag{5}$$

For the sake of brevity spin degrees of freedom are suppressed here. For later use we have expressed the kinetic energy as well as the exchange energy by the corresponding energy densities $\tau(\underline{r})$ and $e_x(\underline{r})$. Since E_{en} and E_{ee} are already functionals of the one particle density it remains to find an approximate representation of E_{kin}, E_{ex} and E_{cor} in terms of the density.

At present there are two regimes, for which approximate forms of these functionals have been discussed. In the first regime we have small, but otherwise arbitrary deviations from a constant density

$$\rho(\underline{r}) = \bar{\rho} + \delta\rho(\underline{r}) \quad , \quad |\delta\rho(\underline{r})|/\bar{\rho} << 1$$

The situation is most appropriate for the discussion of solid state problems. An approximate form of the functional in question can then be derived by the many body methods used for the discussion of the inhomogeneous electron gas.

The second regime corresponds to situations with a slowly varying but otherwise arbitrary density. This condition is usually specified by

$$|\underline{\nabla}\rho|/\rho << K_F$$

This situation is supposed to apply to atomic, molecular and nuclear systems. An approximate form of the functional can be derived by the gradient expansion technique, first introduced by Kirzhnitz.[3] This expansion technique has a strong parallel to the semiclassical expansion of Wigner and Kirkwood.[4]

Once the functionals have been derived within some approximation, the next question to be answered is how they can be applied to atomic and molecular physics problems. There are essentially three options:

Option I: Derive a set of variational equations

$$\delta\rho\left(E_o[\rho] + v_o \int\rho(\underline{r})d^3r\right) = 0 \tag{6}$$

from which the density (and hence the ground state energy and other quantities derived from the density) can be determined.

For example, neglecting the exchange-correlation part and approximating the kinetic energy by the free electron gas expression

$$\tau(\underline{r}) = \frac{3}{10m} (3\pi^2)^{2/3} \rho(\underline{r})^{5/3} \tag{7}$$

one obtains the classical Thomas-Fermi (TF) model.[5] The variational equation (6) can be transformed into the following well known differential equation for the total electrostatic potential V

$$\Delta V = \sigma \cdot (-V)^{3/2} \quad , \quad \sigma = -\frac{8\sqrt{2}}{3\pi} e^2 m^{3/2} \tag{8}$$

(The Lagrangian multiplier V_0 vanishes for neutral systems.) The solution of this equation gives a rough picture of atomic and molecular properties which does not contain any shell effects.

Option II: If a more detailed analysis of the system is required we rely only on the density functional representation of the exchange and correlation energy and use it to set up a Hartree-Fock (HF)-Slater like problem

$$\left[\hat{t} + \hat{v}_n + \frac{\delta E_{ee}}{\delta\rho} + \frac{\delta}{\delta\rho} (E_{ex}+E_{cor})\right] \phi_i = e_i \phi_i \tag{9}$$

from which orbitals and orbital energies are determined in a selfconsistent fashion.

This procedure yields high precision results for atoms and light molecules.[6] In the case of heavier molecules, the selfconsistent calculation need not be carried through for all electrons of the system; the inner electron shells are usually well described by frozen atomic cores since the molecular binding distance is very large compared to the extension of these orbitals. However, if one aims at the quasimolecular description of atomic scattering processes this approximation cannot be maintained since the internuclear separation may become very small during the collision process. Therefore, all electrons have to be included in the selfconsistent solution of Eq. (9). In the case of heavy or superheavy quasimolecules, such as the U-U system, the selfconsistency problem thus becomes extremely tedious if soluble at all. Here a third possibility becomes more attractive which is essentially a combination of the first two options:

Option III: By first solving the variational equation (6) we obtain an approximate density of the system. This density is then inserted in the potential terms

$$\frac{\delta}{\delta\rho} (E_{ee}+E_{ex}+E_{cor})$$

and remains fixed during the subsequent solution of the orbital problem (9). The hope is to obtain reasonably consistent results without tackling the selfconsistency problem. This scheme was first proposed for atoms in 1955 by R. Latter[7] and has later been applied to quasimolecular scattering systems.[8]

For relativistic systems, the main interest has been focussed upon the investigation of higher relativistic orders of the electron-electron interaction such as the Breit or the full transverse interaction.[9] Using option II, important corrections have been found due to the transverse exchange potential. Although these calculations have generally been quite successful the status of the whole relativistic theory cannot be considered completely satisfactory at this stage since there are problems with the kinetic energy representation. These problems date back to the development of the relativistic TF(RTF) model in 1932:[10] If, in parallel to the nonrelativistic TF model, the kinetic energy representation of a homogeneous electron gas

$$\tau_o^{rel} = \frac{m^4}{\pi^2}\left(\frac{1}{8}[x(2x^2+1)(x^2+1)^{1/2} - \text{Arsh } x] - \frac{1}{3}x^3\right) \tag{10}$$

with

$$x = \frac{(3\pi^2)^{1/3}}{m}\,\rho(\underline{r})^{1/3} \tag{11}$$

is used to derive a variational equation (6) for relativistic systems one obtains the RTF equation (for neutral systems)

$$\Delta V_{rel} = \sigma\cdot(-V_{rel})^{3/2}\left[1 - \frac{V_{rel}}{2m}\right]^{3/2} \tag{12}$$

This nonlinear differential equation for the total electrostatic potential V_{rel} (and hence ρ) is to be compared with the nonrelativistic TF equation (8). The relativistic correction

$$\left[1 - \frac{V_{rel}}{2m}\right]^{3/2}$$

shows some reasonable features: It increases with increaing Z and it contributes particularly for small values of r, i.e. for the innermost electrons. On the other hand, however, it overestimates the relativistic effects: Since the right hand sides of equation (8) and (12) are proportional to the density we obtain in the vicinity of the atomic nucleus

$$\rho_{TF} \underset{r \to 0}{\sim} r^{-3/2}$$

$$\rho_{RTF} \underset{r \to 0}{\sim} r^{-3}$$

i.e. the RTF density is not integrable. It can in fact be shown that the RTF equation (12) does not even possess solutions for point-like nuclear boundary conditions.[11]

There is a very hesitating history of attempts to correct this deficiency: The simplest remedy already suggested by Jensen in 1933[12] is the introduction of an extended nucleus instead of the point source. The problem is however: The nuclear dimensions are so small on the atomic scale that the incorrect $1/r^3$ behavior sets in before the cut-off procedure takes over.

Therefore, Ashby and Holzman[13] introduced a hybrid model by using simple relativistic K-shell orbitals to describe the inner part of the density distribution and matched the TF density at a distance r_o from the point charge. As a condition to determine the point r_o they required the kinetic energy functional to be continuous at this point. It turns out that r_o moves inward with increasing Z so that no solution could be found for Z>87.

A look at the nonrelativistic case suggests that the deficiency can be remedied by considering the gradient terms. The first suggestion along these lines is due to Rudkjøbing[14] who proposed an old version of the once iterated Dirac equation as a starting point. In this equation the spin orbit term is made explicit and contributes to the density of states. The resulting correction for the variational equation is then

$$\Delta V_{rel} = -\frac{4e^2}{3\pi}\left[(m-V_{rel})^2 - m^2 - \left(r\,\frac{dV_{rel}}{dr}\right)^2\right]^{3/2}$$

If the spin orbit term is neglected, this equation reduces to the RTF equation (12). In the limit $r \to 0$ the right hand side (i.e. the density) is proportional to $r^{-3/2}$. Unfortunately the solution does not only behave like the nonrelativistic solution at the origin, it agrees with the classical solution practically completely. The correction goes in the wrong direction.

In this paper, we shall present a systematic derivation of relativistic density functionals on the basis of the Dirac equation. The formalism is based on two essential points:

a) the gradient expansion technique first introduced by Kirzhnits[3] which, in principle, yields an exact representation of the HF energy of the system,

b) a precise prescription how to subtract the infinite energy contribution of the filled Dirac sea.

Section 2 will give a detailed description of the Kirzhnits formalism along with some applications to nonrelativistic two-center scattering systems. The relativistic theory is then developed in Section 3. The divergence problems of the traditional RTF-model will be found to be a consequence of an incorrect subtraction of the Dirac sea contribution. On the basis of our formulation we shall finally derive a new relativistic exchange term and a second order gradient correction to the kinetic energy density, i.e. a relativistic Weizsäcker correction. Some preliminary numerical results for relativistic atoms will be given in the final part of Section 3.

2. BASIC FORMALISM AND NONRELATIVISTIC APPLICATIONS

In the following, we shall describe in detail the so-called gradient expansion of the density matrix. Within this paper we restrict ourselves to the HF-limit, i.e. we neglect the correlation energy

$$E_{cor} = 0$$

In this limit, the one particle density matrix can be written as

$$\rho(\underline{r},\underline{r}') = \sum_\alpha n_\alpha \langle \underline{r}|\alpha\rangle\langle\alpha|\underline{r}'\rangle \tag{13}$$

where the one particle orbitals are solutions of the HF variational equation

$$(\hat{t} + \hat{v}_{eff})\,|\alpha\rangle = \varepsilon_\alpha|\alpha\rangle \tag{14}$$

A representation of the one particle density operator can be derived as follows: For the ground state, the occupation numbers n_α (1 for occupied levels, 0 for unoccupied levels) can be represented with a step function involving the Fermi-level ε_F

$$\langle\underline{r}|\hat{\rho}|\underline{r}'\rangle = \sum\!\!\!\!\!\!\int_\alpha \theta(\varepsilon_F-\varepsilon_\alpha)\langle\underline{r}|\alpha\rangle\langle\alpha|\underline{r}'\rangle$$

or in operator form

$$= \sum\!\!\!\!\!\!\int_\alpha \langle\underline{r}|\theta(\varepsilon_F-\hat{t}-\hat{v}_{eff})\,|\alpha\rangle\langle\alpha|\underline{r}'\rangle$$

By use of the completeness of the HF single particle basis we obtain

$$= \langle\underline{r}|\theta(\varepsilon_F-\hat{t}-\hat{v}_{eff})\,|\underline{r}'\rangle$$

For convenience we introduce a "Fermi energy operator" by

$$\hat{E}_F = \varepsilon_F - \hat{v}_{eff}$$

and extract from the simple argument above, the well known representation

$$\hat{\rho} = \theta(\hat{E}_F-\hat{t}) \tag{15}$$

Since the kinetic energy may be calculated from the trace

$$E_{kin} = \text{tr}\ \hat{t}\hat{\rho} = \int d^3r\ \langle \underline{r}|\hat{t}\hat{\rho}|\ \underline{r}\rangle$$

we readily obtain for the kinetic energy density

$$\tau(\underline{r}) = \langle \underline{r}|\hat{t}\hat{\rho}|\ \underline{r}\rangle \tag{16}$$

In order to evaluate the density matrix in a systematic fashion, we first insert the complete system of plane waves $|\underline{k}\rangle$

$$\rho(\underline{r},\underline{r}') = 2\int d^3k\ \langle \underline{r}|\theta(\hat{E}_F-\hat{t})|\underline{k}\rangle\langle\underline{k}|\underline{r}'\rangle \tag{17}$$

Although the plane wave states are eigenfunctions of the kinetic energy operator

$$\hat{t}|\underline{k}\rangle = \frac{k^2}{2m}|\underline{k}\rangle$$

it will in general not be correct to write

$$\Theta(\hat{E}_F-\hat{t})|\underline{k}\rangle = \Theta(\hat{E}_F - \frac{k^2}{2m})\ \underline{k}\rangle \tag{18}$$

since the operators $\hat{E}_F$ and $\hat{t}$ do not commute. Nevertheless, we shall adopt (18) for a while as a lowest order approximation; the effects due to the actual noncommutativity will be discussed below in a systematic way. Further we assume that the HF problem (14) can be approximated to any required accuracy by an appropriate <u>local</u> problem (knowledge of the corresponding local potential is not required as it will be eliminated in the end). Then we obtain for the density matrix

$$\rho_o(\underline{r},\underline{r}') = \frac{2}{(2\pi)^3}\int d^3k\,\theta(E_F(\underline{r}) - \frac{k^2}{2m})e^{i\underline{k}\cdot(\underline{r}-\underline{r}')} \tag{19}$$

Introducing a "position dependent Fermi momentum"

$$K_F(\underline{r}) = \sqrt{2mE_F(\underline{r})}$$

the integral above is readily evaluated yielding

$$\rho_o(\underline{r},\underline{r}') = \frac{K_F^2}{\pi^2}\, j_1(K_F\underline{y})\,\frac{1}{y} \quad \text{with } \underline{y} = \underline{r}-\underline{r}' \tag{20}$$

In the limit $y\to 0$ one obtains

$$\rho_o(\underline{r}) = \frac{1}{3\pi^2}K_F^3(\underline{r}) \tag{21}$$

A similar calculation gives for the kinetic energy density

$$\tau_o(\underline{r}) = \frac{2}{(2\pi)^3} \int d^3k \, \frac{k^2}{2m} \, \theta(E_F(\underline{r}) - \frac{k^2}{2m}) \\ = \frac{1}{10m\pi^2} K_F^{\,5}(\underline{r}) \qquad (22)$$

Eliminating the unknown local Fermi momentum K_F in favor of the density one obtains for the kinetic energy

$$\tau_o(\underline{r}) = \frac{3}{10m} (3\pi^2)^{2/3} \rho_o(\underline{r})^{5/3} \qquad (23)$$

We note one particular point: At the intermediate level (Eqs. (21) and (22)) we have a turning point problem, i.e. the quantity

$$K_F(\underline{r}) = [2m(\varepsilon_F - v_{eff}(\underline{r}))]^{1/2}$$

vanishes on some surface in the asymptotic region (since v_{eff} is purely attractive and $\varepsilon_F<0$) and is not defined outside. With the elimination of K_F in favor of ρ we have implicitly set up a continuation into the (semi-classically forbidden) outer region.

In order to go beyond the lowest order approximation we are left with the problem of calculating the operator function

$$f(\hat{a}+\hat{b})|a\rangle$$

for the case that $|a\rangle$ is an eigenstate of the operator $\hat{a}$, and that $\hat{a}$ and $\hat{b}$ do not commute. The question is resolved by the following argument. First, we use the Laplace representation

$$f(\hat{a}+\hat{b})|a\rangle = \int d\lambda \, c(\lambda) \, \exp[\lambda(\hat{a}+\hat{b})]|a\rangle$$

and the factorisation of the exponential operator

$$\exp[\lambda(\hat{a}+\hat{b})] = \exp[\lambda\hat{b}]\hat{K}(\lambda)\exp[\lambda\hat{a}] \qquad (24)$$

For the operator $\hat{K}$ a differential equation of the form

$$\frac{d\hat{K}}{d\lambda} = [\hat{a},\hat{K}] + \exp(-\lambda\hat{b})[\hat{a},\exp(\lambda\hat{b})]\hat{K}(\lambda) \qquad (25)$$

with the initial condition

$$\hat{K}(0) = 1$$

can be derived. In order to solve this equation one expands $\hat{K}$ into a power series in λ

$$\hat{K}(\lambda) = \sum_{n=0}^{\infty} \lambda^n \hat{O}_n \qquad (26)$$

For the operators $\hat{O}_n$ one obtains from the differential Eq. (25) the recursion formula

$$\hat{O}_n = \frac{1}{n}\left([\hat{a},\hat{O}_{n-1}] + \sum_{i=1}^{n} \hat{C}_i\, \hat{O}_{n-1-i}\right) \qquad (27)$$

with

$$\hat{C}_i = \frac{(-1)^i}{i!}\,\underbrace{[\hat{b},[\hat{b},[\ldots[\hat{b},a]\ldots]]]}_{i \text{ times}}$$

The initial values for the recursion are

$$\hat{O}_o = 1 \qquad \hat{O}_1 = 0$$

Then we proceed as follows: By use of Eq. (24) the exponential operator acting on the eigenfunctions of $\hat{a}$ gives

$$\begin{aligned}\exp[\lambda(\hat{a}+\hat{b})]\,|a\rangle &= \exp[\lambda\hat{b}]\hat{K}(\lambda)\ \exp[\lambda\hat{a}]\,|a\rangle \\ &= \exp[\lambda(\hat{a}+\hat{b})]\hat{K}(\lambda)\,|a\rangle\end{aligned}$$

Here we have recollected the exponential operator with the eigenvalues of $\hat{a}$ at the cost of introducing the operator $\hat{K}(\lambda)$. Inserting the expansion (26) for $\hat{K}$ and using the formal derivative of the operator valued function

$$f^{(n)}\ (\hat{Q}) = \frac{d^n f(\hat{Q})}{d\hat{Q}^n} = \int d\lambda c(\lambda)\lambda^n\ \exp[\lambda\hat{Q}]$$

one obtains the final result

$$f(\hat{a}+\hat{b})\,|a\rangle = \sum_{n=0}^{\infty} f^{(n)}\ (\hat{a}+\hat{b})\ \hat{O}_n\,|a\rangle \qquad (28)$$

For the final exploitation one sets

$$f \to \theta,\ \hat{a} \to -\hat{t} \quad \text{and} \quad \hat{b} \to \hat{E}_F$$

If one again uses the <u>local</u> representation of $\hat{E}_F$, the commutators involved in (27), correspond to gradients of $E_F(\underline{r})$. The commutators can be evaluated in a straightforward manner. If one restricts oneself to the second order in gradient terms ΔE_F, $(\underline{\nabla} E_F)^2$ one has to go to $\hat{Q}_4$, for an expansion to fourth order gradient terms $\Delta^2 E_F$, $(\underline{\nabla} E_F)^4$, etc., one has to go to $\hat{O}_8$. These expansions are then introduced into the representation of density, kinetic energy density and exchange energy density. The integration over momentum space (as well as spatial coordinates in the case of the exchange energy density[15]) can be executed directly, yielding to second order the following expressions:

$$\rho(\underline{r}) = \frac{K_F(\underline{r})^3}{3\pi^2} + \frac{1}{24\pi^2}\frac{\Delta K_F^2}{K_F} - \frac{1}{96\pi^2}\frac{(\nabla K_F^2)^2}{K_F^3} \pm \ldots \qquad (29)$$

$$\tau(\underline{r}) = \frac{K_F(\underline{r})^5}{10m\pi^2} - \frac{1}{48m\pi^2} K_F \Delta K_F^2 - \frac{1}{64m\pi^2}\frac{(\nabla K_F^2)^2}{K_F} \pm \ldots \qquad (30)$$

$$e_x(\underline{r}) = -\frac{e^2}{4\pi^3} K_F(\underline{r})^4 - \frac{e^2}{576\pi^3}\frac{(\nabla K_F^2)^2}{K_F^2} \pm \ldots \qquad (31)$$

The lowest order is again the TF-result. The formal expansion parameter is $\hbar$. Thus we have at this stage again a semiclassical statement, including the WKB like turning point problem discussed before. This problem can be overcome if one eliminates the Fermi momentum in favor of the density (consistently to second order). The result of this process is

$$\tau(\underline{r}) = c_1\, \rho(\underline{r})^{5/3} + c_2 \frac{(\nabla\rho)^2}{\rho(\underline{r})} \qquad (32)$$

$$e_x(\underline{r}) = -\, c_3\, \rho(\underline{r})^{4/3} - c_4 \frac{(\nabla\rho)^2}{\rho(\underline{r})^{4/3}} \qquad (33)$$

with

$$c_1 = \frac{3}{10m}(3\pi^2)^{2/3} \qquad c_2 = \frac{1}{72m}$$

$$c_3 = e^2\,\frac{3}{4}\left(\frac{3}{\pi}\right)^{1/3} \qquad c_4 = \frac{7e^2}{432\pi(3\pi^2)^{1/3}}$$

In the following we shall discuss some applications of these functionals to diatomic scattering systems. Since we eventually plan to treat very heavy systems (for which the relativistic effects are expected to be most relevant) the use of option III is mose adequate, i.e. we solve the variational Eq. (6) in a first step and then treat the orbital problem (9) with fixed potential terms. In order to demonstrate the quality of the results at the level of the first step we take a look on atomic systems, where a full comparison with HF results is possible. Figure 1 shows the density for the Kr atom ($r\sqrt{\rho}$ is plotted as a function of $\sqrt{r}$) calculated from the so-called Thomas-Fermi-Dirac-Weizsäcker (TFDW) functional which contains the first two terms of the kinetic energy density (32) and the lowest order (Dirac) term of the exchange energy density (33). The TFDW density averages beautifully over the shell structure. For exactly this reason it will only be able to reproduce quantities that do not depend sensitively on shell effects.

The total energy is a quantity that does not depend strongly on

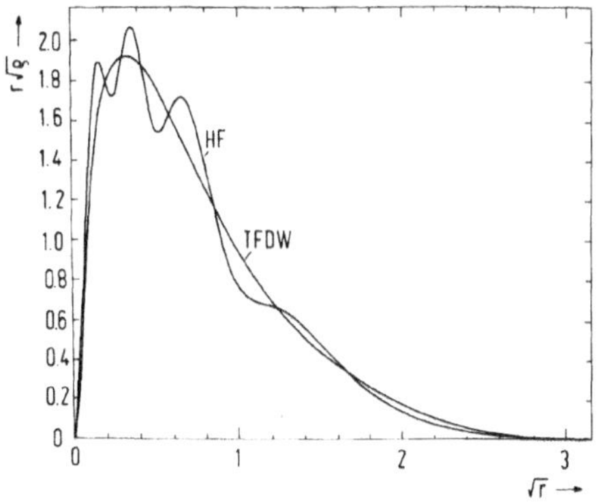

Figure 1. The density of the Kr atom.

the shell fluctuations. This is illustrated in Table 1 which contains the result of a careful numerical solution of the TFDW problem.[16] One notes that the TFDW energies agree with the HF values to better than 4% over the whole periodic table.

A number of authors have discussed the use of an adjustable coefficient c_2 of the gradient term in (32) in order to allow for an effective correction of the terms neglected in the expansion. This can actually be determined from the noninteracting (hydrogen-like) problem. One finds that a factor of 1.8 gives agreement with HF-values to better than 0.4% over the full range of charge numbers.[16]

In the quasimolecular treatment of diatomic scattering problems a basic assumption to be made is the Born-Oppenheimer approximation. This means that the collision is assumed to be slow enough to be well represented by a stationary electronic problem for each fixed internuclear separation. If the process is not adiabatic an alternative approach may be found in time dependent density functional methods.[17] In the following however, we restrict ourselves to the adiabatic case. The solution of th TFDW equation for the system N-N is shown in Figure 2, where the total electronic energy is plotted versus the internuclear separation. The agreement with HF-ground state energies[18] is better than 2%. The corresponding electronic density is then used to carry through step 2 of our programme consisting of the solution of the single particle eigenvalue problem (9) for each fixed internuclear separation. Figure 3 shows a comparison of the correlation diagram with HF results for the 14 electron system N-N. The agreement lies within a few per cent.

It should be emphasized that within this approach we can with a very reasonable increase in computing time generate as accurate correlation diagrammes for heavier systems. As an example Figure 4 shows the correlation diagram for the occupied orbitals of the 70

Table 1. Atomic groundstate energies in atomic units (a.u.): Comparison of TFDW-results with HF- and TF-results. ((1): Calculation with $c_2=1/72m$, (2): with $c_2=1/40m$).

System	$-E_{HF}$	$-E^{(1)}_{TFDW}$	$-E^{(2)}_{TFDW}$	$-E_{TF}$
Ne(10)	128.55	139.89	128.80	165.62
Ar(18)	526.82	561.82	524.75	652.76
Kr(36)	2752.1	2897.2	2744.3	3289.7
Xe(54)	7232.1	7558.7	7309.2	8472.9
Au(79)	17865	18575	17817	20586
Rn(86)	21867	22699	21799	25096
U(92)	25664	26618	25586	29373
Fm(100)	31283	32409	31187	35682
(120)	48203	49838	48069	54578

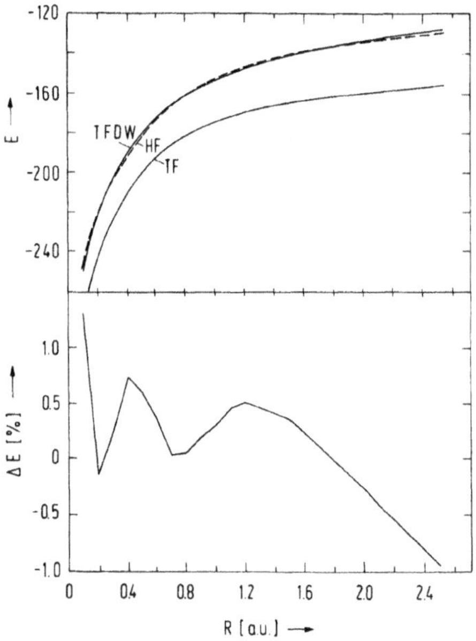

Figure 2. The total electronic energy as a function of the internuclear separation for the N-N system.

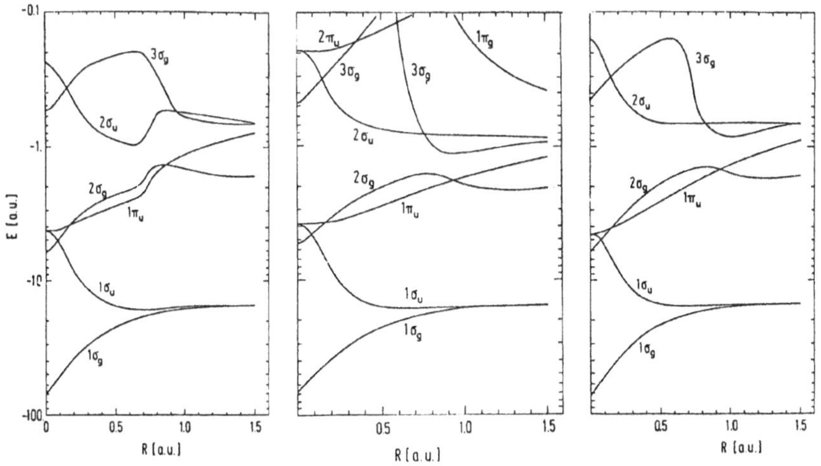

Figure 3. Correlation diagrams for the N-N system. From left to right: HF calculation,[18] result of the variable screening model,[22] correlation diagram based on the TFDW approach.

electron system I-Cl. This system represents a borderline case for the nonrelativistic treatment. For heavier systems relativistic effects have to be taken into account.

As long as the system may be regarded as weakly relativistic, it is possible to set up a gradient expansion[19] in perfect analogy to the nonrelativistic formalism. The basic idea is:

(a) to represent the relativistic density operator as in the nonrelativistic case by

$$\hat{\rho} = \Theta(\varepsilon_F - \hat{h}_{eff})$$

and

(b) to use the Foldy-Wouthuysen (FW)-representation of $\hat{h}_{eff}$ to second order in (p/m)

$$\hat{h}_{eff} = \beta[\hat{p}^2+m^2]^{1/2} + \hat{v}_{eff} + \frac{1}{4m^2}\,\hat{\underline{\sigma}}\cdot(\underline{\hat{\nabla}}v_{eff})\times\hat{\underline{p}} + \frac{1}{8m^2}\,\Delta\, v_{eff}$$

If the gradient expansion is carried through on the basis of this approximation the final result for the kinetic energy density looks as follows:

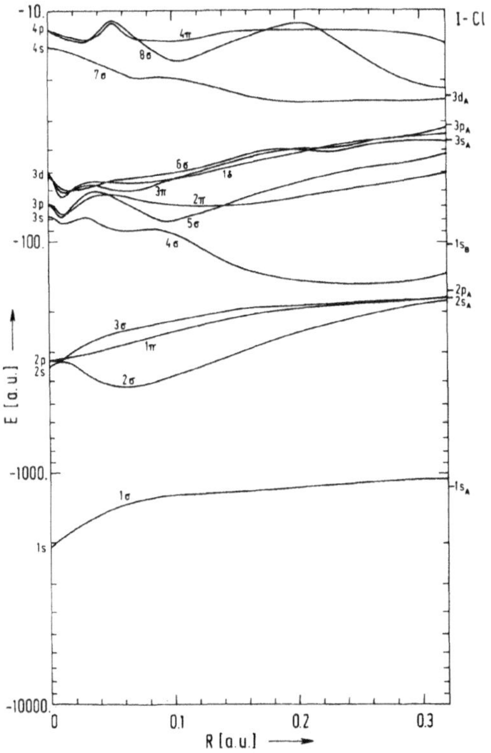

Figure 4. Correlation diagram for the 70 electron system I-Cl.

$$\tau_{FW}(\underline{r}) = \tau_o(\underline{r}) + \frac{1}{72m}\frac{(\nabla\rho)^2}{\rho(\underline{r})} B_{FW}(x) \tag{34}$$

with

$$B_{FW}(x) = \frac{1}{\sqrt{1+x^2}}\left[1 + \frac{x^4}{2} - \frac{x^2}{1+x^2} + \frac{2}{3}\frac{x^4}{(1+x^2)^2}\right] \tag{35}$$

and

$$x = \frac{(3\pi^2)^{1/3}}{m}\,\rho(\underline{r})^{1/3}$$

$\tau_o(\underline{r})$ is given by the lowest order expression (10). For small densities $x \ll 1$ one finds $B(x) \rightarrow 1$ which is the proper nonrelativistic limit. Some results obtained on the basis of this density functional are given in Table 2. The calculations are for neutral atomic systems. One notices (for the corrected coefficient $c_2 = 1.8/72$ m) that the agreement with Dirac Fock results is not as good as in the nonrelativistic case for light to medium atoms, but still acceptable as a first attempt. For heavier systems ($Z \geq 90$) the deviation in-

Table 2. Atomic groundstate energies in a.u.: Comparison of DF-results with results obtained by the density functional methods explained in the text, ((1): FW-version, (2): Simplified fully relativistic version).

	$-E_{HF}$	$-E_{DF}$	$-E^{(1)}_{RTFDW}$	$-E^{(2)}_{RTFDW}$
10	128.55	128.67	129.60	128.53
36	2752.1	2787.3	2862.0	2736.0
70	18392	14052	14202	13608
92	25664	28011	27363	26739
110	39225	44950	42012	42071
120	48203	57387	51773	52742

creases fairly rapidly, i.e. the weakly relativistic approximation inherent in the second order FW representation is not sufficient. Thus, we finally have to face a fully relativistic treatment.

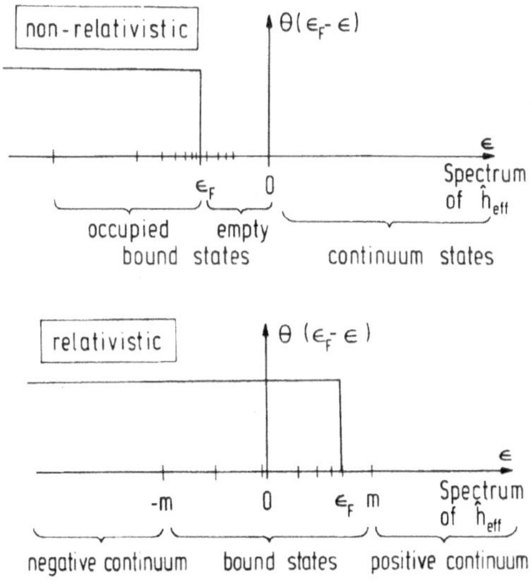

Figure 5. Schematic representation of the ground state occupation numbers for an effective single particle problem.

3. RELATIVISTIC THEORY

When developing a fully relativistic many electron theory an important question is how to subtract the infinite contribution of the filled Dirac sea. In order to find the correct answer, let us first discuss a very simple example: a free noninteracting electron gas. In this case, the single particle orbitals are eigenfunctions

$$\hat{t}\,|\underline{k}sT\rangle = \varepsilon_{kT}\,|\underline{k}sT\rangle$$

of the Dirac operator

$$\hat{t} = \underline{\alpha p} + \beta m \tag{36}$$

The standard representation in configuration space is given by

$$\langle \underline{r}|\underline{k}sT\rangle = \frac{1}{(2\pi)^{3/2}}\, u_s^T(\underline{k})\, e^{i\,\underline{k}\cdot\underline{r}} \tag{37}$$

with

$$u_s^{(+)}(\underline{k}) = \left[\frac{E_k+m}{2E_k}\right]^{1/2} \begin{bmatrix} \chi_s \\ \dfrac{\underline{\sigma}\cdot\underline{k}}{E_k+m}\,\chi_s \end{bmatrix}$$

and

$$u_s^{(-)}(\underline{k}) = \left[\frac{E_k+m}{2E_k}\right]^{1/2} \begin{bmatrix} \dfrac{-\underline{\sigma}\cdot\underline{k}}{E_k+m}\,\chi_s \\ \chi_s \end{bmatrix}$$

The index $s=\pm 1/2$ denotes the spin, while $T=(+)$, $(-)$ classifies positive and negative energy solutions:

$$\varepsilon_{kT} = \begin{cases} +E_k & \text{for } T = (+) \\ -E_k & \text{for } T = (-) \end{cases} \quad \text{with } E_k = \sqrt{k^2+m^2} \tag{38}$$

The density of the system is given by

$$\rho(\underline{r}) = \sum_{sT} \int d^3k\; n_{ksT}\, \langle \underline{k}sT|\underline{r}\rangle\langle\underline{r}|\underline{k}sT\rangle$$

If all the negative energy states (the Dirac sea) and the positive energy states up to a maximum level ε_F are assumed to be occupied, the occupation number n_{ksT} may be represented as in the nonrelativistic case by a step function

$$n_{ksT} = \Theta(\varepsilon_F - \varepsilon_{kT})$$

This leads to

$$\rho(\underline{r}) = 2 \int \frac{d^3k}{(2\pi)^3} \Theta(\varepsilon_F - E_k) + 2 \int \frac{d^3k}{(2\pi)^3} \Theta(\varepsilon_F + E_k)$$

The first step function $\Theta(\varepsilon_F - E_k)$ leads to an upper boundary of the integral, the Fermi momentum

$$K_F = \sqrt{\varepsilon_F^2 - m^2}$$

while the second step function $\Theta(\varepsilon_F + E_k)$ retains the value 1 everywhere. Thus we obtain

$$\rho(\underline{r}) = \frac{K_F^3}{3\pi^2} + \rho_{sea}$$

where the second term resulting from the filled Dirac sea is clearly divergent. The most natural way of eliminating this contribution is to define the occupation numbers from the very beginning as follows

$$n_{ksT} = \Theta(\varepsilon_F - \varepsilon_{kT}) - \Theta(-m - \varepsilon_{kT})$$

which means that only the positive energy states are considered to be occupied. This convention corresponds to the usual definition of the vacuum state for this situation. We then obtain for the density

$$\begin{aligned} \rho(\underline{r}) &= 2 \int \frac{d^3k}{(2\pi)^3} \Theta(\varepsilon_F - E_k) - 2 \int \frac{d^3k}{(2\pi)^3} \Theta(-m - E_k) \\ &+ 2 \int \frac{d^3k}{(2\pi)^3} \Theta(\varepsilon_F + E_k) - 2 \int \frac{d^3k}{(2\pi)^3} \Theta(-m + E_k) \qquad (39) \\ &= \frac{K_F^3}{3\pi^2} \end{aligned}$$

Since $E_k \geq m$ we have $\Theta(-m+E_k)=1$ and $\Theta(-m-E_k)=0$ everywhere. Thus, the last two terms (both divergent) cancel each other while the second term vanishes.

Now let us come back to the case of a nonuniform interacting electron gas in HF-approximation. The total energy of the system is then given by

$$\begin{aligned} E^{(\mathrm{rel})} &= \mathrm{tr}\,[\int d^3r \int d^3r'\, \delta(\underline{r}-\underline{r}')(-i\underline{\alpha}\cdot\underline{\nabla}_r+\beta m)\, \rho(\underline{r},\underline{r}') \\ &+ \int d^3r \rho(\underline{r},\underline{r})\, v_n(\underline{r}) \\ &+ \frac{1}{2}\int d^3r \int d^3r' \rho(\underline{r},\underline{r})\, V(\underline{r},\underline{r}')\, \rho(\underline{r}',\underline{r}') \\ &- \frac{1}{2}\int d^3r \int d^3r' \rho(\underline{r},\underline{r}')\, V(\underline{r},\underline{r}')\, \rho(\underline{r}',\underline{r})] \end{aligned} \tag{40}$$

Again, $v_n(\underline{r})$ is the nuclear Coulomb potential. The particle-particle interaction is given by the series

$$V(\underline{r},\underline{r}') = \frac{e^2}{|\underline{r}-\underline{r}'|} - \frac{e^2}{2}\,\frac{\underline{\alpha}\cdot\underline{\alpha}'}{|\underline{r}-\underline{r}'|} + \frac{\underline{\alpha}\cdot(\underline{r}-\underline{r}')\underline{\alpha}'\cdot(\underline{r}-\underline{r}')}{|\underline{r}-\underline{r}'|^3} + \ldots$$

In the following we restrict ourselves to the lowest order, i.e. the nonrelativistic Coulomb interaction. The inclusion of higher orders, as e.g. the Breit interaction, does not pose principle difficulties.

The density matrix

$$p(\underline{r},\underline{r}') = \sum\!\!\!\!\!\!\int_\nu \; n_\nu\, \phi_\nu(\underline{r})\, \phi_\nu(\underline{r}')^+ \tag{41}$$

consists of the above sum over the occupied states of the corresponding Dirac Fock variational equation

$$\hat{h}_{\mathrm{eff}}\, \phi_\nu(\underline{r}) = \varepsilon_\nu\, \phi_\nu(\underline{r}) \quad , \quad \hat{h}_{\mathrm{eff}} = \underline{\hat{\alpha}}\cdot\underline{\hat{p}}+\hat{\beta}m+\hat{v}_{\mathrm{eff}} \tag{42}$$

It should be noted that each element $\rho(\underline{r},\underline{r}')$ of the density matrix (41), by definition, is a 4x4 matrix. Therefore, the groundstate energy (40) as well as the single particle density are obtained by taking the trace

$$\rho(\underline{r}) = \mathrm{tr}\, \rho(\underline{r},\underline{r}) \tag{43}$$

The HF-basis $\{\phi_\nu(r)\}$ consists of positive continuum states, bound states, and negative continuum states. If all levels with energy less than ε_F are considered to be occupied

$$n_\nu = \Theta(\varepsilon_F-\varepsilon_\nu) \quad \text{(see Fig. 5)}$$

then the density and all other quantities of interest will be divergent due to the infinite contribution of all the negative energy states. Similar to the uniform gas, the natural way of subtracting this contribution is to define the occupation numbers as follows

$$n_\nu = \Theta(\varepsilon_F-\varepsilon_\nu) - \Theta(-m-\varepsilon_\nu) \tag{44}$$

This means that only a window of occupied bound states with energy less than ε_F is extracted from the whole spectrum of $\hat{h}_{\mathrm{eff}}$ (see

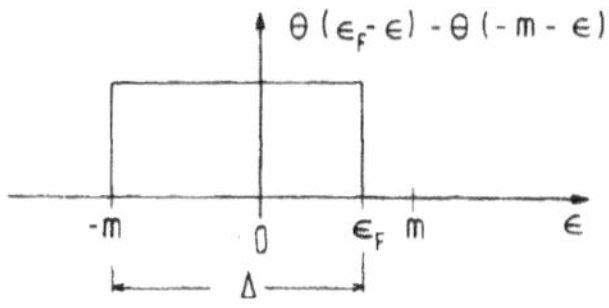

Figure 6. Schematic representation of the occupation numbers chosen for relativistic problems.

Figure 6.) This is, of course, the occupation of states implicitely used in any atomic or molecular Dirac-Fock calculation. From (44) and (41) we readily obtain a representation of the relativistic density operator

$$<\underline{r}\,|\hat{\rho}_{rel}|\,\underline{r}'> = \sum\!\!\!\!\!\!\int_{\upsilon} [\Theta(\varepsilon_F-\varepsilon_\upsilon) - \Theta(-m-\varepsilon_\upsilon)]<\underline{r}|\upsilon><\upsilon|\underline{r}'>$$

$$= <\underline{r}|\ [\Theta(\varepsilon_F-\hat{h}_{eff}) - \Theta(-m-\hat{h}_{eff})]\ \sum\!\!\!\!\!\!\int_{\upsilon}|\upsilon><\upsilon|\underline{r}>$$

This yields, using the completeness of the HF-basis

$$\hat{\rho}_{rel} = \Theta(\varepsilon_F-\hat{h}_{eff}) - \Theta(-m-\hat{h}_{eff})$$

For convenience we define

$$\hat{E}_F := \varepsilon_F - \hat{v}_{eff}$$

$$\hat{G}_F := -m - \hat{v}_{eff}$$

which leads to the finel expression

$$\hat{\rho}_{rel} = \Theta(\hat{E}_F-\hat{t}) - \Theta(\hat{G}_F-\hat{t}) \tag{45}$$

$\hat{t}$ is still the Dirac operator (36). It should be pointed out that the difference between the operators $\hat{E}_F$ and $\hat{G}_F$ is a constant Δ (namely the width of the window) which amounts to approximately 2m.

$$\hat{E}_F - \hat{G}_F = \varepsilon_F + m =: \Delta(\approx 2m) \tag{46}$$

The relativistic gradient expansion of the density matrix is then obtained by insertion of the complete system of free Dirac spinors (37)

$$\rho(\underline{r},\underline{r}') = <\underline{r}|\hat{\rho}_{rel}|\underline{r}'>$$

$$= \sum_{sT} \int d^3k\ <\underline{r}|\theta(E_F(\underline{r})-\hat{t})\,|\underline{k}sT><\underline{k}sT|\underline{r}'> \tag{47}$$

$$- \sum_{sT} \int d^3k\ <\underline{r}|\theta(G_F(\underline{r})-\hat{t})\,|\underline{k}sT><\underline{k}sT|\underline{r}'>$$

If the mathematical theorem (28) is applied twice in the last expression we obtain two formally identical series, one depending on E_F the other depending on G_F. It turns out that all divergent contributions in the two series cancel each other as should be expected from the above consideration.

In order to get some further insight into the features of the relativistic gradient expansion we shall first calculate the density of the system to lowest order, i.e. neglecting the noncommutativity of the operators $\hat{t}$ and $\hat{E}_F(\hat{G}_F)$. From (47) we obtain

$$\begin{aligned}\rho_o(\underline{r}) &= \mathrm{tr}\rho_o(\underline{r},\underline{r}) \\ &= \mathrm{tr}\sum_s \int d^3k\ [\Theta(E_F(\underline{r})-E_k) - \Theta(G_F(\underline{r})-E_k)]\,|\underline{k}s(+)\rangle\langle\underline{k}s(+)|\underline{r}\rangle \\ &+ \mathrm{tr}\sum_s \int d^3k\ [\Theta(E_F(\underline{r})+E_k) - \Theta(G_F(\underline{r})+E_k)]\,|\underline{k}s(-)\rangle\langle\underline{k}s(-)|\underline{r}\rangle \\ &= 2\int\frac{d^3k}{(2\pi)^3}\,\Theta(E_F(\underline{r})-E_k) - 2\int\frac{d^3k}{(2\pi)^3}\,\Theta(G_F(\underline{r})-E_k) \qquad (48) \\ &+ 2\int\frac{d^3k}{(2\pi)^3}\,\Theta(E_F(\underline{r})+E_k) - 2\int\frac{d^3k}{(2\pi)^3}\,\Theta(G_F(\underline{r})+E_k)\ .\end{aligned}$$

The last two (divergent) terms cancel each other. The first term corresponds to the uniform electron gas expression (39) while the second one gives an additional contribution if $G_F>m$. In terms of the effective potential this inequality means

$$v_{eff}(\underline{r}) < -2m \qquad (49)$$

Thus, we obtain a correction to the classical RTF model in the region where the potential is more attractive than -2m.

For the density we obtain

$$\rho_o(\underline{r}) = \rho_1(\underline{r}) - \rho_2(\underline{r}) \qquad (50)$$

with

$$\begin{aligned}\rho_1(\underline{r}) &= \frac{1}{3\pi^2}K_F^3 \\ K_F(v_{eff}) &= \left[E_F^2-m^2\right]^{1/2} = \left[(\varepsilon_F-v_{eff})^2-m^2\right]^{1/2} \qquad (51)\end{aligned}$$

and

$$\rho_2(\underline{r}) = \frac{1}{3\pi^2} q_F^3$$

$$q_F(v_{eff}) = \begin{cases} \left[G_F^2-m^2\right]^{1/2} = \left[(-m-v_{eff})^2-m^2\right]^{1/2} \\ 0 \qquad \text{if } v_{eff} > -2m \end{cases} \tag{52}$$

This means, we have a weakly relativistic domain where the density equals the classical contribution ρ_1 and, in addition, we have a highly relativistic domain, where the "correction" $\rho2$ comes into the game. This is illustrated in Figs. 7 and 8 where the density is shown as a function of the effective potential and as a function of the distance r from the nuclear point charge.

For pointlike nuclei the density is still divergent. But, in contrast to the classical RTF-model, the density is <u>integrable</u>: In the vicinity of the nucleus we obtain

$$\rho_o(r) \underset{v_{eff}<<-2}{\sim} (v_{eff})^2 \sim (\frac{Z}{r})^2$$

It should be noted that the density ρ_2 can be expressed as a simple function of ρ_1 by means of the fundamental equation $G_F = E_F - \Delta$:

$$\rho_2 = \frac{1}{3\pi^2} q_F^3 = \frac{1}{3\pi^2}\left[G_F^2-m^2\right]^{3/2} = \frac{1}{3\pi^2}\left[(E_F-\Delta)^2-m^2\right]^{3/2}$$

$$= \frac{1}{3\pi^2}\left[\left(\left[(3\pi^2\rho_1)^{2/3}+m^2\right]^{1/2} -\Delta^2\right)-m^2\right]^{3/2}$$

For the kinetic energy density we obtain twice the classical RTF-expression (10):

$$\tau_o(\underline{r}) = \tau_o(x) - \tau_o(y) \tag{53}$$

where the first term depends on ρ_1 via

$$x := K_F/m = (3\pi^2\rho_1)^{1/3}/m \tag{54}$$

while the second term depends on ρ_2 via

$$y := q_F/m = (3\pi^2\rho_2)^{1/3}/m \tag{55}$$

The calculation of the exchange energy density is more complicated because a product of two density matrices and an additional integration have to be carried through (see Eq. (4)). The traditional relativistic exchange energy density (to lowest order of the gra-

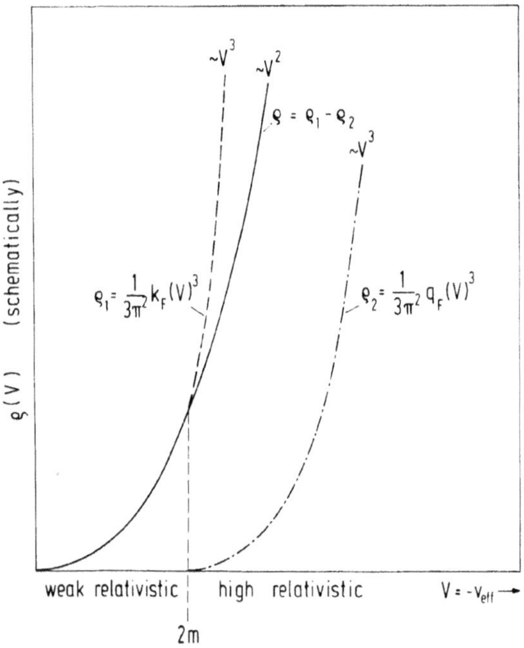

Figure 7. The RTF density for an atomic system as a function of the effective potential.

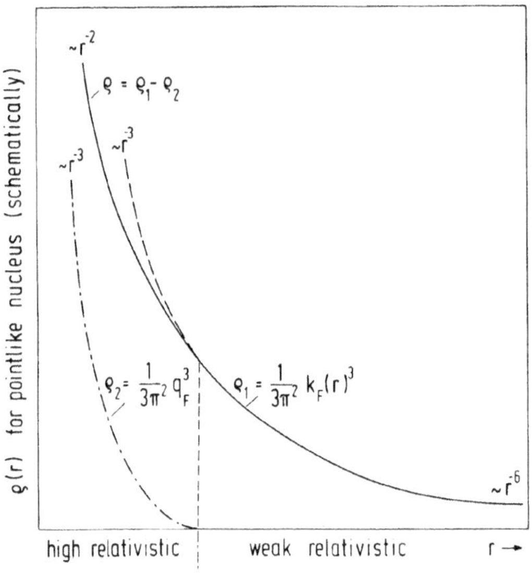

Figure 8. The RTF density for an atomic system as a function of the distance from the nucleus.

dient expansion) which corresponds to the RTF kinetic energy density (10) has first been calculated by Jancovici in 1962.[20] He derived the following expression

$$e_x^{(0)}(\underline{r}) = \frac{e^2 m^4}{(2\pi)^3}\left[x^4 + \frac{2}{3}x^2 E_x^2 - \frac{2}{3}E_x^4 \ln(E_x^2) + \frac{4}{3}x^3 E_x \text{Arsh } x - (xE_x - \text{Arsh } x)^2\right] \tag{56}$$

with

$$E_x = \sqrt{x^2+1}$$

Due to the high relativistic correction ρ_2 we now obtain the following extension

$$\begin{aligned} e_x^{(0)}(\underline{r}) = -\frac{e^2 m^4}{(2\pi)^3}\cdot\Bigg[&(x-y)(x^3-y^3)+\frac{2}{3}(xE_x-yE_y)^2 \\ &-\frac{2}{3}E_x^4\ln(E_x^2) + \frac{2}{3}(E_x^4+E_y^4)\ln\left(\frac{E_xE_y+1+xy}{E_xE_y+1-xy}\right) - \frac{2}{3}E_y^4\ln(E_y^2) \\ &-\left[(x^2-y^2)^2+\frac{1}{6}(E_x-E_y)^4\right]\ln\left(\frac{x-y}{x+y}\right) \\ &+\frac{4}{3}(x^3E_x-y^3E_y)(\text{Arsh } x - \text{Arsh } y) \\ &-\Big([xE_x-yE_y]-[\text{Arsh } x-\text{Arsh } y]\Big)^2\Bigg] \end{aligned} \tag{57}$$

For y=0 this expression reduces to Jancovici's result. An important application of this expression should be the use of the derivative

$$v_x = \frac{\partial e_x^{(o)}}{\partial \rho_o}$$

as an effective exchange potential in relativistic HF-Slater-like calculations (option II).

Let us finally discuss the higher order terms of the gradient expansion. These are obtained by application of the mathematical theorem (28) to Eq. (47). This yields for the density

$$\begin{aligned} \rho(\underline{r}) = &\sum_{sT}\int d^3k \sum_{n=o}^{\infty}\theta^{(n)}(E_F \pm E_K)\ \langle \underline{k}sT|\underline{r}\rangle\langle\underline{r}|\hat{O}_n|\underline{k}sT\rangle \\ - &\sum_{sT}\int d^3k \sum_{n=o}^{\infty}\theta^{(n)}(G_F \pm E_K)\ \langle \underline{k}sT|\underline{r}\rangle\langle\underline{r}|\hat{O}_n|\underline{k}sT\rangle \end{aligned}$$

The operators $\hat{O}_n$ are calculated from (27) by insertion of the Dirac operator

$$\hat{a} = - \hat{t} = - \underline{\alpha} \cdot \hat{\underline{p}} - \beta m$$

As before, we evaluate the operators up to second order gradients of $E_F(\underline{r})$ and $G_F(\underline{r})$. A detailed analysis shows that, in contrast to the nonrelativistic case, each operator $\hat{O}_n$ contains first and second order gradient terms. Therefore, the series over n have to be resummed. This problem is solved by use of the Taylor expansion

$$\theta(x+a) = \sum_{n=o}^{\infty} \frac{1}{n!} \theta^{(n)}(x-a)(2a)^n$$

which is a well defined object in the theory of distributions.[21] The result at the semiclassical stage looks as follows:

$$\rho(\underline{r}) = \rho_1 [E_F] - \rho_2 [G_F] \tag{58}$$

with

$$\rho_1 [E_F] = \frac{1}{3\pi^2} (K_F^3 + f(E_F)(\nabla E_F)^2 + g(E_F)\Delta E_F) \tag{59}$$

$$\rho_2 [G_F] = \frac{1}{3\pi^2} (q_F + f(G_F)(\nabla G_F)^2 + g(G_F)\Delta G_F) \tag{60}$$

and

$$f(E_F) = -\frac{1}{8}\frac{E_F^2}{K_F^3} + \frac{3}{8}\frac{1}{K_F}$$

$$g(E_F) = \frac{1}{4}\frac{E_F}{K_F} + \frac{1}{2}\,\mathrm{Arsh}\left(\frac{K_F}{m}\right)$$

For the kinetic energy density one obtains

$$\tau(\underline{r}) = \tau_o(E_F) + \frac{1}{3\pi^2}(h(E_F)(\nabla E_F)^2 + k(E_F)\Delta E_F)$$
$$- \tau_o(G_F) - \frac{1}{3\pi^2}(h(G_F)(\nabla G_F)^2 + k(G_F)\Delta G_F)$$

with

$$h(E_F) = \frac{1}{4}\frac{E_F}{K_F} - \frac{1}{8}\frac{E_F^3}{K_F^3} - \frac{1}{4}\,\mathrm{Arsh}\left(\frac{K_F}{m}\right)$$

$$k(E_F) = \frac{1}{4}K_F + \frac{1}{4}\frac{E_F^2}{K_F}$$

In the final step we invert the functionals $\rho_1[E_F]$ and $\rho_2[G_F]$ and eliminate in $\tau(\underline{r})$ E_F in favour of ρ_1 and G_F in favour of ρ_2. The final result is

$$\tau(\underline{r}) = \tau_o(x) + \frac{1}{72m} B(x) \frac{(\nabla\rho_1)^2}{\rho_1} - \tau_o(y) - \frac{1}{72m} B(y) \frac{(\nabla\rho_2)^2}{\rho_2} \qquad (61)$$

with

$$B(Z) = \frac{1}{\sqrt{1+Z^2}} + \frac{2Z}{1+Z^2} \text{ Arsh } Z \qquad (62)$$

The quantities x and y are given by (54) and (55).

In contrast to the lowest order, ρ_2 cannot be expressed as a simple function of ρ_1. Instead we have the following differential relation

$$g[\rho_1,\rho_2] := E_F[\rho_1] - G_F[\rho_2] - \Delta = 0 \qquad (63)$$

where the functionals $E_F[\rho_1]$ and $G_F[\rho_2]$ are the inversion (to second order) of (59) and (60).

For the total groundstate energy we thus obtain

$$\begin{aligned} E^{rel} &= \int d^3r\tau[\rho_1,\rho_2] + \int d^3r\ e_x^{(o)}(\rho_1,\rho_2) \\ &+ \int d^3r\ v_n(\underline{r})\ [\rho_1(\underline{r})-\rho_2(\underline{r})] \\ &+ \frac{1}{2} \int d^3r \int d^3r'\ [\rho_1(\underline{r})-\rho_2(\underline{r})] \frac{e^2}{|\underline{r}-\underline{r}'|} [\rho_1(\underline{r})-\rho_2(\underline{r})] \end{aligned}$$

where the functionals (61) and (57) have to be inserted for the kinetic and the exchange energy density. The corresponding variational principle looks as follows

$$\delta \left(E^{rel} + V_o \int d^3r(\rho_1(\underline{r})-\rho_2(\underline{r})) + \int d^3r\ \ell(\underline{r})g[\rho_1,\rho_2] \right) = 0$$

The Lagrangian multiplier V_o guarantees a fixed particle number while the subsidiary condition (63), which relates ρ_1 and ρ_2, is included by means of the "Lagrangian multiplier function" $\ell(\underline{r})$.

The corresponding variational equations have not been solved so far. A qualitative discussion of the solution is shown in Figure 9. As before, we have a dominant contribution ρ_1 to the density and a correction ρ_2 in the highly relativistic region near the nucleus. In contrast to the lowest order, the density is expected to be less singular at the position of the nucleus and decreases exponentially in the asymptotic region.

Numerical results have been obtained so far only for a simpli-

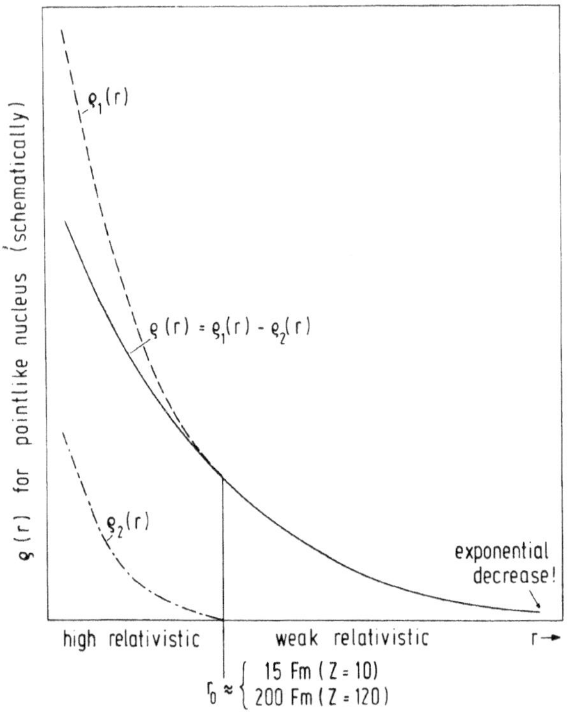

Figure 9. Schematic representation of the solution of the atomic RTFDW variational equations.

fied version of the full set of equations in which we set $\rho_2=\ell\equiv 0$. The results for the total atomic energies are given in the last column of Table 2. It turns out that for the heaviest systems the results are already slightly better than those obtained within the second order FW-approximation. Of course, this is not yet the final answer. The numerical solution of the complete set of variational equations is envisaged as a near future problem.

Finally it should be pointed out that the fully relativistic expression (62) for the kinetic energy density is completely general. It holds for any bound relativistic many Fermion system, independent of the particular external potential and independent of the particle-particle interaction. Therefore, a variety of applications can be expected in many fields of physics.

REFERENCES

1. P. Hohenberg and W. Kohn, Phys.Rev. 136, B864 (1964).
2. A.K. Rajagopal, J.Phys. C11, L943 (1978).
3. D.A. Kirzhnitz, Field Theoretical Methods in Many Body Systems, Pergamon Press (Oxford 1967).
4. E. Wigner, Phys.Rev. 40, 749 (1932);
J.G. Kirkwood, Phys.Rev. 44, 31 (1933);
B. Grammaticos and A. Voros, Ann.Phys. 123, 359 (1979);
5. L.H. Thomas, Proc.Camb.Phil.Soc. 23, 542 (1926);
E. Fermi, Z.Phys. 48, 73 (1928).
6. O. Gunnarsson, P. Johansson, S. Lundqvist, and B.I. Lundqvist, Int.J.Q.Chem.Symp. 9, 83 (1975).
7. R. Latter, Phys.Rev. 99, 510 (1955).
8. E.K.U. Gross and R.M. Dreizler, Phys.Rev. A20, 1798 (1979);
A. Toepfer, E.K.U. Gross, and R.M. Dreizler, Phys.Rev. A20, 1808 (1979);
R.M. Dreizler, E.K.U. Gross, and A. Toepfer, Phys.Lett. A71, 49 (1979);
A. Toepfer, E.K.U. Gross, and R.M. Dreizler, Z.Phys. 298, 167 (1980).
9. A.H. MacDonald and S.H. Vosko, J.Phys. C12, 2977 (1979);
M.P. Das, M.V. Ramana, and A.K. Rajagopal, Phys.Rev. A22, 9 (1980);
M.P. Das, Phys.Rev. A23, 391 (1981);
M.V. Ramana and A.K. Rajagopal, J.Phys. C14, 4291 (1981);
M.V. Ramana and A.K. Rajagopal, Phys.Rev. A25, 96 (1982).
10. M.S. Vallarta and N. Rosen, Phys.Rev. 41, 708 (1932).
11. E.K.U. Gross, A. Toepfer, B. Jakob, and R.M. Dreizler, Proc. XVIIth International Winter Meeting on Nuclear Physics in Bormio, p.84, Instituto Nazionale de Fisica Nucleare, Milano (1979).
12. H. Jensen, Z.Phys. 82, 794 (1933).
13. N. Ashby and M.A. Holzmann, Phys.Rev. A1, 764 (1970).
14. M. Rudkjøbing, Kgl. Danske Videnskab Selskab, Mat.-Fys.Medd. 27, No. 5 (1952).
15. E.K.U. Gross and R.M. Dreizler, Z.Phys. A302, 103 (1981).
16. W. Stich, E.K.U. Gross, and R.M. Dreizler, to be published.
17. M. Horbatsch and R.M. Dreizler, Z.Phys. A300, 119 (1981).
18. W.C. Ermler, R.S. Mulliken, and A.C. Wahl, J.Chem.Phys. 66, 3031 (1977).
19. E.K.U. Gross and R.M. Dreizler, Phys.Lett. A81, 447 (1981).
20. B. Jancovici, Nuovo Cim. 25, 428 (1962).
21. F. Constantinescu, Distributionen und ihre Anwendungen in der Physik, p.59 ff., Teubner, Stuttgart (1974).
22. J. Eichler and U. Wille, Phys.Rev. A11, 1973 (1975).

A PHASE SPACE APPROACH TO ENERGY DENSITIES IN POSITION SPACE

Michael Springborg* and Jens Peder Dahl

Chemistry Department B
The Technical University of Denmark
DTH 301, DK-2800 Lyngby, Denmark

1. INTRODUCTION

The problem we shall consider in the present contribution is that of determining the various (position space) energy densities associated with a stationary state of N electrons moving in the field of one or more fixed nuclei. By an energy density (in position space) we mean a quantity $\varepsilon(\underline{r})$ which only depends on the position $\underline{r}$ of a single electron, and which gives a specific energy contribution when integrated over all space. Let E be the electronic energy of the stationary state considered. We may then partition E as follows

$$E = E_K + E_n + E_{c\ell} + E_{xc} \tag{1}$$

where E_K is the kinetic energy, E_n the potential energy in the field of the nuclei, $E_{c\ell}$ the "classical" Coulomb energy, and E_{xc} the exchange-correlation energy. The corresponding energy densities are then obtained by writing

$$E_K = \int d\underline{r}\ \varepsilon_K(\underline{r}) \tag{2}$$

$$E_n = \int d\underline{r}\ \varepsilon_n(\underline{r}) \tag{3}$$

$$E_{c\ell} = \int d\underline{r}\ \varepsilon_{c\ell}(\underline{r}) \tag{4}$$

$$E_{xc} = \int d\underline{r}\ \varepsilon_{xc}(\underline{r}) \tag{5}$$

*Present address: Max-Planck-Institut für Festkörperforschung, D-7000 Stuttgart 80, Federal Republic of Germany.

Our problem is to determine ε_K, ε_n, $\varepsilon_{c\ell}$, and ε_{xc} from a known N-electron wavefunction.

A density functional theory is obtained when one or more of the energy densities are considered to be functionals of the electron density $\rho(\underline{r})$, and a local density approximation is obtained when these functionals are taken to be such that the corresponding energy density at $\underline{r}$ only depends upon the value of ρ at $\underline{r}$. Obviously it is of a priori interest for density functional theories to determine the energy densities above from first principles.

These energy densities are, however, not uniquely defined, but we shall argue that an appropriate form of the densities can be obtained via the phase space formulation of quantum mechanics.

Since this formulation has been very little used in the study of stationary electronic states, we shall begin by reviewing some of its features in the next section. In the following sections we derive general expressions for various energy densities and show some results for a few simple atoms and molecules. These results are finally analysed within the framework of local density theories.

2. THE PHASE SPACE FORMULATION OF QUANTUM MECHANICS

Phase space representations of quantum mechanics have attracted much attention in the physics literature. The original formulations are due to Weyl[1] and Wigner,[2] but a great number of later investigations have contributed considerably to our understanding of the subject. An introduction to some of the most essential concepts and definitions involved has been given by de Groot and Suttorp.[3] We shall essentially base our presentation on a series of recent papers[4-10] in which we have discussed the phase space formulation per se and also applied it to some simple atomic and molecular systems. The basic concepts are the Wigner function and Weyl's correspondence rule.

2.1. The Wigner Function

Let $|\psi\rangle$ be a normalized state vector in the spin-free Hilbert space associated with a single non-relativistic particle, and let

$$\psi(\underline{r}) = \langle\underline{r}|\psi\rangle \tag{6}$$

and

$$\phi(\underline{p}) = \langle\underline{p}|\psi\rangle \tag{7}$$

be the corresponding position and momentum wavefunctions which are connected by a Fourier transformation

$$\psi(\underline{r}) = h^{-3/2}\int d\underline{p}\ \phi(\underline{p})e^{i\underline{p}\cdot\underline{r}/\hbar} \tag{8}$$

$$\phi(\underline{p}) = h^{-3/2}\int d\underline{r}\ \psi(\underline{r})e^{-i\underline{p}\cdot\underline{r}/\hbar} \tag{9}$$

The quantity $\psi(\underline{r})^*\psi(\underline{r})$ measures the probability density in position space and $\phi(\underline{p})^*\phi(\underline{p})$ is the probability density in momentum space.[11]

The Wigner function $f(\underline{r},\underline{p})$ is then given by either of the two expressions

$$f(\underline{r},\underline{p}) = \left(\frac{2}{h}\right)^3\int d\underline{r}'\psi(\underline{r}-\underline{r}')^*\psi(\underline{r}+\underline{r}')e^{-2i\underline{p}\cdot\underline{r}'/\hbar} \tag{10}$$

and

$$f(\underline{r},\underline{p}) = \left(\frac{2}{h}\right)^3\int d\underline{p}'\phi(\underline{p}-\underline{p}')^*\phi(\underline{p}+\underline{p}')e^{2i\underline{r}\cdot\underline{p}'/\hbar} \tag{11}$$

It satisfies the relations

$$\int d\underline{p}\ f(\underline{r},\underline{p}) = \psi(\underline{r})^*\psi(\underline{r}) \tag{12}$$

$$\int d\underline{r}\ f(\underline{r},\underline{p}) = \phi(\underline{p})^*\phi(\underline{p}) \tag{13}$$

and

$$\iint d\underline{r}\ d\underline{p}\ f(\underline{r},\underline{p}) = 1 \tag{14}$$

Furthermore, if $\hat{a}$ is a quantum mechanical operator acting on the state vector $|\psi\rangle$ we have

$$\langle\psi|\hat{a}|\psi\rangle = \iint d\underline{r}\ d\underline{p}\ f(\underline{r},\underline{p})a(\underline{r},\underline{p}) \tag{15}$$

where $a(\underline{r},\underline{p})$ is the Weyl transform of $\hat{a}$, as discussed in the next section.

The relations (12)-(15) show that the Wigner function plays the rôle of a probability density in phase space, in an operational sense. The values of the function are, however, not restricted to being positive or zero, although they are always real. Hence, one may not *interpret* the function as a probability density. Such an interpretation would of course also be inconsistent with the uncertainty principle. Nevertheless, we may say that $f(\underline{r},\underline{p})$ is a measure of the way the phase space point $(\underline{r},\underline{p})$ supports the given quantum state, and this statement may be made more precise by noting that we may also write

$$f(\underline{r},\underline{p}) = \left(\frac{2}{h}\right)^3 \langle\psi|\hat{\Pi}(\underline{r},\underline{p})|\psi\rangle \tag{16}$$

where $\hat{\Pi}(\underline{r},\underline{p})$ is an inversion operator[5,12,13] with respect to the phase space point $(\underline{r},\underline{p})$. It has the form

$$\hat{\Pi}(\underline{r},\underline{p}) = \left(\frac{1}{2h}\right)^3 \iint d\underline{u}\, d\underline{v}\, \exp\left[\frac{i}{\hbar}(\underline{r}\cdot\underline{u}+\underline{p}\cdot\underline{v})\right] \exp\left[-\frac{i}{\hbar}(\hat{\underline{r}}\cdot\underline{u}+\hat{\underline{p}}\cdot\underline{v})\right] \tag{17}$$

with $\hat{\underline{r}}$ and $\hat{\underline{p}}$ being the quantum mechanical operators associated with $\underline{r}$ and $\underline{p}$ respectively. These operators satisfy the ordinary commutation relation

$$[\underline{e}_1\cdot\hat{\underline{r}},\ \underline{e}_2\cdot\hat{\underline{p}}] = i\hbar\, \underline{e}_1\cdot\underline{e}_2 \tag{18}$$

where $\underline{e}_1$ and $\underline{e}_2$ are arbitrary, but fixed vectors.

The operator $\hat{\Pi}(\underline{r},\underline{p})$ is hermitian and unitary, i.e.,

$$\hat{\Pi}(\underline{r},\underline{p})^{\dagger} = \hat{\Pi}(\underline{r},\underline{p}) \tag{19}$$

$$\hat{\Pi}(\underline{r},\underline{p})^{2} = 1 \tag{20}$$

and may be considered as the operator inducing the transition from a state vector to a Wigner function, in accordance with (16).

The definitions and relations given above all pertain to a single particle without spin. The inclusion of spin and the extension to several particles is straightforward. However, we do not need the full extension for the present purpose. It will be sufficient to consider the description given by the first order spin-free density matrix. For a single particle, with the wavefunction (6), this quantity is defined as

$$\rho_1(\underline{r},\underline{r}') = \psi(\underline{r})\psi(\underline{r}')^* \tag{21}$$

its diagonal elements giving the electron density

$$\rho(\underline{r}) = \rho_1(\underline{r},\underline{r}) \tag{22}$$

In terms of ρ_1 we may evidently write Eq. (10) as

$$f(\underline{r},\underline{p}) = h^{-3}\int d\underline{v}\, \rho_1(\underline{r} + \tfrac{1}{2}\underline{v},\ \underline{r} - \tfrac{1}{2}\underline{v})\, e^{-i\underline{p}\cdot\underline{v}/\hbar} \tag{23}$$

so that the Wigner function is a certain Fourier transform of ρ_1. Inverting (23) gives

$$\rho_1(\underline{r} + \tfrac{1}{2}\underline{v},\ \underline{r} - \tfrac{1}{2}\underline{v}) = \int d\underline{p}\, f(\underline{r},\underline{p})\, e^{i\underline{p}\cdot\underline{v}/\hbar} \tag{24}$$

The definition of the first order spin-free density matrix for an N-electron system should be well known to the reader.[14] The matrix can be written

$$\rho_1(\underline{r},\underline{r}') = \rho_1^{\uparrow}(\underline{r},\underline{r}') + \rho_1^{\downarrow}(\underline{r},\underline{r}') \tag{25}$$

with ↑ and ↓ referring to the spin-up and spin-down parts, respectively. The electron density is still given by (22), and

$$\int d\underline{r}\; \rho(\underline{r}) = N \tag{26}$$

Just as one defines density matrices of various orders it is possible to define Wigner functions of various orders. In particular we may define a first order spin-free Wigner function. It can be written in the form

$$f(\underline{r},\underline{p}) = f^{\uparrow}(\underline{r},\underline{p}) + f^{\downarrow}(\underline{r},\underline{p}) \tag{27}$$

The function itself, and each of its components can be obtained from the corresponding entities in (25) by means of (23), and the relations (23) and (24) are thus also applicable to an N-electron system. A relation similar to (15) also exists. It reads

$$\langle\hat{A}\rangle = \iint d\underline{r}\; d\underline{p}\; f(\underline{r},\underline{p})\; a(\underline{r},\underline{p}) \tag{28}$$

where $\hat{A}$ is a symmetric and spin-free N-electron operator of the form

$$\hat{A} = \hat{a}(1) + \hat{a}(2) + \dots + \hat{a}(N) \tag{29}$$

and $\langle\hat{A}\rangle$ denotes the expectation value of $\hat{A}$ for the N-electron state considered.

2.2. The Weyl Correspondence

The operator $\hat{a}$ and the phase space function $a(\underline{r},\underline{p})$ occurring in Eq. (15) are related via the so-called Weyl correspondence, proposed in 1931[1] before the appearance of the Wigner function.[2] The synthesis of Weyl's and Wigner's approaches, as reflected in Eq. (15), is the result of later investigations (cf. the remarks at the beginning of section 2).

Weyl's correspondence rule implies, that if $a(\underline{r},\underline{p})$ is a function of $\underline{r}$ or $\underline{p}$ alone, then $\hat{a}$ is the same function of the vector operator $\hat{\underline{r}}$ or $\hat{\underline{p}}$. In the more general polynomial case one obtains the operator $\hat{a}$ from the function $a(\underline{r},\underline{p})$ by the replacement of $\underline{r}$ with $\hat{\underline{r}}$ and $\underline{p}$ with $\hat{\underline{p}}$, followed by a proper symmetrization of products of non commuting operators. This symmetrization is such that, if x_i and p_i are corresponding components of $\underline{r}$ and $\underline{p}$, then the operator associated with the function $x_i^n p_i^m$ is

$$\hat{a} = \frac{1}{2^n} \sum_{r=0}^{n} \binom{n}{r} \hat{x}_i^r\, \hat{p}_i^m\, \hat{x}_i^{n-r}$$

$$= \frac{1}{2^m} \sum_{s=0}^{m} \binom{m}{s} \hat{p}_i^s \hat{x}_i^n \hat{p}_i^{m-s} \tag{30}$$

These expressions are due to McCoy.[15]

Weyl's own formulation of the correspondence rule identifies the Fourier components of $\hat{a}$ with respect to the operators $\exp[i(\hat{\underline{r}}\cdot\underline{u} + \hat{\underline{p}}\cdot\underline{v})/\hbar]$ with the Fourier components of $a(\underline{r},\underline{p})$ with respect to the functions $\exp[i(\underline{r}\cdot\underline{u} + \underline{p}\cdot\underline{v})/\hbar]$. It leads to the following relation[4]

$$\hat{a} = \left(\frac{2}{h}\right)^3 \iint d\underline{r}\, d\underline{p}\; a(\underline{r},\underline{p})\hat{\Pi}(\underline{r},\underline{p}) \tag{31}$$

where $\Pi(\underline{r},\underline{p})$ is the inversion operator (17). Thus, we may consider $\hat{\Pi}(\underline{r},\underline{p})$ as generating the Weyl correspondence.

The relation (31) involves an integration over phase space. The inverse relation involves the trace over Hilbert space and is found to be

$$a(\underline{r},\underline{p}) = 2^3\, \mathrm{Tr}[\hat{a}\hat{\Pi}(\underline{r},\underline{p})] \tag{32}$$

When the trace is evaluated in the $\underline{r}$-basis, with $|\underline{r}\rangle$ being the eigenvector of $\underline{r}$ and with

$$\langle\underline{r}|\underline{r}'\rangle = \delta(\underline{r}-\underline{r}') \tag{33}$$

we get

$$a(\underline{r},\underline{p}) = \int d\underline{v}\, \langle\underline{r} + \tfrac{1}{2}\underline{v}|\hat{a}|\underline{r} - \tfrac{1}{2}\underline{v}\rangle\, e^{-i\underline{p}\cdot\underline{v}/\hbar} \tag{34}$$

We shall use this relation later, in evaluating the phase space function corresponding to an exchange operator.

By comparing Eqs. (23) and (34), and by writing Eq. (21) in the form

$$\rho_1(\underline{r},\underline{r}') = \langle\underline{r}|\psi\rangle\langle\psi|\underline{r}'\rangle \tag{35}$$

we observe that the Wigner function for a single particle in the state $|\psi\rangle$ is the Weyl transform of the operator $h^{-3}|\psi\rangle\langle\psi|$. This result, which is well known,[16] unifies Weyl's and Wigner's approaches in a beautiful way. It is readily extended to cover the N-electron case dealt with in Eqs. (25)-(27), namely, if we write $\rho(\underline{r},\underline{r}')$ for an N-electron system as

$$\rho_1(\underline{r},\underline{r}') = \sum_i n_i\, \psi_i(\underline{r})\psi_i(\underline{r}')^* \tag{36}$$

then we obtain the corresponding Wigner function as h^{-3} times the Weyl transform of the density operator

$$\hat{\rho} = \sum_i n_i |\psi_i\rangle\langle\psi_i| \tag{37}$$

Obviously, this result also holds for the spin-up and spin-down parts of ρ_1 separately.

In closing this section it should be pointed out, that other correspondence rules than the one proposed by Weyl have been suggested. For each of these correspondence rules there is a modified Wigner function and hence an equivalent phase space formalism. If, however, some simple requirements about canonical invariance are imposed, then it is found that the Weyl-Wigner formalism is the only acceptable one.[17,4,7] Hence, we may consider this formalism as the canonical phase space formalism.

3. DEFINITION OF POSITION SPACE DENSITIES

The relation (28) suggests that we define the quantity

$$\alpha(\underline{r}) = \int d\underline{p}\ f(\underline{r},\underline{p})\ a(\underline{r},\underline{p}) \tag{38}$$

as the (position space) density associated with the operator $\hat{A}$ and the quantum state under consideration. Integration over all space gives

$$\langle\hat{A}\rangle = \int d\underline{r}\ \alpha(\underline{r}) \tag{39}$$

in analogy with Eqs. (2)-(5).

When $a(\underline{r},\underline{p}) = a(\underline{r})$, a function of $\underline{r}$ alone, we obtain the simple expression

$$\alpha(\underline{r}) = a(\underline{r})\rho(\underline{r}) \tag{40}$$

by virtue of the relation

$$\int d\underline{p}\ f(\underline{r},\underline{p}) = \rho(\underline{r}) \tag{41}$$

In what follows we shall use the definition (38) to calculate the quantities ε_K, ε_n, $\varepsilon_{c\ell}$, and ε_{xc} (Eqs. (2)-(5)). We begin with a discussion of the hydrogen atom and the hydrogen molecular ion. These are one-electron systems and hence only ε_K and ε_n are of relevance. Besides the kinetic energy density defined by (38) we shall also consider kinetic energy densities obtained by other definitions, and for each case we shall study the local interplay between ε_K and ε_n. It turns out that this interplay tends to establish a local virial theorem when the definition (38) is used, and we shall take this as a good reason for considering this definition as the more appropriate one.

4. KINETIC ENERGY DENSITIES. A LOCAL VIRIAL THEOREM

4.1. The Hydrogen-Like Atom

In a previous paper[8] we have performed a comprehensive study of the Wigner function for the ground state of a hydrogen-like atom and displayed its characteristic features by a series of contour maps. We also condidered various definitions of the kinetic energy density and discussed the local virial theorem. The following is a summary of some of the results obtained in Ref. 8. We use atomic units, and hence put m, e and ħ equal to one, m being the electron's mass and -e its charge. Further, we consider the nucleus to be at rest.

With Z being the nuclear charge we have the Hamiltonian

$$\hat{H} = \frac{\hat{p}^2}{2} - \frac{Z}{\hat{r}} \tag{42}$$

The kinetic energy operator is

$$\hat{T} = \frac{\hat{p}^2}{2} \tag{43}$$

and

$$\hat{V}_n = -\frac{Z}{\hat{r}} \tag{44}$$

is the potential energy operator. The ground state wavefunction is

$$\psi(\underline{r}) = (Z^3/\pi)^{1/2}\, e^{-Zr} \tag{45}$$

when we use the position representation, and

$$\phi(\underline{p}) = \frac{2\sqrt{2}}{r}\, Z^{5/2}(p^2 + Z^2)^{-2} \tag{46}$$

when we use the momentum representation. From (44), (45), and the defining equation (40) we immediately find the potential energy density to be

$$\varepsilon_n(\underline{r}) = -\frac{Z}{r}\,\psi(\underline{r})\psi(\underline{r})^* = -\frac{Z^4}{\pi}\frac{1}{r}\, e^{-2Zr} \tag{47}$$

The kinetic energy density becomes

$$\varepsilon_K(\underline{r}) = \int d\underline{p}\; f(\underline{r},\underline{p})p^2/2 \tag{48}$$

By inserting the formal expression (10) for the Wigner function we may derive the more convenient expression

$$\varepsilon_K(\underline{r}) = \frac{1}{2}\left(|\nabla\psi|^2 - \frac{1}{4}\nabla^2|\psi|^2\right)$$

which is due to Ziff et al.[18] and to Cohen.[19] An equivalent expression is

$$\varepsilon_K(\underline{r}) = \frac{1}{2}\left(\frac{1}{2}|\nabla\psi|^2 - \frac{1}{4}\psi^*\nabla^2\psi - \frac{1}{4}\psi\nabla^2\psi^*\right) \tag{50}$$

By inserting the explicit form of the wavefunction, as given by (45), this becomes

$$\varepsilon_K(\underline{r}) = \frac{Z^4}{2\pi}\frac{1}{r}\,e^{-2Zr} \tag{51}$$

The energy density associated with the full Hamiltonian (42), i.e. the total energy density, is obviously

$$\varepsilon(\underline{r}) = \varepsilon_K(\underline{r}) + \varepsilon_n(\underline{r}) \tag{52}$$

We observe that the following relations hold:

$$\varepsilon_K(\underline{r}) = -\frac{1}{2}\,\varepsilon_n(\underline{r}) \tag{53}$$

$$\varepsilon_K(\underline{r}) = -\varepsilon(\underline{r}) \tag{54}$$

and when these are integrated over all space we get

$$E_K = -\frac{1}{2}\,E_n \tag{55}$$

$$E_K = -E \tag{56}$$

Eqs. (55) and (56) express the well-known virial theorem which holds for any eigenfunction of any atom. The virial theorem reflects a certain global balance between the kinetic and potential energies, and Eqs. (53) and (54) show that this balance is also maintained locally for the ground state of a hydrogen-like atom.

The expression (50) may be written

$$\varepsilon_K(\underline{r}) = \frac{1}{2}\left(\varepsilon_{KB}(\underline{r}) + \varepsilon_{KC}(\underline{r})\right) \tag{57}$$

with

$$\varepsilon_{KB}(\underline{r}) = \frac{1}{2}|\nabla\psi|^2 \tag{58}$$

and

$$E_{KC}(\underline{r}) = -\frac{1}{4}\left(\psi^*\nabla^2\psi + \psi\nabla^2\psi^*\right) \tag{59}$$

Both $\varepsilon_{KB}(\underline{r})$ and $\varepsilon_{KC}(\underline{r})$ are proper kinetic energy densities which give the total kinetic energy when integrated over all space, and both expressions are of course well known. Inserting the wavefunction (45) gives

$$\varepsilon_{KB}(\underline{r}) = \frac{Z^5}{2\pi}\,e^{-2Zr} \tag{60}$$

and

$$\varepsilon_{KC}(\underline{r}) = \frac{Z^4}{\pi}\left(\frac{1}{r} - \frac{Z}{2}\right) e^{-2Zr} \qquad (61)$$

None of these expressions satisfy anything resembling a local virial theorem together with $\varepsilon_n(\underline{r})$.

We consider it a virtue when the local quantities reflect a global balance to a maximum degree, and for this reason we like to favour $\varepsilon_K(\underline{r})$ for $\varepsilon_{KB}(\underline{r})$ and $\varepsilon_{KC}(\underline{r})$. But we cannot exclude the possibility that there may be occasions where $\varepsilon_{KB}(\underline{r})$ or $\varepsilon_{KC}(\underline{r})$ have certain advantages. Anyway, our analysis illustrates well the lack of uniqueness in the setting up of energy density functions.

This lack of uniqueness can be related to the possibility of constructing other phase space descriptions than Weyl's and Wigner's (cf. the discussion at the end of section 2.2). As a matter of fact, Cohen[19] has shown that the expressions for $\varepsilon_{KB}(\underline{r})$ and $\varepsilon_{KC}(\underline{r})$ can be derived from alternative phase space descriptions discussed earlier by him.[20] But this in turn provides us with another reason for preferring $\varepsilon_K(\underline{r})$, since we know that such alternative phase space descriptions lack the canonical features possessed by the Weyl-Wigner description.

Accordingly, we consider the definition (48) to be the canonical definition of the kinetic energy density.

The local virial theorem, expressed through Eqs. (53) and (54), is a remarkable one, but we stress that it is only for the ground state of a hydrogen-like atom that it is exactly fulfilled.

4.2. The Hydrogen Molecular Ion

In the present section we shall study the kinetic and potential energy densities for the ground state of H_2^+, with the nuclei fixed at their equilibrium separation, R = 2 atomic units. The wavefunction $\psi(\underline{r})$ describing this state is well known. It is obtained by separating the electronic Schrödinger equation in elliptic coordinates,[21-24] but we prefer to display our results in cylindrical coordinates. Thus, we locate the two nuclei on the z axis at -R/2 and R/2 respectively, and use z, r and ϕ as independent coordinates, where r now is the perpendicular distance to the z axis and ϕ is the angle of rotation about this axis. The ground state wavefunction is independent of ϕ.

Using the wavefunction given by Bates et al.[22] we have calculated the electron density ρ and the kinetic energy densities ε_K, ε_{KB} and ε_{KC}, as defined by Eqs. (50), (58) and (59), respectively. Furthermore, we have calculated the potential energy density

$$\varepsilon_n = \left(-\frac{1}{r_a} - \frac{1}{r_b}\right)|\psi|^2 \tag{62}$$

where r_a and r_b measure the distances to the two nuclei.

In Fig. 1 we display the calculated densities by means of contour maps, after each density has been multiplied by $2\pi r$. Thus, each density, $\varepsilon'(z,r)$, is normalized such that

$$\int_{-\infty}^{\infty} dz \int_0^{\infty} dr\ \varepsilon'(z,r) = 1 \tag{63}$$

It is seen that the potential energy density ε'_n and the two kinetic energy densities ε'_K and ε'_{KC} show similar structures, in the sense that they are all quite localized around the two nuclei. The kinetic energy density ε'_{KB} is more delocalized and has a structure similar to that of the electron density ρ'.

To investigate the possible validity of a local virial theorem we have set up a (z,r) grid of M = 6480 points, and at each grid point calculated the values of $\varepsilon_n/\varepsilon_K$, $\varepsilon_n/\varepsilon_{KB}$, and $\varepsilon_n/\varepsilon_{KC}$. Denoting by x_i the value of any of these three ratios at the i'th grid point we have then evaluated the mean value

$$\mu = \frac{1}{M}\sum_{i=1}^{M} x_i \tag{64}$$

and the standard deviation

$$\sigma = \left\{ \frac{1}{M-1}\left[\sum_{i=1}^{M} x_i^2 - \frac{1}{M}\left(\sum_{i=1}^{M} x_i\right)^2\right]\right\}^{1/2} \tag{65}$$

The results are shown in Table 1, where we have also listed x_{max}

Table 1. An investigation of the local balance between the potential and kinetic energies in H_2^+. For explanation see text.

x	μ	σ	x_{max}	x_{min}
$-\varepsilon_n/\varepsilon_K$	2.87	0.34	4.65	2.06
$-\varepsilon_n/\varepsilon_{KC}$	5.70	4.78	38100	-3210
$-\varepsilon_n/\varepsilon_{KB}$	2.07	16.9	943	0.352

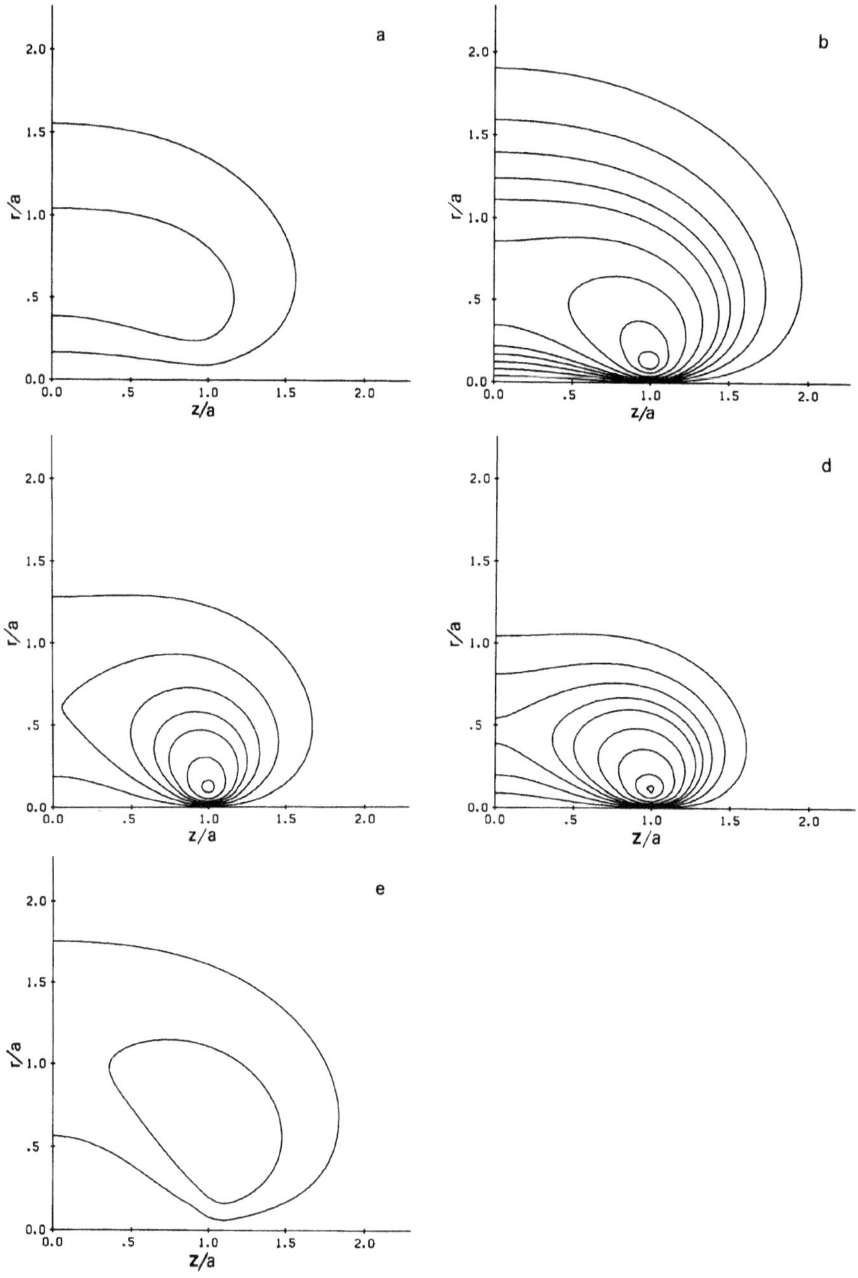
a
b
d
e
r/a
z/a
0.0
.5
1.0
1.5
2.0

and x_{min} which denote the largest and smallest values of x encountered at any grid point.

It is obvious from Table 1 that the distribution of $\varepsilon_n/\varepsilon_K$ is much sharper than the distributions of $\varepsilon_n/\varepsilon_{KB}$ and $\varepsilon_n/\varepsilon_{KC}$. Furthermore, the mean value of $\varepsilon_n/\varepsilon_K$ only deviates 1.5% from the value -2.83 determined by the ratio E_n/E_K, where E_n and E_K are the total potential and kinetic energies as defined by Eqs. (2) and (3), respectively.

Since R is chosen as the equilibrium distance the virial theorem states that

$$(E_n + \frac{1}{R})/E_K = -2 \tag{66}$$

(see, e.g., Ref. 25). The above E_n/E_K ratio of -2.83 is in fact in agreement with Eq. (66). Hence we can state that the virial theorem is locally satisfied to a substantial degree of accuracy, but only when we adapt the definition (48) for the kinetic energy density. So, again we must consider this definition as the canonical one.

4.3. The Hydrogen Molecule

In the preceding two sections we have shown that the interplay between the potential and kinetic energy terms tends to establish a local virial theorem for the ground states of the one-electron systems H and H_2^+. In continuation of this finding we shall now investigate the potential and kinetic energy densities for the ground state of a two-electron system: the hydrogen molecule.

The ground state of H_2 is a singlet state. By eliminating the spins of the two electrons we obtain the spin-free wavefunction $\Phi(\underline{r}_1,\underline{r}_2)$ which is symmetric under interchange of the two particles. Extremely accurate representations of Φ are now known, but for the present purpose we have considered it sufficient to base the analysis on the wavefunction determined by Shull and Ebbing.[26] This function is a linear combination of a covalent and an ionic term constructed

←

Fig. 1. Contour curves in the (z,r) plane, for various ground state densities in H_2^+. Distances in units of a = R/2. (a) Electron density ρ'. Contour values: 0.1, 0.2. (b) Potential energy density ε_n'. Contour values: -0.05, -0.1, -0.15, -0.2, -0.25, -0.35, -0.5, -0.75, -1.0. (c) Kinetic energy density ε_K'. Contour values: 0.05, 0.1, 0.15, 0.2, 0.25, 0.35, 0.5. (d) Kinetic energy density ε_{KC}'. Contour values: 0.05, 0.1, 0.15, 0.2, 0.25, 0.35, 0.5, 0.75, 1.0. (e) Kinetic energy density ε_{KB}'. Contour values: 0.05, 0.1. All contour values in atomic units.

from scaled 1s Slater orbitals slightly displaced from the nuclear positions (floating orbitals). From such a wavefunction it is relatively straightforward to calculate the electronic density

$$\rho(\underline{r}) = 2\int d\underline{r}_2 |\Phi(\underline{r}, \underline{r}_2)|^2 \tag{67}$$

and the kinetic energy densities ε_K, ε_{KB} and ε_{KC}. ε_K is obtained from Eq. (28) via a modified Eq. (49). ε_{KB} and ε_{KC} are obtained from the modified Eqs. (58) and (59). The potential energy density associated with the attraction between the electrons and the nuclei has a form similar to (62):

$$\varepsilon_n = \left(-\frac{1}{r_a} - \frac{1}{r_b}\right) \rho(\underline{r}) \tag{68}$$

In addition to the above mentioned densities we must, however, also consider a density $\varepsilon_e(\underline{r})$ associated with the electron-electron repulsion energy E_e, such that

$$E_e = \int d\underline{r}\ \varepsilon_e(\underline{r}) \tag{69}$$

The energy E_e is, of course, the sum of the two last terms in Eq. (1), but in the present section we are not interested in the partitioning of E_e.

The obvious thing is to define $\varepsilon_e(\underline{r})$ according to the relation

$$\varepsilon_e(\underline{r}) = \int d\underline{r}_2\ \Phi(\underline{r}, \underline{r}_2) \frac{1}{r_{12}} \Phi(\underline{r}, \underline{r}_2) \tag{70}$$

where r_{12} is the distance between the two electrons. This density is then equal to $\rho(\underline{r})$ multiplied by the so-called reduced local energy.[27-30]

Adding $\varepsilon_n(\underline{r})$ and $\varepsilon_e(\underline{r})$ leads to the energy density associated with the total potential energy E_P of the molecule:

$$\varepsilon_P(\underline{r}) = \varepsilon_n(\underline{r}) + \varepsilon_e(\underline{r}) \tag{71}$$

ε_P now takes the role played by ε_n in the previous section.

As in that section we introduce primed quantities satisfying Eq. (63) and measure distances in terms of R/2, where R is the equilibrium bond distance, 1.428 atomic units. Using a grid of M = 2070 points we have then obtained the results shown in Fig. 2 and Table 2.

Fig. 2 shows that the densities ρ', ε_n', ε_K', ε_{KC}', and ε_{KB}' have the same characteristics as for H_2^+. The density ε_e' is low and has a structure similar to that of ρ'.

Table 2. An investigation of the local balance between the potential and kinetic energies in H_2. For explanation see text.

x	μ	σ	x_{max}	x_{min}
$-\varepsilon_P/\varepsilon_K$	2.53	0.26	4.12	1.66
$-\varepsilon_P/\varepsilon_{KC}$	-0.38	35.6	1007	-471
$-\varepsilon_P/\varepsilon_{KB}$	1.52	2.35	28.0	0.48

Table 2 shows, in analogy with Table 1, that the virial theorem is locally satisfied to a considerable degree of accuracy when and only when the kinetic energy density is described by ε_K. The value of μ = 2.53 compares well with the value 2.61 of the global ratio $-E_P/E_K$.

Thus, we have strengthened our previous conclusion, that the most appropriate definition of the kinetic energy density is the one based on the Wigner function according to Eqs. (38) and (48).

5. COULOMB AND EXCHANGE ENERGY DENSITIES

So far we have studied the densities $\varepsilon_K(\underline{r})$ and $\varepsilon_n(\underline{r})$ in Eqs. (2) and (3), and in section 4.3 we have also considered the sum of $\varepsilon_{c\ell}$ and ε_{xc} for a two-electron system, the hydrogen molecule. In what follows we shall now consider the densities $\varepsilon_{c\ell}(\underline{r})$ and $\varepsilon_{xc}(\underline{r})$ separately and calculate them for a few simple atoms, in the restricted Hartree-Fock approximation. Since we adopt the Hartree-Fock approximation we shall write $-k(\underline{r})$ instead of $\varepsilon_{xc}(\underline{r})$ and E_x instead of E_{xc}. Eq. (5) is then replaced by

$$E_x = -\int d\underline{r}\; k(\underline{r}) \tag{72}$$

The form of $\varepsilon_{c\ell}(\underline{r})$ is easily written down, but the problem of specifying $k(\underline{r})$ is not so simple. As with the kinetic energy density we turn to a phase space formulation.

5.1. The Weyl Transform of the Exchange Operator

We consider a closed shell system with 2n electrons for which the Hartree-Fock wavefunction is a Slater determinant with n doubly occupied orbitals:

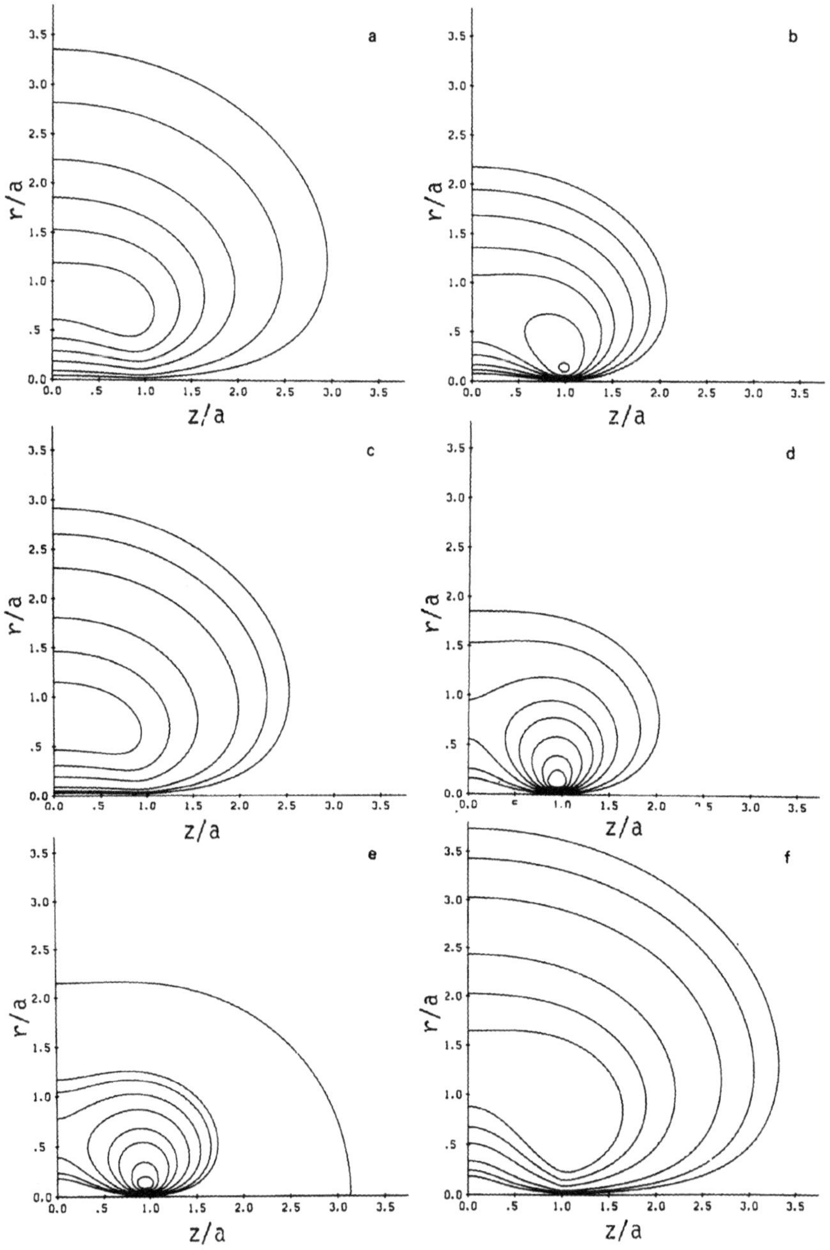
a
b
c
d
e
f
r/a
z/a
0.0
.5
1.0
1.5
2.0
2.5
3.0
3.5

$$\Psi = |\psi_1\bar{\psi}_1 \cdots \psi_n\bar{\psi}_n| \tag{73}$$

The so-called Coulomb operator is defined as

$$\hat{J} = 2\sum_{i=1}^{n} \hat{J}_i \tag{74}$$

with

$$\hat{J}_i\psi(\underline{r}_1) = e^2\int d\underline{r}_2 \frac{\psi_i(\underline{r}_2)^*\psi_i(\underline{r}_2)}{r_{12}} \psi(\underline{r}_1) \tag{75}$$

and the exchange operator is

$$\hat{K} = \sum_{i=1}^{n} \hat{K}_i \tag{76}$$

with

$$\hat{K}_i\psi(\underline{r}_1) = e^2\int d\underline{r}_2 \frac{\psi_i(\underline{r}_2)^*\psi_i(\underline{r}_1)}{r_{12}} \psi(\underline{r}_2) \tag{77}$$

The first order spin-free density matrix of Eq. (25) has the form (36) and is

$$\rho_1(\underline{r}_1,\underline{r}_2) = 2\sum_{i=1}^{n} \psi_i(\underline{r}_1)\psi_i(\underline{r}_2)^* \tag{78}$$

$\rho_1^{\uparrow}$ and $\rho_1^{\downarrow}$ are identical and both equal to

$$\rho_1^{\uparrow}(\underline{r}_1,\underline{r}_2) = \sum_{i=1}^{n} \psi_i(\underline{r}_1)\psi_i(\underline{r}_2)^* \tag{79}$$

In terms of ρ_1 and $\rho_1^{\uparrow}$ we may accordingly write

$$\hat{J}\psi(\underline{r}_1) = e^2\int d\underline{r}_2 \frac{\rho(\underline{r}_2)}{r_{12}} \psi(\underline{r}_1) \tag{80}$$

←

Fig. 2. Contour curves in the (z,r) plane, for various ground state densities in H_2. Distances in units of a = R/2. (a) Electron density ρ'. Contour values: 0.05, 0.1, 0.2, 0.3, 0.4, 0.5. (b) Electron-nucleus potential energy density ε_n'. Contour values: -0.25, -0.35, -0.5, -0.75, -1.0, -1.5, -2.5. (c) Electron-electron potential energy density ε_e'. Contour values: 0.02, 0.03, 0.05, 0.1, 0.15, 0.2. (d) Kinetic energy density ε_K'. Contour values: 0.1, 0.15, 0.25, 0.35, 0.45, 0.6, 0.8, 1.0. (e) Kinetic energy density ε_{KC}'. Contour values: 0.0, 0.2, 0.25, 0.35, 0.45, 0.75, 1.0, 1.5, 2.0. (f) Kinetic energy density ε_{KB}'. Contour values: 0.02, 0.03, 0.05, 0.1, 0.15, 0.2. All contour values in atomic units.

and

$$\hat{K}\psi(\underline{r}_1) = e^2\int d\underline{r}_2 \frac{\rho_1^{\dagger}(\underline{r}_1,\underline{r}_2)}{r_{12}} \psi(\underline{r}_2) \tag{81}$$

The Weyl transform of the Coulomb operator is seen to be

$$J(\underline{r}_1) = e^2\int d\underline{r}_2 \frac{\rho(\underline{r}_2)}{r_{12}} \tag{82}$$

It is just the classical Coulomb potential associated with the electronic density $\rho(\underline{r})$. To find the Weyl transform $K(\underline{r},\underline{p})$ of the exchange operator we use the expression (34), and hence we must first evaluate the matrix elements of $\hat{K}$ in the position representation. We get immediately:

$$\begin{aligned} \langle\underline{r}'|\hat{K}|\underline{r}''\rangle &= \int d\underline{r}_1\ \delta(\underline{r}_1 - \underline{r}')\hat{K}\delta(\underline{r}_1 - \underline{r}'') \\ &= e^2\int d\underline{r}_1 d\underline{r}_2\ \delta(\underline{r}_1 - \underline{r}') \frac{\rho_1^{\dagger}(\underline{r}_1,\underline{r}_2)}{r_{12}} \delta(\underline{r}_2 - \underline{r}'') \\ &= e^2 \frac{\rho_1^{\dagger}(\underline{r}',\underline{r}'')}{|\underline{r}'-\underline{r}''|} \end{aligned} \tag{83}$$

Eq. (34) then gives:

$$\begin{aligned} K(\underline{r},\underline{p}) &= \int d\underline{v}\ \langle\underline{r} + \frac{1}{2}\underline{v}|\hat{K}|\underline{r} - \frac{1}{2}\underline{v}\rangle\ e^{-i\underline{p}\cdot\underline{v}/\hbar} \\ &= e^2\int d\underline{v} \frac{\rho_1^{\dagger}(\underline{r} + \frac{1}{2}\underline{v},\ \underline{r} - \frac{1}{2}\underline{v})}{v} e^{-i\underline{p}\cdot\underline{v}/\hbar} \end{aligned} \tag{84}$$

Next we introduce the Wigner function $f^{\dagger}(\underline{r},\underline{p})$ corresponding to $\rho_1^{\dagger}$. It is just the sum of the n functions which result from Eq. (10) when ψ therein is successively replaced by $\psi_1,\psi_2,\ldots,\psi_n$. We insert the expression (24) and get:

$$K(\underline{r},\underline{p}) = e^2\int\int d\underline{v}\ d\underline{p}' \frac{f^{\dagger}(\underline{r},\underline{p}')}{v} e^{i(\underline{p}'-\underline{p})/\hbar} \tag{85}$$

and since

$$\int d\underline{v} \frac{e^{i\underline{k}\cdot\underline{v}}}{v} = \frac{4\pi}{k^2} \tag{86}$$

this becomes

$$K(\underline{r},\underline{p}) = 4\pi e^2\hbar^2\int d\underline{p}' \frac{f^{\dagger}(\underline{r},\underline{p}')}{|\underline{p}-\underline{p}'|^2} \tag{87}$$

Thus a relatively transparent expression exists for the Weyl transform of the exchange operator.

It is remarkable that this expression already occurs in Dirac's 1930 "Note on exchange phenomena in the Thomas atom",[31] but it was derived there under "the further approximation of neglecting the fact that the coordinates and momenta do not commute, so that the problem reduces to a classical problem". Our derivation shows that the validity of Eq. (87) extends far beyond this approximation. (In comparing with Dirac's paper, note that the $\rho(qp')$ in his Eq. (17) equals h^3 times our $f^{\uparrow}(\underline{r},\underline{p}')$.)

Dirac evaluated $K(\underline{r},\underline{p})$ for the specific case of the Thomas-Fermi atom, where it is assumed that $f^{\uparrow}(\underline{r},\underline{p})$ and $f^{\downarrow}(\underline{r},\underline{p})$ are both equal to $1/h^3$ whenever p is less than a certain value $p_0(r)$, and both are equal to zero when p is larger than this value. Performing the integrations in Eq. (87) then gives the expression

$$K(\underline{r},\underline{p}) = \frac{e^2}{\pi\hbar}\left\{\frac{1}{2}\,\frac{p_0(r)^2 - p^2}{p}\,\ell n\left|\frac{p_0(r) + p}{p_0(r) - p}\right| + p_0(r)\right\} \tag{88}$$

(Note that Dirac's Eq. (19) gives twice this value, because he erroneously put the phase space density equal to $2/h^3$ instead of $1/h^3$. See also Jensen.[32])

To our knowledge Eq. (88) is the only application that has so far been made of Eq. (87), and this is probably due to the approximations assumed by Dirac in his derivation. Here we evaluate the right hand side of Eq. (87) under the assumption that

$$\rho_1^{\uparrow}(\underline{r},\underline{r}') = \phi(\underline{r})\phi(\underline{r}')^* \tag{89}$$

where $\phi(\underline{r})$ is a normalized 1s Gaussian:

$$\phi(\underline{r}) = \left(\frac{2\alpha}{\pi}\right)^{3/4} e^{-\alpha r^2} \tag{90}$$

Introducing atomic units we get for the Wigner function corresponding to $\rho_1^{\uparrow}$:[8]

$$f^{\uparrow}(\underline{r},\underline{p}) = \frac{1}{\pi^3}\, e^{-2\alpha r^2}\, e^{-\frac{p^2}{2\alpha}} \tag{91}$$

and when this is inserted into (87) and the integrations performed, we obatin the result

$$K(\underline{r},\underline{p}) = \frac{16}{\sqrt{\pi}}\,\frac{\alpha}{p}\, e^{-2\alpha r^2}\, F\left(\frac{p}{\sqrt{2\alpha}}\right) \tag{92}$$

where

$$F(x) = e^{-x^2}\int_0^x e^{t^2}\,dt \tag{93}$$

is Dawson's integral (see Ref. 33 p. 298).

It is of course of considerable interest to evaluate the form of $K(\underline{r},\underline{p})$ for more general situations, but in the present context we shall be content with the two examples of Eqs. (88) and (92).

5.2. Formulae for the Coulomb and Exchange Energy Densities

On the basis of the preceding section we shall now derive the expressions for the Coulomb and exchange energy densities associated with the Slater determinant (73).

The total Coulomb energy is

$$E_{c\ell} = \sum_{j=1}^{n} \langle\psi_j|\hat{J}|\psi_j\rangle \tag{94}$$

and by introducing the Coulomb potential (82) and the electron density from (78) this becomes

$$E_{c\ell} = \frac{1}{2}\int d\underline{r}\; J(\underline{r})\rho(\underline{r}) \tag{95}$$

Hence the Coulomb energy density is

$$\varepsilon_{c\ell} = \frac{1}{2}\, J(\underline{r})\rho(\underline{r}) \tag{96}$$

The total exchange energy is

$$E_x = -\sum_{j=1}^{n} \langle\psi_j|\hat{K}|\psi_j\rangle \tag{97}$$

In analogy with Eq. (15) we write it as

$$E_x = -\iint d\underline{r}\; d\underline{p}\; f^{\dagger}(\underline{r},\underline{p})K(\underline{r},\underline{p}) \tag{98}$$

The prescription (38) then gives the exchange energy density as

$$k(\underline{r}) = \int d\underline{p}\; f^{\dagger}(\underline{r},\underline{p})K(\underline{r},\underline{p}) \tag{99}$$

which by insertion of (87) becomes, in atomic units:

$$k(\underline{r}) = 4\pi\iint d\underline{p}_1\, d\underline{p}_2\, \frac{f^{\dagger}(\underline{r},\underline{p}_1)f^{\dagger}(\underline{r},\underline{p}_2)}{|\underline{p}_1-\underline{p}_2|^2} \tag{100}$$

This is a qualitatively satisfactory expression, but from a computational point of view it is advantageous to reduce it by replacing $f^{\dagger}(\underline{r},\underline{p})$ with $\rho_1^{\dagger}(\underline{r}_1,\underline{r}_2)$ according to Eq. (23). By also using Eq. (79) we can then show that

$$k(\underline{r}) = \sum_{i=1}^{n} \sum_{j=1}^{n} k_{ij}(\underline{r}) \tag{101}$$

where

$$k_{ij}(\underline{r}) = 4\int d\underline{r}_1 \frac{1}{r_1} \psi_i(\underline{r}-\underline{r}_1)^* \psi_i(\underline{r}+\underline{r}_1)\psi_j(\underline{r}+\underline{r}_1)^* \psi_j(\underline{r}-\underline{r}_1) \tag{102}$$

We note that

$$\int d\underline{r}\, k_{ij}(\underline{r}) = K_{ij} \tag{103}$$

where K_{ij} is the usual exchange integral between the orbitals ψ_i and ψ_j, i.e.

$$K_{ij} = \iint d\underline{r}_1 d\underline{r}_2\, \psi_i(\underline{r}_1)^* \psi_j(\underline{r}_2)^* \frac{1}{r_{12}} \psi_i(\underline{r}_2)\psi_j(\underline{r}_1) \tag{104}$$

In the following we shall use the expressions (96) and (101) to calculate the Coulomb and exchange energy densities for a few simple atoms.

5.3. Coulomb and Exchange Energy Densities for He, Be, Ne, and Ar

From the expressions (95) and (101) we have calculated the electron density and the Coulomb and exchange energy densities for the ground state of each of the closed shell atoms He, Be, Ne, and Ar. The atomic orbitals were represented as linear combinations of Gaussians, since this greatly simplifies the evaluation of the integrals (102). For He and Be a basis set of 5 s-type Gaussians was used,[34] for Ne a basis set of 5 s-type and 3 p-type Gaussians,[34] and for Ne a basis set of 9 s-type and 5 p-type Gaussians.[35]

The results of the calculations are shown in Figs. 3a-d, where we have plotted the energy densities against the electron density using logarithmic scales on both axes. The range of r-values is for each figure from 0.01 atomic units (highest densities) to 3.00 atomic units (lowest densities).

The four figures look very similar, the relation between energy density and electron density being quite smooth in all cases. The only irregularities appear on the exchange energy curves. These irregularities reflect the shell structure of the atoms, as a comparison between Fig. 3b and Fig. 4 reveals. Thus, Fig. 4 shows the radial electron densities for Be, and the dip on the total density curve of that figure matches quite well with the irregularity on the exchange energy curve of Fig. 3b. For Ar (Fig. 3d) there are two irregularities, one associated with the transition from the n=1 to the n=2 shell, the other with the transition from the n=2 to the n=3 shell, with n denoting the principal quantum number. (Note, however, that the densities in Fig. 3 are ordinary densities, not radial densities.)

We shall now consider the possibility of constructing an analytic fit to the curves of Fig. 3. We treat the Coulomb and exchange curves separately.

The fact that the Coulomb curves are almost straight lines suggests that we should look for an analytic representation to these curves of the form

$$\varepsilon_{c\ell}(\underline{r}) = k\rho(\underline{r})^{\alpha} \tag{105}$$

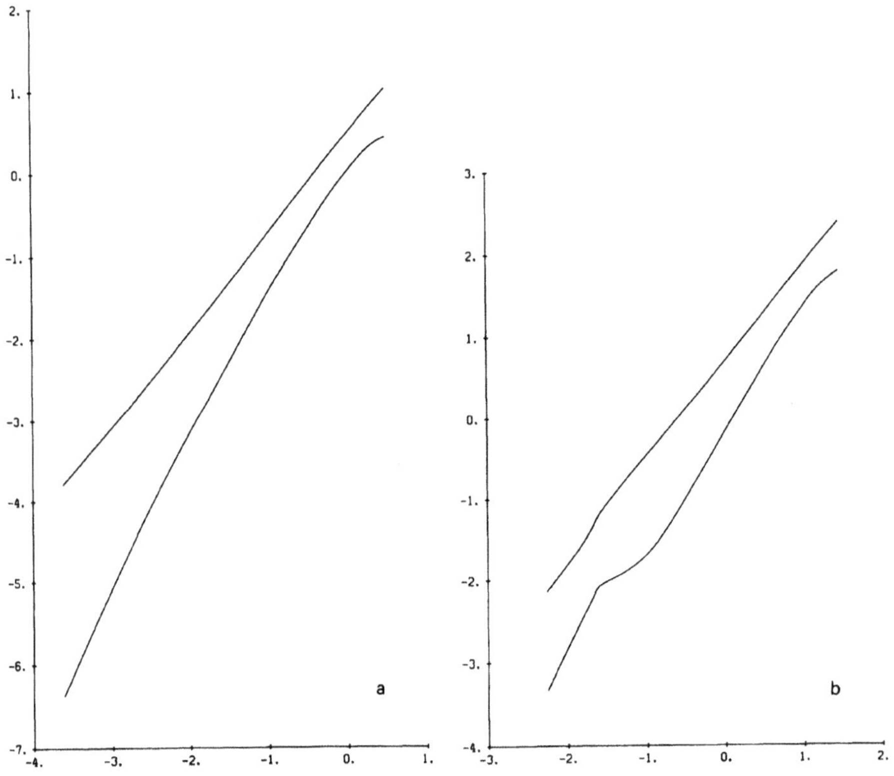

Fig. 3. Coulomb energy densities multiplied by 2 (upper curves) and exchange energy densities (lower curves) vs. electron densities. Atomic units. Logarithmic scales (base 10) on both axes. (a) Helium. (b) Beryllium.

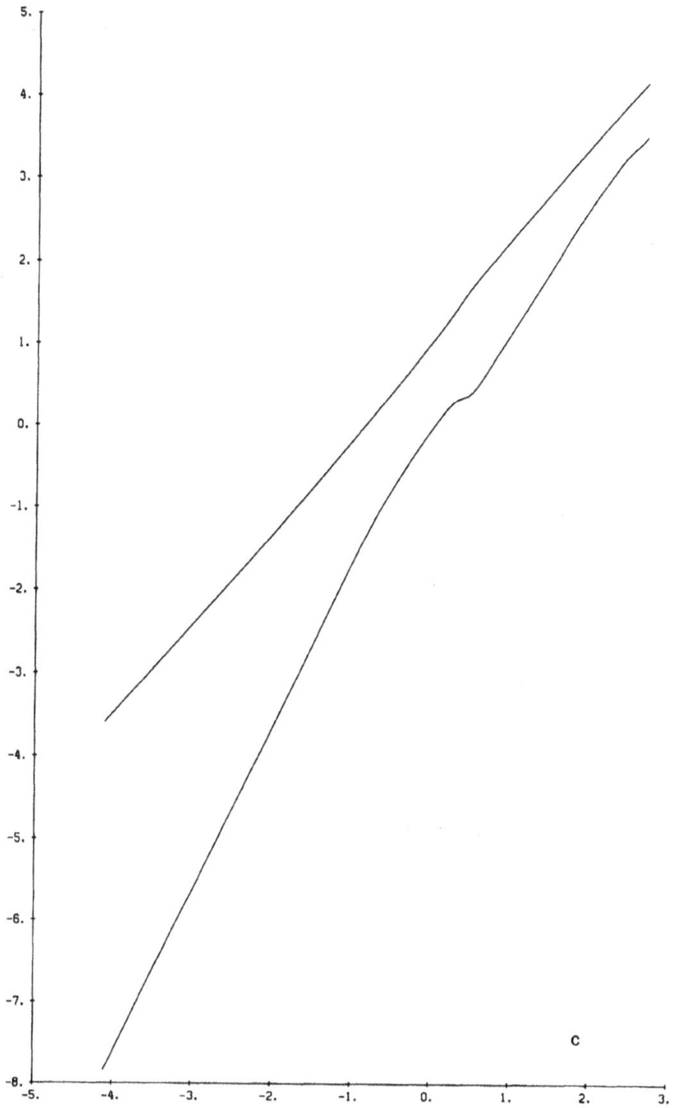

Fig. 3c. Neon Coulomb energy density multiplied by 2 (upper curve) and exchange energy density (lower curve) vs. electron density. Atomic units. Logarithmic scales (base 10) on both axes.

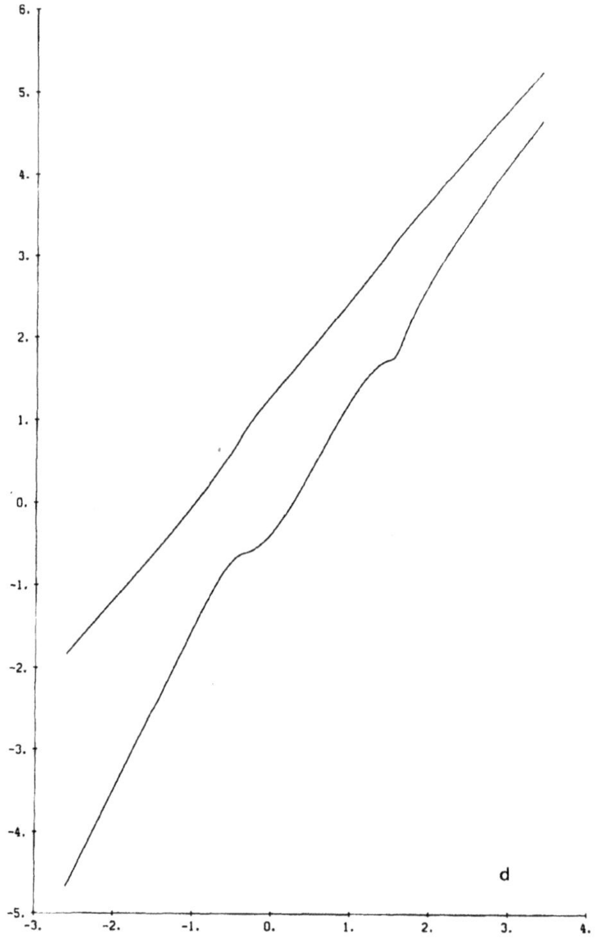

Fig. 3d. Argon Coulomb energy density multiplied by 2 (upper curve) and exchange energy density (lower curve) vs. electron density. Atomic units. Logarithmic scales (base 10) on both axes.

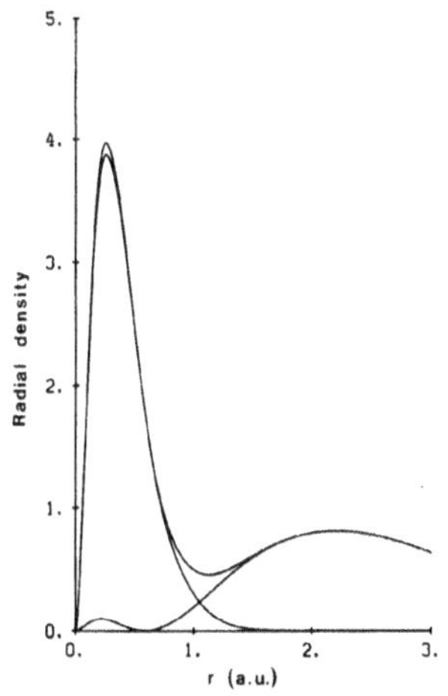

Fig. 4. 1s, 2s, and total electron densities vs. r for beryllium. All densities are radial densities (they include a factor $4\pi r^2$), obtained from the (5s) expansion in Ref. 34. Atomic units.

We have tested this representation by minimizing the quantity

$$R = \sum_{i=1}^{300} [\log e(i) - \log \varepsilon(i)]^2 \tag{106}$$

where i specifies equidistant points between r=0.01 and r=3.00 atomic units. $\varepsilon(i)$ is the value from the exact calculation, e(i) the corresponding value obtained from Eq. (105). Thus, we have obtained the results presented in Table 3. The representation is very satisfactory.

Table 3. Parameters of the analytic fit (105)

	He	Be	Ne	Ar
k	1.390	2.399	5.587	8.867
α	1.183	1.225	1.161	1.215
R	0.09	0.84	0.91	0.61

Table 4. Parameters of the analytic fit (109)

	He	Be	Ne	Ar
k_1	1.248	0.779	0.600	0.633
α_1	1.702	1.341	1.746	1.523
R	8.10	5.05	24.47	15.52

The parameter α is seen to be quite similar for all the atoms considered, whereas k of course varies with the number of electrons. This variation can, however, also be parameterized by choosing a representation of the form

$$\varepsilon_{c\ell}(\underline{r}) = k_0 N^{\beta} \rho(\underline{r})^{\alpha} \tag{107}$$

where N is the number of electrons. We have found that the expression

$$\varepsilon_{c\ell}(\underline{r}) = 0.747 N^{0.868} \rho(\underline{r})^{1.189} \tag{108}$$

gives a good representation of the Coulomb energy density for all four atoms considered.

The exchange energy curves of Fig. 3 are more difficult to describe analytically. We have tried the representation

$$k(\underline{r}) = k_1 \rho(\underline{r})^{\alpha_1} \tag{109}$$

and obtained the parameter values of Table 4. Obviously, this representation is much less satisfactory than the corresponding one for the Coulomb energy, and hence we have not attempted to construct a representation similar to (108). We have tried to improve the representation (109) by adding terms containing $|\nabla\rho|$, but have been unable to construct an analytic expression which would give a good representation of the exchange energy density for all four atoms, with a common set of parameters.

6. LOCAL DENSITY FUNCTIONALS

The analytic representations obtained above for $\varepsilon_{c\ell}(\underline{r})$ and $k(\underline{r})$ express these quantities as local functionals of the electron density $\rho(\underline{r})$. The Coulomb functional (108) is universal in the sense that it is valid for all four atoms considered. The exhange functional (109) is, on the other hand, different for each atom. We shall now try to understand the differences between these exchange functionals by relating them to two simple models: the Thomas-Fermi atom and the Gaussian atom. We shall also compare the Coulomb functional (108) with a similar functional due to Parr and coworkers and, in that connection, comment on a traditional scaling argument.

6.1. The Thomas-Fermi Atom

The Thomas-Fermi model is, as already mentioned in section 5.1, based on the assumption that the Wigner functions $f^{\uparrow}(\underline{r},\underline{p})$ and $f^{\downarrow}(\underline{r},\underline{p})$ both are equal to $1/h^3$ whenever p is less than a certain value $p_0(r)$, and both are equal to zero when p is larger than this value. The total Wigner function $f(\underline{r},\underline{p})$ is the sum of its spin-up and spin-down parts, as in Eq. (27).

The electron density in the Thomas-Fermi atom is

$$\rho(\underline{r}) = \int d\underline{p}\ f(\underline{r},\underline{p}) = \int_0^{p_0(r)} (4\pi p^2 dp)\ \frac{2}{h^3} = \frac{8\pi}{3}\left(\frac{p_0(r)}{h}\right)^3 \tag{110}$$

For the Weyl transform of the exchange operator we have Dirac's expression (88), which we rewrite as

$$K(\underline{r},\underline{p}) = 4e^2\ \frac{p_0(r)}{h}\ G(\eta) \tag{111}$$

with

$$\eta = \frac{p}{p_0(r)} \tag{112}$$

and

$$G(\eta) = \frac{1}{2} + \frac{1-\eta^2}{4\eta}\ \ell n \left|\frac{1+\eta}{1-\eta}\right| \tag{113}$$

Obviously $K(\underline{r},\underline{p})$ can also be written as

$$K(\underline{r},\underline{p}) = 4e^2 \left[\frac{3}{8\pi}\ \rho(r)\right]^{1/3} G(\eta) \tag{114}$$

The function $G(\eta)$ varies smoothly between the values 1/2 and 1. Slater[36] replaced it by its averaged value 3/4 and thus introduced the local exchange potential which bears his name:

$$V_{XS}(r) = 3e^2 \left[\frac{3}{8\pi}\ \rho(r)\right]^{1/3} \tag{115}$$

The exchange energy density of the Thomas-Fermi atom can be obtained from Eq. (99) or (100). It is

$$k(r) = \frac{3}{2}\ e^2 \left(\frac{3}{8\pi}\right)^{1/3} [\rho(r)]^{4/3} \tag{116}$$

It was first obtained by Bloch,[37] and although he actually derived it in a different way he did evaluate the integral in Eq. (100). We see that

$$\frac{dk(r)}{d\rho(r)} = \frac{2}{3} V_{xS}(r) \tag{117}$$

as discussed by Kohn and Sham.[38]

6.2. The Gaussian Atom

Let us define a Gaussian atom as an atom for which $\rho_1^{\uparrow}(\underline{r},\underline{r}')$ and $\rho_1^{\downarrow}(\underline{r},\underline{r}')$ both have the form (89), with $\phi(\underline{r})$ as given by (90). The electron density for such an atom is

$$\rho(r) = 2\left(\frac{2\alpha}{\pi}\right)^{3/2} e^{-2\alpha r^2} \tag{118}$$

It is normalized to 2 electrons, but we could in principle have normalized it to any number.

For the Weyl transform of the exchange operator we have previously derived the expression (92), which we rewrite as

$$K(\underline{r},\underline{p}) = 8\left(\frac{2\alpha}{\pi}\right)^{1/2} e^{-\alpha r^2} \frac{1}{\mu} F(\mu) \tag{119}$$

with

$$\mu = \frac{p}{\sqrt{2\alpha}} \tag{120}$$

By using (118) we get

$$K(\underline{r},\underline{p}) = \frac{2\pi}{\alpha} \rho(r) \frac{1}{\mu} F(\mu) \tag{121}$$

The function $F(\mu)/\mu$ varies smoothly between 1 and 0. By replacing it with its average value 1/2 we get the local exchange potential

$$V_{XG}(r) = \frac{\pi}{\alpha} \rho(r) \tag{122}$$

The exchange energy density of the Gaussian atom can be obtained from Eq. (100) or, more easily, from an integral of the type (102). It is found to be

$$k(r) = \frac{\pi}{2\alpha} [\rho(r)]^2 \tag{123}$$

We see that

$$\frac{dk(r)}{d\rho(r)} = V_{XG}(r) \tag{124}$$

The difference between the two expressions (117) and (124), which manifests itself through the factor 2/3, can be rationalized in the following manner.

Whenever the exchange operator $\hat{K}$ associated with the Slater determinant (73) is replaced by a multiplicative exchange potential $V_X(\underline{r})$, the total exchange energy (97) becomes

$$E_X = -\frac{1}{2}\int d\underline{r}\; V_X(\underline{r})\rho(\underline{r}) \tag{125}$$

The exchange energy density is accordingly

$$k(\underline{r}) = \frac{1}{2} V_X(\underline{r})\rho(\underline{r}) \tag{126}$$

It may happen that $V_X(\underline{r})$ is a local potential of the form

$$V_X(\underline{r}) = A[\rho(\underline{r})]^s \tag{127}$$

where A and s are constants. Then we get that

$$k(\underline{r}) = \frac{1}{2} A[\rho(\underline{r})]^{s+1} \tag{128}$$

and hence

$$\frac{dk(\underline{r})}{d\rho(\underline{r})} = \frac{1}{2}(s+1)V_X(\underline{r}) \tag{129}$$

The relations(117) and (124) are special cases of this result, with s equal to 1/3 and 1 respectively.

The consistent description of exchange which we have obtained for the Thomas-Fermi atom and the Gaussian atom demonstrates that the phase space formulation is a natural intermediate in the construction of exchange energy densities, just as it was a natural intermediate in the construction of kinetic energy densities.

6.3. The Exchange Energy Densities for He, Be, Ne, and Ar Revisited

We may adopt the viewpoint that the "soft" Thomas-Fermi atom and the "hard" Gaussian atom represent two extremes, two limiting cases between which any atom can be placed so far as exchange effects are concerned. If we look at Table 4 in this perspective, it is seen that the "metallic" beryllium atom stands out as an atom which is very close to the Thomas-Fermi limit. α_1 is almost exactly 4/3, and k_1 is approximately 0.7386 which is the value of the constant in Eq. (116), in atomic units. Be is a soft atom.

The α_1 values for He and Ne show that these two atoms are closer to the Gaussian atom than to the Thomas-Fermi atom. They may be characterized as being hard atoms. Obviously Ar is an atom in between.

We believe that this way of interpreting the results in Table 4 is a meaningful one, and it would be interesting to analyse other atoms in a similar way.

6.4. The Coulomb Energy Density Revisited

Parr and his coworkers have been very active in deriving local energy functionals of the electron density.[39] Thus, they have derived the following expression for the Coulomb energy densities of neutral atoms[40,41]

$$\varepsilon_{c\ell}(\underline{r}) = 0.8994\ N^{2/3}\ \rho(\underline{r})^{4/3} \qquad (130)$$

It is of the form (107), like our expression (108). Our α exponent is, however, somewhat smaller than 4/3, and accordingly the Coulomb energy is "softer" in our picture. The sum of the α and β exponents is exactly 2 in Eq. (130), in Eq. (108) it is 2.057.

Parr et al. fixed the $\alpha = 4/3$ value from a scaling argument and the value $\alpha+\beta = 2$ from a normalization argument. The constant 0.8994 was determined by comparison with Hartree-Fock results.

Our analysis is based on detailed calculations, and it is gratifying that the expressions (108) and (130) are so similar. Since our $\alpha+\beta$ value is very close to 2 we may accept the normalization argument as being very strong. The scaling argument appears somewhat weaker.

We agree that a universal functional must have $\alpha = 4/3$, and the idea of Parr et al. was indeed to construct a universal functional, but we want to stress that if the constant k_0 in Eq. (107) is allowed to vary slightly from atom to atom, then we can allow α to differ from 4/3 without destroying the scaling argument. That this is so becomes obvious if one compares the exchange expressions (116) and (123). One of them has $\alpha_1 = 4/3$, the other has $\alpha_1 = 2$, but both have the proper scaling property, because the denominator in Eq. (123) also changes during scaling.

7. CONCLUSIONS

The problem of defining appropriate one-electron energy densities for electronic states of atoms, molecules and solids is not at all straightforward, and it has occupied many authors, most of whom have not been mentioned in our paper. But several of them have given beautiful contributions to the present volume, and by studying those the reader should obtain a good picture of the various approaches. (See also the articles in the recent book by Keller and Gásquez[43] and in the recent book by Lundqvist and March.[44])

We have suggested a new approach, based on the phase space formulation of quantum mechanics. It is new in the sense that it has not been coherently formulated before, but it has deep historical roots in the works of Thomas, Fermi and Dirac, and phase space notions have always been freely used by the workers in the field.

One advantage of our approach is the comprehensive picture it gives of the field, another is that it leads to energy densities with optimal properties. For the kinetic energy density we have illustrated this by focusing on the local virial theorem which is not satisfied for other definitions. For the exchange energy density we have shown that our definition connects this density with an averaged exchange potential in a natural way. Obviously the expression (104) for an exchange integral offers several ways of integrating over the coordinates of one electron such that an energy density results. The integration occurring in our work turns out to be the one described by Eq. (102). Other authors[42] have chosen to integrate directly over say $\underline{r}_2$, in the integrand of Eq. (104). The problem is, however, that the derivative $dk(\underline{r})/d\rho(\underline{r})$ comes out differently in each case, at the same time as this derivative is supposed to play the role of a local potential in the one-electron equations of density functional theory. In other words, there are two conditions to watch, one is that Eq. (72) should be satisfied, the other is that $dk(\underline{r})/d\rho(\underline{r})$ should be physically meaningful. We believe that Eq. (99) controls the situation in our approach.

REFERENCES

1. H. Weyl, "The Theory of Groups and Quantum Mechanics", Dover, New York (1931).
2. E. Wigner, Phys. Rev. 40, 749 (1932).
3. S.R. de Groot and L.G. Suttorp, "Foundations of Electrodynamics", North-Holland Publishing Company, Amsterdam (1972).
4. J.P. Dahl, Physica 114A, 439 (1982).
5. J.P. Dahl, Phys. Scripta 25, 499 (1982).
6. J.P. Dahl, p. 557 in: "Energy Storage and Redistribution in Molecules", J. Hinze, ed., Plenum Press, New York (1983).
7. M. Springborg, J. Phys. A 16, 535 (1983).
8. J.P. Dahl and M. Springborg, Mol. Phys. 47, 1001 (1982).
9. M. Springborg, Theoret. Chim. Acta 63, 349 (1983).
10. M. Springborg and J.P. Dahl, to be published.
11. V.H. Smith, Jr., this volume.
12. A. Grossmann, Commun. Math. Phys. 48, 191 (1976).
13. A. Royer, Phys. Rev. A15, 449 (1977).
14. R. McWeeny and B.T. Sutcliffe, "Methods of Molecular Quantum Mechanics", Academic Press, New York (1969).

15. N.H. McCoy, Proc. Natl. Acad. Sci. U.S.A. 18, 674 (1932).
16. H.J. Groenewold, Physica 12, 405 (1946).
17. J.G. Krüger and A. Poffyn, Physica 85A, 84 (1976).
18. R.M. Ziff, G.E. Uhlenbeck, and M. Kac, Phys. Rep. C32, 1970 (1977).
19. L. Cohen, J. Chem. Phys. 70, 788 (1979).
20. L. Cohen, J. Math. Phys. 7, 781 (1966).
21. Ø Burrau, Mat. Fys. Medd. Dan. Vid. Selsk. 7, No. 14 (1927).
22. D.R. Bates, K. Ledsham and A.L. Stewart, Phil. Trans. Roy. Soc. London A246, 215 (1953).
23. G. Hunter and H.O. Pritchard, J. Chem. Phys. 46, 2146 (1967).
24. E. Teller and H.L. Sahlin, p. 35 in: "Physical Chemistry, an Advanced Treatise, Vol. 5", H. Eyring, ed., Academic Press (1970).
25. J.C. Slater, "Quantum Theory of Molecules and Solids. Vol. 1. Electronic Structure of Molecules", McGraw-Hill, New York (1963).
26. H. Shull and D.D. Ebbing, J. Chem. Phys. 28, 866 (1958).
27. G.F. Thomas, F. Javor, and S.M. Rothstein, J. Chem. Phys. 64, 1574 (1976).
28. F. Javor, G.F. Thomas, and S.M. Rothstein, Int. J. Quantum Chem. 11, 59 (1977).
29. H.H. Grelland, Int. J. Quantum Chem. 19, 847 (1981).
30. H.H. Grelland and J. Almlöf, Int. J. Quantum Chem. 21, 885 (1982).
31. P.A.M. Dirac, Proc. Cambridge Phil. Soc. 26, 376 (1930).
32. H. Jensen, Z. Phys. 89, 713 (1934).
33. M. Abramowitz and I.A. Stegun, "Handbook of Mathematical Functions", U.S. National Bureau of Standards, Washington D.C. (1964).
34. F.B. van Duijneveldt, "Gaussian Basis Sets for the Atoms H-Ne for Use in Molecular Calculations", IBM Technical Report RJ 945 (1971).
35. S. Huzinaga, "Approximate Atomic Functions", Div. of Theor. Chem., Dep. of Chem., Univ. of Alberta (1971).
36. J.C. Slater, Phys. Rev. 81, 385 (1951).
37. F. Bloch, Z. Phys. 57, 545 (1929).
38. W. Kohn and L.J. Sham, Phys. Rev. 140, A1133 (1965).
39. R.G. Parr, this volume.
40. R.G. Parr, S.R. Gadre, and L.J. Bartolotti, Proc. Natl. Acad. Sci. U.S.A. 76, 2522 (1979).
41. S.R. Gadre, L.J. Bartolotti, and N.C. Handy, J. Chem. Phys. 72, 1034 (1980).
42. M. Berrondo and O. Goscinski, Chem. Phys. Letters 62, 31 (1979).
43. J. Keller and J.L. Gázquez, eds., "Density Functional Theory", Lecture Notes in Physics Vol. 187, Springer-Verlag, Berlin 1983.
44. S. Lundqvist and N.H. March, eds., "Theory of the Inhomogeneous Electron Gas", Plenum, New York (1983).

APPLICATIONS

APPLICATION OF THE HARTREE-FOCK-SLATER METHOD IN PHOTOELECTRON SPECTROSCOPY

E.J. Baerends and J.G. Snijders

Department of Theoretical Chemistry, Free University
De Boelelaan 1083
1081 HV Amsterdam, The Netherlands

C.A. de Lange and G. Jonkers

Department of Physical Chemistry, Free University
De Boelelaan 1083
1081 HV Amsterdam, The Netherlands

1. A MOLECULAR LOCAL DENSITY METHOD: THE HARTREE-FOCK-SLATER LCAO METHOD

A. Introduction

Current interest in density functional theory stems from two sources. In the first place the density functional approach to the correlation problem, originating from the famous Hohenberg-Kohn theorem,[1] is of great intrinsic theoretical interest. In the second place, the approximation of the exact (but unknown) exchange-correlation potential with a potential based on the local density, to which density functional theory has given some respectibility albeit no justification, is of great practical interest, as are its predecessors, the Thomas-Fermi model and Slater's $\rho^{1/3}$ exchange potential. In band structure calculations simple local exchange-correlation potentials have always dominated the scene, and in the field of molecular calculations they have stimulated the development of important alternatives to the standard *ab initio* Hartree-Fock LCAO programs. We may mention in particular the multiple scattering $X\alpha$ method,[2-4] the discrete variational $X\alpha$ methods,[5-8], the LCAO-$X\alpha$ methods,[9-10] and the LMTO method.[11] In the present contribution we will discuss the discrete variational $X\alpha$ method (often called LCAO Hartree-Fock-Slater) in its STO-expansion/density-fit form.[5] This as well as the other implementations of local density potentials are useful as they provide relatively efficient methods for obtaining information on the electronic structure of large systems, such as inorganic and organometallic complexes, or clusters. There remains, of course, the important question of the justification of the theoretical model used. The existing computational schemes are valuable as they offer the possibility of explicitly testing present and new local potentials in molecular calculations. As an example we will briefly discuss a comparison between Slater's Xα potential and more recently proposed local (spin) density potentials. In the present paper, however, our emphasis will not be on the comparison between different varieties of (local) density functionals, but rather on an assessment of the reliability of such functionals, in particular Xα, in terms of agreement with experiment, and of their usefulness in interpreting experimental data. We will restrict ourselves here to a discussion of photoelectron spectroscopic data. Photoelectron spectroscopy provides one of the most direct ways of probing molecular electronic structures, and is equally useful for systems composed of first and second row atoms as for transition metal complexes or even systems containing heavy elements such as actinides. One particular advantage of the computational scheme we have developed is its ability to deal with all such systems, without any need for reparameterisation etc. An essential prerequisite, however, for any first principles method applied to heavy elements is the incorporation of relativistic effects. We plan to give a survey in this paper of application of the LCAO-HFS method to systems ranging from light, organic molecules to heavy-atom compounds. This survey is not intended to be complete or comprehensive, but will be focused on the interplay between experimental (UPS) and theoretical (HFS) work in our own laboratory.

The plan of this paper is as follows. In an introductory chapter we will first briefly discuss the DVM computational scheme, and use this scheme to investigate the Xα and LSD functionals in molecular calculations. Next the incorporation of relativistic effects will be dealt with, and examples given for atoms, in particular atomic ionisation potentials.

The second chapter will deal with the application of ionisation potentials. Various systems will be discussed, ranging from small test molecules (N_2,H_2O) to organic systems (the aromatic molecules benzene, pyridine, phosphorine) and a transition metal complex (cobaltocene). In chapter 3 the application of the relativistic HFS-LCAO method will be discussed, and examples given for a variety of iodine containing molecules.

In chapter 4 and 5 we will consider systematically related series of molecules, and compare to appropriate HFS calculations. In chapter 4 halogenides and cyanides of group VI A elements will be discussed. In chapter 5 halogenides of group IV A elements are treated.

B. Computational Aspects

The technical advantages offered by the replacement of the exchange operator of the Hartree-Fock model by a simple local potential dependent on the local electron density have first been exploited in atomic calculations (cf. Herman and Skillman[12]), and next in band structure calculations which by the use of a muffin-tin model for the electronic potentials had been reduced partly to atomic (i.e. spherical potential) problems. The MSXα method is very much in the same spirit. The LCAO-HFS method however is closer to current ab initio Hartree-Fock computational schemes in that it also employs an expansion method. At the same time, however, it takes advantage of the local character of the Xα (or L(S)D) potential by evaluating the Hamiltonian matrix elements by a simple numerical integration procedure. We summarize the essential ingredients of our methods in the following points.

1. The one-electron equation

$$\hat{h}^0 \psi_i^0 = (-1/2\,\nabla^2 + V_N + V_C^0 + V_X^0)\,\psi_i^0 = \varepsilon_i^0\,\psi_i^0 \tag{1}$$

where

$$V_N(1) = \sum_\alpha - Z_\alpha / r_{1\alpha}$$

is the nuclear potential,

$$V_C^0(\vec{r}_1) = \int [\rho^0(\vec{r}_2)/r_{12}]\, d\vec{r}_1$$

the Coulomb potential, and V_X^0 the Xα or L(S)D potential, is solved by expanding ψ_i^0 in a set of basis functions, which we choose to be exponential functions (STOs) centered on the nuclei:

$$\psi_i^0 = \sum_\mu \chi_\mu\, c_{\mu i}^0 \tag{2}$$

Our computational scheme does not pose any limitations on the type of basisfunctions to be used. STOs are convenient as they lead to fairly rapid convergence. We note that basis set deficiency is the most serious drawback of most present day ab initio Hartree-Fock calculations on large molecules. In the calculations to be reported here we have used at least double zeta quality, and often triple zeta for valence electrons. Exponents have been generated by fits to numerical atomic orbitals from Herman-Skillman type calculations.[13]

2. The Hamiltonian matrix elements in the secular equation

$$(\underline{\underline{H}} - \varepsilon \underline{\underline{S}})\underline{c} = 0$$

are evaluated by a direct numerical integration procedure:

$$H_{\mu\upsilon} = \langle\chi_\mu|\hat{h}^O|\chi_\upsilon\rangle$$
$$\simeq \sum_k \chi^*_\mu(\vec{r}_k)\ \hat{h}^O\chi_\upsilon(\vec{r}_k)\ \omega(\vec{r}_k)$$

(the $\underline{\underline{S}}$ matrix is calculated analytically).

The sample points $\vec{r}_k$ and weight factors $\omega(\vec{r}_k)$ are determined according to the Diophantine method introduced by Ellis and Painter.[14,15] The matrix elements are not determined very accurately by this procedure, as we only use a moderate number of sample points (500-1000 per atom). In many cases, however, an accuracy of e.g. ~0.1 eV is quite sufficient, if one realizes that we are dealing with the Hamiltonian matrix elements directly, not with the two-electron integrals individually. We thus get the same accuracy, or even better,[16] in e.g. the one-electron energies, or ionisation potentials. Other error sources, in particular basis set deficiency, are however in practice always more serious than 0.1 eV, whereas the simple local density schemes, or Hartree-Fock, are by no means capable of reproducing experimental ionisation energies to this accuracy. Given these limitations, there does not seem to be any point in improving the numerical accuracy, although this is of course possible if the need arises, by just adding sample points.

The numerical integration is very flexible, in the sense that any local exchange-correlation potential may be used. We will make a comparison between $X\alpha$ and LSD potentials in the next section, but we will use in this paper as in all other work, the $X\alpha$ potential with α=0.70.[17]

3. The calculation of the local potentials V_N and V_X in each sample point is very simple if the latter depends only on the local density. Exact calculation of $V_C(\vec{r}_k)$:

$$V_C(\vec{r}_k) = \int \frac{\rho(\vec{r})}{|\vec{r}-\vec{r}_k|}\, d\vec{r} = \sum_{\mu,\upsilon} P_{\mu\upsilon} \int \frac{\chi^*_\mu(\vec{r})\chi_\upsilon(\vec{r})}{|\vec{r}-\vec{r}_k|}\, d\vec{r}$$

would require the evaluation of many three-center nuclear attraction integrals. Without sacrificing too much accuracy we may simplify the calculations of $V_C(\vec{r}_k)$ by expanding the electron density in a set of functions (STOs) centred on the nuclei of the system:

$$\rho(\vec{r}) \simeq \sum_i a_i\ f_i\ (\vec{r})$$

We split up the total density into one- and two-centre contributions,

$$\rho(\vec{r}) = \sum_{A \le B} \rho_{AB}(\vec{r})$$

and each $\rho_{AB}(\vec{r})$ is fitted with the functions on A and B by a least-squares optimisation of the coefficients a_i, In practice we use as fitting functions the products of basis functions on each nucleus, omitting those that would give rise to linear dependency problems. In this way accurate fits of the density may be obtained. We wish to emphasize that this is an important feature of our scheme, distinguishing it from a method such as MSXα that introduces serious approximations to the electronic potentials.

The Coulomb potential $V_C(\vec{r}_k)$ at each sample point is now calculated from the fitted density

$$V_C(\vec{r}_k) = \int \frac{\rho(\vec{r})}{|\vec{r}-\vec{r}_k|} d\vec{r} = \sum_i a_i \int \frac{f_i(\vec{r})}{|\vec{r}-\vec{r}_k|} d\vec{r}$$

The integrals $\int [f_i(\vec{r})/|\vec{r}-\vec{r}_k|] d\vec{r}$ are calculated at each sample point for all fit functions. This is done once, and on iterating changes in the density are represented by changes in the coefficients a_i.

4. It is possible to freeze the cores, i.e. to assume that the core orbitals would not change in going from atom to molecule. The atomic core densities are used to get the core contribution to the electronic potentials:

$$V_C(\vec{r}_k) = \int \frac{\rho_{core}(\vec{r})}{|\vec{r}-\vec{r}_k|} d\vec{r} + \int \frac{\rho_{val}(\vec{r})}{|\vec{r}-\vec{r}_k|} d\vec{r}$$

$$V_X(\vec{r}_k) = - 3\alpha_{EX} \left[\frac{3}{8\pi} (\rho_{core}+\rho_{val})\right]^{1/3}$$

Collapse of the valence solutions into the core is prevented by expanding them in basis functions that have been explicitly orthogonalized to the core.

We wish to point out that this scheme is asymptotically exact, i.e. we may obtain solutions arbitrarily to the exact ones by enlarging basis and fit sets. and the number of sample points. This is not normally useful. Rather it is our policy to save computation time by maintaining a balance between the approximations introduced in various stages of the calculation. For instance, given a double zeta basis set there is no use in spending more time on the integration and the density fit than is required to get the overall accuracy of 0.1 eV in ionisation energies. However, this flexibility of our scheme, permitting us to obtain highly accurate solutions if desired, is very valuable if one wishes to assess the performance of a specific local potential, such as Xα, or to compare different potentials. We will briefly discuss such a comparison in the next section.

C. Xα and L(S)D results for diatomics

We have used the capability of our method to yield accurate solutions to one-electron equations with a local exchange-correlation potential for a comparison between the standard Xα potential and the L(S)D potential for molecular properties.[18] The LSD potential is the one used in their molecular calculations by Gunnarsson, Harris, and Jones,[19] and has been given by Gunnarsson and Lundquist [20] (cf. also Von Barth and Hedin[21]). The results[18] are displayed in Fig. 1. We have used the accurate (triple-zeta) basis of Ref. 22, with one 3d polarisation function added. This basis is not yet complete, but probably fairly close to the Hartree-Fock-Slater limit. This is substantiated by the very good overall agreement of the Xα calculations with the ones by Dunlap et al.[23] using large gaussian basis sets, and different fitting procedures. The remaining small differences may be due to the different bases, or to computational error sources (e.g. fitting), but are irrelevant for the present comparison of Xα with LSD.

It is interesting to observe from Fig. 1 that both Xα and LSD give quite good agreement with experiment for R_e, ω_e, and D_e of the diatomic molecules investigated. Whereas HF shows the well-known systematic deviations from experiment - too low D_e, too small R_e, too high ω_e - this is not the case for Xα, which rather shows a scattering about the experimental values, and also not for LSD although D_e appears to be systematically larger than experiment. The difference between the Xα and LSD results for D_e is caused by a difference in the Xα and LSD functionals in the case of spin-polarised systems, which get a smaller total energy for LSD. As spin polarisation is most important in the atoms, the difference in D_e's is mostly due to the difference in the reference energy of the free atoms.

In cases where this is a large effect (i.e. small D_e, significant spin polarisation in the atom), and where the Xα D_e is too small, such as H_2,[24] Li_2 (not shown, see 25) and LiH[18], the LSD D_e will be closer to experiment,[18,25,26] but on the average Xα is found to be in better agreement with experiment (see Fig. 1).

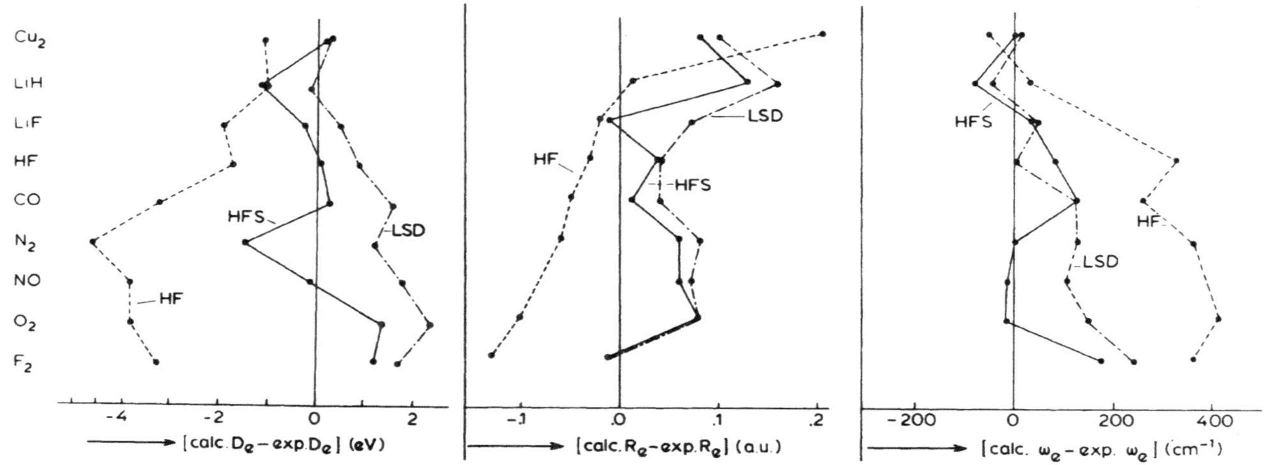

Figure 1. The deviation of calculated spectroscopic constants from experimentsl values, for Hartree-Fock-Slater calculations (——), ab initio Hartree-Fock calculations (·····), and local (spin) density calculations (-·-·-) (see Ref. 18).
[a]Dissociation energies (D_e)
[b]Equilibrium bond lengths (R_e)
[c]Harmonic stretch frequencies (ω_e)

The consistently better agreement with experiment found for the local density functionals compared to Hartree-Fock is interesting, but the lack of a convincing theoretical justification is unsatisfactory. While there is at present much interest in the behaviour of density functionals in atoms and solids, we wish to emphasize the need for investigations of their performance for molecular binding. From the present work we conclude that in spite of theoretical support for LSD, this density functional does not constitute a significant improvement over the simple Xα functional for bond properties.

D. The relativistic HFS method

To include the effects of relativity in the Hartree-Fock-Slater method described above, a scheme based on perturbation theory was developed.[27,28] This method takes advantage of the fact that the largest influence of relativity in molecules is to be found among the core orbitals, which are very similar to their atomic counterparts. The core orbitals are therefore kept frozen and are transferred from fully relativistic Dirac-Fock-Slater calculations on the constituent atoms which can be routinely performed. The smaller relativistic effects on the valence orbitals can then be treated by perturbation theory to first order in the square of the fine structure constant α^2 ($\alpha=e^2/4\pi\varepsilon_0 hc$). We can distinguish four perturbations due to relativity:

a. The mass-velocity correction due to the relativistic increase of the mass of an electron with its velocity:

$$h^1_{MV} = \frac{1}{8}\alpha^2 p^4 = -\frac{1}{8}\alpha^2 \nabla^4$$

where $\vec{p}$ is the linear momentum operator.

b. The Darwin correction. This term has no classical relativistic analogue and is due to the small scale irregular motion of an electron about its mean position, the so-called "Zitterbewegung" (see e.g. Ref. 29)

$$h^1_D = \frac{1}{8}\alpha^2 \nabla^2 V^0_T$$

where V^0_T is the total (non-relativistic) potential. The main contribution to this operator comes from the nuclear potential

$$V_N = \sum_\mu Z_\mu/r_{1\mu} \quad ,$$

$$h^1_D \simeq \frac{1}{8}\alpha^2 \nabla^2 V_N = \frac{1}{8}\alpha^2 \sum_\mu - 4\pi Z_\mu \delta\ (r_1-r_\mu)$$

is almost a sum of Dirac δ-functions on the nuclei. It is therefore only important for orbitals that have a non-zero value at the nuclei, i.e. orbitals that contain atomic s-orbital contributions.

c. Spin-orbit coupling. This is the well-known coupling of the electron-spin to the magnetic field produced by its own orbital motion.

$$h^1_{SO} = \frac{1}{2}\alpha^2\,\vec{s}.(\vec{\nabla}V^0_T\times\vec{p})$$

where $\vec{s}$ is the spin angular momentum operator.

d. The change in electronic potential induced by relativity. This is the indirect effect mentioned in the introduction. The core contribution to this term can be written as just the difference of the relativistic and non-relativistic atomic core potentials. For the valence contribution we need the first order change in the orbitals ϕ^1_i, which are calculated by perturbation theory, to construct the first order change in the valence density ρ^1_{val}

$$h^1_{POT} = \sum_\mu (V^{rel}_{core,\mu}) - (V^{nrel}_{core,\mu}) + \int \frac{\rho^1_{val}}{r_{12}}\,d\tau_2 + \text{exchange terms}$$

Since $h^1 = h^1_{MV} + h^1_D + h^1_{SO} + h^1_{POT}$ depends on ρ^1 and hence the ϕ^1_i which are determined through the perturbation equation

$$(h^1 - \varepsilon^1_i)\,\phi^0_i = (\varepsilon^0_i - h^0_i)\,\phi^1_i$$

which in turn contains h^1, one has to solve this equation iteratively until self-consistency is reached.

The correction to the orbital energies can now be calculated from

$$\varepsilon^1_i = \langle\phi^0_i|h^1|\phi^0_i\rangle = \varepsilon^1_{MV} + \varepsilon^1_D + \varepsilon^1_{SO} + \varepsilon^1_{POT}$$

Apart from computational advantages, one of the important assets of a perturbational approach is the clear cut separation of the relativistic corrections of different physical origin (h^1_{MV}, h^1_D, h^1_{SO}, h^1_{POT}), which allows a more illuminating interpretation of the results.

E. Comparison with atomic Dirac-Fock-Calculations

In order to assess the adequacy of the proposed perturbation scheme, we have compared it with full Dirac-Fock-Slater calcula-

Table 1. Orbital energies in Ev

	nrel(num.)	nrel(anal.)	pert.(dir.)	pert.(dir.+indir.)	quasirel.	rel.
Au						
5d 3/2	7.47	7.48	10.17	7.20	7.24	7.24
5d 5/2			8.67	5.69	5.81	5.70
6s 1/2	3.67	3.68	5.36	4.72	5.13	5.28
Hg						
5d 3/2	11.50	11.52	14.73	10.40	10.37	10.38
5d 5/2			12.93	8.61	8.67	8.52
6s 1/2	4.80	4.80	6.88	5.75	6.16	6.30
Rn						
6s 1/2	16.19	16.21	22.70	19.51	20.90	21.16
6p 1/2	7.12	7.13	10.97	8.70	9.16	9.73
6p 3/2			8.55	6.28	6.27	6.11

tions on several atomic systems.[27] In Table 1 we show the valence orbital energies for three elements (Au, Hg, and Rn) for which relativistic effects are particularly large. In the first two columns we compare the non-relativistic orbital energies from numerical calculation and HFS-LCAO calculation showing that the basisset used[13] is large enough that we have essentially reached the basisset limit. In the third column we show the effect of the direct relativistic corrections (h^1_{MV}, h^1_D, h^1_{SO}). In all these cases there is a sizable stabilization by relativity. In column four we add the indirect potential effect which partly counteracts the direct stabilization of orbital energies. It is quite clear that this effect is large and should be included to reach any kind of agreement with the full relativistic result in the last column. Any remaining discrepancies with the full relativistic result must be due to effects of higher order than α^2. They can come from two sources, however, namely higher order corrections to the relativistic perturbation operators (i.e. h^2 etc.) and energetic effects of h^1 in higher order. The importance of the last effect can be taken into account by solving the so-called quasi relativistic equation:

$$(h^0 + h^1)\ \phi_i = \varepsilon_i\ \phi_i$$

effectively taking h^1 into account up to infinite order while neglecting h^2.[10,11]

The results of this procedure are shown in the fifth column of Table 1. In Au and Hg the influence on the d levels is negligible but the 6s orbital energies are brought into much closer agreement with the full relativistic result. The remaining discrepancy in the Rn 6p1/2 energy must be due to h^2 which is difficult to calculate. It can thus be concluded that first order perturbation theory, possibly improved by the quasi-relativistic procedure can give an accurate description of relativistic effects on orbital energies except in cases where relativistic effects are extremely large as in the spin-orbit splitting of Rn.

Another criterion to assess the accuracy of the method is to compare relativistic changes in the orbitals themselves as given by ϕ^1 in first order, with the changes determined from the difference of a non-relativistic and a fully relativistic numerical calculation. As an example we show in Fig. 2 the orbital densities of the 6s1/2 orbital of Gd^+. It can be seen that the first order result is quite satisfactory. For other orbitals and other elements comparable results have been obtained.[27]

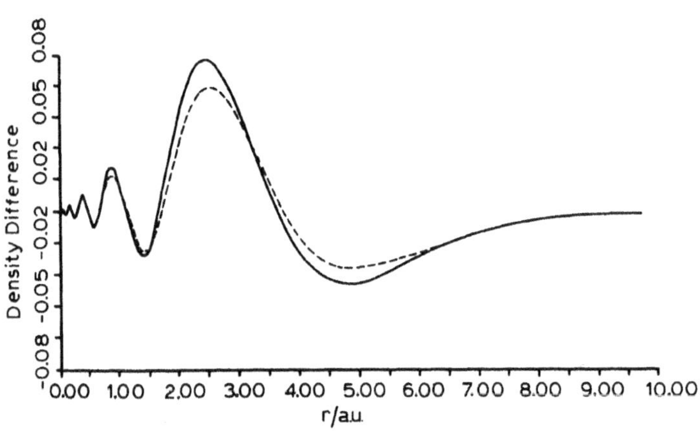

Figure 2. Relativistic correction to the $6s_{1/2}$ density in Gd^+, exact (——) and by perturbation theory (----).

2. APPLICATION OF THE NON-RELATIVISTIC HFS METHOD IN PHOTO-ELECTRON SPECTROSCOPY

A. Introduction: Small molecules

Extensive tests have been carried out on small molecules to establish the reliability of the HFS model for ionisation energies.[5,30-34] In the case of MSXα calculations the results may differ considerably from accurate Xα results (sometimes several eV) due to the muffin tinning of the potential, but typically the effect of muffin tinning on ionisation potentials is in the order of 1 eV.[30,35] In the case of calculations using the expansion method there are many examples in the literature that at least basis sets of double zeta quality are required to approach exact HF results with an accuracy of about 0.5 eV. The same holds for HFS calculations.[30] The calculations to be reported in this paper therefore employ basis sets of at least double zeta, and often better, quality.

As for comparison with experiment we may say that for systems with light atoms only the absolute agreement is usually within 0.5 eV, and for systems with heavy atoms (second transition series and beyond) usually within 1 eV. Relative band positions are predicted with higher accuracy, to within a few tenths of an eV. As an illustration of the comparison between HF, HFS, MSXα, and experiment we show in Table 2 and Fig. 3 results for N_2 and H_2O. The general agreement with experiment is striking. It is well-known that in the case of N_2 Hartree-Fock calculations (both Koopmans' theorem and ΔSCF) lead to the wrong order of the first two ionic

Table 2a. Ionisation potentials of H_2O

Orb.	EXP.[a] (eV)	HFS	MSXα[b]	HF[c] (KT)	HF[c] (ΔSCF)
$1b_1$	12.6	12.5	15.0	13.8	11.1
$3a_1$	14.7	14.7	15.5	15.8	13.3
$1b_2$	18.4	18.9	15.7	19.5	17.6
$2a_1$	32.2	30.9	30.1	36.8	34.3

[a]K. Siegbahn et al. "ESCA applied to free molecules", Amsterdam, 1969.
[b]J.W.D. Connolly, J.R. Sabin, J.Chem.Phys. 56 (1972) 5529.
[c]M.F. Guest, V.R. Saunders, Molec.Phys. 29 (1975) 873.

Table 2b. Ionisation potentials of N_2

Orb.	EXP.[a] (eV)	HFS	MSXα[b]	HF[c] (KT)	HF[c] (ΔSCF)	GF[d]
$3\sigma_g$	15.5	15.5	14.1	17.3	16.0	15.5
$1\pi_u$	16.8	17.4	18.2	16.8	15.7	16.8
$2\sigma_u$	18.6	18.3	18.3	21.2	19.9	19.0
$2\sigma_g$	37.3	37.3	31.5	40.1		

[a]C.R. Brundle, D.B. Brown, Spectrochim. Acta 27A (1971) 2491.
[b]J.W.D. Connolly, U. Gelius, C. Nordling, H. Siegbahn, J.Chem. Phys. 58 (1973) 4265.
[c]P.E. Cade, K.D. Sales, A.C. Wahl, J.Chem.Phys. 44 (1966) 1973.
[d]W. von Niessen, G. Diercksen, L.S. Cederbaum, J.Chem.Phys. 67 (1977) 4124.

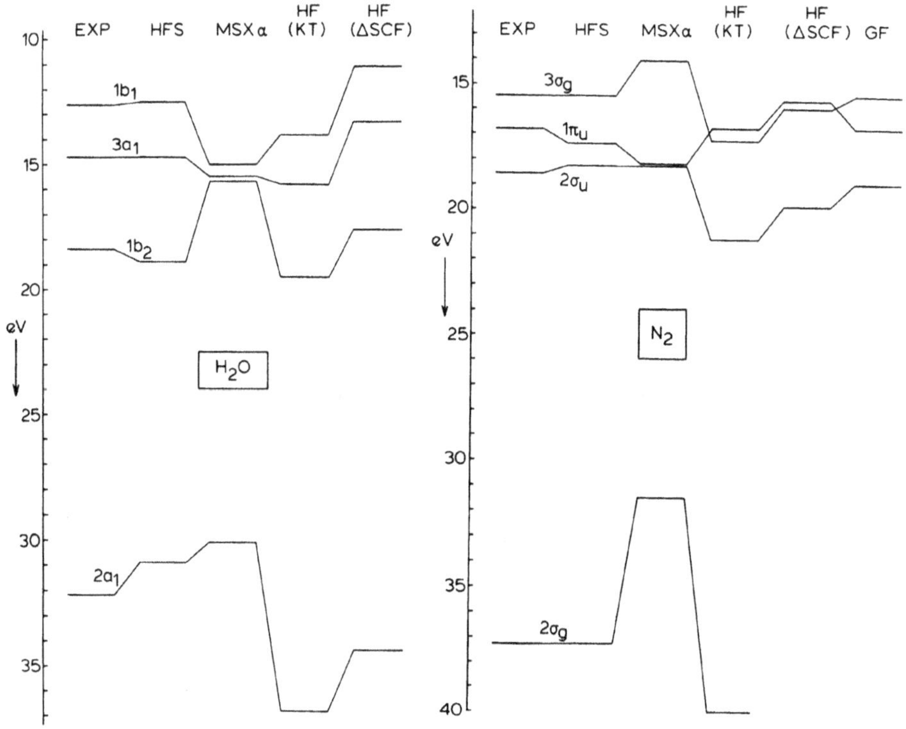

Figure 3a. Experimental and various theoretical ionisation potentials of H_2O.
3b. Experimental and various theoretical ionisation potentials of N_2.

states. This does not occur in the HFS calculations. In a recent study De Alti et al.[35] documented a number of cases where wrong orderings in HF (KT) calculations are absent in HFS (gaussian based LCAO-Xα) calculations. Although the HFS model also has its failures (cf. next section) it is clear that it nevertheless is in excellent agreement with experiment in most cases, and for that reason a very valuable assignment tool.

B. Benzene, pyridine, phosphoridine

Considerable controversies have existed on the assignment of the photoelectron spectra of, in particular, benzene and pyridine (see [36,37] and Refs. therein). In the case of benzene the discussions have mainly - but by no means exclusively ! - concerned the assignment of the first three bands, in particular whether the position of the first σ peak ($3e_{2g}(\sigma)$) was below[38] the two π peaks ($1e_{1g}(\pi)$ and $1a_{2u}(\pi)$), as assumed in elementary Hückel theory, or in between.[39] In the case of pyridine again the assignment of the first ionisation was controversial, in particular the position of the nitrogen lone pair peak. This IP has sometimes been assigned to the third experimental peak, although the general inclination has been to place it under the first band, which corresponds to two IPs. It was not clear, however, whether the first ionisation of these two should be π or n. We refer to the papers by Von Niessen, Cederbaum, and coworkers[36,37] for extensive discussion of both the experimental and theoretical work. It is somewhat disconcerting that in spite of a huge theoretical effort the assignments could not be established before the detailed work, including correlation corrections, by the quoted authors. As for the ab initio HF calculations the inconclusiveness originated both from the diverse Koopmans' results obtained in different basis sets (it is hard to get close to the HF limit for systems of this size), and from the need to include relaxation effects, i.e. to perform expensive SCF calculations for all relevant ionic states.

We present in Tables 3-5 the results of LCAO-HFS calculations on the ionisation spectra of benzene, pyridine, and phosphoridine,[40] and compare to HF and many-body Green's function calculations,[36,37] and to MSXα calculations (for benzene,[41]). We have used a double-zeta basis of STOs, which is comparable to the Gaussian basis sets used in Refs. 36, 37.

For benzene we note that all calculations give the now preferred π σ π ordering of the first three levels. The HF Koopmans IPs are however quantitatively rather far off the experimental ones (e.g. 2 eV for $2e_{2g}(\sigma)$). The Green's function calculations give much better agreement, although the $2e_{2g}(\sigma)$ is still 0.6 eV off experiment. The HFS calculations give the first three IPs very close to experiment (~0.1 eV), and also in the π σ π order. If we go to higher IPs, the HFS results are in general satisfactory, except for the $1b_{2u}$, $1b_{1u}$ pair of levels. Although the $2a_{1g}$ is below the $1b_{1u}$ in all calculations, including HFS, resolving one of the assignment controversies, and in agreement with the intensity arguments put forward by Gelius,[42] there is a discrepancy between the HFS (and MSXα) calculations and the other ones concerning the ordering of the $1b_{2u}$, $1b_{1u}$ pair of levels. As pointed out recently[35] this seems to be a genuine deficiency of Xα methods, which also consistently misplace the $1b_1$ π-orbital in furan and its homologues.

Experimental and several theoretical results for pyridine are given in Table 4. Again the first three IPs agree very well with experiment. It is clear that $7a_1(n)$ (the N lone pair) and $1a_2(\pi)$ both belong to the first band in the photoelectron spectrum, as

Table 3. Ionisation potentials of benzene

MO	HF[a] (Koopmans)	Relax./Corr. (ΣR)[a]	EXP.[b]	HFS (trans.state)	MXSα[c] (overl. spheres)
$1e_{1g}(\pi)$	9.31	9.10	9.3	9.4	10.08
$2e_{2g}(\sigma)$	13.47	11.95	11.4	11.3	11.48
$1a_{2u}(\pi)$	13.79	12.26	12.1	12.1	12.85
$2e_{1u}$	16.14	14.46	13.8	13.4	13.56
$1b_{2u}$	17.02	14.83	14.7	14.4	16.09
$1b_{1u}$	17.42	15.75	15.4	13.9	13.53
$2a_{1g}$	19.47	17.48	16.9	15.9	16.07
$1e_{2g}$	22.43	20.01	19.2	17.6	18.90
$1e_{1u}$	27.59[c]		22.8	21.4	22.83
$1a_{1g}$	31.26[c]		26.5	24.7	26.12

[a] Ref. 36.
[b] Ref. 36.
[c] Ref. 41.

predicted also by Green's function calculations, but the HFS calculations clearly indicate the first IP under the band to be the $7a_1(n)$, in agreement with the very first assignment based on electron impact, by Higasi and Omura et al.[43] The splitting between the $7a_1(n)$ and $1a_2(\pi)$ is however too large (0.5 eV) in the HFS calculation. Going to higher IPs we observe again very nice agreement with experiment, even more satisfactory in fact than in the Green's functions calculations. There is, however, again a problem with the 8th and 9th IPs (as in benzene), the $5a_1$ and $3b_2$. Whereas the Green's function calculations give these two degenerate, at 16.3 eV, and assign them to the experimental IP at 15.7 eV, we find the $3b_2$ at 15.5, close to the experimental 15.7, but the $5a_1$ at 14.4 eV, degenerate with $4b_2$, and right on the experimental IP of 14.4 eV. As the absolute agreement with experiment is excellent in this case, and over the whole spectrum this leaves some doubt about the correct assignment in this region of the spectrum. We also wish to draw attention to the large difference between the HF (Koopmans) ordering π π n, for the first levels, and the ordering with relaxation etc. taken into account (n π π). This is an example of the large relaxation energy for nitrogen lone pair levels, connected to the strongly localised character of the lone pair orbital. This may also, if symmetry equivalent N lone pairs are present, as

Table 4. Ionisation potentials of pyridine

MO	HF[a] (Koopmans)	Relax./Corr.[a] (Σ^R)	EXP.[a]	HFS (trans. state)
$7a_1(n)$	11.29	9.59)		(9.7
		)	9.7	(
$1a_2(\pi)$	9.81	9.57)		(10.0
$2b_1(\pi)$	10.65	10.24	10.5	10.5
$5b_2$	14.25	12.87	12.5	12.1
$1b_1(\pi)$	14.98	13.43	13.2	13.1
$6a_1$	15.89	14.18	13.7	13.5
$4b_2$	16.61	15.11	14.4	(14.4
				(
$5a_1$	17.84	16.32)		(14.4
		)	15.7	
$3b_2$	18.26	16.33)		15.5
$4a_1$	19.93	18.00	17.1	16.5
$2b_2$	23.45	21.17	19.6	18.8
$3a_1$	23.50	21.15	20.0	19.1
$1b_2$			23.3	22.5
$2a_1$			24.2	23.6

[a] Ref. 37.

in pyrazine, lead to symmetry-breaking in calculations on the ion, as discussed for the N lone pairs of $Se(CN)_2$, in Ref. 44 (cf. chapter 4).

Finally, we give, in Table 5, the IPs of phosphoridine. We agree with the Green's function calculation on the first ionisation being from a π level ($2b_1$). The second band contains the P lone pair IP, and another π level. We find in fact an ordering of these two levels ($1a_2(\pi)$ and $7a_1(n)$) that is reversed compared to the Green's function calculation, but they are so close in all calculations (0.1 - 0.2 eV) that the theoretical results are inconclusive

Table 5. Ionisation potentials of phosphoridine

MO	HF[a] (Koopmans)	Relax./Corr.[a] (Σ^R)	EXP.[a]	HFS (trans. state)
$2b_1(\pi)$	8.97	8.74	9.2	9.1
$1a_2(\pi)$	9.74	9.20	9.8	9.7
$7a_1(n)$	10.53	9.44	10.0	9.6
$5b_2$	12.92	11.72	11.5	11.1
$1b_1(\pi)$	13.61	12.15	12.1	11.9
$6a_1$	14.73	13.37	~13.1	12.8
$4b_2$	16.23	14.77	~14.5	13.9
$3b_2$	16.84	15.18	~14.9	14.6
$5a_1$	16.90	15.38	~15.3	13.8
$4a_1$	19.03	17.33		15.8
$3a_1$	21.76	19.20		17.6
$2b_2$	22.26	20.17		18.1

[a]Ref. 37.

on this detail. Going to higher IPs there are similar discrepancies with the Green's function calculations as in pyridine (note in particular the assignment of the 8th and 9th IP) but experimental data are imprecise in that region.

It is noteworthy that the effect of relaxation does not change the Koopmans' ordering of the π and n levels, although the Koopmans' correction is evidently largest for the P lone pair orbital. This is probably related to the much more diffuse character of the P lone pair than the N lone pair, a fact with some chemical significance. A brief comment on those features of these photoelectron spectra that convey chemically useful information is in order here. We observe in Fig. 4, comparing the upper σ and π levels of benzene, pyridine, and phosphoridine, that the π levels of pyridine are considerably lowered compared to benzene. The π levels of phosphori-

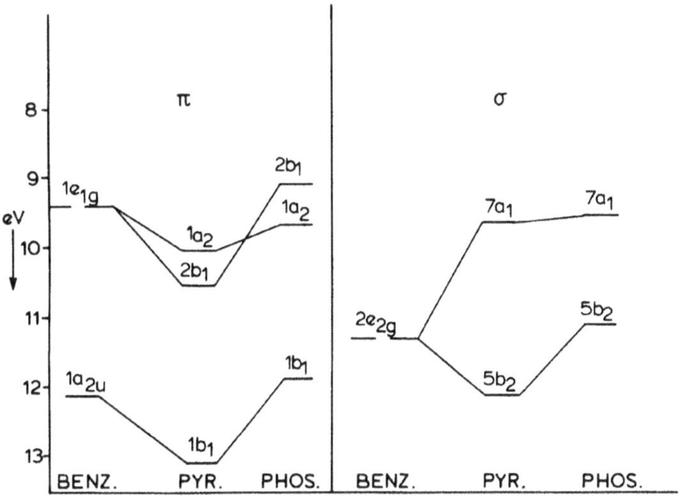

Figure 4. Calculated ionisation potentials of the first three MOs in benzene, pyridine, and phosphoridine.

dine, however, stay close to those of benzene (although there is a splitting of $1e_{1g}$ (benzene) into $1a_2$ and $2b_1$ (phosphoridine)). This is in accordance with the often quoted chemical resemblance between carbon and phosphorus. In the σ system we observe the strong destabilization of the upper σ level of pyridine and phosphoridine with respect to benzene; these are the lone pairs, representing the major difference between benzene and the hetero-aromatics. The $5b_2$ is much closer to the parent $2e_{2g}$ (benzene) in phosphoridine then in pyridine.

C. Cobaltocene

The relative efficiency of the various Xα methods has made the field of transition metal chemistry a particularly fruitful area of application. Here again we find in general excellent agreement between calculated and experimental ionisation potentials.[45-49] Many photoelectron experiments have been considered in the light of Xα calculations, where it is to be noted that the muffin-tin approximation is intuitively expected to be less severe here, in particular in highly symmetric, strongly ionic complexes, than in organic systems with highly directional, covalent bonds. Organometallic systems, such as metallocenes, may also be less suitable for the muffin-tin approximation. As an example of the application of the LCAO-HFS method to transition metal complexes we will discuss the photoelectron spectrum of cobaltocene.[50,51] The level ordering

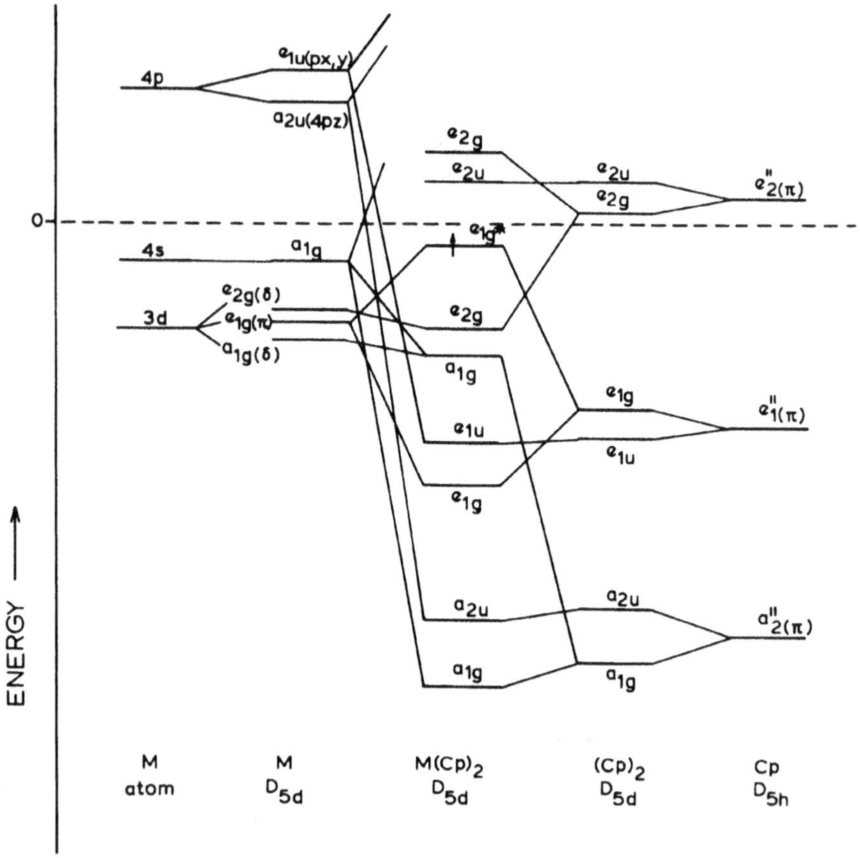

Figure 5. Qualitative level ordering diagram of a metallocene. In $Co(Cp)_2$ the e^*_{1g} is singly occupied.

diagram (Fig. 5) shows that the π orbitals of the rings will slightly split in $(Cp)_2$, forming combinations of different symmetries, suitable for formation of $\sigma(a_{1g}$ and $a_{2u})$, $\pi(e_{1g}$ and $e_{1u})$, and $\delta(e_{2g})$ bonds with the metal. The π bond resulting from interaction of the $e_{1g}(Cp\text{-}\pi)$ with $e_{1g}(d_\pi)$ is generally believed to be strong, and to play a leading role in metal-ring bonding. The $e_{1g}(Cp\text{-}\pi)$ is accordingly stabilized, and ends up below $e_{1u}(Cp\text{-}\pi)$, although in $(Cp)_2$ it is above. The $e_{1u}(Cp\text{-}\pi)$ may however also be stabilized, by interaction with $4p_{x,y}$, and the actual ordering of the e_{1u}, e_{1g} levels, and the contribution of the $4p_{x,y}$ orbitals to the bonding, has been a subject of debate.[52,53] The antibonding e^*_{1g} is undoubtedly sufficiently destabilized to be well above the $e_{2g}(d_\delta)$ and $a_{1g}(d_\sigma)$ orbitals, but the relative ordering of the latter, which are anyway close, has been difficult to establish. The $a_{1g}(d_\sigma)$ may be destabilized by interaction with a lower lying $a_{1g}(Cp\text{-}\pi)$, but is possibly stabilized by interaction with a_{1g}(4s). The $e_{2g}(d_\delta)$ is stabilized by interaction with the e_{2g}(Cp-π), which is the only interaction resulting in δ-bonding.

The ordering of the $a_{1g}(d_\sigma)$ and $e_{2g}(d_\delta)$ in ferrocene has been the subject of some controversy. Whereas Hartree-Fock calculations (both Koopmans' theorem and ΔSCF) put e_{2g} above a_{1g},[52] but by 1.8 eV instead of the experimental 0.3 eV, the MSXα calculations by Rösch and Johnson[54] lead to the reverse ordering, and the LCAO-HFS calculations by Baerends and Ros[45] resulted in degenerate a_{1g} and e_{2g} orbitals. The two Xα calculations, however, suffered from the neglect of proper treatment of the multiplet states resulting from these orbital ionisations. In the case of cobaltocene the multiplet splittings have now been taken into account according to the method described in [55] and [49]. We will consider the calculations on cobaltocene in some detail (cf. [49]).

The calculated splitting between the $e^*_{1g}(d_\pi)$ and $e_{2g}(d_\delta)$ is 2.17 eV, and the ground state of cobaltocene is quite unambiguously $^2E_{1g}$, in agreement with experiment. The splitting between $a_{1g}(d_\sigma)$ and $e_{2g}(d_\delta)$ is only 0.3 eV, and in order to interpret the UPS data we list the ionic states that may arise:

Configuration		States
$(a_{1g})^2\ (e_{2g})^4\ (e^*_{1g})^1$	(ground state)	$^2E_{1g}$
$(a_{1g})^2\ (e_{2g})^4$	$(e^*_{1g} \to \infty)$	$^1A_{1g}$
$(a_{1g})^2\ (e_{2g})^3\ (e_{1g})^1$	$(e_{2g} \to \infty)$	$^3E_{1g}(\delta)$, $^3E_{2g}(\delta)$, $^1E_{1g}(\delta)$, $^1E_{2g}(\delta)$
$(a_{1g})^1\ (e_{2g})^4\ (e_{1g})^1$	$(a_{1g} \to \infty)$	$^3E_{1g}(\sigma)$, $^1E_{1g}(\sigma)$

In their UPS measurements Evans et al.[50] have labelled the bands in the photoelectron spectrum a, b, c, d, and they consider four possible assignments (see also [51]):

	a	b	c	d
	5.56 eV	7.18	7.63	8.01
Assignment				
I	$^1A_{1g}$	$^3E_{2g}(\delta)$, $^3E_{1g}(\delta)$	$^1E_{2g}(\delta)$	$^1E_{1g}(\delta)$
II	$^1A_{1g}$	$^3E_{2g}(\delta)$, $^3E_{1g}(\delta)$	$^1E_{2g}(\delta)$, $^1E_{1g}(\delta)$	$^3E_{1g}(\sigma)$
III	$^1A_{1g}$	$^3E_{2g}(\delta)$, $^3E_{1g}(\delta)$	$^3E_{1g}(\sigma)$	$^1E_{2g}(\delta)$, $^1E_{1g}(\delta)$
IV	$^1A_{1g}$	$^3E_{2g}(\delta)$, $^3E_{1g}(\delta)$	$^1E_{2g}(\delta)$, $^1E_{1g}(\delta)$	$^3E_{1g}(\sigma)$, $^1E_{1g}(\sigma)$

All these assignments agree on putting the triplet states due to $e_{2g}(d_\delta)$ ionisation under band b, and may therefore be considered to favour an $e_{2g}(d_\delta)$ one-electron level above $a_{1g}(d_\sigma)$, as in ferrocene. Of course the situation is not completely unambiguous, because the singlets arising from d_δ ionisation may come at higher energy than the triplet of d_σ ionisation, as in possibility III. Possibilities II and IV put the d_σ ionisations under the high energy band d, whereas in assignment I they are assumed so high in energy that they fall under the ligand ionisations. There is no doubt that at least a full multiplet splitting of the states is required in order to make a meaningful assignment. We have obtained the following order for the triplets and singlets:

$$^1A_{1g} < {}^3E_{2g}(\delta), {}^3E_{1g}(\delta) \lesssim {}^3E_{1g}(\sigma) < {}^1E_{2g}(\delta), {}^1E_{1g}(\delta) < {}^1E_{1g}(\sigma).$$

The actual values of the computed ionisation potentials are given in Table 6 and compared to experiment, and to MSXα calculations by Weber et al.[56] The overall agreement with experiment is quite satisfactory. We note that our order of IPs is the same as in assignment III of Evans et al., but we obtain better correlation with the experimental peak positions when we assign $^3E_{1g}(\sigma)$ also to band b, and the d_δ singlets, $^1E_{2g}(\delta)$,$^1E_{1g}(\delta)$ to band c. We then also obtain an assignment of the $^1E_{1g}(\sigma)$, namely to band d. Bands e and f agree rather well with simple orbital ionisations of the ligand e_{1u} and e_{1g} levels, computed without further multiplet splitting. This assignment remains tentative for two reasons. First, it is not certain that this assignment can be made to agree with the experimental relative intensities. It is clear that band b is by far the most intense one, as we would predict, but in particular band d may be more intense than would be expected on basis of our assignment. Secondly, we have not included configuration interaction. This might be important for the position of $^3E_{1g}$ terms, which are very close in our calculations. Elementary ligand field considerations give for the interaction element $2\sqrt{6}$ B, which is 0.25 eV with a reasonable B value of 0.05 eV. We can thus not rule out the possibility that $^3E_{1g}(\sigma)$ falls under band c.

3. APPLICATION OF THE RELATIVISTIC HFS METHOD IN PHOTOELECTRON SPECTROSCOPY

A. Introduction: Atoms, I_2, HgI_2

To get a feeling of the accuracy that can be achieved in predicting ionisation potentials (I.P.) by the HFS method, we present in Fig. 6 the I.P. of the rare gases and the Hg atom and compare with experiment. The I.P. have been calculated using Slater's transition state concept to account for relaxation.[57] In general the agreement between theory and experiment is quite reasonable

Table 6. Ionisation potentials of cobaltocene (all values in eV)

Orbital ionisation	States	HFS	EXP.[a]	MSXα[b]
$4e^{*}_{1g}(d_{\pi})$	$^{1}A_{1g}$	4.73	5.56(a)	$4.87(d_{\pi})$
$3e_{2g}(d_{\delta})$	$^{3}E_{1g}(\delta)$	6.80		
	$^{3}E_{2g}(\delta)$	6.80	7.18(b)	$7.75(d_{\sigma})$
	$^{3}E_{1g}(\sigma)$	7.08		
$5a_{1g}(d_{\sigma})$	$^{1}E_{1g}(\delta)$, $^{1}E_{2g}(\delta)$	7.48	7.63(c)	
	$^{1}E_{1g}(\sigma)$	7.72	8.01(d)	$8.12(d_{\delta})$
$4e_{1u}(Cp\text{-}\pi)$		8.37	8.66(e)	8.34
$3e_{1g}(Cp\text{-}\pi)$		9.14	9.88(f)	9.05

[a]Ref. 50.
[b]Ref. 56.

and in particular the spin-orbit splittings seen in the heavier elements are reproduced quite well, a relativistic treatment being of course essential in these cases.

As an example of a simple molecule we show in Fig. 7 the non-relativistic, relativistic and experimental photoelectron spectrum of I_2.[28] Again a similar agreement with experiment is reached as in the rare gases. The main effect of relativity is a spin-orbit splitting of the π_u and π_g I.P.

As a second example we show in Fig. 8 the calculated and experimental photoelectron spectra of the linear HgI_2.[28] The spectrum divides into two parts. Below 14 eV we find a σ_g bonding level consisting mainly of I5p_z and Hg6s, a non-bonding σ_u level concentrated mainly on I and the π_u and π_g non-bonding levels which are only slightly split, because there is no low lying orbital of the proper symmetry on the intervening Hg with which they could interact in this energy range. Relativity stabilizes the σ_g MO which is not unexpected in view of the stabilization of Hg6s orbital mainly through the mass-velocity mechanism. The π orbitals are spin-orbit

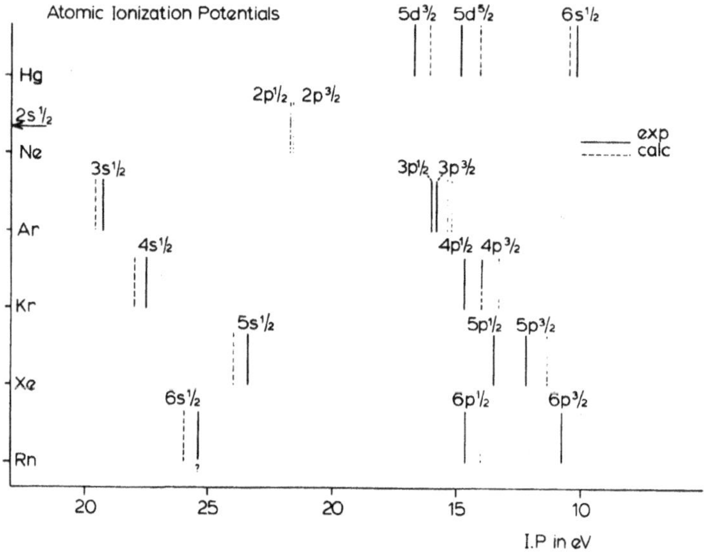

Figure 6. Experimental and theoretical ionisation potentials of Hg and the rare gases Ne-Rn.

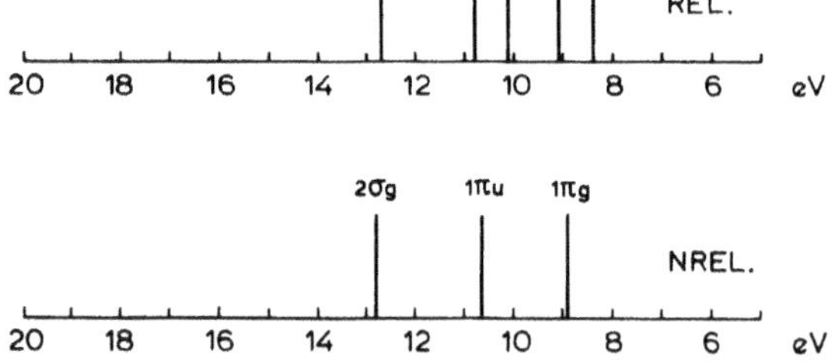

Figure 7. Experimental and theoretical photoelectron spectrum of I_2.

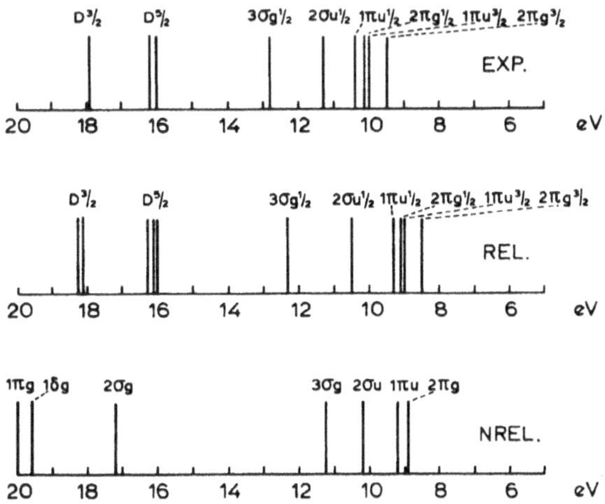

Figure 8. Experimental and theoretical photoelectron spectrum of HgI_2.

split by an amount that is actually larger than the π_u/π_g saparation so that the two level systems interpenetrate as is found experimentally.

Above 14 eV we find the ionisations due to the Hg5d electrons. The experimental spectrum shows essentially two bands separated by an amount close to the spin-orbit splitting of the 5d level in the Hg atom. This has been interpreted at first as an indication that the d electrons in HgI_2 are essentially atomic in character, the levels being shifted down by the increased charge on Hg and slightly split by a weak ligand field of axial symmetry. Calculations, however, give a somewhat different interpretation. The non-relativistic treatment shows that in fact the ligand field is quite strong, separating a level of σ_g symmetry ($5d_{z^2}$), from the two remaining ones of π_g and δ_g symmetry by a considerable amount, mainly due to (antibonding) interaction with the I5s orbitals. The direct relativistic effects, however, partly cancel this interaction and bring the levels much closer together again, while off-diagonal spin-orbit interactions between them separate two groups in an atomic-like pattern, in agreement with experiment. It is seen that it is quite important to take into account off-diagonal spin-orbit interactions between levels that are slosely spaced.

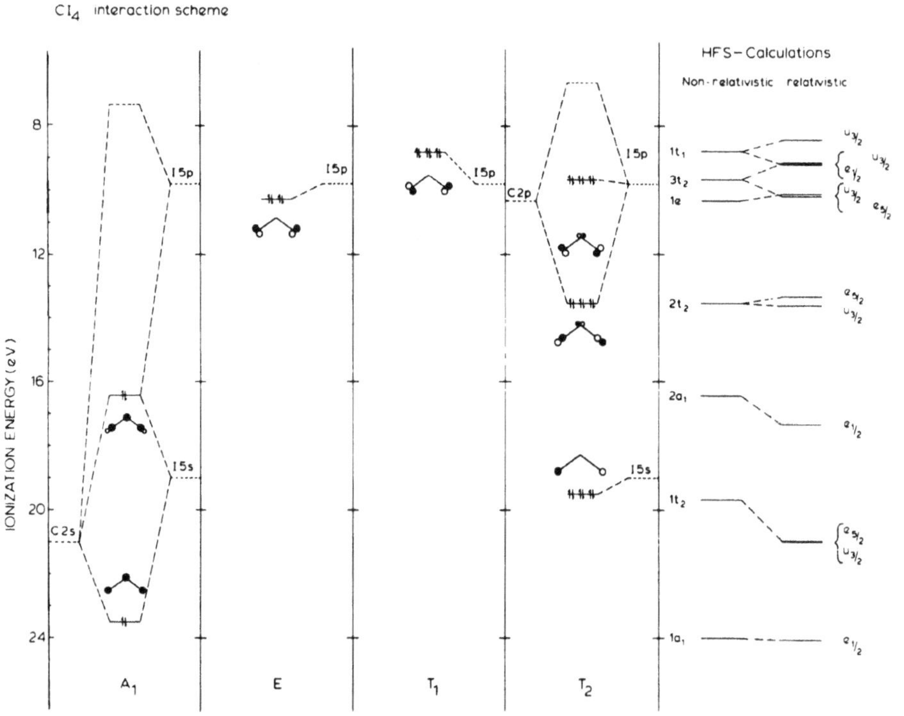

Figure 9. Ionisation energies for CI_4, obtained from non-relativistic and relativistic HFS-calculations. A pictorial representation of the approximate character of the occupied orbitals is given in a CI_2 plane.

B. CI_4

The chemical bonding and the P.E. spectrum[58] of CI_4 can be qualitatively understood from the interaction of the I5s and 5p with the C2s and 2p orbitals. In Fig. 9 we present an interaction diagram for the various molecular symmetry species. In the a_1 symmetry we have the deep lying $1a_1$ orbital which is a bonding combination of C2s and I5s, and a $2a_1$ orbital which is predominantly an antibonding combination of C2s and I5s but has some (~10%) radial I5p mixed in. The $3a_1$ orbital which has a dominant radial I5p character is empty in this molecule. In the e and t_1 symmetries we find two bonding tangential I5p lone pair orbitals, these being no low lying orbitals of the correct symmetry on C. The relative ordering of these orbitals can be understood from the fact that the 1e is slightly I-I bonding while the $1t_1$ is somewhat antibonding. Finally in the t_2 symmetry we have a $1t_2$ orbital which is a nonbonding combination of I5s, the $2t_2$ orbital which is a strongly bonding combination of radial I5p with C2p and the $3t_2$ orbital which is essentially a nonbonding lone pair orbital made up from the tangential I5p's, which overlap very little with central C2p's.

The most important modification of this picture by relativity is due to spin-orbit coupling, which is important in this molecule as is to be expected considering the large spin-orbit splitting in the I atom. The largest splittings are found among the lone pair orbitals. The $1t_1$ gives rise to $u_{3/2}$ and $e_{1/2}$ in double group notation, with a separation of 0.70 eV. The splitting of the $3t_2$ into $u_{3/2}$ and $e_{5/2}$ is even larger amounting to 1.03 eV. In the relativistically calculated ionisation spectrum the $1t_1(e_{1/2})$ is virtually degenerate with $3t_2(u_{3/2})$ as are $3t_2(e_{5/2})$ and $1e(u_{3/2})$. The bonding $2t_2$ orbital is split to a lesser extent (0.26 eV) as can be expected, since it has considerable C2p character. Moreover the $u_{3/2}$ and $e_{5/2}$ components are reversed compared to those of $3t_2$. All these features are borne out in the experimental spectrum (see Table 7). Experimentally the $1t_1(u_{3/2})$ and $3t_2(u_{3/2})$ are slightly Jahn-Teller split. The reversal of components between $2t_2$ and $3t_2$ is experimentally supported by intensity considerations. There is good overall agreement between theory and experiment.

In Table 7 we also analyze the relativistic correction to the IE's into the various physical effects in operation. It can be seen that the Darwin correction is only important for those orbitals that contain an appreciable I5s contribution ($1a_1$, $1t_2$, $2a_1$ and to a lesser extent $2t_2$). Also the mass-velocity correction is largest for these orbitals since the I5s orbital penetrates deeply into the core where electrons move with relativistic speed.

C. C_2I_4

Another example of a P.E. spectrum which has been analyzed by the current method is that of C_2I_4.[59] Calculations were performed using basis sets of double zeta Slater type function quality for the carbon 2s and 2p and for the iodine 4d, 5s, and 5p atomic orbitals (AO's) with the atomic cores kept frozen.[13] Because the behaviour of the iodine 4d AO's turned out to be core-like, they were excluded from the valence orbital numbering. The geometry used in the calculations was taken from electron diffraction data: r(C=C) = 1.365 Å, r(C-I) = 2.106 Å, and (I-C-I) = 114.2°.[60] In the following discussion the molecule with D_{2h} symmetry is placed in the xy-plane with the carbon atoms on the x-axis. It should be realized that the molecular geometry is such that the iodine atoms are arranged approximately in a square.

The planar C_2I_4 molecule is imagined to be constructed from

Table 7. Summary of experimental band positions and of non-relativistic and relativistic HFS-calculations for CI_4 (in eV). Given are non-relativistic ionisation energies (Nrel) and corrections due to the relativistic increase of mass with velocity (Δ_{MV}), Darwin correction due to Zitterbewegung (Δ_{DAR}), direct (diagonal) spin-orbit interaction (Δ_{SO}), indirect (off-diagonal) spin-orbit interaction (Δ_{SO}^{ind}) and relativistic change in the electronic potential (Δ_{POT}). In the final columns the total relativistic energies are assigned and compared to experiment.

	Nrel	Δ_{MV}	Δ_{DAR}	Δ_{SO}	Δ_{SO}^{ind}	Δ_{POT}	Rel.	Relat. state	Exp.
$1t_1$	8.84	0.88	0.02	-0.17	-0.04	-1.05	8.48	$u_{3/2}$	9.10±0.02
									9.19±0.02
		0.88	0.02	0.36	0.11	-1.03	9.18	$e_{1/2}$	9.58±0.02
$3t_2$	9.67	0.80	0.00	-0.31	-0.01	-0.99	9.16	$u_{3/2}$	9.78±0.02
									9.89±0.02
		0.80	0.00	0.62	0.07	-0.97	10.19	$e_{5/2}$	10.76±0.02
1e	10.25	0.74	0.01	0.00	0.05	-0.94	10.11	$u_{3/2}$	10.75±0.05
$2t_2$	13.58	1.35	-0.56	-0.15	-0.02	-0.84	13.36	$e_{5/2}$	13.09±0.05
		1.35	-0.56	0.07	0.03	-0.85	13.62	$u_{3/2}$	13.45±0.05
$2a_1$	16.46	4.12	-2.32	0.00	0.13	-1.05	17.34	$e_{1/2}$	
$1t_2$	19.68	5.19	-2.96	-0.01	0.34	-1.27	20.97	$e_{5/2}$	
		5.19	-2.96	0.01	0.33	-1.27	20.98	$u_{3/2}$	
$1a_1$	23.99	1.73	-0.94	0.00	0.20	-0.91	24.07	$e_{1/2}$	

two different fragments, viz. C_2 and I_4. HFS-LCAO non-relativistic calculations were performed for the separate fragments, allowing a Mulliken population analysis of the HFS-LCAO ground state of C_2I_4 in terms of the fragment orbitals (FRO's). In the overall C_2I_4 molecule the iodine $5p_x$ and $5p_y$ in-plane valence orbitals form radial and tangential symmetry combinations. The radial combinations interact with the dicarbon FRO's of the correct symmetry to form bonding and antibonding molecular orbitals (MO's), while the tangential combinations essentially constitute the iodine lone pair orbitals.

Fig. 10 gives a pictorial representation of the molecular orbital ordering of C_2I_4. The scheme can be understood in terms of interactions between the dicarbon FRO's and the iodine 5s and 5p valence AO's in the different D_{2h} symmetry irreducible representations. The interaction pattern is based on the non-relativistic HFS transition state results (solid lines), while the (dotted) dicarbon and iodine levels "before interaction" are estimates in the final molecular field. The figure is almost self-explanatory when one realizes that interaction lines are only drawn when the dicarbon FRO or iodine valence AO contribution to the resulting MO exceeds 15%. On the right in the figure the shifts of the transition state energies under the influence of relativistic effects are shown.

In Table 8 the non-relativistic and relativistic HFS-LCAO results are summarized together with experimental results.[59] The values obtained for the different relativistic corrections are listed separately. In the table heading Δ_{POT} represents the correction due to relativistic contraction of the orbitals, Δ_{MV} is the mass velocity term, and Δ_{DAR} the correction due to the "Zitterbewegung" of the electrons. The sum of the non-relativistic results and the above relativistic corrections is found under the heading Δ_{SUM}. In D_{2h} double group symmetry the orbitals belong to either the gerade representation $e_{1/2g}$ or to the ungerade $e_{1/2u}$. The spin-orbit operator cannot remove the Kramers' degeneracies of these orbitals, but provides an interaction mechanism which may lead to spectral shifts. Separate diagonalization of the $e_{1/2g}$ and $e_{1/2u}$ blocks including the effects of the spin-orbit operator leads to corrections to the IE's shown under the heading Δ_{SO}^{ind}. The final relativistic IE's agree well with the experimental ones, apart from an overall shift of about 1 eV.

The relativistic results in Table 8 merit some final remarks. The $1b_{1u}$ MO which is fairly well described by the C=C π_u FRO (~75%), is readily recognized from the small Δ_{POT} and Δ_{MV} contributions, illustrating the well-known fact that relativity is much less important for carbon than for iodine. The large Darwin corrections to $2b_{2u}$, $3a_g$, and $2b_{3u}$ indicate significant I 5s contributions to these MO's which is confirmed by the Mulliken population analysis. The indirect spin-orbit contribution is especially

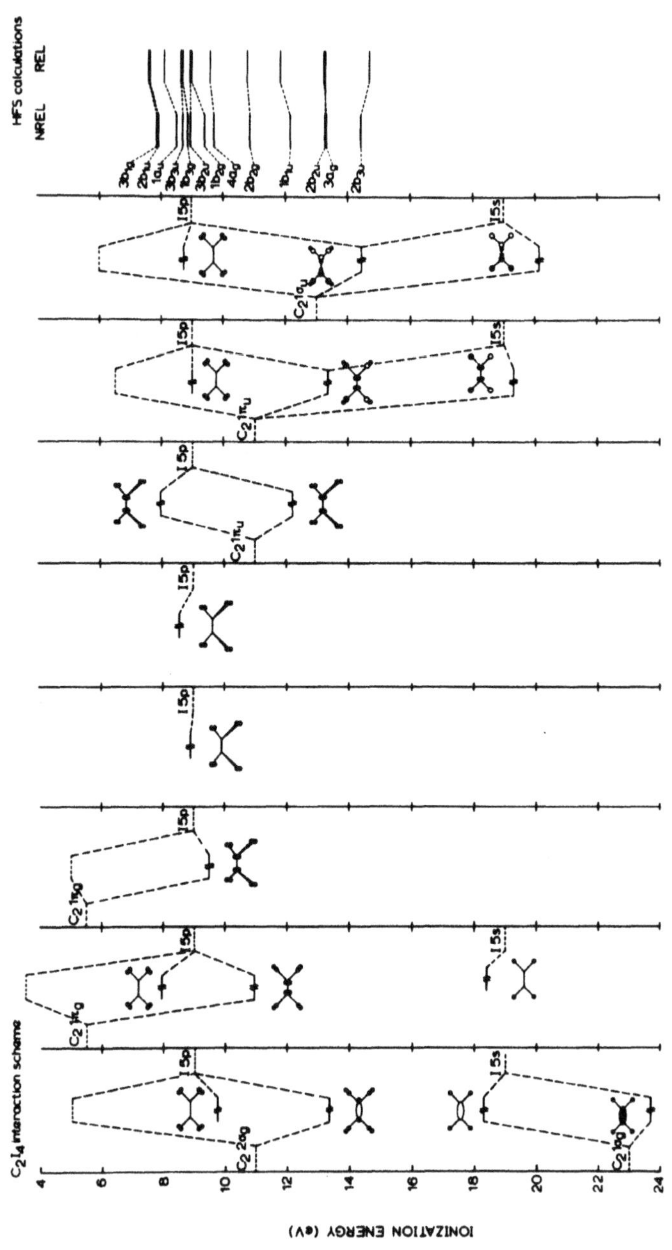

Figure 10. Ionisation energies for C_2I_4 from non-relativistic and relativistic HFS calculations. The dominant character of each occupied orbital is indicated schematically.

Table 8. Summary of non-relativistic and relativistic HFS calculations for C_2I_4 and experimental band positions and assignments (in eV).

	Nrel	Δ_{POT}	Δ_{MV}	Δ_{DAR}	Δ_{SUM}	Δ_{SO}^{ind}	Rel.	EXP.	
$3b_{1g}$	7.92	-1.13	0.91	0.06	7.76	-0.10	7.66	8.65 ± .02	$e_{1/2g}$
$2b_{1u}$	7.97	-0.84	0.62	0.02	7.77	-0.11	7.66	8.82 ± .02	$e_{1/2u}$
$1a_u$	8.57	-1.07	0.86	0.02	8.38	-0.24	8.14	9.58 ± .02	$e_{1/2u}$
$3b_{3u}$	8.74	-1.06	0.83	0.02	8.53	0.20	8.73	10.08 ± .02	$e_{1/2u}$
$1b_{3g}$	8.89	-1.00	0.81	0.02	8.72	0.05	8.77	10.21 ± .02	$e_{1/2g}$
$3b_{2u}$	9.02	-1.03	0.80	0.05	8.84	0.19	9.03	10.33 ± .02	$e_{1/2u}$
$1b_{2g}$	9.47	-0.94	0.69	0.02	9.24	-0.18	9.06	10.58 ± .02	$e_{1/2g}$
$4a_g$	9.74	-0.99	0.73	0.05	9.53	0.10	9.63	10.92 ± .02	$e_{1/2g}$
$2b_{1g}$	10.92	-1.00	0.88	-0.11	10.69	0.16	10.85	11.72 ± .02	$e_{1/2g}$
$1b_{1u}$	12.21	-0.48	0.15	0.00	11.88	0.02	11.90	12.62 ± .02	$e_{1/2u}$
$2b_{2u}$	13.33	-0.81	1.53	-0.68	13.37	-0.01	13.36	14.00 ± .05	$e_{1/2u}$
$3a_g$	13.34	-0.79	1.43	-0.60	13.29	0.00	13.29	14.00 ± .05	$e_{1/2g}$
$2b_{3u}$	14.46	-0.91	2.71	-1.45	14.81	-0.04	14.77	15.80 ± .05	$e_{1/2u}$

important for those MO's with significant I 5p character ($3b_{1g}$ to $2b_{1g}$).

D. $BiCl_3$, $BiBr_3$, and BiI_3

The photoelectron spectra of the bismuth trihalides were analyzed with HFS calculations using basissets of double zeta quality.[61] The geometries used are $BiCl_3$(R(Bi-Cl) = 2.48 Å, < Cl-Bi-Cl = 100°), $BiBr_3$(R(Bi-Br) = 2.63 Å, < Br-Bi-Br = 100°), and BiI_3(R(Bi-I) = 2.87 Å, < I-Bi-I = 98°). In the following the z-axis is the symmetry axis. The bonding picture that emerges from the calculations is shown in Fig. 11 for $BiCl_3$. We show the ionisation energies, together with the character of the corresponding orbitals in the ground state. Atomic energies are estimates in the final molecular field and are indicated for ease of interpretation only. The form of the different orbitals is indicated schematically, some in a plane through Bi and one Cl, bisecting the angle between the other two Cl's. Others are drawn in the plane of the halogens. The percentages indicate the result of a Mulliken population analysis on the basis of the actual atomic ground state orbitals, i.e. not the basis functions used.

In the a_1 symmetry we find the deep lying $2a_1$ orbital which consists predominantly of Bi6s, stabilized by some Cl3s and radial Cl3p admixture. The $3a_1$ is a bonding orbital basically made up out of the Bi6p_z and tangential Cl3p orbitals which have an almost π-like bonding overlap. The $4a_1$ orbital consists of the Cl3p_z orbitals interacting in an antibonding way with an almost sp hybridized orbital on Bi. It is interesting to note that this highest occupied molecular orbital is not a pure Cl lone pair orbital but has considerable Bi character (25%), accounting for the donor properties of this type of molecule. Since there are no high lying orbitals of a_2 symmetry on Bi, we find in this symmetry a pure Cl lone-pair orbital, made from the tangential Cl3p orbitals in the plane of the three Cl's. The orbital is slightly Cl-Cl antibonding. Finally, in the e symmetry we have a strongly bonding 2e orbital arising from a σ-like overlap of the Bi6p_x and 6p_y orbitals with suitable combinations of the radial Cl3p's. The more σ-like overlap explains the ordering of the 2e and $3a_1$ bonding levels. The 3e and 4e are essentially (99 and 98%) Cl lone-pair orbitals. The 3e is somewhat Cl-Cl bonding, while the 4e is slightly Cl-Cl antibonding like the $1a_2$ with which it is almost degenerate.

For $BiBr_3$ and BiI_3 the picture is very similar. We only give the ionisation energies in Fig. 12. Apart from an expected shift to lower I.E.'s going to the heavier halides, we note that the 4e and $1a_2$ are relatively destabilized as we go down the periodic system, becoming almost degenerate with the $4a_1$ in BiI_3. This can be understood if one realizes that the antibonding halogen-halogen

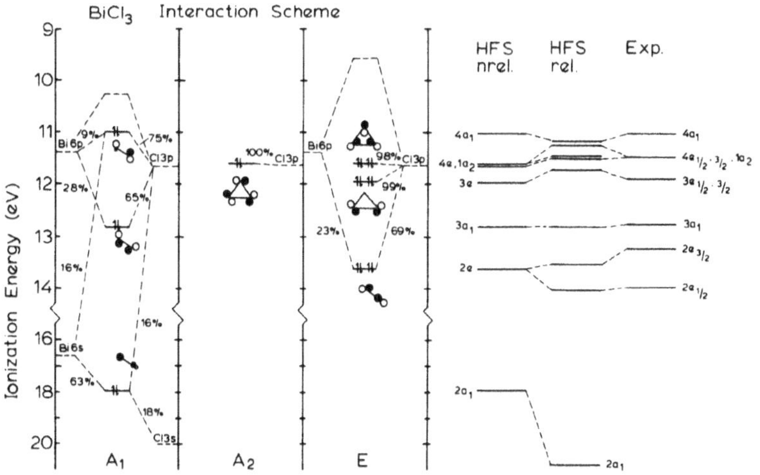

Figure 11. Bonding pattern for $BiCl_3$ (see text). Non-relativistic, relativistic, and experimental ionisation energies in eV. Calculated IE's are shifted by 1.26 eV.

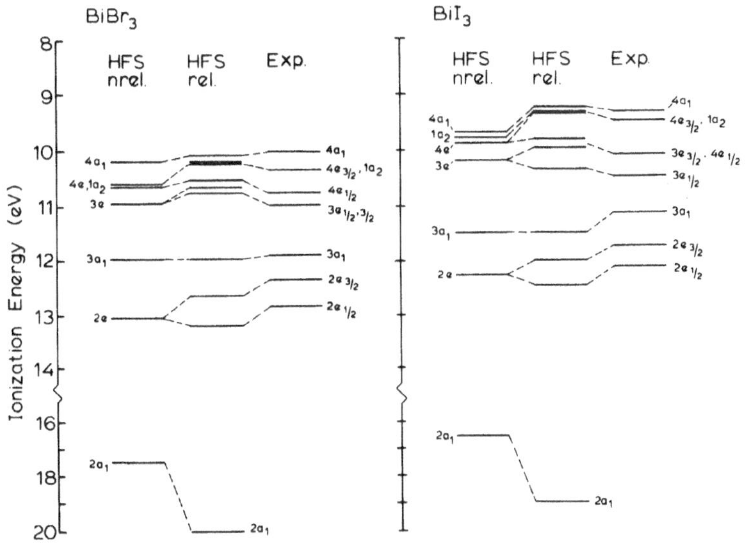

Figure 12. Non-relativistic, relativistic, and experimental ionisation energies of $BiBr_3$ and BiI_3 in eV. Calculated IE's are shifted by 0.96 eV ($BiBr_3$) and 1.20 eV (BiI_3).

overlap increases in these orbitals as the corresponding p orbitals grow in size in the heavier halogens.

In view of the heavy Bi nucleus and of the heavier halogens it is essential to also consider the modification of the above picture due to relativistic corrections. In the case of BiI_3 where relativistic effects are largest, we show the results of the perturbative calculations in Fig. 12 and Table 9. In Table 9 we separate the different contributions to the relativistic effect in IE's according to their physical origin. For all levels there is a considerable stabilization due to the negative definite mass-velocity effect, especially in those MO's that have an appreciable density on the heavy Bi atom (2e, $3a_1$, and $4a_1$). For the $2a_1$, which consists largely of the deeply penetrating Bi6s, the stabilization is enormous (7.67 eV !). The destabilizing Darwin interaction is only important for MO's that contain some Bi6s ($2a_1$, $3a_1$, $4a_1$) or I5s ($2a_1$, 2e) admixture, these being the only valence AO's with non-vanishing density on the nuclei. The change in electronic potential destabilizes all levels, even overcompensating for the mass-velocity effect in the lone pair orbitals. This can be explained by the general tendency of s and p orbitals to contract due to relativity, confining the electrons to a smaller space and hence increasing their repulsion. We have separated the spin-orbit effect into two contributions. First there is a direct diagonal splitting of the orbitals of e symmetry into two components $e_{1/2}$, $e_{3/2}$. It should be realized that the $e_{3/2}$ representation is reducible in the C_{3v}^* double group, so that the degeneracy of the $e_{3/2}$ levels is not due to the molecular symmetry. Rather we have here a case of Kramers' degeneracy due to time-reversal symmetry. This direct splitting is important in the 2e and 4e levels, but quite negligible in the 3e. Apart from this direct spin-orbit effect we have to take into account that orbitals of the same double group symmetry (but not necessarily the same single group symmetry) can have an off-diagonal indirect spin-orbit interaction between them, which can be of the same order of magnitude as the direct effect, especially when they are close in energy, as is the case with the halogen lone pair levels. This interaction will tend to push the interacting levels apart, stabilizing the deeper lying ones, while destabilizing the higher levels. This effect is quite pronounced in BiI_3, where the $4a_1$ and $1a_2$ (both of $e_{1/2}$ symmetry) are pushed up while the $3e_{1/2}$ is stabilized considerably as is the $2e_{1/2}$ to some extent. The $4e_{1/2}$, lying in between these levels, is relatively unaffected. A similar but much smaller effect operates between the $4e_{3/2}$ and $3e_{3/2}$. It is seen that it is mainly this indirect interaction that is responsible for the splitting of the 3e level. Note that in all cases the $e_{1/2}$ component of the e levels is the deeper lying one as is the case in atomic p-levels. Comparing the final relativistic IE's to experiment in Fig. 12 we see that there is excellent agreement apart from an overall shift of 1.2 eV. The $2a_1$ IP calculated at 18.94 eV is probably too deep to be easily accessible experimen-

Table 9. Analysis of relativistic corrections to the IE's of BiI_3. All energies in eV.

	non-[a] relat.	mass-velocity	Darwin	potential correction	spin-orbit (direct)	spin-orbit (indirect)	relat.	exp.
$2a_1$	16.55	7.67	-3.67	-2.05	0.0	0.44	18.94	
$2e_{1/2}$	12.32	1.50	-0.22	-1.48	0.14	0.24	12.49	12.12
$2e_{3/2}$	12.32	1.50	-0.22	-1.52	-0.14	0.09	12.02	11.75
$3a_1$	11.48	1.68	-0.29	-1.41	0.0	0.04	11.50	11.17
$3e_{1/2}$	10.20	0.77	0.01	-1.05	0.02	0.40	10.35	10.49
$3e_{3/2}$	10.20	0.77	0.01	-1.08	-0.02	0.09	9.97	10.11
$4e_{1/2}$	9.88	0.82	0.01	-1.12	0.21	0.02	9.82	10.11
$4e_{3/2}$	9.88	0.82	0.01	-1.10	-0.21	-0.07	9.33	9.49
$1a_2$	9.80	0.84	0.00	-1.17	0.00	-0.10	9.37	9.49
$4a_1$	9.71	1.61	-0.34	-1.29	0.00	-0.46	9.23	9.32

[a]Shifted by 1.20 eV.

tally. At 12.12 and 11.75 eV we have the spin-orbit split bonding 2e, at 11.17 eV the bonding $3a_1$ and at 10.49 eV the $3e_{1/2}$ component of the spin-orbit split 3e lone pair orbital.

The band at 10.11 eV can be assigned to a superposition of the $3e_{3/2}$ and $4e_{1/2}$ IE's, while those at 9.49 and 9.32 eV contain the $4a_1$, $4e_{3/2}$, and $1a_2$, the $4a_1$ probably being the most weakly bound level. It should be noted that the designation $4a_1$ and $1a_2$ loses much of its meaning due to the above mentioned off-diagonal spin-orbit interaction that mixes the orbitals of non-relativistic $4a_1$ and $1a_2$ symmetry. For $BiCl_3$ and $BiBr_3$ the results are shown in Fig. 11 and Fig. 12, respectively. The bonding part of these spectra ($3a_1$,2e) is very similar to that of BiI_3, the spin-orbit splitting in the 2e being due to the Bi atom. In $BiBr_3$ the spin-orbit splitting in the 3e has become too small to be resolved, but from the calculations we assign the IE's at 10.71 and 10.33 eV to the spin-orbit split 4e, the $4e_{3/2}$ being very close to the $1a_2$ and unresolved. Finally we again find the $4a_1$ as the most weakly bound level at 10.02 eV. Overall agreement between theory and experiment is again satisfactory. Finally in $BiCl_3$ the splitting of the 4e has also virtually disappeared, the 3e and 4e being assigned to the bands at 11.87 and 11.46 eV, respectively. The $4a_1$ is again calculated to be the least stable and is assigned to the band at 10.99 eV. In the calculation we now find the $1a_2$ to be almost degenerate with the $4a_1$. Nevertheless from intensity considerations it seems more likely that the $1a_2$ IE is hidden under the relatively broad band at 11.46 eV. The assignments given here agree well with those given by Lee et al.[7] for the other group Va trihalides.

In conclusion we can say that these calculations provide a good understanding of the experimental photo-electron spectra of the bismuth-trihalides, while at the same time illuminating the various physical effects in operation in these systems.

4. DIVALENT GROUP VI A COMPOUNDS

A. Introduction

In this section we shall be concerned with molecules which consist of a divalent atom belonging to group VI A (O,S,Se,Te) and from halogen atoms (F,Cl,Br,I) or pseudohalogens (CN) as ligands. For many molecules of this type He(I) photoelectron (PE) spectra have been obtained and spectral assignments have been published. We shall limit our discussion to those systems for which HFS calculations have been performed.

Since the majority of the molecules belonging to the above class are unstable, the measurement of their PE spectrum has in many cases not been a simple and straightforward matter. In order to generate the desired compounds in sufficient steady-state concentration in the gas phase, in situ preparation techniques in the spectrometer ionization chamber have often proved necessary. In the course of the experimental study of reactive systems spectrometer contamination is a severe problem. Although the effects of contamination on spectrometer sensitivity and resolution can never be eliminated completely, a sensible spectrometer design and careful experimental procedures can go a long way in keeping the problems within acceptable limits. When the PE spectra of unstable species are studied, their bands must often be disentangled from those arising from stable precursor materials or unreactive end products. In order to eliminate these unwanted signals and to make the detection of spectra of transient species feasible, microprocessor-based modulation methods or spectrum stripping techniques have been developed and proved successful. In spite of such experimental difficulties an appreciable number of compounds based on the divalent group VI A elements have been studied in detail. We shall start our description with various group VI A dihalides, followed by a series of compounds in which one halogen atom has been replaced by a CN group. Finally an example of a dicyano compound will be discussed. The CN containing compounds with oxygen or sulphur as the group VI A atom are rather intriguing in the sense that the OCN or SCN moieties are ambidentate, bonding at either end and forming either the (thio)cyanate or iso(thio)cyanate structure. Recent PE studies indicate that molecules such as FSCN,[63] ClSCN,[64] BrSCN[64] or $S(CN)_2$ [65] have a distinct preference for the "normal" structure whereas OCNCl,[66] OCNBr,[66] OCNI,[66] and OCNCN[67] prefer the "iso" form.

When the valence region of the bent AX_2(A=O,S,Se,Te; X=F,Cl, Br,I) is probed with He(I) radiation of 21.2 eV a limited number of spectral bands is observed and an unambiguous assignment is usually possible. For both group VI A atom and ligand atoms the valence s AO's lie much deeper than the valence p AO's (see Table 10). Hence the valence orbitals which correspond to the PE bands in the He(I) spectrum are mainly built up from valence p AO's while occupied s and empty AO's play a minor role.

Although the inner valence region can in principle be studied using He(II) radiation of 40.8 eV, the low intensity of presently available He(II) sources in combination with the difficulties associated with the usually low steady-state concentration obtainable in the case of unstable products, have proved to be rather prohibitive in most cases. However, the orbitals situated in the inner valence region should contain important contributions from valence s AO's on both central atom and ligands. Moreover, sizable interactions between central atom and ligand valence s AO's should be expected. For the pseudohalogen CN with its σ-type fragment orbitals far below the π-type fragment and nitrogen lone pair levels the situation is similar to the halogen case.

Table 10. Group VI A and halogen free atom valence subshell binding energies,[68] values in eV

atom	n	ns	np
O	2	28.48	13.62
F	2	37.85	17.42
S	3	20.20	10.36
Cl	3	24.54	12.97
Se	4	20.15	9.75
Br	4	23.80	11.85
Te	5	17.84	9.01
I	5	20.61	10.45

B. Dihalides

The dihalides have a bent structure and possess C_{2v} symmetry. In the following discussion the molecules are assumed to lie in the xz plane with the z-axis along the C_2-axis. In order to understand the common features in the PE spectra of these compounds, we show in Fig. 13 an orbital interaction diagram for $TeBr_2$ obtained on the basis of HFS calculations. As becomes clear from the figure, the level ordering can be understood in terms of a simple MO scheme involving only the interaction of Te 5p and Br 4p orbitals. We consider the various symmetry species separately.

a_1. From the in-plane Br $4p_x$ and $4p_z$ orbitals two combinations of a_1 symmetry can be constructed. In principle these combinations can interact with the Te based $5p_z$ AO. The interaction between these three symmetry fragment orbitals results in three MO's, of which the combination of $4p_z$ Br AO's acquires very little Te character and remains essentially Te-Br nonbonding. The Br $4p_x$ combination of a_1 symmetry interacts strongly with Te $5p_z$ to give a bonding and an anti-bonding MO.

a_2. Since there are no low-lying orbitals of a_2 symmetry on Te, the a_2 combination of the out-of-plane Br $4p_y$ AO's forms a nonbonding MO.

b_1. For this symmetry our starting point is a radial and a tangential combination built from in-plane Br 4p AO's and the Te $5p_x$ AO. Again three MO's are obtained of which the tangential Br 4p combination remains Te-Br non-bonding. From the strong interaction between the radial Br 4p combination and Te $5p_x$ a bonding and an anti-bonding MO result.

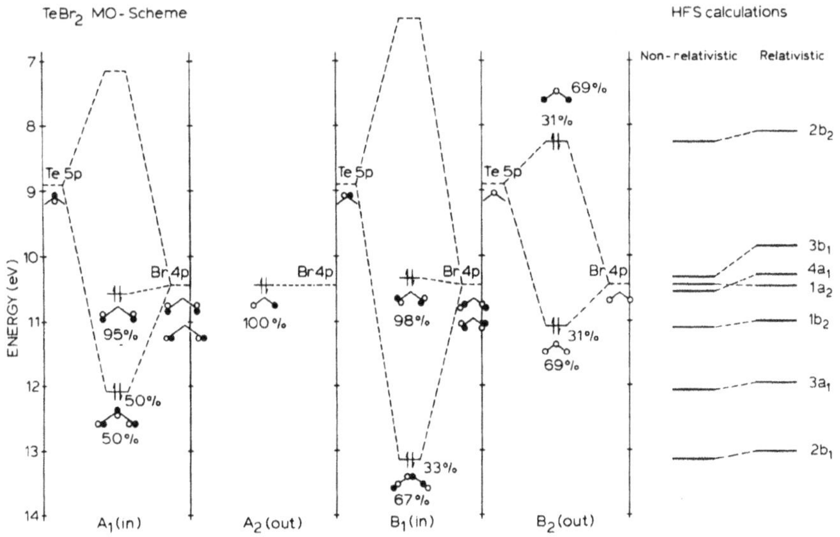

Figure 13. Orbital interaction diagram and ionisation energies for $TeBr_2$ obtained from non-relativistic and relativistic HFS calculations.

b_2. The b_2 combination of the out-of-plane Br $4p_y$ orbitals has a π-type interaction with Te $5p_y$, giving rise to a bonding and an anti-bonding b_2 MO.

With regard to the anti-bonding orbitals, the π-type overlap in the b_2 symmetry block is expected to be smaller than the more favourable overlaps in a_1 and b_1 symmetry. When we consider the bonding orbitals, the overlap between the AO combinations of b_2 symmetry is again smaller than the overlaps in the a_1 and b_1 blocks. Since we have to accomodate 20 valence electrons in all, 14 of which are valence p-electrons, we expect a highest occupied anti-bonding orbital $2b_2$, followed by three non-bonding orbitals $3a_1$, $1a_2$, and $3b_1$ whose relative order is uncertain a priori. Finally three bonding orbitals in the order of increasing IE, viz. $1b_2$, $3a_1$, and $2b_1$ should occur. Valence orbital numbering is used throughout. The corresponding He(I) PE spectrum should show an isolated first band, next three close-lying bands, followed by three distinct features due to the bonding orbitals in the valence region. As we shall see these general features will return in the PE spectra of all the dihalides studied.

(i) OF_2

The He(I) spectrum of the relatively stable compound OF_2 has been recorded by two separate groups.[69,70] Some dispute has arisen about the number of experimental bands in the 16-17 eV region. In both [69] and [70] an assignment is preferred with two PE bands between 16 and 17 eV and the feature corresponding to the $1a_2$ MO outside this region. This interpretation was based on ab initio calculations using Koopmans' theorem[69,70] and semi-empirical calculations.[69] However, in [70] it is mentioned that an alternative interpretation of the spectrum which places $1a_2$ together with $4a_1$ and $3b_1$ in the 16-17 eV region cannot be ruled out. On the basis of this alternative assignment the PE spectrum of OF_2 would take the general form described above for the group VI A dihalides. Subsequently this point of view has been fully confirmed by various good-quality calculations such as those based on ordinary third-order Rayleigh-Schrödinger perturbation theory,[71] HFS calculations[72] and ab initio many-body Green's function calculations.[73] A summary of the experimental data and the results of these calculations is presented in Table 11. In addition it should be mentioned that orbital correlations for various difluoro molecules[74] as well as simple sum-rule considerations[75] also place $1a_2$ in the 16-17 eV region. In contrast, a recent SCF CI double zeta quality calculation suggests an assignment which, although not satisfactory in every respect, places $1a_2$ at 18.50 eV.[76]

(ii) SF_2

The PE spectrum of this rather elusive species has been recorded in an experiment in which atomic fluorine was reacted with flowers of sulphur.[32] Apart from some photoionization features due to other species, the spectrum shows a well-resolved first band at 10.35 eV with clear vibrational components. Next a broad feature has been observed which on the basis of the intensity distribution has been ascribed to one band at 15.4 eV and two superimposed bands at 16.2 eV. Finally features at 18.3 and 19.3 eV have been found. Due to the rising background which occurs with the type of spectrometer used, no more bands have been located in the He(I) region (see Fig. 14). The observed spectrum is in agreement with the expected general form of the PE spectra of divalent group VI A dihalides.

An assignment of the spectrum based on HFS calculations has been given. For SF_2 an STO basis set of double zeta quality has been used. As a test of the importance of d-orbital participation HFS calculations have been performed with and without a 3d polarization function in the sulphur basis set. Inclusion of the polarization function had a small effect on the valence IE's and introduced no change in the calculated ordering. However, agreement between calculated and experimental electric dipole moments was much

Table 11. Summary of experimental data and theoretical results for the vertical ionisation energies (in eV) of OF_2 in the outer valence region

Symmetry	Exp.[a]	RS[b]	GF[c]	HFS[d]
$2b_2$	13.25	13.29	13.06	12.83
$4a_1$	16.10	16.33	16.37	16.02
$3b_1$	16.44	16.54	16.46	15.90
$1a_2$	~16.44	16.69	16.95	15.95
$1b_2$	18.50	18.55	18.95	17.91
$3a_1$	19.55	18.90	19.49	19.27
$2b_1$	20.7	19.99	20.57	20.73

[a]The "alternative" assignment of Ref. 70 is taken.
[b]Rayleigh-Schrödinger perturbation theory.[71]
[c]Green's function method.[73]
[d]Hartree-Fock-Slater, double zeta quality.[72]

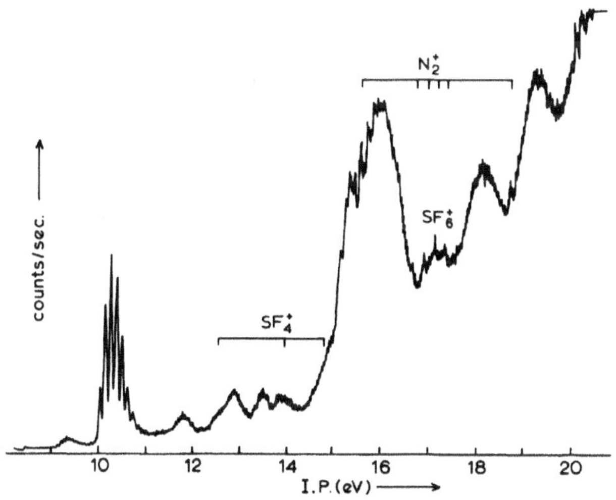

Figure 14. The He(I) photoelectron spectrum of SF_2.

improved by including a sulphur 3d function.[32]

In the HFS-based assignment the first band is ascribed to $2b_2$, with the strong progression of 935 ± 40 cm^{-1} arising from the symmetrical stretch in the ion. The increased vibrational frequency compared to that in the neutral species is in accordance with the calculated strongly anti-bonding S-F overlap. The transition at 15.4 eV is assigned to $4a_1$ which is calculated to be delocalized over the entire molecule and somewhat S-F anti-bonding. The two overlapping bands at 16.2 eV correspond to $3b_1$ and $1a_2$ which according to HFS should be very close. The remaining two bands in the valence region at 18.3 and 19.3 eV have been assigned to the bonding $1b_2$ and $3a_1$ orbitals respectively.

A detailed ab initio study using a large gaussian basis set and including d polarization functions has been reported.[77] When Koopmans' theorem is assumed to be valid the resulting calculated IE's agree less well with the experimental PE spectrum than the HFS results in which relaxation effects are included.

A subsequent Green's function calculation on SF_2[73] leads for the bands corresponding to the orbitals $2b_2$, $4a_1$, $3b_1$, and $1a_2$ to the same assignment as the HFS results. Differences occur deeper in the valence region where the experimental feature at 18.3 eV on the basis of the Green's function calculation should be ascribed to both $1b_2$ and $3a_1$, and the band at 19.3 eV to $2b_1$. Despite this discrepancy in the deeper valence region the overall agreement between HFS and Green's function results is satisfactory.

(iii) SBr_2

Evidence for this short-lived molecule has been obtained for the first time in a PE experiment on the reaction products of bromine atoms and flowers of sulphur.[33] Two bands were detected, the first PE band assigned to $2b_2$ at 9.36 eV with resolved vibrational structure with a spacing of 400 ± 50 cm^{-1} and a rather sharp feature at 11.4 eV ascribed to $1a_2$. The vibrational progression on the first band has been assigned to excitation of the symmetrical S-Br stretch vibration. HFS calculations of double zeta quality were performed in this study.

Subsequently the full He(I) PE spectrum of SBr_2 has been obtained via a spectrum stripping technique which removed overlapping signals of unwanted products.[78] The stripped spectrum shows the now familiar picture of an isolated first band, three close-lying bands and three well-separated features arising from three bonding orbitals. As an aid in the spectral assignment the VEOMP molecular orbital method[79,80] based on the all-electron Gaussian 70 ab initio programme[81] with a minimum valence STO-3G basis[82] has been employed.

A comparison between HFS and VEOMP calculations shows that in the spectral region where $4a_1$, $3b_1$, and $1a_2$ are situated both methods do not arrive at the same ordering. However, in view of the closeness of these IE's neither method is expected to produce definite results. In the deeper valence region the HFS calculations agree definitely better with the experimental IE's than the VEOMP results.

A slight difficulty remains in the position of the out-of-plane $1b_2$ orbital. Experience shows that especially for π-type orbitals simple semi-empirical ZDO or sum-rule considerations[75] are useful in predicting the positions of π-type bands when results for related molecules are available. The predicted and measured values for $1b_2$ in SBr_2 disagree by a rather large amount of 0.4 eV. It may be that inaccuracies inherent in the spectrum stripping technique are at the root of this problem.

(iv) SeF_2

Evidence for the short-lived molecule SeF_2 has been obtained in the reaction between fluorine atoms, generated in a microwave discharge in CF_4, with elemental selenium. A rather weak first band of the PE spectrum of SeF_2 has been observed at 10.2 eV.[34] This band is shown on an expanded scale in Fig. 15 and exhibits vibrational structure with an average spacing of 780 ± 50 cm^{-1}. This progression has been assigned to excitation of the symmetrical stretch vibration. The first band is assigned to ionization from the π-type $2b_2$ MO which on the basis of HFS calculations is characterized by a strong anti-bonding Se-F overlap. HFS calculations predict the other photoionization phenomena above 14 eV. Due to experimental difficulties no bands in this region have been observed until now.

(v) $SeCl_2$

The He(I) PE spectrum of the unstable $SeCl_2$ has been detected by passing atomic or molecular chlorine over selenium powder.[34,83] The spectrum is presented in Fig. 16. Although only six bands are observed in the spectrum, the intensity distribution indicates that under the structure at 12.30 eV two photoionization phenomena are hidden. The first band at 9.52 eV exhibits vibrational structure with a regular spacing of 450 ± 50 cm^{-1}.[34] The band is assigned to the π-type $2b_2$ MO and the progression arises from the symmetrical stretch in the ion. The vibrational frequency in the ion is increased compared to that of the symmetrical stretch in the neutral $SeCl_2$, in accord with the HFS calculated strong anti-bonding Se-Cl overlap. Since accurate geometrical data for $SeCl_2$ are lacking, the HFS calculations have been performed for various values of the bond angle between 90° and 109.5°. The calculated IE's were found

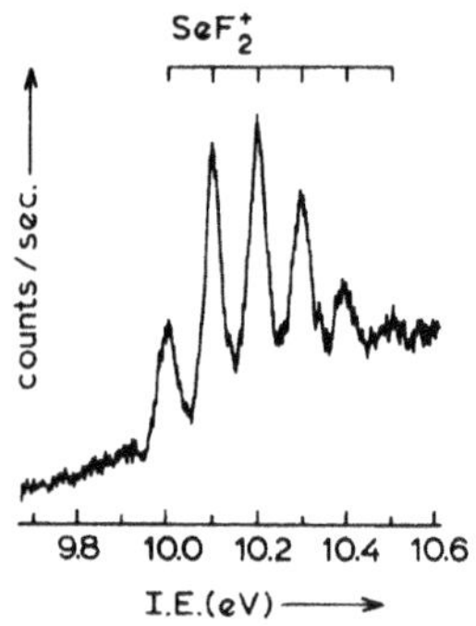

Figure 15. Expanded band associated with the first ionisation energy of SeF_2.

to remain constant within 0.3 eV for bond angle variations in this range. The HFS based assignment shows, apart from the isolated $2b_2$ band, three close-lying features corresponding to three essentially non-bonding MO's with mainly chlorine character, and three well-separated deeper-lying Se-Cl bonding MO's. The experimental IE's appear to be in better agreement with the HFS results[34] than with those of VEOMP calculations[83] in which the three IE's corresponding to the three bonding MO's are computed to be almost degenerate.

A correlation diagram of the assigned vertical IE's of $SeCl_2$, SCl_2[84,85] and OCl_2[69] versus the first IE's of the central atom (see Fig. 17) shows the behaviour of the vertical IE's. The fact that

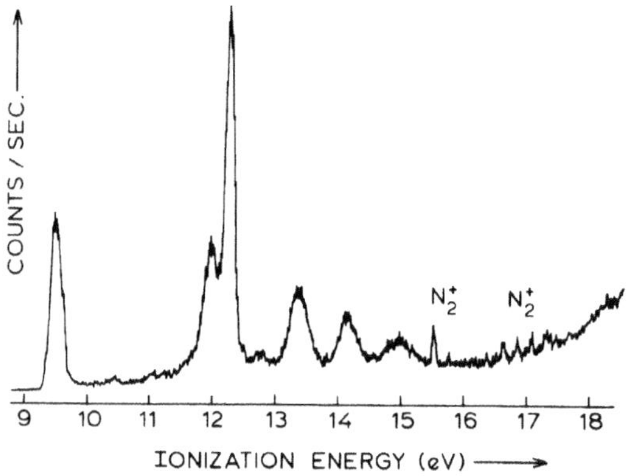

Figure 16. The He(I) photoelectron spectrum of $SeCl_2$.

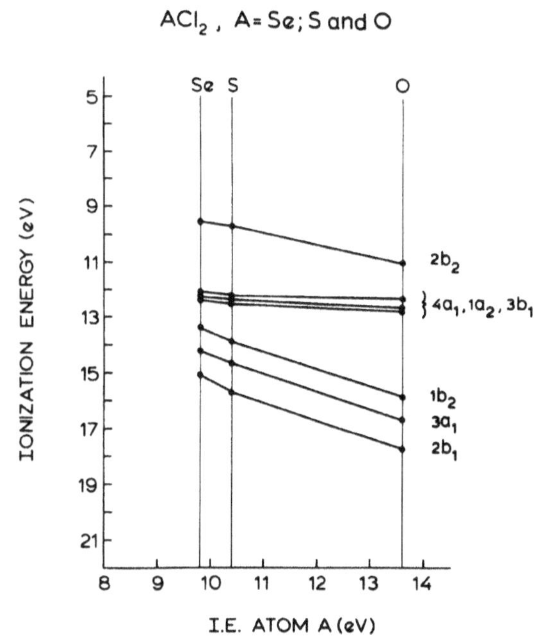

Figure 17. Correlation diagram of the observed vertical ionisation energies of $SeCl_2$, SCl_2, and OCl_2 versus the first atomic ionisation energies of their central atoms.

the positions of $4a_1$, $1a_2$, and $3b_1$ hardly depend on the central atom is once more an indication that the three MO's around 12 eV have very little central atom character and are essentially non-bonding.

(vi) $SeBr_2$

The He(I) PE spectrum of the short-lived $SeBr_2$ has been detected by passing atomic or molecular bromine over elemental selenium.[34,78] The spectrum is presented in Fig. 18 and consists of the expected seven bands with two of them, at 11.31 and 11.40 eV, almost coincident. No vibrational structure is observed on any of the bands. The experimental spectrum, the results of HFS calculations and hence the spectral assignment[34] are in close agreement with those obtained for $SeCl_2$. Since $3b_1$, $4a_1$, and $1a_2$ are calculated to be almost coincident, the relative order of these three orbitals cannot be assigned conclusively. However, if 4d orbital participation is excluded the $1a_2$ MO cannot have any selenium character on symmetry grounds. Because of the dominant non-bonding character calculated for $1a_2$ the corresponding PE transition is ascribed to

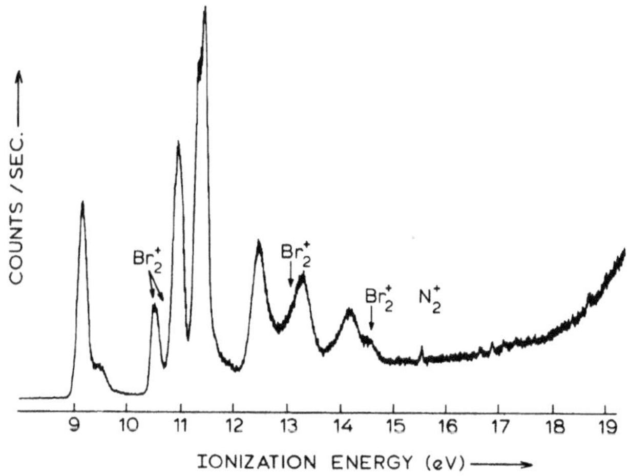

Figure 18. The He(I) photoelectron spectrum of $SeBr_2$.

the sharpest peak in this region, at 11.40 eV.

As in the case of $SeCl_2$, the HFS results[34] show much better agreement with experimental IE's than the VEOMP calculations.[78]

(vii) $TeCl_2$

He(I) PE spectra of the unstable species $TeCl_2$ have been detected by leading molecular chlorine over elemental tellurium heated to about 400°C.[86] If the tellurium powder is not heated, no detectable reaction occurs. The spectrum is presented in Fig. 19 and shows the usual pattern with one isolated band with low IE, a number of almost coinciding features at higher IE and finally three well-separated bands corresponding to bonding MO's. The observed pattern is well reproduced by double zeta STF quality HFS calculations. Spectrum and spectral interpretation show strong similarities to those of $TeBr_2$.[86]

Since tellurium is a rather heavy atom, one might expect the occurrence of relativistic effects in the valence PE spectrum of a molecule such as $TeCl_2$. HFS calculations in which relativistic effects have been taken into account through perturbation theory have been performed for this compound.[86] Computed relativistic corrections to the non-relativistic IE's are of the order of 0.1 eV and mainly show up as a uniform shift of all the bands. Since a shift between HFS calculated and experimental IE's tends to be

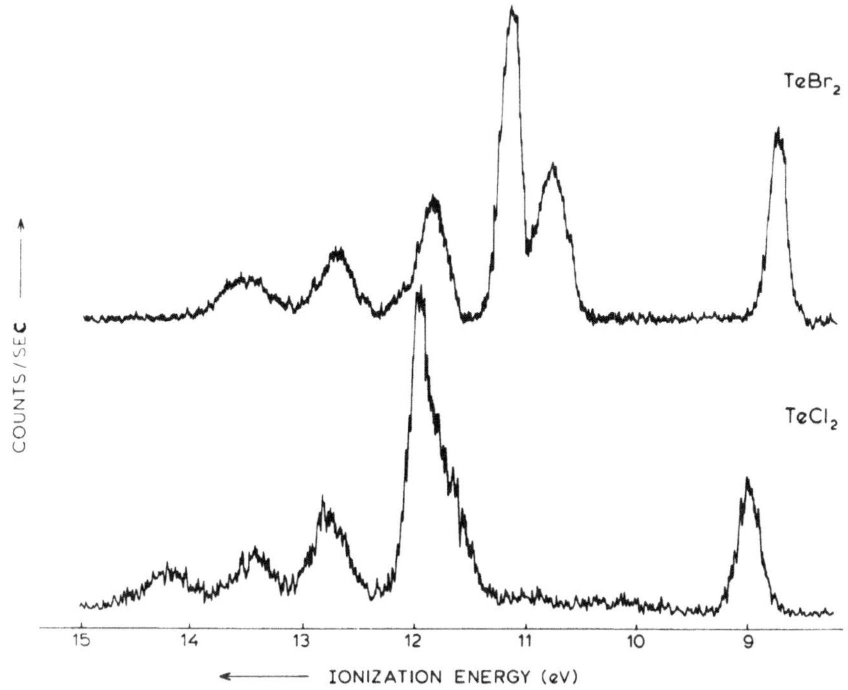

Figure 19. The He(I) photoelectron spectra of $TeCl_2$ and $TeBr_2$.

present anyway, it can be concluded that relativistic effects have a minor influence on the valence PE spectrum of $TeCl_2$.

(viii) $TeBr_2$

The valence PE spectrum of this short-lived compound (see Fig. 19) has been observed in the reaction between molecular bromine and heated tellurium powder.[86] The analysis of the general features of this spectrum under the neglect of relativistic corrections has been described in an earlier paragraph. However, for this molecule with fairly heavy nuclei a detailed understanding of the PE spectrum can only be obtained when relativistic corrections to the IE's are considered. HFS calculations including the effects of relativity have been performed on this compound.[86]

In Fig. 20 we show how the various relativistic contributions shift the IE's in the case of $TeBr_2$. It is seen that a considerable stabilization of all levels occurs through the mass-velocity correction, while those orbitals which have considerable Te character

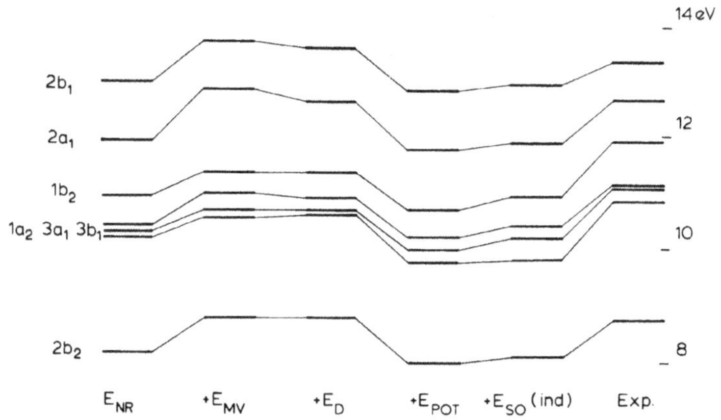

Figure 20. Relativistic corrections to the IE's of $TeBr_2$.

are affected most, Te being the heaviest centre. The Darwin term is only important for those orbitals that have some Te or Br s character ($3a_1, 2b_1$) as expected from the considerations given before. The potential corrections which have to be taken into account overcompensate the relativistic stabilization so that the relativistic IE's are actually somewhat smaller than the non-relativistic ones. Among the three non-bonding MO's ($4a_1, 1a_2, 3b_1$) an interesting additional effect appears. Since all three are of different nondegenerate symmetry there is no direct spin-orbit splitting (the diagonal elements of h^1_{SO} vanish identically), but since these levels are so close in energy, they can in fact be coupled due to off-diagonal elements of the spin-orbit operator between them (they all belong to the same double group symmetry $e_{1/2}$ in C^*_{2v}). This separates one of the non-bonding MO's from the other two as found experimentally in $TeBr_2$ (Fig. 19). The effect is calculated to be very small in $TeCl_2$, which is not surprising since the relevant orbitals are almost entirely of halogen character and relativity is expected to be of less importance for Cl than for Br. Indeed, experimentally the splitting is not resolved in $TeCl_2$ (Fig. 19). It seems therefore rather well established that we are here dealing with a spin-orbit splitting which is not predicted by symmetry as such.

(ix) TeI_2

Experiments are presently underway to observe the He(I) PE spectrum of TeI_2. In all the methods attempted until now to generate this unstable compound in the gas phase, the TeI_2 PE bands were heavily masked by iodine bands. The off-diagonal spin-orbit effect apparent in $TeBr_2$ is calculated to be even more important in TeI_2.[87]

(x) Related compounds

Two more members belonging to the class of group VI A dihalides have been studied by means of He(I) PE spectroscopy, viz. the rather stable OCl_2[69] and the stable compound SCl_2.[84,85] The spectra show the general features of those of the group VI A dihalides treated above. Since no HFS calculations have been performed on these molecules we shall refrain from further discussion.

C. Molecules containing one CN group

When in the group VI A dihalides one of the halogen ligands is replaced by the pseudohalogen CN group, planar molecules of C_s symmetry result. These compounds can occur either with the "normal" structure or in the "iso" form. For molecules in the "normal" form the replacement of a halogen atom by CN affects the valence PE spectrum in a distinct manner. Instead of the valence p AO's of the halogen the relevant fragment orbitals in the CN group are two filled π_{CN} orbitals, one in-plane and one out-of-plane, one fully occupied σ-type or nitrogen lone pair orbital and one half-filled σ orbital. The latter orbital plays a role in the bonding of the CN group to the group VI A atom. In the He(I) region of the PE spectra of such compounds we expect on these simple grounds eight photoionization bands. An extra complication arises from the fact that the IE's corresponding to $\pi_{CN}(a',a'')$ and the $n_N(a')$ lone pair orbital tend to lie close together in the spectrum. This situation can lead to an assignment problem. As with the dihalides, for a general understanding of the valence PE spectra of tbe compounds belonging to this class no valence s or d orbital participation is required. In the calculations the molecules with C_s symmetry are located in the xy plane.

For compounds which occur in the "iso" form the valence PE spectra cannot be compared in a simple way to those of the dihalides since the molecular bonding is entirely different.

(i) FSCN

When molecular chlorine at low pressure is passed through a column of silver thiocyanate (AgSCN), the compound $(SCN)_2$ is generated in the gas phase.[64] The reaction of the latter molecule with molecular fluorine in a flow system produces a novel unstable compound, which has recently been detected using PE spectroscopy and which has been shown to be FSCN.[63]

On the basis of simple MO considerations eight photoionization bands are expected in the He(I) region of the PE spectrum which is presented in Fig. 21. This expectation is supported by the results of HFS calculations which also predict eight IE's below 21.1 eV.

The first band in the FSCN spectrum at a vertical IE of 10.93 eV corresponds to the 3a" out-of-plane MO. The vibrational progression associated with this IE (840 ± 50 cm^{-1}) is clearly observed. A reasonable estimate for an S-F stretching frequency of 810 cm^{-1} can be obtained from the matrix isolated infrared spectrum of SF_2.[88] When the progression in the first band of FSCN is assigned to the symmetrical S-F stretch, the slight increase of the frequency in the ion indicates electron removal from a slightly anti-bonding MO. This is in accord with the HFS results which describe the 3a" MO as an orbital mainly located on sulphur (~70%) with weakly anti-bonding S-F character.

By comparing the position of the first IE's of the closely related SF_2,[32] FSCN and $S(CN)_2$[65] molecules, the first IE of SF_2 has the lowest value, despite the two strongly electronegative fluorine atoms bonded to the central atom. HFS calculations on these molecules show about the same degree of localization on the sulphur atom (~70%). Apparently the observed trend in the position of the first IE in these molecules arises from a conjugative interaction due to overlap between the sulphur out-of-plane 3p and the fluorine out-of-plane 2p orbitals. This is illustrated by the different characters of the first IE in these compounds: strongly anti-bonding in SF_2 (eight vibrational components), weakly anti-bonding in FSCN (four vibrational components) and almost non-bonding in $S(CN)_2$ (two vibrational components with a strong o-o transition).

The relative intensity of the band system observed between 13 and 14 eV indicates that it could well contain three ionization phenomena. This is in full accord with the HFS calculations which predict little separation between the three bands arising from the cyano group (a'(π_{CN}), a"(π_{CN}), a'(n_N)). Because of the localized character of n_N (~80%) this orbital should give rise to the sharp feature at 13.59 eV. This value corresponds closely to n_N positions in related compounds.[64] On the basis of the HFS calculations, the in-plane π_{CN} is located at 13.5, the out-of-plane π_{CN} at 13.9 eV.

The fairly sharp band at 15.20 eV is assigned to an ionization from an in-plane (a') MO which can roughly be viewed as a fluorine lone pair orbital. Around 17.5 eV a rather broad feature is supposed to be composed from two bands, at 17.4 and 17.6 eV, ascribed to the out-of-plane fluorine lone pair and a strongly S-F bonding MO

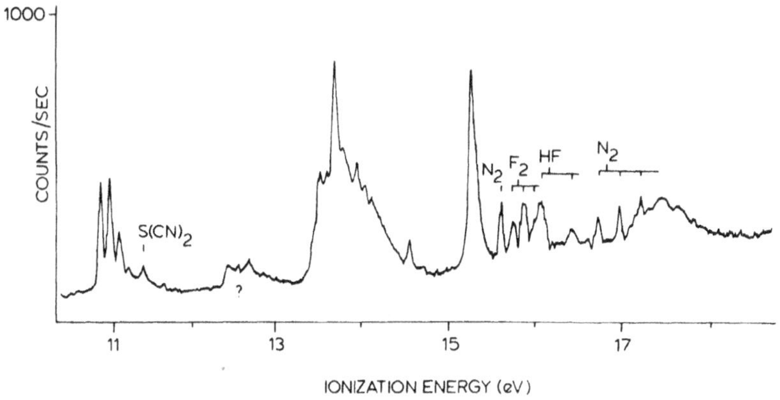

Figure 21. The He(I) photoelectron spectrum of FSCN. The weak structure seen on the 13-14 eV band system is probably due to $(CN)_2$ and FCN impurities.

respectively. Indications for a band corresponding to 4a', an orbital delocalized over the entire molecule, have been found at ~19.4 eV. Results are summarized in Table 12.

(ii) ClSCN

By studying the gas phase equilibrium between $(SCN)_2$ and molecular chlorine PE spectra of the unstable compound ClSCN have been obtained.[64] In the He(I) region eight photoionization bands have been detected which could be observed without appreciable spectral overlap. The spectral assignment has been arrived at using ab initio calculations with an STO-3G minimal basis set internal to the Gaussian 70 programme.[81] Subsequently HFS calculations have been performed for this compound.[89] These results generally confirm the assignment given before, although an interchange of 2a" (π_{CN} out-of-plane) and 7a' (π_{CN} in-plane) is preferred on the basis of the latter calculations. HFS results and assignments given in [63] and [64] are summarized in Table 12.

Further confirmation of the assignments of the PE spectra of both ClSCN and FSCN is given by a correlation diagram (Fig. 22). It is apparent that the positions of the CN-based orbitals (a'(π_{CN}), a"(π_{CN}), a'(n_N)) show little variation in going from one compound to the other. The orbitals which possess halogen character, in particular those associated with the halogen lone pair, undergo much stronger changes.

Table 12. Experimental and HFS calculated IE's and assignments of FSCN and ClSCN (all values in eV).

HFS ion state	FSCN[63] HFS	FSCN[63] EXP.	ClSCN[64,89] HFS	ClSCN[64,89] EXP.
3a"	10.6	10.93 ± 0.02	9.85	10.52
8a'(n_N)	13.5	13.59 ± 0.02	12.02	12.67
7a'(π_{CN})	13.7	13.4 ± 0.1	12.95(π_{CN})	13.05
2a"	13.8	13.9 ± 0.1	12.48	13.28
6a'	14.7	15.20 ± 0.02	13.47(n_N)	13.66
1a"	17.5	17.4 ± 0.1	13.94	14.45
5a'	17.6	17.6 ± 0.1	14.99	15.44
4a'	19.4	19.4 ± 0.1	17.22	17.55

In the HFS calculations on ClSCN d-orbital participation is expected to be small and should have its largest effect for atoms with a net positive charge.[90] Inclusion of a d function on sulphur did not change the overall pattern of the calculated PE spectrum very much, but turned out to be important to properly reproduce the observed splitting between the bands corresponding to 1a" and 5a'.

(iii) ClSeCN

He(I) spectra of the transient species ClSeCN have been obtained by passing molecular chlorine over silver selenocyanate (AgSeCN).[91] The PE spectrum which contains only small contributions from unwanted products, is presented in Fig. 23. In the He(I) region the expected number of eight photoionization bands has been observed and an assignment based on HFS calculations has been given.[91] Both PE spectrum and spectral analysis are similar to the ClSCN case.

The first band at 10.19 eV originates from 3a" which is fairly well described by an out-of-plane combination of the p_z orbitals on the halogen and selenium atoms (~85%). The band shows a progression of 440 ± 50 cm^{-1} ascribed to the Se-Cl stretch vibration. This vibrational frequency appears to be somewhat larger than an estimated value for the same normal mode in the neutral species, thus indicating ionization from an MO with anti-bonding Se-Cl character and confirming the HFS assignment. The second IE is assigned to 8a'

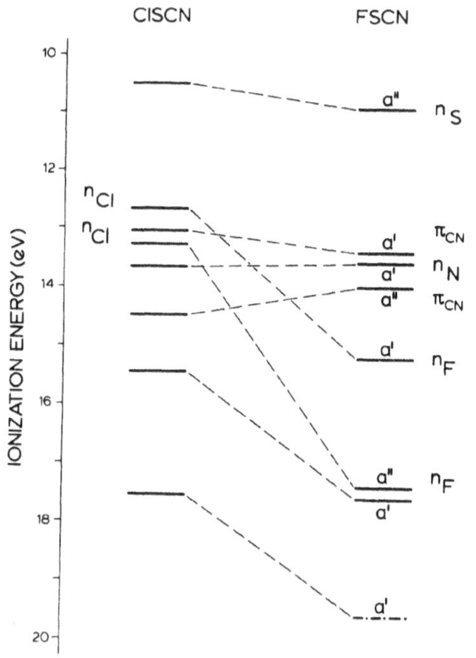

Figure 22. Correlation diagram of the experimental IE's of ClSCN and FSCN, assigned according to [63] and [64].

which can be described approximately as a halogen lone pair orbital (~70%). The third and fourth IE in ClSeCN have merged into one PE feature at ~12.9 eV which is also reproduced by the HFS transition state calculations. The sharp peak at 12.83 eV which is probably accompanied by a CN stretch vibrational component (~1900 cm^{-1}) is assigned to 7a', an in-plane π_{CN} type orbital. The 2a" IE, which can be described approximately as the chlorine out-of-plane lone pair, is then located at 12.9 eV. The sharp fifth IE at 13.39 eV is assigned to the strongly localized non-bonding nitrogen lone pair orbital 6a'.

It should be pointed out that orbital analyses on the neutral molecule show the 6a' and 7a' to be respectively π_{CN} (in-plane) and nitrogen lone pair (n_N). However, in the Slater transition state the character of both orbitals interchanges. This unusual state of affairs is confirmed by the transition state calculations performed with the molecular orbitals of the neutral molecule as a basis set. Since in the transition state relaxation effects are taken into account to a large extent, the transition state charac-

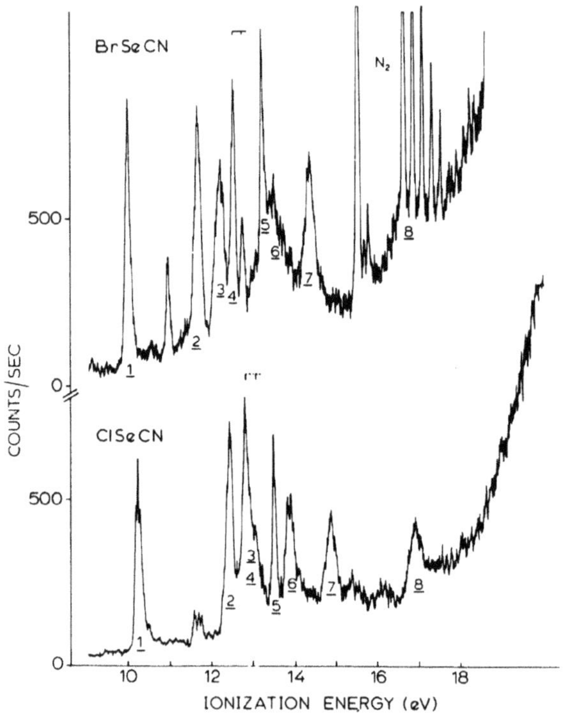

Figure 23. He(I) photoelectron spectra of BrSeCN and ClSeCN. In the BrSeCN spectrum the band at 10.96 eV is due to $Se(CN)_2$, while N_2 bands are also evident. In the ClSeCN spectrum the band at 11.6 eV is due to Cl_2.

ters are to be preferred over orbital characters derived from the HFS neutral ground state calculations.

The sixth IE corresponds to 1a" which can be viewed as the π_{CN} out-of-plane fragment orbital. The seventh IE is assigned according to HFS to 5a', an orbital which is rather delocalized over the entire molecule. Without a 4d polarization function on Se in the basis set the orbitals 1a" and 5a' are calculated to be almost degenerate. Inclusion of a 4d function separates 1a" and 5a' by 0.6 eV compared to an observed splitting of ~1 eV. Finally the IE observed at 16.95 eV corresponds to the 4a' bonding MO. This is the only orbital in which the Se 4s AO participates significantly in the bonding.

(iv) BrSeCN

The He(I) PE spectrum of this novel unstable species has been observed (Fig. 23) and assigned on the basis of HFS calculations.[91] The spectrum is very similar to that of ClSeCN. The third and fourth band, which almost coincide in the ClSeCN spectrum, are well separated in the case of BrSeCN. The first band shows a vibrational progression of 360 ± 50 cm^{-1} which is attributed to the Se-Br stretch. Since the analysis of the BrSeCN spectrum is very similar to that of ClSeCN we shall only mention those features of the assignments which differ in the two compounds. In the calculations on 5a', corresponding to the seventh IE at 14.41 eV in BrSeCN it appears that this orbital is predominantly located on the CN fragment, thus explaining the almost degenerate transition state energies obtained for 1a" and 5a'. Inclusion of a 4d polarization function on Se in the BrSeCN calculations does not lift this calculated degeneracy although experimentally a separation of almost 1 eV is observed.

(v) Related compounds

For a number of compounds belonging to the present class PE information has been obtained but no HFS calculations have been performed. In this context we should mention a study on BrSCN[64] which shows a PE spectrum quite similar to that of ClSCN. Also the measurement of the first band of FSeCN presented in Fig. 24[92] should be reported. Position and structure of the first PE bands in the series SeF_2, FSeCN, $Se(CN)_2$[91] shows the same trend as that described for the related SF_2, FSCN, and $S(CN)_2$. The first vertical IE of SeF_2 occurs at 10.20 eV and shows about 6 vibrational components. The corresponding orbital is calculated to be strongly Se-F antibonding. The first vertical IE of FSeCN occurs at 10.44 ± 0.02 eV and has a progression consisting of about 4 components separated by 800 ± 100 cm^{-1} associated with it. The first IE in $Se(CN)_2$ at 10.96 eV shows a dominant o-o transition, corresponding to an almost nonbonding orbital. Finally we mention an interesting PE study on the molecules OCNCl, OCNBr, and OCNI[66] which have been observed in the "iso" form only.

D. Molecules containing two CN groups

In this section we shall concentrate on molecules which contain a divalent group VI A atom and two cyano groups. When small nonlinear molecules which contain one cyano group are studied, this moiety gives rise to three ionization phenomena, two of π_{CN}, one of n_N type, in the 13-15 eV region. The band corresponding to ionization from the non-bonding nitrogen lone pair orbital is generally sharp. The π_{CN} bands often show vibrational structure arising from the CN stretch. Quantum chemical calculations of various degrees

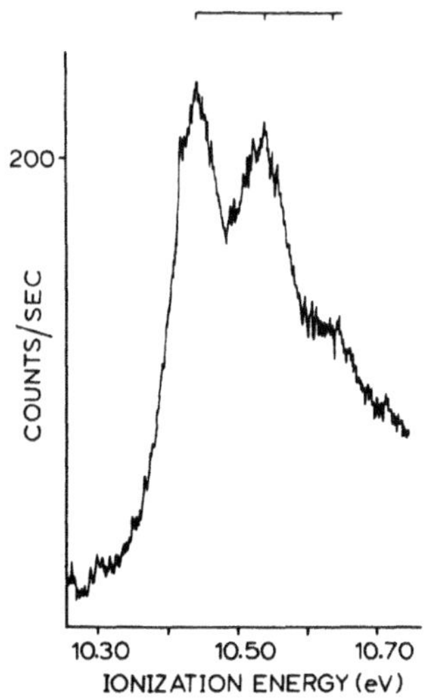

Figure 24. Expanded first band associated with the first IE of FSeCN.

of sophistication are usually utilized in the assignment of the experimental PE spectra, but the close proximity of the three bands arising from the CN group causes problems. For small molecules containing two or more symmetry-equivalent CN groups the situation is more complicated. Experimentally a large number of bands is concentrated in a narrow spectral region. Theoretically both HF-Koopmans and symmetry-restricted HF-ΔSCF calculations often show up with an ordering which deviates from the experimentally preferred one,[93] suggesting that correlation energy contributions to the various CN type ionization energies are different. It is well-known that "symmetry breaking", in particular when strongly localized orbitals are concerned, may influence the calculated IE's and bring them better into line with observed values. In a recent paper[44] it has been shown how for molecules with C_{2v} symmetry which contain two cyano groups, even in the outer valence region breaking the symmetry may have appreciable effects in the case of strongly localized orbitals.

(i) $Se(CN)_2$

The He(I) PE spectrum of the fairly stable $Se(CN)_2$ molecule has been obtained and an assignment of the spectrum, mainly based on an analysis of experimental spectral features and on the correspondence with the spectra of closely related compounds has been published.[91] For a theoretical interpretation of the PE spectra of small molecules the HFS method has been shown to be very reliable for a rather extensive series of compounds. However, application of the HFS scheme to $Se(CN)_2$ assuming a C_{2v} symmetry constraint predicted an order of spectral bands which was at variance with the experimentally preferred assignment. This behaviour was rather unexpected in view of the good agreement between experiment and HFS-calculated spectra within a C_{2v} symmetry constraint for the related compounds $SeCl_2$ and $SeBr_2$.[34]

For HF calculations it has been emphasized by Löwdin[94] that an important part of the correlation error can be due to the symmetry constraint which is usually imposed. Excitation and ionization energies computed without this symmetry constraint often show better agreement with experimental results than the symmetry-restricted ones. These "broken symmetry" solutions, specifically lower the energies of excitations involving orbitals which can be viewed as localized, whereas excitation energies involving delocalized orbitals are hardly affected. The importance of the "broken symmetry" approach has been demonstrated for core holes, for deep valence holes and for electronic excitation spectra. However, for ionizations from outer valence levels "broken symmetry" considerations are scarce.

In the above mentioned paper[44] it is shown how the difficulties encountered in the assignment of the PE spectrum of $Se(CN)_2$ (see Fig. 25) disappear in a "broken symmetry" description. In particular, the positions of the levels associated with the nitrogen lone pairs are effected in the "broken symmetry" scheme. The excellent agreement obtained between the "broken symmetry" results and the experimentally favoured assignment of the outer valence $Se(CN)_2$ PE spectrum as exemplified in Fig. 26, suggests that the considerations in this paper are generally applicable even in the outer valence region whenever localized orbitals play a role. For more details we refer to [44].

(ii) $S(CN)_2$

The He(I) PE spectrum of the related compound $S(CN)_2$ has been recorded before[65,95] and shows strong similarities with that of $Se(CN)_2$. The assignments presented for the $S(CN)_2$ spectrum[65,95] do not agree with our preferred assignment as far as the three π_{CN} (a_1,a_2,b_1) and n_N (a_1,b_1) type MO's are concerned.[91] Our preferred orbital ordering for $S(CN)_2$ would be identical to the one for $Se(CN)_2$. It would be of interest to perform HFS calculations on $S(CN)_2$ and to investigate whether "broken symmetry" effects play a similar role in the outer valence PE spectrum of this compound.

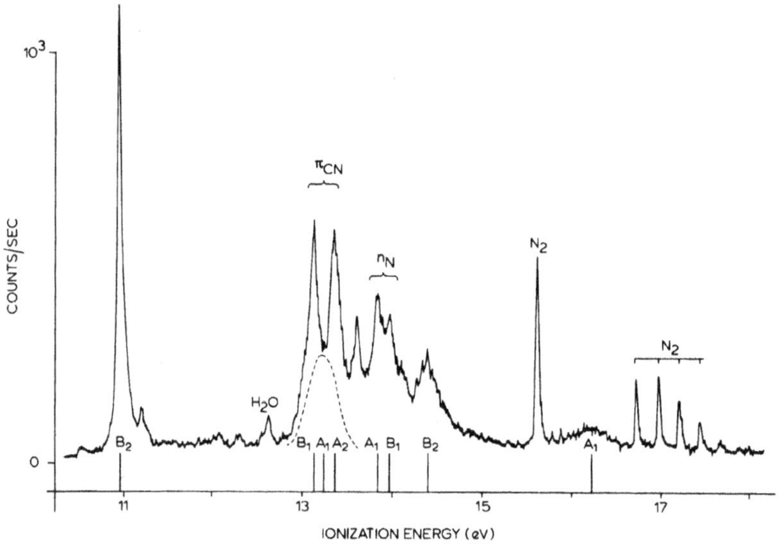

Figure 25. He(I) photoelectron spectrum and experimentally preferred assignment of $Se(CN)_2$ assuming C_{2v} symmetry.

(iii) OCNCN

The He(I) PE spectrum of the compound OCNCN (cyanogen isocyanate) has been recorded and the molecule has been found to occur in the "iso" form.[67] In view of the different structure no direct comparison with other compounds treated in this section can be made. The PE spectrum of OCNCN has been assigned on the basis of experimental considerations and semi-empirical calculations.[67] Subsequently this assignment has been confirmed by HFS calculations.[96]

5. GROUP IV A DIHALIDES

As shown in the previous section, for molecules which contain a divalent group VI A atom an abundance of PE data is available. In contrast, for compounds with a divalent group IV A atom the PE information obtained until now is rather limited. Apart from applications to a number of chalcogenides,[97-102] PE spectroscopy has been used to study the following group IV A compounds: CF_2 [103], SiF_2,[104,105] SnX_2 and PbX_2 (X=Cl,Br),[106-108] PbI_2,[107], PbX_2 (X=Cl,I),[109] and recently GeX_2 (X=F,Cl,Br,I).[110-112] All these molecules belong to the class of the dihalides. The Pb and Sn based compounds are rather stable and their PE spectra can there-

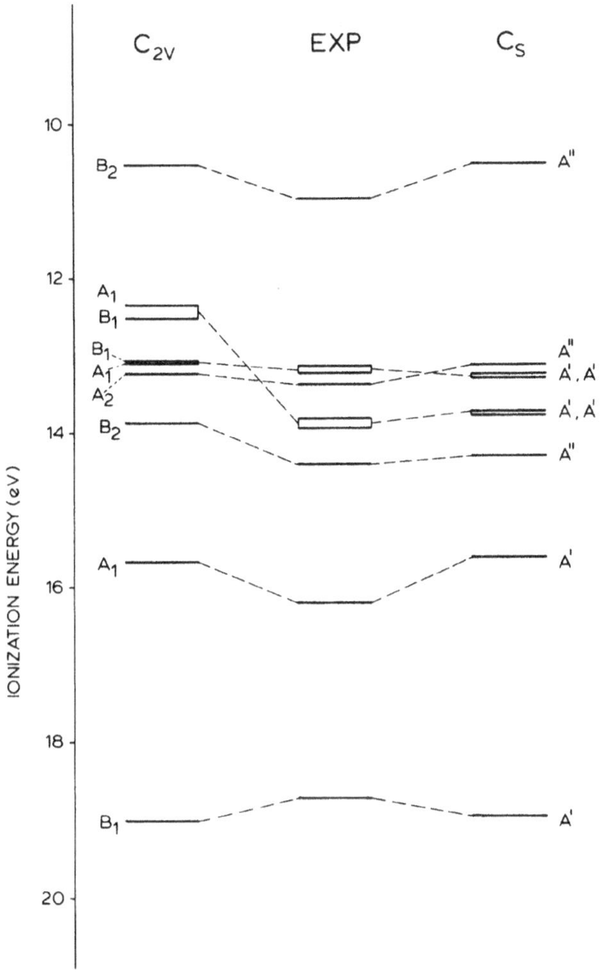

Figure 26. Comparison of experimental ionisation energies and results of HFS calculations with C_{2v} and C_s symmetry constraints for $Se(CN)_2$.

fore be measured without too many complications. However, the Ge, Si, and C dihalides are rather elusive species which tend to be short-lived under room temperature conditions in the gas phase. In order to study the latter molecules by means of the PE technique, careful spectrometer design and sensible experimental procedures are again required. In this section we shall concentrate on those compounds for which HFS calculations have been performed. We shall therefore limit our discussion to the difluorides of carbon[103] and silicon[104,105] and to the germanium dihalides.[110-112]

In the dihalides of the group VI A elements the valence s AO's in both the group VI A atoms and the halogen atoms lie much deeper than the corresponding valence p levels (see Table 10). Therefore the chemical bonding in these compounds can be understood in simple terms by taking only MO's constructed from valence p AO's into consideration. This very characteristic feature is at the root of all the PE results of the group VI A compounds described before and greatly simplifies the interpretation of their spectra. In the group IV A elements the situation is quite different. The valence s and p AO's are generally fairly close in energy (see Table 13) and hence participation of the valence s group IV A AO's in the chemical bonding can by no means be neglected.

As an illustration the valence MO diagram of $GeCl_2$ (situated in the xz plane) is considered. To a first approximation only valence p electrons are taken into account. Each halogen atom contributes two "lone pairs" and one unpaired electron, while the germanium atom contributes two unpaired electrons. The germanium out-of-plane $4p_y$ orbital remains empty. On the basis of this picture these twelve electrons give rise to six bands in the He(I) region of the PE spectrum. In C_{2v} symmetry the halogen valence p orbitals form tangential, radial, and out-of-plane combinations which transform as a_1, a_2, b_1, and b_2. The two germanium unpaired electrons combine with those on the halogen atoms to form two bonding MO's of a_1 and b_1 symmetry. Hence we expect four close-lying halogen lone pair features (a_1, a_2, b_1, and b_2), followed at higher IE's by two bonding MO's (a_1 and b_1). However, this simple MO scheme has to be modified, because the germanium 4s orbital which is energetically close to the halogen valence p-orbitals can participate in the interactions. As the germanium 4s transforms as a_1, the two above-mentioned a_1 type MO's are expected to undergo conjugative shifts. Orbitals of other symmetries should remain unchanged. In summary seven ionization bands are expected in the He(I) region, with the highest IE mainly germanium 4s character.

Fig. 27 gives a pictorial representation of the orbital interaction scheme and the MO ordering in $GeCl_2$. The pattern can be understood in terms of interactions between germanium 4s and 4p and chlorine 3p AO's. The scheme is based on the results of HFS calculations (solid lines), while the (dotted) 4s, 4p germanium, 3p chlorine and unoccupied $GeCl_2$ levels are estimates in the final molecular field. This figure is almost self-explanatory when one realizes that interaction lines are only drawn when the AO contribution to the resulting MO exceeds 15%.

From these considerations the expected PE spectrum should show a first band ($4a_1$) followed by two fairly close-lying bands with strong halogen lone pair character ($3b_1$, $1a_2$) and next a

Table 13. Group IV A free atom valence subshell binding energies,[68] values in eV

Atom	n	ns	np
C	2	16.59	11.26
Si	3	13.46	8.51
Ge	4	14.3	7.9
Sn	5	12	7.34
Pb	6	10	7.42

number of bands corresponding to orbitals with slight ($1b_2$, $3a_1$) and progressively stronger bonding character ($2b_1$, $2a_1$). Valence orbital numbering is used throughout. The character of the MO's in $GeCl_2$ is shown schematically in Fig. 27.

In the following we shall treat the various compounds separately.

(i) CF_2

The He(I) and He(II) spectra of the unstable compound CF_2 have been observed by means of a microwave discharge in gaseous tetrafluoroethylene. An assignment of the spectrum has been presented on the basis of semi-empirical and ab initio calculations.[103] The first band corresponds to $4a_1$ and shows a very extensive progression of 650 ± 40 cm^{-1}, corresponding to the symmetric bending mode in the ion and indicative of the strong C-F anti-bonding character of this orbital. The remaining bands lie below 16 eV. There is only approximate agreement between calculated and observed band positions.[103]

Subsequently double-zeta quality HFS calculations have been carried out for this compound,[111] the results being in reasonable agreement with the observed PE spectrum. However, the discrepancies between calculated and experimental IE's are larger for this compound than is normally experienced. The underlying reasons are not known in detail, although HFS calculations on fluorine containing compounds have a tendency to reproduce experimental IE's less well than usual. The HFS results on CF_2 are summarized in Table 14.

(ii) SiF_2

The PE spectrum of the unstable compound SiF_2 has been ob-

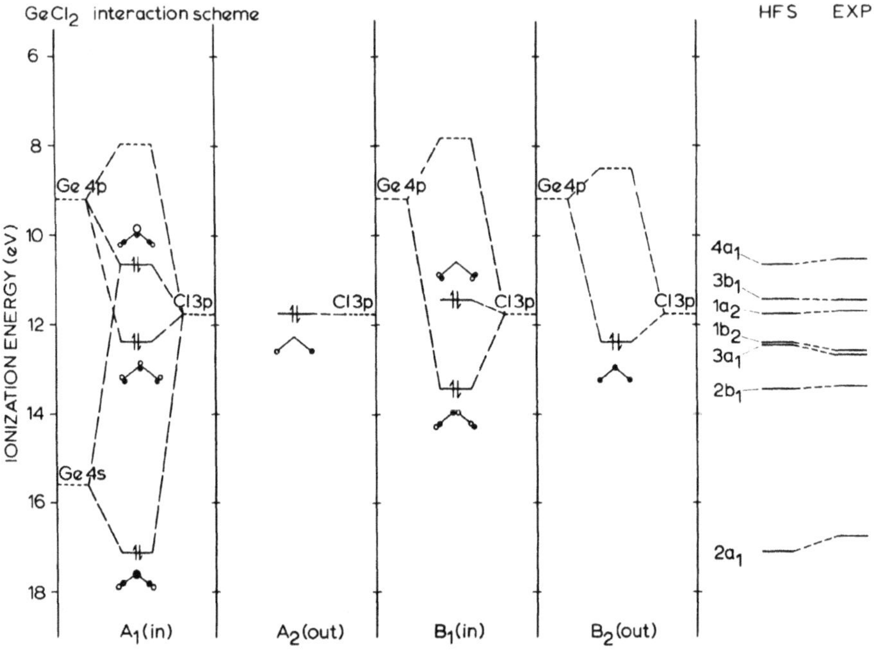

Figure 27. Experimental IE's and orbital interaction scheme based on the results of HFS LCAO calculations (shifted by 0.7 eV) for $GeCl_2$.

served by flowing gaseous SiF_4 over pure silicon at high temperature[104,105] or by heating a mixture of CaF_2 and silicon.[105] A spectral assignment has been presented using results from semi-empirical calculations. Again an isolated first band corresponding to $4a_1$ is found, followed by a number of rather close-lying features with vertical IE's between 15.5 and 17.8 eV. A careful study of the first band has suggested a progression of 350 ± 100 cm^{-1} arising from the symmetric bending mode in the ionic state. The corresponding orbital is predominantly located on silicon and possesses both silicon 3s and 3p character.[105]

Later double-zeta quality HFS calculations have been performed for this compound.[111] The calculated IE's show fair agreement with the experimental ones and confirm the somewhat tentative previous assignment. A summary of the SiF_2 results is given in Table 14.

Table 14. Experimental IE's, results of HFS LCAO calculations for CF_2, SiF_2 (double zeta) and GeF_2 (triple zeta) and spectral assignments (all values in eV). In addition experimental and calculated electric dipole moments are given.

MO	CF_2 HFS(DZ)	CF_2 EXP.	SiF_2 HFS(DZ)	SiF_2 EXP.	GeF_2 HFS(TZ)	GeF_2 EXP.
$4a_1$	12.60	12.27	10.84	11.18	11.30	11.98
$3b_1$	17.20	16.4	14.60	15.5	13.54	~14.4
$1a_2$	17.81	17.4	15.30	15.8	13.87	~14.4
$3a_1$	19.03	19.2	15.60	16.8	14.46	15.55
$1b_2$	19.68	20.8	15.99	17.06	14.44	~15.9
$2b_1$	20.54	22.2	16.44	17.8	15.05	~16.2
$2a_1$	23.20	24.0	18.30	-	17.62	18.70
dip.mom.(D)	0.39	0.46	1.71	1.28	2.75	2.61

(See Ref. 111 for more details).

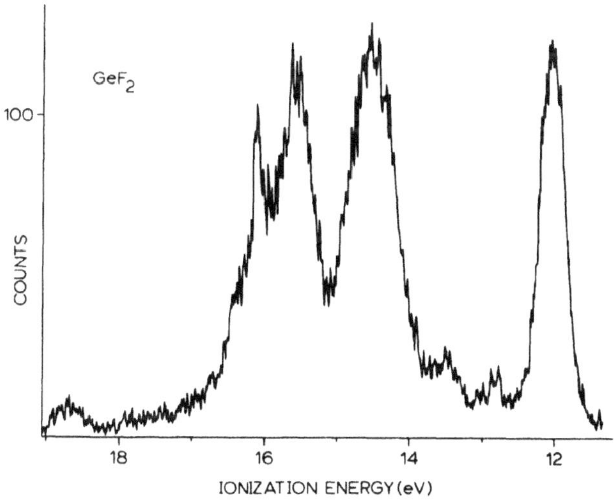

Figure 28. He(I) photoelectron spectrum of GeF_2. The rather sharp feature at 16.05 eV is due to HF.

(iii) GeF_2

The PE spectrum of the unstable compound GeF_2 has recently been obtained from the solid state reaction between GeS and PbF_2 at ~300°C,[111] and has been assigned using triple-zeta quality HFS calculations. The spectrum is presented in Fig. 28. It shows a first band at 11.98 eV, a feature suggestive of two photoionization phenomena at ~14.5 eV, a structure of probably three overlapping bands between 15.5 and 16.1 eV and finally a weak band at 18.70 eV. The calculated and experimental IE's and the preferred assignment which is tentative as far as the close-lying bands are concerned, are presented in Table 14.

The calculations show that in many ways the chemical bonding in CF_2, SiF_2, and GeF_2 is similar. In each PE spectrum we can distinguish a fluorine lone pair region defined by $3b_1$ (100%), $1a_2$ (100%), and $3a_1$ (90%) ionization phenomena. While for GeF_2 and SiF_2 the $1b_2$ and $2b_1$ IE's are fairly close to this lone pair region, in CF_2 these two IE's are much more stabilized due to strong bonding interaction of the carbon 2p AO's with the fluorine 2p AO symmetry combinations. The location of the $4a_1$ and $2a_1$ IE's is, roughly speaking, determined by the interaction between the central atom valence s AO and the radial fluorine 2p combination of a_1 symmetry. This might explain the unusual trend observed in the positions of the first band ($4a_1$) in CF_2, SiF_2, and GeF_2 which is parallelled by the valence s orbital IE's of the free central atoms (see Table 13).

Moreover, the calculations show the trend towards ionic bonding character throughout this series of molecules, which is illustrated by the charges on the central atom, i.e. +0.5 e (C), +1.0 e (Si), and +1.1 e (Ge). Also the contribution of the valence p AO of the central atom to the $4a_1$ MO is increased in going from germanium/silicon to carbon, in line with the well-known tendency of the carbon atom to form s-p hybrids.

(iv) $GeCl_2$

The He(I) PE spectrum of $GeCl_2$ (see Fig. 29) has recently been measured by carrying out the solid state reaction between GeS and $PbCl_2$ in the spectrometer.[110]

The first IE is assigned to $4a_1$ which corresponds to a radial combination of chlorine p AO's (60%) destabilized through an antibonding interaction with a germanium sp hybrid orbital.

The second band system is assigned to two ionizations, from $3b_1$ and $1a_2$, which are both completely localized on the chlorine atoms. The $3b_1$ MO, constructed from tangential chlorine p AO's,

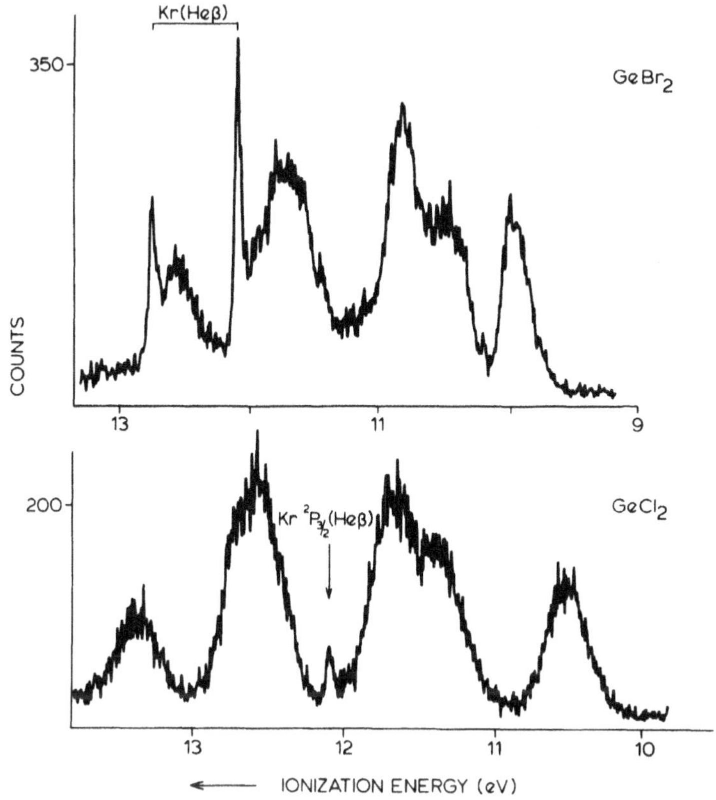

Figure 29. Part of the He(I) photoelectron spectra of $GeCl_2$ and $GeBr_2$.

has a slightly anti-bonding chlorine-chlorine interaction which is probably responsible for the slight destabilization relative to $1a_2$ which is constructed from two hardly interacting out-of-plane chlorine p AO's.

In the third band system intensity considerations indicate two closely spaced ionization features. One band arises from the chlorine out-of-plane p combination with b_2 symmetry which is stabilized through interaction with the unoccupied germanium $4p_y$ orbital (~20%). The second band arises from a bonding $3a_1$ MO described as an in-phase combination of germanium $4p_z$ (~15%) with a tangential chlorine p AO combination.

The IE at 13.41 eV is assigned to the $2b_1$ bonding MO, de-

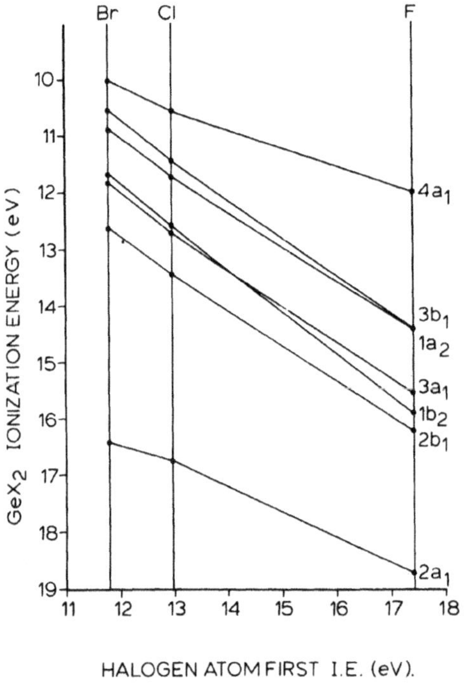

Figure 30. Correlation diagram of the experimental valence IE's of GeF_2, $GeCl_2$, and $GeBr_2$ versus the first atomic IE's for the halogen atoms.

scribed by an in-phase combination of germanium $4p_x$ (~20%) and radial chlorine p contributions. Finally the IE at 16.7 eV is ascribed to $2a_1$ which is mainly germanium 4s in character but stabilized through interaction with the radial chlorine p orbital a_1 symmetry combination (~30).

HFS calculations of triple-zeta quality have been performed for $GeCl_2$. The calculated IE's agree very well with the experimental ones apart from an overall shift of 0.7 eV (see Fig. 27).

(v) $GeBr_2$

The He(I) PE spectrum of $GeBr_2$ (see Fig. 29) has recently been obtained in the solid state reaction between GeS and $PbBr_2$.[110] The spectrum is very similar to that of $GeCl_2$ and the spectral interpretation is almost analogous to that given above for $GeCl_2$. Hence we shall refrain from a detailed description.

HFS calculations of triple-zeta quality have been performed for $GeBr_2$. The calculated IE's agree again very well with the experimental band positions when an overall shift of 0.7 eV is applied. Relativistic effects are found to be unimportant in the He(I) spectrum of this compound.[110]

When the experimental PE band positions for GeF_2, $GeCl_2$, and $GeBr_2$ are plotted against the first IE's of fluorine, chlorine, and bromine atoms (17.42, 12.97, and 11.81 eV respectively) approximately linear correlations are obtained, thus providing confirmation for our assignments (fig. 30). The relatively gentle slopes of the $2a_1$ and $4a_1$ correlation lines are due to the significant germanium 4s participation in these orbitals, the steeper slopes of the remaining lines indicate predominant halogen character.

(vi) GeI_2

The solid state reaction between GeS and PbI_2 generates GeI_2 in sufficient concentration in the gas phase to allow detection of its He(I) PE spectrum.[112] Good spectra have been obtained and an analysis on the basis of HFS calculations including relativistic corrections which are required for this compound, is in progress.[112]

REFERENCES

1. P. Hohenberg, W. Kohn, Phys.Rev. 136 (1964) B864.
2. K.H. Johnson, J.Chem.Phys. 45 (1966) 3085.
3. J.C. Slater, Adv. Quantum Chem. 6 (1972) 1.
4. K.H. Johnson, Adv. Quantum Chem. 7 (1973) 143.
5. E.J. Baerends, D.E. Ellis, P. Ros, Chem.Phys. 2 (1973) 41.
6. A. Rosén, D.E. Ellis, H. Adachi, F.W. Averill, J.Chem.Phys. 65 (1976) 3629.
7. B. Delley, D.E. Ellis, J.Chem.Phys. 76 (1982) 1949.
8. G.L. Gutsev, A.A. Levin, Chem.Phys. 51 (1980) 459.
9. H. Sambe, R.H. Felton, J.Chem.Phys. 62 (1975) 1122.
10. B.I. Dunlap, J.W.D. Connolly, J.R. Sabin, J.Chem.Phys. 71 (1979) 3396
11. O. Gunnarsson, J. Harris, R.O. Jones, Phys.Rev. B 15 (1977) 3027.
12. F. Herman, S. Skillman, Atomic Structure Calculations, Prentice-Hall, Englewood Cliffs, N.J. (1963).
13. J.G. Snijders, P. Vernooijs, E.J. Baerends, Atomic Nucl. Data Tables, 26 (1981) 483.
14. D.E. Ellis, Int.J. Quantum Chem. 2 S (1968) 35.
15. D.E. Ellis, G.S. Painter, Phys.Rev. B 1 (1970) 4747.
16. S.F. Boys, Proc.Roy.Soc. A 309 (1969) 195.

17. E.J. Baerends, P. Ros, Chem.Phys. 2 (1973) 52.
18. E.J. Baerends, P. Ros, Int.J. Quantum Chem. 12 S (1978) 169.
19. O. Gunnarsson, J. Harris, R.O. Jones, J.Chem.Phys. 67 (1977) 3970.
20. O. Gunnarsson, B.I. Lundquist, Phys.Rev. B 13 (1976) 4274.
21. U. von Barth, L. Hedin, J.Phys. C 5 (1972) 1629.
22. P.S. Bagus, T.L. Gilbert, C.C.J. Roothaan, J.Chem.Phys. 56 (1972) 5195.
23. B.I. Dunlap, J.W.D. Connolly, J.R. Sabin, J.Chem.Phys. 71 (1979) 4993.
24. W. Heijser, A.Th. v. Kessel, E.J. Baerends, Chem.Phys. 16 (1976) 371.
25. D. Post, E.J. Baerends, Surf. Science 109 (1981) 167.
26. R.O. Jones, J.Chem.Phys. 76 (1982) 3098.
27. J.G. Snijders, E.J. Baerends, Mol.Phys. 36 (1978) 1789.
28. J.G. Snijders, E.J. Baerends, P. Ros, Mol.Phys. 38 (1979) 1909.
29. R.E. Moss, Advanced Molecular Quantum Mechanics (Chapman and Hall, London, 1973).
30. E.J. Baerends, P. Ros, Chem.Phys. 8 (1975) 412.
31. D.M. de Leeuw, R. Mooyman, C.A. de Lange, Chem.Phys. 63 (1979) 57.
32. D.M. de Leeuw, R. Mooyman, C.A. de Lange, Chem.Phys. 34 (1978) 287.
33. D.M. de Leeuw, R. Mooyman, C.A. de Lange, Chem.Phys.Lett. 61 (1979) 191.
34. D.M. de Leeuw, R. Mooyman, C.A. de Lange, Chem.Phys. 38 (1979) 21.
35. G. de Alti, P. Decleva, A. Lisini, Chem.Phys. 66 (1982) 425.
36. W. von Niessen, L.S. Cederbaum, J.Chem.Phys. 65 (1976) 1378.
37. W. von Niessen, G.H.F. Diercksen, L.S. Cederbaum, Chem.Phys. 10 (1975) 345.
38. A.W. Potts, W.C. Price, D.G. Streets, T.A. Williams, Faraday Discussion 54 (1972) 168.
39. F. Lindholm, Faraday Discussion 54 (1972) 200.
40. W. Oosterbeek, E.J. Baerends, unpublished results.
41. N. Rösch, W.G. Klemperer, K.H. Johnson, Chem.Phys.Lett. 23 (1973) 149.
42. U. Gelius, J. Electron Spectrosc. Rela.Phenom. 5 (1974) 985.
43. K. Higasi, I. Omura, H. Baba, J.Chem.Phys. 24 (1956) 623.
44. G. Jonkers, C.A. de Lange, L. Noodleman, E.J. Baerends, Mol. Phys. (1982), in press.
45. E.J. Baerends, P. Ros, Chem.Phys.Lett. 23 (1973) 391.
46. E.J. Baerends, P. Ros, Mol.Phys. 30 (1975) 1735.
47. E.J. Baerends, Ch. Oudshoorn, A. Oskam, J.Electr.Spectrosc. Relat.Phenom. 6 (1975) 259.
48. T. Ziegler, A. Rauk, E.J. Baerends, Chem.Phys. 16 (1976) 209.
49. C. Famiglietti, E.J. Baerends, Chem.Phys. 62 (1981) 407.
50. S. Evans, M.L.H. Green, B. Jewitt, G.H. King, A.F. Orchard, J.C.S. Faraday II, 70 (1974) 356.

51. C. Cauletti, J.C. Green, M. Kelly, P. Powell, J. von Tilborg, J. Electron Spectrosc.Relat.Phenom. 19 (1980) 327.
52. M.M. Rohmer, A. Veillard, Chem.Phys. 11 (1975) 349.
53. D.N. Hendrickson, Y.S. Sohn, D.M. Duggan, H.B. Gray, J.Chem. Phys. 58 (1973) 4666.
54. N. Rösch, K.H. Johnson, Chem.Phys.Lett. 24 (1974) 179.
55. T. Ziegler, A. Rauk, E.J. Baerends, Theoret.Chim.Acta 43 (1977) 261.
56. J. Weber, A. Goursot, E. Pénigault, J.H. Ammeter, J. Bachmann, J.Am.Chem.Soc. 104 (1982) 1491.
57. J.C. Slater, The Self Consistent Field for Molecules and Solids, Vol. 4, (McGraw-Hill, New York, 1974).
58. G. Jonkers, C.A. de Lange, J.G. Snijders, Chem.Phys. (1982) in press.
59. G. Jonkers, S.M. van der Kerk, R. Mooyman, C.A. de Lange, J.G. Snijders, Chem.Phys. (1982) in press.
60. T.G. Strand, Acta Chem.Scand. 21 (1967) 2111.
61. R.G. Egdell, M. Hotokka, L. Laaksonen, P. Pyykkö, J.G. Snijders, Chem.Phys. (1982) submitted for publication.
62. T.W. Lee, J.W. Rabalais, J.Chem.Phys. 60 (1974) 1172.
63. G. Jonkers, O. Grabandt, R. Mooyman, C.A. de Lange, J.Electron Spectrosc.Relat.Phenom. 26 (1982) 147.
64. D.C. Frost, C.B. MacDonald, C.A. McDowell, N.P.C. Westwood, J.Am.Chem.Soc. 103 (1981) 4423.
65. P. Rosmus, H. Stafast, H. Bock, Chem.Phys.Lett. 34 (1975) 275.
66. D.C. Frost, C.B. MacDonald, C.A.McDowell, N.P.C. Westwood, Chem.Phys. 47 (1980) 111.
67. D.C. Frost, H.W. Kroto, C.A. McDowell, N.P.C. Westwood, J. Electron Spectrosc.Relat.Phenom. 11 (1977) 147.
68. Data are taken from: T.A. Carlson, Photoelectron and Auger Spectroscopy (Plenum, New York, 1976) p. 337. Original sources are quoted therein.
69. A.B. Cornford, D.C. Frost, F.G. Herring, C.A. McDowell, J. Chem.Phys. 55 (1971) 2820.
70. C.R. Brundle, M.B. Robin, N.A. Kuebler, H. Basch, J.Am.Chem. Soc. 94 (1972) 1451.
71. D.P. Chong, F.G. Herring, D. McWilliams, J.Chem.Phys. 61 (1974) 78; J.Chem.Phys. 61 (1974) 958.
72. D.M. de Leeuw, Ph.D.thesis, Free University, Amsterdam, 1979.
73. W. von Niessen, J. Electron Spectrosc.Relat.Phenom. 17 (1979) 197.
74. D. Chadwick, A.B. Cornford, D.C. Frost, F.G. Herring, A. Katrib, C.A. McDowell, R.A.N. McLean, in Electron Spectroscopy, edited by D.A. Shirley (North-Holland, Amsterdam, 1972) p. 453-469.
75. D.M. de Leeuw, C.A. de Lange, Chem.Phys. 61 (1981) 109.
76. K.E. Valenta, K. Vasudevan, F. Grein, J.Chem.Phys. 72 (1980) 2148.
77. C. Thomson, Chem.Phys.Lett. 44 (1976) 475.
78. E. Nagy-Felsobuki, J.B. Peel, Chem.Phys. 45 (1980) 189.

79. E. Nagy-Felsobuki, J.B. Peel, Australian J.Chem. 31 (1978) 2571.
80. E. Nagy-Felsobuki, J.B. Peel, J.Chem.Soc. Faraday Trans. II, 74 (1978) 2204.
81. W.J. Hehre, W.A. Lathan, R. Ditchfield, M.D. Newton, J.A. Pople, QCPE 11 (1973) 236.
82. W.J. Hehre, R.F. Stewart, J.A. Pople, J.Chem.Phys. 51 (1969) 2657.
83. E. Nagy-Felsobuki, J.B. Peel, J.Chem.Soc. Faraday Trans. II, 76 (1980) 148.
84. R.J. Colton, J.W. Rabalais, J. Electron Spectrosc. Relat. Phenom. 3 (1974) 345.
85. B. Solouki, P. Rosmus, H. Bock, Chem.Phys.Lett. 26 (1974) 20.
86. G. Jonkers, C.A. de Lange, J.G. Snijders, Chem.Phys. 50 (1980) 11.
87. G. Jonkers, W. v.d. Meer, C.A. de Lange, J.G. Snijders, unpublished results.
88. A. Haas, H. Willner, Spectrochim. Acta 34A (1978) 541.
89. G. Jonkers, C.A. de Lange, unpublished results.
90. C.A. Coulson, Nature 22 (1969) 1106.
91. G. Jonkers, R. Mooyman, C.A. de Lange, Mol.Phys. 43 (1981) 655.
92. G. Jonkers, R. Mooyman, C.A. de Lange, unpublished results.
93. S. Bell, Chem.Phys.Lett. 67 (1979) 498.
94. P.O. Löwdin, Adv.Chem.Phys. 14 (1969) 283.
95. H. Stafast, thesis, University of Frankfurt (1974).
96. G. Jonkers, R. Mooyman, C.A. de Lange, Chem.Phys. 57 (1981) 97.
97. D.W. Turner, C. Baker, A.D. Baker, C.R. Brundle, Molecular photoelectron spectroscopy (Wiley Interscience, New York, 1970) p. 49.
98. N. Jonathan, A. Morris, M. Okuda, K.J. Ross, D.J. Smith, Faraday Disc.Chem.Soc. 54 (1972) 48.
99. E.A. Colbourn, J.M. Dyke, E.P.F. Lee, A. Morris, I.R. Trickle, Mol.Phys. 35 (1977) 873.
100. E.A. Colbourn, J.M. Dyke, A. Fackerell, A. Morris, I.R. Trickle, J.Chem.Soc. Faraday Trans. II, 73 (1978) 2278.
101. M. Wu, T.P. Fehlner, J.Am.Chem.Soc. 98 (1976) 7578.
102. M.G. White, R.A. Rosenberg, S.-T. Lee, D.A. Shirley, J.Electron Spectrosc.Relat.Phenom. 17 (1979) 323.
103. J.M. Dyke, L. Golob, N. Jonathan, A. Morris, M. Okuda, J.Chem. Soc. Faraday Trans. II, 70 (1974) 1828.
104. T.P. Fehlner, D.W. Turner, Inorg.Chem. 13 (1974) 754.
105. N.P.C. Westwood, Chem.Phys.Lett. 25 (1974) 558.
106. S. Evans, A.F. Orchard, J.Electron Spectrosc.Relat.Phenom. 6 (1975) 207.
107. A.W. Potts, W.C. Price, Physica Scripta 16 (1977) 191.
108. A.W. Potts, M.L. Lyus, J.Electron Spectrosc.Relat.Phenom. 13 (1978) 327.
109. J. Berkowitz, in: Electron spectroscopy, ed. D.A. Shirley, (North-Holland Publishing Company, Amsterdam, 1972), p. 391-9.

110. G. Jonkers, S.M. van der Kerk, C.A. de Lange, Chem.Phys. (1982) in press.
111. G. Jonkers, S.M. van der Kerk, R. Mooyman, C.A. de Lange, submitted for publication.
112. G. Jonkers, S.M. van der Kerk, R. Mooyman, C.A. de Lange, J.G. Snijders, manuscript in preparation.

SCF-Xα-SW CLUSTER MOLECULAR-ORBITAL MODELS OF CATALYSTS, ALLOYS, MAGNETIC MATERIALS, AND SUPERCONDUCTORS

K.H. Johnson

Department of Materials Science and Engineering
Massachusetts Institute of Technology
Cambridge, Massachusetts 02139, U.S.A.

Recent self-consistent-field Xalpha scattered-wave (SCF-Xα-SW) molecular-orbital studies of the electronic structures, chemical bonding, and related (e.g. catalytic, magnetic, and superconducting) properties of metal and alloy clusters are reviewed, including results for: (1) Li clusters, (2) Cu, Ni, Pd, and Pt clusters, (3) relativistic contributions to cluster electronic structure, (4) the interaction of hydrogen with Ni, Pd, and Pt clusters, (5) metal-support ($Pt-SiO_2$ and $Ru-SiO_2$) interactions, (6) magnetism of Fe clusters, (7) cluster models of transition-metal impurities and their magnetic properties in copper and aluminum hosts, (8) cluster models of intermetallic compounds and solid-solution alloys, (9) cluster models of amorphous alloys, and (10) cluster molecular-orbital models of superconductors.

1. INTRODUCTION

The electronic structures and related properties of small transition- and noble-metal clusters are of considerable current interest and importance. For example, the active centers of commercial heterogeneous catalysts often consist of small metallic, bimetallic, or multimetallic clusters, typically based on Group-VIII and IB elements, supported on a refractory material such as silica or alumina.[1-3] The electronic structures of small metallic and bimetallic clusters are also intrinsically interesting to the extent they mimic the bulk metals, and in their utility as models for alloy electronic structure.

In comparison with the large number of band-structure calculations for crystalline metals and alloys,[4] relatively little fundamental work has been directed to the electronic structures of small metallic clusters until recently. Band theory, in its conventional form, is based on the assumptions of long-range crystalline order. Bloch's theorem, and reciprocal- or $\vec{k}$-space representation, which do not apply to small clusters, where there is at most only short-range order.

Molecular-orbital theory, on the other hand, is well suited for describing metal-cluster electronic structure, although applications to transition-metal clusters were originally limited largely to semiempirical molecular-orbital methods.[5] With the development of the self-consistent-field-X-alpha scattered-wave (SCF-Xα-SW) approach to molecular-orbital theory,[6] based on the combined use of Slater's Xα density-functional approximation to exchange and correlation[7] and the multiple-scattered-wave method of solving the one-electron Schrödinger equation,[8] it has become practical to calculate the electronic structures of complex polyatomic molecules and clusters from first principles, using only moderate amounts of computer time.[9] The principal objectives of such studies are:

1. the systematic calculation of the electronic structures of metal and alloy clusters by the SCF-Xα-SW method, with emphasis on clusters containing transition and noble elements;

2. the use of the calculated electronic structures as a basis for understanding the electronic, magnetic, superconducting, and catalytic properties of small clusters and larger particles including, where necessary, the effects of supporting environments;

3. the use of the cluster results as a basis for describing local electronic, magnetic, superconducting, and other physical properties of bulk crystalline transition metals and their alloys.

2. APPLICATIONS

A. Li Clusters

Li clusters containing up to 13 atoms were the first metallic aggregates investigated by the SCF-Xα-SW method, including effects of spin polarization.[10] Results of these studies that are of general significance in regard to the physical properties of small metal clusters are: (1) there is a gradual increase of cohesive energy per atom with increasing cluster size and coordination number, a 13-atom cluster yielding approximately 60% of the cohesive energy per atom characteristic of the bulk solid; (2) there is a gradual decrease of first ionization potential (as calculated by the SCF-Xα "transition-state" procedure[6,7]) with increasing cluster size, approaching the bulk work function; (3) a cluster with icosahedral geometry (see Fig. 1(b)) is the most energetically stable of all the clusters considered; (4) where the highest occupied cluster molecular orbital is degenerate, a Jahn-Teller distortion can remove the degeneracy, lower the cluster symmetry, and lower the energy of the cluster.

B. Cu, Ni, Pd, and Pt Clusters

Following the work on Li clusters, SCF-Xα-SW studies of Cu, Ni, Pd, and Pt clusters containing up to 13 atoms were carried out and compared with the results of similar calculations by the extended-Hückel and CNDO methods. Various phases of this work are presented in Refs. 11-13. One is at first struck by the systematic similarities, rather than the differences, between the SCF-Xα electronic structures of the clusters and those of the corresponding bulk metals. For example, the manifold of orbital energy eigenvalues for 13-atom clusters having the cubo-octahedral nearest-neighbor coordination (see Fig. 1(a)) characteristic of the fcc lattice exhibit all the principal features of the bulk band structures, e.g., overlap of the "d band" by the "s,p band", a sharp peak in the density of states around the Fermi energy for the Ni_{13}, Pd_{13}, and Pt_{13} clusters, increasing band width through this series, and magnetic spin polarization in the case of Ni_{13} (see Fig. 2). These results have more recently been corroborated by SCF-Xα-LCAO calculations on small nickel clusters,[14] and the inadequacy of symmetry-restricted Hartree-Fock theory in determining the electronic structures of such clusters has been demonstrated.[15] The theoretical study of a finite metallic cluster of reasonably high symmetry, such as the 13-atom cubo-octahedron, has the further advantage that, by virtue of the component molecular-orbital symmetries and charge distributions, its electronic structure can be partitioned into eigenstates which are delocalized throughout the cluster

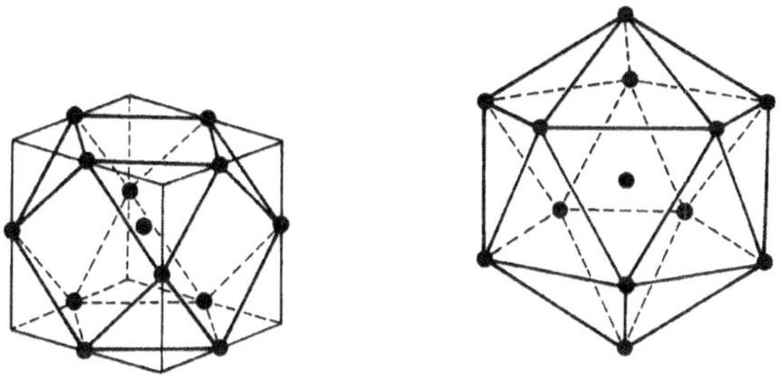

(a) (b)

Figure 1. Cubo-octahedral cluster containing 13 atoms.

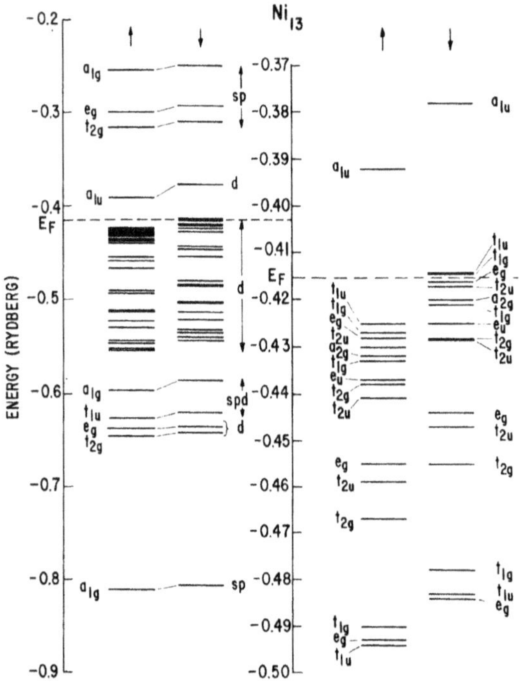

Figure 2. Spin-polarized SCF-Xα-SW electronic energy levels for a cubo-octahedral Ni_{13} cluster with near-crystalline nickel. Levels are labeled according to the irreducible representations of the O_h symmetry group. "Fermi level" E_F separates the occupied levels from the unoccupied ones.

(the "bulk-like" states) and eigenstates that are primarily localized on atoms at the cluster boundary (the "surface-like" states), as shown in Fig. 3. The electronic structures of such clusters can therefore be used as analogues of both the bulk and surface electronic structures of the corresponding crystalline metals, as well as models for small metallic and bimetallic clusters of catalytic importance.

C. Relativistic Effects

In order to determine the effects of special relativity on the electronic structures of small transition-metal clusters, we have applied the relativistic version of the SCF-Xα-SW method, developed by Yang and Rabii,[16] to tetrahedral Ni_4, Pd_4, and Pt_4 clusters, octahedral Ni_6, Pd_6, and Pt_6, and cubo-octahedral Pt_{13} clusters.

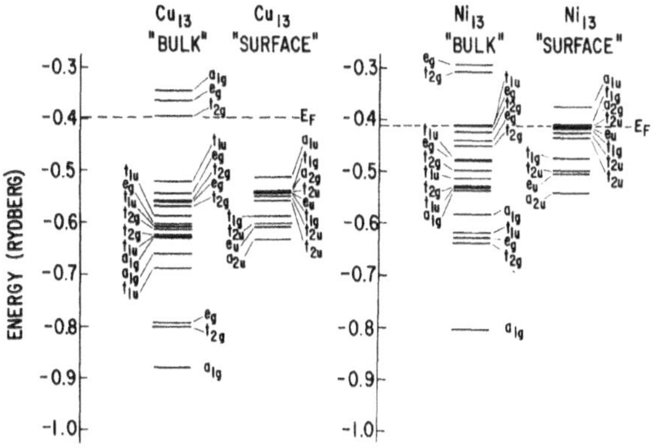

Figure 3. Separation of the SCF-Xα-SW electronic energy levels for Cu_{13} and Ni_{13} clusters into "bulk" and "surface" components.

The relativistic energy-level shifts and spin-orbit splittings are, as anticipated, significant for platinum clusters. relatively small for palladium clusters, and negligible for nickel clusters. The principal relativistic effects on the electronic structure of a platinum cluster are to widen its "d band" and lower its center of gravity on the energy scale in comparison with the nonrelativistic limit. Because SCF-Xα orbital energy eigenvalues, by virtue of their equivalence to derivatives of the total energy with respect to occupation number,[7] can be identified with "orbital electronegativities,"[17,18] the main effect of relativity is to increase the effective group electronegativity of a platinum aggregate as compared with the nonrelativistic limit, thereby affecting its chemisorptive and catalytic properties. For example, the relativistic increase of orbital electronegativity for a platinum aggregate is crucial to explaining why platinum if effectively an electron acceptor, rather than a donor, with respect to atomic hydrogen (see following Sec. 2D). Further details of this work, including a comparison of relativistic and nonrelativistic electronic structures for Pt, Pd, and Ni clusters are published in Ref. 13.

D. Interaction of Hydrogen with Ni, Pd, Pt Clusters

The interaction of atomic hydrogen with small Ni, Pd, and Pt clusters has been studied by the SCF-Xα-SW method, including relativistic effects, as models for hydrogen chemisorption on and interstitial solubility in these transition metals. The nature of the interaction is governed principally by the proximity in energy (or corresponding "orbital electronegativity")[17,18] and concomi-

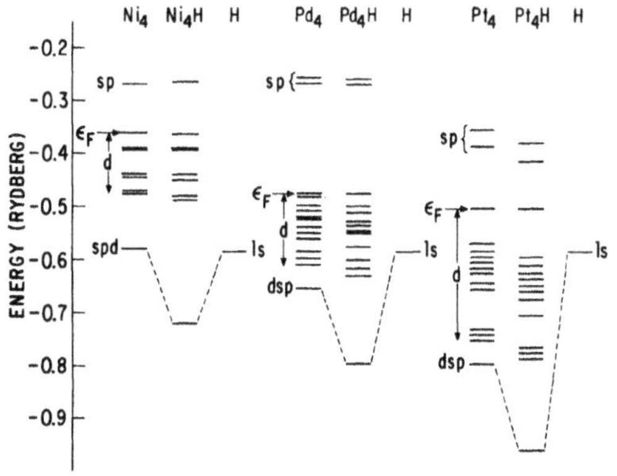

Figure 4. SCF-Xα orbital energies of tetrahedral group-VIII transition-metal clusters with and without interstitial atomic hydrogen, including relativistic corrections for the Pd_4, Pd_4H, Pt_4, and Pt_4H clusters.

tant overlap of symmetry-conserving orbitals near the bottom of the cluster d bands with the H 1s orbital. The main result is the splitting off in energy of a hydrogen-metal bonding orbital from the bottom of the d band of each cluster, accompanied by much smaller level shifts within the d band, as illustrated in Fig. 4 for hydrogen interacting with 4-atom clusters of Ni, Pd, and Pt (relativistic effects included). Because the H 1s orbital (as calculated by the SCF-Xα method) lines up in energy with the nickel cluster spd-hybrid orbital at the bottom of the d band, the relative contributions of Ni atomic orbitals to hydrogen bonding are approximately 60% Ni(4s), 30% Ni(3d), and 10% Ni(4p). In contrast, for palladium and platinum clusters, where the H 1s orbital lines up with the center and top of the respective d bands, the Pd 4d and Pt 5d orbitals account for over 90% of the hydrogen-metal bonding orbital character, thereby dominating the interaction of atomic hydrogen with these metals. The above findings are essentially unaltered for hydrogen interacting with larger Ni, Pd, and Pt clusters. The calculated electronic structures, especially the differences in s- and d-orbital contributions to hydrogen bonding, have been used as a basis for explaining why hydrogen solubility and catalytic activity vary among Ni, Pd, and Pt. Further details are discussed in Ref. 13.

Following the above SCF-Xα-SW studies of atomic hydrogen interacting with Ni, Pd, and Pt clusters, we have investigated the precursor to such interaction, namely, the dissociation of molecular hydrogen (H_2) at low-coordination cluster sites. The previously

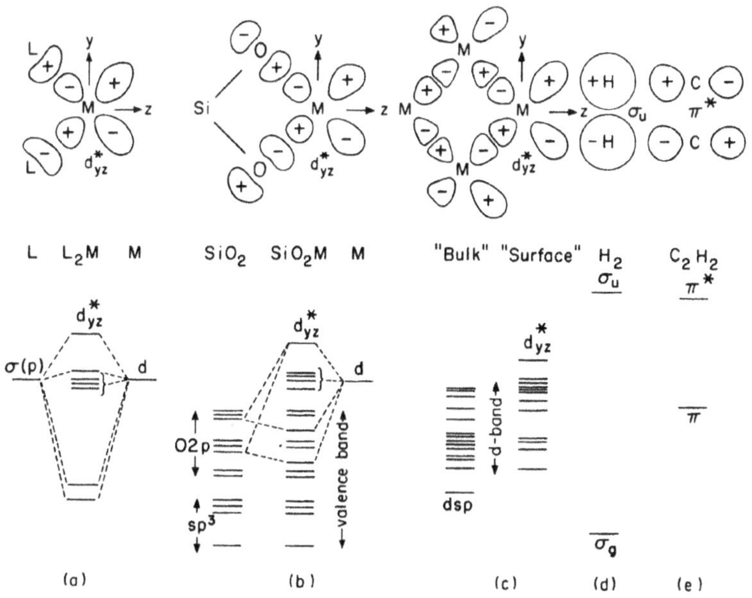

Figure 5. Comparison of the electronic structures associated with coordinatively-unsaturated transition-metal (e.g., M = Pt) sites: (a) in a transition-metal coordination complex (L = PH_3); (b) on a silica (SiO_2) support; (c) at the periphery or "surface" of a transition-metal cluster. Also shown, for comparison, are the HOMO and LUMO of the reactant molecules; (d) H_2 and (e) C_2H_2. All the orbital energy levels shown in the lower half of the diagram are displayed schematically on the same relative scale and for M = Pt in (a), (b), and (c) span an energy range of approximately 0.5 Ry. Symmetry-conserving orbital wavefunctions involved in reactivity are mapped in the y-z plane in the upper half of the diagram.

discussed (see Sec. 2B and Fig. 3) partitioning of the molecular-orbital eigenstates of a 13-atom cubo-octahedral transition-metal cluster into "bulk" and "surface" components is reproduced schematically in Fig. 5(c). In close analogy to the electronic structure of a coordinatively unsaturated transition-metal complex shown in Fig. 5(a), in which a d_{yz}* orbital is split off in energy from the nonbonding d orbitals by the strong antibonding interaction with the ligand p-like lone-pair orbitals, the direct interaction of the d orbitals of a metal site on the cluster periphery with the d orbitals on neighboring metal sites (viewed as "ligands") leads to a splitting off in energy of a strongly antibonding d_{yz}* orbital

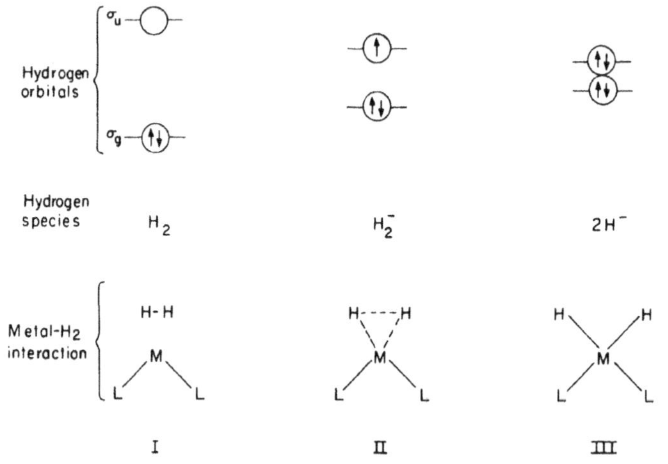

Figure 6. Proposed orbital mechanism for H_2 dissociation at a coordinatively-unsaturated transition-metal site.

from the top of the main d-orbital manifold. For cluster geometries other than the one implied in Fig. 5(c), the most strongly antibonding orbital can have a different local symmetry, e.g., d_{z^2}*, and it is found that for more general cluster configurations or surface morphologies a range of orbital symmetries will be encountered. Such orbitals can be interpreted as cluster analogues of "surface states" that are split off in energy from the top of the d band of a bulk crystalline transition metal. In analogy to the electronic structure of the coordinatively unsaturated transition-metal complex in Fig. 5(a), the d_{yz}* cluster orbital in Fig. 5(c) is symmetry-conserving with respect to the lowest unoccupied orbitals (σ_u and π*) of molecules such as H_2 and C_2H_2, thereby facilitating electron flow between the metal cluster and these molecules. This suggests a mechanism for H_2 bond cleavage at a single metal site on the cluster periphery or "surface" analogous to that known in Fig. 6 for a coordinatively unsaturated metal complex.

E. Metal-Support (Pt-SiO_2 and Ru-SiO_2) Interactions

As prototype models for catalyst-support interactions, we have carried out SCF-Xα-SW molecular-orbital studies of Pt and Ru atoms and aggregates supported on silica (SiO_2). It has already been shown that one can accurately model the electronic structure of pure SiO_2 by performing SCF-Xα-SW molecular-orbital calculations on representative clusters.[19] Although definitive structural data are not yet available, a possible metal-support configuration, consistent with recent EXAFS data,[20] is the one illustrated schemati-

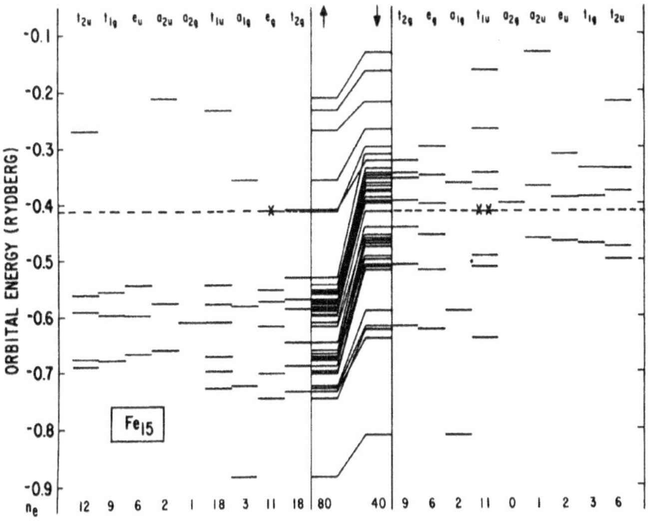

Figure 7. Orbital eigenvalues from the spin-polarized calculation for Fe_{15}. The dashed line indicates the position of the Fermi level, which is coincident with the $6e_g^{\uparrow}$ (one electron) and $4t_{1g}^{\downarrow}$ (two electrons) levels. The bottom line shows the number of electrons of each spin and symmetry type (point group O_h) as well as the total number of majority- and minority-spin electrons.

cally in Fig. 5(b) in which the oxygen atoms at the SiO_2-M interface serve as "ligands" of the metal atom in a fashion analogous to the coordinatively unsaturated L_2M complex (e.g., L = PH_3: M = Pt) considered in Fig. 5(a).

The primarily covalent interaction between the metal d orbitals and oxygen nonbonding 2p orbitals that constitute the top portion of the SiO_2 valence band lead to metal-oxygen bonding orbitals embedded in the valence band and to nonbonding and antibonding d orbitals split off in energy from the top of the valence band and positioned within the SiO_2 band gap. The highest of these antibonding orbitals is principally metal d_{yz}* in character for the chosen coordinate system and has a one-to-one correspondence with the HOMO of the coordinatively unsaturated L_2M (M = Pt) Complex (cf. Fig. 5(a)) which has been shown in Fig. 6 to be responsible for activation of H_2 dissociation through symmetry-conserving overlap with the H_2 antibonding σ_u orbital. These theoretical results suggest that there are strong analogies among catalytically active sites of (a) molecular transition-metal coordination complexes, (b) supported metal catalysts, and (c) transition-metal clusters and particles. The orbital symmetries and orderings are qualitatively similar for SiO_2Pt and SiO_2Ru, although in the latter system the d_{yz}* antibonding orbital is empty. This difference in orbital occupancy suggests that the bonding and catalytic activity of Ru atoms on silica should be significantly different from those of Pt atoms on silica. Recent EXAFS, electron microscopy and chemisorption studies[20,21] indicate that ruthenium interacts strongly with a silica substrate, forms "raft-like" structures, and chemisorbs oxygen (O_2) molecularly, whereas such effects have not yet been observed for Pt supported on silica. SCF-Xα-SW cluster models for ruthenium aggregates on SiO_2 suggest that the SiO_2-Ru bonding perpendicular to the silica surface actually enhances Ru-Ru bonding parallel to the surface, thereby providing a theoretical justification for the formation of raft-like structures.

F. Electronic Structure and Magnetic Properties of Fe Clusters

SCF-Xα-SW molecular-orbital calculations have also been carried out for iron clusters ranging in size from 4 atoms to 15 atoms.[22,23] The calculations were carried out in spin-unrestricted form, i.e., different orbitals for different spins, in precisely the same fashion as described for nickel clusters in Ref. 12. The resulting spin-orbital energy eigenvalues for a 15-atom cluster having the local coordination up to second-nearest neighbors of bcc crystalline α-iron are shown in Fig. 7, labeled according to the irreducible representations of the O_h symmetry group. The two highest occupied spin orbitals, $6e_g^{\uparrow}$ and $4t_{1u}^{\downarrow}$, are nearly degenerate, partially occupied with one and two electrons, respectively, and determine the cluster Fermi energy (indicated by the dashed line).

From Fig. 7 one can extract the following characteristics which are directly comparable with the band structure of ferromagnetic crystalline α-iron calculated by Tawil and Callaway:[24]

(1) The cluster electronic structure can be characterized in terms of manifolds of predominantly d-like spin orbitals bracketed in energy by orbitals which have significant s,p character, in analogy to the overlap of the d band by the s,p band in bulk crystalline iron.

(2) The Fermi level of the cluster passes through the top of the "spin-up" (↑) d-orbital manifold and through the center of the "spin-down" (↓) d-orbital manifold in a fashion similar to the intersection of the Fermi level with the "majority-spin" and "minority-spin" d bands of ferromagnetic iron (see Fig. 2 of Ref. 24).

(3) The exchange splitting and width of the cluster d band are comparable with the values for bulk ferromagnetic iron.

(4) The cluster total density of states shows a pronounced two-

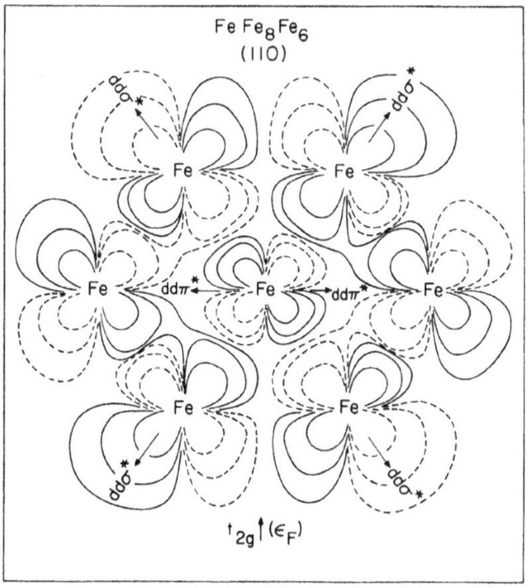

Figure 8. Contour map of the lowest unoccupied "majority-spin" molecular orbital, $7t_{2g}\uparrow$ (see Fig. 7), of an Fe_{15} cluster. The solid and dashed contours represent positive and negative values, respectively, of the wave function.

peak structure, which is primarily a consequence of the exchange splitting, consistent with the density of states for ferromagnetic iron (see Fig. 5 of Ref. 24).

(5) The highest occupied and lowest unoccupied cluster "majority-spin" molecular orbitals, $6e_g\uparrow$ and $7t_{2g}\uparrow$, respectively, have antibonding hybrid dsp character, as exemplified by the $7t_{2g}\uparrow$ wavefunction contour map in Fig. 8, plotted in a (110) plane. The solid and dashed contours, respectively, represent positive and negative values of the wavefunction.

The cluster molecular orbitals can be further utilized as a basis for understanding the transition between collective ferromagnetism (or "super-paramagnetism")[25] and local paramagnetism with decreasing particle size or increasing temperature. For example:

(1) The average paramagnetic magneton numbers per atom for the Fe_9 and Fe_{15} clusters are 2.9 and 2.7, respectively, approaching the 2.2 value for ferromagnetic crystalline α-iron, whereas SCF-Xα results for smaller iron clusters, such as Fe_4, indicate

magneton numbers per atom as large as 3.0. The predicted increase of magneton number per atom with decreasing iron cluster size is supported experimentally by Mössbauer studies of small iron aggregates isolated in inert matrices.[26]

(2) The partial-wave decomposition of the Fe_9 and Fe_{15} cluster orbitals indicates that the contribution of the 4s-like components to spin polarization, although relatively small in magnitude, is opposite in direction (antiparallel to) that of the dominant 3d-like components, in good agreement with the analysis of magnetic form factors in polarized neutron scattering data for crystalline iron.[27]

(3) Somewhat more spin density in the Fe_9 and Fe_{15} clusters is concentrated in the e_g orbitals than in the t_{2g} orbitals, providing a local model for the concentration of spin density along the [100] direction (the direction of easy magnetization) in crystalline α-iron, as deduced from neutron-diffraction measurements.[27,28]

(4) Neutron-diffraction studies of crystalline iron at temperatures greater than the Curie temperature T_c[29-31] suggest that short-range ordering of spins in the form of "spin clusters" (~ 10 Å in size) persists well into the paramagnetic region, and that the paramagnetic magneton number is approximately 25% greater than the value for ferromagnetic iron ($T < T_c$). It is possible, therefore, that the electronic structure of crystalline iron in the paramagnetic region is more appropriately represented statistically by a local spin-polarized cluster model, than by the extended $\vec{k}$-space band-structure description, since the cluster electronic structure is not dependent on the assumption of long-range crystalline ordering of spins. Utilizing the bcc Fe_{15} cluster as a prototype model for a "spin cluster", we find that the majority-spin d orbitals, $7t_{2g}\uparrow$ and $6e_g\uparrow$ (see Fig. 7), are the highest partially occupied and lowest unoccupied orbitals near the Fermi energy at $T = 0^oK$. Thermally induced "spin-flip" electronic excitations from the highest occupied orbitals of the minority-spin d manifold to the unoccupied $7t_{2g}\uparrow$ and $6e_g\uparrow$ orbitals for $T > T_c$ results in an effective 30% increase of cluster magnetic moment and magneton number, thereby providing a good model for the observed 25% increase in magnetic moment of crystalline iron in the paramagnetic region. Such excitations only slightly alter the average exchange splitting and total density of states of the cluster, consistent with the apparent persistence of the local density of states for crystalline iron at $T > T_c$ (as measured, for example, by photoemission), an experimental observation which heretofore has been inexplicable within the framework of conventional ferromagnetic energy-band theory.

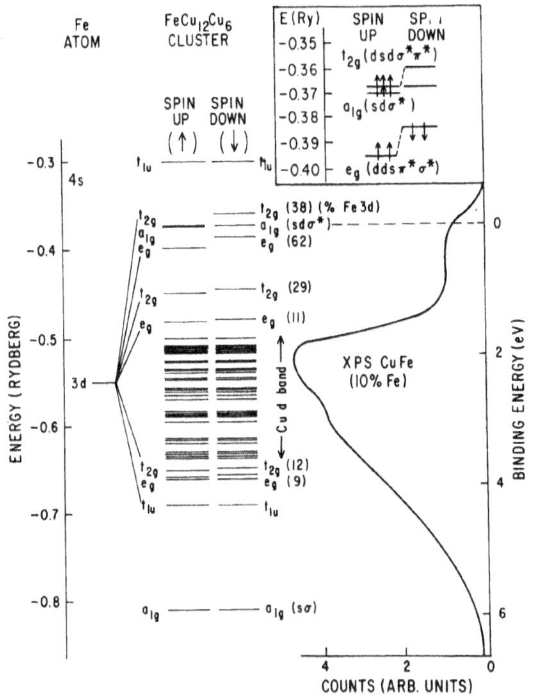

Figure 9. SCF-Xα-SW spin-polarized molecular-orbital energy levels of 19-atom cluster representing the local environment of an Fe impurity in a copper crystal up to second-nearest neighbors. The calculated levels are compared with the measured X-ray photoelectric spectrum (XPS) of bulk copper containing a dilute concentration of substitutional Fe impurities.

G. Cluster Molecular-Orbital Model for the Local Electronic Structure, Chemical Bonding, and Magnetic Moments of Transition-Metal Impurities in Copper

Spin-unrestricted self-consistent-field X-alpha scattered-wave (SCF-Xα-SW) cluster molecular-orbital models have been constructed for transition-metal impurities (Ni, Fe, Mn, and V) alloyed in crystalline copper, in order to gain insight into the occurrence of local magnetic mements and the Kondo effect,[32,33] traditionally viewed as many-body effects arising from the interaction of the transition-metal impurities with the conduction electrons of the host metal. A 19-atom cluster representing the local molecular environment (12 nearest neighbors and 6 second-nearest neighbors) of an isolated transition-metal impurity in an otherwise perfect crystalline copper lattice yields manifolds of molecular-orbital energy levels (such as those shown in Fig. 9 for iron in copper) corresponding to the copper "d band", bracketed by impurity d-orbital energy levels of e_g and t_{2g} symmetries that are bonding with respect to the bottom of the copper d band and antibonding with respect to the top of the d band. The antibonding cluster orbitals (see inset of Fig. 9) in conjunction with Hund's rules, (1) are discrete analogues of Friedel-Anderson[34,35] virtual impurity states, (2) satisfy a "sum rule" analogous to the Friedel sum rule, (3) have spin occupancies and polarization in accord with measured magnetic moments, (4) are consistent with measured trends of residual electrical resistivity, and (5) describe the "renormalizing" effects of the local crystalline environment on the transition-metal impurity atom. The distribution of occupied $FeCu_{12}Cu_6$ cluster spin-orbital energy levels is in good agreement with the X-ray photoelectron spectrum (XPS) recently measured by Höchst et al.[36] for a dilute alloy of iron in copper, as shown in Fig. 9 (Fermi level indicated by the dashed line).† In particular, the broad "shoulder" in the XPS spectrum between the Cu d-band maximum and the Fermi energy, is explained in the present theoretical model as arising from the e_g and t_{2g} molecular orbitals above the Cu d band which have significant Fe 3d σ^* and π^* antibonding components (the Fe 3d percentages are indicated in parentheses in Fig. 9). The most striking result of these theoretical studies is the discovery that local coordination chemical bonding dominates the interaction of the transition-metal impurity with its crystalline environment, yielding a large splitting ($\sim$ 4-4.5 eV) between bonding and antibonding impurity molecular-orbital energy levels (see Fig. 9) and a spin splitting ($\sim$ 0.2-0.6 eV) of the highest occupied antibonding t_{2g} and e_g orbitals (the Friedel-Anderson-like impurity states) that is an order of magnitude smaller than values assumed in previous theoretical models (e.g., the Anderson[35] model).† The near degeneracy of the latter t_{2g} and e_g spin orbitals (for Fe and Mn impurities) with cluster a_{1g} molecular-orbital analogues of the sd conduction-band eigenstates of crystalline copper near the Fermi energy (see inset of Fig. 9) suggests that the "spin-compensation cloud" often associated with the Kondo effect is considerably smaller in magnetization and more spatially localized around the impurity than assumed in previous formal theories[33] and, in conjunction with the SCF-Xα transition-state concept, also provides a discrete orbital basis for discussing the onset of "localized spin fluctuations" around the Kondo temperature. The possible contributions of spin-orbit coupling and Jahn-Teller

†A recently published density-functional model for magnetic impurities in copper by Podloucky et al. (Phys.Rev. B22, 5777 (1980)), which does not properly include the chemical bonding effects of the local crystalline environment on the impurity site is in qualitative disagreement with the present theoretical model and with the XPS measurements of Höchst et al.[36] and merely reproduces the Anderson model.

effects to impurity electronic structure have been determined, with detailed examples of how the former effect can lead to the quenching of the impurity magnetic moment below the Kondo temperature. Further details of this work are presented in Ref. 37.

H. Cluster Molecular-Orbital Model for the Local Electronic Structure, Chemical Bonding, and Magnetic Moments of Transition-Metal Impurities in Aluminum

SCF-Xα-SW cluster molecular-orbital studies of the type described in the preceding Sec. G have been extended to the problem of transition-metal impurities in aluminum. First-row transition-metal (Ti, V, Cr, Mn, Fe, Co, and Ni) atoms were successively substituted for the central Al atom in a cluster representing the local environment of bulk fcc crystalline aluminum, in a fashion similar to that described in Sec. G for a copper host, except that in the present case a 43-atom cluster ($AlAl_{12}Al_6Al_{24}$) representing the aluminum host up to third nearest neighbors was chosen to model the somewhat more delocalized electronic structure of this material. Indeed, previous SCF-Xα-SW studies of a 43-atom aluminum cluster by Salahub and Messmer[38] have demonstrated that this cluster yields a

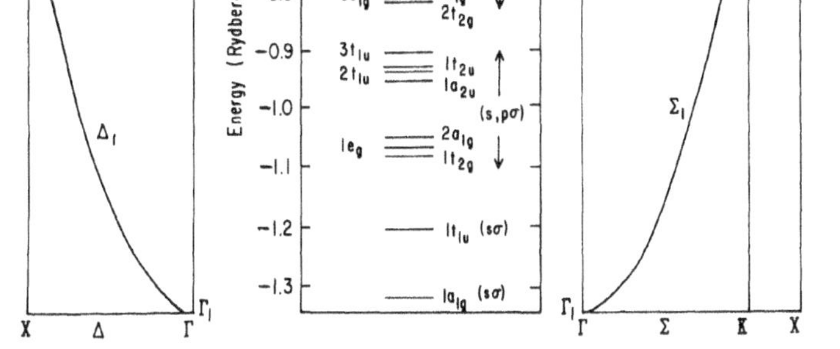

Figure 10. Comparison of the SCF-Xα-SW molecular-orbital energy levels of a 43-atom aluminum cluster, representing the local environment of fcc crystalline aluminum up to third nearest neighbors, with the bulk energy bands.

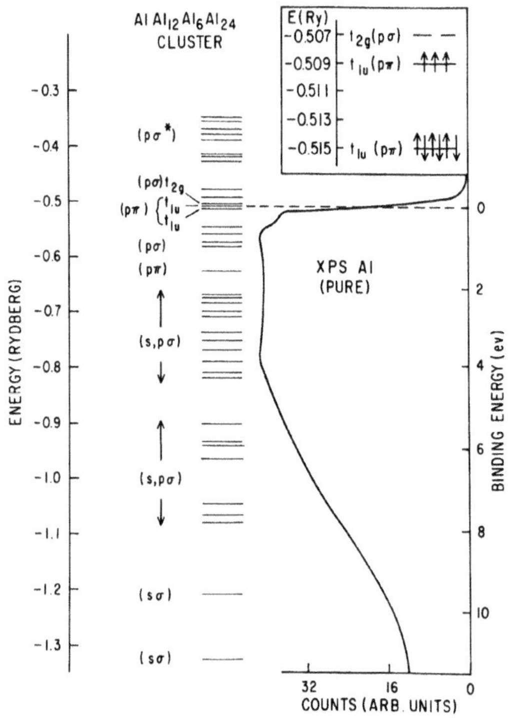

Figure 11. Comparison of the SCF-Xα-SW molecular-orbital energy levels of a 43-atom aluminum cluster, representing the local environment of crystalline aluminum up to third-nearest neighbors, with the measured X-ray photoelectron spectrum (XPS) of an aluminum crystal.

manifold of s,p σ- and π-bonding molecular orbitals having a total band width and general character consistent with the band structure of crystalline aluminum, as shown in Fig. 10, and in good agreement with measured XPS data for crystalline aluminum, as shown in Fig. 11.

Perhaps the most striking feature of the SCF-Xα-SW results for the 43-atom pure aluminum cluster is the π-bonding character of the highest occupied molecular orbital ($8t_{1u}$ in Fig. 10), i.e. the molecular orbital that coincides with the Fermi level, the wavefunction contour map of which is shown in Fig. 12, plotted in a (200) plane up to second-nearest-neighbor atoms. As will be shown in Sec. 2K, the spatially extended π-bonding topology of this orbital is a precursor to superconductivity in pure aluminum. Thie "real-space" character of aluminum molecular orbitals at the Fermi energy cannot

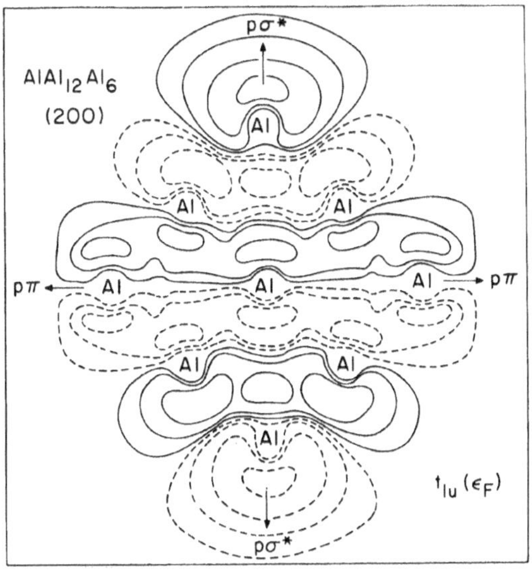

Figure 12. Contour map of the wave function for the highest occupied molecular orbital, $t_{1u}(\varepsilon_F)$, of a 43-atom aluminum cluster, plotted in the (200) crystallographic plane up to second-nearest neighbors. The solid and dashed contours represent positive and negative values, respectively, of the wave function.

be extracted easily from the conventional "reciprocal ($\vec{k}$)-space" Bloch description of the crystal band structure (see Fig. 10), except in hindsight as an appropriate combination of Σ_1, Σ_3, and Δ_1 Bloch eigenstates, and thus has eluded solid-state physics.

The SCF-Xα-SW molecular-orbital energy levels for a Mn impurity substituted for the central atom of the 43-atom aluminum cluster are shown in Fig. 13, where they are compared with the 3d energy level of an isolated free Mn atom (and with the 3d atomic energy levels of the other first-row transition elements). The cluster results are also compared directly with XPS data for a dilute AlMn alloy recently obtained by Steiner et al.[39] Comparison of Fig. 13 with the results for pure aluminum in Fig. 11 reveals that the main effects of alloying Mn impurities in an aluminum host, neglecting impurity-impurity interactions, are: (1) to empty out locally the otherwise spatially extended t_{1u} pπ-bonding orbital mapped in Fig. 12, and (2) for the Mn 3d orbitals to overlap and hybridize with the surrounding Al 3p orbitals to produce a narrow manifold of occupied spatially localized bonding and antibonding t_{2g} and e_g molecular orbitals near the Fermi energy (indicated by the dashed line in

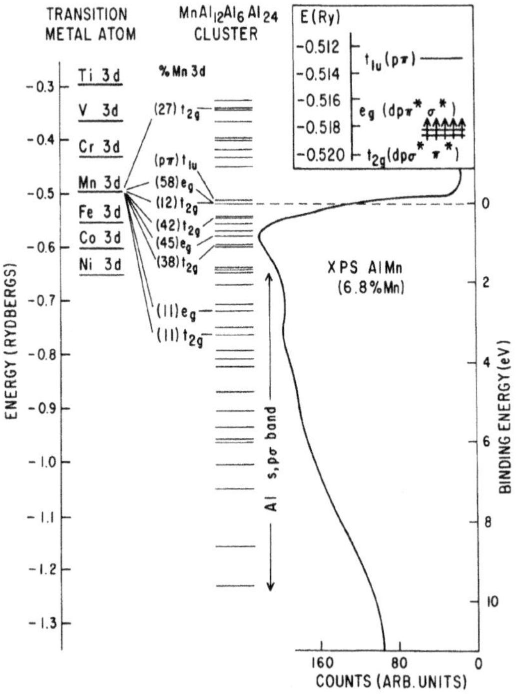

Figure 13. SCF-Xα-SW molecular-orbital energy levels of a 43-atom cluster representing the local environment of a Mn impurity of an aluminum crystal up to third-nearest neighbors. The calculated levels are compared with the measured X-ray photoelectron spectrum (XPS) of bulk aluminum containing a dilute concentration of substitutional Mn impurities.

Fig. 13), which are responsible for the peak in XPS intensity just below the Fermi energy. The t_{2g} and e_g bonding and antibonding orbitals containing 38% or more Mn 3d character (indicated in parantheses in Fig. 13) are mapped in Figs. 14 through 17. The highest occupied molecular orbitals are nearly degenerate t_{2g} and e_g $d p\sigma^*$-$d p\pi^*$ antibonding orbitals, as indicated in the inset of Fig. 13 (cf. inset of Fig. 11). The $d p\sigma^*$-$d p\pi^*$ antibonding topology is clearly evident in the wavefunction contour maps of Figs. 16 and 17. These molecular orbitals are analogous to the t_{2g} and e_g $dsd\sigma^*$-$dds\pi^*$ antibonding orbitals shown in the inset of Fig. 9 and in Sec. G to be associated with the local spin magnetic moment of an Fe impurity in a copper host above the Kondo temperature. However, for Mn impurities in aluminum, the fluctuations in occupancy

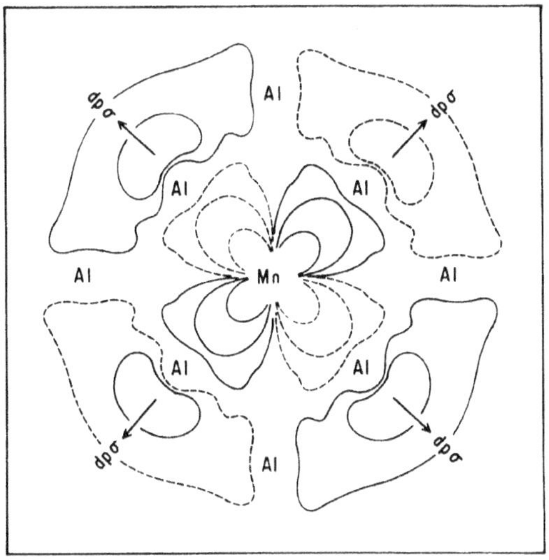

Figure 14. Contour map of the highest occupied bonding t_{2g} cluster molecular-orbital wavefunction for a Mn impurity in an aluminum crystal.

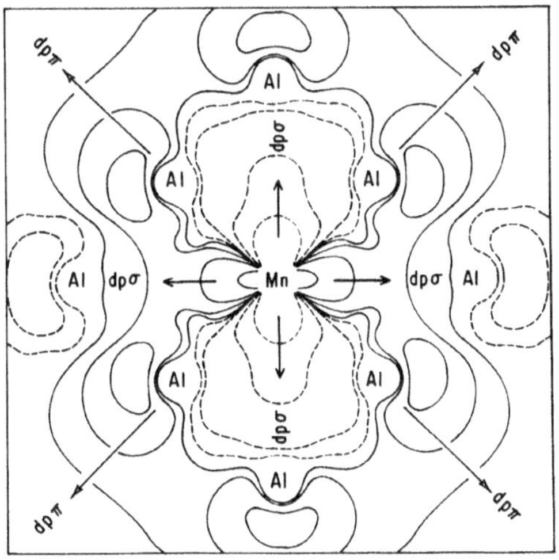

Figure 15. Contour map of the highest occupied bonding e_g cluster molecular-orbital wavefunction for a Mn impurity in an aluminum crystal.

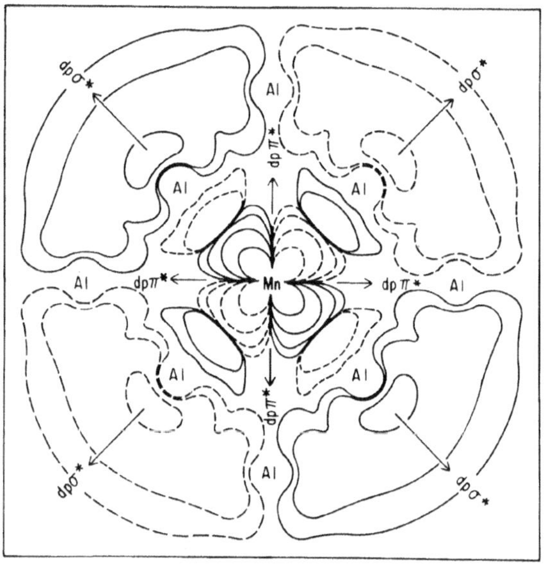

Figure 16. Contour map of the highest occupied antibonding t_{2g} cluster molecular-orbital wavefunction for a Mn impurity in an aluminum crystal.

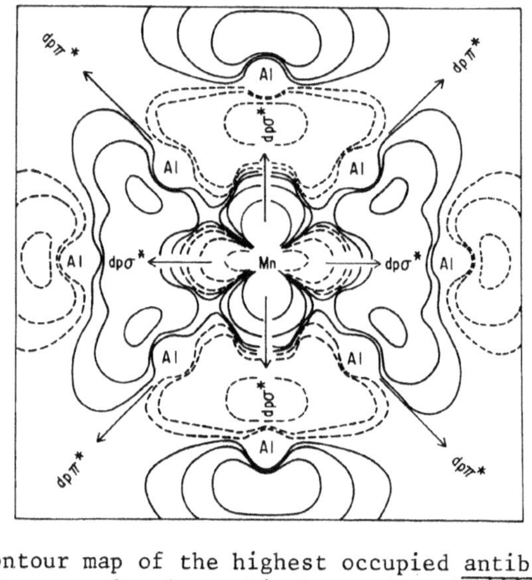

Figure 17. Contour map of the highest occupied antibonding e_g cluster molecular-orbital wavefunction for a Mn impurity in an aluminum crystal.

of the nearly degenerate t_{2g} and e_g antibonding orbitals at the Fermi energy by five electrons (see inset of Fig. 13), coupled with a relatively small exchange splitting of the orbital energies, prevent stable magnetic moment formation and are responsible for "spin fluctuations" in the AlMn system.[33] These antibonding orbitals also provide a straightforward explanation of the destructive effect of the Mn impurities on the superconductivity of aluminum in terms of the emptying out of the t_{1u} ($p\pi$) orbital.

SCF-Xα-SW cluster molecular-orbital models for Fe, Co, and Ni impurities in aluminum show a systematic downward shift in energies and full electron occupancy of the bonding and antibonding impurity-derived orbitals with respect to the Fermi energy. This trend is consistent with the relative energies of the corresponding atomic 3d orbitals in Fig. 13, explains the nonmagnetic ground states of these impurities in aluminum,[33] is in good agreement with the positions of the peaks in XPS intensity for these alloys,[39] and suggests that the effective cancellation of occupied bonding orbitals by occupied antibonding orbitals is largely responsible for the very low solubility of these impurities in aluminum.

Further details of the above results will appear in a later publication.[40]

I. Local Electronic Structures and Chemical Bonding of Intermetallic Compounds and Solid-Solution Alloys

The SCF-Xα-SW cluster molecular-orbital method has been used to determine the local electronic structures of the intermetallic compound Ni_3Ti and solid-solution alloy αCuZn. It has been found that relatively small clusters which preserve local compositional and structural order yield local densities of states comparable to those obtained in more elaborate band-structure theories, provide a realistic model for local chemical bonding effects neglected in traditional $\vec{k}$-space alloy theories, and lead to a "real-space" chemical description of alloy phase stability. For the example of Ni_3Ti, the cluster electronic structure provides a simple yet remarkably accurate interpretation of photoemission spectra for this compound recently measured by Fischer and Kelemen. Details are published in Ref. 41.

J. Local Electronic Structure and Chemical Bonding of Amorphous Alloys

A unified conceptual basis for understanding and predicting the electronic structures and glass-forming capabilities of amorphous alloys has been developed, using the results of first-principles self-consistent X-alpha (SCF-Xα-SW) cluster molecular-orbital calculations. Prototype cluster models have been constructed for Cu-Zr, Ni-Ti, and Pd-Si alloys which exemplify two major classes of binary (A-B) glass-forming systems, namely:

(1) Metallic glasses based on noble or transition elements (e.g., A = Cu, Ni) toward the right of the periodic table and transition elements (e.g., B = Zr, Ti) toward the left of the periodic table.

(2) Metalloid glasses based on transition elements (e.g., A = Pd) toward the right of the periodic table and nonmetallic elements (e.g., B = Si) toward the middle of the periodic table.

The calculated electronic structures are in excellent quantitative agreement with and provide an interpretation of published photoelectron spectra for the above amorphous alloys. All three systems investigated are characterized by an occupied d-band associated principally with the A (e.g., Cu, Ni, Pd) components and a partially occupied band lying at somewhat higher energies and associated mainly with the B (e.g., Zr, Ti, Si) components, although there is significant A-B bonding and B-A antibonding character within these respective bands and some electron transfer from B to A.

The most striking feature of the calculated electronic structures is the nature of the electronic states coinciding with or just below the Fermi energy, i.e., the highest occupied molecular orbitals. These states can be described as forming highly delocalized and spatially directed B-B (e.g., Zr-Zr d-orbital or Si-Si p-orbital) chemical bonds which are promoted, in part, by the composite bonding-antibonding interaction with the d-orbitals localized on neighboring A (e.g., Cu or Pd) components. These bonds are clearly visible in the molecular-orbital wavefunction contour maps shown in Fig. 18 for the Cu-Zr and Pd-Si systems, respectively, where the solid and dashed contours represent positive and negative phases of the wavefunction, respectively.

The existence of directed Zr-Zr and Si-Si bonds in amorphous Cu-Zr and Pd-Si alloys, respectively, may, at first sight, seem surprising since the Zr and Si atoms are not nearest neighbors in these alloys, and the Zr-Zr and Si-Si bond distances are significantly larger than those in elemental solid Zr and Si. Nevertheless, the effects of the nearest-neighbor Cu and Pd d-orbitals in acting as "bridges" for the formation of Zr-Zr and Si-Si bonds are clearly evident in the above contour maps.

On a scale spanning many atoms, the Zr-Zr and Si-Si bonds form a type of spatially extended "electron network" or "electron superlattice" in the vicinity of the Fermi energy, which we believe is related to the glass-forming capabilities of these and similar alloys. In the case of the Cu-Zr system, the electron network orbital topology at the Fermi energy is precisely of the type re-

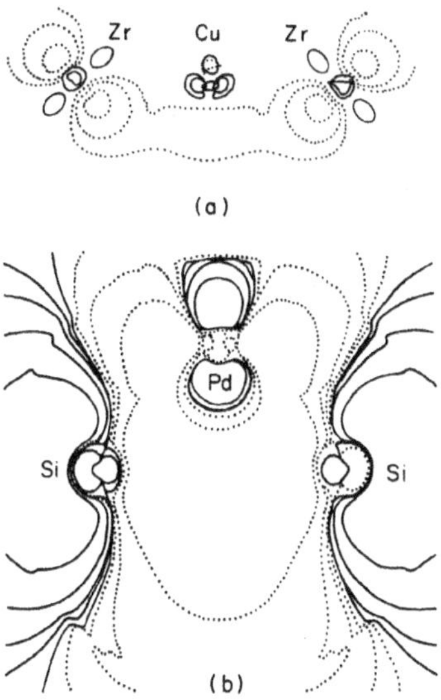

Figure 18. (a) Contour map of the wavefunction for the cluster molecular orbital at the Fermi energy of amorphous Cu-Zr.
(b) Contour map of the wavefunction for the cluster molecular orbital at the Fermi energy of amorphous Pd-Si.

cently found by this author (see the following Sec. K) to be necessary for the occurrence of superconductivity in a material. Indeed, amorphous Cu-Zr alloys have recently been observed to be superconducting.[42]

K. Molecular Orbitals and Superconductivity

A real-space molecular-orbital description of electronic wavefunctions which are postulated to be the precursors of the superconducting state in metals, alloys, compounds, amorphous, and organic solids is presented, based on self-consistent-field X-alpha scattered-wave (SCF-Xα-SW) molecular-orbital calculations for clusters representing the local molecular environments in these materials. It is shown that there is a persistent correlation between the occurrence of superconductivity in a material and the existence of spatially delocalized molecular orbitals at the Fermi energy which are bonding within and antibonding between "layers" or "tubes" of overlapping atomic orbitals that span many atoms, forming a type of "electron superlattice" or "network" at the Fermi energy, as exemplified by pπ "layered" molecular-orbital

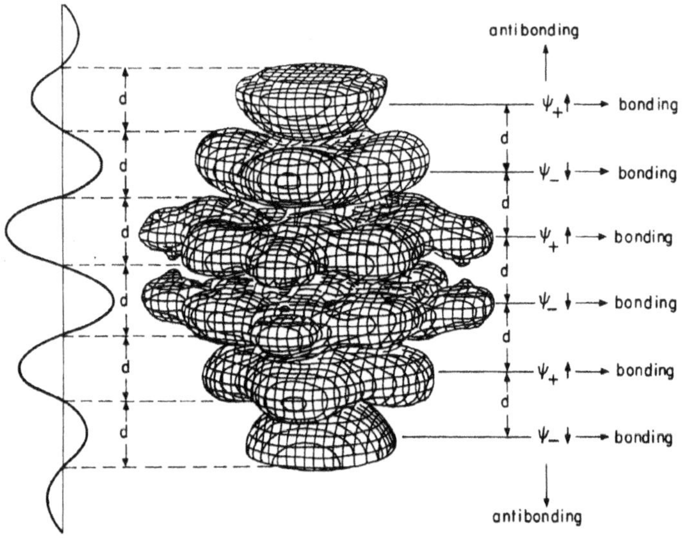

Figure 19. Three-dimensional contour map of the wavefunction for the highest occupied molecular orbital, $t_{1u}(\varepsilon_F)$, of a 43-atom aluminum cluster.

components ψ_+ and ψ_- in Al and $(TMTSF)_2PF_6$ (see Fig. 19, the three-dimensional version of Fig. 12), and by dδ "tubular" molecular-orbital components ψ_+ and ψ_- in Nb and Nb_3Sn (See Figs. 20 and 21, respectively). This description of the precursor superconducting state is consistent with the original conjectures of London[43] that the superconducting-state wavefunction is "molecular" in nature, "rigid" in character, and of wide spatial extent, from which observed physical properties (e.g., diamagnetism and nondissipative electrical currents) of the superconducting state logically follow. The molecular-orbital model is further shown to be consistent with Cooper's concept of electron pairing in the superconducting state through a net attractive electron-electron interaction but differs from the Bardeen-Cooper-Schrieffer (BCS)[44] theory of superconductivity in attributing the pairing to "valence-bond-like" electron occupation of layered or tubular molecular orbitals at the Fermi energy, rather than to the absorption and emission of virtual phonons. However, recent simple physical arguments presented by Weisskopf[45] for the formation of Cooper pairs suggest that the two points of view are not incompatible. Approximate formulae for calculating the electron pair binding energy and superconducting transition temperature, analogous to the BCS formulae but expressed entirely in terms of two simple molecular-orbital parameters, have been derived and applied to various superconductors. Moreover, through an elementary phase-space argument, the molecular-orbital description of Cooper pairs is shown to be qualitatively consistent

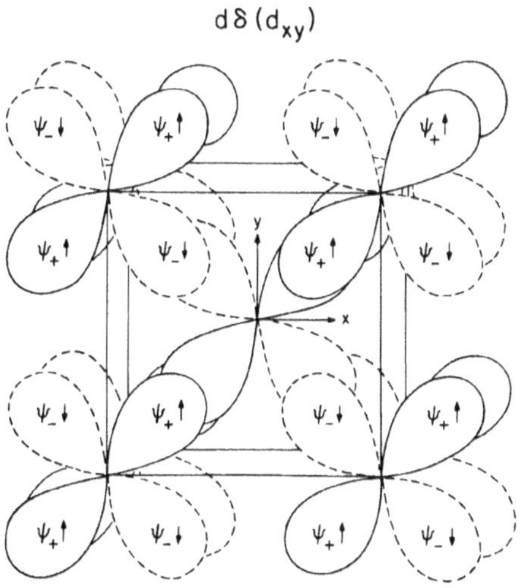

Figure 20. Schematic perspective drawing of "tubular" dδ-bonding cluster molecular-orbital components in a bcc crystal arising from the overlap of d_{xy} atomic orbitals.

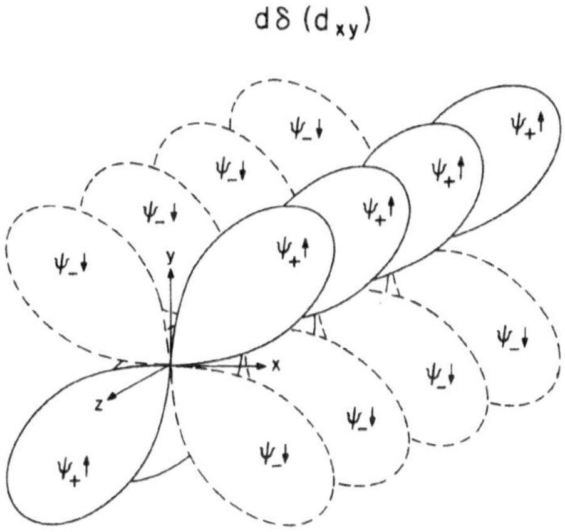

Figure 21. Schematic perspective drawing of "tubular" dδ-bonding molecular-orbital components along a line of atoms arising from the overlap of d_{xy} atomic orbitals along the z-axis.

with Schafroth's[46] description of electron pairs as "quasi-molecules" undergoing Bose-Einstein-like condensation to the superconducting state. The above molecular-orbital criteria for superconductivity are diametrical to those for the occurrence of local magnetic moments and ferromagnetism, namely, the existence of spatially localized, exclusively antibonding (e.g., dσ*, dπ*, or dδ*) molecular orbitals near the Fermi energy. Furthermore, in contrast to the ordering of electron spins on atomic sublattices in conventional antiferromagnetism, the valence-bond-like correlation of Cooper-Schafroth electron pairs at the Fermi energy in composite bonding-antibonding layered or tubular molecular-orbital components, spatially delocalized between the atoms, corresponds to a type of conduction-electron "antiferromagnetism" or spin density wave, thus offering an explanation for the occurrence of this antiferromagnetism in superconductors such as $(TMTSF)_2PF_6$[47] and Nb_3Sn.[48] These molecular-orbital criteria therefore provide a conceptual basis for understanding the generally mutually exclusive incidence of superconductivity and magnetism among the elements of the periodic table, although they can also be used to explain the occasional coexistence of superconductivity and ferromagnetism or antiferromagnetism in some materials. The molecular-orbital model for superconductivity complements BCS theory in that it permits the prediction of which materials are likely to be superconductors and which are not entirely on the basis of the molecular-orbital topology at the Fermi energy. It offers an explanation of the vanishing isotope effect in certain superconductors. It contributes to the clarification of the issue of non-phonon mechanisms of superconductivity. It provides insight into the correlations of superconductivity with other physical properties, such as lattice instabilities. It is readily applicable to superconductors lacking long-range crystalline order, such as amorphous alloys and small particles. Finally, the molecular-orbital approach can be used to explain in simple terms why some materials (e.g., Cu, Ag, and Au) are neither superconducting nor magnetic, why certain quasi-one-dimensional organic solids, such as TTF-TCNQ, are not superconductors while others, such as $(TMTSF)_2PF_6$, are superconductors,[47] and to assess critically the likelihood of superconductivity in certain other types of substances (e.g., other types of organic solids; metallic hydrogen at attainable high pressures), and to suggest ways of systematically improving existing classes or synthesizing novel classes of superconducting materials. Further details will appear in other publications.[49]

ACKNOWLEDGMENTS

I am grateful to the Office of Naval Research and the National Science Foundation for supporting this research.

REFERENCES

1. M. Boudart, in Proceedings of the Robert A. Welch Foundation Conference on Chemical Research. XIV Solid State Chemistry, edited by W.O. Milligan (The Robert A. Welch Foundation, Houston, Texas, 1970), p. 299.
2. J.H. Sinfelt, in Annual Review of Materials Science, edited by R.A. Huggins, R.H. Bube, and R.W. Roberts (Annual Reviews, Palo Alto, California, 1972), V. 2, p. 641.
3. J.F. Anderson, Structure of Metallic Catalysts (Academic, New York, 1975).
4. Electronic Density of States, edited by L.H. Bennett (National Bureau of Standards Spec. Pub. 323, Washington, D.C. 1971).
5. R.C. Baetzold, J.Chem.Phys. 55, 4363 (1971); J.Catal. 29, 129 (1974); R.C. Baetzold and R.E. Mack, J.Chem.Phys. 62, 1513 (1975); A.B. Anderson and R. Hoffman, J.Chem.Phys. 61, 4545 (1974); G. Blyholder, J.Chem.Phys. 62, 3193 (1975).
6. J.C. Slater and K.H. Johnson, Phys.Rev. B5, 844 (1972); K.H. Johnson and F.C. Smith, Jr., Phys.Rev. B5, 831 (1972); J.C. Slater and K.H. Johnson, Physics Today 27, 34 (1974).
7. J.C. Slater, in Advances in Quantum Chemistry, Vol. 6, edited by P.-O. Löwdin (Academic, New York, 1972), p. 1; J.C. Slater, The Self Consistent Field for Molecules and Solids, Vol. 4 of Quantum Theory of Molecules and Solids (McGraw-Hill, New York, 1974).
8. K.H. Johnson, J.Chem.Phys. 45, 3085 (1966); K.H. Johnson, in Advances In Quantum Chemistry, Vol. 7, edited by P.-O. Löwdin (Academic, New York, 1973), p. 143.
9. K.H. Johnson, in Annual Review of Physical Chemistry, Vol. 26, edited by H. Eyring, C.J. Christensen, and H.S. Johnston (Annual Reviews, Palo Alto, California, 1975), p. 39.
10. J.G. Fripiat, K.T. Chow, M. Boudart, J.B. Diamond, and K.H. Johnson, J.Molec,Catal. 1, 59 (1975); C.Y. Yang and K.H. Johnson (unpublished work).
11. R.P. Messmer, C.W. Tucker, Jr., and K.H. Johnson, Chem.Phys. Lett. 36, 423 (1975).
12. R.P. Messmer, S.K. Knudson, K.H. Johnson, J.B. Diamond, and C.Y. Yang, Phys.Rev. B13, 1396 (1976).
13. R.P. Messmer, D.R. Salahub, K.H. Johnson, and C.Y. Yang, Chem. Phys.Lett. 51, 84 (1977).
14. R.P. Messmer and S.H. Lamson, Chem.Phys.Lett. (in press).
15. R.P. Messmer, C.M. Kao, and T.C. Caves, Chem.Phys.Lett. (in press).
16. C.Y. Yang and S. Rabii, Phys.Rev. A12, 362 (1975); C.Y. Yang, Chem.Phys.Lett. 41, 588 (1976).
17. K.H. Johnson, Intern.J. Quantum Chem. 11S, 39 (1977).
18. K.H. Johnson, Crit.Rev. Solid State and Mat.Sci. 7, 101 (1978).
19. J.A. Tossell, D.J. Vaughan, and K.H. Johnson, Chem.Phys.Lett. 20, 329 (1973); J.A. Tossell, J.Am.Chem.Soc. 97, 4840 (1975); J.Phys.Chem. Solids 36, 1723 (1975).

20, F.W. Lytle, G.H. Via, and J.H. Sinfelt, J.Chem.Phys. 67, 3831 (1977).
21. E.B. Prestidge, G.H. Via, and J.H. Sinfelt, J.Catal. 50, 115 (1977).
22. C.Y. Yang, Ph.D. Thesis, Department of Materials Science and Engineering, Massachusetts Institute of Technology, June, 1977.
23. C.Y. Yang, K.H. Johnson, D.R. Salahub, J. Kasper, and R.P. Messmer, Phys.Rev. B24, 5673 (1981).
24. R.A. Tawil and J. Callaway, Phys.Rev. B7, 4242 (1973).
25. B. Abeles, in Applied Solid State Science (Academic, New York, 1976), p. 90.
26. P.A. Montano, P.H. Barrett, and Z. Stanfield, Solid State Comm. 15, 1675 (1974); J.Chem.Phys. 64, 2896 (1976).
27. C.G. Shull and Y. Yamada, J.Phys.Soc. Japan 17S, BIII, 1 (1962).
28. C.G. Shull and H.A. Mook, Phys.Rev.Lett. 16, 184 (1966).
29. M.K. Wilkinson and C.G. Shull, Phys.Rev. 103, 516 (1956).
30. H.A. Gersch, C.G. Shull, and M.K. Wilkinson, Phys.Rev. 103, 103 (1956).
31. S. Spooner and B.L. Averbach, Phys.Rev. 142, 291 (1966).
32. A.J. Heeger, in Solid State Physics, edited by F. Seitz, D. Turnbull, and H. Ehrenreich (Academic, New York, 1969), Vol. 23, p. 283.
33. G. Grüner, Adv. in Phys. 23, 941 (1974).
34. J. Friedel, Can.J.Phys. 34, 1190 (1956); J.Phys. Radium 19, 573 (1958); Suppl. Nuovo Cimento VII, 287 (1958).
35. P.W. Anderson, Phys.Rev. 124, 41 (1961).
36. H. Höchst, P. Steiner, and S. Hüfner, Z.Phys. B38, 201 (1980).
37. K.H. Johnson, D.D. Vvedensky, and R.P. Messmer, Phys.Rev. B19, 1519 (1979).
38. D.R. Salahub and R.P. Messmer, Phys.Rev. B16, 2526 (1977).
39. P. Steiner, H. Höchst, W. Steffen, and S. Hüfner, Z.Phys. B38, 191 (1980).
40. K.H. Johnson, Phys.Rev. (manuscript in preparation).
41. T.E. Fischer, S.R. Kelemen, K.P. Wang, and K.H. Johnson, Phys. Rev. B20, 3124 (1979).
42. O. Rapp, B. Lindberg, H.S. Chen, and K.V. Rao, J.Less-Common Metals 62, 221 (1978).
43. F. London, Superfluids, Vol. 1 (Wiley, New York, 1950).
44. J. Barden, L.N. Cooper, and J.F. Schrieffer, Phys.Rev. 108, 1175 (1957); L.N. Cooper, Phys.Rev. 104, 1189 (1956).
45. V.F. Weisskopf (to be published and private communication).
46. M.R. Schafroth, J.M. Blatt, and S.T. Buther, Helv.Phys.Acta 30, 93 (1957); M.R. Schafroth, Phys.Rev. 130, 1244 (1958).
47. Physics Today 34, 17 (1981).
48. S.N. Ekbote and A.V. Narlikar, Mat.Res.Bull. 15, 827 (1980).
49. K.H. Johnson and R.P. Messmer, Synth. Metals (in press).

SOME APPLICATIONS OF THE DENSITY-FUNCTIONAL FORMALISM TO CHEMISORPTION

B.I. Lundqvist

Institute of Theoretical Physics
Chalmers University of Technology
S-412 96 Göteborg, Sweden

Some recent applications of the density-functional formalism using the local-density approximation for exchange-correlation effects are briefly reviewed.

1. INTRODUCTION

Among the many areas, where the density-functional formalism,[1] in particular with the local-density approximation for exchange-correlation effects,[2] has developed to a useful tool, is chemisorption. The usefulness in this context is not only the possibility of calculating numbers that characterize the chemisorption with a reasonable accuracy. At least as useful and certainly more stimulating is the fact that the tool contributes to the development of new concepts and ideas and the identification of key factors in the chemisorption process.

Chemisorption is the type of sorption, where atoms and molecules are attracted to a solid or liquid surface due to forces that emerge from the sharing of electrons between the adsorbate and substrate.[3] At focus is thus the electronic structure of the combined adsorbate-substrate system, or rather, still more focused, the changes in the electron structure caused by the presence of the extra adparticle. This distinction is made, as it is reflected in the theoretical approaches to chemisorption.[3] Among the methods that aim at a description of chemisorption by solving the Schrödinger-like equations of the density-functional formalism there is one group that directly attack the electron structure of the combined adsorbate-substrate system, approximating the latter by, e.g., a cluster[4] or a slab.[5] There is also another group of methods, in which the starting point is the solution of the electron-structure problem of the substrate, described as a semi-infinite jellium, a semi-infinite crystal or a slab, into which the adsorbated is then embedded.[6-8] This paper will be devoted to the application of two embedding schemes, one developed by Gunnarsson and Hjelmberg for the numerical solution of the Kohn-Sham equations,[6] and another one, due to Nørskov and others,[9-11] which is particularly useful for qualitative and semi-quantitative considerations.

In practice, a brief account is given of an ongoing program aiming at the theoretical description of the static and dynamic aspects of the adsorption of atoms and molecules on metal surfaces.[9] This program is addressed to such questions as: Why are molecules attracted to surfaces? How can molecules become trapped on surfaces? Why do molecular bonds break at surfaces? What is the dynamics of these phenomena? In addition, it provides data for an interplay with experiment.

The presentation will cover the following points: Section 2 describes the main features of the calculational method and the models used. Some characteristic properties of and trends in the adsorbate-induced electron-structure are described in Section 3. Potential-energy curves of simple atomic molecular species on free-electron-like metals are reviewed in Section 4, together with a conceptual picture of the bond-weakening of molecules on metal surfaces. The effective-medium scheme is described in Section 5, and in Section 6 applications to transition-metal substrates are described. The paper is concluded with some perspectives on this kind of approach to chemisorption theory.

2. CALCULATIONAL METHODS AND MODELS

The theoretical description of the chemisorption problem is made complex by the (N+1)-atom nature of the problem, with N being a macroscopic number, and by its low symmetry. The Kohn-Sham (KS) scheme[12] has the advantages of (1) being simple, providing a basis for a one-electron molecular orbital (MO) description, (2) including exchange and correlation effects, and (3) having shown to give results with a useful accuracy for atoms, molecules, and solids, even in approximate versions.[12]

The crucial quantity in this scheme is the exchange-correlation (XC) energy functional $E_{xc}[\rho]$. As an exact knowledge of this functional is equivalent to knowing the exact solution to the whole many-body problem, the KS scheme is useful only through the introduction of approximate functional forms for it. The most commonly used approximation, <u>the local density (LD) approximation</u>,

$$E_{xc}[\rho] \approx \int \rho(r)\varepsilon_{xc}(\rho(r))d^3r \qquad (1)$$

where $\varepsilon_{xc}(\rho)$ is the XC energy per electron of a homogeneous electron liquid with density ρ, means that one replaces the true XC hole locally by that for the homogeneous electron liquid at the local density. For spin polarized systems, Eq. (1) can be generalized to a local spin density (LSD) approximation.[12,14] These approximations and approximations with non-local dependences on the density are discussed in the paper by Jones and Gunnarsson in this volume.[13]

Traditionally, a single adsorbate on a semi-infinite substrate has been viewed as the central chemisorption configuration. This geometry has a very low symmetry, with the periodicity of the surface destroyed by the adsorbate. If the ion-lattice structure of the substrate is ignored, a cylindrical symmetry simplifies the problem, although the strong variations along the rotational axis make the problem far from trivial. This is the approximation of the so-called semi-infinite jellium or the planar, uniform positive background model. This model for the bare metal surface has been shown to represent accurately the ground state properties of free-electron-like metals,[15] and should describe well the interference, charge transfer and screening effects occuring in the adsorbate-substrate system and, with a good XC energy functional, also the Coulomb repulsion for such metals.

Lang and Williams (LW)[16] have designed an efficient method of solving the equations of the Kohn-Sham scheme, tailor-made for the adatom-jellium system. For the combined system, the only major simplification that can be utilized is the fact that the metal electrons screen out the effects of the adatom on the charge density and potential (though not on the wave functions) except in the vicinity of the adatom. Lang and Williams exploit this fact, when solving the Lippman-Schwinger equation equivalent to the one-electron equation of the KS-scheme for the continuum states. They have calculated local densities of states, densities, dipole moments, equilibrium bond lengths, and chemisorption energies for H, Li, O,[16] Si, and Cl[17] on a high density metallic substrate.

Gunnarsson and Hjelmberg[7] have suggested another method that has been applied to the adatom-jellium problem, too, but which has a more general relevance as an embedding scheme. This scheme is inspired by Grimley's[6] method to treat the overcompleteness problem. It utilizes the fact that screening by the substrate conduction electrons makes the changes in the electron density localized around the adsorbate. To describe the density in this local region as well as possible a *finite* set of localized functions, $\{\phi_n\}_{n=1}^{N}$, is introduced, and the problem of solving the Kohn-Sham one-electron equation in the adsorbate-substrate geometry is formulated as an NxN matrix problem. Far away from the adsorbate, on the other hand, the solutions for the clean surface, $\{\phi_k\}$, give a convenient description. The MO solutions for the combined system are thus suitably expanded in terms of this combined basis,

$$|\chi_\nu\rangle = \sum_{n=1}^{N} c_n^\nu |n\rangle + \sum_k c_k^\nu |k\rangle \tag{2}$$

The overcompleteness of (2) is remedied by imposing the subsidiary conditions

$$\sum_k c_k^\nu \langle n|k\rangle = 0, \quad n = 1,\ldots,N \tag{3}$$

Eqs. (2) and (3) can be systematically used in the Kohn-Sham equations to generate the equation for the adsorbate-induced Green function and approximations thereto.[7]

In terms of the projection operator $P = \sum_{n=1}^{N} |n\rangle\langle n|$ the assumption that the induced density can be expressed in terms of the local basis set only can be written as[18]

$$\Delta\rho(\vec{r}) \simeq -\frac{1}{\pi}\int_{-\infty}^{\varepsilon} F d\varepsilon \ \mathrm{Im}\langle\vec{r}|P(G - G^M)P|\vec{r}\rangle \tag{4}$$

where the pure-metal Green function G^M satisfies

$$(\varepsilon - H^M)G^M = 1 \tag{5}$$

and the corresponding equation for the combined system can be written

$$(\varepsilon - H)G = 1 \tag{6}$$

where $H = H^M + V$, V being defined by

$$\delta v_{eff}(\vec{r}) \equiv v_{eff}[n;\vec{r}] - v_{eff}[n_M;\vec{r}] \tag{7}$$

By assuming approximate completeness, $P \simeq 1$, in Eqs. (5) and (6) one gets

$$G_A = G_A^M (1 - V_A G_A^M)^{-1} \tag{8}$$

In practice three approximations have been made[7,18] that all improve, when the set of functions $\{\phi_n(\vec{r})\}$ gets more complete in the perturbed region, which usually occurs by increasing the number N of these functions. The practical usefulness of the method, on the other hand, hinges upon the matrices, and thus N, being not too big.

A prerequisite for the method is the Green function G^M for the clean substrate. A relatively simple but yet physically reasonable evaluation of G^M is obtained in the semi-infinite jellium model.[19] Effects of the ion lattice have so far only been accounted for in a perturbative way.[19] Originally designed for atomic substrates, the method has also been used for molecular hydrogen, defining the local basis set on a sphere with center on the midpoint of the molecular axis.[20,21,22]

3. ADSORBATE-INDUCED ELECTRON STRUCTURE

When an atom or molecule comes close to a surface, its electron structure undergoes significant changes. The sharing of electrons between the adsorbate and the substrate leads both to interference (resonance, hopping or mixing) between the two kinds of states and to increased Coulomb repulsion due to the penetration of electrons. The interference causes shifts of the adsorbate levels. The quasi-continuum of substrate one-electron states makes the adsorbate levels to broaden, corresponding to a decay of the adsorbate state into bulk states, unless their energy lies in an energy gap.[23] It also causes the level shift to vary with the distance from the surface in a way that is characteristic for both substrate and adsorbate electron features. In accounting for adsorbate-level variation, one has to distinguish between ionization and affinity levels. Already at large distances, the image charge effect makes an ionization level move upwards in energy upon approaching the surface, the ionization potential being effectively $I_{eff} = I - e^2/4d$, and an electron affinity level downwards, with the affinity energy $A_{eff} = A + e^2/4d$,[24] the distance d here being measured from the image plane.[15] The classical value for the image energy does not apply very close to the image plane, however.

In this region the calculations on adsorbates on jellium (Je) surfaces (Fig. 1) for atomic adsorbates (O, Si and Cl)[16] show both a dependence on the individual character of the adsorbate state and a certain correlation with the variations of the effective electron potential V^o_{eff}, i.e., the total potential felt by the electron in the clean substrate.

The correlation has thoroughly been established in calculations for H on Je[18] and H_2 on Je[20,21] by varying the electron density (the bulk electron density $n_o = (4\pi r_s^3/3)^{-1}$ is commonly expressed in terms of the radius r_s in atomic units). The resulting H-induced density of states $\Delta n(\varepsilon)$ (Fig. 2) has a mean peak position that follows $V^o_{eff}(d)$ closely.

The results of these kinds of calculations have been condensed into a simple principle:[25] The interaction with metals shifts affinity energy levels downwards and increases the population of them. On metals, the latter can occur with little energy cost, as the Coulomb repulsion between two "adsorbate" electrons is reduced considerably, due to the screening by the conduction electrons. This mechanism is intimately tied to the fundamental property of metals of having a quasi-continuous electron excitation spectrum. Therefore, the simple principle is expected to have a general applicability.

The principle has further illustrations from results for H_2 on jellium.[20-22] Fig. 3 presents some features of the H_2-induced electron structure for a hydrogen molecule on an Mg-surface. A free H_2

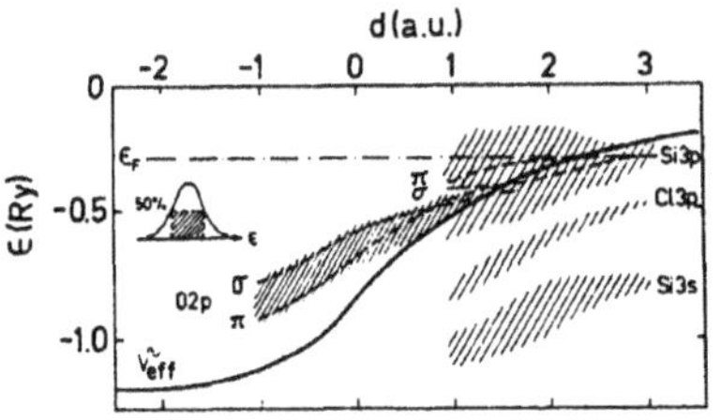

Figure 1. Characteristic behavior of O, Si and Cl resonances as a function of metal-adatom separation d (measured from the jellium edge) for adsorption on a jellium substrate with $r_s = 2$. The shaded region indicates the energy values within the full width at half the maximum for each d value. The fulldrawn curve gives the effective one-electron potential V^o_{eff} of the bare metal, and the dash-dotted line indicates the Fermi level ε_F of the substrate. The dashed curves show peak positions for $p_z(\sigma)$ and $p_{x,y}(\pi)$ components of O-2p and Si-3p resonances. The material is from Ref. 16 and N.D. Lang (private communication).

molecule has a doubly occupied, bonding 1σ molecular orbital and one empty, antibonding $2\sigma^x$ molecular-orbital resonance.

As the molecule comes close to the surface, the energies of the split-off $1\tilde{\sigma}$ orbital (the tilde indicating a chemisorbed state) and the $2\tilde{\sigma}^x$ molecular-orbital resonance, as given by the Σ_u-projection of the H_2-induced density of states, are shifted downwards. Lying below the bottom of the substrate conduction band, the $1\tilde{\sigma}$ level is sharp. The broad $2\tilde{\sigma}^x$ resonance gets a still larger shift downwards in mean energy together with an increased width. The bonding-antibonding separation thus gets reduced, as the molecule is pressed into the jellium. Concomitantly the antibonding $2\tilde{\sigma}^x$ resonance gets an increased occupation.

Figures 1-3 illustrate that the adsorbate-induced electron densities of states have resonance peaks, whose energy vary with the distance from the surface in a characteristic way. Certain rules can be set up for this variation and how it depends on substrate properties. In some cases it just follows the variation of $V^o_{eff}(d)$.[16] More generally, proper projections of the clean-substrate density of states appear to affect the variation in an important way.[27] An adsorbate-induced state or resonance closely follows the projection with the same symmetry. The atomic H resonance has s character and should therefore follow $V^o_{eff}(d)$ (cf. Figs. 1 and 2) rather closely. For H_2 parallel (perpendicular) to the surface the antibonding resonance has $\pi(\sigma)$ character. This makes the antibonding resonance shifted down less in the perpendicular than in the parallel case. Similarly, one can trace the difference in energy between the π- and σ-resonances

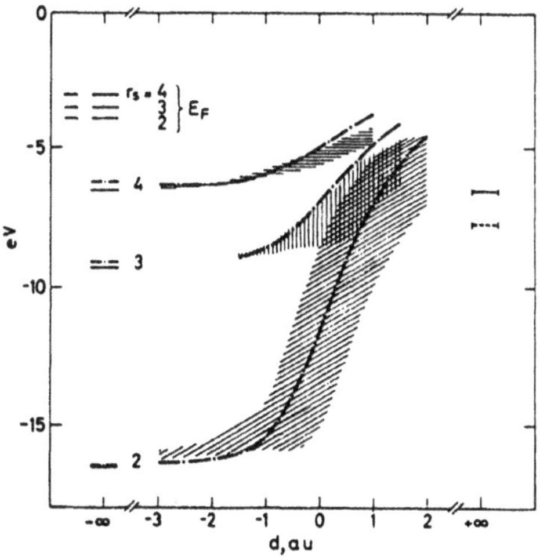

Figure 2. The peak position (shaded area) in the H-induced density of states[19], as a function of the distance d between the H atom and jellium edge, is shown for different r_s values, corresponding to the conduction electron densities of Na, Mg and Al, respectively. The dash-dotted curves give the effective electron potential of the clean surface, $V^o_{eff}(d)$. To the left ($d = -\infty$), the free-atom results are given both for a spin-polarized (dashed line) and an unpolarized (full line) calculation.

for O and Si adsorbed on Al (Fig. 1) to the difference in symmetry.

The adsorbate-induced electron structure also depends on the interatomic distance of the molecular adsorbate. Figure 3 shows that for a fixed distance d but for an increased H-H-separation R, the split-off $1\tilde{\sigma}$ -level is shifted upwards in energy and the antibonding $2\sigma^*$ resonance downwards. This means that with larger R value the intramolecular bonding and antibonding resonances come closer in energy and also closer to the H-induced resonance for adsorbed atomic hydrogen. This is all in agreement with what a simple consideration over the decreasing atomic overlap with increasing R would imply.

4. POTENTIAL-ENERGY SURFACES

The above mentioned scheme has been applied to the calculation of potential-energy (PE) surfaces of light atoms[16-19] and molecule[20-22] on free-electron-like metal surfaces. So far, H_2 is the only molecule, for which self-consistent calculations have been performed. Results for H_2 on the Mg (0001) surface will be used here as

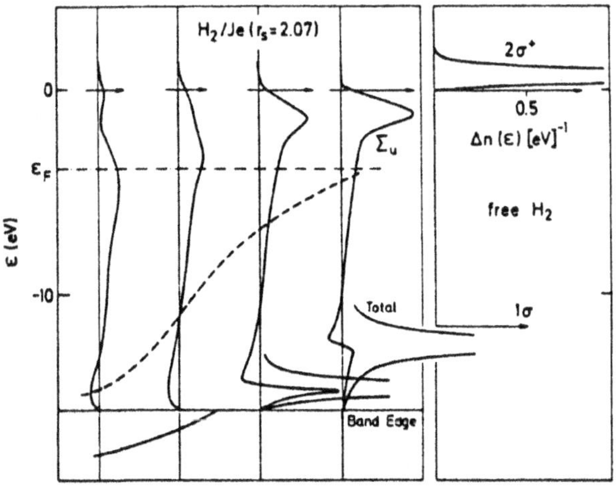

Figure 3. H_2-induced density of states (DOS), $\Delta n(\varepsilon)$, for H_2 standing upright on a jellium surface, as a function of the distance d, between the surface and the midpoint of H_2, with r_s = 2.07, as for Al. The $2\tilde{\sigma}^x$ MO resonance is identified with the Σ_u projection of $\Delta n(\varepsilon)$ (the $1\tilde{\sigma}$ MO, denoted "total" in the figure, contains all symmetries). The internuclear distance is kept at R = 1.4 a.u. The dashed curve denotes $V^o_{eff}(d)$, the effective electron potential of the clean jellium surface. In the upper right corner the free H_2 DOS is drawn as comparison.[20]

an illustration.[22] It should be stressed that the numbers given for the potential energies in Fig. 5 should not be read literally. Limited accuracy of the local-density approximation, the jellium model, the incomplete set of local basis functions, the first-order expression for the lattice effects etc. introduce an uncertainly also in relative PE-values of at least a few tenths of an eV. Being the first of its kind, however, this study helps to develop concepts like activation barrier, precursor state and associative versus dissociative adsorption.

With these reservations the following picture of the adsorption process emerges from the calculations: (1) Upon approaching the surface, the molecule feels an activation barrier. The barrier has a pass outside the atop position and the molecules will be directed towards this site. (2) A molecule passing the outer barrier cannot dissociate directly over the atop site. A second activation barrier prevents the dissociation into two separated H atoms in the energetically more favourable bridge (Fig. 5) or center sites. (3) For dissociation from the bridge site into the adjacent center site the ac-

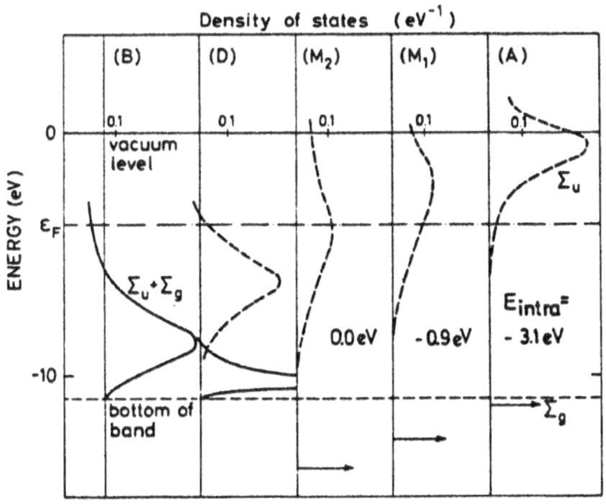

Figure 4. Development in the one-electron density of states along the reaction path in Fig. 5 for an H_2 molecule approaching an Mg (0001) surface. First the antibonding (Σ_u) and bonding (Σ_g)H_2 states are shifted down as the distance d to the first Mg layer decreases for an almost constant H-H distance R. Two spectra are shown corresponding to d = 3.5 a.u. (M_1) and d = 2.5 a.u. (M_2) in the molecular minimum (M) of Fig 5. As the antibonding resonance gets filled, the H-H interaction E_{intra} goes to zero. As the molecule dissociates, M→ D→ B, the bonding-antibonding splitting decreases and finally two noninteracting H resonances result.

tivation energy is much lower. (4) If the molecule does not enter the surface in the particular geometry mentioned in (3) (which is unlikely) and therefore cannot dissociate directly, it can be adsorbed into a molecular state (the local minimum M in Fig. 5) and then migrate along the surface easily.

The features of the PE-surface can be understood in terms of the behaviour of the H-induced electron structure (Fig. 4). The initial increase in energy seen in Fig. 5, when the molecule approaches the surface, is analogous to the kinetic-energy repulsion found, when other closed-shell adsorbates, such as the isoelectronic He, approach a surface.[28] For H_2, the repulsion levels off to an activation barrier because the antibonding orbital of H_2 is shifted down and gradually filled, as the molecule approaches the surface (Fig. 4). This filling of the antibonding H_2 orbital, on the other hand, weakens the H-H interaction E_{intra}, as indicated in Fig. 4. The activation barrier for molecular adsorption can therefore also be thought of as

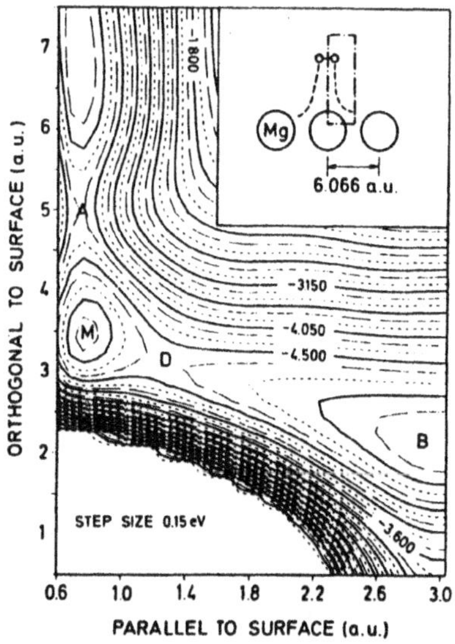

Figure 5. Equipotential-energy curves for H_2 on a Mg (0001) surface, with H_2 parallel to the surface dissociating over the atop site into the bridge sites (see geometry in inset). The energies in electron volts are relative to those of the free atoms. The molecular adsorption energy in M is thus only 0.4 eV. The distance from the surface is measured from the first atomic layer, and the distance parallel to the surface, which equals one-half of the H-H separation R, is measured from the atop site towards the bridge site.[22]

being due to the fact that the H-H interaction (E_{intra}) starts weakening before the H-surface bonds (E_{extra}) have started to build up. In this E_{intra}/E_{extra} picture the activation energy E_a^d for dissociation is due to the existence of the residual H-H interaction even close to the surface. As the molecule dissociates, M → D → B in Figs. 4 and 5, the bonding-antibonding splitting decreases, and finally two interacting H resonances result. Since the filling of the antibonding resonance and thereby the size of the H-H interaction depend on the H_2-surface distance, there is a distribution of E_a^d's that depends on where on the surface the dissociation takes place. E_a^d is smallest when the molecule dissociates from the twofold bridge site, where the minimum in M is close to the surface, into the threefold center sites, which are the most stable sites for atomic adsorption.[19] This site specificity of E_a^d is very important in an interpretation of experimental kinetic data.

Finally, the repulsion from E_{intra} prevents the H atoms from getting close enough to the substrate atoms to really feel the difference between the different sites. This makes the activation energy for molecular diffusion all over the surface very small. Similarly, it is the kinetic-energy repulsion that makes atomic adsorbates migrate easily, however, only along certain diffusion paths.[19]

The bond of the precursor state M is chemical, although weak, resulting from cancellations beteen the simultaneous buildup of the H-surface bond (E_{extra}) and weakening of the H-H bond (E_{intra}). The large electronic rearrangements behind it are shown in Fig. 4. A physisorbed state should exist outside the outer activation barrier as it does for the rare gases.

5. EFFECTIVE MEDIUM SCHEMES AND CHEMISORPTION ON TRANSITION METHODS

The calculation described so far has been self-consistent, what concerns the interaction of the atom or molecule with substrate electrons described in the jellium model. The lattice structure has then been introduced by treating the pseudopotentials due to the ions in first-order perturbation theory. From such calculations it has become apparent that the local density of the host material is a key parameter for the determination of several adsorbate/impurity-induced properties,[9,10,29,30] such as the chemisorption energy. A description where such properties are calculated, as for a homogeneous system at the local density, gives a good zeroth-order estimate but has to be corrected for local variations, due to inhomogeneities in the electrostatic potential and in the electron structure due to d-electrons etc.

The Effective-Medium Theory is an attempt to account for the effects of the local interactions in a simple but quantitative way.[9-11] The basic idea is calculationally to replace the true, low-symmetric host by an "effective" host, having a higher symmetry and thereby being simpler to describe theoretically.[9-11] The most symmetric effective host is the homogeneous electron gas. The chemisorption or, more generally, embedding energy ΔE, defined as the difference in energy between the combined atom and host system and the separated atom and host, would in the simplest approximation be

$$\Delta E^{(o)}(\underline{r}) \simeq \Delta E^{hom}(\bar{n}_o(\underline{r})) \tag{9}$$

where $\Delta E^{hom}(n)$ is the embedding energy of the atom in a homogenous electron gas of electron density n, $n_o(\underline{r})$ is the host electron density at the site $\underline{r}$ of the atom, and $\bar{n}_o(\underline{r}) \simeq n_o(\underline{r})$ in this approximation. The host is characterized by only one quantity, $n_o(\underline{r})$. The properties of the atom are given by $E^{hom}(n)$, the so-called immersion energy that can be calculated once and for all for each atom[31] or molecule.

One can account for effects of, e.g. d electrons and local variations in the electrostatic potential by correcting Eq. (9) in a straightforward manner. The starting point is the Hohenberg-Kohn density functional[1] for the total energy in the Kohn-Sham form[2], using the local-density approximation for the exchange-correlation energy.[3]

If deviations of the host density $n_o(\underline{r})$ from homogeneity is considered in perturbation theory, one finds that the average density $n_o(\underline{r})$ in Eq. (9) should be obtained with the atom-induced electrostatic potential in the homogeneous electron gas with the density $n_o(\underline{r})$ as a weight function.[9-10] One also finds the first-order correction as an electrostatic interaction integral between the bare substrate electrostatic potential and the atom-induced charge density. The resulting approximation for ΔE is justified, when the substrate-electron density and electrostatic potential do not vary too strongly over a characteristic atomic volume and when the adsorbate is not particularly polarizable. The basis for the perturbation theory is the screening that makes the atomic charge density localized. This atom-induced density has Friedel oscillations that only slowly decay with the distance from the atom, however. In cases with strong scattering centers the first-order electrostatic integral is likely to give an unrealistically large contribution.

This problem can be solved by restricting the spatial region, where the deviation of the host density from homogeneity is assumed to be small, to the close vicinity of the atom.[11] Outside this region, where the host potential is treated to infinite order, the atom can be regarded as the small perturbation. With such a mixed perturbation theory, Eq. (9) and the first-order term are recovered, however with all integrals over the atomic region only. The outside region gives rise to an additional term that represents the covalent interaction between the atom and the host. When choosing the homogeneous electron gas as the effective medium, the chemisorption energy can be written[11]

$$\Delta E \simeq \Delta E^{(o)} + \int_a \phi_o(\underline{r})\Delta\rho_a(\underline{r})d\underline{r} + \Delta E^{cov} \qquad (10)$$

where $\phi_o(\underline{r})$ is the host electrostatic potential, $\Delta\rho_a(\underline{r})$ the atom-induced charge density within region a close to the atom, as calculated in the homogeneous host, and $\Delta E^{cov} = \delta\Delta\ (\Sigma_j\varepsilon_j)$ describes the covalent aspects of the binding. The latter interpretation can be given to the difference in atom-induced shifts in the one-electron energy parameters, $\delta\Delta(\Sigma_j\varepsilon_j)$, since it is governed mainly by the possibility of finding a resonance between the atom- and host-derived one-electron energies.[11] The covalent term ΔE^{cov} requires a calculation of the one-electron spectrum for the atom in the real host for fixed potentials, i.e. not self-consistently. Together with the fact that ΔE^{cov} can often be treated more approximately than the first two terms of Eq. (10), when the latter give the major contribution of ΔE, as for hydrogen, this makes ΔE^{cov} accessible with rather simple methods. For chemisorbed hydrogen or hydrogen as an interstitial impurity in a transition metal, the 1s level interacts with the valence electrons of the host. The covalent term ΔE^{cov} measures the difference in hybridization of this level, when going from the homogeneous effective medium to the transition-metal host. As the effective medium has only s- and p-electrons, and as in the transition metals the s- and p-wave phase shifts are rather free-electron like, the major difference then is the effect of the host d-electrons. With a rather crude description of ΔE^{cov}, the potential-energy curves for H on Ni (100) and Ni (111) in Fig. 6 have been calculated, and the observed trends in the hydrogen chemisorption energy and heat of solution along the transition-metal series can be accounted for, the main variation in the covalent term coming from the increased filling of the host d-band upon going to the right in the series.[11-32] The covalent term is particularly important in, e.g., explaining the differences between Cu and Ni. In Cu it is zero, because both the bonding and antibonding levels, resulting from the 1s-d-hybridization, are filled, while the higher energy of the d-band in Ni results in a dominance of bonding effects. In an extensive calculation for the three transition-metal series, chemisorption energy, equilibrium position, and vibrational frequency of H, calculated in this way[32], agree well with experimental data, where such are available.

In a recent series of papers, Upton, Goddard, and coworkers[4] have performed systematic studies of a sequence of atomic adsorbates

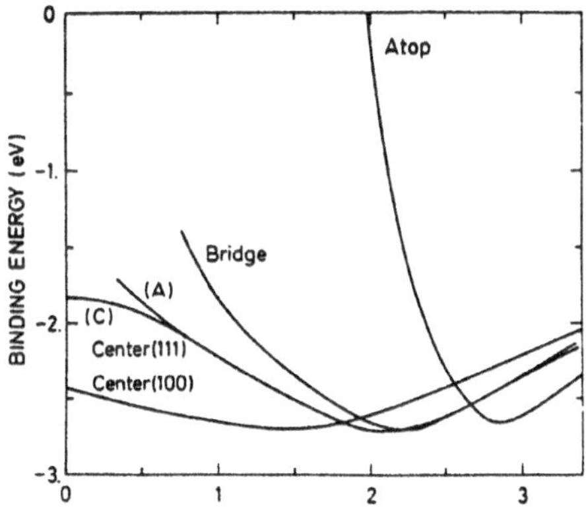

Figure 6. Binding energy of H as a function of distance outside the atop, bridge and center sites on the Ni (100) and (111) surfaces. The atop and bridge site potential-energy curves are indistinguishable for the two surfaces. For the (111) surface, the difference between center sites with and without an atom underneath [center (A) and center (C), respectively] only shows up very close to the surface. The distance is relative to the first Ni layer.[11]

on nickel surfaces, modelled by small clusters. Characteristic features of these calculations are that the conduction electrons are described as moving in an effective potential from the Ni-electrons in an averaged d^9-configuration.[4] The electron structure is then calculated from first principles on a Hartree-Fock level and, if found necessary, with electron correlation effects included via the Generalized Valence-Bond method. On a 20-atom Ni cluster bulk properties and properties at four-coordinate, two-coordinate and terminal sites have been modelled. The main results of the H calculations are that the chemisorption energies increase with ligancy of the binding site, i.e. with the number of nearest neighbours, that the Ni-H bond distance increases with ligancy, and that the vibrational frequency of the H atom relative to the surface decreases markedly with H ligancy.[4]

To indicate the status of our present quantitative knowledge about potential-energy surfaces, a comparison between calculated and measured data for H adsorbed on Ni (100) can be made. The adsorption site is judged to be a center position in both theory and practice. The Effective-Medium Scheme, the Upton-Goddard method, and experiment give 2.7, 3.0 and 2.74 eV,[33] respectively, for the chemisorption energy and 79, 73 and 74 meV,[34] respectively, for the frequency of the vibration normal to the surface. On Ni (111) the corresponding energies and frequencies are 2.7, 3.2 and 2.74 eV[33] and 134, 155 and 139 meV,[35] respectively.

6. CONCLUSIONS

Some applications of the density-functional formalism[1] to the chemisorption problem have been briefly reviewed. In particular, two methods for calculating binding energies in the Kohn-Sham scheme[2] for localized species (atoms or molecules) in extended conducting media have been mentioned, the Gunnarsson-Hjelmberg embedding scheme[7] and the Effective-Medium Scheme of Nørskov.[11] Qualitative aspects on chemisorption have been stressed, such as the behaviour of the adsorbate-induced electron structure and the mechanism for breaking molecular bonds at surfaces. Quantitative results have been referred to, such as the potential-energy surfaces of H_2 on an Mg surface and of H on transition-metal surfaces.

There are many ways, by which these calculations can be extended and improved. For instance, calculations on other adsorbates, in particular strongly electronnegative or-positive ones, are desirable. Immersion energies for molecules would be valuable as inputs into Effective-Medium-Scheme calculations. As an example of a desirable improvement, the development of the embedding scheme with multi-center basis functions can be mentioned.

Due to their relative simplicity and good account of the actual physics, the two schemes should also in the future be valuable, by themselves and as supplements to the first-principle calculations on more realistic model systems, from which now potential-energy curves are beginning to result.[5]

ACKNOWLEDGEMENT

This work has been supported by the Swedish Natural Science Council.

REFERENCES

1. P. Hohenberg and W. Kohn, Phys.Rev. 136, B 864 (1964).
2. W. Kohn and L.J. Sham, Phys.Rev. 140, A 1133 (1965).
3. See, e.g., B.I. Lundqvist, H. Hjelmberg and O. Gunnarsson, in: Photoemission and the Electronic Properties of Surfaces, Eds. B. Feuerbacher, B. Fitton and R.F. Willis, Wiley, New York (1978); J.P. Muscat and D.M. Newns, Progr.Surface Sci. 9, 1 (1978).
4. See, e.g., T.H. Upton and W.A. Goddard, III, Phys.Rev.Lett. 42, 472 (1979); CRC Crit.Revs. in Solid State and Materials Sciences 10, 261 (1981).
5. C. Umregar, Bull.Am.Phys.Soc. 28, 538 (1983). For some of the ingredients of this calculation, see e.g., O.K. Andersen, Phys.Rev. B 12, 3060 (1975), H. Krakauer, M. Pasternak and A.J. Freeman, Phys.Rev. B 19, 1706 (1979) and M. Weinert, E. Wimmer and A.J. Freeman, Phys.Rev. B 26, 4571 (1982).
6. T.B. Grimsley and C. Pisani, J.Phys. C 7, 2831 (1974).
7. O. Gunnarsson and H. Hjelmberg, Physica Scripta 20, 192 (1975).
8. J. Inglesfield, J.Phys. C 14, 3795 (1981); J.Phys. F 11, L 287 (1981).
9. J.K. Nørskov and N.D. Lang, Phys.Rev. B 21, 2136 (1980).
10. M.J. Stott and E. Zaremba, Phys.Rev. B 22, 1564 (1980).
11. J.K. Nørskov, Phys.Rev. B 26, 2875 (1982); Phys.Rev.Lett. 48, 1620 (1982).
12. See, e.g., O. Gunnarsson and B.I. Lundqvist, Phys.Rev. B 13, 4274 (1976) and O. Gunnarsson, in: NATO, Advanced Study Institute Series, Vol. 42, Electrons in disordered metals and at metallic surfaces, Eds. P. Phariseau and B.L. Gyorffy.
13. U. von Barth and L. Hedin, J.Phys. C 5, 1629 (1972).
14. R.O. Jones and O. Gunnarsson, in this volume.
15. N.D. Lang, Solid State Phys. 28, 225 (1973); In: Theory of the Inhomogeneous Electron Gas, Eds. S. Lundqvist and N.H. March, Plenum Press, in print.
16. N.D. Lang and A.R. Williams, Phys.Rev. B 18, 616 (1978) and references therein.
17. N.D. Lang and A.R. Williams, Phys.Rev.Lett. 37, 212 (1976).
18. H. Hjelmberg, Physica Scripta 18, 481 (1978).
19. O. Gunnarsson, H. Hjelmberg and B.I. Lundqvist, Phys.Rev.Letters 37, 292 (1976); Surface Sci. 63, 348 (1977).

H. Hjelmberg, O. Gunnarsson and B.I. Lundqvist, Surface Sci. 68 158 (1977); H. Hjelmberg, Surface Sci. 81, 539 (1979).
20. H. Hjelmberg, B.I. Lundqvist, and J.K. Nørskov, Phys.Scripta 20, 192 (1979).
21. P.K. Johansson, Surf.Sci. 104, 510 (1981).
22. J.K. Nørskov, A. Houmøller, P.K. Johansson and B.I. Lundqvist, Phys. Rev.Lett. 46, 257 (1981).
23. R.W. Gurney, Phys.Rev. 47, 479 (1935).
24. A.C. Hewson and D.M. Newns, Jap.J.Appl.Phys.Suppl. 2, 2:121 (1974).
25. B.I. Lundqvist, O. Gunnarsson, H. Hjelmberg and J.K. Nørskov, Surf.Sci. 89, 196 (1979).
26. P.K. Johansson, B.I. Lundqvist, A. Houmøller and J.K. Nørskov, in: Recent Developments in Condensed Matter Physics, Ed. J.T. Devreese, Plenum Publ.Corp., p. 605 (1981).
27. O. Gunnarsson, H. Hjelmberg and J.K. Nørskov, Phys.Scripta 22, 165 (1980).
28. N. Esbjerg and J.K. Nørskov, Phys.Rev.Lett. 45, 807 (1980).
29. B.I. Lundqvist, J.K. Nørskov and H. Hjelmberg, Surface Sci. 80, 441 (1979).
30. J.K. Nørskov, Solid State Commun. 24, 691 (1977); Solid State Commun. 25, 995 (1978); Phys.Rev. B 20, 446 (1979).
31. M.J. Puska, R.M. Nieminen and M. Manninen, Phys.Rev. B 24, 3037 (1980).
32. P. Nordlander, S. Holloway and J.K. Nørskov, to be published.
33. K. Christmann, O. Schober, G. Ertl and M. Neumann, J.Chem.Phys. 60, 472 (1974).
34. S. Andersson, Chem.Phys.Lett. 55, 185 (1978).
35. W. Ho, J. DiNardo and E.W. Plummer, J.Vac.Sci.Technol. 17, 134 (1980).

DENSITY FUNCTIONAL CALCULATIONS FOR ATOMIC CLUSTERS

J. Harris and J. E. Müller

Institut für Festkörperforschung
der Kernforschungsanlage Jülich
D-5170 Jülich 1, W.-Germany

1. INTRODUCTION

a. The HKS-Functional

The calculation of the adiabatic ground state energy $E[\{\vec{R}_i\}]$ of molecules, a long-standing problem in chemistry, is traditionally handled in one of two ways. Quantum chemists employ ab initio methods (meaning one or another brand of CI) which lead in principle to an exact solution of the many-electron wave-equation. Because the practical difficulties in achieving such a solution are formidable for all but the smallest systems, chemists over the years have developed empirical or semiempirical methods, such as extended-Hückel, which involves parameterization of a model Hamiltonian. These methods are not based on an attempt to solve the many-electron wave-equation, but rely on the transferability of chemical experience, i.e. the known behaviour of a given radical in a series of different environments is used to construct a model that can predict the behaviour of the same radical in a new environment. While this is an entirely legitimate scientific method giving useful results for very complex systems, it has obvious limitations. One might say that the two types of methods traditionally favoured by chemists are limited by placing either too much or too little emphasis on the many-electron wave-equation. This highlights the role that density functional methods might play in chemistry i.e. the many-electron problem is taken seriously, but not too seriously!

The idea of using a density functional to evaluate $E[\{\vec{R}_i\}]$ goes back to the era of Thomas and Fermi. However, the Thomas-Fermi functional is extremely limited in application because it fails to describe correctly the balance between kinetic and potential energy near the nuclei - the kinetic energy cost of a build up of charge near the nuclei is underestimated. This obviates the method for use in atomic structure calculations, and it was not until 1965 that Hohenberg, Kohn and Sham (HKS)[1] showed how to correct the flaw by separating off the dominant part of the kinetic energy and treating it exactly. In HKS's scheme, a Thomas-Fermi-like approximation is applied only to the exchange-correlation term in the energy functional.

In spite of this improvement, HKS, in their original papers, did not envisage applications of their functional in chemistry because the conditions that would justify the approximate treatment of exchange and correlation, namely that the electron density should be everywhere slowly varying, are blatantly violated. Only gradually, over the years, has it emerged that, for reasons only partially understood, the local approximation for exchange and correlation gives results of useful accuracy even where it should not. The most striking example is the case of the H_2 molecule, where the HKS-local density functional reproduces the exact binding energy curve with remarkable accuracy.[2] Results for larger diatomic molecules, while somewhat sobering,[3] nevertheless demonstrate a useful and consistent accuracy and we will show presently that this persists even in as delicate a case as the inversion curve of the ammonia molecule. We believe, therefore, that density functional calculations will play an important role in the next decade in complementing traditional methods based directly on the wave-equation.

An illustration of this role might be the case of the Be_2 molecule for which a He_2-like ground state energy curve was anticipated. The first evidence of a substantial bond energy was provided by an HKS local density calculation,[4] which gave a binding energy curve with a minimum at about 4.9 a.u., a separation clearly at variance with a Van der Waals binding mechanism. The physical origin of the increased binding, a hybridization effect involving the Be(2p) level, is automatically included in the density functional formalism, but is sufficiently obscure in a CI description that it was missed entirely until the density functional result suggested it ought to be there. This example, which of course should not be overinterpreted, points nevertheless to a central problem in CI. For systems where the ground state energy is not known experimentally, it is difficult to establish just when all the configurations of relevance have been included. There is, perhaps, some parallel with many-body perturbation theory, also in principle exact, but only if one is able to sum an infinite number of infinite series!

It is not our purpose in this article to give a detailed critique of other methods, or even to describe in detail the intricacies of the density functional formalism. The subject has been extensively reviewed,[5] and the original papers of HKS[1] are clear and accessible. Our main aim is to outline the way we implement the HKS scheme, and to illustrate its usefulness by presenting some results recently obtained for the H_2O and NH_3 molecules.

To establish notation, and to set up a discussion of a few essential points of special relevance to chemists, we describe briefly the main features of the HKS method. The functional is separated into three parts:

$$E[n] = T_o[n] + E_{es}[n] + E_{xc}[n] \tag{1.1}$$

where E_{es} is the total electrostatic energy (including interaction of electrons with nuclei and of nuclei with each other), T_o is the one-electron part of the kinetic energy, and E_{xc} by definition the remainder, is the exchange-correlation energy. The electrostatic energy is given by

$$E_{es}[n] = \int d\vec{r}\, n(\vec{r}) \left[V_N(\vec{r}) + \frac{1}{2}\phi(\vec{r})\right] + \sum_{i \neq j} \frac{Z_i Z_j}{2R_{ij}} \tag{1.2}$$

where $\phi(\vec{r})$ is the electrostatic potential due to the electron density $n(\vec{r})$, $V_N(\vec{r})$ that of the nuclei, and Z_i the nuclear charge of the ith atom. This energy term can be evaluated directly for an arbitrary electron density $n(\vec{r})$. The kinetic term T_o cannot be written explicitly in terms of the density but is defined with respect to a system of N non-interacting electrons in some external potential $V_{eff}(\vec{r})$. To evaluate T_o it is necessary to introduce a set of single particle orbitals $\psi_n(\vec{r})$, from which a determinantal state Ψ for the independent-electron system can be constructed.* The density $n(\vec{r})$ and T_o are then related via

$$n(\vec{r}) \equiv \langle\Psi|\hat{n}(\vec{r})|\Psi\rangle = \sum_n f_n |\psi_n(\vec{r})|^2 \tag{1.3}$$

and

$$T_o \equiv \langle\Psi| -\frac{1}{2}\sum_{i=1}^{N} \nabla_i^2 |\Psi\rangle = \sum_n f_n \int d\vec{r}\, \psi_n^*(\vec{r}) \left[-\frac{1}{2}\nabla^2\psi_n(\vec{r})\right] \tag{1.4}$$

where the f_n are occupation numbers characterizing the state Ψ. This feature of the HKS functional - that one cannot associate a T_o with a given $n(\vec{r})$, but must generate the two quantities in pairs via (1.3) and (1.4) - is unavoidable, and indeed, is the central new idea of the theory. The feature is at once responsible for the qualitative improvement in accuracy over Thomas-Fermi, and for the fact that the HKS functional cannot be minimized simply. That is, although we work with a <u>density functional</u>, we cannot avoid the

---

* Symmetry requirements sometimes demand the inclusion of more than one determinant.

introduction of one-electron orbitals. In practice, therefore, the HKS-scheme is an orbital theory. (We comment on the meaning of these orbitals presently.)

Once an approximation for E_{xc} has been agreed on, (1.1) can be minimized either by varying the coefficients of a suitably large basis of functions in which the $\psi_n(\vec{r})$ are expanded, or by converting the extremal condition to the solution of a one-electron equation, as in the Hartree-Fock method (the parallel is not accidental, as we will see.) The former, more direct route involves expanding each ψ_n in a basis, and evaluating E by minimizing in a multi-parameter space. As anyone knows who has tried it, this is very tricky because the variation of E in such a parameter space is neither smooth nor simple - one's minimizer tends to run up and down valleys and hop across saddle points ad infinitum. For this reason, minimization is almost universally achieved via the second, indirect route, i.e. the differential equation is used as an efficient and reliable guide to the minimum of the functional.

The equation that the ψ_n satisfy at the extremum[1] is

$$\{-\frac{1}{2}\nabla^2 + V_{eff}(\vec{r}) - \varepsilon_n\}\,\psi_n(\vec{r}) = 0 \tag{1.5}$$

with the auxiliary self-consistency condition

$$V_{eff}(\vec{r}) = V_N(\vec{r}) + \phi(\vec{r}) + \mu_{xc}(\vec{r}) \tag{1.6}$$

$$\mu_{xc}(\vec{r}) = \frac{\delta E_{xc}}{\delta n(\vec{r})} \tag{1.7}$$

Though it would be nice to be able to solve these equations exactly, this is neither easy, nor necessary if we bear in mind that the functional is stationary. The crucial element of an HKS calculation is not the solution of the differential equation but the evaluation of the functional. If this is done sufficiently accurately, the stationarity property of E will guarantee that we get a good energy from a solution of (1.5)-(1.7) that would be inadequate, say, for accurate eigenvalue determination.* The bulk of this article describes how we attempt to achieve a solution of (1.5)-(1.7) for atomic clusters - i.e. with localized boundary conditions - that is accurate enough, in the above sense, yet remains attainable for relatively large systems. To do this we expand the ψ_n in a somewhat unusual set of basis functions. The choice of basis is motivated by the requirement that it give reasonable accuracy with only a few functions per atom (typically one more than the number of bound valence levels in the free atom). We describe this basis in Section 2,

*The eigenvalues do not obey a variational principle because the differential equation is non-linear.

and show in Section 3 how the bits and pieces of the calculation can be efficiently programmed. Results demonstrating the relative accuracy of the resulting scheme are then presented in Section 4. In the remainder of this introduction we address a few general points of relevance to chemists, that we feel are not always adequately emphasized.

b. The Role of the One-Electron Equation: HKS vs. HF

In Hartree-Fock (HF) theory the determinant Ψ_{HF}, constructed from orbitals which solve the HF one-electron equations exactly, has a very precise significance. It is the best possible determinantal representation of the exact many-electron wave-function, in the sense that Ψ_{HF} minimizes the expectation value of the many-electron Hamiltonian in the space spanned by all wave-functions representable by a single determinant. This fact is largely responsible for the popularity of Hartree-Fock. However, the HF energy functional $\langle\Psi_D|\hat{H}|\Psi_D\rangle$, where Ψ_D is an arbitrary determinant, can also be thought of as a special case of the density functional, i.e. one in which E_{xc} in (1.1) is taken to be the "Fock" or "exchange" term of the HF expression. This being so, if some other approximation is used for E_{xc}, the determinantal function Ψ constructed from the solutions of (1.5)-(1.7) cannot be an optimal (in the above sense) representation of the many-electron wave-function.

One might ask, what is the relationship between the HKS one-electron determinant Ψ and the true ground state wave-function? This question is rather difficult to answer because the relationship is an extremely tenuous one. In spite of its orbital flavour, HKS theory makes no statement whatsoever about the wave-function of an interacting system. The function Ψ refers strictly to a non-interacting system and is introduced merely to relate density and one-particle kinetic energy. The most that can be said is that if $\mu_{xc}(\vec{r})$ were exactly known and (1.5)-(1.7) solved exactly, then Ψ would refer to an eigenstate of an independent electron system that corresponds to an electron density equal to the ground state density of the interacting system. If Ψ is used to evaluate matrix elements of operators not simply related to the electron density then the justification must be found (if it is to be had at all) outside the framework of density functional theory. It is important to note, however, that a similar statement can be made about Ψ_{HF}. This is the "best determinant" only in the sense that it minimizes $E_D = \langle\Psi_D|\hat{H}|\Psi_D\rangle$. However, again assuming μ_{xc} to be known exactly, the HKS wave-function Ψ, though it gives a worse E_D, gives the best attainable value for $n(\vec{r}) = \langle\Psi|\hat{n}(\vec{r})|\Psi\rangle$. The question as to which of the two, Ψ_{HF} or Ψ, gives the best approximation to the true groundstate wave function is therefore meaningless. For some quantities Ψ_{HF} is better, for others Ψ may be better. In all probability, neither is very good, in general.

As a rider to the above discussion, we point out that an HKS calculation cannot be "improved" by doing CI on the function Ψ, or by trying to defeat the Pauli principle by admitting non-integral occupation numbers.[6] In the HKS approach, Ψ is the wave-function of a non-interacting system, i.e. it belongs to a single configuration. It should be noted, however, that HKS's way of separating the energy functional into one-particle kinetic, electrostatic, and exchange-correlation terms is by no means unique. Other ways of grouping the terms have been suggested,[7] though, so far as we are aware, these have not yet been made the basis of practical computational schemes.

c. The Approximation for E_{xc}: Local Density vs. Xα

Although the density functional scheme admits many alternatives - as we have seen, Hartree-Fock can be regarded as one - HKS's original proposal of using a local density approximation* for E_{xc} has gained a special significance. Devious schemes too numerous to cite have been devised to improve on this approximation, but none seems to be used by others than its proponents - a crucial test!

The exchange-correlation energy may be written,

$$E_{xc} \equiv \int d\vec{r}\, n(\vec{r})\, \varepsilon_{xc}(\vec{r}) \tag{1.9}$$

and the local approximation amounts to writing

$$\varepsilon_{xc}(\vec{r}) \simeq \varepsilon^{h}_{xc}(n(\vec{r})) \tag{1.10}$$

for the energy density, where $\varepsilon^{h}_{xc}(n)$ is the exchange-correlation energy of a homogeneous electron gas having density n.** While this is recognizably the first term in a gradient expansion, the reader will be pardonned for wondering whether such an expansion converges for molecules and what the exchange-correlation energy of a homogeneous gas, which has only extended states and a continuous excitation spectrum, might have to do with a system as small as a few $(\text{Bohr})^3$, for which the energy levels are separated by Hartrees. Clearly, the gradient expansion does not converge in atoms. If it did one could also expand the kinetic energy, ending up with a simple generalization of Thomas-Fermi-Dirac theory. Equally clearly, the exchange-correlation energy density of an atom has no obvious physical connection with that of a homogeneous electron gas.

---

*In fact, this was only one of several possibilities envisaged by HKS.

**To avoid unnecessary complication, we consider only the spin-restricted formalism.

The success of a simple formula like (1.10) in general applications, though still far from understood in detail, can be made plausible by the following series of arguments. Firstly, it was known to Hartree fifty years ago that the essential energy balance in atoms, particularly in regions near the nuclei, is determined primarily by T_0 and E_{es}. While important, E_{xc} is not critical i.e. it is the right term in the functional approximate. Secondly, the exchange-correlation energy density $\varepsilon_{xc}(\vec{r})$ seems to be insensitive to details of the electronic structure. It may be written, suggestively, as

$$\varepsilon_{xc}(\vec{r}) = -\frac{1}{2R_{xc}(\vec{r})} \tag{1.11}$$

where $R_{xc}(\vec{r})$ is the radius of the hole an electron at $\vec{r}$ creates around itself as a consequence of Coulomb interaction with the other electrons (the so-called exchange-correlation hole). $R_{xc}(\vec{r})$ can be defined precisely in terms of the pair-correlation function,[8,9] and this definition suggests that most of the orbital structure of the atom or molecule is averaged out i.e. that $R_{xc}(\vec{r})$ is a relatively smooth function of $\vec{r}$.* In fact, dimensional arguments having nothing to do with the rate of variation of the density $n(\vec{r})$, indicate that at least over a limited region of space $R_{xc}(\vec{r})=\alpha\, n^{-1/3}(\vec{r})$,[10] with the constant α depending only on the <u>shape</u> of the exchange-correlation hole. This shows that a "local" relation between $\varepsilon_{xc}(\vec{r})$ and $n(\vec{r})$ holds even when the latter is rapidly varying, and fails when the hole shape varies rapidly from point to point within the molecule. The success of local approximation for $\varepsilon_{xc}(\vec{r})$ seems to imply that the shape of the hole is rather constant over those regions which contribute most to the change in E_{xc} when the nuclear coordinates change.

Chemists are more familiar with the Hartree-Fock-Slater-Xα (HFS-Xα) method than with the HKS-LD scheme. In practice, the differences are slight. In HFS-Xα one takes the dimensional argument $R_{xc}(\vec{r}) = \alpha\, n^{-1/3}(\vec{r})$ literally - HFS is a "fixed hole shape" theory. In HKS-LD the hole shape is taken to be that of a homogeneous electron gas. This is an "almost constant hole shape" theory** and is equivalent to using an α-value that has a weak but uniquely prescribed density dependence. In a spin-restricted formalism, and for the α-values commonly used in HFS-Xα, the two approximations lead to similar results.*

---

*A very complete and illuminating discussion of the exchange-correlation hole of the H-atom can be found in Reference 9.

**The hole radius of a homogeneous electron gas depends on two lengths, λ_F, λ_{FT}, the Fermi and Fermi-Thomas wave-lengths.

In spite of this formal similarity, the difference in historical development of the two schemes should not be overlooked. HFS-Xα was conceived, and is apparently still viewed in some circles, as an approximation to Hartree-Fock. The HFS-Xα wave-function was regarded as an approximation to Ψ_{HF}, which in turn was thought of as an approximation to the many-electron wave-function. Similarly, the HFS-Xα energy was thought of as being an approximation to E_{HF}, the lowest expectation value of the many-electron Hamiltonian taken with determinantal functions. By contrast, the HKS scheme points to the density rather than the wave-function as a basic variable, interprets Ψ in terms of an equivalent system of independent electrons, and determines the energy by evaluating a density functional rather than a wave-function functional (as the expectation value of the Hamiltonian is occasionally termed).

In our work, we prefer to use the LD** rather than Xα approximation for three reasons. (i) Where spin-flip energies are involved, the corresponding spin-unrestricted approximation has a clear advantage. (ii) It is naturally satisfying to be able to use the same energy expression for systems with rapidly and slowly varying densities. (iii) The total absence of a free parameter eliminates the temptation to play with it when results do not meet up to expectations! In this last point, we differ from many chemists, who seem to appreciate the freedom of allowing α to vary from application to application.

As a final point, we remark that no local approximation for $\varepsilon_{xc}(\vec{r})$ can reproduce correctly the changes in E_{xc} that accompany the formation of chemical bonds. Furthermore, the inevitable errors will accumulate as the nuclei are drawn far apart to form isolated atoms. For this reason, we should not be surprised if the theory gives worse results for dissociation energies than for intra-molecular properties such as internuclear separations and vibration frequences. Similarly, we are right to be suspicious of the approximation whenever an energy difference involves states having substantially different densities.

---

*The same is not true of the spin-unrestricted theories, where the Xα approximation grossly overestimates the energy lowering on adopting parallel spin configurations. This is because the Xα exchange-correlation hole relates only to electrons of the same spin i.e. allowance is made only for Pauli principle correlations, or exchange.

**Specifically, $\varepsilon^{h}_{xc}(n)$ in (1.10) is taken from the latest and presumably best calculations for the homogeneous electron gas.[11]

2. BASIS FUNCTIONS FOR DENSITY FUNCTIONAL CALCULATIONS

a. General Considerations

The criteria we should bear in mind when constructing basis functions for use in HKS-LD calculations are illustrated by the case of the H_2 molecule, where the single trial orbital*

$$\psi(\vec{r}) = \left(\frac{\lambda^3}{2\pi}\right)^{1/2} \left(\frac{1}{1+S_\lambda}\right)^{1/2} [\exp(-\lambda|\vec{r}-\vec{R}_1|) + \exp(-\lambda|\vec{r}-\vec{R}_2|)] \quad (2.1)$$

with $\vec{R}_1, \vec{R}_2$ the nuclear positions, S_λ the usual overlap integral, and λ a variation parameter, reproduces the exact HKS-LD binding energy curve within a few percent.[8] This example shows that a small basis can give good results, and also brings to light the crucial advantage that occurs when orbital exponents can be varied. A much larger basis of LCAO's with fixed exponents would be necessary to achieve comparable accuracy. This is the reason why we prefer a variational procedure that includes variation of orbital exponents.

The basis we employ is totally unfamiliar to chemists. For this reason, it may be useful to include a few remarks on the historical development of our method. The first serious attempt to solve problems of the HKS-LD type for molecules was via the so-called scattered-wave (SW) method.[12] This method, a direct transcription of the KKR band-structure method employed in solid state physics, solves (1.5)-(1.7) for the case that the potential $V_{eff}(\vec{r})$ has muffin-tin form. That is, V_{eff} was taken to be spherically symmetric inside touching spheres surrounding the atoms and in a concave sphere of infinite extent whose surface circumscribes the entire cluster. In the interstitial region between concave and convex spheres the potential was taken to be a constant. The SW-Xα method turned out to give very bad results (amongst them was the prediction that the water molecule is straight[13]) and this seems to have influenced chemists sufficiently profoundly that the proponent of local potential calculations tends to be greeted with considerable scepticism. It is important to understand, however, that the main flaw in the SW-Xα method as practised, was the failure to recognize the point mentioned in the introduction, namely that the energy expression must be evaluated accurately. If the proponents of SW-Xα had not made the crucial mistake of "muffin-tinizing" the energy functional they would have obtained excellent results for the total energy, in spite of the muffin-tin potential used to generate the ψ_n.

This was demonstrated by Gunnarsson, Harris and Jones,[3] who adapted an alternative band-structure method - the LMTO method of Andersen[14] - which at the time was establishing itself in solid

---

*Throughout this article, unless otherwise stated, Hartree units are used.

state physics. In spite of a simplifying restriction imposed on the exponents of the orbital tails - the so-called common-κ approximation - results of useful accuracy were obtained for a wide variety of diatomics spanning the periodic table.[3] Subsequent work on first-row diatomics[15,16] using extensive LCAO basis sets revealed an inadequacy of the common-κ approximation for these cases (the approximation is worst when the valence eigenvalues are widely spread), but led to results that were of even better quality, and consistently better than Hartree-Fock. This holds for Xα as well as for LD, and is an obvious source of mystery to those who regard HFS-Xα as an approximation to Hartree-Fock!

While the extension of the LCAO basis to systems larger than diatomics involves essentially no drastic changes, the same has turned out not to be true of the original LMTO method. Limitations imposed by the nature of the basis - in particular, the division of space into convex and concave spheres as in the SW-method - and the way in which integrations were carried out, led to the abandonment of the method as unsuited for general application. The method we use currently, in all essential features that of Harris and Painter,[17] while leaning somewhat towards the LCAO philosophy in that all orbitals decay exponentially away from their associated atomic centres, nevertheless retains certain key features of the LMTO method. In particular, the $(\phi,\dot{\phi})$ construction of the orbitals inside the atomic spheres plays a crucial role. For this reason we will refer to it as a "localized" muffin-tin-orbital basis and denote it by the original notation, LMTO. Since the original basis - the "linear" MTO basis - has been abandoned and we do not anticipate its adoption by other workers, this should not lead to confusion.

b. The localized MTO Basis

One motivation for using muffin-tin orbitals (MTO's) as basis functions is that they are exactly orthogonal to all the core states of the cluster. In conjunction with the frozen core approximation, the MTO basis allows us to forget about the core altogether, no matter how complicated this is.* A second important feature of the MTO's is that they are very specific to the actual potential of the system. This means that only a minimal number of basis functions is needed which gives not only computational speed but also an appealing simplicity to the results. The price we have to pay is that the basis functions must be recalculated each time the potential is changed (for example, on each iteration in self-consistent calculations).

*Recall that in the conventional LCAO scheme it is necessary to include at least one orbital for each core state, a burden which makes calculations for systems containing heavier atoms increaseingly cumbersome.

To construct a MTO basis we separate the full potential into two parts: (i) a muffin-tin part, spherically symmetric inside atomic spheres of radii S_i centred on each atomic site, and constant between these spheres; and (ii) a non-muffin-tin part that includes the rest of the potential. No outer sphere is needed in the present approach. We construct basis functions χ_L^i for each partial wave $L\equiv(\ell,m)$ and each site i. Inside its own atomic sphere

$$\chi_L^i(\vec{r}_i) = [\phi_\ell^i(\varepsilon_\nu,r_i) + \omega_\ell^i\dot{\phi}_\ell^i(\varepsilon_\nu,r_i)]\ Y_L(\hat{r}_i)\ ,\ r < S_i \qquad (2.2)$$

where $\vec{r}_i=\vec{r}-\vec{R}_i$, with $\vec{R}_i$ the nuclear positions. The function $\phi_\ell^i(\varepsilon_\nu,r)$ is a solution of the radial wave-equation for the muffin-tin part of the potential with energy ε_ν, and $\dot{\phi}_\ell^i(\varepsilon_\nu,r_i)$ is its energy derivative $\partial\phi_\ell^i(\varepsilon_\nu,r)/\partial\varepsilon$. As can be seen on noting that (2.2) has the form of a Taylor series, χ_L^i inside its own sphere represents an approximate solution of the radial wave-equation with energy $\varepsilon_\nu+\omega_\ell^i$, valid within a range of one principal quantum number about ε_ν. Since ϕ_ν and $\dot{\phi}_\nu$ are orthogonal to all states of the potential inside sphere i that have negligible amplitude at the sphere boundary, χ_L^i is automatically orthogonal to all core states corresponding to site i.[14]

Outside the atomic spheres we define

$$\chi_L^i(\vec{r}_i) = G_L^i[K_\ell(\kappa_\ell^i,r_i) + A_\ell^i K_\ell(\tilde{\kappa}_\ell^i,r_i)]\ Y_L(\hat{r}_i)\ ,\ r_i > S_i \qquad (2.3)$$

where

$$K_\ell(\kappa,r) \equiv i(i\kappa)^{\ell+1}\ h_\ell^{(1)}(i\kappa r) \qquad (2.4)$$

are decaying Hankel functions. The parameters G_L^i and ω_ℓ^i are determined by the requirement that χ_L^i be continuous and differentiable at the atomic sphere boundary. In the other spheres the orbital tail (2.3) is again augmented via the $(\phi,\dot{\phi})$ construct (2.2), which ensures that χ_L^i is orthogonal to the core states of the sites $j\neq i$. Although in principle any function could be chosen for the orbital tail (2.3), the use of Hankel functions simplifies the matching of the tails to $(\phi,\dot{\phi})$ constructs on the other sites because these functions obey simple expansion theorems. At least two Hankel functions are needed to properly represent the details of the orbitals tails. The actual shape of the tails is determined by the parameters $\kappa_\ell^i,\tilde{\kappa}_\ell^i$, and A_ℓ^i,* which in principle could be regarded as variational parameters. In practice, however, three parameters per inequivalent site is much more variational freedom than one actually needs, and it is convenient to restrict this freedom by determining a linkage between the parameters. We distinguish between the orbitals which are bound in the isolated atom and those which are not bound.

*We shall assume $\tilde{\kappa}_\ell^i > \kappa_\ell^i$.

For the atom-bound orbitals we have taken the orbital tails to be atom-like. We perform separate atomic calculations for each constituent atom for a range of values of the corresponding nuclear charge Z^*. The resulting contracted ($Z^*>Z$) or expanded ($Z^*<Z$) atomic orbitals are then fitted to a function of the form (2.3). The quantities $\kappa_\ell^i(Z^*)$, $\tilde{\kappa}_\ell^i(Z^*)$, and $A_\ell^i(Z^*)$ are then determined by a single variational parameter for each orbital. In general, we choose to vary the leading exponent κ_ℓ^i, to which we link the remaining parameters.

For the atom-unbound orbitals this linkage procedure cannot be used. In order to generate a family of orbital tails which cover all possible situations two independent parameters are needed: (i) κ_ℓ^i to determine the spatial extent of the orbital and (ii) $\tilde{\kappa}_\ell^i$ or A_ℓ^i to fix the logarithmic derivative at the atomic sphere radius. In practice, however, we have found that provided reasonable parameters* are chosen, the variation of the atom-unbound orbitals has very little effect on the total energy and need not be carried out. This means that the entire basis depends on only a few exponent parameters, for example 3 for systems like H_2O and NH_3, which gives a variational problem of manageable proportions.

To minimize the functional within the space spanned by the LMTO's we expand the eigenfunctions in (1.5) as

$$\psi_n(r) = \sum_{L,i} C_{iL}^n \, \chi_L^i(\vec{r}_i) \tag{2.5}$$

where the C_{iL}^n are linear parameters and the orbital exponents κ_ℓ^i are held constant. The best linear parameters are determined by converting (1.5) to a matrix equation, using the eigenvectors to construct a density and new potential, and iterating towards self-consistency. As the iterations progress, we find that the value of the functional falls steadily and becomes constant in a very few iterations,** long before e.g. the eigenvalues settle at their final values. This is a key sign that the variational principle for the energy is operating, and is a very critical test of the accuracy with which the functional is evaluated. Even a small error in the energy shows up immediately in the convergence during iteration of the potential.

*Typically $\kappa_\ell^i \simeq (2S_i)^{-1}$ and a logarithmic derivative $S_i \chi_L^{i\prime}/\chi_L^i \simeq \ell$ at the atomic sphere radius leads to a reasonable unbound orbital tail.

**2 to 6, depending on how far the start potential is from its final self-consistent value. Iterations are carried out with a feedback of about 50% of the old Coulomb potential. Feedback is not applied to the exchange-correlation potential, which is calculated using the new density in each iteration.

Once the energy is converged to a given accuracy, the orbital exponents κ_ℓ^i are changed and the procedure repeated. So far we have not attempted to automate the variation of the κ_ℓ^i but use our experience to vary them individually in the order of their importance in the bonding. Once the exponents are determined for one geometry, new values for neighbouring geometries can be found very easily and a map of the energy as a function of the nuclear coordinates determined.

From the numerical point of view, there are three non-trivial computational steps: (i) Evaluation of the matrix elements of the one-electron Hamiltonian with the χ_L^i; (ii) construction of the potential; and (iii) evaluation of the energy. These we discuss in Section 3, which the reader uninterested in such details can skip without loss of continuity.

3. COMPUTATIONAL ASPECTS

a. Matrix Element Evaluation

We consider the Hamiltonian and overlap matrices

$$H_{LL'}^{ij} = \langle\chi_L^i|-\frac{1}{2}\nabla^2 + V_{eff}(\vec{r})|\chi_{L'}^j\rangle$$

$$O_{LL'}^{ij} = \langle\chi_L^i|\chi_{L'}^j\rangle \qquad (3.1)$$

with χ_L^i the orbitals detailed in Section 2. The one-electron potential $V_{eff}(\vec{r})$, is separated into muffin-tin and non-muffin-tin parts. All contributions to $H_{LL'}^{ij}$ arising from the muffin-tin part can be evaluated analytically, as detailed in Reference 17. In addition, the non-muffin-tin terms arising from within the spheres can be handled easily with the aid of a partial wave expansion of the potential. Only the integrals $\langle\chi_L^i|V_{eff}(\vec{r})|\chi_L^j\rangle_{\Omega I}$, where the integration region Ω_I, is the space outside all of the spheres, pose calculational problems. These can be performed using the smoothed integrand technique of Reference 17. The integrand is continued smoothly into the spheres and all-space integrals performed by Gaussian quadrature. Contributions arising from inside the spheres are then subtracted. We currently use a modified procedure whereby variables for Gaussian quadrature are defined that automatically exclude two of the spheres from the integral, thus reducing the subtraction error substantially.[18] Although Gaussian quadrature is efficient, these integrals represent the time limiting step in the computation because they must be performed in each iteration. Once the matrix elements are calculated diagonalization is achieved via standard techniques to give the coefficients C_{iL}^n in (2.5) and the corresponding eigenvalues, ε_n, which satisfy

$$\sum_{L',j} (H^{ij'}_{LL'} - \varepsilon_n O^{ij}_{LL'}) C^n_{jL'} = 0 \qquad (3.2)$$

b. The Coulomb Potential

For a given set of variation parameters, κ^i_ℓ, C^n_{iL}, and occupation numbers f_n, the electron density is

$$n(\vec{r}) = \sum_n f_n |\psi_n(\vec{r})|^2 = \sum_{ij} n_{ij}(\vec{r}) \qquad (3.3)$$

with

$$n_{ij}(\vec{r}) = \sum_n f_n \sum_{L,L'} C^n_{iL} C^n_{jL'} \chi^i_L(\vec{r}) \chi^j_{L'}(\vec{r}) \qquad (3.4)$$

Outside the spheres this must be evaluated directly. Inside sphere i, however, $n(\vec{r})$ can be expanded conveniently in partial waves[17]

$$n(\vec{r}) = \sum_L n^i_L(r_i) Y_L(\hat{r}_i) \qquad (3.5)$$

Because the exchange-correlation potential μ_{xc} depends only on the local density its evaluation presents no problems. Similarly, the Coulomb potential due to the electron density inside the spheres can be determined directly from (3.5) as a partial wave expansion. However, the Coulomb potential due to the remaining part of the density, namely

$$\phi_I(\vec{r}) = \int_{\Omega_I} d\vec{r}' \frac{n(\vec{r}')}{|\vec{r}-\vec{r}'|} \qquad (3.6)$$

cannot be determined easily because a separate three dimensional integral over an awkward region would be required for each point $\vec{r}$ for which the Coulomb potential is needed. To circumvent this difficulty, we follow other workers[16,19] and introduce an auxiliary set of basis functions, for which Poisson's equation can be integrated analytically. The interstitial electron density $n(\vec{r})$ is approximated by $\tilde{n}(\vec{r})$, given by

$$\tilde{n}(\vec{r}) = \sum_{\kappa iL} n_{\kappa iL} K_\ell(\kappa_{i\ell} r_i) Y_L(\hat{r}_i) \qquad (3.7)$$

We have here again chosen the decaying Hankel functions (2.4) as basis functions because they are eigenfunctions of the Laplacian and obey simple expansion theorems. The coefficients $n_{\kappa iL}$ in (3.7) are determined by performing a constrained least squares fit of (3.7) to the exact electron density (3.3). This involves the calculation of integrals

$$t_{\kappa iL} = \int_{\Omega_I} d\vec{r}\, n(\vec{r}) K_\ell(\kappa_{i\ell} r_i) Y_L(\hat{r}_i) \qquad (3.8)$$

in each iteration and these are performed using Gaussian quadrature as in secular matrix element evaluation. This fit procedure was found adequate for molecules like H_2O, NH_3, but would become increasingly difficult to carry out for larger systems, where it is preferable to carry out separate fits of the partial densities $n_{ij}(\vec{r})$ in (3.3).[14] Note that we rely on a density fit only over the region Ω_I, where $n(\vec{r})$ is relatively smooth. By contrast, conventional LCAO methods fit the density over all space, even in the region close to the nuclei where $n(\vec{r})$ varies very rapidly. We are able to avoid this because of the $(\phi,\tilde{\phi})$ construction and the simple expansion properties of the orbital tails (2.3).

On using the fit density $\tilde{n}(\vec{r})$ instead of the exact density $n(\vec{r})$ to calculate the Coulomb potential component $\phi_I(\vec{r})$, we make an error $\delta\phi(\vec{r})$. This error contributes in two ways. Firstly, the self-consistency condition (1.6) will only apparently be obeyed because the Coulomb potential used does not follow exactly from the corresponding density. Since we know from experience that the functional becomes stationary long before the potential is self-consistent, this error is obviously negligible. More serious is the fact that $\phi(\vec{r})$ is needed in evaluating the functional. Dunlap et.al.[16] have stressed the importance of treating this error very carefully and we too have found that any error that is linear in $\delta\phi(\vec{r})$ is unacceptable. Fortunately, such errors can be avoided, as we now demonstrate.

c. Evaluation of the Energy Functional

This is greatly facilitated if we make the frozen-core approximation, where the core density is assumed independent of the nuclear positions. Here we ignore the polarization of the core that accompanies chemical bonding. This has been shown to have a negligible effect on the energy even when the core is fairly extended.[20] Once this approximation is made a large part of the total energy is independent of the nuclear positions and need never be evaluated. The part that is dependent on $\vec{R}_i$ may be written[3]

$$E_v = T_o^v + \int d\vec{r}\, n_v(\vec{r}) \left\{\frac{1}{2}\phi_v(\vec{r}) + \varepsilon_{xc}^h(n(\vec{r})) + \phi_c(\vec{r})\right\} + \int d\vec{r}\, n_c(\vec{r}) \left\{\varepsilon_{xc}^h(n(\vec{r})) - \varepsilon_{xc}^h(n_c(\vec{r}))\right\} + \frac{1}{2}\sum_{i\neq j} \frac{Z_v^i Z_v^j}{R_{ij}} \quad (3.9)$$

Here Z_v^i is the charge of core i, R_{ij} are the internuclear separations, T_o^v, $n_v(\vec{r})$ are valence kinetic energy and density given by (1.4), (1.3) with the n-sum over only valence orbitals, $n_c(\vec{r})$ and $n(\vec{r})\equiv n_c(\vec{r})+n_v(\vec{r})$, are core and total electron densities, respectively, and $\phi_c(\vec{r})$ is the Coulomb potential of nuclei plus core electrons. The core density appears in (3.9) because the exchange-correlation energy density $\varepsilon_{xc}^h(n)$ is a non-linear function of n.

We wish now to evaluate (3.9) with single-particle orbitals (2.4) that give the density (3.3). The final two terms present no problems. The first two can be simplified on relating T_o^v to the eigenvalues of the secular equation (3.2). That is,

$$T_o^v \equiv \sum_n^v f_n \sum_{\substack{i,j\\L,L'}} c_{iL}^n c_{jL'}^n \langle \chi_{L'}^j | -\frac{1}{2}\nabla^2 | \chi_L^i \rangle \tag{3.10}$$

can be written

$$T_o^v = \sum_n^v f_n \varepsilon_n - \int d\vec{r}\, n_v(\vec{r})\, V_{eff}(\vec{r}) \tag{3.11}$$

where ε_n are the eigenvalues defined by (3.2) and $V_{eff}(\vec{r})$ the potential used in constructing the Hamiltonian matrix (3.1). The first two terms of eqn. (3.9) can now be combined to give

$$\sum_n^v f_n \varepsilon_n + \int d\vec{r}\, n_v(\vec{r})\, \{\frac{1}{2}\phi_v(\vec{r}) + \varepsilon_{xc}^h(n(\vec{r})) - V_{eff}^v(\vec{r})\} \tag{3.12}$$

where

$$V_{eff}^v(\vec{r}) = V_{eff}(\vec{r}) - \phi_c(\vec{r})$$

Next, we need to take account of the fact that $\phi_v(\vec{r})$ is known only approximately. That is, we can calculate only $\tilde{\phi}_v(\vec{r})$, the Coulomb potential corresponding to the fit density, $\tilde{n}(\vec{r})$. Following Dunlap et.al.,[6] we make the approximation

$$\frac{1}{2}\int d\vec{r}\, n_v(\vec{r})\phi_v(\vec{r}) \simeq -\frac{1}{2}\int d\vec{r}\, \tilde{n}_v(\vec{r})\tilde{\phi}_v(\vec{r}) + \int d\vec{r}\, n_v(\vec{r})\tilde{\phi}_v(\vec{r}) \tag{3.13}$$

which is in error only to second order in $\delta n(\vec{r})=n_v(\vec{r})-\tilde{n}_v(\vec{r})$. Thus with second order errors (3.12) can be written in a form convenient for computation:

$$\sum_n^v f_n \varepsilon_n + \int d\vec{r}\, n_v(\vec{r})\, \{\tilde{\phi}_v(\vec{r}) + \varepsilon_{xc}^h(n(\vec{r})) - V_{eff}^v(\vec{r})\}$$
$$- \frac{1}{2}\int d\vec{r}\, \tilde{n}_v(\vec{r})\tilde{\phi}_v(\vec{r}) \tag{3.14}$$

The first term is trivial and with the choice of density basis (3.7) the last can be evaluated analytically.[18] Only the second term requires serious computation, and is evaluated by separating off the part within the spheres and smoothing the remainder to give an all-space integral.[17] This can be evaluated by Gaussian quadrature and again, a mesh can be used that automatically eliminates up to two of the spheres. The precise way this is done will depend on the cluster under investigation. For example, for H_2O and NH_3 it is obviously advantageous to use a mesh that eliminates the heavy atom sphere. In practice, we have found that a mesh of ~1000 points gives a relative error of ~10^{-4} which is sufficient because this contribution to the energy is only about 1 Hartree. Note that when $V^v_{eff}(\vec{r})$ is close to self-consistency the second term in (3.14) approaches

$$\int d\vec{r}\; n_v(\vec{r})\; \{\varepsilon^h_{xc}(n(\vec{r})) - \mu_{xc}(n(\vec{r}))\} \tag{3.15}$$

which behaves roughly like $\int d\vec{r}\; n_v^{4/3}(\vec{r})$. As an independent check of the error made in evaluating this integral, we use the same mesh to evaluate the integral $\int d\vec{r}\; n(\vec{r})$, which is, of course, known exactly.

d. Discussion

In any calculational scheme it is important to keep apart the physical approximation - in our case the local density approximation for exchange and correlation - and further approximations of a mathematical nature. Only when this is done properly and the mathematical errors are under control can we hope to test the accuracy of the physical approximation by comparing with measured energies. For example, early work using the scattered-wave Xα approximation was interpreted by some as implying the failure of the local Xα-approximation i.e. the physical approximation for exchange. In fact, the bad results reported using this scheme reflected merely a poor mathematical approximation of the energy expression.

In our case, the factors governing the mathematical accuracy are as follows: (i) Errors due to the numerical integrations that are needed in evaluating matrix elements $H^{ij}_{LL'}$, in (3.1),* the density coefficients in (3.8) and the final energy integral. These errors can all be made smaller by increasing the number of points in the Gaussian meshes, and therefore are subject to a straightforward convergency test. (ii) Incompleteness of the density basis. This too is under control to a considerable extent since the addition of new functions is straightforward and results in an improved fit and a reduction in the quadratic error in the energy. In the calculations reported in Section 4, we used 3 Hankel functions per partial wave with widely spread exponents, and included s, p, and d partial waves on all sites. This fit basis has been found flexible enough to handle all the geometries considered so far. Typically, the root mean square fit error was about 10^{-3}. A precise value for the concomitant error in the energy cannot be made easily. However, the smoothness of the energy as a function of the nuclear coordinates on a scale ~0.01 eV suggests that this error cannot be very large. (iii) Incompleteness of the orbital basis. As we have stated, our basis is constructed so as to provide maximum accuracy with a minimum number of functions. For example, the results of Section 4 for H_2O and NH_3 were achieved with s, p, and d orbitals on the heavy atom site, and s and p orbitals on the H sites, giving unsym-

*The overlap integrals can be evaluated analytically.

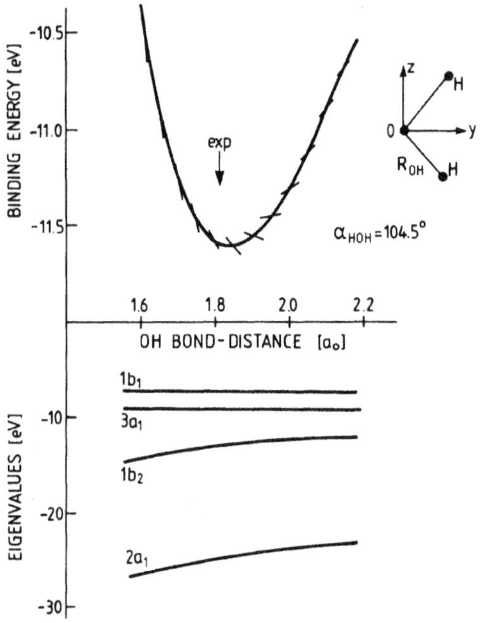

Fig. 1. HKS-LD binding energy of H_2O as function of the OH distance for fixed bond angle. The bars through the points represent the internuclear radial force. Lower half: Variation of eigenvalues for the molecular orbitals involved in the bond.

metrized secular matrices of rank 17, 21, respectively. Adding d-waves on the H sites changed the energy of H_2O by a negligible amount, while f-waves on the O site brought about a lowering of ~0.1 eV and did not alter the equilibrium geometry. Obviously, the basis is essentially converged in ℓ. However, there is still an error, as is clear in Figs. 1 and 2, where the energy of the H_2O molecule as a function of the OH bond-length and HOH bond angle is plotted, together with the corresponding force component. The force is calculated using a modified Hellman-Feynmann formula whose derivation assumes that the density functional has been minimized exactly.[21] The deviations apparent in the figure between the magnitude of the force and the gradient of the energy are a sign that an exact minimization has not been achieved. This is probably due to the fact that the $(\phi,\dot{\phi})$ construct (2.2) for the oxygen s-orbital has been used over an energy range comparable with the spacing between principle quantum numbers 2 and 3. In such a situation one might expect an improved correspondence of energy and force if hybridization of O_{2s} with a O_{3p}-like orbital is included, a feature that we are

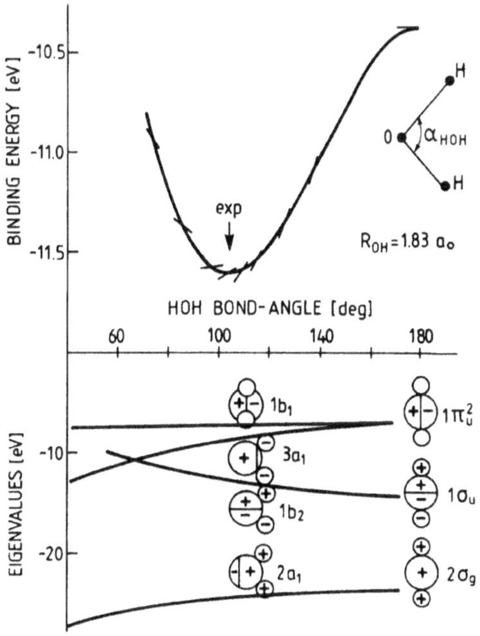

Fig. 2. HKS-LD binding energy of H_2O as function of the bond angle for fixed OH distance. The bars through the points represent the internuclear angular force. Lower half: Variation of eigenvalues for the molecular orbitals involved in the bond. The shape of the molecular orbitals is schematically shown.

currently building into our code.* The effect is negligible for systems with a smaller eigenvalue spread, for example the Be_2 dimer, where an excellent correspondence of energy and force results with only a minimal basis.[21]

This discussion shows that all mathematical errors that arise in our calculational scheme are essentially under control. While

*To assess correctly the probable basis error, it is important to bear in mind that the balance of forces is very delicate. For example, an error of only a few percent in the gradient of the electronic Coulomb potential at the nuclear positions is sufficient to give the deviations apparent in Figs. 1 and 2. So far as we are aware, no other worker performing similar calculations has yet shown a comparison of simultaneously calculated energies and forces.

the orbital basis is not quite extensive enough, we expect only a minimal change in the calculated geometry on carrying out the improvement mentioned in point (iii) above. Thus we believe the results of Section 4 give a reliable indication of the quality of the results that can be expected in HKS-LD calculations of bond lengths and angles in polyatomic molecules.

4. NUMERICAL RESULTS

a. Preamble: The HKS Scheme and Molecular Orbital Chemistry

The molecular orbital theory of chemistry as set out in standard chemistry texts[22,23] makes use of concepts that do not obviously have relevance in many-electron systems. What are the molecular orbitals into which electrons are put, and how are the energies, that they have when they are there, determined? The physicist searches frustratedly for precise definitions of these objects. If he is persistent and has a feeling for nuance, he will conclude that the orbitals and eigenvalues of the Hartree-Fock equations are meant, though rarely used because they are hard to determine. However, since he will be aware that this approximation can be nonsensical in an extended system,* and that virtually the entire industry of quantum chemistry justifies its existence via the need to go "beyond Hartree-Fock", this is unlikely to impress him very much. How can we build a theory of chemistry on concepts that have meaning only for wave-functions that cannot possibly represent many-particle states?

The chemist, on the other hand, accepts the molecular orbital or one-electron picture because he knows it explains much of chemistry rather well. He does not deal with extended systems, has no reason to doubt the qualitative accuracy of Hartree-Fock, and so is hard put to recognize any advance as a result of HKS's approach to the density functional. To the physicist interested in ground state energy calculation, however, the HKS approach gives a new view of the problem. In essence it amounts to a formal legitimization of molecular orbital chemistry. The molecular orbitals are the functions $\psi_n(\vec{r})$ that make up the ground state wave-function of the equivalent system of independent electrons. The orbital energies are their eigenvalues. The equivalent system can be precisely defined,** and the correspondence with the real system visualized in terms of an adiabatic switching-on of the electron-electron interaction and the concomitant switching-off of an external field in such a way

---

*e.g. simple metal densities of states show singular behaviour at the Fermi level.

**If it exists!

as to leave the density unchanged. This process, which has the same flavour as a popular approach to Fermi-liquid theory,[24] can even be mathematized and is, indeed, the background to the discussion in Section 1.c. concerning the exchange-correlation energy.

If we accept the formal existence of the equivalent independent-electron system, the energetic behaviour of the physical system can be discussed and understood rigorously in a one-particle language. It can be demonstrated that when the electron energies of the equivalent system go down as distant nuclei approach so does the energy of the physical system. It can be shown that orbitals of the equivalent system whose energies go down or up are, respectively, bonding or antibonding in the usual chemical sense. In this respect HKS have helped the physicist to understand molecular orbital chemistry in his own way, and the chemist, though he may find this development bewildering, can at least be happy that the physicist has finally seen the light and learned the language!

b. The Water Molecule

To our knowledge, only one local potential calculation has been reported for this molecule.[25] The local density approximation was used and the functional minimized with the aid of an LCAO 4-31 G basis. Figures for the bond-distance and bond-angle of $R^o_{OH}=1.85$, $\alpha^o_{HOH}=111^o$ were quoted. The results we report below suggest that the orbital basis used in this calculation was not quite converged.

Using an LMTO basis with s- and p-waves on all sites and d-waves on the O-site - in all, 5 independent radial functions - we find the energy variation shown in Figs. 1 and 2. The energy is plotted against bond-distance for $\alpha_{HOH}=104.5^o$ (Fig. 1), and against bond-angle for $R_{OH}=1.83$ a.u. (Fig. 2).* The values $R^o_{OH}=1.83$ a.u., $\alpha^o_{HOH}=104.5^o$ correspond to the calculated equilibrium geometry, and are in excellent agreement with the experimental values $R^{expt}_{OH}=$ 1.809 a.u., $\alpha^{expt}_{HOH}=104.52^o$.[26] By comparison, the text book Hartree-Fock values are $R^{HF}_{OH}=1.78$ a.u., $\alpha^{HF}_{HOH}=106.6^o$.[27] The binding energy with respect to isolated atoms, was found to be $E_B=11.6$ eV, somewhat larger than the measured value of 10.09 eV,[23] but better than $E^{HF}_B=6.9$ eV.[27]** Polynomial fits to the curves shown in Figs. 1 and 2 gave frequencies for the totally symmetric stretch and bend modes of 3420 and 1530 cm^{-1}, respectively. The corresponding experimental values are 3556.65 and 1594.6 cm^{-1}.[23]

*The bars drawn on these figures refer to the calculated force. If the orbital basis were exactly converged these would be exactly tangential to the energy curves. See Section 3.d. for details.

**The Hartree-Fock energy attained in these calculations was estimated to be within 0.1 eV of the HF limit.

In addition to giving this quantitatively correct energy variation, HKS-LD yields a simple picture of the forces that determine the non-linear geometry. Shown in Figs. 1 and 2 are the eigenvalues corresponding to the molecular orbitals $2a_1$, $3a_1$, $1b_2$, and $1b_2$ of the water molecule. The structure of these orbitals is well-known, and most easily visualized in the linear configuration, where $1b_1$ and $3a_1$ are degenerate and correspond simply to the oxygen $2p_x$ and $2p_y$ orbitals oriented perpendicular to the molecular axis. In the linear configuration the OH bonds are due to $2a_1$ and $1b_2$, which are linear combinations of O_{2s} and O_{2p_z} with the hydrogen s-orbitals.* As we draw the O atom out of line with the H atoms along, say, the y-axis (Fig. 2), the degeneracy of $1b_1$ and $3a_1$ is lifted. $1b_1$ remains essentially pure O_{2p_x}, while $2a_1$, $3a_1$ become hybrid O_{2s}, O_{2p_y} orbitals with admixture of a reflection-symmetric combination of H_s orbitals. The $2a_1$ orbitals are weighted predominantly on the side of the oxygen closest to the hydrogens, while $3a_1$ becomes a "lone-pair". Fig. 1 shows the variation of the orbital eigenvalues with R_{OH} for a non-linear geometry and illustrates that the $3a_1$ lone-pair is OH non-bonding, as in the linear geometry.

When the bond-angle decreases from 180° at fixed R_{OH} (Fig. 2), however, the $3a_1$ orbital moves to lower energy because of (i) the admixture in O_{2p_y} of a lower-lying oxygen orbital, and (ii) the HH bonding nature of the orbital. $2a_1$ is more strongly HH bonding, but moves down rather less than $3a_1$ because the two effects (i) and (ii) above compete.** The downward movement of the a_1 orbital energies accounts for the lowering of the total energy as the bond angle increases, but is partially compensated for by the upward movement of $1b_2$, which is HH anti-bonding (see the diagrams drawn on the figure).

It is tempting to regard the equilibrium position as determined by the competition between these two effects. However, this is not correct as can be seen on noting that the sum of the eigenvalues in Fig. 2 decreases steadily with decreasing bond angle. The competition is in fact between a net "orbital energy" force tending to decrease the bond angle and a "nuclear de-screening" force that arises from an imperfect cancellation of the inter-nuclear repulsion as the charge at the midpoint of the bond between the hydrogens builds up. Because the internuclear repulsion does not influence eigenvalue behaviour it is never possible to determine

---

*In this qualitative discussion we ignore the atom-unbound orbitals. These are, however, very important for a quantitative description of the energy variation. See also Section 4.c.

**Note that it is not easy to determine a priori how these orbital energies move. For example, Herzberg,[23] following Walsh,[27] has the $3a_1$ energy moving <u>upwards</u> with decreasing bond angle.

equilibrium geometries from eigenvalue movement alone. At best, all we can do is relate roughly the strength of the binding that results when a given geometric parameter changes to the rate at which the eigenvalue sum decreases when the nuclei are far enough apart for us to neglect the de-screening effect.

c. The Ammonia Molecule

The ground-state energy of the ammonia molecule has been the subject of many Hartree-Fock calculations, the most recent of which show good correspondence with experimentally determined parameters. For example, there seems to be general agreement that the Hartree-Fock inversion barrier height is 0.225 eV,[29] as against the measured 0.25 eV,[30] and that the NH distance in the equilibrium geometry is R_{NH}^{eq}=1.89 a.u.,[29] compared with the measured 1.912 a.u. The text book figure of α_{HNH}=107.2^o [29] for the bond-angle also compares favourably with the measured 106.7^o, but has been challenged recently by Rodwell and Radom,[31] who obtained a lower energy at α_{HNH}= 107.7^o. This notwithstanding, it is clear that the Hartree-Fock approximation contains all the essential elements of the energy balance.

Aside from an SW-Xα calculation in which the barrier height was found to vary from -1.2 eV to +0.2 eV according to how the muffin-tin radii were chosen,[32] only Kitaura et.al.[24] have reported local density calculations for NH_3. Using an LCAO 4-31 G orbital basis they found the planar geometry to be favoured. The addition of 3d-functions on the N-site and 2p-functions on the H-sites was then found to stabilize the out-of-plane geometry with parameters R_{NH}^{eq}=1.93 a.u., α_{HNH}^{eq}=106°, and a barrier height of 0.17 eV.

These figures are in good correspondence with calculations we have performed using the LMTO basis with s- and p-partial waves on all sites, and d-waves on the nitrogen site. The results are preliminary and we have not yet investigated the influence of higher partial wave components, which we expect to be greater than for H_2O because of the more delicate energy balance. A complete set of results for this and other molecules showing interesting inversion behaviour will be published in due course. Figure 3 shows our calculated energy as a function of Z_N, the normal distance of the N-atom from the plane of the three H-atoms. The NH distance was kept fixed at 1.925 a.u., a value intermediate between the separations R_{NH}=1.90, 1.94 that minimized the calculated energy for bond angles α_{HNH}=120° (in-plane) and α_{HNH}=106°, respectively.* As can be

*The Hartree-Fock bond lengths, though somewhat smaller, also show a contraction of R_{NH} as the nitrogen atom enters the plane of the hydrogens.[31]

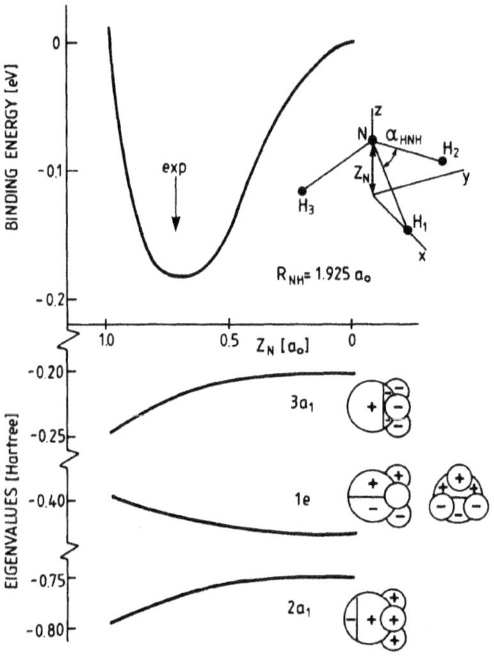

Fig. 3. HKS-LD energy of NH_3 as the N atom moves out of plane for fixed NH distance. Lower half: Variation of eigenvalues for the molecular orbitals involved in the bond. The shape of the molecular orbitals is schematically shown.

seen in the figure, the energy falls as the N atom moves out of plane, goes through a rather flat minimum and then rises steeply. The minimum region coincides with the experimentally determined bond angle, and the barrier height is about 0.18 eV. The lowest energy obtained, when compared with atom calculations using the same functional (but spin unrestricted), corresponds to a binding energy of 14.6 eV for NH_3, a 14% overestimate. The flatness of the energy in the minimum region implies a substantial zero-point motion i.e. the energy varies by <5 meV, or 40 cm^{-1} between bond-angles of $\alpha_{HNH}=103^o$ and 110°.* A polynomial fit to the energy curve for $0.4<Z_N<0.9$ gave a "single well" vibration frequency of 913 cm^{-1},

*This, and the too low barrier height, may be a sign of orbital basis nonconvergence. We note, however, that the uncertainty in the Hartree-Fock value of α_{HNH}^{eq} also points to a rather flat minimum.

somewhat smaller than the "centre-of-gravity" of the measured symmetric and anti-symmetric "bend" vibration energies of 932.5 and 968.3 cm^{-1}.[23] This is consistent with the underestimate of the well depth. The variation of the energy with R_{NH} for bond-angle $\alpha_{HNH}=106^o$ was found to correspond to a vibration frequency of 3300 cm^{-1}, which should be compared with the measured value of 3336.2 cm^{-1},[23] for the totally symmetric NH stretch vibration.

A picture of the energy balance of NH_3 in terms of molecular orbitals can be given that is a precise parallel to that of the water molecule. The eigenvalues are shown in Fig. 3 and the competition between the a_1 orbitals, which tend to favour the out-of-plane geometry, and the two-fold degenerate 1e orbitals, which are HH anti-bonding and so favour the in-plane configuration, is clear. Since there are equal numbers of electrons in the orbitals favouring and disfavouring the in-plane geometry, the result of this competition is not easy to guess.* In Hartree-Fock, for instance, and in the LCAO-LD calculations mentioned earlier, the result was found to depend on whether unbound-atom functions are included in the basis. We find that the eigenvalue sum does show a weak downward trend as the nitrogen moves out of plane, indicating that though equal numbers of "bonds" and "anti-bonds" are present, the a_1 bonds are slightly stronger than the 1e anti-bonds.

This discussion accentuates the delicacy of the energy balance in the ground state of NH_3. In view of this it is rather surprising that one-electron approximations like Hartree-Fock and HKS-LD describe this balance essentially correctly. This can only mean that the precise details of exchange and correlation have relatively little influence on the variation of the energy with the nuclear coordinates.

5. CONCLUDING REMARKS

In this article, we have tried to address some of the issues involved when local potential equations are used to calculate the adiabatic energy of many-electron systems. We have emphasized that the formal justification of this approach does not involve an attempt to approximate the many-electron wave-function. The one-electron equation refers not to the physical system but to a fictitious system of non-interacting electrons - the equivalent system. The significance of the equivalent system is that in principle it has the same electron density as the physical system. The link between the real and equivalent systems is provided by the term E_{xc} in the

*In the ground state of CH_3 there is one fewer electron in $3a_1$ and so one net electron favouring the in-plane geometry, which is stable experimentally.

energy functional, through its functional derivative μ_{xc}, which, via the self-consistency condition, determines when the densities of the two systems are the same. If E_{xc} and μ_{xc} were exactly known (and if we ignore some questions concerning mathematical rigour), the energy of the real system could be determined exactly by solving the one-particle equations.

These ideas were put forward in the two classic papers of Hohenberg, Kohn and Sham.[1] They constitute a new way of looking at the problem of energy calculation that is physically and mathematically appealing. Whereas standard quantum chemical methods seek to approximate the many-electron wave equation, many details of which do not contribute to the energy, HKS address the question, "what is the minimum amount of information we need to calculate the energy ?" Their answer is that we need the exchange-correlation energy density, $\varepsilon_{xc}(\vec{r})$, and its functional derivative with respect to the electron density.

These theoretical ideas are interesting in their own right - they provide, for example, a new way of viewing the Hartree-Fock approximation. However, this interest would be academic were it not for the fact that $\varepsilon_{xc}(\vec{r})$ seems to be astonishingly easy to approximate. For a wide variety of systems, the simplest possible local approximation has been found to give excellent results. This we have demonstrated for the case of the water molecule, for which the local approximation leads to a more accurate description of the energy variation than Hartree-Fock. Although the reason why the local approximation is so accurate is still unclear, we have given very general arguments based on the relation of $\varepsilon_{xc}(\vec{r})$ to the form of the exchange-correlation hole surrounding an electron at $\vec{r}$, that lead to a local relationship irrespective of the size of the density gradients in the system. That is, the reason why the local approximation works in molecules has little or nothing to do with the fact that it becomes exact in the limit of slowly varying density.

The second focus of this article has been to describe and demonstrate the efficiency of an alternative basis to LCAO for performing density functional calculations. We have shown that our localized muffin-tin orbital basis (LMTO) gives good accuracy with a minimum number of radial functions and is so constructed as to simplify certain aspects of the energy calculation. For example, the use of the $(\phi,\dot{\phi})$ combinations of Andersen coupled with the frozen core approximation virtually eliminates the core from the calculation. In addition, reliance on a fitting of the electron density to a separate basis is reduced considerably because the main part of the electronic Coulomb potential can be determined via only one-dimensional integration. On the other hand, programming the LMTO scheme has posed practical problems because special integration techniques are needed that have no analogue in LCAO. This means that we have not been able to make use of any of the standard programmes available in the various quantum chemistry libraries that have been built up over the years. However, we believe that these problems have been overcome to a large extent, and our results for H_2O and NH_3 are an encouraging sign that the LMTO basis is quite accurate enough to give useful results.

In conclusion, we apologize to the reader who feels that over large stretches of this article he has been told what he already knows. However, we feel that many of the points we have discussed have evolved through the literature in a rather roundabout way and that the interested non-expert, particularly if he is a chemist, might appreciate our attempt to collect them together in a unified way.

REFERENCES

1. P. Hohenberg and W. Kohn, Phys. Rev. 136, B864 (1964); W. Kohn and L.J. Sham, Phys. Rev. 140, A1133 (1965).
2. O. Gunnarsson and P. Johansson, Int. J. Quantum Chem. 10, 307 (1976).
3. O. Gunnarsson, J. Harris, and R.O. Jones, Phys. Rev. B 15, 3027 (1977). See also, J. Chem. Phys. 67, 3970 (1977); 68, 1190 (1978); 70, 830 (1979); as well as Refs. 15 and 16.
4. R.O. Jones, J. Chem. Phys. 71, 1300 (1979).
5. See, for example, W. Kohn and P.D. Vashishta, General Density Function Theory, in: "Theory of the Inhomogeneous Electron Gas", S. Lundqvist and N.H. March, eds., Plenum Press, New York (1982).
6. J. Harris, Int. J. Quantum Chem. 13, 189 (1979).
7. M. Levy, Proc. Natl. Acad. Sci. USA 76, 6062 (1979).
8. O. Gunnarsson and B.I. Lundqvist, Phys. Rev. B 13, 4274 (1976).
9. O. Gunnarsson, B.I. Lundqvist, and J.W. Wilkins, Phys. Rev. B 10, 1319 (1974).
10. J. Harris, NATO Summer School on "The Electronic Structure of Complex Systems", W. Temmerman and P. Phariseau, eds., Plenum Press, New York (to be published).
11. S.H. Vosko, L. Wilk, and M. Nusair, Can. J. Phys. 58, 1200 (1980).
12. See, K.H. Johnson, Adv. Quantum Chem. 7, 143 (1973).
13. J.W.D. Connolly and J.R. Sabin, J. Chem.Phys. 56, 5529 (1972).
14. O.K. Andersen, Phys. Rev. B 12, 3060 (1975).
15. E.J. Baerends and P. Ros, Int. J. Quantum Chem. S 12, 169 (1978).
16. B.I. Dunlap, J.W.D. Connolly, and J.R. Sabin, J. Chem. Phys. 71, 3396 (1979).
17. J. Harris and G.S. Painter, Phys. Rev. B 22, 2614 (1980).
18. Calculational details will be presented in J.E. Müller, R.O. Jones, and J. Harris (to be published).

19. E.J. Baerends, D.E. Ellis, and P. Ros, Chem. Phys. 2, 41 (1973).
20. U. von Barth and C.D. Gelatt, Phys. Rev. B 21, 2222 (1980).
21. J. Harris, R.O. Jones, and J.E. Müller, J. Chem. Phys. 75, 3904 (1981).
22. J.C. Slater, "Quantum Theory of Molecules and Solids", Vol. 1, McGraw Hill, New York (1963).
23. G. Herzberg, "Electronic Spectra of Polyatomic Molecules", van Nostrand Reinhold, New York (1966).
24. D. Pines and P. Nozières, "Theory of Quantum Fluids", Benjamin, New York (1968).
25. K. Kitaura, C. Satoko, and K. Morokuma, Chem. Phys. Lett. 65, 206 (1979).
26. Quoted in "The Structure and Properties of Water" by D. Eisenberg and W. Kauzmann, Oxford University Press, London (1969).
27. T.H. Dunning, R.M. Pitzer, and S. Aung, J. Chem. Phys. 57, 5044 (1972).
28. A.D. Walsh, J. Chem. Soc., 2260 (1953).
29. A. Rank, L.C. Allen, and E. Clementi, J. Chem. Phys. 52, 4133 (1970).
30. J.D. Swalen and J.A. Ibers, J. Chem. Phys. 36, 1914 (1962).
31. W.R. Rodwell and L. Radom, J. Chem. Phys. 72, 2205 (1980).
32. U. Wahlgren, Chem. Phys. Lett. 20, 246 (1973).

COHESIVE ENERGY, SPECTROSCOPIC, AND MAGNETIC PROPERTIES OF SMALL CLUSTERS

D. E. Ellis and B. Delley

Department of Physics and Astronomy and
Materials Research Center
Northwestern University, Evanston
Illinois 60201, U.S.A.

Applications of the self-consistent one-electron local density theory are described within the molecular orbital framework. The discrete variational method is used to determine cohesive energy, spectroscopic, and magnetic properties. We discuss data representative of recent progress in three areas: size effects in small metal particles, magnetic impurities in transition metal hosts, and the effects of hydrogen (or muon) impurities in transition metals.

1. INTRODUCTION

Local density theory had its beginnings and early successes in solid state physics. The development of position-dependent exchange potentials by Slater and his co-workers was driven by the necessity of finding a single-particle Hamiltonian which was simple enough to be applied to crystalline solids, without sacrificing the essential physics.[1] The local exchange approximation turned out to be better than its inventors could have supposed in their wildest dreams. This in turn led to extraordinary theoretical efforts which produced a solid conceptual foundation for the local density approach to electronic structure. This work is not yet finished, and the interplay between applications and fundamental studies is one of the interesting aspects of the present-day scene.

In this paper, we discuss recent applications to properties of small clusters of atoms which have relevance to some areas of current interest in metal physics. First, size effects on cohesive energy and bond distance for copper particles are considered. These data indicate not only the predictive power of self-consistent local density methods, but also point the way for future improvements in the underlying theory. Second, the classic problem of a single magnetic impurity in a metallic host is considered. Passing from the FeAg case where the host is magnetically inert, to the superparamagnetic host in FePd, we explore the range of response of the metal to the embedded moment. The results suggest that a more chemical picture, of bonding and wavefunction mixing between host and impurity, is required in order to understand the nature of observed induced moments. Finally, effects of hydrogen or muon impurities on electronic structure and magnetic hyperfine fields are considered. Measurements and theoretical calculations for hydrogenic impurities are of importance not only because of their value as spatially localized probes of internal fields, but also because of technical problems of hydrogen storage and retrieval in metals, and the embrittlement of reactor vessels.

2. THEORETICAL MODEL AND COMPUTATIONAL APPROACH

The theoretical basis of self-consistent local density models is described in detail elsewhere in this volume, and in the literature.[1,2] Here we will merely sketch salient features of our particular approach in describing structure of both free molecules and embedded clusters.[3] The non-relativistic spin dependent Hamiltonian operator is written as

$$h_\sigma = t + V_{Coul} + V_{xc,\sigma} \tag{1}$$

where t is the kinetic energy, V_{Coul} the Coulomb potential due to both electrons and nuclei, and $V_{xc,\sigma}$ is some local approximation to the fundamentally nonlocal exchange and correlation potential. For many applications, the simplest form of $V_{xc}=K\rho_\sigma^{1/3}$ where K is a known constant gives an entirely satisfactory representation of spectroscopic and total energies.

The Schrödinger equation can be solved variationally, to determine cluster eigenfunctions

$$\Psi_{i\sigma} = \sum_j \chi_j(\vec{r})C_{ji}^{\sigma} \tag{2}$$

in terms of an expansion over symmetry-adapted basis functions $\{\chi_j\}$. The charge and spin densities can then be formed from the individual spin components,

$$\rho_\sigma = \sum_i f_{i\sigma}|\Psi_{i\sigma}|^2 \tag{3}$$

and iterations are carried out to achieve a self-consistent density. Details of the numerical integration and iteration procedure have been presented elsewhere.[3,4] In a typical application 200-500 integration points per distinct atom and 20-30 iterations are required to obtain reasonably accurate well-converged eigenvalues and eigenvectors.

The basis functions used in the calculations reported here were constructed from numerical free atom wavefunctions. In some cases additional quasi-bound states were generated by embedding the atom in a potential well in order to extend the variational freedom.

Molecular Coulomb and exchange potentials used in the self-consistency loop were generated using a model charge and spin density, since solution of the Poisson equation for the eigendensity of Eq. (3) is awkward. In the self-consistent-charge (SCC) approximation,[4] the model density is given by

$$\tilde{\rho} = \sum_{\nu n \ell} f_{n\ell}^{\nu} |R_{n\ell}^{\nu}(r_\nu)|^2 \tag{4}$$

where $\{f_{n\ell}^{\nu}\}$ are Mulliken atomic orbital populations found by analysis of the eigenvectors, and $\{R_{n\ell}^{\nu}\}$ are the corresponding radial wavefunctions used in the variational basis. The SSC approximation to ρ thus consists of spherical overlapping atom-like densities, and is found to produce reasonably good results for a variety of molecular and solid state systems. An improved model density can be obtained by using a multicenter-multipolar expansion,

$$\tilde{\rho} = \sum_{\nu n \ell m} \rho_{n\ell m}^{\nu}(r_\nu)\, Y_{\ell m}(\hat{r}_\nu) \tag{5}$$

with radial amplitudes determined by the least squares criterion: minimize $<(\rho-\tilde{\rho})^2>$. This leads to a Self-Consistent Multipolar (SCM) iteration procedure with controllable precision and a higher degree of accuracy which is needed for close study of molecular binding energies and other sensitive properties. The SCM procedure is clearly more time-consuming and thus costly, to execute. However, for a discussion of bonding in low symmetry complexes, and for molecules with significant covalent bonding, inclusion of $\ell=1$ (dipolar) and $\ell=2$ (quadrupolar) density and potential components is quite important.[5]

Development of total energy algorithms for multicenter systems, which are capable of "chemical accuracy" has proved to be quite a challenge. However, three or four distinct schemes have now emerged which have demonstrated the necessary precision and stability. The method that we use in the present study of small metal particles is described elsewhere.[6] The chief point to note here is that the cohesive energy of interest is of the order of 2 eV/atom and is thus a tiny part of the total energy. In order to calculate the binding between fragments A and B, $\varepsilon = E_{tot}(A)-E_{tot}(B)$ it is necessary to arrange terms in the total energy expression[1,6]

$$E_{tot}(\rho) = <t> + <V_{Coul}^{ne}> + \frac{1}{2} <V_{Coul}^{ee}> + \frac{3}{4} <V_{xc}> + V_{Coul}^{nn} \tag{6}$$

to obtain maximum error cancellation. The brackets in Eq. (6) denote summation over occupied levels making up the density ρ_σ, as well as integration over spatial co-ordinates. The notation n,e refers to nuclear and electronic components, respectively. In practice we find that very stable values for ε result from using the model density $\tilde{\rho}$ to determine the potential energy, with integration carried out on the <u>same</u> mesh as in the variational calculation.

3. BINDING ENERGIES AND EQUILIBRIUM GEOMETRY VERSUS PARTICLE SIZE

Electronic properties of small particles are of interest in many different fields; to mention only one: catalysis of hydrocarbon synthesis reactions. To begin our study of size effects, we have chosen Cu and Ag clusters where the presence of a filled d-band simplifies theoretical studies. Good experimental information is also available, especially for small Cu particles. Comparisons between the $3d^{10}$ and $4d^{10}$ systems are interesting especially since the silver clusters are catalytically active.

Experimental X-ray diffraction studies on small noble metal particles and thin plates, which started 30 years ago,[7] have suggested that the lattice becomes contracted with decreasing particle size. A recent extended X-ray absorption fine structure (EXAFS) study[8] supports these conclusions. Vibrationally resolved optical band spectroscopic measurements have allowed the extraction of information about the binding energy curves of Cu dimers;[9,10] the bond length is considerably shorter than the bulk nearest neighbor distance and the binding energy is greatly reduced.

We have performed self-consistent calculations of the electronic structure and binding energy for high symmetry small particles including: Cu_2, Cu_4 (with T_d as well as $D_{\infty h}$ symmetry), Cu_{13}, Ag_{13}, and Cu_{79} (with O_h symmetry). An extended discussion of our results will be given elsewhere;[6] here we only want to summarize the main findings.

The metallic bond is seen to be more delocalized than that of covalent molecules; this is hardly surprising. However, it is remarkable that in the interatomic region and along bond lines, charge densities comparable to a bulk metal are obtained already for the Cu <u>dimer</u>. We find that on formation of the particles mentioned before, the electronic charge density becomes slightly reduced inside the atomic core region when compared to the free atom, with significant consequences for core-level binding energies.

Small particles have a discrete energy spectrum of electronic states. If such discrete spectra are suitably broadened, some resemblance to bulk spectra becomes recognizable. For the Cu_{79} particle, the position and width of the d-band distribution in the central part of the cluster are within 0.2 eV of the bulk values. Since the density of states is a "long range property", it would not be expected that any of the presently studied clusters should come close to the bulk results. The local densities of states for the larger clusters show a d-band narrowing and shift for the surface layer of atoms, quite analogous to that found in thin film studies.

Before we discuss results for nearest neighbor distance and cohesive energy as a function of particle size, it is useful to define a size parameter. Cluster size dependent quantities can be considered as surface effects. If N is the number of atoms in the cluster, then the volume goes like N, the diameter goes like $N^{1/3}$ and the surface area like $N^{2/3}$. In discussing surface effects on the cohesive energy, which is defined on a per atom basis, we find the surface-to-volume ratio $N^{-1/3}$ as the appropriate parameter.

We show in Fig. 1 calculated equilibrium nearest neighbor distances for the Cu_2, Cu_4, and Cu_{13} particles. The calculated values show a rather regular reduction of nearest neighbor distance on going to smaller particle size. We have also included in the figure the experimental bulk and dimer bond distance together with experimental

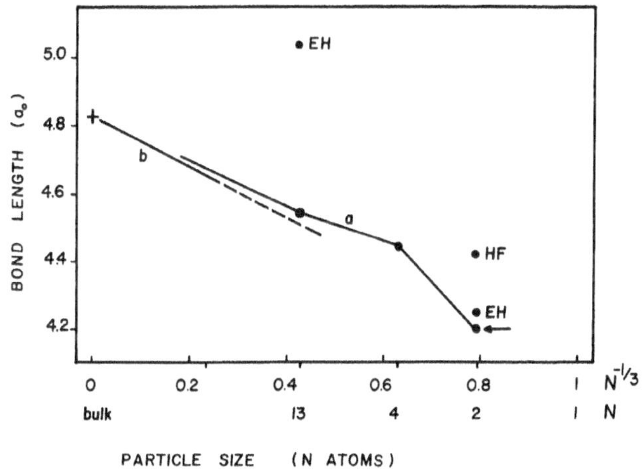

Fig. 1. Bond length versus particle size from Ref. 6. Calculated bond lengths are connected by line and extrapolated linearly to the theoretical bulk value from Ref. 11. Curve b from EXAFS data, Ref. 8 and connected with experiment bulk value. The arrow indicates the experimental value for the dimer - in excellent agreement with the calculated local density value and the value from a configuration interaction calculation (P. Joyes and M. Leleyter, J. Phys. B6, 150 (1973)). Results of other calculations are denoted as EH for extended Hückel.[12] The result of a Hartree-Fock calculation is denoted by HF.[13]

results from EXAFS[8] measurements. Recalling that local density theory has also been used to predict bulk lattice constants,[11] which agree with experiment, we can see that this theory is very consistent with observed experimental trends. Other theories, as indicated in the figure, have been less successful. For Ag_{13} we find a 2.2% reduction in nearest neighbor distance, as compared to the experimental bulk value.

We have found cohesive energies of 1.05 eV/atom for Cu_2, 1.26 eV/atom for tetrahedral Cu_4, 1.23 eV for linear Cu_4, 2.19 eV for Cu_{13}, and 3.03 eV for Cu_{79} (the latter with a 4.80 a_o fixed nearest neighbor distance). Fig. 2 shows these values plotted against our size parameter $N^{-1/3}$. For $N^{-1/3}=1$ we have the isolated atom. A calculation of the total energy for the atom with the local-spin density theory provides the reference state with which cohesive energies have to be compared. Spin contributions to the cohesive energy vanish for Cu_2, which is a closed shell molecule. For the larger Cu clusters, spin effects are negligibly small, as we have demonstrated in the Cu tetramer case.

The cohesive energies versus particle size show a surprisingly linear trend when scaled against the 'surface' parameter. Comparison with experiment can be made for the dimer and agreement is excellent. On the bulk side of the figure, cohesive energies are overestimated by ca. 15-20% by local density theory. Surface tension values derived from a fit through present cluster cohesive energies, overestimate again experimental data by 15-20%. While these discrepancies are somewhat less than entirely satisfactory, present results still suggest that the trend of cohesive energies versus particle size can be obtained correctly by local density theory.

We also have included results obtained by extended Hückel[12] and by Hartree-Fock theory[13] in the figure. Both of these theories appear to have difficulty in reproducing the trends of bond length and cohesive energy with particle size.

4. LOCALIZED MAGNETIC MOMENTS IN METALS: Fe<u>Ag</u> and Fe<u>Pd</u>

The existence and behaviour of localized magnetic moments as a function of atomic impurity as well as host properties has been studied both experimentally as well as theoretically for a long time. In the case of Fe impurities the response of the host can lead to an apparently nonmagnetic state in Fe<u>Aℓ</u>; in Fe<u>Ag</u> the impurity moment is of comparable magnitude to an Fe atom; i.e. ~2.9μ_B and in giant moment systems, such as Fe<u>Pd</u>, a moment of ~12μ_B is found.

Previous theories have mostly considered single site models where the host has been approximated as a free electron gas,[14] or by a crystal field.[15] More recent theories take the response of the

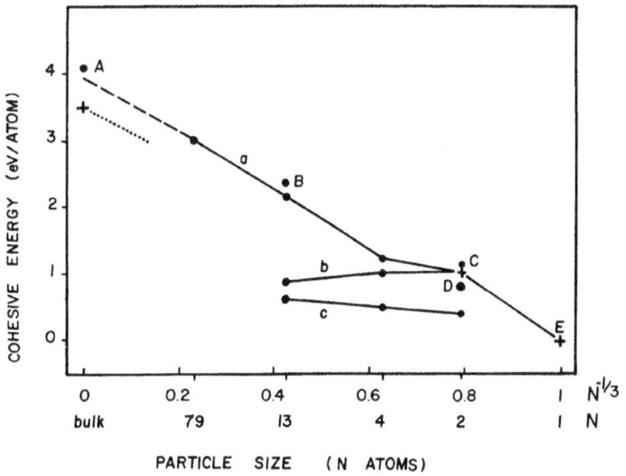

Fig. 2. Cohesive energy versus particle size from Ref. 6. N=number of atoms in the cluster and $N^{-1/3}$ is an effective surface/volume ratio. The experimental values for the atom, dimer, and bulk are indicated by a cross (+). The cross for the dimer coincides with the present α=0.7 result and the extended Hückel (EH) result.[12] Point A is the bulk value obtained in Ref. 11 with the Hedin-Lundqvist potential; B is the cohesive energy for the 13 atom cluster with the same potential; C is the value for the dimer by Harris and Jones[22] using the Gunnarsson-Lundqvist (GL) potential, it coincides with the present value for the GL potential; D is the value for the dimer by configuration interaction theory; E is the atomic binding energy, which is the zero reference for each method. The present result uses the spin-polarized atom $3d{\uparrow}^5 3d{\downarrow}^5 4s{\uparrow}$ configuration as the reference state. A line (a) connects the present α=0.7 results. The extrapolation to bulk is discussed in the text. The line (b) connects extended Hückel results[12] and the line (c) connects Hartree-Fock results.[13] The dotted line is taken from surface tension data: Yu. V. Naidich and V.N. Eremenko, Fiz. Metal. Metalloved, 11, 883 (1961).

host into account by a Green's function method;[16] until very recently, however, self-consistent treatment has been restricted to the impurity site.[17]

Cluster theories allow for self-consistent treatment of several neighbor shells. Two varieties are possible; in the supercell approach,[18] the cluster is repeated periodically, which allows the study of interacting impurities at higher concentrations. A finite cluster approach can treat one or more impurities and long-range effects can be taken into account by suitable boundary conditions or a Green's function embedding.

In the present study we have treated clusters of 43 and 55 atoms containing a single impurity with free molecule boundary conditions. In view of the magnetic properties of interest we have solved the local spin-density equations self-consistently with a spin dependent potential,[19] allowing the orbitals for the two spin directions to have different shapes.

Before considering the impurity results it is useful to describe what is obtained for the pure host metals. We need to recall that the spectrum of a finite cluster is discrete. The discrete level density of states (DOS) for the cluster is made continuous by suitable broadening of the line spectrum, and a volume-integrated partial density of states can be obtained as

$$D_{\nu\sigma}(E) = \int_{\Omega\nu} \rho_\sigma(\vec{r},E)d^3r \quad (7)$$

In the pure Ag_{43} and Pd_{43} clusters, a total bandwidth for the d-band is obtained which is within 0.1-0.2 eV of the bulk band structure result. The local DOS for the surface shell of atoms is narrowed as might be expected. For the center of the cluster local DOS gives gross features such as position of peaks and edge of d-band, which are similar to bulk results.

In the $FeAg_{42}$ cluster the spin-dependent partial DOS in the iron Wigner-Seitz sphere shows a sharp majority spin peak, which is bound by 1.6 eV with respect to E_F, plus a small amount of additional DOS at higher binding energy. The Fe minority spin partial DOS shows also a sharp peak at E_F and a small additional DOS inside the Ag d-band. These partial Fe DOS peaks arise from e_g and t_{2g} one electron states, which have a large Fe d-population and some s and p population on the Ag atoms. As the minority peak states (at E_F) are only partially filled a spin density results which integrates to ~$3\mu_B$ in the Fe Wigner-Seitz sphere. On the Ag atoms the spin density is rather small and oscillating. These results are consistent with previous interpretations, e.g. Refs. 16,17,20, and therefore they help to establish the finite cluster scheme as a useful model for a magnetic bulk system.

In the $FePd_{42}$ and $FePd_{54}$ clusters the partial DOS for majority spin in the Fe Wigner-Seitz cell is band-like, with a width comparable to the Pd d-band width. The Fe minority spin partial DOS exhibits a band-like part as well as a peak above E_F, as shown in Fig. 3. The states contributing to the Fe DOS are e_g and t_{2g} states with Fe d-character and are hybridized strongly with Pd d-states.

Contour plots of orbital densities show that the magnetic states

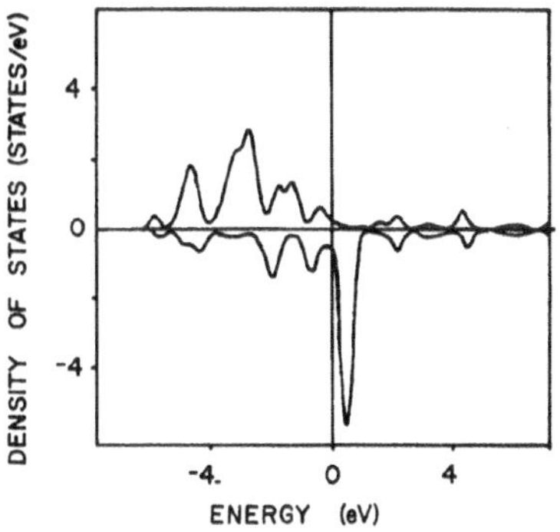

Fig. 3. Partial density of states inside the Fe "atomic" sphere for the $FePd_{54}$ cluster.

are rather delocalized throughout the cluster. It is not possible to predict the shape of the spin-density by picking just one or two orbital densities. Rather a summation and integration over all orbital densities has to be performed. The result for a (100) plane is shown in Fig. 4. There is a large spin-density on the Fe atom, which integrates to $3.2\mu_B$ in the W-S sphere. The neighboring Pd atoms are clearly polarized with a moment of $0.1\text{-}0.2\mu_B$. The total moment for the interior 43 atoms of the 55 atom cluster integrates to $6.8\mu_B$.

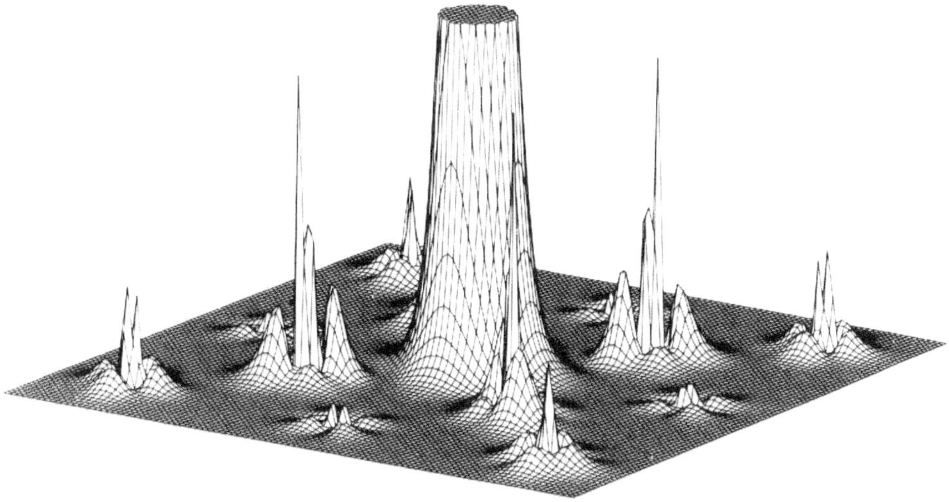

Fig. 4. Spin density for $FePd_{54}$ cluster in a 100 plane, truncated at 0.25 e/a_o^3.

It is especially interesting to observe that in the Fe<u>Pd</u> clusters a part of the moment is due to orbitals that have zero Fe d-population. Such orbitals, which are mostly Pd d-like, describe host exchange effects that give rise to the enhanced magnetic susceptibility of Pd. This exchange enhancement is a local property of the Pd host, which is due in the first place to the high DOS of Pd at the Fermi energy.

While this exchange enhancement of the Pd host is an important factor determining the magnitude of the giant moment, it is not what causes the giant moment to arise in the first place. We find that spin-polarization is transmitted from Fe to Pd mainly by hybridization effects. In order to clarify the role of hybridization effects, we have performed a model calculation in which we set the Hamiltonian matrix elements between Fe d- and Pd d-states to zero. This leaves the neighbor Pd atoms with tails that overlap with the Fe impurity site and sample its potential; i.e. the direct exchange mechanism remains operative. The result of such a calculation is a normal moment on Fe ($3\mu_B$) but no giant moment. This clearly shows that the occurrence of the host magnetic halo is due to the nonlocal susceptibility arising from hybridization effects.

We have thus shown that small clusters can be useful models for magnetic impurities in bulk systems, and have explored the essential differences between inert hosts like Ag and the superparamagnetic host Pd. The fully self-consistent treatment of the first few neighbor shells is indispensable for gaining a useful modeling of the response of systems like Fe<u>Pd</u>. More complete analyses are given in a separate publication.[21]

5. HYDROGENIC IMPURITIES IN TRANSITION METALS

Certain electronic properties depend sensitively upon the wavefunction character of states near the Fermi energy. Transport coefficients, and electric and magnetic hyperfine fields in metals are suitable examples. One can attempt a direct calculation of such properties using the discrete level structure of a finite cluster model with "free molecule" boundary conditions. However, it is clear that the results can be unstable, in the sense that motion of a single level in passing above or below E_F, can produce a major change in say, the hyperfine fields. The predictions will then be strongly dependent upon cluster size, truncations in potential or wavefunction expansions, and other computational artifacts.

The discrete levels of a cluster can be broadened to form a band structure which will hopefully asymptotically approach the band structure of the extended solid. A simple approach, which has been used extensively to describe spectral densities, consists simply of broadening cluster levels with a Lorenzian line shape. An embedding potential is included in the cluster Hamiltonian to represent cluster-environment interactions.[20] An energy-independent broadening is, however, not a very good model for the transition metals; we know that localized d-electron states form rather narrow bands compared to more diffuse s,p states which occupy a similar energy interval.

A better model can be found by imposing suitable boundary conditions on the cluster wavefunctions. The general idea of matching cluster functions onto continuum waves, which approximately represent propagating states of the host medium, has been explored in the context of the multiple-scattering Xα scheme[22] and of the discrete variational Xα scheme.[23] These wavefunction matching schemes provide a physically motivated model with quite reasonable energy band distributions for both localized and diffuse states. The computational effort is, of course, greater than that of simple level-broadened models, since the spectral densities have to be iteratively determined as part of the self-consistency procedure. On the other hand, this method is very much simpler than the rigorous lattice Green's function approach, which typically requires knowledge of the unperturbed host Green's function prior to construction of cluster states.[16,17] In our simplified approach, both cluster and host can be brought simultaneously to self-consistency, and the choice of which atoms belong to "cluster" and "host" can be freely varied.

As a single example, we discuss briefly the magnetic moments and contact hyperfine fields in transition metals, and the effect of hydrogen impurities on these quantities. A detailed description is published elsewhere.[23]

The von Barth-Hedin potential[19] was adopted to describe the spin-dependent exchange and correlation interactions, and DV-Xα self-consistent calculations were carried out on clusters containing up to 38 atoms. The cluster wavefunctions were expanded in the form

$$\Psi\,(\varepsilon_k,\vec{r}) = a_k\Theta_k + \sum_{j=1}^{N} C_j\Phi_j \tag{8}$$

where Θ_k is a chosen asymptotic ($r\to\infty$) function and the $\{\Phi_j\}$ form a cluster-localized expansion basis. An inhomogeneous Schrödinger equation is solved at each energy ε_k. A particularly simple form is obtained if we choose $\{\Phi_j\}$ as the orthonormalized eigenfunctions to the homogeneous problem of the isolated cluster, with Θ_k also chosen to be an eigenstate of some model Hamiltonian associated with potential $V_o(\vec{r})$. Then the amplitudes are related by

$$(\varepsilon_i - \varepsilon_k)C_i = -\langle\Phi_i|u|\Theta_k\rangle a_k \tag{9}$$

where $u = V_{cluster}(\vec{r}) - V_o(\vec{r})$.

The eigenstates of Eq. (8) can be interpreted as built up from orthogonal cluster-localized and propagating parts; this leads naturally to an energy-dependent line profile $d_i(\varepsilon)$ or density of states (DOS) for each cluster level. If the matrix element of u appearing in Eq. (9) is slowly varying with energy, the DOS approaches a Lorenzian function with width $|<\Phi_i|u|\Theta>/<\Theta|\Theta>^{\frac{1}{2}}|$.

A typical result is given in Fig. 5 for the octahedral interstitial site occupied by a proton or muon in fcc nickel. The discrete levels, which have finite density at the H-site for a 38 atom cluster, are indicated in the bottom panel. The total (spin ↑ and ↓ shown separately) density of states, H-site DOS, and H-site spin density distribution are displayed. We find here, and in related Fe, Co, and Ni systems, that the hyperfine fields result from a delicate balance between negatively exchange-polarized "bound-paired" states (not shown on the figure) and positive "unpaired band" contributions. The rather good agreement found for calculated moments (and reduction in moment for metal sites adjacent to the impurity)

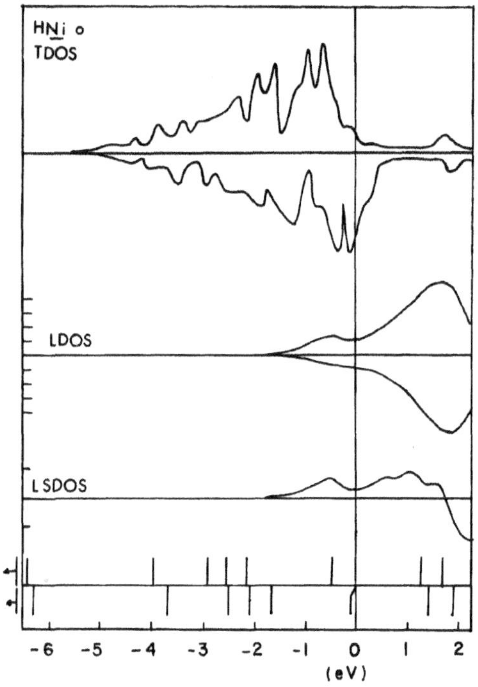

Fig. 5. Total density of states (TDOS) for spin ↑,↓, local density of states (LDOS) and local spin density (LSDOS) for octahedral position H impurity in nickel (from Ref. 23). Discrete energy levels of 38 atom cluster with nonzero density at the H-site are indicated in the lower panel.

and hyperfine fields in comparison with experiment is quite encouraging. These results suggest that physically motivated cluster level broadening schemes are capable of making quantitative predictions of properties which depend upon the metallic band structure. This opens possibilities for the study of disordered and defective systems which are hardly amenable to periodic band structure approaches.

ACKNOWLEDGEMENTS

This research was partially supported under the NSF-MRL program through the Materials Research Center of Northwestern University (Grant No. DMR79-23573). We acknowledge the essential contribution of B. Lindgren concerning the hydrogenic impurity problem.

REFERENCES

1. J.C. Slater, The Self-Consistent Field for Molecules and Solids (McGraw-Hill, NY, 1974).
2. For example, U. von Barth and L. Hedin, J. Phys. C 5, 1629 (1972); O. Gunnarsson and B.I. Lundqvist, Phys. Rev. B 13, 4274 (1976); also see article by B.I. Lundqvist in this volume.
3 E.J. Baerends, D.E. Ellis, P. Ros, Chem. Phys. 2, 41 (1973); F.W. Averill and D.E. Ellis, J. Chem. Phys. 59, 6412 (1973).
4. A. Rosén and D.E. Ellis, J. Chem. Phys. 65, 3629 (1976).
5. B. Delley and D.E. Ellis, J. Chem. Phys. 16, 1949 (1982); B. Delley (unpublished).
6. B. Delley, D.E. Ellis, and A.J. Freeman (unpublished); ibid Bull. Amer. Phys. Soc. 27, 345 (1982).
7. F.W.C. Boswell, Proc. Phys. Soc. London A 64, 465 (1951).
8. G. Apai, J.F. Hamilton, T. Stohr, and A. Thompson, Phys. Rev. Lett. 43, 165 (1979).
9. B. Kleeman and S. Lundqvist, Ark. Phys. Sv 8, 333 (1954).
10. N. Ashlund, R.F. Barrow, W.G. Richards, and D.N. Travis, Ark. Phys. Sv. 30, 171 (1965).
11. V.L. Moruzzi, J.F. Janak, and A.R. Williams, Calculated Electronic Properties of Metals (Pergamon, New York, 1978).
12. A.B. Anderson, J. Chem. Phys. 68, 1744 (1978).
13. C. Bachmann, J. Demuynck, and A. Veillard, Faraday Symp. 14, 170 (1980).
14. P.W. Anderson, Phys. Rev. 124, 41 (1961).
15. L.L. Hirst, Phys. Kondens. Mat. 11, 255 (1970).
16. R. Podloucky, R. Zeller, and P.H. Dederichs, Phys. Rev. B 22, 5777 (1980).
17. R. Zeller and P.J. Braspenning, Sol. Sta. Commun. 42, 70 (1982).
18. T. Jarlborg, B. Delley, D.E. Ellis, and A.J. Freeman, Bull. Amer. Phys. Soc. 27, 214 (1982).

19. We use the form of von Barth-Hedin, J. Phys. C 5, 1629 (1972), with constants C_p=0.225, r_p=21 according to L. Hedin and B.I. Lundqvist, J. Phys. C 4, 2064 (1971) and $C_F=C_p/2$, $r_F=2^{4/3}r_p$.
20. Previous DV-Xα studies of magnetic impurities in an embedding potential include: G.A. Benesh and D.E. Ellis, Phys. Rev. B 24, 1603 (1981); D.E. Ellis, G.A. Benesh, and E. Byrom, Phys. Rev. B 20, 1198 (1979). The MS-Xα approach, with free molecule boundary conditions, has also been used; for example, K.H. Johnson, D.D. Vvedensky, and R.P. Messmer, Phys. Rev. B 19, 1519 (1979).
21. B. Delley, D.E. Ellis, and A.J. Freeman (unpublished); ibid, Bull. Amer. Phys. Soc. 27, 214 (1982).
22. V.I. Anisimov, V.A. Gubanov, D.E. Ellis, and E.Z. Kurmaev, J. Phys. F 11, 405 (1981).
23. B. Lindgren and D.E. Ellis, Phys. Rev. B 26 (July, 1982).

CALCULATIONS OF HYPERFINE INTERACTIONS IN TRANSITION METAL COMPOUNDS IN THE LOCAL DENSITY APPROXIMATION

Diana Guenzburger

Centro Brasileiro de Pesquisas Físicas - CBPF/CNPq
Rua Xavier Sigaud, 150
22290 Rio de Janeiro, RJ, Brasil

A survey is made of some theoretical calculations of electrostatic and magnetic hyperfine interactions in transition metal compounds and complex ions. The molecular orbital methods considered are the Multiple Scattering and Discrete Variational, in which the local Xα approximation for the exchange interaction is employed. Emphasis is given to the qualitative informations, derived from the calculations, relating the hyperfine parameters to characteristics of the chemical bonds.

1. INTRODUCTION

Hyperfine interactions are electromagnetic interactions between a nucleus and surrounding charges, other than the Coulomb interaction between the external charge distribution, and the nucleus considered as a point charge. These small interactions are of the order of 10^{-8} eV, and may be detected through their effect on the nuclear or electronic energy levels.

The experimental techniques employed to measure hyperfine interactions are, most commonly, spectroscopic in nature. Some techniques that involve only the ground state of the nucleus are Nuclear Magnetic Resonance,[1] Electron Paramagnetic Resonance,[2] Nuclear Quadrupole Resonance[3] etc. Other techniques, such as Mössbauer spectroscopy[4] and Perturbed Angular Correlation,[5] involve nuclear excited states as well.

Hyperfine interactions provide information on the nucleus, and on its electronic environment. As such, they have become an important tool in the investigation of solid state electronic properties, and in the description of chemical bonds in molecules.

Investigations of electronic properties through hyperfine interactions have outstanding importance for transition metal compounds and complex ions. Many such compounds have unpaired electrons in open shells and are suitable for Electron Paramagnetic Resonance studies. Others, due to the characteristics of the metal nucleii, are convenient probes for Mössbauer spectroscopy studies. Among these last, Fe has been extensively employed, through the 14.4 keV transition of ^{57}Fe, providing wealthy data which can be related to ionic or covalent bonding to ligands.

In relating experimental measurements to electronic properties, wave functions of good quality are needed to calculate the hyperfine parameters. On the other hand, wave functions of full Hartree-Fock level are still difficult to obtain for transition metal compounds. The Hartree-Fock-Slater method,[6] which employes the local density approximation known as Xα for the exchange interaction, constitutes an alternative which is less costly computationally. The success achieved with this method in calculating other properties of transition metal molecules has encouraged its use in hyperfine interaction studies, and now a sufficient number of investigations have been reported to allow an assessment of some of its successes and failures.

2. ELECTROSTATIC HYPERFINE INTERACTIONS

The energy of the electrostatic interaction of a nuclear charge distribution $\rho(\vec{r})$ placed in a potential $\phi(\vec{r})$ due to external charge is:

$$W = \int \rho(\vec{r})\phi(\vec{r})d\tau \tag{1}$$

Expanding $\phi(\vec{r})$ in a Taylor series around $\vec{r} = 0$, and defining the total nuclear charge as $Ze = \int \rho(\vec{r})d\tau$, we obtain:

$$\begin{aligned} W = Ze\phi(0) &+ \sum_{\alpha=1}^{3} \left(\frac{\partial \phi}{\partial x_\alpha}\right)_o \int \rho(\vec{r})\, x_\alpha d\tau \\ &+ \frac{1}{2} \sum_{\alpha\beta}\sum \left(\frac{\partial^2 \phi}{\partial x_\alpha \partial x_\beta}\right)_o \int \rho(\vec{r}) x_\alpha x_\beta d\tau + \ldots \end{aligned} \tag{2}$$

The first term is the Coulomb point charge interaction, and the second (dipole) term vanishes because of parity conservation of the nuclei. Rewriting the third term by introducing $r^2 = \sum_\alpha x_\alpha^2$, we find

$$W' = \frac{1}{6}\sum_{\alpha}\left(\frac{\partial^2\phi}{\partial x_\alpha^2}\right)_o \int\rho(\vec{r})r^2 d\tau$$
$$+ \frac{1}{6}\sum_{\alpha\beta}\sum\left(\frac{\partial^2\phi}{\partial x_\alpha \partial x_\beta}\right)_o \int\rho(\vec{r})(3x_\alpha x_\beta - r^2\delta_{\alpha\beta})d\tau \qquad (3)$$

From the first term in W' is derived the monopole shift δ of a nuclear level. Making use of Poisson's equation $\nabla^2\phi(\vec{r}) = -4\Pi\rho_{e\ell}(\vec{r})$ at $\vec{r} = 0$, we find:

$$\delta = \frac{2\Pi}{3} Ze^2|\Psi(0)|^2 \langle r^2\rangle_N \qquad (4)$$

where $\rho_{e\ell} \equiv -e|\Psi(0)|^2$ is the electronic charge density within the nuclear volume and $\langle r^2\rangle_N$ is the mean square radius of the nuclear charge distribution.

In a nuclear resonance (Mössbauer) experiment, if $\rho_{e\ell}(0)$ is different at the emitting (S) and absorbing (A) nuclei, because of different electronic environments, an isomer shift (IS)[7] of the resonance line is observed, given by:

$$IS = \frac{2\Pi}{3} Ze^2[\langle r^2\rangle_e - \langle r^2\rangle_g][|\Psi(0)|^2_A - |\Psi(0)|^2_S]S'(Z) \qquad (5)$$

where $\langle r^2\rangle_e$ and $\langle r^2\rangle_g$ are the mean square nuclear radii in the excited and ground nuclear levels of the Mössbauer transition, and S'(Z) a factor to account for relativistic effects.

The second term of W' in Eq. (3) expresses classically the interaction between the electric field gradient tensor and the nuclear quadrupole moment tensor.[8] The quantum mechanical treatment of this interaction involves the evaluation of the matrix elements of the electrostatic Hamiltonian $H_{e\ell}$ defined as:

$$H_{e\ell} = \sum_{i=1}^{N}\sum_{p=1}^{Z} \frac{eK_i}{|\vec{r}_i - \vec{r}_p|} \qquad (6)$$

where $\vec{r}_p$ are the position vectors of the Z nuclear protons surrounded by N exterior charges K_i. Expanding $(1/|\vec{r}_i - \vec{r}_p|)$ in Legendre polynomials and making use of the addition theorem of the spherical harmonics, we find:

$$H_{e\ell} = \sum_{\ell,m}\left[\sum_p \sqrt{\frac{4\Pi}{2\ell+1}}\, e\, r_p^\ell\, Y_\ell^m(\theta_p\phi_p)\right]\left[\sum_i \sqrt{\frac{4\Pi}{2\ell+1}} \frac{K_i Y_\ell^{m*}(\theta_i\phi_i)}{r_i^{\ell+1}}\right]$$

or

$$H_{e\ell} = \sum_{\ell,m} [\varepsilon_{\ell m}][v^*_{\ell m}] \tag{7}$$

The quadrupolar Hamiltonian H_Q is the $\ell=2$ term of the expansion. This interaction may be considered as a perturbation that will mix the substates of a nuclear level of spin I. Making use of the Wigner-Eckart theorem and the definition of the nuclear quadrupole moment Q,

$$Q = \langle II|\sum_p (3z_p^2 - r_p^2)|II\rangle \tag{8}$$

the matrix elements of H_Q are expressed as

$$\langle Im_j|H_Q|Im_i\rangle = \frac{eQ}{2}\,\frac{\langle I2m_i m|Im_j\rangle}{\langle I2I0|II\rangle}\,v^*_{2m} \tag{9}$$

where m_j and m_i are projections of the nuclear spin I on the z-axis and the coefficient on the right side includes Clebsch-Gordon coefficients, with the usual selection rule $m=m_j-m_i$. The electric field gradient tensor obtained from the field at the nucleus due to the external charges K_i has as its components:

$$V_{k\ell} = \left\{\frac{\partial^2}{\partial x_k\,\partial x_\ell}\left[\sum_i \frac{K_i}{|\vec{r}_i-\vec{r}_n|}\right]\right\}_{\vec{r}_n\to 0} = \sum_i K_i\,\frac{3x_{ik}x_{i\ell}-\delta_{k\ell}r_i^2}{r_i^5} \tag{10}$$

The matrix representing this tensor is symmetric and, due to Laplace's equation, traceless, so that only five independent components remain. These may be related to the elements v_{2m} by a simple transformation of coordinates. In a coordinate system in which $V_{k\ell}$ is diagonal, we have then:

$$\begin{aligned} v_{20} &= V_{zz}/2 \\ v_{2\pm1} &= 0 \\ v_{2\pm2} &= \frac{1}{2\sqrt{6}}\,(V_{xx}-V_{yy}) \end{aligned} \tag{11}$$

Two special cases may be considered. In the first case, the charge distribution around the nucleus has axial symmetry, and the quadrupole interaction (Eq. (9)) becomes simply:

$$E_Q = \langle IM|H_Q|IM\rangle = eQ\left[\frac{3M^2-I(I+1)}{4I(2I-1)}\right]V_{zz} \tag{12}$$

In the second case, I=3/2 and the matrix with elements defined by Eq. (9) may be diagonalized to give two doubly degenerate levels corresponding to $|M|=3/2$ and $|M|= 1/2$. The splitting of a I=3/2 level is seen, for example, in Mössbauer experiments of Fe, Ru and

Au isotopes, of which many compounds have been investigated.

Defining further:

$$\eta = \frac{V_{xx}-V_{yy}}{V_{zz}} \tag{13}$$

with the convention that $|V_{zz}|\geq|V_{yy}|\geq|V_{xx}|$, so that $0\leq\eta\leq1$, we have finally:

$$\Delta EQ = \frac{eQ}{2} V_{zz}(1+\frac{\eta^2}{3})^{1/2} \tag{14}$$

which is the energy splitting of the two sublevels $|M|=3/2$ and $|M|=1/2$. In a Mössbauer experiment, this is manifested by a splitting of the resonance line. In the case of axial symmetry, Eq. (14) reduces to:

$$\Delta EQ = \frac{eQ}{2} V_{zz} \tag{15}$$

For higher spin states, the matrix $\langle H_Q \rangle$ may be diagonalized; for I<1, no splitting occurs.

3. MAGNETIC HYPERFINE INTERACTIONS

The magnetic moment of a nucleus $\vec{u}$ interacts with an external magnetic field, which, when arising from the surrounding electrons, is called the hyperfine field.[2] We have then, for the magnetic dipole interaction Hamiltonian:

$$H_{HF} = -\vec{u}\cdot\vec{H} = -g_N \, \mu_N \, \vec{I}\cdot\vec{H} \tag{16}$$

where g_N and μ_N are, respectively, the nuclear g factor and nuclear magneton.

The calculation of magnetic hyperfine interactions consists in evaluating the first order correction on the energy levels of the system considered

$$E_{HF} = \langle \Psi | H_{HF} | \Psi \rangle \tag{17}$$

with the hyperfine Hamiltonian defined as

$$H_{HF} = \sum_i (H_F^i + H_D^i + H_L^i)$$

where

$$H_F^i = \frac{8\Pi}{3} g_e g_N \mu_B \mu_N \ \delta(\vec{r}_i)\ \vec{I}\cdot\vec{S}_i$$

$$H_D^i = g_e g_N \mu_B \mu_N \left[3\ \frac{(\vec{I}\cdot\vec{r}_i)(\vec{S}_i\cdot\vec{r}_i)}{r_i^5} - \frac{\vec{I}\cdot\vec{S}_i}{r_i^3}\right] \qquad (18)$$

$$H_L^i = g_e g_N \mu_B \mu_N \frac{\vec{I}\cdot\vec{L}_i}{r_i^3}$$

In Eq. (18), g_e is the electron g factor and μ_B the Bohr magneton. H_F is the Fermi (contact) interaction, H_D the spin dipolar, and H_L the orbital dipolar interactions. This last, in many cases, may be considered negligible.

To establish a link with experimental measurements, a spin Hamiltonian is defined, which allows the description of experimental results through a limited number of parameters (the "hyperfine constants"). For this purpose, Eq. (18) are used to accomplish the integration in Eq. (17) over the electronic spacial coordinates leaving the nuclear and electronic spin operators in their explicit form.

We have then, for the Fermi interaction

$$H_F = a\ \vec{I}\cdot\vec{S}$$

where

$$a = \frac{8\Pi}{3} g_e g_N \mu_B \mu_N \sum_i n_i |\psi_{i\uparrow}(0)|^2$$

for the direct interaction of unpaired s-symmetry electrons with the nucleus, or

$$a = \frac{8\Pi}{3} g_e g_N \mu_B \mu_N \ \sum_i n_i \left[|\psi_{i\uparrow}(0)|^2 - |\psi_{i\downarrow}(0)|^2\right] \qquad (19)$$

for the interaction of the spin density at the nucleus which arises through the polarization of s-symmetry electrons in closed shells by unpaired electrons of different symmetry.[9]

The spin dipolar interaction may be expressed in terms of a symmetric and traceless tensor $\tilde{B}$, such that

$$H_D = \vec{I}\cdot\tilde{B}\cdot\vec{S}$$

The diagonal components of the tensor $\tilde{B}$ may be expressed in terms of the parameters:

$$b = \frac{1}{2} g_e g_N \mu_B \mu_N \sum_i n_i \int \psi_i^*(\vec{r}) \left[\frac{3z^2-r^2}{r^5}\right] \psi_i(\vec{r}) d\tau$$
$$b' = \frac{3}{2} g_e g_N \mu_B \mu_N \sum_i n_i \int \psi_i^*(\vec{r}) \left[\frac{x^2-y^2}{r^5}\right] \psi_i(\vec{r}) d\tau \quad (20)$$

The summations in Eq. (20) include all unpaired electrons of appropriate symmetry, as well as polarized electrons in closed shells; in cases of axial symmetry, b' is equal to zero.

In a Mössbauer experiment the magnetic hyperfine interaction is seen to lift the degeneracy of the |IM> nuclear levels.

4. CALCULATIONS OF HYPERFINE INTERACTIONS

The Hartree-Fock-Slater[6] approximation for the calculation of molecular orbitals ψ_i of transition metal compounds and complex ions is based on the local approximation for the exchange potential, expressed as (in Hartrees):

$$V_{x\alpha\uparrow}(\vec{r}) = -3\alpha\left[\frac{3}{4\pi} \sum_i n_i |\psi_{i\uparrow}(\vec{r})|^2\right]^{1/3} \quad (21)$$

with a similar expression for spin down. The parameter α, as derived rigorously from a variational treatment,[10] is equal to 2/3. However, different values have been used, as derived by fitting the total energies obtained for free atoms to the Hartree-Fock energies.[6]

In this approximation, the one-electron equations to be solved are of the form

$$\left[-\frac{1}{2}\nabla^2 + V_c(\vec{r}) + V_{x\alpha\uparrow}(\vec{r})\right] \psi_{i\uparrow}(\vec{r}) = \varepsilon_i \psi_{i\uparrow}(\vec{r}) \quad (22)$$

where $V_c(\vec{r})$ is the Coulomb potential due to both electrons and nuclei.

The methods of calculation which have been employed are the Multiple Scattering (MSXα)[11] and Discrete Variational (DVM).[12] The description of these methods will not be given here, since this has been done in detail in the original literature. It will be sufficient to remind that in the MSXα, a "muffin-tin" approximation to the potential is employed, with spheres around each nucleus defining regions of spherically-symmetrical potentials. In these regions, the wave functions are then an expansion of spherical harmonics multiplied by a numerical radial function, this last being quite flexible to adapt to the molecular potential. In the DVM method no such approximation to the potential is made, and the molecular orbitals are expanded in a basis of atomic orbitals; numerical integrations are performed to calculate the matrix elements of the secular equations.

4.a. Isomer Shifts

The problem posed in the calculation of isomer shifts as in Eq. (5) is the evaluation of the electronic densities at the nuclei. In the orbital approximation, we have then:

$$IS = \alpha \left[\sum_i n_i |\psi_i(0)|_A^2 - \sum_j n_j |\psi_i(0)|_S^2 \right] \qquad (22)$$

where the summations are over the molecular orbitals of s-type symmetry. Since the early days of Mössbauer spectroscopy, many attempts have been made to evaluate the so-called calibration constant α of ^{57}Fe, in order to obtain the value of the nuclear parameter $\Delta\langle r^2\rangle_{e,g}$. For this purpose, Hartree-Fock wave functions of free ions were related to the IS values of ionic compounds. Besides this quantitative interest, questions arouse as to what bonding mechanisms the different values of IS of different Fe compounds were to be ascribed. The main effects that seemed relevant were the screening of 3s and 4s electrons by 3d electrons, differences in the number of 4s electrons in different configurations, and Fe orbital contractions when orthogonality to ligand orbitals was considered ("overlap distortion"). But no answer could be given as to the relative roles of these effects in an actual molecule.

Walsh and Ellis[13] studied the variations of the electron density at the nucleus of the $FeAr_{12}$ cluster to study the IS of ^{57}Fe in a solid argon matrix. The DVM method was used, but the potential was not made self-consistent. With these calculations it was shown that overlap distortion and potential distortion (relaxation) have opposite effects on the electronic density at the nucleus, and partially cancel.

The first attempts to study ^{57}Fe isomer shifts in complex ions with the MSXα method were quite discouraging. Ellis and Averill[14] investigated the tetrahedral anions $[FeCl_4]^{-1}$ and $[FeCl_4]^{-2}$ in the spin-restricted approximation, and obtained a calibration constant $\alpha \sim -0.8$, very far from the range of "reasonable" values which had been obtained from free-ion calculations. A detailed study of the IS of several high-spin halogen compounds of Fe was made with the MSXα method by de Siqueira et al.[15] Both octahedral and tetrahedral clusters were considered in spin-unrestricted calculations. This work showed that electron densities at the Fe nucleus were dependent on the radii chosen in forming the "muffin-tin" model potential, and so no definite value of the calibration constant could be derived.

The subsequent continuation of this work to include covalent complexes of Fe, has, however, changed this outlook to a certain extent. Complexes of Fe^{+2} and Fe^{+3} with CN^- ligands have a much more covalent character than halogen complexes, and this is made evident by their low-spin configurations. The octahedral complexes $[Fe(CN)_6]^{-4}$ and $[Fe(CN)_6]^{-3}$ were studied with the MSXα method and values of $|\Psi(0)|^2$ derived.[16,17] The tetrahedral ferrate ion FeO_4^{-2}, in which Fe has the unusual formal charge +6, is present in compounds having the lowest values of isomer shifts of all Fe compounds measured. This ion, which is in a $^3A_2$ ground state, with two electrons in the last occupied e orbital, was investigated with the spin-polarized MSXα method.[18]

A compilaticn of these results is given in Table 1 and Figure 1. Although the dependence on the chosen "muffin-tin" scheme is evident, this model does provide useful information, if reasonable empirical criteria are used in defining the model potential. For example, in the case of the CN^- complexes, the choice of the "muffin-tin" tangent spheres radii was very limited, due to the short C-N distance, which allowed only small radii associated to the C and N atoms. One may also make use of Pauling's ionic or covalent radii to help in defining a "muffin-tin" potential.

In Table 1 and Fig. 1, some trends in $|\Psi(0)|^2$ may be observed. The values of $|\psi(0)|^2$ for the 1s and 2s orbitals have not been considered because they show differences among different compounds which are very small. For the Fe(3s) and valence orbitals, the fluorides show an increase in the electron density at the origin with an increase in the Fe formal oxidation state. The increase in the valence density (Fe(4s)) plays an important role in this trend. As for $[Fe(CN)_6]^{-4}$ and $[Fe(CN)_6]^{-3}$, one may explain the considerably lower values of the isomer shifts, compared to the high-spin (ionic) complexes of same formal charge on Fe, by the much larger valence (Fe(4s)) density at the origin present in the covalent ions, as a result of the strong σ-donation property of the CN^- ligands.

Besides increase in 4s charge densities, another mechanism which has been evoked to explain differences in IS values are changes in 3d populations, which would result in differences in 3s densities at the origin, since the 3d electrons screen the 3s electrons from the nucleus. Examination of Table 1 shows small differences in $|\psi(0)|^2_{3s}$, and so this mechanism is seen to be less important.

Populations of an atom in a molecule may be approximately defined in the MSXα model by considering the amplitude of the radial functions inside the atomic sphere, as compared to the free atom.[17,19] Comparison of the Fe(3d) populations, calculated in this manner for the covalent complexes in Table 1, show small differences, which is consistent with the small differences in $|\psi(0)|^2_{3s}$. Particularly striking is the case of FeO_4^{-2}: formally, Fe has a $3d^2$ configuration, but in the molecule the 3d population increases

Table 1. Values of $|\Psi(0)|^2$ in a_0^{-3} (1s and 2s electrons excluded) calculated with the MSXα method for various complexes of Fe. Fluorine anions from Ref. (15), $[Fe(CN)_6]^{-4}$ from Ref. (16), $[Fe(CN)_6]^{-3}$ from Ref. (17), and FeO_4^{-2} from Ref. (18).

	FeF_6^{-3}		FeF_6^{-4}		FeF_6^{-5}	$[Fe(CN)_6]^{-4}$	$[Fe(CN)_6]^{-3}$	FeO_4^{-2}
	R_{Fe}=1.76[a] Fe-F=1.80[b]	R_{Fe}=1.76 Fe-F=1.91	R_{Fe}=2.04 Fe-F=2.06	R_{Fe}=1.76 Fe-F=2.06	R_{Fe}=1.76 Fe-F=2.18	R_{Fe}=2.48 Fe-C=1.92	R_{Fe}=2.47 Fe-C=1.90	R_{Fe}=2.00 Fe-O=1.65
3s	140.25	140.58	140.22	139.63	138.73	140.45	140.51	141.17
Valence	4.12	3.07	2.51	1.94	1.30	6.40	6.94	7.49
Total	144.37	143.65	142.73	141.57	140.03	146.85	147.45	148.66

[a]Radii in a.u.
[b]Distances in Å.

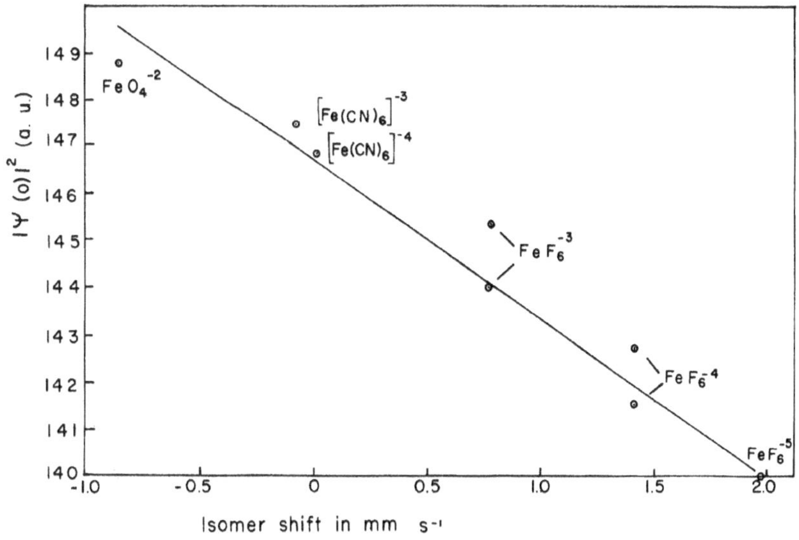

Figure 1. Values of $|\Psi(0)|^2$ calculated with the MSXα method (1s and 2s orbitals excluded), plotted against isomer shifts of ^{57}Fe. See Refs. (15)-(18) for details of calculations and references to isomer shifts values. The two values of $|\Psi(0)|^2$ for FeF_6^{-4} and FeF_6^{-3} refer to two different muffin-tin radii.

to 5.8.[18] The very low IS value may thus be ascribed mainly to a large "invasion" of valence electrons, caused by the very attractive potential of Fe^{+6}: the population analysis results in ~1.3 Fe(4s) electrons.

One must bear in mind, however, that Coulomb correlation is not included in this approximation; if it were considered, these results might change somewhat, in that correlation would tend to oppose the accumulation of electronic charge, especially in orbitals of limited spacial extent, such as Fe(3d).

In Fig. 1, the IS and electron densities are related by a straight line, which gives the value $\alpha = -0.29$ for the IS calibration constant, similar to the value $\alpha = -0.30$ obtained from Hartree-Fock "ab-initio" calculations of CN^- and F^- complexes.[20] Although the present number must be viewed with caution, the results do indicate that a consistent trend among different Fe compounds is obtained.

The development of low-temperature matrix isolation techniques has made possible the measurement of hyperfine interactions in isolated molecules and clusters of transition elements. Since the inert-gas matrices are expected to interact very little with the clusters, it is an ideal situation to study chemical bonds without exterior interference. Hyperfine parameters have been calculated with the XαDVM method for Fe_2 and other Fe dimers, namely FeMn, FeCo, FeNi, and FeCu, and related to ^{57}Fe Mössbauer measurements in noble gas matrices.[21] This task met with several difficulties. These molecules present a dense band of valence one-electron levels, and assigning the ground-state configuration based solely on the molecular orbital levels schemes is not possible. In these circumstances, Coulomb correlation effects are also expected to be important, and Configuration Interaction calculations would be required. However, certain general informations on the electronic densities at the Fe nucleus were obtained. For example, it is observed that the sum of the contributions $|\psi_i(0)|^2$ of the valence orbitals, if occupied by the same number of electrons, decreases slightly when going from FeMn to FeCu. This is related to an increase in the Fe(3d) Mulliken population and decrease in 4s. It is curious to notice that this same trend may explain the increase of the IS values of Fe impurities in transition metal hosts Mn to Cu, as well as those of first-neighbour Fe atoms in Fe metal containing Mn, Co, Ni, and Cu impurities.[21] This suggests that isomer shifts of systems involving transition elements metallic bonds depend on fundamental features that may be represented by a single bond between two atoms.

In the case of Fe_2, the effect of different interatomic distances on $|\Psi(0)|^2$ was also investigated. The results are shown in Fig. 2. There is a dramatic increase of $|\Psi(0)|^2$ at shorter distances, not followed by any significant changes in 3d or 4s populations. This may be interpreted in terms of an orthogonality ("overlap distortion") effect, which, with the LCAO basis used in the DVM method, is manifested by a greater participation of the inner atomic orbitals of the basis (mainly 3s) in the valence molecular orbitals. Again, the trend shown in Fig. 2 may be related to the trend observed for isomer shifts of Fe metal at different pressures:[21] at higher pressures, the isomer shifts are smaller.

Another interesting research area in Fe compounds is the investigation, in inert matrices, of molecules that undergo photochemical reactions. The photochemical decomposition of $Fe(CO)_5$ is known to produce species in which one or more CO ligands have been detached. Molecular orbital calculations of the fragments $Fe(CO)_n$, $5 \geq n \geq 1$, have been performed recently with the DVM method, in the spin-unrestricted form for the case of open-shell fragments.[22] Hyperfine interactions were studied and related to bonding characteristics. For the case of $Fe(CO)_5$ and $Fe(CO)_4$, the IS values are known: for $Fe(CO)_4$, the quadrupole splittings ΔEQ and IS values were measured in recent experiments of Mössbauer spectroscopy of

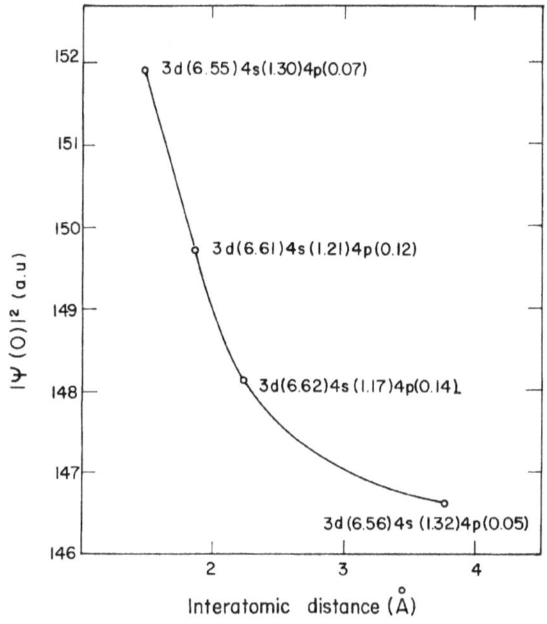

Figure 2. Values of $|\Psi(0)|^2$ (1s and 2s orbitals excluded) at an Fe nucleus of Fe_2, at different interatomic distances, for configuration $(\ldots 1\delta_g^2\ 1\delta_u^2\ 6\sigma_u^2\ 3\pi_g^2)$.

$Fe(CO)_5$ absorbed in a polyethylene film at room temperature, and irradiated with UV light[23] (see Table 2).

The calculated values of $|\Psi(0)|^2$ for ion pentacarbonyl and its fragments are very similar, and so similar IS values are predicted. As the number of carbonyls decreases, a small decrease in $|\Psi(0)|^2$ is observed; for the smaller fragments $Fe(CO)_2$ and FeCO, the trend is reversed. This is explained as follows: in $Fe(CO)_5$, a large valence contribution to $|\Psi(0)|^2$ is calculated. This is not due to a large Fe(4s) population, since this is quite negligible (see Table 2), but to participation of the Fe(3s) orbitals of the basis in the valence molecular orbitals ("contraction"). This "contraction" is decreased slightly in $Fe(CO)_4$, resulting in a slightly smaller $|\Psi(0)|^2$ and, accordingly, higher value of the IS (see Table 2). This result shows that, in this case, an analysis based solely on the 3d and 4s populations would be misleading, as seen in Table 2. As the number of CO ligands decrease further, an increase in Fe(4s) populations is observed, and this compensates for the smaller "contraction" of valence orbitals.

Table 2. Values of $|\Psi(0)|^2$ in a_0^{-3}, experimental and calculated isomer shifts and quadrupole splittings, and Fe populations for $Fe(CO)_5$ and $Fe(CO)_4$. The IS for $Fe(CO)_4$ was calculated by fitting the experimental value of the IS for $Fe(CO)_5$, and using $\alpha = -0.29$. ΔEQ for $Fe(CO)_4$ was obtained with Q(Fe) = 0.158 barn, which reproduces the experimental ΔEQ of $Fe(CO)_5$. The IS values are referred to Fe metal

$Fe(CO)_5$ D_{3h}		$Fe(CO)_4$ C_{2v}	
$8a_1(3s)$	140.46	$7a_1'(3s)$	140.41
$9a_1$–$19a_1$	6.99	$8a_1'$–$13a_1'$	6.63
Total	147.45	Total	147.04
3d population 6.65 4s population 0.01		3d population 6.61 4s population 0.07	
IS measured:	−0.174±0.005	IS measured:	−0.124±0.008
IS calculated:	−0.174 (mm/s)	IS calculated:	−0.060 (mm/s)
ΔEQ measured:	+2.52±0.01	ΔEQ measured:	1.83±0.02
ΔEQ calculated:	+2.52 (mm/s)	ΔEQ calculated:	+1.62 (mm/s)

In concluding this section, it would be worthwhile to mention that little systematic investigation has been done to evaluate the effect of the local exchange approximation on the sensitive hyperfine parameters. Comparisons of $|\Psi(0)|^2$, calculated in the Xα approximation, with Hartree-Fock values, were made for Ru atoms and ions, in several configurations.[24] Several values of α were investigated, since, as mentioned before, not always the value $\alpha=2/3$ is employed. Similar studies for other elements would certainly be useful.

4.b. Quadrupole Interaction

The electronic contributions to the electric field gradient

that originates the quadrupole splitting ΔEQ of a nuclear level with spin I = 3/2, as in Eq. (14), are given by:

$$V_{zz} = -e \sum_i n_i < \psi_i \left| \frac{3\cos^2\theta - 1}{r^3} \right| \psi_i > \qquad (23)$$
$$\frac{V_{xx} - V_{yy}}{e} = \sum_i n_i < \psi_i \left| \frac{3\sin^2\theta \cos 2\phi}{r^3} \right| \psi_i >$$

where ψ_i are molecular orbitals.

Considerable less attention has been given to the theoretical study of quadrupole splittings in transition metal compounds with Xα methods, as compared to isomer shifts. Early MSXα calculations [14] of tetrahedral $FeCl_4^{-1}$ and $FeCl_4^{-2}$ allowed the comparison of $<r^{-3}>$ values of 3d-like orbitals with atomic and ionic values. More recently, the MSXα method has been applied[25] in the overlapping-spheres version, to calculate ΔEQ values in the very covalent complexes $[Fe(CN)_5NO]^{-2}$, $[Fe(CN)_5CO]^{-3}$, and $[Fe(CN)_5N_2]^{-3}$. Special attention was given to analyse the relative roles of σ and Π bonding of the various ligands in producing the field gradients. For calculating ΔEQ, an atomic model was employed, using the Fe 3d and 4p populations obtained in the molecular calculations. The agreement with the experimental values was found to be good.

The Xα-DVM method was recently employed[26] to calculate quadrupole splittings of linear Au(I) compounds, for which experimental values were available from ^{197}Au Mössbauer spectroscopy. Special numerical integration schemes were employed, both for the DVM variational calculation as for the calculation of the matrix elements of Eq. (23). These last were fully evaluated, including multicenter terms and core orbitals. Fairly good agreement with experimental values was found for $[Au(CN)_2]^{-1}$ and $[AuCl_2]^{-1}$. The large negative value of ΔEQ measured for $K[Au(CN)_2]$ was seen to derive mainly from large one-center contributions given by σ_μ^+ valence orbitals with significant core atomic orbital (Au(5p)) character or, in other words, contracted towards the Au atom in the molecular axis direction. The inclusion of the core 5p orbital in the LCAO expansion greatly magnifies the electric field gradient of a σ_μ^+ orbital, due to the large $<r^{-3}>_{5p}$ value.

Finally, it may be mentioned that the effect of the Xα local density approximation on $<r^{-3}>$ values of ions and atoms, as compared to Hartree-Fock values, was studied for two transition elements,[24] namely Fe and Ir, in several configurations and several values of α. The best agreement is found for $\alpha = 2/3$; the d and f orbitals compare better with the Hartree-Fock values than the p orbitals, the larger discrepancies being observed for the outer p orbitals.

4.c. Magnetic Hyperfine Interactions

The calculation of the Fermi (contact) hyperfine interaction (Eq. (19)) in the local exchange approximation presents no difficulties when the spin density at the nucleus is the direct contribution of an unpaired electron in an s-type orbital. However, when the spin density at the origin is the result of exchange polarization of s-type orbitals by unpaired p or d electrons, it has been known for some time that the Xα approximation gives poor results, compared to Hartree-Fock.[6] An extensive study reported for Mn^{+2} ion ($3d^5$) indicates that the spin densities at the nucleus of the 1s, 2s, and 3s orbitals are systematically smaller than those obtained by the Hartree-Fock method.[27] This was ascribed to a slightly larger expansion of the 3d orbital wave functions obtained in the Xα calculation. A systematic study of the Xα spin densities of s shells for the atoms B, N, F, Al, P, and Cl (with unpaired p electrons) was reported.[28] Values of radial spin densities and spin densities at the origin were compared to spin-polarized Hartree-Fock values. The differences noticed in the radial spin densities are larger near the nucleus than in the valence region. The best agreement found was for B and Al, which may be explained by the fact that the electronic density is more slowly varying for these atoms. A slowly varying density is a condition for the applicability of the statistical Xα approximation.

In spite of these shortcomings, the Xα approximation can still be made useful in calculations of Fermi hyperfine fields in transition metal compounds. Two cases of Fe complexes were studied [18,29] in detail with the MSXα method, the tetrahedral ions FeO_4^{-2} and FeS_4^{-5}. The former is known to have two electrons in the last occupied orbital (e^2) and a $^3A_2$ ground state.[18] The latter is in a high-spin $3d^5$ configuration, with a $^6A_1$ ground state.[29] In both cases, the experimental hyperfine fields are much lower than what would be expected from the spin-polarization due to two (FeO_4^{-2}) and five (FeS_4^{-5}) unpaired electrons in an Fe 3d shell.[6,30] This was ascribed, on the basis of the MSXα calculations, to covalency effects. One of the features noticed is that bonding with the ligands produces an inflow of spin ↓ electrons into the Fe(3d) orbital, such that the number of effective unpaired electrons in that orbital is reduced, as compared to the free ions Fe^{+6} and Fe^{+3}, thus producing a smaller polarization of the inner s shells.

Another study of contact hyperfine terms in an ionic cluster of transition element is the MSXα calculation of Co^{+2} in a LiF crystal, represented by the octahedral $[CoF_6]^{-4}$ cluster.[31] The spin density was calculated both at the Co^{-2} and F^- nuclei.

An investigation of the spin-Hamiltonian parameters has been performed for $[CrOCl_4]^{-1}$ with wave functions obtained from MSXα calculations.[32] The agreement with Electron Spin Resonance data was found to be good. Molecular orbital coefficients, as defined in the MSXα method, and values of $\langle r^{-3}\rangle_{3d}$, were needed for the calculations. Similar treatments of the spin hyperfine parameters were given in the theoretical study of ESR spectra of organometallic compounds with the MSXα method. Copper porphine[33] and Cobaltocene Cp_2Co (CP=η^5-cyclopentadienyl)[34] are among the organometallics investigated. Several refinements in the method for calculating the hyperfine constants were tested, such as spin-polarization, inclusion of excited states that mix with the ground state via spin-orbit coupling, and use of $\langle r^{-3}\rangle$ values derived from the molecular functions, instead of atomic Hartree-Fock values.

5. CONCLUSIONS

In analysing the results reviewed here of calculations of hyperfine interactions in transition metal compounds in the local Xα approximation, some conclusions may be extracted. First, it may be observed that, where molecular or cluster calculations are concerned, approximations other than the local exchange have also been employed, such as the muffin-tin potential, or numerical three-dimensional integrations. In this case, it is very difficult to extract from the results the shortcomings which are due solely to the local density approximation. It seems that more systematic investigations would be necessary. Another unfortunate feature is the very small number of "ab-initio" calculations of hyperfine interactions of such systems with which to make comparisons.

Regarding the calculations mentioned here, correlations with experimental results seem to indicate that, as far as trends and qualitative aspects are concerned, the Xα Molecular Orbital methods have led to some understanding of the bonding mechanisms which explain the influence of the different electronic environments on the hyperfine parameters. Much understanding has been gained in this aspect. However, quantitative results at the present moment seem to be requiring more accurate calculations. Moreover, too little is actually known on the influence of many-body effects, which might have an outstanding role in explaining such small interactions.

REFERENCES

1. A. Abragam, "The Principles of Nuclear Magnetism", Oxford University Press, London (1961).
2. B. Bleaney, "Hyperfine Structure and Electron Paramagnetic Resonance", in "Hyperfine Interactions", ed. A.J. Freeman and R.B. Frankel, Academic Press, New York (1967); A. Abragam and B. Bleaney, "Electron Paramagnetic Resonance of Transition Ions", Clarendon Press, Oxford (1970).
3. T.P. Das and E.L. Hahn, "Nuclear Quadrupole Resonance Spectroscopy", Academic Press, New York (1958).

4. G.M. Bancroft, "Mössbauer Spectroscopy", McGraw-Hill, London (1973); N.N. Greenwood and T.C. Gibb, "Mössbauer Spectroscopy", Chapman and Hall, London (1971); J.M. Friedt and J. Danon, "Mössbauer Spectroscopy: Principles and Examples of Chemical Applications", in "Modern Physics in Chemistry", vol. 2, ed. E. Fluck and V.I. Goldanskii, Academic Press, London (1979).
5. R.D. Gill, "Gamma-ray Angular Correlations", Academic Press, New York (1975); H. Frauenfelder and R.M. Steffen, in "Alpha-Beta - and Gamma-ray Spectroscopy", vol. 2, ed. K. Siegbahn, North-Holland, Amsterdam (1965).
6. J.C. Slater, "The Self-consistent Field for Molecules and Solids", vol. 4 of "Quantum Theory of Molecules and Solids", McGraw-Hill, New York (1974).
7. B.D. Dunlap and G.M. Kalvius, "Theory of Isomer Shifts", in "Mössbauer Isomer Shifts", ed. G.K. Shenoy and F.E. Wagner, North-Holland, Amsterdam (1978).
8. E.N. Kaufmann and R.J. Vianden, Rev.Mod.Phys. 51, 161 (1979), and references therein.
9. R.E. Watson and A.J. Freeman, "Hartree-Fock Theory of Electric and Magnetic Hyperfine Interactions in Atoms and Magnetic Compounds", in "Hyperfine Interactions", ed. A.J. Freeman and R.B. Frankel, Academic Press, New York (1967).
10. W. Kohn and L.J. Sham, Phys.Rev. A 140, 1133 (1965).
11. J.C. Slater and K.H. Johnson, Phys.Rev. B 5, 844 (1972); J.C. Slater, Advan.Quant.Chem. 6, 1 (1972); K.J. Johnson, Advan.Quant.Chem. 7, 143 (1973).
12. D.E. Ellis, Int.J.Quant.Chem. S2, 35 (1968); D.E. Ellis and G.S. Painter, Phys.Rev. B 2, 2887 (1970); A. Rosén, D.E. Ellis, H. Adachi, and F.W. Averill, J.Chem.Phys. 85, 3629 (1976).
13. P.F. Walch and D.E. Ellis, Phys.Rev. B 7, 903 (1973).
14. D.E. Ellis and F.W. Averill, J.Chem.Phys. 60, 2856 (1974).
15. M.L. de Siqueira, S. Larsson, and J.W.D. Connolly, J.Phys. Chem. Solids 36, 1419 (1975).
16. D. Guenzburger, B. Maffeo, and M.L. de Siqueira, J.Phys.Chem. Solids 38, 35 (1977).
17. D. Guenzburger, B. Maffeo, and S. Larsson, Int.J.Quant.Chem. 12, 383 (1977).
18. D. Guenzburger, D.M.S. Esquivel, and J. Danon, Phys.Rev. B 18, 4561 (1978).
19. S. Larsson, Theoret.Chim. Acta 49, 45 (1978).
20. W.C. Nieuwpoort, D. Post, and P.Th. Van Duijnen, Phys.Rev. B 17, 91 (1978).
21. D. Guenzburger and E.M.B. Saitovitch, Phys.Rev. B 24, 2368 (1981).
22. D. Guenzburger, E.M.B. Saitovitch, M.A. de Paoli, and H. Manela, to be submitted for publication.
23. M.A. de Paoli, E.M.B. Saitovitch, S. Oliveira, and D. Guenzburger, to be submitted for publication.

24. S.K. Lie, D.M.S. Esquivel, and D. Guenzburger, Chem.Phys. Letters 57, 458 (1978).
25. M. Braga, A.C. Pavão, and J.R. Leite, Phys.Rev. B. 23, 4328 (1981).
26. D. Guenzburger and D.E. Ellis, Phys.Rev. B. 22, 4203 (1980).
27. T.M. Wilson, J.H. Wood, and J.C. Slater, Phys.Rev. A. 2, 620 (1970).
28. R. Ahlberg and S. Larsson, Chem.Phys. Letters 49, 491 (1977).
29. C.A. Taft and M. Braga, Phys.Rev. B. 21, 5802 (1980).
30. R.E. Watson and A.J. Freeman, Phys.Rev. 123, 2027 (1961).
31. E.L. Albuquerque, B. Maffeo, H.S. Brandi, and M.L. de Siqueira, Solid State Comm. 18, 1381 (1976).
32. K.K. Sunil, J.F. Harrison, and M.T. Rogers, J.Chem.Phys. 76, 3078 (1982).
33. D.A. Case and M. Karplus, J.Am.Chem.Soc. 99, 6182 (1977).
34. J. Weber, A. Goursot, E. Pénigault, J.H. Ammeter, and J. Bachmann, J.Am.Chem.Soc. 104, 1491 (1982).

LOCAL DENSITY APPROACH TO SURFACES AND ADSORBED LAYERS

E. Wimmer*, A.J. Freeman, and M. Weinert

Northwestern Univ., Dept. of Physics and Materials
Research Center
Evanston, IL 60201, USA

We show that the local density problem for the thin film geometry can be solved with high accuracy by employing our all-electron full-potential linearized augmented-plane-wave (FLAPW) method. This is achieved by removing all shape approximations in the charge density and the potential and by using a highly flexible variational basis set. Its utility even for treating molecules on surfaces is demonstrated for the severe test case of the (nearly) free O_2 molecule where the FLAPW method is found to be at least as accurate as state-of-the art localized-basis molecular methods. The high accuracy and precision of the FLAPW approach for metallic surfaces is shown by our finding that for the interpretation of surface sensitive photoemission measurements of the Al-2p core levels both a surface induced core level shift of 120 meV to reduced binding energies and a 38 meV crystal field splitting are important. Based on all-electron local density calculations we explain the lowering of the work function of a transition metal surface [W(001)] upon deposition of a Cs over-layer by a multiple dipole formation. Finally, we demonstrate for a graphite monolayer that local density total energies give excellent descriptions of equilibrium geometries and discuss the overestimation of local-density cohesive energies due to an incomplete treatment of correlation effects in the free atom.

*permanent address: Inst. f. Phys. Chem., University of Vienna, Währingerstr. 42, A-1090 Vienna, Austria.

1. INTRODUCTION

Over the decades, while the need for a rigorous quantum mechanical treatment of surfaces has been strongly felt, the task has seemed monumental. Traditional approaches such as Hartree-Fock theory and determinantal multiconfiguration interaction expansions, successful in the treatment of small molecules, are impracticable and incapable of handling the electronic structure of surfaces. On the other hand, the local density approximation (LD) to the density-functional theory of Hohenberg, Kohn, and Sham[1,2] has been highly successful in the theoretical-computational treatment of crystalline solids. The collective approach to the many-body problem intrinsic in density-functional theory results in a radical computational simplification: only effective one-particle Schrödinger or Dirac equations have to be solved. In bulk calculations, the solution of the LD one-particle equations has been traditionally simplified by assuming a muffin-tin form of the density and the potential. This approximation may be justifiable in closed packed bulk systems, but it becomes more questionable in open structures and certainly is unacceptable in surface calculations when high accuracy is desired. Shape approximations to the charge density and the effective potential also cloud possible limitations of the LD theory.

2. COMPUTATIONAL ASPECTS

Currently one of the most successful methods for ab initio surface calculations consists in modelling the surface by a slab (or thin film) either in the form of a superlattice geometry with appropriate space between slabs or, more naturally, by a single slab. The slab thickness is chosen such that in the center the local quantities such as local density of states, charge densities and spin densities approach the bulk values. For transition metals usually 5-7 layers and about 9 layers for s-p metals are found to be sufficient.

One of the most accurate and powerful schemes to solve the local (spin) density (LSD) one-particle equations for the film (single slab) geometry is the all-electron full-potential linearized-augmented-plane-wave (FLAPW) method.[3-5] The basic idea in this variational method is the partition of real space into three different regions, namely spheres around the nuclei, vacuum regions on both sides of the film, and the remaining interstitial region. In each of these regions the "natural" form of the variational basis functions is adopted, i.e. plane waves in the interstial region, a product of radial functions and spherical harmonics inside the spheres, and in the vacuum a product of functions which depend only on the coordinate normal to the film and 2-dimensional plane waves. Each of these basis functions is continuous in value and derivative across the various boundaries. This is possible because inside the spheres (and analogously in the vacuum) two radial functions for each ℓ-value are used, namely the solution of the radial Schrödinger equation for the current potential and its energy derivative.

In the FLAPW method no shape approximations are made to the charge density and the potential. Both the charge density and the effective one-electron potential are represented by the same analytical expansions, i.e. a Fourier representation in the interstitial region, an expansion in spherical harmonics inside the spheres, and a 2-dimensional Fourier series in each vacuum plane parallel to the surface. The generality of the potential requires a method to solve Poisson's equation for a density and potential without shape approximations. This is achieved by a new scheme which goes beyond the Ewald method.[5,6] The key idea in this new scheme is the obvervation that the potential outside a sphere does not depend on the actual charge density inside the sphere but only on the multipole moments of the sphere. Now, Poisson's equation is solved straightforwardly when the charge density is given in a Fourier representation. Because of the sharp structure of the charge density in the core region (including the nuclear charge), a Fourier expansion of the total density would be extremely slowly convergent. However, since the potential outside the sphere depends on the charge inside only through the multipole moments, the true charge density can be replaced by a smooth density which has a rapidly converging Fourier series and the same multipole moments as the true density. With this replacement of the density inside the spheres, we have a Fourier expansion of a charge density which gives the correct potential outside and also on the sphere boundaries. To find the potential inside the spheres we are faced in a final step with a standard boundary-value problem of classical electrostatics which can be solved from the original charge densities inside the spheres and the potential on the sphere boundaries by a Green's function method.

Thus, the FLAPW method allows a fully self-consistent solution of the one-particle LSD equations for the film geometry and yields charge densities close to the LSD limit. Besides the total charge density the key quantity in density-functional theory is the total energy corresponding to the ground state charge density. Recently we have presented a new scheme[15] to calculate accurate and stable all-electron density functional total energies and have applied it within the FLAPW method. The capability of total energy calculations for various geometrical arrangements provides us with a powerful theoretical tool to study the energetics and consequently dynamics of surfaces and overlayers.

In the following section we will discuss selected and characteristic results of local density FLAPW calculations for clean surfaces and surfaces with overlayers.

3. APPLICATIONS

A. Precision of the FLAPW Method: O_2 Molecule

The quality of the local density approximation for the treatment of surfaces can be assessed only if our solutions of the LD one-particle equations have a high degree of accuracy. In achieving this goal the FLAPW method is designed to remove approximations not inherent in local density theory such as shape approximations to the charge density and potential. Since the FLAPW method is a variational method it has to be demonstrated that the FLAPW basis sets allow enough variational freedom. This is best tested by pushing the FLAPW method - inherently a method for condensed systems - to the dilute molecular limit and making detailed comparison with accurate molecular LCAO type calculations.

This test has been done by forming a 2-dimensional hexagonal array of O_2 molecules and separating the molecules far enough so that the interaction between them becomes negligible. For this geometry we have performed a fully self consistent FLAPW calculation[5] and found our eigenvalues and charge densities for the O_2 molecule (Fig. 1) in disquieting disagreement with available LCAO results. Further investigation revealed - quite to our surprise - that the LCAO basis sets in the molecular calculations did not have sufficient variational freedom. In particular the bonding $2\sigma_g$ state (see Fig. 1) which originates from the atomic 2s state and which has a sharp double-peak structure and a high amplitude in the bonding axis, was found to be difficult to describe adequately with a standard LCAO basis set. The inclusion of off-site function in the LCAO basis set greatly improved its quality and led to excellent agreement with the FLAPW results. Inspired by the PLAPW method, Delley et al.[7] showed recently that an improvement over the use of on-site virtual (unoccupied) atomic orbitals (as in the standard LCAO description) is found by adding additional on-site basis functions obtained from positive ions. It is interesting to note that the self-consistent energy of the $2\sigma_g$ eigenvalue increases with an improved basis, whereas the total energy of course decreases as is required by the variational principle.

From this severe test of the FLAPW method in the dilute limit we can conclude that the accuracy of the FLAPW method in its proper domain, namely condensed systems, is at least as good as in the dilute case. In the following sections we will show applications to condensed systems, namely a clean s-p metal surface and a d-metal surface with an alkali overlayer.

B. Core Level Shifts on the Al (001) Surface

Surface sensitive photoemission experiments by Eberhardt et al.[8]

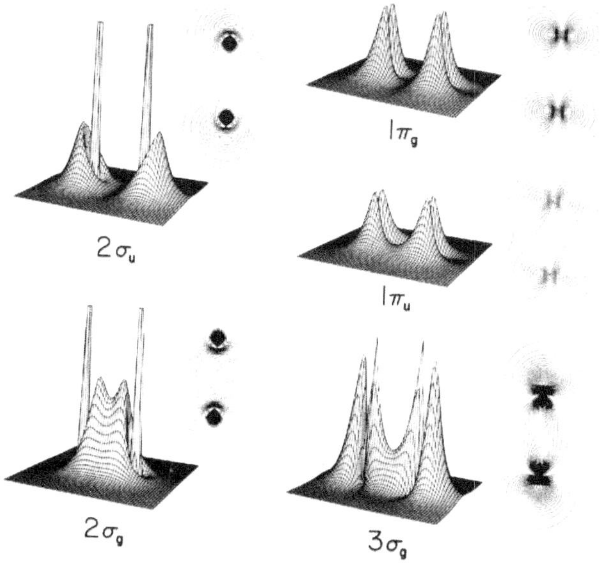

Figure 1. One-particle densities for the occupied valence states in an O_2 molecule obtained by the FLAPW method. The value of the lowest contour and the contour spacings are 0.016 electrons/(a.u.)^3 [after Ref. 5].

using synchrotron radiation suggested a surface induced broadening (crystal field splitting) of the Al-2p core level signal by 100-200meV but no detectable core level shift within an uncertainty of ± 40meV. However, the determination of these small effects is close to the experimental resolution and thus a challenge to accurate ab initio surface calculations. For this reason we undertook an FLAPW calculation of the Al (001) surface represented by a 9 layer slab.[9] Our results suggested a 2p-core level shift of 120meV to smaller binding energies for the surface atoms and 50meV for the subsurface atoms. For the surface atoms we also found a crystal field splitting of 38meV. The prediction of a surface induced 2p-core level shift in addition to the crystal field splitting was somewhat contradictory to the interpretation of Eberhardt et al. Independently, Chiang and Eastman[10] concluded from their photoemission experiments with synchrotron radiation a 2p-core level shift, but they did not consider crystal field splitting effects. Our work showed that both crystal field splitting and core level shifts must be considered in interpreting the data. Remarkably, the local density functional FLAPW calculation clarifies the interpretation of state-of-the-art photoemission data even on a scale of milli-volts.

Further, our theoretical work function of 4.5 eV is in very good agreement with the experimental value of 4.41 ± 0.03 eV.[11] It should be pointed out that the comparison of the local-density value for the work function, i.e. the energy difference between the Fermi energy and the vacuum zero, and the experimental work function is meaningful[12] and gives a quantitative statement of the quality of the local density approach to the many-body problem - provided no other approximations cloud the accuracy.

C. Cs Overlayer on W(001)

An interesting case of overlayers is that of alkali metals deposited on transition metal surfaces: upon alkalization the work function of the surface is reduced dramatically leading to a high electron emission rate. At higher temperatures and lower coverages alkali metals desorb mostly as positive ions. Therefore alkali overlayers on transition metal surfaces have important technological applications such as photoemitters and thermionic converters. The prototype of this class of systems is Cs on W, for which we have performed the first rigorous, all-electron local density functional study.[13]

The chemisorption process is characterized by comparing the results of independent self-consistent calculations for (1) a five-layer W(001) slab, (2) an unsupported Cs monolayer, and (3) Cs in a c(2×2) coverage on W(001) at distances ranging from 2.6 to 2.9 Å. We find that the reduction in the work function upon cesiation arises from a multiple dipole formation at the surface and interface layers. A characteristic feature of the clean W(001) surface is the pronounced spill-out of electrons from the surface layer into the vacuum which leads to the formation of a strong surface dipole (Fig. 2). This surface dipole is reduced in the cesiated system: Cs forms a metallic overlayer with its valence electrons polarized toward the W surface, thus reducing the spill-out into the vacuum. Correlated with the polarization of the Cs valence electrons, the Cs semicore 5p electrons are markedly counter-polarized (Fig. 3). Because of the strong metallic screening, changes in the electronic environment are localized to the Cs overlayer and the Cs/W interface layer. The net result of these polarizations is a reduction of the effective electrostatic surface barrier. As a result, all W states, and also the Fermi level, are shifted to lower binding energies with respect to the vacuum and consequently the work function is reduced by 2.0-2.5 eV, depending on the Cs-W distance used.

The work function of 4.6 eV found for a 7-layer W(001) film is in excellent agreement with the experimental value of 4.63 eV.[14]

D. Local Density Total Energies: Graphite Monolayer

A monolayer of carbon atoms lends itself ideally as an illu-

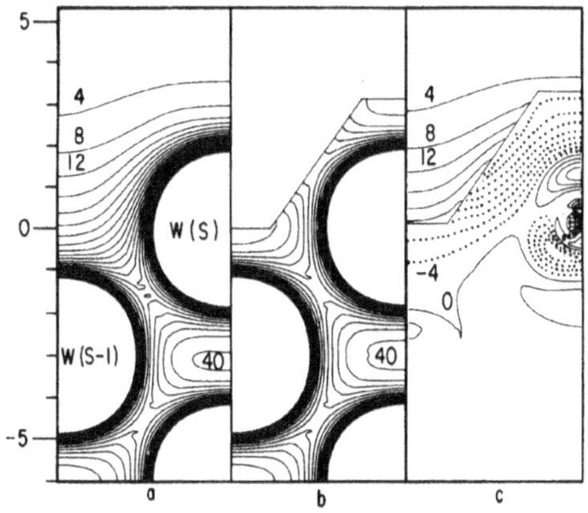

Figure 2. Total charge density on (a) a clean W(001) surface, (b) and "ideal" surface (cut-off of bulk) and (c) the difference between (a) and (b), showing the spill-out of electrons into the vacuum. The units are 10^{-3} electrons $/(a.u.)^3$. Dotted contour lines indicate loss of electrons. The vertical scale at the left gives the distance from the surface W atoms in atomic units [after Ref. 13].

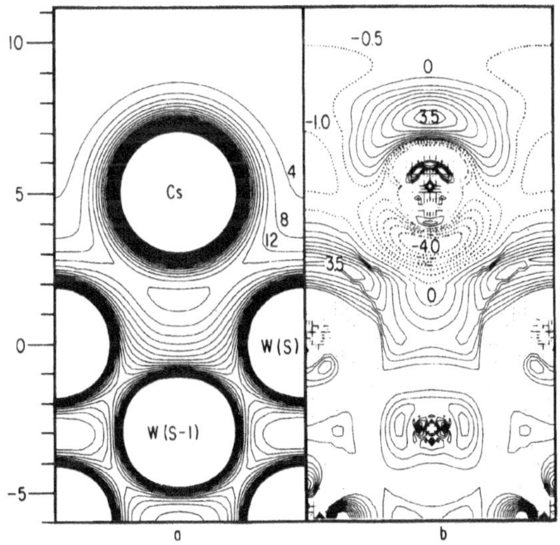

Figure 3. (a) Total electronic charge density for c(2×2) Cs on W(001) and (b) the difference charge density between that of Cs/W and the superposed density of clean W(001) and a Cs monolayer (in units of 10^{-3} electrons$/(a.u.)^3$). The vertical scale at the left gives the distance from the surface W atoms in atomic units [after Ref. 13].

strative and severe test case of the accuracy of local density FLAPW total energy calculations. For this reason we have performed calculations for carbon atoms in a 2-dimensional hexagonal graphite lattice and an artificial square lattice for variable bond distances.[15] As expected, the hexagonal lattice is energetically favored (by ~ 3.3 eV/atom) compared to the square lattice (Fig. 4). The calculated bond length for the hexagonal graphite monolayer of 2.450 Å is contracted by 0.4% compared to the experimental value of 2.461 Å for bulk graphite. Negative thermal expansion coefficients below 400°C for the basal plane of graphite indicate that this contraction may be a real physical effect. Thus, the structural information obtained from local density calculations seems to be highly accurate and reliable. While we find that the structural information is rather insensitive to the variational quality of the basis, this is not the case for the cohesive energy which is sentitive on the quality of the basis used. For a converged variational basis set, the energy difference between the graphite monolayer and the free (spin-polarized) carbon atoms overestimates the cohesive energy by about 1.3 eV per carbon atom. This tendency of local density functional theory to overestimate the cohesive energy was also pointed out for bulk metals.[16] The error appears to originate from an underestimation of the correlation effects in the LD free atoms compared to the condensed phase[15] (where the electron-gas approxi-

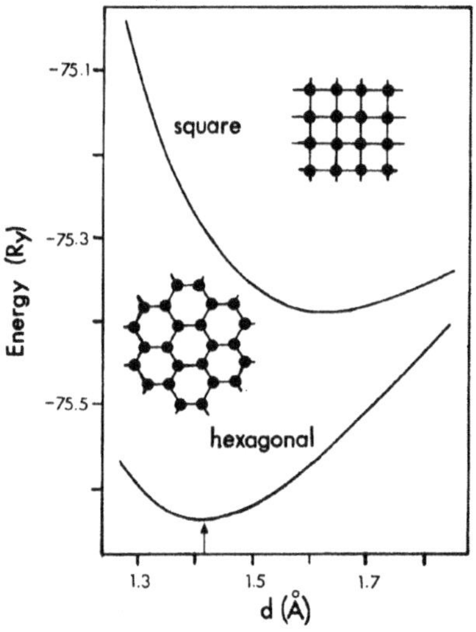

Figure 4. Total energy per atom vs. nearest neighbor distance for a square and a hexagonal monolayer of carbon atoms. The arrow indicates the experimental value for bulk graphite [after Ref. 15].

mation is obviously more appropriate) and not to the neglect of the self-interaction correction. Additional calculations performed using the frozen core approximation indicate that there is a cancellation of systematic local density errors (in particular the self-interaction correction) for the core electrons, when comparing the free atom with condensed phases. This cancellation of LD errors for the core electrons is certainly even more rigorously fulfilled when comparing two different condensed phases.

4. SUMMARY AND CONCLUSIONS

We have discussed and shown how the full-potential linearized-augmented plane-wave (FLAPW) method for thin films provides an highly accurate and reliable scheme to solve the local (spin) density functional one-particle equations and allows the computation of accurate LSD total energies. In particular, we have discussed applications of this method to the dilute molecular limit (O_2 molecule), a clean Al(001) surface, a W(001) surface with a Cs overlayer, and total energies for a carbon monolayer in a graphite and an artificial square lattice. The calculation on the O_2 molecule demonstrated that the FLAPW method is at least as accurate, if not more so, than standard local density molecular LCAO type calculations, although the dilute limit is a severe test for a plane-wave method which is really designed for condensed systems. The study of the 2p-core levels on an Al(001) surface revealed that the local density one-particle energies are accurate down to the meV range in establishing relative core level positions and crystal field splitting effects on surfaces. Theoretical local density work functions seem to be accurate within about $\pm$ 0.1 eV compared to experiment, which we have demonstrated for the case of a s-p metal surface, namely Al(001), and the d-metal surface of a 7-layer W(001) slab. Total density difference maps are shown to be an extremely helpful tool in studying changes in the electronic structure due to overlayers as demonstrated by the cesiation of a W(001) surfaces. In the case of Cs on W(001) we found that the lowering of the work function upon cesiation is due to a multiple dipole formation. The simple ionic picture of Cs^+ on W is clearly inadequate within a quantum mechanical picture. The density and difference density maps give a much better physical picture of the charge distribution and rearrangement on a surface, as well as being rigogously defined from a theoretical point of view.

The calculation of all-electron local density total energies is demonstrated to be a highly accurate tool in determining the structure of 2-dimensional systems. The internal accuracy is sufficient to discriminate between different phases as shown for the example of a carbon monolayer in a graphite and an artificial square lattice. The cohesive energy obtained from local density functional calculations are found to be overestimated due to an underestimate of the correlation in the free atom.

Future work will certainly go in the direction of systematic studies of the electronic structure of clean surfaces and surfaces with atomic and molecular overlayers employing accurate all-electron methods based on local density functional theory. Local density total energies will provide accurate structural and vibrational information for surfaces. A requirement of total energy calculations is certainly a high degree of self-consistency in the charge density such as 10^{-5} electrons per $(\text{Bohr})^3$ average difference between the final input and output densities. The next logical step after the successful treatment of itinerant cases such as clean surfaces and dense overlayers consists in the fully atomistic treatment of single atoms and molecules interacting with a surface (i.e. solution of the 2-dimensional impurity problem).

The local density functional approach proves to be a highly accurate and powerful tool in studying the electronic structure of surfaces and overlayers and is particularly promising in questions of phase stabilities, surface energetics and surface dynamics.

ACKNOWLEDGEMENT

We are grateful to our colleagues B. Delley, H. Krakauer, and S. Ohnishi for informative discussions. This work was supported by the National Science Foundation (Grant No. DMR 77-23776 and under the NSF-MRL program through the Materials Research Center of Northwestern University, Grant No. DMR 79-23573), the U.S. Office of Naval Research (Grant No. N00014-81-K-0438), and the U.S. Department of Energy.

REFERENCES

1. P. Hohenberg and W. Kohn, Phys.Rev. 136, B864 (1964).
2. W. Kohn and L.J. Sham., Phys.Rev. 140, A1133 (1965).
3. H. Krakauer, M. Posternak, and A.J. Freeman, Phys.Rev. B19, 1706 (1979).
4. M. Posternak, H. Krakauer, A.J. Freeman, and D.D. Koelling, Phys.Rev. B21, 5601 (1980).
5. E. Wimmer, H. Krakauer, M. Weinert, and A.J. Freeman, Phys. Rev. B24 , 864 (1981).
6. M. Weinert, J.Math.Phys. 22, 2433 (1981).
7. B. Delley, A.J. Freeman, M. Weinert, and E. Wimmer, Phys.Rev. B (to be published).
8. W. Eberhardt, G. Kalkoffen, and C. Kunz, Solid State Commun. 32, 901 (1979).
9. E. Wimmer, M. Weinert, A.J. Freeman, and H. Krakauer, Phys.Rev. B24, 2292 (1981).

10. T.-C. Chiang and D.E. Eastman, Phys.Rev. B23, 6836 (1981).
11. T.K. Grepstad, P.O. Gartland, and B.T. Slagsvold, Surf.Sci. 57, 348 (1976).
12. N.D. Lang and W. Kohn, Phys.Rev. B3, 1215 (1971).
13. E. Wimmer, A.J. Freeman, M. Weinert, H. Krakauer, J.R. Hishes, and A.M. Karo, Phys.Rev.Lett. 48, 1128 (1982).
14. P.L. Billington and T.N. Rhodin, Phys.Rev.Lett. 41, 1602 (1978).
15. M. Weinert, E. Wimmer, and A.J. Freeman, Phys.Rev. B (to be published).
16. W.L. Moruzzi, J.F. Janak, and A.R. Williams, Calculated Electronic Properties of Metals (Pergamon, 1978).

ELECTRON DENSITIES IN SOLID COMPOUNDS

Karlheinz Schwarz and Peter Blaha

Institut für Technische Elektrochemie
Technische Universität Wien
A-1060 Vienna, Austria

Density functional theory is a more recent approach for treating a many-particle system. In practice this theory is applied by using local density approximations. Applications to atoms provide ionization potentials or electron densities which agree well with experiments or with other theories. Different exchange-correlation potentials and their effects on properties of atoms, molecules, and solids have been discussed by several authors in literature, thus only some important results are summarized here. Then we emphasize applications to solids and focus on electron densities in various compounds (TiC, TiN, TiO, MgO, and NaF), where we discuss differences in the chemical bonding. An attempt is made to interpret band-structure results in a molecular-orbital language in order to bridge the gap between the physicist's band picture and the chemist's bond picture.

1. LOCAL DENSITY APPROXIMATIONS

The Xα-method is one of the frequently used schemes for incorporating electron density distributions into calculations on many-electron systems. In the one-particle Schrödinger equation the essential part is the exchange potential (in Rydberg atomic units)

$$V_{X\alpha}(r) = -3\alpha\left[\frac{3}{\pi}\rho(r)\right]^{1/3} = \alpha\, V_{XS}(r) \qquad (1)$$

which is proportional to the cube root of the electron density $\rho(r)$. In the Xα-method this potential can be expressed as the original exchange potential V_{XS} suggested by Slater[1] times α an exchange parameter.[2,3]

In the more general density functional formalism correlation contributions can be included explicitly in the exchange-correlation potential according to Kohn and Sham[4]

$$V_{xc} = d(\rho\varepsilon_{xc})/d\rho \tag{2}$$

Within the local density approximation (LDA) we can write all these exchange-correlation potentials in the form

$$V_{xc}(r) = \alpha(r)\ V_{XS}(r) \tag{3}$$

in order to show the similarity to the Xα-method where $\alpha(r)$ is a constant which is denoted by α_{vt} when it is determined by the virial theorem criterion.[2,3]

Another way to conveniently write the exchange-correlation potential is the form

$$V_{xc}(r) = \beta(r_s)\frac{2}{3}\ V_{XS}(r) \tag{4}$$

where the interelectronic separation parameter r_s is locally determined by the requirement $(4\pi/3)r_s^3\ \rho(r)=1$.

Wigner[5,6] gave a very simple interpretation of correlation. In the low-density limit the electrons in an electron gas will avoid each other most efficiently, if they are arranged as in a closed-packed crystal, a configuration for which the energy in (2) is known. Then an interpolation of the correlation energy between the high- and the low-density limit can be made.

For the paramagnetic case two examples for the correlation enhancement factor β in Eq. (4) are given below: i) the one following Wigner[5-7]

$$\beta_W(r_s) = 1 + [0.9604\ r_s(r_s+5.85)/(r_s+7.8)^2] \tag{5}$$

ii) the other parametrization which is frequently used is due to Hedin and Lundqvist[8]

$$\beta_{HL}(r_s) = 1 + 0.0316\ r_s\ \ln(1+24.3/r_s) \tag{6}$$

A comparison of these two representative exchange-correlation potentials with that of the Xα-method is made for the titanium atom: in Fig. 1a the radius dependent prefactor $\alpha(r)$ defined in (2) is shown for both cases; these two curves behave similarly in the sense that they start near α=2/3 and increase monotonically with distance from the nucleus. The radial charge density $\sigma(r)=4\pi r^2\rho(r)$ shown in Fig. 1b clearly shows the K-, L-, and M-shell in titanium. By combining these results we find that the function $\alpha(r)$ is smaller than α_{vt} for the core electrons, but greater than α_{vt} for the

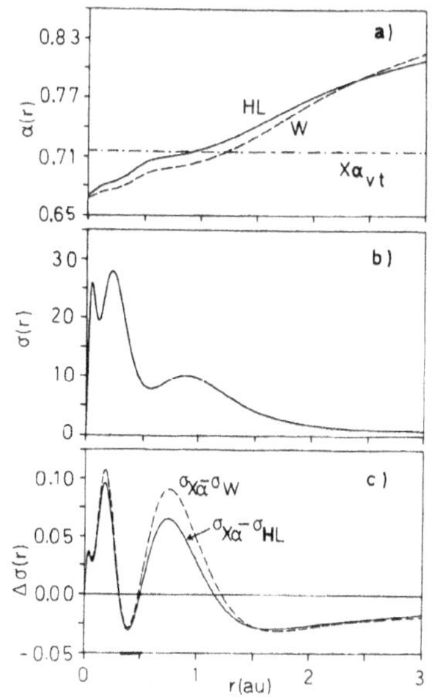

Figure 1. Results for the titanium atom obtained by the $X\alpha_{vt}$-,[2] the Hedin-Lundqvist (HL)-method[8] and the Wigner (W) exchange correlation potential:[5-7]
[a]the function $\alpha(r)$ defined by Eq. (3),
[b]the radial charge density $\sigma(r) = 4\pi r^2 \rho(r)$ of the $X\alpha$ calculation,
[c]the difference $\Delta\sigma(r)$ between two radial charge densities.

outer electrons. In Table 1 it is shown how this deviation affects the orbital energies. For example, why is it that the 1s orbital energy is raised, when V_{HL} is used instead of the $X\alpha_{vt}$-potential. This comes about, because in the relevant radial range the effective $\alpha(r)$ is smaller than α_{vt} leading to a less attractive V_{xc} potential; by an analogous argument the valence states, whose wave functions dominate at large radii, are lowered in energy with respect to the $X\alpha$-orbital eigenvalues. For a similar discussion see Wimmer et al.[7]

The exchange parameter α_{vt} can be interpreted as weighted radial average over $\alpha(r)$ and thus the $X\alpha$-potential should contain some correlation effects, although this is not the way it has been constructed.

The orbital energies of LDA schemes can be used to calculate ionization potentials, but by means of Slater's transition state method rather than by Koopmans' theorem. Such calculations have

Table 1. Orbital energies ε (Ryd) and their differences for the Ti atom using three different V_{xc}.

$n\ell$	$\varepsilon_{X\alpha}$	ε_{HL}	ε_{W}	$\varepsilon_{X\alpha}-\varepsilon_{HL}$	$\varepsilon_{X\alpha}-\varepsilon_{W}$	$\varepsilon_{HL}-\varepsilon_{W}$
1s	-355.326	-354.523	-354.469	-0.803	-0.857	-0.054
2s	-39.040	-38.902	-38.877	-0.138	-0.163	-0.025
2p	-32.706	-32.555	-32.528	-0.151	-0.178	-0.027
3s	-4.491	-4.510	-4.502	0.019	0.011	-0.008
3p	-2.817	-2.841	-2.833	0.024	0.016	-0.008
3d	-0.297	-0.337	-0.335	0.040	0.038	-0.002
4s	-0.288	-0.335	-0.344	0.047	0.056	-0.009

been reported by Schwarz,[9] who compared them with Hartree-Fock results. The different interpretation of the orbital energies in LDA calculations has important consequences; for example in connection with stable negative ions which are found unstable in LDA.[10] This is one of the shortcomings of the LDA schemes originating from the wrong asymptotic form of V_{xc} at large radii. This problem is closely related to the self-interaction correction which has found much attention recently.[11]

Since the electron density plays a central role in density functional theory, it is of interest to know how much the density is affected by the choice of V_{xc}. The difference in $\sigma(r)$ with respect to the radial density $\sigma_{X\alpha}(r)$ of Fig. 1b is illustrated in Fig. 1c using a 200 times enlarged scale. These $\Delta\sigma$ curves show that $\sigma(r)$ is not extremely sensitive to V_{xc} and its relative change with respect to the Xα results is of the order of one per cent or less. These variations of $\sigma(r)$ are again consistent with Fig. 1a, since in the inner core region the effective $\alpha(r)$ is smaller than α_{vt} leading to a less attractive V_{xc} which causes the corresponding wave functions to be more diffuse and $\Delta\sigma$ to be positive. The opposite argument holds for the outermost electrons at large radii.

There are several other LDA for V_{xc} than the ones discussed here, but most of them are quite close to V_W or V_{HL}. Since these two potentials lead to very similar results (Table 1 and Fig. 1), we can conclude that at least in the paramagnetic case the particular choice of V_{xc} should be less important as long as the potential has a form similar to V_{HL}. A convenient way to judge the similarity between potentials is the analysis by means of $\alpha(r)$ according to Eq. (3) as illustrated in Fig. 1a.

2. APPLICATION TO SOLIDS

In solid state theory LDA schemes are widely used especially in energy band calculations, where one should specify the exchange-correlation potential V_{xc} and also the method which is used to solve the one-electron Schrödinger equations. Since often different methods yield accurate solutions, it is mainly V_{xc} which determine the results and then the choice of V_{xc} is responsible for the success or the failure of a calculation. In this context the reader is referred to a work by Wimmer et al.[7] who among others have investigated the influence of different V_{xc} on energy bands.

In the following we present special applications and discuss some aspects of chemical bonding in solids by studying electron densities of a series of compounds all of which crystallize in the rock salt structure.

Several refractory metal compounds including TiC, TiN, and TiO have been investigated by Neckel et al.[12] They used the Xα-method for V_{xc} and performed self-consistent energy band calculations by the augmented plane wave (APW) method. Recently this work was extended by the present authors[13] who calculated the electron densities of these compounds by the linearized APW method. Some of these results will be discussed below.

These compounds exhibit an unusual combination of properties such as ultra-hardness, high melting points, and metallic conductivity and thus are of great technological and theoretical interest. The bonding mechanism consists of ionic (small charge transfer from titanium to the non-metal atom) metallic (conductivity of the compounds is comparable to that of pure titanium) and "covalent" contributions. It seems that it is this mixture in the bonding which is responsible for their properties. How can we learn about the relation between bonding and properties ? Many experimental and theoretical investigations are needed to achieve this goal. Here we want to illustrate - for a few examples - what can be learned from electron densities.

The band structure of the refractory metal compounds TiX discussed here (X denotes one of the non-metal atoms C, N, or O) consists of a low lying band originating from the X-2s states, the next lowest bands are mainly derived from X-2p states (termed p-band) while the bands further above are formed predominantly from Ti-3d states (called d-band). In TiC and TiN the p- and d-band overlap, but in TiO they are separated by a gap.

In these compounds a certain amount of covalent bonding is involved as discussed by Neckel et al.[12] In TiC for example it is the C-p Ti-d interaction which is very important. This can be seen indirectly from the energy bands or the partial densities of states

In order to make this point more evident, however, electron densities have proven to be extremely useful.[13]

In the following we show that energy band results can be interpreted in a very simple molecular orbital (MO) picture. For that purpose we select one representative $\vec{k}$-point inside the first Brillouin zone, namely Δ half-way between Γ and X. In the range of the C-p Ti-d bands there are six energy states of various symmetries, the lower three of which are occupied in TiC. For these three states the corresponding electron densities in the (100) plane are shown in Fig. 2. (Actually the sum over the star of $\vec{k}$ is taken in order to obtain the required cubic symmetry). The d-orbitals centered on a titanium site can be classified according to e_g or t_{2g} symmetry corresponding to the octahedral ligands (or the cubic environment). Ti d-orbitals of e_g symmetry can form pd_σ bonds with the neighbouring C-p orbitals. If the Ti d-orbitals have t_{2g} symmetry either a metal-metal dd_σ bond can form (as for Δ_2, with hardly any admixture from carbon wavefunctions) or pd_π bonds are possible (see bottom of Fig. 2).

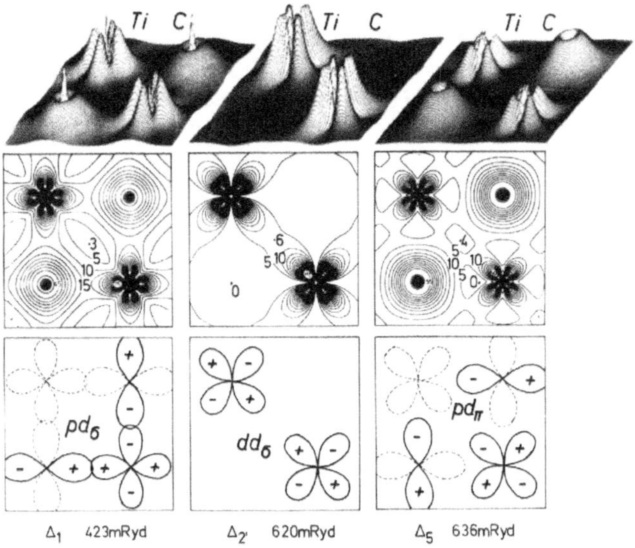

Figure 2. Charge densities corresponding to the occupied band states $E_{n\vec{k}}$ (for star $\vec{k} = \Delta = (1,0,0)\pi/a$ of the p- and d-band) in the (100) plane for TiC; each density is normalized to 2 e in the unit cell (cut-off at 1.0 eÅ^{-3}); contour intervals 0.05 eÅ^{-3} (numbers are in units of 0.01 eÅ^{-3}). The bottom panel shows schematically the corresponding bonding types.

The fact that the density around the C site is nearly spherically symmetric - in both cases for Δ_1 (pd_σ) or Δ_5 (pd_π) - comes about from the cubic symmetry, i.e. we have effectively added with the same weights the densities of p_x, p_y, and p_z orbitals and the sum is spherically symmetric. In the case of the Δ_1 state there is (on both sites at C and Ti) also a small s contribution which can be seen from the corresponding spikes near C and Ti in the top of Fig. 2. This feature is not present in the other two cases.

It is important to note that the relative densities near C and Ti are comparable in magnitude in both pd bonds and that the orbitals involved are similar in size. Since the corresponding C-p and Ti-d partial densities of states[12] are both large in this energy range, all essential criteria are satisfied which guarantee an efficient covalent bond between C and Ti.

In simple Mo theory two orbitals of appropriate symmetry can form a bonding and an antibonding MO. This picture can also be used in connection with energy band calculations as is shown in Fig. 3. In TiC the two Δ_5 states correspond to such bonding and antibonding states and their spectral weights are balanced indicating a relatively strong covalency. Since the O-p orbital energy is lower than that of C-p, the energy separation between the X-p and the Ti-d state is greater in TiO than it is in TiC. Consequently in TiO the bonding MO has predominantly O-p character while the

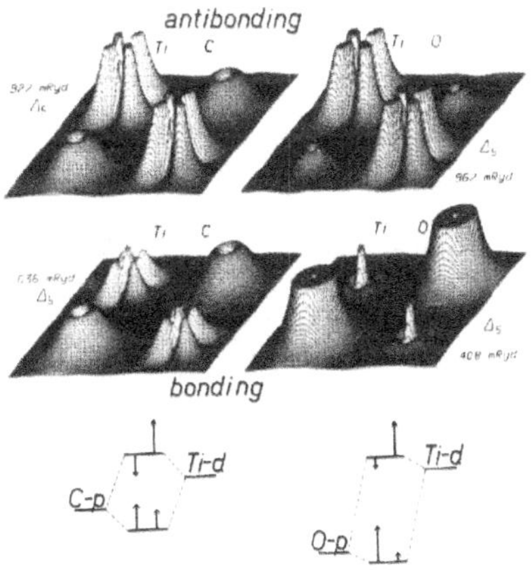

Figure 3. Charge densities corresponding to two states of Δ_5 symmetry of TiC (left) and TiO (right); for definitions see Fig. 2. At the bottom the schematic MO diagram is shown in which the spectral weight of the wave functions is indicated by arrows pointing up or down corresponding to the sign of the coefficients.

antibonding state has a large coefficient for the Ti-d wave functions as is obvious from Fig. 3. Such a change in weights is also present in TiC but to a much lesser extent.

The sum of such orbital densities (as shown in Figs. 2 and 3) taken over all occupied $E_{n\vec{k}}$ leads to the valence electron densities which are discussed in detail elsewhere.[13] But here we summarize that in the series TiC, TiN, and TiO the densities around the non-metal sites and their localization increase, while around the Ti site the valence density deviates from spherical symmetry changing from e_g symmetry for TiC to t_{2g} for TiO.

An analysis of the band structure results shows that in TiC the covalent component is an important factor in the chemical bonding. Since essentially just those states are occupied which correspond to bonding MOs, while all antibonding states remain empty, this covalency is likely to contribute to the ultra-hardness and to the high melting point observed. In TiO the situation is different; the covalent component is greatly reduced (see the discussion above) and, since TiO has two valence electrons in addition to TiC, antibonding states must be occupied. This fact could be the reason why about 15 per cent of vacancies are found on both the Ti- and the O-sublattice. If we describe TiO to be more ionic in the sense that the covalent component is reduced, it is of interest to compare TiO with a more typical ionic solid. For that purpose we have calculated the valence electron densities of MgO and NaF, since for these compounds similar self-consistent APW calculations were available.[14] We have used the same LAPW procedure as for the calculations on the refractory transition metal compounds.[13]

The total valence electron densities of TiO and MgO are shown in Fig. 4. In TiO we see the pronounced t_{2g} symmetry component in the density around Ti as mentioned above and as discussed elsewhere.[13] The other densities (around O or Mg) are essentially spherically symmetric. The comparison between TiO and MgO yields a very similar density around O, but as expected more charge near Ti than near Mg. The valence charge density in MgO practically drops to zero relatively close to the Mg nucleus, a fact which might indicate a very small ionic radius, but we will show that this argument is somewhat misleading. In order to illustrate this situation we show in Fig. 5 the density of MgO in the (100) direction between O and Mg, i.e. between the next nearest neighbours (NN); also displayed are the densities between second NN in the (110) directions. The valence density vanishes at about 0.5 Å from the Mg nucleus. If we add the electron density corresponding to the Mg core electrons ($1s^2 2s^2 2p^6$), then the minimum in the total density moves outward from the Mg nucleus to almost 1 Å.

After the effect of the core densities has been demonstrated we can focus again on just the valence electron densities. A com-

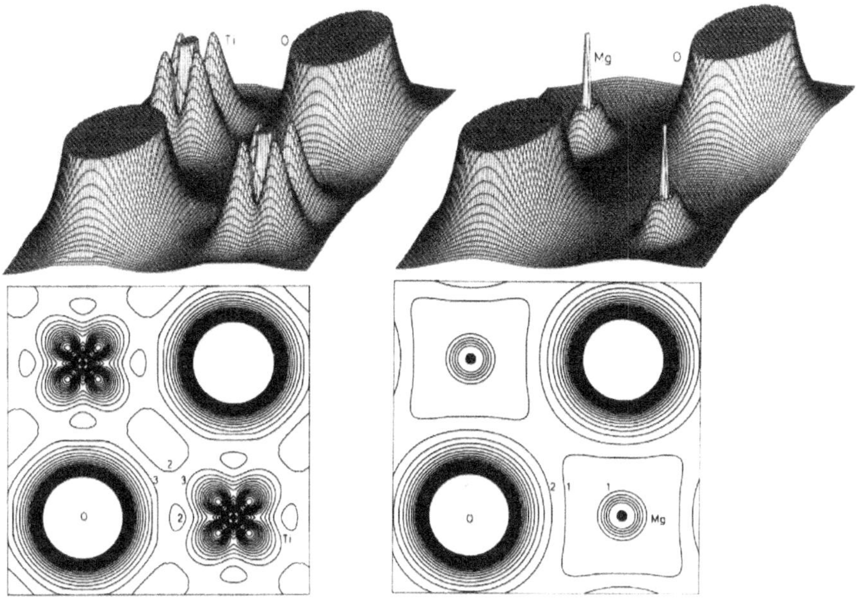

Figure 4. Valence charge densities of TiO and MgO in the (100) plane. Contour intervals 0.1 eÅ^{-3} (numbers are in these units), cut-off at 1.7 eÅ^{-3}.

parison of such densities is made in Fig. 6 and an interpretation is attempted on the following simple assumption: if the density between two atoms is high, it indicates a large overlap and consequently a (covalent) interaction between these atoms. By using this simple rule we find that in TiC the density between Ti and C in (100) is significantly higher than in the (110) direction between second NN. This observation suggests a strong C-Ti interaction. In TiO, however, the minimum in $\rho(r)$ has about the same value along (100) and (110), but here we must keep in mind that the density increases to a local saddle point away from the O-Ti internuclear connection (Fig. 4). From just this valence density a strong pd_π bond might be deduced, but an analysis in terms of contributions from different valence bands has shown that this is not the case.[13]

The minimum value of the density between the oxygen atoms is about the same in TiO and in MgO and it is by no means very small. At least in MgO one usually interpretes this situation as anion contact. A comparison of MgO with the isoelectronic NaF clearly shows that in NaF the density between the ions is really small and

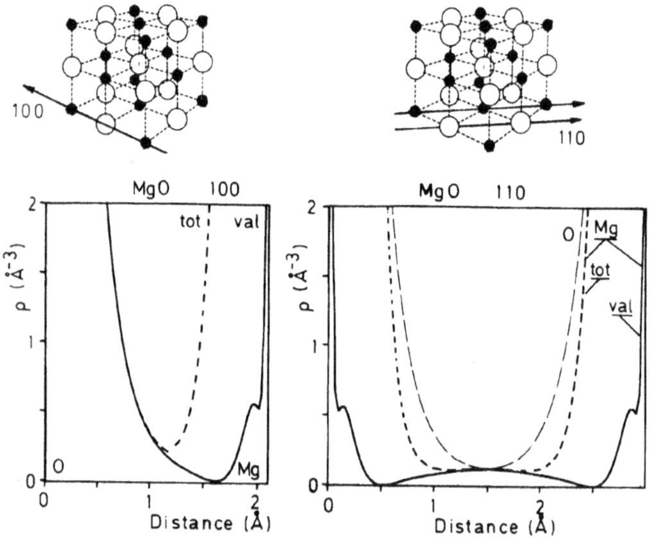

Figure 5. Electron densities in MgO along the (100) and (110) direction: for Mg both densities, the valence and the total, are shown, where the latter contains the Mg core density ($1s^22s^22p^6$).

it is justified to call it a typical ionic solid. In the ionic MgO, however, a different bonding situation occurs, which originates in the fact that in contrast to F^-, which is a stable anion, O^{2-} is unstable as a free ion. This distinct behaviour of F^- and O^{2-} leads, for example, to a qualitative difference in the polarizability.[15] It also causes the O^{2-} density to be very sensitive to the crystalline environment[14] and it had lead Pendry[16] to speak about new elements in ionic bonding.

With the present apllications we wanted to demonstrate that the electron densities provide a powerful tool to illustrate bonding effects and that they can help to bridge the gap between the physicist's band picture (in reciprocal $\vec{k}$-space) and the chemist's bond picture (in direct $\vec{r}$-space). We hope that we were able to show that electron densities can supplement other methods of studying chemical bonding in solids.

ACKNOWLEDGEMENTS

This work was supported by the "Hochschuljubiläumsstiftung der Stadt Wien".

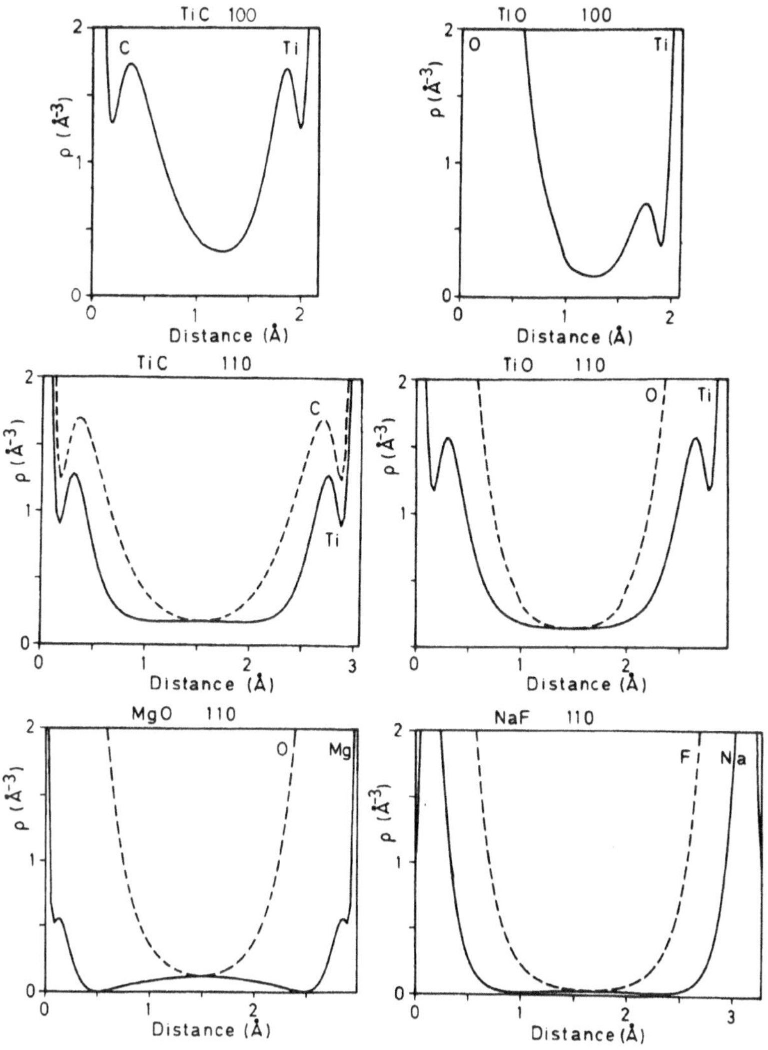

Figure 6. Valence electron densities in TiC, TiO, MgO, and NaF in the (100) or (110) direction.

REFERENCES

1. J.C. Slater, A simplification of the Hartree-Fock method, Phys.Rev. 81, 385 (1951).
2. K. Schwarz, Optimization of the statistical exchange parameter α for the free atoms H through Nb, Phys.Rev. B 5, 2466 (1972).
3. K. Schwarz, Optimized exchange parameter α for the atoms with higher Z, Theoret.Chim. Acta 34, 225 (1974).
4. W. Kohn and L.J. Sham, Self-consistent equations including exchange and correlation effects, Phys.Rev. 140A, 1133 (1965).
5. E.P. Wigner, On the interaction of electrons in metals, Phys. Rev. 46, 1002 (1934).
6. E.P. Wigner, Effects of the electron interaction on the energy levels of electrons in metals, Trans. Faraday Soc. 34, 678 (1938).
7. E. Wimmer, A. Neckel, and K. Schwarz, The influence of different exchange-correlation potentials on the energy band structure of ScP, J.Phys.C. 12, 5441 (1979).
8. L. Hedin and B.I. Lundqvist, Explicit exchange-correlation potentials, J.Phys.C. 4, 2064 (1971).
9. K. Schwarz, First ionization potentials of atoms obtained with local density functional schemes, J.Phys.B. 11, 1339 (1978).
10. K. Schwarz, Instability of stable negative ions in the Xα method or other local density functional schemes, Chem.Phys. Lett., 57, 605 (1978).
11. J.P. Perdew and A. Zunger, Self-interaction correction to density functional approximations for many-electron systems, Phys.Rev.B. 23, 5048 (1981).
12. A. Neckel, P. Rastl, R. Eibler, P. Weinberger, and K. Schwarz, Results of self-consistent band structure calculations for ScN, ScO, TiC, TiN, TiO, VC, VN, and VO, J.Phys.C. 9, 579 (1976).
13. P. Blaha and K. Schwarz, Electron densities and chemical bonding in TiC, TiN, and TiO derived from energy band calculations, Int.J. Quantum Chem. (in press).
14. J. Redinger and K. Schwarz, Electronic charge distribution of the polarizable O^{2-} ion in MgO and CaO in contrast to the F^- ion in NaF, Z.Physik B. 40, 269 (1981).
15. A. Bussmann, H. Bilz, R. Roenspiess, and K. Schwarz, Oxygen polarizability in ferroelectric phase transitions, Ferroelectrics 25, 343 (1980).
16. J.B. Pendry, Incomplete orbitals - new elements in ionic bonding, J.Phys.C: Solid State Phys., 10, 809 (1977).

LOCAL DENSITY APPROXIMATIONS FOR RELATIVISTIC EXCHANGE ENERGIES

A.H. MacDonald

Division of Physics
National Research Council of Canada
Ottawa, Canada K1A 0R6

The use of local density approximations to approximate exchange interactions in relativistic electron systems is reviewed. Particular attention is paid to the physical content of these exchange energies by discussing results for the uniform relativistic electron gas from a new point of view. Work on applying these local density approximations in atoms and solids is reviewed and it is concluded that good accuracy is usually possible provided self-interaction corrections are applied. The local density approximations necessary for spin-polarized relativistic systems are discussed and some new results are presented.

1. INTRODUCTION

The use of local exchange potentials to simplify electronic structure calculations has been common since Slater's original proposal for a statistical $\rho^{1/3}$ approximation (Slater, 1951). Later on, the early relativistic calculations for atoms generally took over the non-relativistic form when local approximations for the Dirac-Fock exchange operator were adopted (e.g., Liberman et al. 1965). It was not until quite recently that Ellis (1977) pointed out the need for relativistic corrections to the local approximation for the Coulomb exchange operator. Shortly thereafter, in the context of the relativistic density functional formalism (MacDonald 1978, MacDonald and Vosko 1979, Rajagopal 1978), it was suggested that a local density approximation could also be made for quantum-electrodynamical corrections to the Coulomb interaction, effects corresponding to the virtual exchange of transverse photons between the Dirac electrons. It is these relativistic local exchange approximations which are the subject of this article.

In Sec. 2, we examine the exchange energy of the relativistic electron gas and, in doing so, attempt to bring out the physical content of the exchange energies being discussed. Recent work based on the applications of relativistic exchange approximations in atoms and in solids is reviewed in Sec. 3. The spin-polarized version of these exchange approximations is discussed in Sec. 4 and some new results are presented. Finally in Sec. 5, we conclude with some suggestions for future work.

2. EXCHANGE ENERGY OF THE UNIFORM RELATIVISTIC ELECTRON GAS

In the Coulomb gauge the Hamiltonian for interacting Dirac and Electromagnetic fields may be written as

$$H = H_{oD} + H_{oem} + \frac{e^2}{2}\int d\vec{r}\int d\vec{r}\,' \frac{\hat{n}(\vec{r})\hat{n}(\vec{r}\,')}{|\vec{r}-\vec{r}\,'|} - e\int d\vec{r}\hat{\psi}^{+}\vec{\alpha}\hat{\psi}(\vec{r})\cdot\hat{\vec{A}}(\vec{r}) \quad (1)$$

where H_{oem} and H_D are the Hamiltonian for a free electromagnetic field, and a non-interacting Dirac field (possibly including an external potential), $\hat{n}(\vec{r}) = \hat{\psi}^{+}(\vec{r})\hat{\psi}(\vec{r})$ and $\hat{\psi}(\vec{r})$ and $\hat{\vec{A}}(\vec{r})$ are the Dirac and electromagnetic field operators. (See MacDonald, 1978 or Rajagopal, 1979 for details.) The third term in Eq. (1) if the instantaneous Coulomb interaction and the fourth is the interaction between the Dirac current and the quantized transverse gauge radiation field. By exchange-energy we shall mean the interaction energy to leading order in $\alpha_{f.s.} \equiv e^2/\hbar c$, i.e. when the third term in Eq. (1) is treated by first order perturbation theory and the fourth-term by second order perturbation theory. In the case of a constant external potential, the electron system is uniform and the Dirac field becomes

$$\hat{\psi}(\vec{r}) = \sum_{s,\vec{k}} \frac{1}{\sqrt{V}} [\hat{b}(\vec{k},s)u(\vec{k},s) + \hat{d}^{+}(-\vec{k},s)v(-\vec{k},s)]\exp(i\vec{k}\cdot\vec{r}) \quad (2)$$

where $\hat{b}(\vec{k},s)$ and $\hat{d}(\vec{k},s)$ are electron and positron annihilation operators and $u(\vec{k},s)$ and $v(\vec{k},s)$ are the corresponding free-particle Dirac spinors. In this case the exchange energy can be readily evaluated and equivalent analytic expressions were given by several workers some time ago (Akhiezer and Peletminskii 1960, Zapolsky 1960, Jancovici 1962).

In discussing these results, for reasons which will become apparent in the next section, we separate the contribution from the third term in Eq. (1), which we refer to as the Coulomb exchange energy (E_X^C), and the contribution from the fourth term in Eq. (1) which we refer to as the transverse exchange energy E_X^T. For E_X^C we follow Ellis (1977) in noting that just in the non-relativistic case it may be written as

$$E_X^C = \frac{V}{2} \int \frac{d\vec{r}}{r} \bar{\rho}\, \rho_x(r) \tag{3}$$

where $\bar{\rho} = Q/V$ is the average charge density of the electron gas and $\rho_x(r)$, the exchange charge density, is defined by

$$\rho_x(r) = \bar{\rho}[\langle\Phi_o|n(\vec{x})n(\vec{x}+\vec{r}) - n(\vec{x})\delta(\vec{r})|\Phi_o\rangle - 1] \tag{4}$$

where $|\Phi_o\rangle$ is the non-interacting ground state. It can be shown (Ellis 1977, MacDonald 1978) that

$$\rho_x(r) = -\frac{q\bar{\rho}}{4}\left[\frac{j_1^2(k_F r)}{(k_F r)^2} + (1-\lambda)A^2(k_F r) + \lambda B^2(k_F r)\right] \tag{5a}$$

where

$$A(x) = \sum_{k=0}^{\infty} \frac{j_{k+1}(x)\lambda^k(2k+1)!!}{(2k+1)x^{k+1}} \tag{5b}$$

$$B(x) = \sum_{k=0}^{\infty} \frac{j_{k+2}(x)\lambda^k(2k+1)!!}{(2k+1)x^{k+1}} \tag{5c}$$

$$x = (\lambda\!\!\!^{-}_c k_F)^2/(1+(\lambda\!\!\!^{-}_c k_F)^2) \tag{5d}$$

k_F is the Fermi radius of the electron gas, and $\lambda\!\!\!^{-}_c$ is the electron Compton wavelength divided by 2π. Plots of this exchange charge density for several values of $\beta \equiv \lambda\!\!\!^{-}_c k_F$ are shown in Fig. 1. We see that relativistic effects tend to increase the size of the exchange-hole charge around each electron although, as is readily demonstrated using Eq. (4), the total charge in the exchange hole is still e. The movement of exchange charge away from each electron would be expected to reduce the exchange energy in the relativistic case, and this is exactly what occurs.

The result for the relativistic Coulomb exchange energy may be written in the form (MacDonald 1978, MacDonald and Vosko 1979)

$$E_{XC} = \left[-\frac{2e^2 V k_F^4}{(2\pi)^3}\right] \phi_c \tag{6}$$

where the first term is the non-relativistic expression for the exchange density and ϕ_c may be regarded as a relativistic correction factor

$$\phi_c \equiv \frac{5}{6} + \frac{\beta^{-2}}{3} + \frac{2\beta^{-1}\eta \ln(\xi)}{3} - \frac{2\beta^{-4}}{3}\eta^4 \ln(\eta) - \frac{1}{2}\left[\frac{\beta\eta - \ln(\xi)}{\beta^2}\right]^2 \tag{7a}$$

$$= 1 - \frac{\beta^2}{9} + \frac{13\beta^4}{180} + \ldots \qquad \beta << 1 \tag{7b}$$

$$= \left[\frac{1}{3} + \frac{2}{3}\ln 2\right] + \beta^{-2}\left[\frac{4}{3}\ln 2 - \frac{1}{3}\right] + \ldots \qquad \beta >> 1 \tag{7c}$$

where $\eta = (1+\beta^2)^{1/2}$ and $\xi = \beta + \eta$. ϕ_c is a monotonically decreasing function of β which approaches one in the non-relativistic limit and $1/3 + 2/3 \ln 2 \simeq 0.795$ in the ultrarelativistic limit. The fact that it changes so little in passing to the relativistic regime may be understood in terms of the preservation of the total charge in the exchange hole.

To leading order in β^2 the transverse exchange energy may be included by adding an additional two-particle interaction term to the Dirac Hamiltonian and dropping the coupling to the electromagnetic field. This additional interaction term may be written as

$$H_{TB} = -\frac{1}{2c^2}\int d\vec{r} \int d\vec{r}\,' \frac{\hat{\vec{J}}(\vec{r})\cdot\hat{\vec{J}}(\vec{r}\,')}{|\vec{r}-\vec{r}\,'|} + \frac{1}{2c^2}\int d\vec{r} \int d\vec{r}\,' \frac{(\vec{J}(\vec{r})\cdot(\vec{r}-\vec{r}\,'))(\vec{J}(\vec{r}\,')\cdot(\vec{r}-\vec{r}\,'))}{|\vec{r}-\vec{r}\,'|^2} \tag{8}$$

This interaction was first introduced by Breit (1929, 1930, 1932) and has received considerable attention in atomic calculations (Grant, 1965; Kim, 1967; Smith and Johnson, 1967; Mann and Johnson, 1971; Huang et al., 1976). The first term, the Gaunt (1932) interaction (H_{TM}) is simply the unretarded magnetic interaction between two Dirac currents while the second term (H_{TR}) is the leading retardation correction to the Coulomb interaction

$$(\vec{J}(\vec{r}) = -ec\hat{\psi}^+(\vec{r})\,\vec{\alpha}\,\hat{\psi}(\vec{r}\,'))$$

The contribution of the magnetic current-current interaction to the energy may be expressed in terms of a current-current correlation function just as the Coulomb energy is expressed in terms of a charge correlation function:

$$E_{XTM} = \langle\phi_o|H_{TM}|\phi_o\rangle \tag{9a}$$

$$= \frac{-N^2}{2V c^2} \int \frac{d\vec{r}}{r} g_J^{(o)}(\vec{r}) \tag{9b}$$

where

$$g_J^{(o)}(\vec{r}) = \langle\phi_o|\hat{\vec{J}}(\vec{x}+\vec{r})\cdot\hat{\vec{J}}(\vec{x}) - \delta(\vec{r})\, J^2(\vec{x})|\phi_o\rangle \tag{9c}$$

An expression for $g_J^{(o)}$ can be determined following the same line as was used to obtain $\rho_x(r)$

$$g_J^{(o)}(r) = \frac{-9}{2\beta^2}\left[3\left(\frac{j_1^2(k_F r)}{(k_F r)^2} - (1-\lambda)A^2(k_F r)\right) - \lambda B^2(k_F r)\right] \tag{10a}$$

$$= \frac{-9}{4}\left[\frac{3j_1^2(k_F r)}{(k_F r)^2} - \frac{2j_1(k_F r)j_2(k_F r)}{(k_F r)^3} - \frac{j_2^2(k_F r)}{(k_F r)^2}\right] + \theta(\beta^2) \tag{10b}$$

(Plots of this quantity for several values of β are presented in Fig. 2).

Similarly, the retardation energy may be interpreted in terms of an effective additional contribution to the exchange charge density:

$$\rho_{XR}(r) = \frac{9}{4}\bar{\rho}\left[(1-\lambda)A^2(k_F r) - \frac{j_1^2(k_F r)}{(k_F r)^2}\right] \tag{11}$$

Comparing Eqs. (5a) and (11) we see that the retardation terms has a tendency to cancel the relativistic corrections to $\rho_x(r)$. In fact, in the ultrarelativistic limit $\bar{\rho}^{-1}(\rho_x(r{=}0) + \rho_{XR}(r{=}0))$ is unchanged from its non-relativistic value. This cancellation between retardation effects and the relativistic expansion of the Dirac-Fock exchange hole manifests itself again it we calculate the exchange energy to leading order in β^2 using the Breit interaction. From Eq. (11) it follows that the retardation contribution is

$$E_{XTR} = \left(-\frac{2e^2 V k_F^4}{(2\pi)^3}\right)\cdot\frac{\beta^2}{9} \tag{12}$$

exactly cancelling the leading relativistic correction to the Dirac-Fock Coulomb exchange energy. The magnetic exchange energy, which may be thought of as representing the repulsive interaction between the current of one Dirac electron and the net current of all other electrons, is found to be

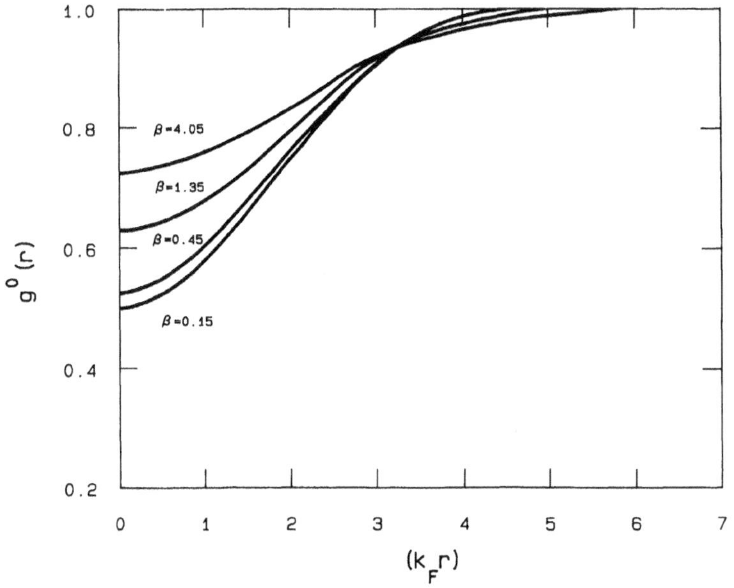

Figure 1. Pair-correlation function, $g^{(o)}(r)$, for a non-interacting relativistic electron gas for various values of $\beta \equiv \bar{\lambda}_c k_F$. The exchange charge-density, $\rho_x(r)$, is related to $g^o(r)$ by $\rho_x(r) = \bar{\rho}(g^o(r)-1)$.

$$E_{XTM} = \left(-\frac{2e^2 V k_F^4}{(2\pi)^3}\right) \cdot \frac{-2\beta^2}{3} \tag{13}$$

This is the dominant leading order relativistic exchange correction.

To calculate the transverse exchange energy beyond leading order, it is necessary to return to Eq. (1) and use quantumelectrodynamical methods. The result is (e.g. Jancovici, 1962)

$$E_{XT} = \left(-\frac{2e^2 V k_F^4}{(2\pi)^3}\right) \phi_T \tag{14a}$$

$$E_{XR} \equiv E_{XC} + E_{XT} = \left(-\frac{2e^2 V k_F^4}{(2\pi)^3}\right) \phi_R \tag{14b}$$

$$\phi_T = \frac{1}{6} - \frac{\beta^{-2}}{3} - \frac{2\beta^{-1}\eta \ell n(\xi)}{3} + \frac{2\beta^{-4}\eta \ell n \eta}{3} - \left(\frac{\beta\eta - \ell n\xi}{\beta^2}\right)^2 \tag{14c}$$

$$\phi_R = 1 - \frac{3}{2}\left(\frac{\beta\eta - \ell n\xi}{\beta^2}\right)^2 = 1 - \frac{2}{3}\beta^2 + \frac{2}{5}\beta^4 \ldots \quad \beta \ll 1 \tag{14d}$$

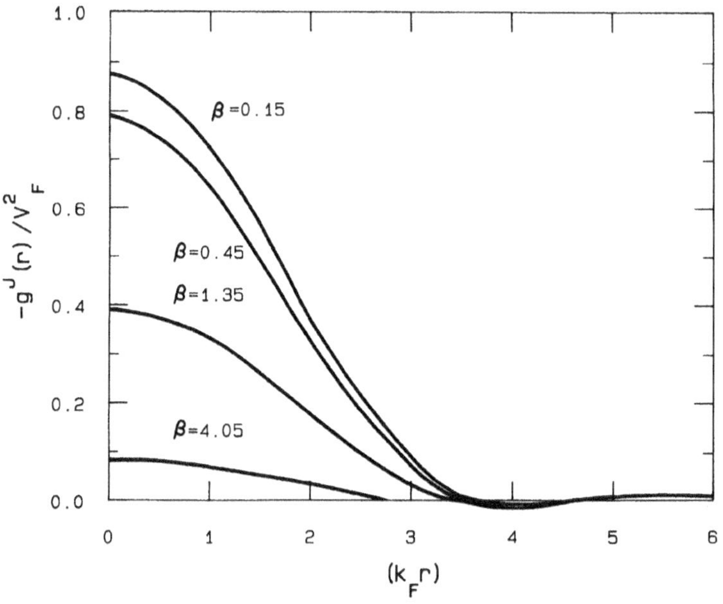

Figure 2. Current pair-correlation function, $g_J^{(o)}(r)$, in units of $(\hbar k_F/m)^2 \equiv V_F^2$ for various values of $\beta \equiv \lambda_c k_F$. V_F is the non-relativistic Fermi velocity. Note that $g_J^{(o)}(r)$ increases more slowly with increasing density than V_F^2 mainly because of the relativistic mass-velocity effect.

$$= -\frac{1}{2} + \frac{3\ln\beta}{\beta^2} + \ldots \qquad \beta >> 1 \tag{14e}$$

Note that at all densities, the transverse exchange energy is ~6 times larger in magnitude than the relativistic shift in the Coulomb Dirac-Fock exchange energy. In the ultrarelativistic limit the total exchange energy is reversed in sign and half as large in magnitude as the non-relativistic exchange energy. Fig. 3 shows how ϕ_C, ϕ_T, and ϕ_R vary in going from non-relativistic to ultrarelativistic limits.

3. RELATIVISTIC EXCHANGE ENERGIES AND POTENTIALS IN ATOMS AND SOLIDS

The discussion in this section will be in the context of the relativistic density-functional formalism (MacDonald 1978, Rajagopal 1978, and MacDonald and Vosko 1979, see also the review in this volume by Gross 1983). In the more standard implementation of this

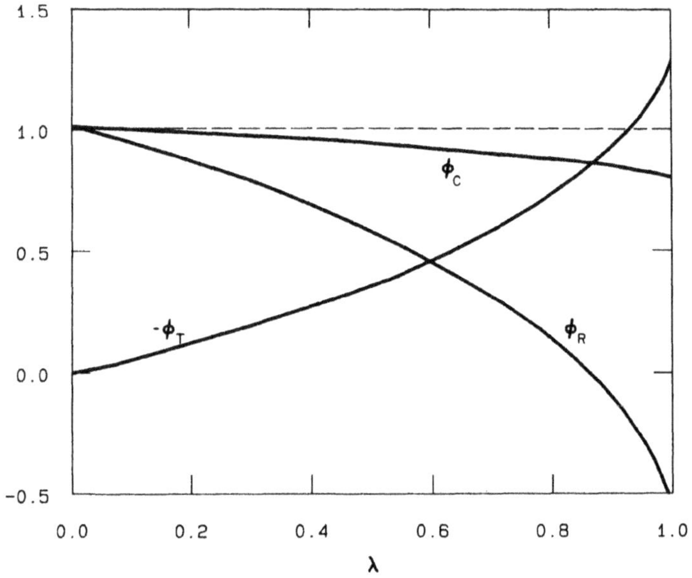

Figure 3. Relativistic correction factors for Coulomb (ϕ_C), transverse (ϕ_T) and total ($\phi_R = \phi_C + \phi_T$) relativistic exchange energies. Results are plotted against $\lambda \equiv \beta^2/(1+\beta^2)$ so that $\lambda = 0$ corresponds to the non-relativistic limit and $\lambda = 1$ corresponds to the ultrarelativistic limit.

formalism, the total energy and density of a relativistic electron system are to be determined by self-consistently solving the following single-particle equations

$$(-i\hbar c\vec{\alpha}\cdot\vec{\nabla} + \beta mc^2 + v_{eff}(\vec{r}))\ \psi_i(\vec{r}) = \varepsilon_i\psi_i(\vec{r}) \tag{15a}$$

$$n(\vec{r}) = \sum_i \psi_i^+(\vec{r})\ \psi_i(\vec{r})\theta(\mu-\varepsilon_i) \tag{15b}$$

$$v_{eff}(\vec{r}) = v_{ext}(\vec{r}) + e^2 \int\frac{d\vec{r}\,'n(\vec{r}\,')}{|\vec{r}-\vec{r}\,'|} + \frac{\delta E_{XC}[n]}{\delta n(\vec{r})} \tag{15c}$$

and then using the following expression for the energy

$$E = \sum_i \varepsilon_i(\mu-\varepsilon_i) - \frac{e^2}{2}\int d\vec{r}\int d\vec{r}\,'\ \frac{n(\vec{r})n(\vec{r}\,')}{|\vec{r}-\vec{r}\,'|} \tag{16}$$

$$+ E_{XC}[n] - \int d\vec{r}\ n(\vec{r})\ \frac{\delta E_{XC}[n]}{\delta n(\vec{r})}$$

In Eq. (15), $v_{ext}(\vec{r})$ is the external (normally nuclear) potential and the functional $E_{XC}[n]$ is defined as the difference between the total internal energy and the sum of kinetic energy and classical Hartree energy contributions. When the local density approximation is adopted

$$
\begin{aligned}
E_{XC}[n] &= \int d\vec{r}\; n(\vec{r})\; \varepsilon_{XC}(n(\vec{r})) \\
&\equiv \int d\vec{r}\; n(\vec{r})(\varepsilon_{XR}(n(\vec{r})) + \varepsilon_{CR}(n(\vec{r}))
\end{aligned}
\tag{17}
$$

where in the second form of the right hand side we have explicitly separated exchange contributions, defined to be those of leading order in the fine structure constant α_{fs}, and the higher order correlation contributions. In Eq. (17) $\varepsilon_{XR}(n)$ and $\varepsilon_{CR}(n)$ are the exchange and correlation energies per electron for the relativistic electron gas.

The relativistic changes in $\varepsilon_{XR}(n)$ were discussed in the previous section. The relativistic correlation energy, which has been numerically evaluated by Ramana and Rajagopal (1981a) (See also Ramana and Rajagopal (1982)), is not discussed here. We content ourselves with the observation that for densities high enough for relativistic effects to be important $\varepsilon_{CR}(n)$ is small compared to $\varepsilon_{XR}(n)$. In the local density approximation, the relativistic exchange potential

$$
\begin{aligned}
v_{XR}(n)(\vec{r}) &= \frac{\delta E_{XR}[n]}{\delta n(\vec{r})} = \frac{d}{dn}(n\varepsilon_{XR}(n)\Big|_{n=n(\vec{r})} \\
&\equiv \mu_X^{NR}(n(\vec{r}))\; \psi_R(n(\vec{r}))
\end{aligned}
\tag{17a}
$$

where

$$
\psi_R(n) = -\frac{1}{2} + \frac{3\ln\xi}{2\beta\eta} = 1 - \beta^2 + \frac{4\beta^4}{5} + \ldots \qquad \beta << 1 \tag{17b}
$$

and $\mu_x^{NR}(n) = 4/3\; \varepsilon_X^{NR}(n)$ is the $\alpha = 2/3$ exchange potential which appears in the non-relativistic density-functional formalism. $\psi_R(n)$ is a relativistic correction factor for the total exchange potential, analogous to the correction factor $\phi_R(n)$ for the exchange energy density in Eq. (14d). $\psi_R(n)$ and $\phi_R(n)$ have very similar behavior, and both approach the value -1/2 in the ultrarelativistic limit.

A second possible implementation of the relativistic density-functional formalism (MacDonald 1978, MacDonald and Vosko 1979) is to make a local-density-approximation only for the transverse contribution to the exchange-energy. In this case the single particle equations would be Dirac-Fock like rather than Dirac-Slater like as in Eq. (15a). For this implementation a transverse-only local exchange potential is required:

$$V_{XT}(n) \equiv \frac{d}{dn}(n\varepsilon_{XT}(n)) \equiv \mu_X^{NR}(n(\vec{r}))\ \psi_T(n) \tag{18a}$$

where

$$\psi_T(n) = -\frac{5}{6} + \frac{\beta^{-1}\ln\xi}{2\eta} - \frac{2\beta\ln\xi}{3\eta} + \frac{2\eta^2\beta^{-2}\ln\eta}{3} \tag{18b}$$

The best opportunity for testing the accuracy of local density approximations for relativistic exchange energies, seems to be by comparing with atomic calculations of transverse exchange contributions to total atomic energies (Mann and Johson 1971) and atomic ionization energies (Huang et al. 1976). Mann and Johson's atomic total energy calculations were performed using a quantum-electrodynamics based generalization of the Breit interaction which describes the full transverse interaction energy. In these atomic calculations E_{XT} must be treated perturbatively, and so the consistent local-density approximation for E_{XT} is

$$E_{XT} = \int d\vec{r}\ n^{(o)}(\vec{r})\ \varepsilon_{XT}(n^{(o)}(\vec{r}')) \tag{19}$$

where $n^{(o)}(\vec{r})$ is the atomic density calculated withour including the transverse exchange term in the potential. As was first pointed out by MacDonald and Vosko (1979), and subsequently verified by other workers (Das et al. 1980, Ramana et al. 1982), the local density approximation seriously overestimates the transverse exchange energy in atoms, by ~150% for He and decreasing to by ~50% for heavy atoms. However, as has recently been demonstrated by Perdew and Cole (1982), this error is largely removed when a self-interaction correction (e.g., see Perdew and Zunger 1981 and Perdew 1983) is applied to the local density approximation, i.e. when

$$E_{XR}[n] \simeq \int d\vec{r}\ n(\vec{r})\ \varepsilon_{XR}(n(\vec{r})) - \sum_\alpha \int d\vec{r}\ n_a(\vec{r})\ \varepsilon_{XR}(n_\alpha(\vec{r})) \tag{20a}$$

and

$$V_{XR}(n(\vec{r})) = \mu_X^{NR}(n(\vec{r}))\ \psi_R(n(\vec{r})) - \psi_X^{NR}(n_\alpha(\vec{r}))\psi_R(n_\alpha(\vec{r})) \tag{20b}$$

where $n_\alpha(\vec{r})$ is the contribution to the density from orbital "α". (The functional for the transverse exchange energy only would be corrected similarly. Note that the contributions to the single-particle potential become orbital dependent.) For the total transverse exchange energy, the self-interaction corrected local-density approximation, is in error by ~25% for He (Z = 2) and this error decreases to ~ several % for heavy atoms. The presence of substantial self-interaction corrections for E_X^T might have been expected, since most of this energy comes from interactions between the two 1s electrons.

For individual orbitals, comparisons of shifts in orbital

energy levels with shifts in ionization energies suggests that the local density approximation is fairly accurate for p, d, and f orbitals but not for s orbitals (MacDonald et al. 1982, Koelling and MacDonald 1982). For 1s orbitals, Perdew and Cole (1982) have shown that self-interaction corrections improve the local density approximation, although neither the error of the local density approximation nor the degree of improvement from self-interaction corrections, are as great as in the total energy case. For 2s orbitals, Breit interaction calculations find that E_X^T comes predominantly from interactions between 2s and 1s orbitals (Mann and Johnson 1971). Consistent with this, Perdew and Cole (1982) find that self-interaction corrections do not significantly modify the 2s removal energies in atoms. The failure of the uncorrected local density approximation is probably due not to self-interaction overcounting but rather due to the fact that the non-locality of the actual current-current interaction is important for interactions between orbitals localized in different regions of space.

In summary, it seems that relativistic exchange effects can generally be described fairly well by the local density approximation provided, at least the cases of total atomic energies and of 1s orbital removal energies, that self-interaction corrections. For atoms, the most accurate approach would probably be to use Dirac-Fock single-particle equations and a self-interaction corrected local density approximation for the transverse exchange energy. These local density approximations make it possible to accurately include quantum-electrodynamical corrections to the Coulomb interaction in electronic structure calculations without a significant increase in complexity. For that reason we suggest that, even though these corrections are often minor as we discussed before, they ought to be routinely included in calculations for heavy systems.

We now turn to a discussion of the trends of shifts in orbital eigenvalues caused by relativistic exchange corrections. A fairly complete investigation has been presented by Koelling and MacDonald (1982). For the inner core levels they find shifts due to relativistic exchange corrections which increase rapidly with atomic number and decrease rapidly with principle quantum number. The 1s orbital shifts increase approximately as $Z^{3.2}$, the 2s levels as $Z^{3.7}$, the $2p^{1/2}$ levels as $Z^{4.2}$, and $2p^{3/2}$ levels as $Z^{4.1}$. Of more importance for the chemical properties of molecules and solids are the smaller shifts in the valence orbital eigenvalues. Koelling and MacDonald (1982) report results for the shifts in valence band-center positions for solids from the 3d transition series to the actinides. The relativistic exchange correction is repulsive and so its direct effect is to increase orbital energy levels. However, because it is more repulsive at higher density, it tends to shift charge away from the nucleus and this can create an indirect attractive effect. For higher ℓ-states which have little weight near the nucleus, the indirect effect tends to increasingly dominate. For transition metals the most relevant energy-shift is the relative change in s and d band positions which is ~3.5mRy, 5mRy, and 10mRy in the 3d, 4d, and 5d series respectively. These shifts should be compared with typical d-bandwidths of ~400mRy. For the actinides and rare earths the relative s-f shifts are ~22mRy and ~16mRy respectively.

4. THE RELATIVISTIC SPIN-DENSITY FUNCTIONAL FORMALISM

The non-relativistic spin-density functional formalism (von Barth and Hedin 1972, Gunnarsson and Lundqvist 1976) is based on considering the Hamiltonian operator

$$\hat{H} = \hat{H}_{INT} + \int d\vec{r}\{\hat{n}(\vec{r})v_{ext}(\vec{r}) - \hat{\vec{m}}(\vec{r})\cdot\vec{B}_{ext}(\vec{r})\} \tag{21}$$

where $\hat{H}_{INT}$ is the internal Hamiltonian for a non-relativistic system of interacting electrons, $\hat{n}(\vec{r}) = \hat{\psi}^+(\vec{r})\hat{\psi}(\vec{r})$ and $\hat{\vec{m}}(\vec{r}) = \mu_B\hat{\psi}^+(\vec{r})\vec{\sigma}\hat{\psi}(\vec{r})$ are the relativistic number and spin-magnetization density operators, and v_{ext} and $\vec{B}_{ext}$ are external fields. (μ_B is the electron Bohr Magneton.) The Hamiltonian of Eq. (21) is not complete for atoms, molecules or solids in a magnetic field since it does not include the diamagnetic term. However this is of little consequence since the main purpose for going to a spin-density functional formalism is not to calculate response to, or electronic structure in the presence of, a magnetic field but rather to provide a basis for the description of naturally spin-polarized systems such as ferromagnets or non-closed-shell atoms. In such systems the density-functional theorems, which apply only for non-degenerate ground states, do not strictly hold unless an infinitesimal symmetry-breaking field $\vec{B}_{ext}$ is applied. As emphasized by Gunnarsson and Lundqvist (1976) this results in a much better description because the electron gas data for $E_{XC}[n,\vec{m}]$ in the local-spin-density approximation is taken from a spin-polarized electron gas which provides a much better approximation for the many-body effects in a spin-polarized system.

In the relativistic case the most natural generalization of the density-functional is to a formalism in which quantities are expressed as a functional of the Dirac four current density $(n(\vec{r}), \vec{J}(\vec{r}))$. Unfortunately, this approach leads to serious difficulties associated with diamagnetic effects and, in the case of solids, Landau levels. (For a more complete discussion see MacDonald 1983.) Moreover, this approach would not lead to the familiar spin-density functional formalism in the non-relativistic limit. As suggested by MacDonald and Vosko (1979), a more practical approach is to use a relativistic Hamiltonian of the same form as Eq. (21) but with $\hat{H}_{INT}$ now given as in Eq. (1), $\hat{\psi}$ becoming a Dirac field operator and $\vec{\sigma}$ becoming its four-component Dirac generalization,

$$\vec{\Sigma} \equiv \begin{pmatrix} \vec{\sigma} & 0 \\ 0 & -\vec{\sigma} \end{pmatrix} \quad (22)$$

When this is done the Kohn-Sham equations follow in the usual way with the ground state becoming a functional of the expectations of the density operator, $\hat{n}(\vec{r}) = \hat{\psi}^+(\vec{r})\hat{\psi}(\vec{r})$, and the magnetization density operator $\hat{\vec{m}}(\vec{r}) = \psi^+(\vec{r})\beta\vec{\Sigma}\psi(\vec{r})$. It is this approach which we refer to as the relativistic spin-density-functional formalism.

In order to make this scheme useful it is necessary to calculate the exchange energy of a relativistic spin-polarized electron gas. This question has been addressed previously by Ramana and Rajagopal (1979, 1981b, 1982) and we examine it in some detail below. The functional $E_{XR}[n,\vec{m}]$ is defined to be the exchange-energy for the state of the system which is the ground state $|\Phi_o\rangle$ in some external fields $v_{ext}(\vec{r})$ and $\vec{B}_{ext}(\vec{r})$, and which has

$$\langle\Phi_o|n(\vec{r})|\Phi_o\rangle = n(\vec{r}) \quad (22a)$$

and

$$\langle\Phi_o|\hat{\vec{m}}(\vec{r})|\Phi_o\rangle = \vec{m}(\vec{r}) \quad (22b)$$

The local density approximation is

$$E_{XR}[n,\vec{m}] \simeq \int d\vec{r}\; n(\vec{r})\; \varepsilon_{XR}(n(\vec{r}),\vec{m}(\vec{r})) \quad (23)$$

In Eq. (23), $\varepsilon_{XR}(n,\vec{m})$ is the exchange energy in the <u>ground</u> state of the electron gas in uniform external fields V_{ext} and $\vec{B}_{ext}$ which has (uniform) densities n and $\vec{m}$. Thus $\varepsilon_{XR}(n,\vec{m})$ should be evaluated by finding the ground state of the <u>non-interacting</u> electron gas for some specified V_{ext} and $\vec{B}_{ext}$ and using perturbation theory away from that state to determine ε_{XR}. At the same time n and $\vec{m}$ can be determined for given V_{ext} and $\vec{B}_{ext}$ and chemical potential by taking the expectation in the interacting ground state.

For the uniform non-interacting system the ground state has all positive energy single-particle eigenstates of the equation

$$[c\vec{\alpha}\cdot\vec{p} + \beta mc^2 + \mu_B\, B_{ext}\, \Sigma_z]\; \psi(\vec{p},s) = \varepsilon(\vec{p},s)\psi(\vec{p},s) \quad (24)$$

occupied up to the Fermi level. We assume a gap between positive-energy (electron) and negative-energy (positron) solutions and hence require $\mu_B\, B_{ext} < mc^2$. For a uniform system our results cannot depend on the direction of $\vec{B}_{ext}$ and we have arbitrarily chosen this to be the Z-direction. For $\vec{B}_{ext} = 0$ the solutions to Eq. (24) are the fimiliar Dirac plane-wave spinors (e.g., Bjorken and Drell

1964) which we denote by $u(\vec{p},s)$ for the electron solutions. Here the index s refers to the spin of the spinor in the frame of reference in which the momentum, $\vec{p}$, is zero. Ramana and Rajagopal (1981b) have evaluated $\varepsilon_{XC}(n,|\vec{m}|)$ and $\varepsilon_{XT}(n,|\vec{m}|)$ by starting from a non-interacting state which differs from the $B_{ext} = 0$ ground state only through changes in the occupation numbers for the $u(\vec{p},s)$ spinors. This approach has the advantage that the calculation can be greatly simplified using trace identities but has the disadvantage, as can be seen from the discussion above, that for $m \neq 0$ the results do not become exact even in the uniform system limit.

The positive energy eigenvalues of Eq. (24) are given by

$$\varepsilon(\vec{p},s) = mc^2\left(1+H^2+(\bar{\lambda}_c k)^2 + 2s\left[H^2(1+\bar{\lambda}_c(k^2-k_z^2)\right]^{1/2}\right)^{1/2} \tag{25}$$

where $\vec{k} = \hbar^{-1}\vec{p}$ and $H = \mu_B B_{ext}/mc^2$. Thus in the presence of an applied field, B_{ext}, the Fermi surface of the homogeneous system is no longer spherical. Nevertheless, given H and a Fermi energy, ε_F, it is still possible to obtain an expression for the electron density as a function of ε_F and H

$$\begin{aligned} n(s) &= \Omega^{-1} \sum_{\vec{p}} \theta(\varepsilon_F - \varepsilon(\vec{p},s)) \\ &= \frac{1}{6\pi^2}\,[\varepsilon_F^2 - (1+sH)^2]^{3/2} \\ &- \frac{sH\varepsilon_F^2}{4\pi^2}\left[\cos^{-1}\left(\frac{1+sH}{\varepsilon_F}\right) - \left(\frac{1+sH}{\varepsilon_F}\right)\left(1 - \left(\frac{1+sH}{\varepsilon_F}\right)^2\right)^{1/2}\right] \end{aligned} \tag{26}$$

where ε_F is in units of mc^2 and the density is in units of $\bar{\lambda}_C^{-3}$. We have not been able to obtain a similar expression for the magnetization density in the non-interacting ground state but a numerical evaluation for given ε_F and H is readily carried out. Since the total number operator commutes with the full Hamiltonian n(s) is unchanged by including interactions. On the other hand $|m|$ in the interacting ground state will differ from $|m|$ in the non-interacting ground state. (This is not true in the non-relativistic case.) However, since we want a relationship between ε_{XT}, ε_{XC} and n and m correct to leading order in $\alpha_{f.s.}$ we can ignore this correction.

Given the non-interacting ground state, expressions for ε_{XT} and ε_{XC} nay be derived which are of exactly the same form as in the non-polarized cases. The complications of non-spherical Fermi surfaces and the absence of the simplifying trace identities related to the matrix elements between $u(\vec{p},s)$ spinors has forced us to evaluate these expressions by a brute-force numerical approach. The details of this calculation are discussed elsewhere (MacDonald 1983b). We have fit our numerical results to a simple analytic form and obtained the following results.

$$\varepsilon_{XC}(n,\xi) = -\frac{e^2 k_F^4}{n(2\pi)^3} \cdot \left[(1+\xi)^{4/3} + (1-\xi)^{4/3} + 2(\phi_c(n)-1)(1+0.58\xi^2)\right] \tag{27a}$$

$$\varepsilon_{XT}(n,\xi) = -\frac{2e^2 k_F^4}{n(2\pi)^3} \cdot \phi_T(n) \cdot \left(1+\xi^2(\frac{5}{9} - 0.16\lambda)\right) \tag{27b}$$

Here $\xi \equiv \mu_B^{-1}|m|/n$. Note that, unlike Ramana and Rajagopal (1981b) we find that at a given density the relativistic correction factors (i.e. ϕ_T or ϕ_C^{-1}) are larger in magnitude at finite ξ than at $\xi = 0$. To emphasize this we have plotted the value of $\beta \equiv \bar{\lambda}_c(3\pi^2 n)^{1/3}$ at which $\varepsilon_{XC}(r_s\xi)+\varepsilon_{XT}(r_s,\xi)$ is zero as a function of ξ in Fig. 4. (Ramana and Rajagopal show similar curves which bend the other way.) This result might have been expected since at a given density, the polarized gas has more electrons moving at higher volocities.

5. CONCLUSIONS

Experience to date demonstrates that relativistic exchange energies can usually be well described by local density approximations. For the inner core levels of atoms however, self-interaction corrections are required. We would like to emphasize that a self-interaction corrected local density approximation for the transverse part of the exchange energy would fit very naturally into the framework of Dirac-Fock atomic calculations. Since these quantum electrodynamical corrections to the Coulomb interaction are very difficult to include in a direct way, we believe that this practice should be encouraged. For non-closed shell atoms or ferromagnets a spin-polarized local density approximation is required. Approximate formulas are given for the exchange energy of the uniform relativistic gas in Sec. 4 (eqs. 27). These formulas now make it possible to see if a better description of transverse exchange is possible by going to a spin density functional just as in the case of the non-relativistic exchange energy. We believe these calculations should be undertaken.

ACKNOWLEDGEMENTS

The author would like to acknowledge many stimulating conversations with Professor S.H. Vosko on the topics discussed above. Thanks are due to Professor J.P. Perdew and Professor A.K. Rajagopal for useful discussions and for providing pre-publication copies of some articles. The influence of collaborations with Dr. D.D.

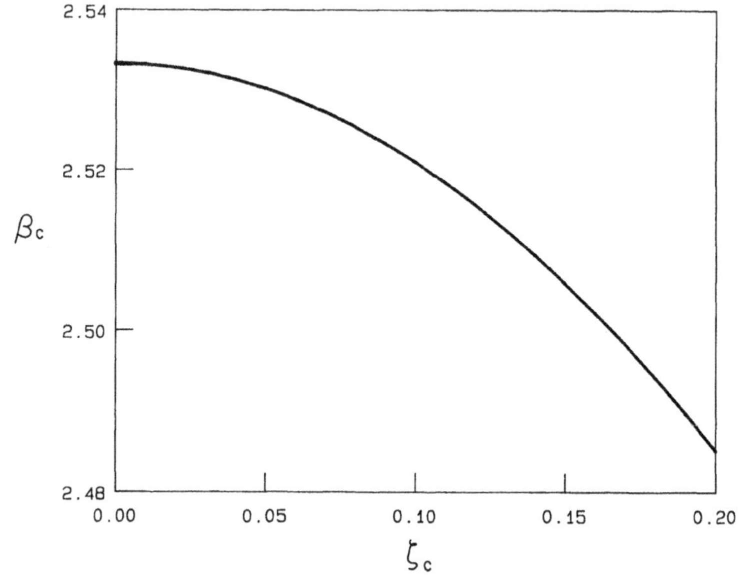

Figure 4. Locus of solutions to $\varepsilon_{XC}(n_c,\xi_c) + \varepsilon_{XT}(n_c,\xi_c) \equiv \varepsilon_{XR}(n_c,\xi) = 0$ ($\beta_c = \lambda_c(3\pi^2 n_c)^{1/3}$). At $\xi = 0$ the relativistic exchange corrections decrease $\varepsilon_{XR}(n_c,\xi)$ in magnitude relative to its non-relativistic value as density is increased. Eventually $\varepsilon_{XR}(n_c,\xi)$ reaches zero and changes sign.

Koelling on the ideas discussed in this review is also gratefully acknowledged.

REFERENCES

Akhiezer, I.A. and Peletminski, 1960, Sov.Phys. JETP 11, 1316.
Bjorken, J.D. and Drell, S.D., 1964, Relativistic Quantum Mechanics (McGraw-Hill, New York).
Breit, G., 1929, Phys.Rev. 34, 553.
Breit, G., 1930, Phys.Rev. 36, 383.
Breit, G., 1932, Phys.Rev. 39, 616.
Das, M.P., Ramana, M.V., and Rajagopal, A.K., 1980, Phys.Rev. A 22, 9.
Ellis, D.E., 1977, J.Phys.B 10, 1.
Gaunt, J.A., 1929, Proc.Roy.Soc. (London), A122, 513.
Grant, I.P., 1965, Proc.Roy.Soc. (London), 86, 523.
Gross, E., 1983 (this volume).
Gunnarsson and Lundqvist, 1976, Phys.Rev. B 13, 4274.
Huang, K.N., Aoyagi, M., Chen, M.H., Crasemann, B., and Mark, H., 1976, At. Data and Nuc. Data 18, 243 (1976).

Jancovici, 1962, Nuovo Cimento 25, 429.
Kim, Y., 1967, Phys.Rev. 154, 17.
Koelling, D.D., and MacDonald, A.H., 1982, in "Relativistic Effects in Atoms, Molecules, and Solids", Malli ed., Plenum, New York.
Liberman, D.A., Waber, J.T., and Cromer, D.T., 1965, Phys.Rev. 137, A27.
MacDonald, A.H., 1978, Ph.D. Thesis, U. of Toronto.
MacDonald, A.H. and Vosko, S.H., 1979, J.Phys. C 12, 2977.
MacDonald, A.H., Daams, J.M., Vosko, S.H., and Koelling, D.D., 1982, Phys.Rev. B 25, 713.
MacDonald, A.H., 1983a, in "Relativistic Quantum Chemistry", S.N. Datta ed. (to be published).
MacDonald, A.H., 1983b, to be submitted for publication.
Mann, J.B. and Johson, W.R., 1971, Phys.Rev. A 4, 41.
Perdew, J.P. and Cole, Lee A., 1982, J.Phys. C (to appear).
Perdew, J.P. and Zunger, Alex, 1981, Phys.Rev. B 23, 5048.
Perdew, J.P. 1983 (this volume).
Rajagopal, A.K., 1978, J.Phys. C 11, L943.
Ramana, M.V. and Rajagopal, A.K., 1981, Phys.Rev. A 24, 1689.
Ramana, M.V. and Rajagopal, A.K., 1982, Adv. in Chem.Phys. (to appear).
Ramana, M.V. and Rajagopal, A.K., 1979, J.Phys. C 12, L845 (Erratum 14, L111).
Ramana, M.V. and Rajagopal, A.K., 1981b, J.Phys. C 14, 4291.
Ramana, M.V., Rajagopal, A.K., and Johson, W.R., 1982, Phys.Rev. A 25, 96.
Smith, F.C. and Johnson, W.R., 1967, Phys.Rev. 160, 136.
Slater, J.C., 1951, Phys.Rev. 81, 385.
von Barth, U. and Hedin, L., 1972, J.Phys. C 5, 1629.
Zapolsky, H.S. and Cornell, U., LNS report, unpublished, September 1960; quoted by E.E. Salpeter, Ap.J. 134, 669, 1961.

RELATIVISTIC MULTIPLE SCATTERING Xα CALCULATIONS

H. Chermette

Institut de Physique Nucléaire (et I N2 P3)
Université Claude Bernard Lyon-1
43 Bd du 11 Novembre 1918
69622 Villeurbanne Cedex, France

A. Goursot

ERA N° 386 - Ecole Nationale Supérieure
de Chimie de Mulhouse
68093 Mulhouse Cedex, France

A one component relativistic theory has recently been developed and tested on isolated atoms[1] and on molecules[2] through the molecular scattered-wave formalism of Johnson,[3] while its application to energy-band calculations (through a relativistic augmented-plane-wave program) also has been considered.

In this theory, the Dirac-Slater problem is solved through the same approximations used by Darewich et al. or Cowan and Griffin[4] for solving the Dirac-Hartree-Fock equations. The approximations lie in dropping the spin-orbit term from the exact equation and in neglecting the contribution of the small component orbitals to the self-consistent central field. This neglect is fully justified since only about 1% of the total electronic charge is described by the small components. The spin-orbit effects are evaluated once self-consistency has been reached by diagonalizing the matrix of the spin-orbit operator among the various states associated with a given eigenvalue. This procedure is very satisfactory for core levels, but cannot be so accurate for (valence) closely spaced eigenvalues. In that case, the interaction of the molecular orbitals (M.O.) associated to these eigenvalues can be taken into account by diagonalizing the matrix of the spin-orbit operator among all these M.O.'s.

The differential equation for the large component $G_{n\ell}$ radial function is:

$$[H_{NR} + H_m + BH_D]\, G_{n\ell} = E_{n\ell}\, G_{n\ell} \tag{1}$$

where H_{NR} is the standard non relativistic Hamiltonian (including Slater $X\alpha\ \rho^{1/3}$ exchange), H_m is the mass-velocity operator and H_D is the standard Darwin operator; the operator B is given by

$$B = \left\{1 + \frac{\alpha^2}{4}\,[E_{n\ell} - V(r)]\right\}^{-1} \tag{2}$$

where $\alpha \simeq 1/137$ is the fine structure constant and V(r) is the one-electron central field potential function. The standard Darwin operator is recovered by setting B equal to unity, according to Pauli's theory.

Similarly the spin-orbit operator to be used in perturbation theory when applied to approximate levels which are j degenerate is given by:

$$\zeta(r) = 2B\,\frac{\alpha^2}{4}\,\frac{1}{r}\,\frac{dV}{dr} \tag{3}$$

The usual (Pauli) form of the spin-orbit operator is again obtained by setting B equal to unity.

The one-electron structure of a molecule, when calculated with relativistic corrections, may be quite different from the non relativistically calculated one: both core and valence energy levels exhibit large shifts which may be considered as being composed of two parts. The first part, the so-called "direct" relativistic effect, is obtained by calculating the expectation value of the relativistic portion of the one-electron Hamiltonian (i.e. the sum of the mass-velocity and Darwin operators) over the associated self-consistent relativistic orbitals. The remaining portion of the energy shift is evidently to be attributed to the changes in the SCF potential, and is the so-called "indirect" effect. While the direct effect always produces a stabilization of the orbitals (decrease of the eigenvalues) the indirect effect leads to a destabilization of these orbitals which can be easily understood as resulting from a larger shielding of the nucleus by the relativistically contracted inner orbitals. However the relative importance of these two effects may be quite different for each orbital, leading to significant changes in the eigenvalues diagram. In general, one observes strong direct (downward) shifts for s and p levels, strong indirect (upward) shifts for f levels, while for d orbitals, both effects are more or less balanced. These effects have been studied in detail by Boring and Wood[2,5,6] for the uranium atom, the UF_6 molecule and the UO_2^{x+} ions. They performed test calculations in which only the core uranium orbitals were treated relativistically, while the remaining ones were handled non relativistically. Then, they introduced, one by one, these last orbitals in the relativistic set: when an orbital became handled relativistically, a downward shift of its eigenvalue occured (direct effect), but it was more or less attenuated by indirect effects when the other orbitals were introduced in the relativistic set.

These trends have been observed in all the molecular relativistic calculations performed up to now. For instance, in a calculation of the XPS spectra of WO_3 and A_xWO_3 bronzes, Chermette et al.[7] obtained the wrong order for the 4f and 5p energy levels without relativistic corrections, while a relativistic calculation leads to the correct order; moreover the calculated spin-orbit splittings lead to predicted peak positions which are in good agreement with the experimental spectra.

The necessity of introducing the relativistic corrections self-consistently in the Schrödinger equation has also to be emphasized. The failure to this process explains the discrepancy found between the $KTaO_3$ XPS spectrum and non relativistic MSXα calculations on the TaO_6^{7-} cluster, in which (only) direct relativistic effects were added after self-consistency and estimated from atomic relativistic shifts.[8]

In Table 1 are reported the eigenvalues of the $IrCl_6^{3-}$ cluster in both relativistic and non-relativistic calculations.[9] One can note the already outlined trend of contraction of the s and p orbitals and the importance of the indirect effects (destabilization) on the d and f levels. Although the 1s energy shift is very large (400 eV), it can be pointed out that the relative energy stabilization of the s levels increases from the 1s (10%) to the 5s (30%). Similarly the relative stabilization of p levels increases from 5% to 10%.

A similar trend has been observed for all other 5d or 6d compounds, sometimes with a larger indirect effect on f levels (e.g. 20% in WO_6^{6-}),[10] so that the necessity of a self-consistent relativistic calculation is still again emphasized. Relativistic calculations on 4d compounds exhibit similar features, although some slight differences are to be pointed out: as expected, the stabilization of the s and p levels of the 4d elements are relatively smaller than in heavier atoms (e.g. 2% for Mo 1s), but the relative destabilization of the 4d levels is similar, or even slightly larger, than for 5d or 6d elements.[11,12]

The relaxation of molecular orbitals in the ionization processes is sometimes defined as the difference between the ground state Xα

Table 1. Xα eigenvalues (core levels) in $IrCl_6^{3-}$ cluster (Ry)

Non relativistic calculation		Relativistic calculation	Increase in binding energy (%)
Ir 1s	-5088.9	-5549.3	9
2s	-844.5	-971.7	15
3s	-195.6	-227.2	16
4s	-40.27	-48.27	20
5s	-5.48	-7.02	28
2p	-811.3	-850.9	5
3p	-180.0	-191.4	6
4p	-33.61	-36.39	8
5p	-3.39	-3.75	11
3d	-150.8	-150.2	-0.4
4d	-21.39	-21.29	-0.5
4f	-5.13	-4.49	-12

eigenvalues and the Slater's transition state values. In fact, the transition state procedure takes into account the removal of the self-interaction term (which is necessary to compare with Hartree-Fock eigenvalues) as well as the true relaxation of orbitals resulting from changes in the nuclear shieldings through the ionization processes. Nevertheless if one admits the first definition, the MO's relaxation in relativistic calculations may be compared with their relaxation in non relativistic calculations: the d orbitals, which get similar binding energies in both kind of calculations, exhibit quite identical relaxation energies; on the contrary, when the orbitals are stabilized by relativistic effects (s and p orbitals), their relaxation increases simultaneously.[13]

Spin-orbit (S.O.) parameter values are obtained from Eq. 3 and are used to evaluate the S.O. interactions assuming the MO's are restricted to the main atomic orbitals of the heavy elements. Then, for close valence eigenvalues, it is necessary to diagonalize the matrix of the S.O. operator with respect to the associated M.O.'s. However, it has been shown[2,9,10] that such S.O. parameter values could be quite different for the corresponding M.O.'s. For example, in the case of the $IrCl_6^{3-}$ cluster, the ζ_{Ir5d} parameter decreases from the value of 520 Ry for the H.O.M.O., i.e. the valence band (V.B.) edge to 260 Ry at the V.B. bottom.[9] Such a variation underlines the limitations of the ligand field model in which a single metal S.O. parameter is used for all the metal d MO's. On the other hand, it is worth noting that these S.O. parameter values can be quite different from the atom (or free ion) values.

The use of the multiple scattering Xα wave functions for the calculation of one electron properties has become popular during these last years since Case and Karplus' work on copper porphyrin.[12] Let us recall that the MSXα radial wave functions are determined by numerical integration of the radial Schrödinger equation, being reoptimized at each step of the self-consistent process. This procedure has the advantage of avoiding the crucial problem of selecting an adequate LCAO basis set and gives full flexibility to the atomic orbitals of a given ℓ value. The incidence of the relativistic effects on the shape of the radial wave functions is illustrated on Fig. 1 where some iridium s, p and d core orbitals of $IrCl_6^{2-}$ are plotted in both relativistic and non relativistic cases. If spin polarized

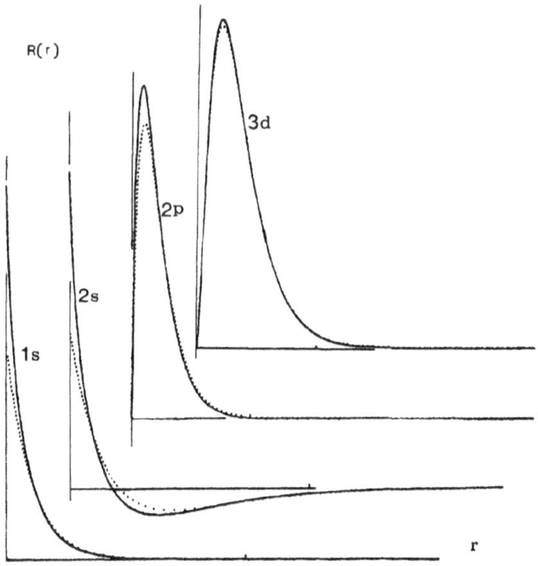

Fig. 1. Iridium radial wave function in $IrCl_6^{2-}$ cluster (from ref. 14).
1 r unit = 0.25 a.u. for 1s, 2s orbitals.
= 0.40 a.u. for 2p, 3d orbitals.
dots = non relativistic wave functions.

Table 2. Comparison between relativistic and non relativistic results with experiment ($IrCl_6^{2-}$ cluster, from ref. 14)

	Non Relativistic	Relativistic	Experiment
Fermi Term $A_F(Ir)$	-10.7	-31.9	-
P(Ir)	29.2	30.8	-
$\|A_Z\|_{Ir}$	32.3	27.1	26.5 ± 1.0
g	-1.762	-1.790	-1.786

calculations are considered, the spin up and spin down radial components of the wave functions exhibit slight differences, due to the effect of the unpaired spin upon the overall potential of the same orientation. The polarization of the MO's having s-type components is responsible for the existence of the isotropic (Fermi contact) hyperfine interaction:

$$A_F = \frac{8\pi}{3} g_e \beta_e g_N \beta_N \left(\rho\uparrow(o) - \rho\downarrow(o)\right)$$

$\rho\uparrow(o)$ and $\rho\downarrow(o)$ being the spin up and spin down electronic densities at the nucleus. The influence of the relativistic correction on the electronic spin densities is emphasized by results taken from $IrCl_6^{2-}$ calculations[14] and reported on Table 2. The relativistic value of the Fermi contact term is almost three times larger than the non relativistic one, and leads to a much better agreement with experiment for the calculated magnetic properties (see below). This increase of the isotropic contribution has already been underlined in previous atomic calculations[15] where relativistic calculations have also led to about three times larger contact term. In this paper we shall not discuss in detail the calculation of magnetic properties since it is done in D. Guenzberger's contribution to this book.[16] However, we will just focus on the incidence of the relativistics effects in the computation of magnetic properties like the isotropic (Fermi term), anisotropic (P factor) contributions to the hyperfine interactions tensor of a cluster. As just said, the results recently obtained for $IrCl_6^{2-}$ cluster[13] and summarized in Table 2 do not deserve large comments since it is obvious from these results that the much better agreement with experiment is obtained from the relativistic calculation.

CONCLUSION

In this article, the necessity to include self-consistent relativistic corrections in molecular calculations has been pointed out for all compounds involving heavy atoms. Most of the changes in the electronic properties are due to the mass-velocity and the so-called Darwin terms so that the use of Wood and Boring's Hamiltonian is very convenient for this purpose as it can be easily included in MSXα programs. Although the spin orbit operator effects are only obtained by perturbation theory, the results compare fairly well with experiment and with other relativistic calculations, namely Hartree-Fock-Slater calculations,[17] Dirac-Slater calculations,[18] or Koelling and Harmon calculations.[19] The comparison of these methods on typical examples is developed in Boring and Wood's contribution to this book.[20] Finally, although most of the relativistic calculations which have been published up to now have been performed through the Xα approximation, the same relativistic scheme can be used with other local spin density approximations.[21]

REFERENCES

1. J.H. Wood and A.M. Boring, Phys. Rev. B 18, 2701 (1978).
2. A.M. Boring and J.H. Wood, J. Chem. Phys. 71, 32 (1979).
3. K.H. Johnson, Adv. Quant. Chem. 7, 143 (1973).
4. J.W. Darewych, A.E.S. Green, and D.L. Sellin, Phys. Rev. A 3, 502 (1971); R.D. Cowan and D.C. Griffin, J. Opt. Soc. Am. 66, 1010 (1976).
5. A.M. Boring and J.H. Wood, J. Chem. Phys. 71, 392 (1979).
6. J.H. Wood, A.M. Boring, and S.B. Woodruff, J. Chem. Phys. 74, 5225 (1981).
7. H. Chermette, G. Hollinger, and P. Pertosa, Chem. Phys. Letters 86, 170 (1982).
8. H. Chermette, P. Pertosa, F.M. Michel-Calendini, and J. Weber, Chem. Phys. Lett. 65, 385 (1979).
9. A. Goursot and H. Chermette, Chem. Phys. (to appear, 1982).
10. H. Chermette, P. Pertosa, A. Goursot, and E. Penigault, Int. J. Quant. Chem. Symp. (to appear, 1982).
11. A. Lebeuze, M.A. Makhyoun, R. Lissilour, and H. Chermette, J. Chem. Phys. (to appear, 1982).
12. H. Chermette and C. Pedrini, unpublished results on $NaCl:Ag^+$ optical properties.
13. D.A. Case and M. Karplus, J. Am. Chem. Soc. 99, 6182 (1977).
14. A. Goursot, C. Daul, and H. Chermette, (to be published).
15. P. Pyykkö, Phys. Letters 35 A, 53 (1971) and 38 A, 218 (1972).

16. D. Guenzberger (in this book).
17. See J.G. Snijders and E.J. Baerends's contributions to this book.
18. D.A. Case and C.Y. Yang, J. Chem. Phys. 72, 3443 (1980).
19. D.D. Koelling and B.N. Harmon, J. Phys. C 10, 3107 (1977).
20. A.M. Boring and J. Wood (in this book).
21. See, for example, O. Gunnarsson, B.I. Lundqvist or J. von Boehm's contributions to this book.

DIRAC SCATTERED WAVE CALCULATIONS

Cary Y. Yang

Surface Analytic Research, Inc.
465A Fairchild Drive, Suite 128
Mountain View, CA 94043, U.S.A.

David A. Case

Department of Chemistry, University of California
Davis, CA 95616, U.S.A.

The self-consistent-field-Xα-Dirac-scattered wave (SCF-Xα-DSW) method is briefly reviewed. The method treats relativistic effects fully within the one-electron local exchange framework. Examples of DSW results illustrate the versatility of the method, and its potential impact on studies of transition metal complexes and surfaces.

1. INTRODUCTION

During the past decade, the Xα method has become increasingly popular as an approximate approach to determining molecular and cluster electronic structures, especially in the areas of inorganic chemistry and solid-state and surface physics. It is descendent of the Thomas-Fermi-Dirac theory, which postulates an exchange potential proportional to the cube root of the electron density.[1] In 1951, Slater[2] proposed using such a term as effective potential, yielding a standard non-relativistic one-electron differential equation (in atomic units)

$$[-\frac{1}{2}\nabla^2 + V_c + V_{X\alpha\uparrow}]\ \phi_{i\uparrow} = \varepsilon_{i\uparrow}\ \phi_{i\uparrow} \qquad (1)$$

Here $\phi_{i\uparrow}$ is an orbital for an electron with $M_S = 1/2(\uparrow)$, and $\varepsilon_{i\uparrow}$ is its one-electron energy, V_C is the classical Coulomb potential (including electron self-interaction terms), and $V_{X\alpha\uparrow}$ is an approximate potential representing the effect of electron exchange. In Slater's model, this is related to $\rho\uparrow$, the local density of electrons of the same spin

$$V_{X\alpha\uparrow} = -3\alpha(3\rho\uparrow/4\pi)^{1/3} \tag{2}$$

Here α is an adjustable parameter, which is usually determined by comparing Hartree-Fock and $X\alpha$ atomic calculations. Eqs. (1) and (2) define what is usually termed the $X\alpha$ method.

One popular method of solving Eq. (1) approximately is to use multiple scattering, or scattered wave (SW) theory. The first derivation of energy band structure using this technique was given by Korringa.[3] Subsequently Kohn and Rostoker[4] employed a variational approach with the Green's function technique and arrived at identical results. This so-called KKR method for calculating electronic energy band structure of crystalline solids is similar to the augmented-plane-wave (APW) scheme reported earlier by Slater.[5] In fact, their equivalence has been extensively studied by Ziman,[6] Lloyd,[7] Slater,[8] and Johnson.[9]

The generalization of the SW technique for finite polyatomic systems with no translational invariance was suggested by Slater[10] and subsequently developed by Johnson.[11] Numerous applications to molecules and clusters followed, which can be found in review articles by Johnson,[12-14] Connolly,[15] Weinberger and Schwartz,[16] Rösch,[17] Messmer,[18] and most recently Case.[19] The self-consistent-field-$X\alpha$-scattered-wave (SCF-$X\alpha$-SW) method, as it is popularly called, probably will not replace traditional quantum chemical methods for computing molecular orbitals. However, because of its computational tractability, both in basis representation and in numerical procedures, the SCF-$X\alpha$-SW method is more suitable in establishing bonding trends for elements across the rows as well as down the columns of the periodic table. Furthermore, molecular properties calculations requiring knowledge of continuum or scattering states can be accomplished with reasonable accuracy and with little extra effort.[20,22]

To implement the scattered wave method, the potential is assumed to be spherically symmetric inside a cell surrounding each atom, and outside an "outer sphere", that surrounds the entire molecule. In the intersphere region, a constant potential is assumed. With these assumptions, the wavefunction may be determined essentially exactly, subject only to errors in the radial numerical integrations inside the atomic spheres, and to truncation of the angular momentum expansions. In practice, it is usually feasible to include enough mesh points and spherical harmonic expansion functions to obtain convergence, even for sizeable molecules.

In view of the promise the $X\alpha$ method holds for practical calculations on molecules and clusters containing heavy atoms, it is not surprising that several methods have been developed to include relativistic effects. Two of these use the scattered wave formalism. The first is a quasirelativistic theory in which the radial functions inside each atomic sphere satisfy an average Dirac equation which includes the mass-velocity and Darwin corrections, but not the spin-orbit term.[22-24] Such calculations are no more difficult or expensive than non-relativistic multiple scattering calculations, yet include important direct and indirect level shifts.[24-28] A second step may follow in which the spin-orbit operator is diagonalized in the space of the valence orbitals.[25] In this way complex wavefunctions that transform according to the molecular double point group are obtained.

A more direct as well as more accurate approach is to use the full four-component Dirac formalism and double group symmetry throughout. For solids, the Dirac extension of the KKR method was reported by Onodera and Okazaki[29] and by Takada.[30] Yang and Rabii,[31] Cartling and Whitmore,[32] and Rosicky et al.[33] presented formulations of the SW secular equations for molecules. Yang[34] also developed the complete double point group symmetrization scheme. The SCF-Xα-SW computer codes were adapted to yield a parallel Dirac-scattered wave (DSW) method, which was then applied to a few systems containing heavy elements.[35-40] More recently, the self-consistent version of the DSW scheme has been coded,[41,42] and further calculations[42-46] have been performed using the resulting SCF-Xα-DSW method. Some of these results will be highlighted in this paper.

2. DIRAC SCATTERED-WAVE FORMALISM

In this section, the derivation of the DSW secular equations will be briefly reviewed. Details of the formulation can be found in Yang and Rabii[31] and Yang.[34]

(a) The Secular Problem

We start by replacing Eq. (1) with the one-electron Dirac equation[47]

$$(c\vec{\alpha}\cdot\vec{p} + \beta mc^2 + VI_4)\ \psi = W\psi \tag{3}$$

where $V = V_c + V_{X\alpha}$, and the average of the spin-up and spin-down charge densities is used in calculating the exchange potential. This can be transformed into an integral equation for any given volume v in space

$$\int_v \widetilde{G}(\vec{r},\vec{r}')V(\vec{r}')\psi(\vec{r}')d\vec{r}' - ic\int\widetilde{G}(\vec{r},\vec{r}')\vec{\alpha}\cdot\hat{n}'\psi(\vec{r}')dS' = \begin{cases} \psi(\vec{r}) & \vec{r}\in v \\ 0 & \vec{r}\notin v \end{cases} \tag{4}$$

where $\hat{n}$ is the unit outward normal on the surface enclosing v, and $\tilde{G}(\vec{r},\vec{r}')$ is the Dirac free electron Green's function.[47] Within the general SW framework, the solution to Eq. (4) is reduced to that of a boundary-value problem. The derivation can proceed once v is specified and boundary conditions are established. If we choose our zero of energy to equal the constant potential in the intersphere region, Eq. (4) becomes

$$\psi(\vec{r}) = ic\int\tilde{G}(\vec{r},\vec{r}')\vec{\alpha}\cdot\hat{n}'\psi(\vec{r}')dS' \quad (\vec{r}\varepsilon v) \tag{5}$$

where the intersphere region is the chosen volume. The secular equations will emerge from the matching of trial wave functions across its boundaries.

One assumes a trial wave function in the spherical regions (atomic and outer) with the form

$$\psi^{\alpha} = \sum_{Q} A_Q^{\alpha} \begin{bmatrix} g_{\kappa}^{\alpha}\ \chi_Q^{\alpha} \\ if_{\kappa}^{\alpha}\ \chi_{\bar{Q}}^{\alpha} \end{bmatrix} \tag{6}$$

where α runs over the outer and atomic spheres, and $Q\equiv(\kappa,\mu)$ and $\bar{Q}\equiv(\kappa,\mu)$. The functions g_{κ}^{α} and f_{κ}^{α} are numerical solutions of the Dirac radial equation. They can be determined by integrating outward from the nuclei (in the atomic spheres) and inward from infinity (in the outer sphere). The starting conditions for the numerical integration satisfy the known asymptotic forms for solutions of the Dirac equation in the vicinity of a nucleus and at infinity. This procedure is analogous to that used in numerical atomic Dirac-Slater calculations,[48] and serve to prevent contamination from states of negative energy, which satisfy the differential equation but are not of physical interest for chemical problems. By directly imposing the proper boundary conditions, we avoid the stability problems that can arise when basis-set expansion methods are applied to the Dirac equation.[49-51] Furthermore, scattering (continuum) boundary conditions in the outer sphere are almost as easy to impose as bound state boundary conditions. These two advantages make the SW formalism attractive for a variety of investigations of the Dirac equation. Although all of the calculations reported here are for bound states, we are in the process of preparing computer codes to handle continuum states.

For the matching problem, there are two types of boundaries for the intersphere region. The first, S_1, is the outersphere surface, while S_2 represents the conglomeration of all atomic sphere surfaces. The specific boundary conditions corresponding to Eq. (5) are given by

$$\psi(\vec{r})|_{S_1} = \psi^o(\vec{r}_o)|_{S_1} - \; ; \quad \psi(\vec{r}')|_{S_1} = \psi^o(\vec{r}_o{}')|_{S_1} + \qquad (7)$$

$$\psi(\vec{r})|_{S_2} = \psi^\alpha(\vec{r}_\alpha)|_{S_2} + \; ; \quad \psi(\vec{r}')|_{S_2} = \psi^\alpha(\vec{r}_\alpha{}')|_{S_1} -$$

The final secular equations have been given in detail elsewhere.[31] Only the important features are highlighted here. The secular matrix M consists of a diagonal term, $[T_\kappa^\alpha]^{-1}$, or the inverse t-matrix, which, in terms of ordinary scattering theory, is the cotangent of the phase shift of the κth partial wave scattering by the αth sphere. Hence T_κ^α is potential-dependent. The off-diagonal terms, $G_{QQ'}^{\alpha\alpha'}$, or structure factors, on the other hand, are purely geometrical entities and are only functions of atomic positions and sphere radii. The structure factors, as in all scattered-wave theories, represent the configuration of the boundary-value problem, while the t-matrices describe the nature of the force fields acting within the given configuration. The complete separation of the two quantities facilitates their computation and in turn the solution of the secular equations.

The t-matrices and structure factors reduce exactly to their usual Schrödinger SW counterparts[31,11] in the limit $c \to \infty$. The solution of the secular equations $\underset{\sim}{M}\,\underset{\sim}{C} = \underset{\sim}{0}$ for bound states gives rise to a set of one-electron orbitals, each of which has single-center partial-wave expansion in angular momentum in the spherical regions (as in Eq. (6)), and a multi-center expansion in the intersphere region. The orbital wave functions can be normalized numerically in the spherical region and analytically in the intersphere region.[34] The proper occupation of these orbitals and subsequent generation of a new potential energy represent the onset of the SCF cycle, for which a convergence criterion for the potential is customarily set.

3. SELECTED APPLICATIONS

Iodine

The initial applications of the DSW method were not self-consistent ones, i.e., the potentials were generated from nonrelativistic scattered wave calculations and then used as input for "one-shot" application of the Dirac equation.[35-40] In spite of this drawback, computed ionization potentials for I_2 were in good qualitative agreement with experimental results.[36] More recently, we have prepared self-consistent versions of the DSW computer codes,[41-42] and it is of interest to see how well this method works in simple test cases. Here we present for the first time self-consistent DSW results for the iodine molecule.

Table 1 compares calculated and experimental ionization potentials in the valence region. The ionization potentials were calculated using the Slater transition state method, and the α factors, molecular geometry and sphere radii are the same as in the earlier, non-self-consistent calculations.[36] The overall agreement with experiment is quite good: the π ionization potentials are all within 0.3 eV of experiment, while the σ_g ionization potential is underestimated by 1.3 eV. The spin orbit splittings of the π orbitals are in excellent agreement with experiment, showing that the π_g splitting is smaller than that of π_u. These two splittings would be the same in a minimum basis set theory in which only one-center matrix elements are considered. The DSW results in Table 1 illustrate the ability of this method to go beyond the simplest descriptions, predicting finer details about relativistic effects.

We can compare the results in Table 1 to earlier ones[36] that used a non-relativistic transition state potential. The differences are minor: the levels deriving from the non-relativistic π_g and π_u orbitals are 0.9 eV higher in the new self-consistent calculation than in the earlier, "one-shot" results, while the level arising from the σ_g orbital increases by 1.1 eV. These changes have the effect of going from calculated ionization potentials slightly higher than experiment in the old results, to the values slightly lower than experiment shown in Table 1. The largest discrepancy is clearly with the σ_g orbital, whose calculated ionization potential is in error by 1.3 eV, causing it to be too close to the π levels.

It has been known for some time, from non-relativistic multiple scattering calculations,[15,19] that errors of this sort in σ-π separations can be removed to a large extent by allowing the atomic spheres to overlap. This is also the case for relativistic calculations, as shown in Table 2, where we compare ground state results for I_2 using touching spheres and with spheres whose radius is 15% larger. As the final column demonstrates, this inclusion of overlap raises the valence π levels by about 0.2 eV, while <u>lowering</u> the $2\sigma_g$ level by about 0.5 eV. This change increases the σ-π separation in iodine by about 0.7 eV, bringing it into much closer agreement with experiment. We expect in most cases that the effects of sphere overlap in relativistic scattered wave calculations will be similar to those found in non-relativistic calculations, and that the results including overlap will give improved representations of the molecular electronic structures.

<u>Xenon Fluoride</u>

One of our main goals in developing a four-component theory has been to gain new insight into relativistic contributions to spin-dependent properties. For example, such effects can be important in the hyperfine splittings observed in electron spin resonance (esr) spectra. When spin-orbit mixings are important, it is no longer

Table 1. I_2 Ionization Potentials[a]

Orbital	Non-rel.[b]	Rel.[c]	Δ[d]	Exp.[e]	Δ[d]
$\pi_g(3/2)$	9.1	8.92	0.66	9.02	0.65
$\pi_g(1/2)$	9.1	9.58		9.67	
$\pi_u(3/2)$	10.6	10.33	0.80	10.61	0.80
$\pi_u(1/2)$	10.6	11.13		11.41	
$\sigma_g(1/2)$	11.2	11.62		12.95	

[a]Values in eV.
[b]Touching-spheres results, from Ref. 36, using the Slater transition state.
[c]This work, with touching spheres, using the Slater transition state.
[d]Spin-orbit splittings of the π-orbitals.
[e]See Ref. 36.

possible to separate the observed hyperfine tensors into an isotropic contact term and a traceless dipolar term, since the additional "orbital" contributions are neither isotropic nor traceless. In the case of XeF, early interpretations of the esr spectra did not take this complication into account,[52] but later analyses have shown that the orbital contributions may be large.[53-55] Here we present results of Dirac scattered wave calculations on the hyperfine interactions in radicals with significant spin-orbit coupling.

The traditional theory of such hyperfine interactions begins with a non-relativistic (approximate) wavefunction and adds spin-orbit coupling and the magnetic fields arising from the nuclei as perturbations. One must go to second order in perturbation theory to describe the interaction of orbital motion (induced by the spin-orbit operator) with the magnetic field of the nuclei. This requires a sum over all the excited states of the system, including (at least in principle) an integration over the continuum states. Even in cases where it is reasonable to truncate this sum to a few terms, the inevitable errors in calculation of the matrix elements are compounded by uncertainties in the estimation of excitation energies. In view of this, it is not too surprising that only attempts tc calculate orbital effects in XeF have been empirical ones, making many simplifying assumptions and incorporating other experimental information (the observed g-tensors) into the theory.

Using a Dirac calculation as our zeroth-order wavefunction,

Table 2. Effect of Sphere Overlap in I_2 Orbital Energies[a]

Orbital	Touching spheres	15% overlap	Δ[b]
$2\pi_g(3/2)$	-5.80	-5.59	0.21
$2\pi_g(1/2)$	-6.45	-6.25	0.20
$2\pi_u(3/2)$	-7.24	-7.08	0.16
$2\pi_u(1/2)$	-8.02	-7.80	0.22
$2\sigma_g(1/2)$	-8.46	-8.92	-0.46
$1\sigma_u(1/2)$	-17.00	-16.38	0.62
$1\sigma_g(1/2)$	-18.68	-18.43	0.25

[a]Values in eV, from self-consistent ground state calculations.
[b]Difference between column 3 and column 2.

we can revert to first order perturbation theory for the description of hyperfine effects, in which the only perturbation is the hyperfine energy

$$H_{hf} = e\boldsymbol{\alpha}\cdot\mathbf{A} \tag{8}$$

Here $\boldsymbol{\alpha}$ is a vector of 4 x 4 Dirac matrices, and $\mathbf{A}$ is the vector potential arising from the nuclear magnetic dipole moment $\boldsymbol{\mu}$

$$\mathbf{A} = (\boldsymbol{\mu} \times \mathbf{r})/r^3 \tag{9}$$

By rearranging the triple vector product, Eq. (8) becomes

$$H_{hf} = eg_N\beta_N\mathbf{I}\cdot(\mathbf{r} \times \boldsymbol{\alpha})/r^3 \tag{10}$$

where we have used the relation $\boldsymbol{\mu} = g_N\beta_N\mathbf{I}$. This can be compared to the spin-Hamiltonian representation of the hyperfine interaction

$$H_{spin} = \mathbf{S}\cdot\mathbf{A}\cdot\mathbf{I} \tag{11}$$

In spin-space the basis functions are kets with $M_S = +1/2$ and $-1/2$, while for the operator of Eq. (10), the basis functions are the two components of the Kramers' doublet, which are degenerate in the absence of magnetic fields. By identifying the matrix elements of H_{hf} with those of H_{spin} we can express the results of our hyperfine calculations in terms of the elements of the usual spin-hamiltonian hyperfine tensor.

The matrix elements of H_{hf} in a molecular orbital basis reduce (using the method of Case and Karplus[56]) to those over atomic spinors labelled by quantum number κ and μ. these may be shown to have the following form:[57]

$$\langle\kappa\mu| \left(\frac{\underset{\sim}{r} \times \underset{\sim}{\alpha}}{r^3}\right)_M |\kappa'\mu'\rangle = (\kappa+\kappa')C(1j'j;M\,\mu'\mu)C(1jj';0\tfrac{1}{2}\,\tfrac{1}{2})R_{\kappa\kappa'} \quad (12)$$

where

$$R_{\kappa\kappa'} = \int(f_\kappa g_{\kappa'} + f_{\kappa'} g_\kappa)dr \quad (13)$$

Here g(r) and f(r) are the radial parts of the "large" and "small" components, and the C's are Clebsch-Gordan coefficients. Since the $\underset{\sim}{\alpha}$ matrices are off-diagonal or "odd", Eq. (13) involves products of the large and small components. Hence this method could not be used with pseudorelativistic schemes that deal only with the large components, (unless the small components were to be generated in a subsequent step.) It is this feature that magnetic interactions involve (weighted) overlaps of large and small components that makes it attractive to learn more about the qualitative features of small components in molecules.

The single expression in Eq. (12) includes all of the contact, dipolar and orbital terms, i.e., there is no need to use different operators to represent these, as in non-relativistic theories. The relation between the non-relativistic and relativistic Hamiltonians has been discussed elsewhere.[57] As $c \to \infty$, the small components approach zero at the same rate as the Bohr magneton ($\beta_e = e\hbar/2mc$). Hence the elements of the hyperfine tensor divided by β_e approach a definite non-relativistic limit. We have verified numerically that the hyperfine tensors derived from a DSW calculation with $c = 10^{15}$ a.u. using Eq. (12) agree exactly with the results of a non-relativistic Xα-SW calculation in which the usual Fermi contact and dipolar operators are employed, as of course they should.

Eq. (12) is written for the case that both atomic spinors are on the same center. The analogous two-center matrix elements are much more complicated, and we have not yet included them into our computer codes. However, because of the inverse powers of r, the one-center terms are expected to dominate, and this is indeed verified in non-relativistic calculations. A further problem with the present calculations is that core spin-polarization effects have not been included, and these may also contribute to the observed hyperfine interactions. For these reasons, the relativistic calculations presented here should be considered preliminary. Further calculations including two-center contributions and core polarization effects are planned.

Hyperfine results for XeF are given in Table 3. The unpaired electron resides in an orbital predominantly of σ-character (formed from the Xe5p and F2p orbitals) with about 5% π-character induced by spin-orbit mixing. This unquenched orbital angular momentum about the molecular axis is expected to significantly affect $A_{\perp}$ while having little effect on $A_{\parallel}$.[53-55] This is indeed what is observed in comparing the relativistic and non-relativistic values shown in Table 3. However, the Xα results suggest that both the Xe and the F tensors are affected by orbital motion, in contrast to earlier theories that applied orbital corrections only to the Xe tensor. Many further calculations will be required to establish the general level of accuracy to be expected from this level of theory. Nevertheless, this general approach appears very promising for a wide range of studies on spin-dependent properties of molecules.

Small Metal Clusters

The demand for a computationally tractable MO method in electronic structure studies of systems containing heavy transition and noble metal atoms has been growing along with increased sophistication in spectral measurements for particles and surfaces. Improved understanding of the latter systems will stimulate further advances in the important area of catalysis. State-of-the-art electronic structure calculations on models of these systems[14,18,58,59] have been focusing on qualitative description of bonding trends and interpretation of photoelectron spectra. Despite the qualitative usage of such calculations, relativistic effects remain crucial to correct descriptions of valence as well as core electronic structures of heavy-atom-containing systems. This point is illustrated by the following results on transition metal clusters.

In order to establish the importance of relativistic effects in heavy d-electron systems, a comparison of the Pt_2 valence orbitals calculated with the non-relativistic SCF-Xα-SW and the relativistic SCF-Xα-DSW schemes is presented in Figure 1. The non-relativistic results show a d^{10} configuration, with virtually no sd-hybridization. The relativistic orbitals, on the other hand, reveal a d^9s^1 configuration and sd-hybrids. Aside from the large spin-orbit splitting of the d-complex into $d_{3/2}$ and $d_{5/2}$ sub-complexes and the admixture of the s with the d orbitals, there are no substantial reordering of d orbital energies. However, because of this sd-hybridization, the nature of the chemical bond described relativistically (which is more accurate) differs considerably from that of its non-relativistic counterpart. Hence, in studies of systems involving heavy transition metals, one must take into consideration relativistic effects to obtain realistic bonding information.

Table 3. Hyperfine Parameters for XeF[a]

Atom	non-relativistic Xα-SW	relativistic DSW	experiment[b]
^{129}Xe: $A_{\parallel}$	2093	1987	2366
$A_{\perp}$	-190	946	1221
^{19}F: $A_{\parallel}$	2047	2209	2649
$A_{\perp}$	-299	-748	540

[a]All values in MHz.
[b]Ref. 53; only the absolute magnitudes of the experimental values are known.

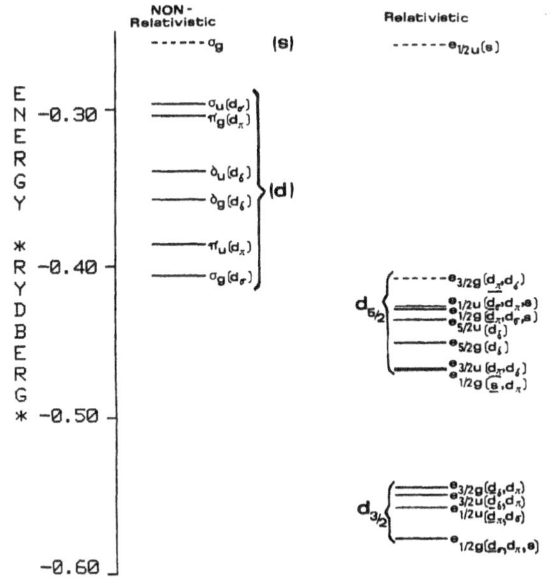

Figure 1. Scattered-wave calculations of Pt_2 valence electronic structure. Broken lines indicate unoccupied orbitals. The dominant non-relativistic character for each relativistic orbital is underlined.

Further relativistic studies of the electronic structure of metal clusters are ongoing. Some examples are given in Figure 2.[45] The valence orbitals for each of the tetrahedral clusters, Ag_4, Au_4, Pt_4, and Pd_4 exhibit band-like complexes. As one would expect, the noble metals have filled d-bands and an unfilled s-band while the $d_{5/2}$- and s-bands are both unfilled in the transition metals. The sd-hybridization increases considerably going from Ag_4 to Au_4 and from Pd_4 to Pt_4. This is primarily a relativistic effect, manifested by stabilization of the s-orbitals and destabilization of the d-orbitals as the element goes down the column of the periodic table.[60] Increased overlap of the s and d-orbitals results.

Because of their unfilled $d_{5/2}$-bands and larger d bandwidths (implying more delocalization), the transition metals Pd and Pt are more likely than the noble metals Ag and Au to have participation of d-electrons in chemisorptive bonds. For the transition metal clusters, the more diffuse and lower s-band in Pt_4 allows most of the charge exchange with an adsorbate through Pt s-electrons. In comparison, for Pd_4, significantly more charge transfer is effected through the Pd d-orbitals. Depending on the adsorption geometry, this fundamental difference in metal-adsorbate bonding between Pd and Pt could give rise to different preferred sites and different

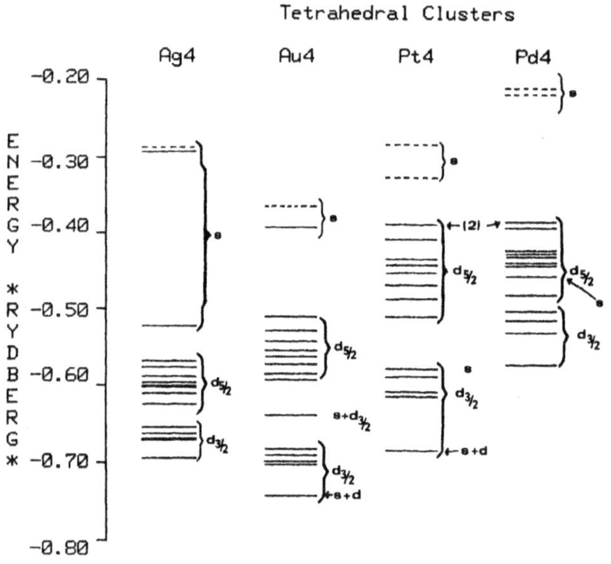

Figure 2. Comparison of heavy transition and noble metal clusters via SCF-Xα-DSW calculations of their tetrahedral valence electronic structure.

dissociation probabilities for the adsorbed species on their respective surfaces.

Calculations on the six-atom octahedral and thirteen-atom cubo-octahedral clusters have also been performed for Pd and Pt. Figures 3 and 4 show the results of the clusters M_n (n = 4,6,13), with the highest occupied molecular orbitals aligned. We note that these 3 clusters represent the smallest non-planar units that are compatible with the face-centered-cubic crystal structure. Hence, it is not unreasonable to assume that these clusters are present in the very early growth stages of small particles or films. The overall d-band features for either metal are similar for all three clusters, the bandwidths increasing with increasing cluster size. For M_{13}, the

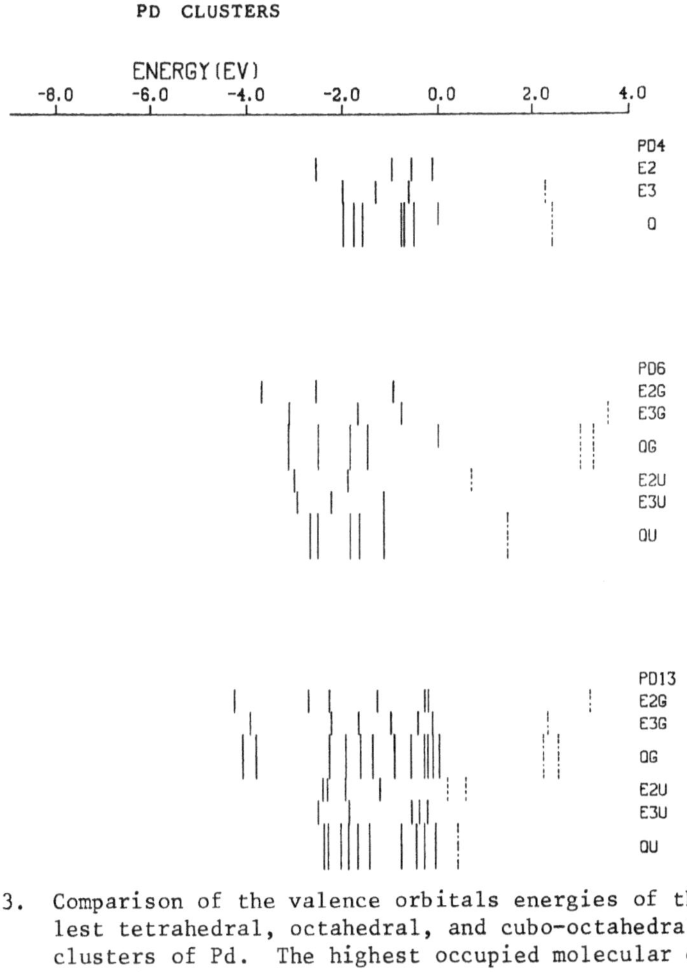

Figure 3. Comparison of the valence orbitals energies of the smallest tetrahedral, octahedral, and cubo-octahedral clusters of Pd. The highest occupied molecular orbitals are aligned.

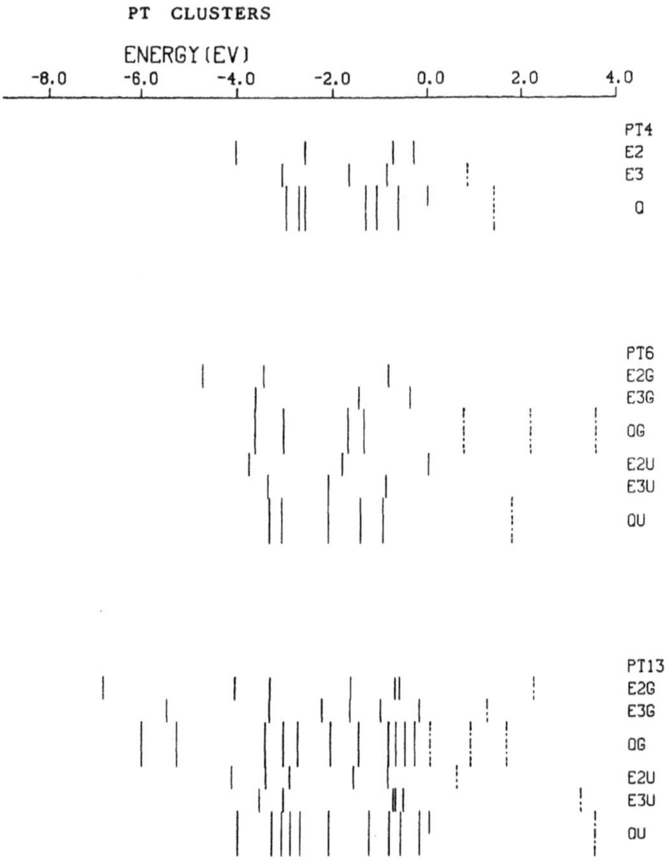

Figure 4. Comparison of the valence orbital energies of the smallest tetrahedral, octahedral, and cubo-octahedral clusters of Pt. The highest occupied molecular orbitals are aligned.

four-level complex with the lowest energies arises principally from the central atom. Since this atom has no unsaturated bonds, its electrons have higher binding energies than those belonging to the peripheral atoms. As a result, the s-d orbitals due to the central atom appear as a separate entity about 1.5 eV below the bottom of the peripheral-atom sd band. This energy separation could be a useful fingerprint in the interpretation of photoemission of EXAFS data for very small particles.

One peculiar aspect of the M_{13} results lies in the densities of states near the HOMO or Fermi level, E_F. In the case of Pt, both Pt_4 and Pt_{13} are open-shell systems, while Pt_6 is clearly closed-shell. Even though Pd_{13} is a closed-shell system, unlike Pd_4 and Pd_6, the density of unoccupied states near E_F is rather high. Hence one would expect similar chemistry for the 3 Pd_n clusters, while Pt_6 is probably somewhat different from the other two. One should bear in mind that for clusters in this size range, any resemblance to the bulk density-of-states does not necessarily imply approaching other bulk properties. These clusters should be treated as individual molecules and possible models of larger clusters of the same or similar geometry. Further analysis of the six- and thirteen-atom results will appear in a later publication.

With the increased availability of synchrotron radiation for surface science experiments, the eventual payoff of calculations such as these lies in the establishment of a far-reaching yet practical theoretical tool to be utilized in conjunction with surface spectroscopic experiments. The increased degree of understanding obtained from the calculations will help shape the basis for better interpretations of experimental data. While no such theoretical tools will be designed with the purpose of replacing experimental measurements, we believe that the proper synergism of experimental and theoretical techniques in the area of suface science is the key to expanding the utilities of most experiments. Moreover, such synergism, if successfully executed, will have an immeasurable impact on surface chemistry and physics.

Quasi One-Dimensional Conductors

Compounds containing the square-planar tetracyanoplatinate ion, $Pt(CN)_4^{2-}$, have been the subject of considerable experimental and theoretical study in recent years.[61,62] In these compounds the $Pt(CN)_4^{2-}$ units are generally parallel stacked in a staggered conformation forming linear chains with a Pt backbone in each chain. Although the compounds are insulating in the pristine form, they acquire anisotropic conductivity (along the Pt back-bone) when they are doped with oxidizing agents such as halogens. Experiments have shown that two phenomena occur simultaneously upon doping: a partial oxidation of each stack and a decrease of the distance between the stacks.[61,62]

The current theoretical interpretation of these results is that the important interaction between stacks must involve the dz^2-like molecular orbitals of the individual $Pt(CN)_4$ units, with the z-axis being the Pt backbone. In the infinite chain the dz^2-like orbitals broadening into a dz^2-like band, the top of which is at the Fermi level. When electrons are depleted from this band by oxidizing dopants, a directional conductivity is created. Simplified band calculations that incorporate these ideas have been presented by Whangbo and Hoffmann,[63] by Messmer and Salahub,[64] and by Yersin et al.[65]

Interrante and Messmer[66] have presented non-relativistic SCF-XαSW calculations for the monomer $Pt(CN)_4^{2-}$ and the staggered dimer $[Pt(CN)_4^{2-}]_2$. For the dimer they performed calculations for a Pt-Pt distance of 2.89 Å, which corresponds to the one found in the bromine-doped compound KCP(Br), $K_2Pt(CN)_4Br_{0.3}\cdot 3H_2O$.[82] They found for the monomer that the dz^2-like orbital is 0.8 eV below the highest occupied molecular orbital (HOMO). For the dimer, however, one of the dz^2-like orbitals becomes the HOMO. They have not reported calculations for larger inter-stack separations, but they suggest that for larger separations, the HOMO may not be dz^2-like. Viswanath and coworkers[67] have performed Extended Huckel MO calculations for the clusters $[Pt(CN)_4^{2-}]_n$ with n = 1, 2 and 3, for the purpose of interpreting optical data. Both eclipsed and staggered conformations have been used for Pt-Pt separations of 3.27 and 3.48 Å. They found that for all cases the HOMO is dz^2-like.

Hoping to obtain a better understanding of the electronic structure of tetracyanoplatinate complexes, we have initiated relativistic SCF-Xα-DSW calculations for clusters up to the trimer.[68] A representative MO diagram is shown in Figure 5.

Results for the monomer unit (or infinite Pt-Pt-separation) calculated with a basis set equivalent to that used for the trimer are also shown for comparative purposes. Total populations for the relativistic and non-relativistic monomers are shown in Table 4. These show that the overall relativistic effects are small in situations such as this, with strong ligand fields. This is in contrast to the metal clusters discussed above, which correspond to very weak ligand fields. Although the relativistic effects cause little net charge transfer between the metal and the ligand, some density shifts are evident in the population ratios shown in Table 4. For the non-relativistic results the ratio of the population in the two components of the same ℓ is always $(\ell+1)/\ell$, whereas the relativistic results are not constrained to this value. As expected, the relativistic ratios are generally smaller, indicating an enchanced population of the lower-energy $j = \ell - 1/2$ component. Hence the Pt p and d populations show substantial deviations from the non-relativistic ratios, indicative of important spin-orbit mixings. This may also be seen by examining the individual orbitals, which we turn to next.

We focus our attention on two aspects of the molecular orbitals, namely, the position of the dz^2-like orbital and the nature of the HOMO. The nature of the HOMO changes with decreasing separation between stacks. While for the monomer and for the trimer at 3.5 Å it is of $(d_\delta + d_\pi)$ character, for the trimer at 2.89 Å it is clearly dz^2-like. Particularly interesting is the examination of the $(d_\delta + d_\pi)$-like orbital at 3.5 Å. Orbital density contour maps show

Table 4. Total Population for $[Pt(CN)_4)]^{2-}$

Orbital		Non-relativistic population[a]	ratio[b]	Relativistic population[a]	ratio[b]
Pt:	$s_{1/2}$	0.935		1.048	
	$p_{1/2}$	0.353	2.000	0.442	1.604
	$p_{3/2}$	0.706		0.709	
	$d_{3/2}$	3.633	1.500	3.677	1.422
	$d_{5/2}$	5.450		5.227	
	$f_{5/2}$	0.153	1.333	0.150	1.320
	$f_{7/2}$	0.204		0.198	
Total Pt:		11.434		11.451	
C:	$s_{1/2}$	1.122		1.107	
	$p_{1/2}$	0.870	2.000	0.878	1.982
	$p_{3/2}$	1.740		1.740	
	$d_{3/2}$	0.085	1.500	0.084	1.524
	$d_{5/2}$	0.128		0.128	
Total C:		3.945		3.937	
N:	$s_{1/2}$	1.372		1.368	
	$p_{1/2}$	1.233	2.000	1.236	2.000
	$p_{3/2}$	2.467		2.472	
	$d_{3/2}$	0.050	1.500	0.049	1.530
	$d_{5/2}$	0.074		0.075	
Total N:		5.196		5.200	

[a]The intersphere charge has been assigned to the atoms following the procedure of Ref. 56. Non-relativistic results are from the DSW program with $c=10^{15}$ a.u.

[b]Ratio of population for the $j = \ell + 1/2$ and $j = \ell - 1/2$ partial waves. For the non-relativistic orbitals these are required to have the values $(\ell+1)/\ell$, whereas the relativistic results may have arbitrary values; see text.

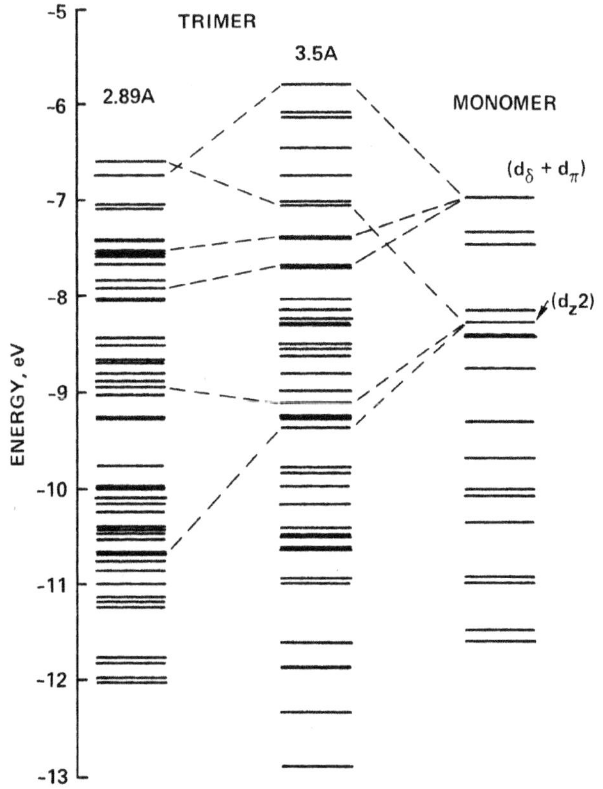

Figure 5. Molecular orbital energy diagram for the monomer $Pt(CN)_4^{2-}$ and for the trimer $[Pt(CN)_4^{2-}]_3$ at two different interstack separations.

that all the electronic charge is localized at the central unit of the cluster. However, strong metal-ligand mixing is observed within the central unit. At 2.89 Å, the HOMO is dz^2-like. This dz^2-like orbital is delocalized throughout the entire cluster, and is of antibonding character between Pt atoms. The dz^2 character at the central atom is 30.1%, and the total dz character of both peripheral Pt atoms is 27.0%. In general this orbital is much more metal-like compared with the $(d_\delta + d_\pi)$ orbital. It is seen from Figure 5 that the energy level position of the dz^2-like orbitals depends critically on the separation between the stacks, and this is reasonable since they are concentrated along the Pt backbone.

These results lead to the following scenario for the insulator-conductor transition. The oxidizing dopant removes electrons from a $(d_\delta + d_\pi)$-like orbital of the insulator, generating an unstable partially oxidized structure. This structure is stabilized by the process which results in a decrease of the inter-stack distance accompanied by the reordering of the dz^2 and $(d_\delta + d_\pi)$ bands at the Fermi level, resulting then in the partial occupation of the dz^2-like band. The stabilization mechanism is suggested below.

Partial oxidation or removal of electrons from the highly localized $(d_\delta + d_\pi)$ band will create a heterogeneous distribution of charges along the $Pt(CN)_4$ units. This decrease in negative charges must in turn produce a decrease in the inter-stack electrostatic repulsion and consequently a shortening of the inter-stack distance. Hence, the antibonding dz^2-like orbital will be pushed up in energy, becoming the new valence band edge. The resulting valence band is partially filled as before. Two competing factors resulting from the partial oxidation of the dz^2-band are principally responsible for maintaining the equilibrium prescribed above. Since the dz^2 valence band is antibonding between neighboring Pt atoms, its partial oxidation results in a strengthening of the Pt-Pt bond, However, further bond-lengths contraction is prevented by the net positive charge created around each Pt atom.

To summarize, our results show that the oxidation process is responsible for drastic changes in the electronic structure of the insulator, particularly in the nature of the bands near the Fermi level. Based on the cluster model, we infer that at the valence band edge of the insulator, there exists a set of nonbonding states highly localized about each stack of the linear chain. Each of these states represents substantial metal-ligand mixings. Partial depopulation of these states leads to a structure where the partially filled valence band has antibonding states delocalized along the Pt backbone. These states have much less-metal-ligand mixings, since they consist essentially of dz^2 character along the Pt backbone. Hence, electron transport occurs principally through the metal atoms. Further details are given elsewhere.[68]

4. CONCLUSIONS

The applications we have discussed here demonstrate the potential utility of the SCF-Xα-DSW method. Further development and applications of this method should lead to better understanding of bonding trends and to more concrete interpretation of experimental spectra for molecules and clusters containing heavy atoms. We are currently working on extensions of the method in three directions: (i) methods to handle continuum orbitals are being developed, which will enable calculations to be made on photoionization intensities and their angular distributions; (ii) a further study of the nature of the small components is being made, with an eye towards improved calculations of spin-dependent properties, such as those outlined above for XeF; (iii) we are preparing computer codes to calculate two-electron matrix elements over the DSW orbitals, allowing us to study multiconfigurational wavefunctions. While it is unreasonable to expect precise quantitative information from scattered-wave calculations, or to argue that they can correctly describe all types of bonding, the relative success of the DSW scheme and the corresponding non-relativistic method[14-19] recommends their use for molecules and clusters too large or too heavy for high quality ab initio computations.

ACKNOWLEDGMENTS

We thank R. Arratia-Perez, J.P. Lopez, and S. Rabii for assistance with the calculations and for helpful discussions. CYY was supported by NASA contracts NAS2-10735 and 10789 and DOE contract DE-AC03-82ER13011. DAC received support from NSF grant CHE81-17101.

REFERENCES

1. P.A.M. Dirac, Proc. Cambridge Phil.Soc. 26, 376 (1930).
2. J.C. Slater, Phys.Rev. 81, 385 (1951).
3. J. Korringa, Physica 13, 392 (1947).
4. W. Kohn and N. Rostoker, Phys.Rev. 94, 1111 (1954).
5. J.C. Slater, Phys.Rev. 51, 846 (1937).
6. J.M. Ziman, Proc.Phys.Soc. (London) 86, 337 (1965).
7. P. Lloyd, Proc.Phys.Soc. (London) 86, 825 (1965).
8. J.C. Slater, Phys.Rev. 145, 599 (1966).
9. K.H. Johnson, Phys.Rev. 150, 429 (1966).
10. J.C. Slater, J.Chem.Phys. 43, S228 (1965).
11. K.H. Johnson, J.Chem.Phys. 45, 3085 (1966).
12. K.H. Johnson, Adv. Quantum Chem. 7, 143 (1973).
13. K.H. Johnson, Ann.Rev.Phys.Chem. 26, 39 (1975).
14. K.H. Johnson, C.R.C. Critical Reviews in Solid State and Materials Science 7, 101 (1978).
15. J.W.D. Connolly, p. 105, in: "Semiempirical Methods of Electronic Structure Calculation, Part A: Techniques,", G.A. Segal, ed., Plenum, New York (1977).
16. P. Weinberger, K. Schwarz, p. 257, in: "International Review of Science, Physical Chemistry, Series Two, Vol. 1, Theoretical Chemistry," A.D. Buckingham, C.A. Coulson, eds., Butterworth, London (1975).
17. N. Rösch, p. 1, in: "Electrons in Finite and Infinite Structures," P. Phariseau, L. Scheire, eds., Plenum, New York (1977).
18. R.P. Messmer, Surf.Sci. 106, 225 (1981).
19. D.A. Case, Ann.Rev.Phys.Chem. (1982), in press.
20. J.L. Dehmer, D. Dill, p. 225, in: "Electro-Molecule and Photon-Molecule Collisions," T. Rescigno, V. McKoy, B. Schneider, eds., Plenum, New York (1979).

21. N.F. Lane, Rev.Mod.Phys. 52, 29 (1980).
22. R.D. Cowan and D.C. Griffin, J.Opt.Soc.Am. 66, 1010 (1976).
23. D.D. Koelling and B.N. Harmon, J.Phys.C. 10, 3107 (1977).
24. J.H. Wood and A.M. Boring, Phys.Rev. B18, 2701 (1978).
25. M. Boring and J.H. Wood, J.Chem.Phys. 71, 32 (1979); ibid. 71, 392 (1979); J.H. Wood, M. Boring, and S. Woodruff, ibid. 74, 5225 (1981).
26. F.A. Cotton, J.Molec.Struct. 59, 97 (1980).
27. J.W.D. Connolly, Chem.Phys.Lett. 68, 549 (1979).
28. G. Thornton, N. Edelstein, M. Rösch, R.G. Egdel, and D.R. Woodwark, J.Chem.Phys. 70, 5218 (1979); G. Thornton, N. Rösch, and N. Edelstein, Inorg.Chem. 19, 1304 (1980).
29. Y. Onodera and M. Okazaki, J.Phys.Soc.Jpn. 21, 1273 (1966).
30. S. Takada, Prog.Theor.Phys. 36, 224 (1966).
31. C.Y. Yang and S. Rabii, Phys.Rev. A12, 362 (1975).
32. B.G. Cartling and D.M. Whitmore, Int.J. Quantum Chem. 10, 393 (1976).
33. F. Rosicky, P. Weinberger, and F. Mark, J.Phys. B9, 2971 (1976).
34. C.Y. Yang, J.Chem.Phys. 68, 2626 (1978).
35. C.Y. Yang, Ph.D. dissertation, University of Pennsylvania (1975).
36. C.Y. Yang, Chem.Phys.Lett. 41, 588 (1976).
37. C.Y. Yang and S. Rabii, Int.J. Quantum Chem. S10, 313 (1976).
38. R.P. Messmer, D.R. Salahub, K.H. Johnson, and C.Y. Yang, Chem. Phys.Lett. 51, 84 (1977).
39. C.Y. Yang, K.H. Johnson, and J.A. Horsley, J.Chem.Phys. 68, 1001 (1978).
40. C.Y. Yang and S. Rabii, J.Chem.Phys. 69, 2497 (1978).
41. D.A. Case and C.Y. Yang, Int.J. Quantum Chem. 18, 1091 (1980).
42. D.A. Case and C.Y. Yang, J.Chem.Phys. 72, 3443 (1980).
43. C.Y. Yang, H.L. Yu, and D.A. Case, Chem.Phys.Lett. 81, 170 (1981).
44. C.Y. Yang and D.A. Case, Surf.Sci. 106, 523 (1981).
45. C.Y. Yang, in: "Proceedings of Electrocatalysis Symposium," Vol. 82-2, eds. W.E. O'Grady, R.N. Ross Jr., and Z.G. Will, Electrochemical Society (1982); S. Rabii and C.Y. Yang, submitted to Chem.Phys.Lett.
46. C.Y. Yang, in: "NATO Advanced Study Institute Proceedings on Relativity Effects of Atoms, Molecules, and Solids," ed. G. Malli, Plenum Press (1982).
47. For a review of the background formalism, see, for example, M.E. Rose, in: "Relativistic Electron Theory," Wiley, New York (1961).
48. I.P. Grant, Adv.Phys. 19, 747 (1970).
49. F. Mark and F. Rosicky, Chem.Phys.Lett. 74, 562 (1980).
50. Y. Ishikawa and G.L. Malli, Chem.Phys.Lett. 80, 111 (1981).
51. W.H.E. Schwarz and E. Weschel-Trakowski, Chem.Phys.Lett. 85, 94 (1982).
52. J.R. Morton and W.E. Falconer, J.Chem.Phys. 39, 427 (1963).
53. R.S. Eachus and M.C.R. Symons, J.Chem.Soc. A304 (1971).
54. R. Lefebvre, Mol.Phys. 12, 417 (1967).

55. F.J. Adrian and A.N. Jette, J.Chem.Phys. 68, 4696 (1978).
56. D.A. Case and M. Karplus, Chem.Phys.Lett. 39, 33 (1976).
57. I. Lindgren and A. Rosén in: "Case Studies in Atomic Physics IV," E.W. McDaniel and M.R.C. McDowell, eds., North-Holland, Amsterdam (1975).
58. R.P. Messmer, Vol. 8, Chapter 6, in: "Semiempirical Methods of Electronic Structure Calculations," G.A. Segal, ed., Plenum, New York (1977).
59. G.A. Ozin, Catal.Rev.-Sci.Eng. 16, 191 (1977).
60. For discussions of this effect, see P. Pyykkö and J.P. Desclaux, Accts.Chem.Res. 12, 276 (1979).
61. J.S. Miller and A.J. Epstein, Progress in Inorg.Chem. 20, 1 (1976).
62. J.M. Williams and A.J. Schultz, p. 377; in "Molecular Metals," W.E. Hatfield, ed., Plenum, New York (1979).
63. M. Whangbo and R. Hoffmann, J.Am.Chem.Soc. 100, 6093 (1978).
64. R.P. Messmer and D.R. Salahub, Phys.Rev.Lett. 35, 533 (1975).
65. H. Yersin, G. Gliemann, and U. Rossler, Solid State Commun. 21, 915 (1977).
66. L.V. Interrante and R.P. Messmer, Chem.Phys.Lett. 26, 225 (1974); p. 382, in: "Extended Interactions Between Metal Ions," L.V. Interrante, ed., American Chemical Society Symposium Series No. 5, American Chemical Society, Washington, D.C. (1974).
67. A.K. Viswanath, M.B. Krogh-Jespersen, J. Vetuskey, C. Baber, W.D. Ellenson, and H.H. Patterson, Molecular Phys. 42, 1431 (1981).
68. J.P. Lopez, C.Y. Yang, and D.A. Case, Chem.Phys.Lett., in press.

THEORETICAL VALENCE CHARGE DENSITIES

FOR GRAPHITE, DIAMOND, AND LiC_6

N.A.W. Holzwarth

Corporate Research-Science Laboratories
Exxon Research and Engineering Company
Linden, New Jersey 07036, U.S.A.

Steven G. Louie

Department of Physics
University of California
Berkeley, California 94720, U.S.A.

Sohrab Rabii

Moore School of Electrical Engineering
University of Pennsylvania
Philadelphia, Pennsylvania 19104, U.S.A.

The valence charge density of graphite was determined from first-principles, self-consistent electronic structure calculations and found to be in excellent agreement with values derived from experimental X-ray form factors. Similar calculations were carried out for the valence charge densities of diamond and of a first stage intercalation compound, LiC_6.

1. INTRODUCTION

Recent advances in numerical techniques have made it possible to carry out first principles calculations of the ground state properties of solids within density functional theory in the local density approximation.[1] The numerical accuracy of these techniques have now enabled workers to address the question of the accuracy of the local density approximation itself. One of the immediate outputs of these calculations is the variationally determined valence charge density.

In the present paper, we discuss some recent results for the calculated valence charge density of graphite[2] and diamond and their comparisons with experimental charge densities derived from the analysis of high resolution X-ray diffraction data.[3,4] For purposes of comparisons, we also discuss results for an intercalation compound, LiC_6.

2. METHOD OF CALCULATION

The first-principles band structure calculations were carried out within density functional theory in the local density approximation using several well-established numerical techniques and approximations, suitably modified to treat the relatively strong electron-core interactions of carbon.[2]

In order to take advantage of the numerical efficiency of a Fourier space evaluation of the matrix elements, the calculations were formulated in terms of pseudopotentials and pseudo-wavefunctions. The electron-ion interactions were represented by norm-conserving pseudopotentials as developed by Hamann, Schlüter, and Chiang.[5] This form of pseudopotential has several advantages. The pseudo-wavefunctions and pseudo-charge converge to the actual wavefunction and actual charge outside spheres of specified radii about each atom and the integrated pseudo-charge is equal to the integrated actual charge inside the core regions. Because of these normalization constraints, the pseudopotentials are energy independent over a reasonable energy range. The electron-electron interactions were represented self-consistently in the local density approximation, using the Hedin-Lundqvist[6] form for the exchange-correlation contribution.

The electronic wavefunctions were presented in terms of a mixed basis set consisting of plane waves and linear combinations of atomic orbitals (LCAO's) as developed by Louie, Ho, and Cohen.[7] The atomic functions were taken to be the numerical s- and p-wave atomic pseudo-wavefunctions. A sufficient number of plane waves was included in the basis set in order to converge the eigenvalues to within 0.2 eV. This relatively large error was permitted in order to be able to treat higher stage graphite intercalation compounds. The self-consistency and the relative positions among levels were converged to well within 0.05 eV. The matrix elements between LCAO's and the local part of the pseudopotential were evaluated using the fast Fourier transform technique suggested by Louie, et al.[7] The matrix elements between LCAO's and the non-local part of the pseudopotential were evaluated using a separable form approximation.[2]

3. VALENCE ELECTRON DISTRIBUTION IN GRAPHITE

Graphite has the Bernal[8] structure with D_{6h}^{4} symmetry and 4 atoms, 2 very slightly inequivalent types, per unit cell. The atoms are arranged in layers of hexagonal lattices with lattice constant a = 2.46 Å; the separation between layers is c/2 = 3.35 Å. Adjacent layers are shifted in an ABAB stacking such that half of the carbon atoms (type "a") are directly above and below carbon atoms in the adjacent layers, while half (type "b") are directly above and below centers of carbon hexagons in the adjacent layers.

The calculated valence charge density for graphite is shown in Fig. 1a. Contours for two different planes are shown, a carbon plane and a plane containing the c axis. The edge at which the two planes meet is shown with a thick line. This figure illustrates the highly

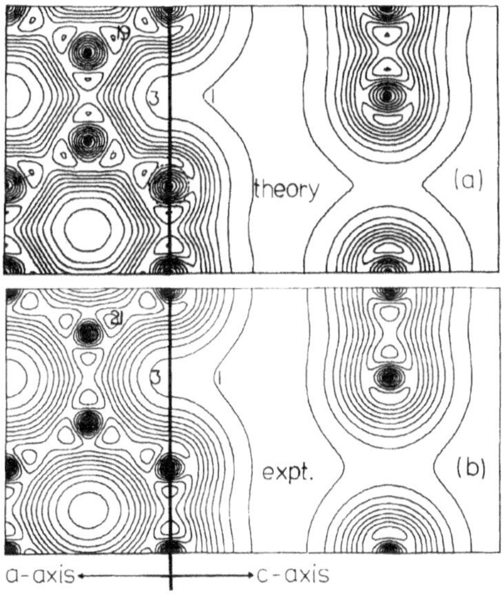

Fig. 1. Contour plots of valence electronic charge density for graphite: (a) present results and (b) results from analysis of X-ray data by Chen, Trucano, and Stewart (Ref. 3). Contour values are given in units of 0.1 e/Å^3. Atomic positions are denoted by filled circles. Two planes are shown - one containing an a-axis and the other containing the c-axis and both intersecting at 90° along a C-C bond. In (a), the dashed circle denotes the pseudopotential radius. In (b), the authors quote a standard deviation of ± 0.1 e/Å^3.

anisotropic structure of the graphite valence density and the large concentration of charge that constitutes the C-C bonds. The density along the bond exhibits a double-humped feature. The peak density is 2.1 e/Å^3, 10% higher than the density at the bond center. By contrast, the charge density between carbon planes, drops to less than 0.1 e/Å^3.

For comparison, the valence charge density generated by the most refined fit to the X-ray form factors by Chen, Trucano, and Stewart,[3] is shown in Fig. 1b. The overall agreement between the two density distributions is excellent. The experimental standard deviation is quoted to be ± 0.1 e/Å^3. In addition, the set of parameters used to generate Fig. 1b overestimates the total density by 0.034 e/Å^3. Therefore, the agreement of the two distributions to within ± 0.15 e/Å^3 throughout the valence region is an encouraging result. Since our density is a "pseudo-charge" density, the shape of the density within the pseudopotential radius (indicated in Fig. 1a by a dashed circle) is not simply related to the real electron density. However, the integrated pseudo-charge within the pseudopotential sphere is equal to the integral of the real charge within the same region.

4. VALENCE ELECTRON DISTRIBUTION IN DIAMOND

Diamond is on average a more dense form of carbon, although the nearest neighbor distance is 10% larger. It has the O_h^7 space group and two equivalent atoms per unit cell with cubic lattice constant a = 3.57 Å. Fig. 2a shows the calculated valence density distribution in a plane containing C-C bonds. The distinctive double humped feature of the bond charge is also exhibited for diamond. Comparing the bond charge of diamond with that of graphite, we see that the peak density for diamond is reduced by 10% of the peak density of graphite. The charge at the bond center in diamond is roughly 75% of its peak density.

The calculated density shown in Fig. 2a agrees well with that calculated by several previous workers.[9-11]

Stewart[4] analyzed that valence density of diamond using powder X-ray diffraction data and the equivalent of only 60% of the reflection information used for the analysis of graphite. The results of his analysis are shown in Fig. 2b. In general the experimentally fit densities are more contracted than the calculated densities. The differences between theory and experiment are greater for diamond than for the case of graphite. This is partially due to the fact that the diamond X-ray data is of slightly lower quality than that of graphite. However, the important qualitative features, namely, the double hump of the density along a C-C bond, and the decrease in bond charge relative to that of graphite are well represented by both the experiment and the calculation.

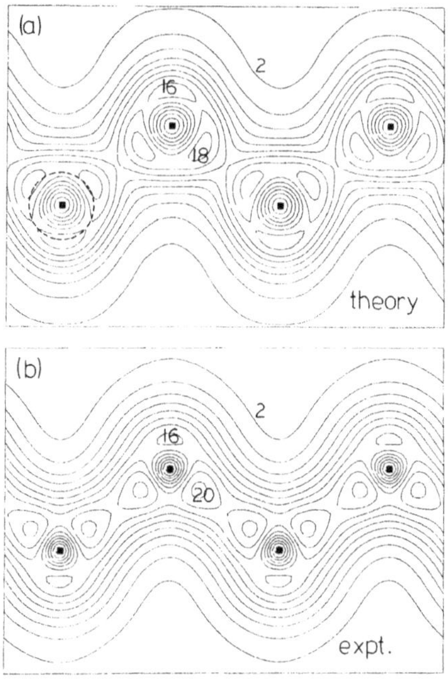

Fig. 2. Contour plots of valence electronic charge density for diamond: (a) present results and (b) results from analysis of X-ray data by Stewart (Ref. 4). Contour values are given in units of 0.1 e/Å^3. Atomic positions are denoted by filled squares. The plane shown contains a C-C bond. In (a), the dashed circle denotes the pseudopotential radius.

5. VALENCE ELECTRON DISTRIBUTION IN LiC_6

First stage LiC_6 has been reported to have the D^1_{6h} space group, having 6 carbon atoms and 1 Li per unit cell.[12] The carbon layers are in AA registry with respect to each other and the Li atoms align themselves between the layers in the center of one third of the carbon hexagons. The hexagonal unit cell parameters are $\sqrt{3}$ a, where a = 2.485 Å and c = 3.706 Å. This represents a 1% dilation of the C-C bond length with respect to that of graphite and a 10% increase in the distance between carbon layers. The calculated density for LiC_6 [13] is shown in Fig. 3 in a plane containing the c-axis and a C-C bond. On the scale shown in the figure, the contours for LiC_6 are barely distinguishable from that of graphite itself, shown in Fig. 1a. In order to consider the effects of charge transfer, therefore, it is useful to define a difference density. We have

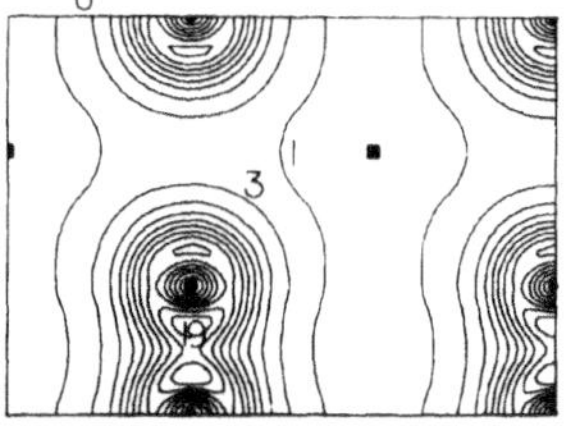

Fig. 3. Contour plot of total valence electron charge density of LiC_6. Contour values are given in units of 0.1 e/Å^3. Atomic positions are denoted by filled squares for Li and circles for C. Plane shown contains c-axis and passes through Li atoms and C-C bonds.

defined a total difference density as the self-consistent valence density of LiC_6 (Fig. 3) minus the self-consistent valence density of graphite modified to the LiC_6 structure. This total difference density is presented in Fig. 4. By means of this difference density we are able to see the build up of π-like charge near the carbon planes, as well as polarization effects due to the Li ions. Negative contours in the regions of the C-C bonds are indicative of bond weakening due to intercalation.

6. DISCUSSION AND CONCLUSIONS

The successful comparison of calculated and experimental valence densities for graphite presented in Fig. 1 and diamond in Fig. 2 is

Fig. 4. Contour plots of the total difference density for LiC_6. Contour values are given in units of 0.001 e/Å^3. Atomic positions are denoted by filled squares for Li and circles for C. Plane shown is the same as for Fig. 3.

an encouraging result which establishes a high degree of confidence in the local density pseudopotential methods.

The opportunity to make a detailed comparison of calculated valence densities with the experimental valence densities derived from X-ray data is relatively rare because there are relatively few experimentally determined valence electron density maps available. Part of the experimental problem (such as in the case of diamond) is the availability of high quality single crystals. For heavier materials, such as Si, the core contributions to the X-ray scattering become more important and the analysis of the valence electron density becomes less accurate.

The charge density itself is important for determining the ground state properties of the materials. The contour plots of the charge density contain important qualitative information about the bonding in these materials. In particular, the double-humped shape of the C-C bond is a distinctive feature that is not exhibited for other Group IV materials such as Si, Ge, and Sn. It is very likely that the double-humped feature can be explained in terms of the relative spatial extents of the s- and p-components of the valence wave function. For carbon, the p-wave component is strongly attracted to the C core, while the s-wave component is attracted less strongly as exhibited by the relative strengths of the ionic pseudopotentials. For the heavier materials, the existence of p core states decreases the p-wave attraction to the core, so that the p-wave components of the wave function are able to have larger extent into the bonding region.

Having established credible results for graphite and diamond, we can then address more detailed questions such as charge transfer in graphite intercalation compounds. From our detailed calculations for LiC_6 we find the effects to be on the order of 5% of the peak density in graphite. However, by defining the difference density shown in Fig. 4, the charging and relaxation effects are clearly seen.

ACKNOWLEDGEMENTS

We would like to thank Professor R.F. Stewart and Dr. J.R. Chelikowsky for helpful discussions.

This work was partly supported by National Science Foundation MRL Grant No. DMR79-23647, ARO Contract DAAG-29-77-C400, and National Science Foundation Grant No. DMR78-22465. S.G. Louie would like to acknowledge support from a Sloan Foundation Fellowship.

REFERENCES

1. P. Hohenberg and W. Kohn, Phys. Rev. 136, B 864 (1964); W. Kohn and L.J. Sham, Phys. Rev. 140, A 1133 (1965).
2. N.A.W. Holzwarth, S.G. Louie, and S. Rabii, to be published.
3. R. Chen, P. Trucano, and R.F. Stewart, Acta Crystallogr. Sec. A 33, 823 (1977).
4. R.F. Stewart, J. Chem. Phys. 58, 4430 (1973).
5. D.R. Hamann, M. Schlüter, and C. Chiang, Phys. Rev. Lett. 43, 1494 (1979).
6. L. Hedin and B.I. Lundqvist, J. Phys. C 4, 2064 (1971).
7. S.G. Louie, K.M. Ho, and M.L. Cohen, Phys. Rev. B 19, 1774 (1979).
8. J.D. Bernal, Proc. Roy. Soc. London, Ser. A 106, 749 (1924).
9. M.T. Yin and M.L. Cohen, Phys. Rev. B 24, 6121 (1981).
10. G.B. Bachelet, H.S. Greenside, G.A. Baraff, and M. Schlüter, Phys. Rev. B 24, 4745 (1981).
11. A. Zunger and A.J. Freeman, Phys. Rev. B 15, 4716 (1977).
12. D. Guerard and A. Herold, Carbon 13, 337 (1975).
13. N.A.W. Holzwarth, S.G. Louie, and S. Rabii, Phys. Rev. Lett. 47, 1318 (1981) and also to be published.

DEPENDENCE OF THE SELF-CONSISTENT ONE-ELECTRON DENSITY OF ANISOTROPIC SEMICONDUCTORS ON LOCAL DENSITY APPROXIMATIONS

J. von Boehm and H.M. Isomäki

Electron Physics Laboratory and
Department of General Sciences
Helsinki University of Technology
SF-02150 Espoo 15
Finland

The valence one-electron densities ρ_v are calculated for 1T $TiSe_2$ and trigonal Se with the self-consistent orthogonalized-plane-wave (SCOPW) method using the most common local density approximations (Dirac-Gáspár-Kohn-Sham, Slater, Hedin-Lundqvist). The resulting SCOPW ρ_v's become smoother (or more metallic) in the sequence: Slater, Dirac-Gáspár-Kohn-Sham, Hedin-Lundqvist, in agreement with the calculation by Wimmer et al. for ScP, but in disagreement with the calculation by Zunger and Freeman for diamond.

1. INTRODUCTION

Most local density approximations for the one-electron exchange or exchange-correlation potentials (in the non-spin-polarized case studied in this paper) are based on the uniform electron gas approximation. Slater[1] obtained his one-electron exchange potential V_S, proportional to $\rho^{1/3}$ (ρ is the one-electron density), by using this approximation for the (averaged) exchange term in the Hartree-Fock (HF) equation. If one makes the uniform electron gas approximation for the exchange term in the HF total energy expression, one obtains the approximate Kohn-Sham[2-4] exchange potential V_{KS} which equals $(2/3)V_S$.[4-6] In the X_α method one uses a parametrized exchange potential αV_S where the parameter α is constant. α is chosen between 2/3 and 1 [e.g., from the requirement that the HF expectation values of the kinetic and potential energies for atoms, calculated with the X_α orbitals, satisfy the virial theorem[7] or by minimizing the atomic HF total energy, calculated with the X_α orbitals, with respect to α (Ref. 8)].

The most general approach starts from the fundamental Hohenberg-Kohn[9] theorem which states that the expression for the ground state energy E of the electron system is a universal functional of ρ and that E has as its minimum value the correct ground state energy which is reached with the ρ of the ground state. When the exchange-correlation energy is approximated by the energy of the uniform electron gas the Hohenberg-Kohn variational principle yields an effective Schrödinger equation.[4] Hedin and Lundqvist[10] have given a parametrized expression of the form $V_{HL} = \beta(\rho)V_{KS}$ for the exchange-correlation potential of this equation. If the correlation is neglected, $\beta(\rho) \equiv 1$.[4] $\beta(\rho)$ varies mainly between 1 (for large ρ) and 1.5 (for low ρ).

The effect of the various exchange or exchange-correlation approximations on the energy bands has been studied by several authors.[6,11-22] However, the effect of the various approximations on ρ has been studied much less in the literature.[16,17,19,20] The aim of this paper is to study the effect of the most common exchange or exchange-correlation approximations (S, KS, HL) on the valence one-electron density ρ_v of semiconductors. This is calculated for the anisotropic semiconductors trigonal Se and 1T $TiSe_2$ with the self-consistent (SC) orthogonalized plane-wave (OPW) method. In the SCOPW method the effective crystal potential and ρ_v are of the general (non-muffin-tin) form. This is important for the proper description of the bonding properties of anisotropic semiconductors. The SC ρ_v was obtained using 235 and 705 symmetry adapted OPW at each of the symmetry points Γ, A, K, H, M and L for Se and $TiSe_2$, respectively.[21,22]

2. RESULTS AND DISCUSSION

Trigonal Se crystallizes into parallel helical chains arranged in a hexagonal array. Each atom has two nearest neighbours belonging to the same chain. Three adjacent atoms of the same chain belong to one primitive unit cell. Therefore the valence bands appear as triplets which are separated from each other. The electrons associated with the two lowest triplets form strong s- and p-type bonds between adjacent atoms of the chains. The p-electrons associated with the uppermost triplet form non-bonding lone-pair ρ_v maxima. The interchain bonding is much weaker than the intrachain bonding but by no means negligible.

The calculated SCOPW ρ_v of Se is shown in the nearest neighbour (NN) direction (within one chain) in Fig. 1 and in the next nearest neighbour (NNN) direction (between the chains) in Fig. 2 (thick and thin lines represent ρ_v's calculated with V_S and V_{KS}, respectively). The double maximum between the NN Se atoms in Fig. 1 consists of s- and p-bonding valence electrons. The maxima at a distance of ~ $1.5a_o$ from the Se atoms in Fig. 2 are associated with the p-non-bonding lone-pair electrons.

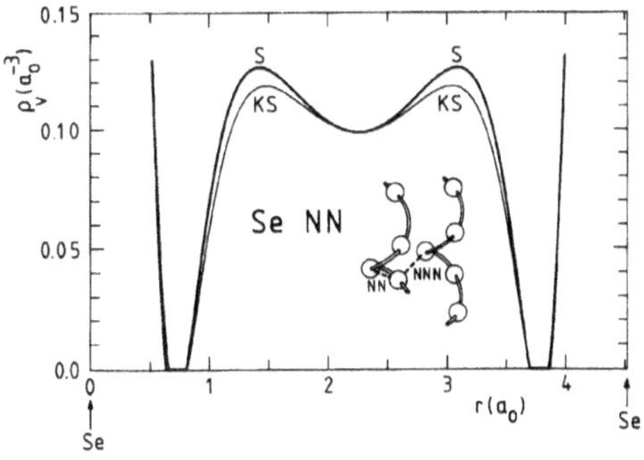

Fig. 1. The SCOPW ρ_v of Se in the nearest neighbour (NN) direction. S and KS denote Slater and Kohn-Sham, respectively. a_o is the Bohr radius. Two nearest chains of the Se crystal are shown as an insertion where the NN and the next nearest neighbour (NNN) directions are marked with dashed lines.

A part of the KS ρ_v has, compared with the S ρ_v, shifted from the bonding regions and the non-bonding lone-pair regions of high ρ_v (Figs. 1 and 2) to the large interchain regions of low ρ_v (Fig. 2). This shift may be associated with the fact that (the negative) V_S is more negative (and thus attractive) than $V_{KS} = (2/3)V_S$, especially in the regions of high ρ_v.

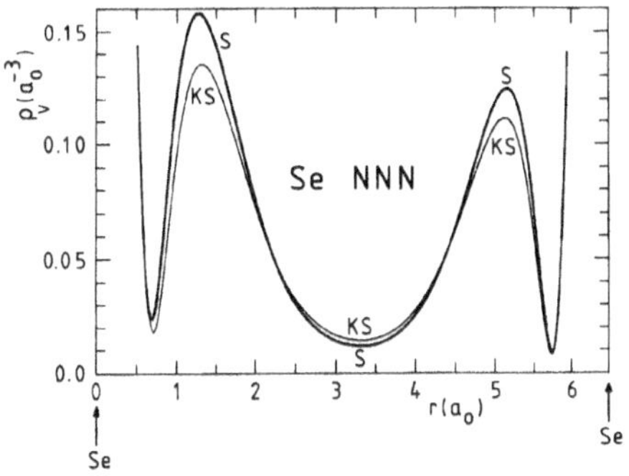

Fig. 2. The SCOPW ρ_v of Se in the next nearest neighbour (NNN) direction. For further information see the insertion and caption of Fig. 1.

1T $TiSe_2$ consists of parallel layers. Each layer consists of one hexagonal Ti plane and two hexagonal Se planes (see the insertion in Fig. 3). The intralayer bonding is strong whereas the interlayer bonding is very weak. One Ti atom and two Se atoms (of the same layer) belong to one primitive unit cell. According to a simplified linear-combination-of-atomic-orbitals (LCAO) description the two lowest valence bands are formed from the Se 4s states whereas the six uppermost valence bands originate from the Se 4p states. The Ti 3d and 4s electrons are transferred to the Se 4p states to complete the occupation of the valence bands.

The calculated SCOPW ρ_v of $TiSe_2$ is shown in the direction between the NN Ti and Se atoms in Fig. 3 and between the nearest Se atoms along the hexagonal plane and between the layers in Fig. 4 (the left and right part of Fig. 4, respectively, see also the insertion in Fig. 3). The thick and thin lines in Figs. 3 and 4 represent ρ_v's calculated with V_{HL} and V_{KS}, respectively.

The HL and KS ρ_v's for $TiSe_2$ differ from each other by one order of magnitude less than the KS and S ρ_v's for Se (cf. Figs. 1-4). Thus, the correlation (in the sense of the local density approximation) has a very minor effect on ρ_v. We find further that a part of the HL ρ_v has, compared with the KS ρ_v, shifted from the neighbourhood of the atoms and the NN Ti-Se bond regions (Figs. 3 and 4) to the larger regions of low ρ_v (the left and right edge of Fig. 4, the difference between the HL and KS ρ_v's falls here within the line width).

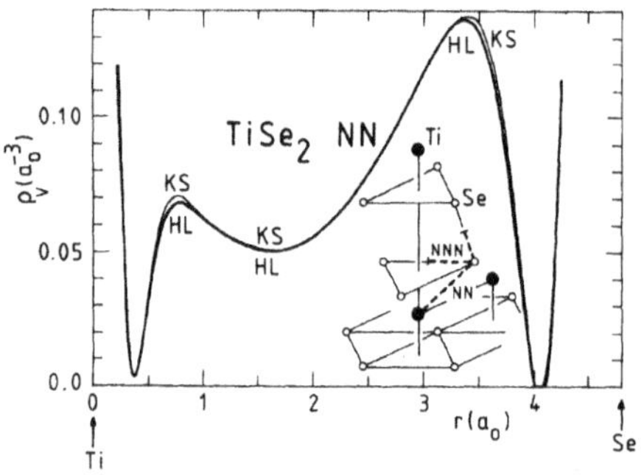

Fig. 3. The SCOPW ρ_v of $TiSe_2$ in the nearest neighbour (NN) Ti-Se direction. KS and HL denote Kohn-Sham and Hedin-Lundqvist, respectively. a_0 is the Bohr radius. The crystal structure of $TiSe_2$ is shown as an insertion where the NN and next nearest neighbour (NNN) directions are marked with dashed lines.

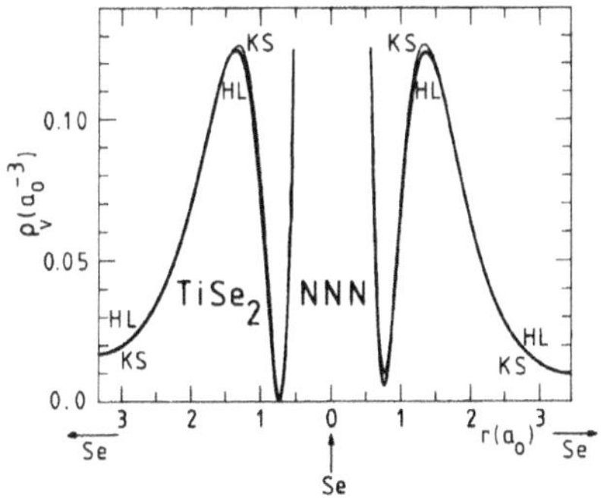

Fig. 4. The SCOPW ρ_v of $TiSe_2$ in the next nearest neighbour (NNN) directions. The left part represents the direction in the hexagonal Se plane from the middle of the nearest Se atoms (the left edge of the figure) to the Se atom. The right part represents the direction from the Se atom to the middle of the nearest Se atoms of the adjacent layers (the right edge of the figure). For further information see the caption and the insertion of Fig. 3.

The SCOPW result is similar to that found by Wimmer et al.[19,20] for ScP with the SC augmented-plane-wave (APW) method. A part of the HL ρ_v by Wimmer et al. has, compared with ρ_v calculated with the virial-theorem $X\alpha$ potential, shifted outside the atomic muffin-tin spheres. The above shift in the HL ρ_v (with respect to the KS ρ_v) may be associated with the fact that $\beta(\rho)$ multiplying (the negative) V_{KS} is larger in the regions of low ρ_v than in the regions of high ρ_v. However, we notice that the SC numerical-basis-set LCAO calculation by Zunger and Freeman[16] for diamond gives a result which is opposite to the SCOPW one, i.e. a part of the HL ρ_v has, compared with the KS ρ_v, shifted to the covalent bond regions and near the atoms. The reason for this discrepancy is not clear to us.

We found above that the SCOPW ρ_v becomes smoother (i.e. more metallic) in the order S, KS and HL. One can therefore expect that the bandwidths will grow and the gaps will decrease in the same order. The SCOPW bands behave precisely in this way[21,22] in agreement with several other band calculations,[6,11-13,17,19,20] but it should be stressed that the differences between the KS and HL bands are extremely small.

ACKNOWLEDGEMENTS

The authors would like to thank Professors T. Stubb and M.A. Ranta for their support.

REFERENCES

1. J.C. Slater, Phys. Rev. 81, 385 (1951).
2. P.A.M. Dirac, Proc. Camb. Phil. Soc. 26, 376 (1930).
3. R. Gáspár, Acta Acad. Sci. Hung. 3, 263 (1954).
4. W. Kohn and L.J. Sham, Phys. Rev. 140, A 1133 (1965).
5. R.D. Cowan, A.C. Larson, D. Liberman, J.B. Mann, and J. Waber, Phys. Rev. 144, 5 (1966).
6. D.J. Stukel, R.N. Euwema, and T.C. Collins, Phys. Rev. B 1, 779 (1970).
7. K. Schwarz, Phys. Rev. B 5, 2466 (1972).
8. J.C. Slater, T.M. Wilson, and J.H. Wood, Phys. Rev. 179, 28 (1969).
9. P. Hohenberg and W. Kohn, Phys. Rev. 136, B 864 (1964).
10. L. Hedin and B.I. Lundqvist, J. Phys. C 4, 2064 (1971).
11. A.M. Boring, Phys. Rev. B 3, 3093 (1971).
12. G.S. Painter, D.E. Ellis, and A.R. Lubinsky, Phys. Rev. B 4, 3610 (1971).
13. L.L. Boyer, D.A. Papaconstantopoulos, and B.M. Klein, Phys. Rev. B 15, 3685 (1977).
14. B.M. Klein, L.L. Boyer, and D.A. Papaconstantopoulos, J. Phys. F 8, 617 (1978).
15. B.M. Klein, L.L. Boyer, D.A. Papaconstantopoulos, and L.F. Mattheiss, Phys. Rev. B 18, 6411 (1978).
16. A. Zunger and A.J. Freeman, Phys. Rev. B 15, 5049 (1977).
17. A. Zunger and A.J. Freeman, Phys. Rev. B 16, 2901 (1977).
18. A. Zunger and A.J. Freeman, Phys. Rev. B 17, 4850 (1978).
19. E. Wimmer, A. Neckel, and K. Schwarz, J. Phys. C 12, 5441 (1979).
20. E. Wimmer, A. Neckel, K. Schwarz, and R. Eibler, J. Phys. C 12, 5453 (1979).
21. J. von Boehm and H. Isomäki, J. Phys. C 13, 4953 (1980).
22. H. Isomäki and J. von Boehm, J. Phys. C 14, L 75 (1981).

γ-RAY BRAGG AND COMPTON SCATTERING EXPERIMENTS USED AS TESTS OF THE VALIDITY OF THE LOCAL DENSITY APPROXIMATION

J.R. Schneider, G.E.W. Bauer
N.K. Hansen and H. Kretschmer

Hahn-Meitner-Institut für Kernforschung
Glienicker Str. 100
D-1000 Berlin 39, FRG

1. INTRODUCTION

Within the one-electron approximation the calculation of the electronic properties of a solid involves the solution of the one-electron Schrödinger equations

$$[-\frac{1}{2}\nabla^2 + V_{cryst}(\vec{r})]\ \psi_{\nu\vec{k}}(\vec{r}) = E_{\nu\vec{k}} \cdot \psi_{\nu\vec{k}}(\vec{r})$$

The vast majority of current band calculations are made with a local crystal potential

$$V_{cryst}(\vec{r}) = V_H(\vec{r}) + V_{xc}(\vec{r})$$

where V_H is the Hartree or Coulomb potential from all charges in the crystal and V_{xc} is an "exchange-correlation" potential taken to depend only on the electron density $\rho(r)$. In many actual calculations V_{xc} is written as

$$V_{xc}(\vec{r}) = \alpha \cdot V_s[\rho(\vec{r})] \equiv \alpha \cdot \frac{3}{2}\left(\frac{3\rho(\vec{r})}{\pi}\right)^{1/3}$$

where α is a parameter which may be density dependent and V_S is the Slater exchange potential.[1] Empirically it has been found that for metals often some α values between 2/3 and 1 can be chosen which give reasonable agreement with experiment. Hohenberg, Kohn and Sham[2-4] developed a rigorous density functional theory and they showed that in principle one can calculate only the charge density and the total energy within this formalism. Therefore it is interesting to compare the calculated charge density with experimental data, e.g. by comparing the measured and the calculated structure factors

$$F_{\vec{H}} = \int_{cell} \rho(\vec{r}) e^{i2\pi\vec{H}\cdot\vec{r}} \, d\vec{r}$$

$\vec{H}$ is the reciprocal lattice vector. From the direct space pseudo one-electron wave-functions $\psi_{\nu\vec{k}}(\vec{r})$ obtained within the local density approximation one can calculate a momentum density

$$\rho(\vec{p}) = \sum_{occ} |\psi_{\nu\vec{k}}(\vec{r}) e^{i\vec{p}\cdot\vec{r}} \, d\vec{r}|^2$$

where the sum is over all occupied states. In a Compton scattering experiment one measures the projection of the total electron momentum density, $\rho(\vec{p})$, onto a line chosen to be parallel to the experimental scattering vector, $\vec{K}$,

$$J_{\vec{K}}(p_z) = \iint \rho(\vec{p}) dp_x dp_y; \quad \vec{K} || p_z$$

which is called the Compton profile. In the density functional theory a correction term has to be added which takes into account the fact that the eigenvectors of the Kohn-Sham equations are not rigorous one-electron states.[5] In the local density approximation this correction term is isotropic and is shown to be very small for Cu, as an example.[6] Therefore Compton profile measurements should provide another powerful test on ground state band structure calculations.

In recent years the accuracy of the measurements of structure factors in imperfect single crystals as well as of Compton profiles was improved substantially. In addition the interpretation of Compton profiles in terms of their Fourier transforms revealed some new aspects of how experiment and theory are connected. The example discussed in the present paper is crystalline copper which for many years provided a testing ground for the theoretical understanding of the electronic structure of the non-simple elements, i.e. metals involving d-bands.

Some years ago it was already found that the anisotropy of the structure factors measured in V, Cr and Fe is much larger than predicted by theory. For Cu it turns out that even recent band structure calculations using various local approximations for the exchange-correlation potential do not agree with the measured structure factors and predict too large anisotropies in the Compton profiles. This disagreement between experimental and theoretical Compton profile anisotropies is also observed for V, Cr and Ni and we interpret this finding as breakdown of the local density approximation for these first row transition metals. A discussion of the measurements and calculations of the charge, spin and momentum density of the metallic elements Ti to Ni was presented recently by Weiss and Mazzone.[7]

The aim of the present paper is to demonstrate the degree of accuracy which can be reached in γ-ray Bragg and Compton scattering experiments, to compare the experimental structure factors and Compton profiles with theory and to start a discussion on the implications of the disagreement between experiment and theory which we found.

2. CHARGE DENSITIES

During the last 15 years most experimental charge density studies were carried out in single crystals with relatively large unit cells so that a larger number of Bragg reflections at low values of $\sin\theta/\lambda$ could be measured. This is an important aspect because changes in the valence charge density affect most strongly the low order reflections. In general these unit cells contain a larger number of atoms or molecules and the bonding between the molecules is considered negligible in most cases. The results of such charge density studies are presented as difference or valence density maps and they are compared with theoretical charge densities of the molecule whenever available. The comparison is complicated due to the thermal motion of the molecule. More recently there is a trend to study smaller systems and to map electrostatic properties.

Actually the limited accuracy of the x-ray structure factor measurements presents probably the most important obstacle for a further development of the field of charge density studies. Therefore a number of groups started to develop new diffractometers making use of the unique properties of Synchrotron radiation as there are the linear polarisation of the high intensity beam and its tunable wavelength extending to very short wavelengths. In a more special approach to measure absolute structure factors in imperfect single crystals γ-radiation of energies between 100 and 500 keV is used on diffractometers installed at the Hahn-Meitner-Institut in Berlin[8-10] (γ=0.03 Å) and at the University of Missouri in Columbia, U.S.A.[11-12] (γ=0.12 Å). Our aim in Berlin is to check the theoretical ground-state charge density $\rho(\vec{r})=|\psi_0(\vec{r})|^2$ of a crystalline solid by measuring the structure factor $F_{\vec{H}}$. The example we want to discuss in this paper is copper.

Before designing an experiment with the above goal we have to find out the necessary accuracy which has to be reached in the measurements so that decisions between various theoretical models are rendered possible. Snow[13-14] has performed self-consistent APW Xα calculations for copper. He used three different exchange parameters α=1, 5/6 and 2/3 and he obtained best agreement between experimental and theoretical energy band gaps for α=5/6. In particular Snow com-

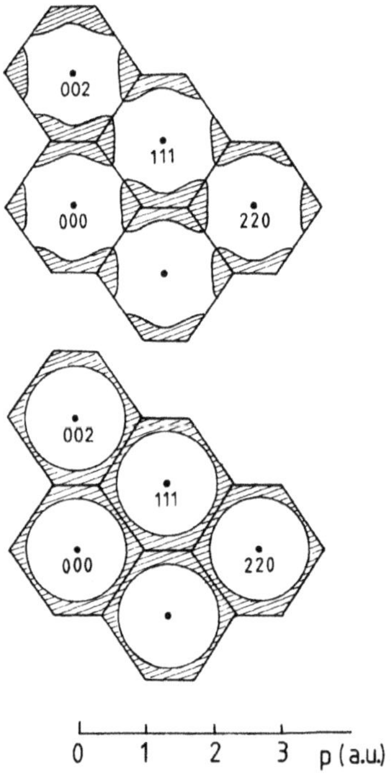

Figure 1. A cut through the reciprocal lattice in copper showing the occupied volume for electrons in the top band. In the lower part of the figure we show the degree of approximation in replacing the Fermi surface by spheres.

pares the experimental value of the radius of the neck of the Fermi surface along <111> of $r_N = 0.28$ Å^{-1} with theoretical results obtained for the 3 choices of α. The necks of the Fermi surface of copper are visible in Fig. 1 which shows a cut through the k-space, the hatched areas are the unoccupied regions between the Fermi surfaces centered at the various reciprocal lattice points. For α = 1 Snow finds no neck at all, for α = 5/6 or 2/3 he gets $r_N = 0.29$ Å^{-1} and 0.53 Å^{-1} respectively. In conclusion, the diameter of the neck of the copper Fermi surface is very sensitive to the choice of the exchange parameter α. Additionally it is known that the calculated value of r_N also depends critically on the numerical procedure applied in the calculation (see e.g. Ref. 25). Next it is interesting to know how much the choice of α affects the wavefunctions and the form factors which can be measured. Fortunately Snow also calculates atomic scattering factors (see Table 3 in Ref. 15) for three values of α. In Fig. 2 we have plotted the difference between his form factors for the various Bragg reflections and those calculated by

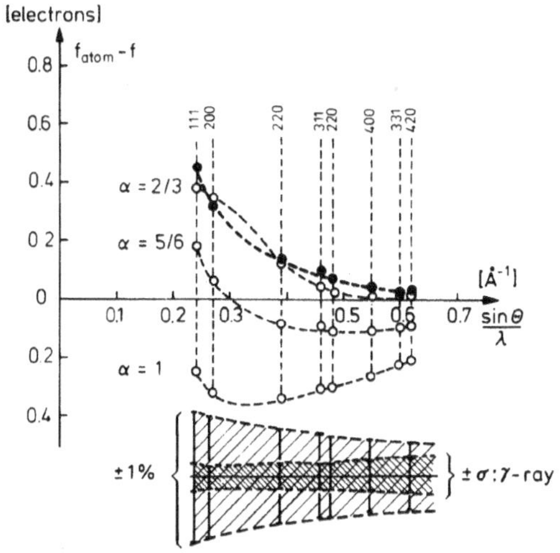

Figure 2. Comparison of the free-atom form factors f_{atom} due to Doyle and Turner[16] with the results of two different band-structure calculations for Cu. The open circles correspond to the SCF Xα APW calculation by Snow[13-14] for three values of the parameter α. The full circles correspond to the results of the SCF LCAO-Gaussian basis calculation of Bagayoko et al.[23-24] In the lower part of the figure we compare the σ = ± 1% error band with the error in the γ-ray diffraction measurements by Schneider et al..[9]

Doyle and Turner[16] in a relativistic Hartree-Fock calculation for the free atom. Additionally we have indicated the error band which one would obtain if the measured structure factors are affected by an error of 1%, which is the accuracy that many people believe can be achieved in an accurate x-ray diffraction study. By comparing the three theoretical data sets with the 1% error band we have to conclude that the error of the experimental copper structure factors should be of the order of ±0.5%. This demand is fulfilled in our γ-ray measurements as demonstrated by the second, much narrower error band shown in Fig. 2, which corresponds to the error in our absolute structure factors.[9]

To measure structure factors as a test of the theoretical crystal wavefunctions is by far not a new idea. Copper, for example, was studied by various authors using x-ray diffraction techniques. In Fig. 3 we have plotted the difference between the experimental results obtained by four groups on one hand and the free-atom form factor due to Doyle and Turner[16] on the other hand. The error bars on the individual x-ray form factors are rather large, but worse,

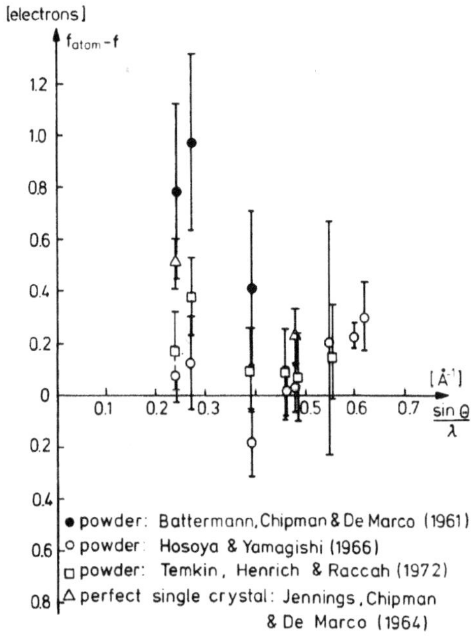

Figure 3. Comparison of the copper free-atom form factors f_{atom} due to Doyle and Turner[16] with the experimental scattering factors determined by various groups using different X-ray diffraction techniques.

the scatter of the experimental data obtained by the various groups for a given reflection is in general so large that only little can be learnt from a comparison between experiment and theory. This very unsatisfactory situation concerning the x-ray experiments results from the large number of problems related to the sample preparation and the measurements themselves, as well as to the assumptions one has to make in calculating structure factors from Bragg scattered x-ray intensities. The improvement in accuracy of the structure factor measurements by means of γ-ray diffractometry compared to x-ray diffraction data is mainly due to the fact that most of the sources of error which affect the x-ray data simply do not exist in γ-ray diffraction experiments. The polarisation of the incident γ-ray beam is well known and each Bragg reflection is measured on an absolute scale. The scattering geometry is simple and large samples can be used so that surface effects are of minor importance and the sample thickness can be measured accurately. The absorption of 412 keV γ-radiation in matter is weak and cryostats can be used without introducing additional experimental difficulties. The attenuation coefficient for beam absorption in the sample must not be determined explicitly. Extinction is much easier to handle than in x-ray diffraction experiments and anomalous dispersion does not occur. In Table 1 atomic scattering form factors measured at perfect and imperfect Cu single crystals are compared with the results of various band structure calculations including the most recent ones.

One of the earliest complete band calculations for Cu is due to Burdick.[17] In his non self-consistent, muffin-tin AΓW calculation he used a semi-empirical potential originally suggested by Chodorow[18] in order to account for exchange in a simple way. It is defined as that potential which when used in a single Hartree equation leads to the Hartree-Fock one-electron wavefunctions for d-electrons, that is, it gives the right result including exchange interaction without actually constructing an appropriate antisymmetric wavefunction. The band structure calculated by Burdick[17] is in excellent agreement with recent angle-resolved photoemission measurements performed by Thiry, Chandesris, Lecante, Guillot, Pinchaux and Pētroff[19] which were analysed within a free-electron-like model for the final state bands. Arlinghaus[20] recalculated the band structure of Cu along the lines given by Burdick[17], but he extended his calculation to provide theoretical atomic form factors which we present in Table 1. From Snow's self-consistent APW Xα calculation[13-14] we give only the results for $\alpha = 1$ and $\alpha = 2/3$. Wakoh and Yamashita[21] calculated form factors from wavefunctions obtained by the Green's function method using the Chodorow potential as well as a self-consistent potential constructed by means of a modified Hartree-Fock-Slater approximation.[22] The authors state that the two metallic potentials are similar to each other and therefore the difference between the two sets of theoretical atomic scattering factors is only of the order of 0.06 electrons, the use of the Chodorow potential yields smaller form factors. The band structure calculation due to Bagayoko, Laurent, Singhal and Callaway[23] is self-consistent. A local, density dependent exchange-correlation potential and the method of linear combination of Gaussian orbitals with a large basis set is used. The authors calculate the energy-band dispersions, the atomic form factors[24] and the Compton profiles. The most recent calculation of the electronic structure of copper which includes a calculation of structure factors is due to MacDonald, Daams, Vosko and Koelling.[25] In this self-consistent calculation the authors introduced relativistic kinematics and no muffin-tin or other type of "shape" approximation to the crystal potential is used. The exchange-correlation potentials employed are those implied by the local density approximation of the density functional formalism. The authors compare a few of their calculated scattering form factors with experimental data due to Jennings, Chipman and DeMarco,[26] Freund[27] and Schneider, Hansen and Kretschmer.[9] Unfortunately the data given in Table 1 of their paper[25] quoted due to Schneider et al.[9] are wrong. MacDonald et al.[25] must have used an earlier draft of our paper, later we discovered a mistake in our calculation of the low temperature lattice parameter which affected our earlier data set significantly. The numbers quoted in ref.[9] are

Table 1. Comparison of Experimental and Theoretical Scattering Form Factors for Cu

(hkl)	Jennings et al.[26] (1964) a = 3.6148 Å	Schneider et al.[9] (1981) a = 3.6035 Å	Arlinghaus[20] (1967) (Chodorow) a = 3.6147 Å	Snow[13-14] (1968) $\alpha=1$ a = 3.6148 Å	Snow[13-14] (1968) $\alpha=2/3$ a = 3.6148 Å
(111)	21.52 (10)	21.51 (5)	21.54	22.33	21.63
(200)		20.22 (4)	20.25	21.04	20.40
(222)	14.01 (10)	14.07 (5)	13.90	14.53	14.10
(333) } (511) }	9.41 (10)		9.51		
(333)		9.49 (6)			
(511)		9.53 (6)			

(hkl)	Wakoh et al.[21] (1971) (Chodorow)	Wakoh et al.[21] (1971) Modified HFS approximation	Bagayoko et al.[23-24] (1980) a = 3.6032 Å	MacDonald et al.[25] (1982) Exchange only a = 3.6032 Å	MacDonald et al.[25] (1982) Non relativistic, exch.-correl. a = 3.6148 Å
(111)	21.67	21.72	21.68	21.65	21.73
(200)	20.39	20.46	20.35	20.32	20.39
(222)	14.04	14.10	14.17	14.19	14.25
(333)			9.58	9.63	9.66
(511)			9.59	9.64	9.67

correct.* Although the data sets due to Jennings et al.[26] and Freund[27] are both measured on perfect single crystals the results show a large discrepancy which is probably due to the different approaches the authors chose to solve the problems caused by the long tail of the diffraction pattern and by thermal diffuse scattering. TDS is particularly serious in the measurements of Bragg intensities in perfect single crystals. Because Freund's values for the 111 and 222 reflecions, as an example, are much higher than the experimental results of other groups, we shall consider only the structure factor measurements by Jennings et al.[26] and by Schneider et al.[9] in the further discussion. Concerning the X-ray powder measurements we have the feeling that the problems inherent in that technique[28] were not fully recognized by the authors at that time.

From the selection of Bragg reflections considered in Table 1, which we took over from the paper by MacDonald et al.,[25] one has to conclude that there is very good agreement between the two experimental data sets and the theoretical scattering from factors calculated by Arlinghaus.[20] All the other calculations provide values which are higher than the low order experimental data. MacDonald et al.[25] have found that the charge density of Cu is not very sensitive to improvements in the calculations like the inclusion of relativistic kinematics and non muffin-tin effects. The same holds for a more carefully sampling of the Brillouin zone. The net effect of the relativistic corrections and the inclusion of the correlation potential is a small contraction of the charge density, which would lead to larger scattering factors, however, MacDonald et al.[25] conclude that these two effects cannot fully explain the difference between their results and those obtained by Arlinghaus.[20] They presume that the phenomenological local potential due to Chodorow[18] might be including exchange-effects more reliable than the standard local density approximation.

Fig. 4 shows the atomic scattering factors of Cu as determined by γ-ray diffraction together with the results of four band structure calculations. We have plotted the difference between the experimental and theoretical values on one hand and the free-atom results of Doyle and Turner[16] on the other hand. The lattice parameters used in the various solid-state calculations are slightly different and, except for the calculation by Bagayoko et al.,[23] clearly differ from our 50K lattice parameter. In comparing the measured structure factors with the theoretical ones, one must take into account that the $\sin\theta/\lambda$ value for a given reflection does change with the lattice parameter; the value of the free-atom form factor for the low order reflections is about 0.05 electrons higher at room temperature than at 50K owing to a contraction of the reciprocal lattice of about 0.3%. As a consequence, one should compare the difference between superimposed

*See erratum Phys.Rev.B. 26, 3473 (1982).

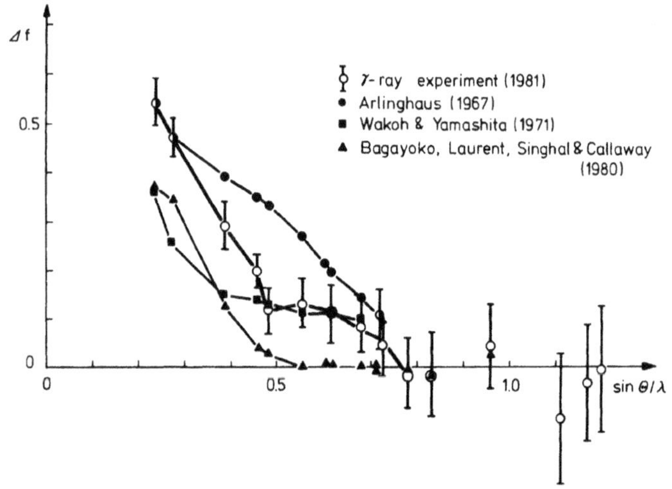

Figure 4. Comparison of the experimental γ-ray form factors of Cu due to Schneider et al.[9] with theoretical data from different band structure calculations. The plot shows the different $\Delta f=f_{DT}-f$, where f_{DT} is the free-atom value of Doyle and Turner.[16] f represents either the experimental or one of the theoretical form factor values.

free-atoms and the experiment with the difference between free-atom model and theory, always with the appropriate lattice constants for the free-atom models. We now see that the Arlinghaus calculation[20] fits perfectly well the experimental values for the reflections 111 and 200, whereas there is a net disagreement for the reflections in the range of $0.3 \leq \sin\theta/\lambda \leq 0.6$ Å^{-1}. Concerning the calculation due to Wakoh and Yamashita[21] we have plotted only the results obtained with their modified HFS potential which fit well our experimental data in the range of momentum transfer of $0.5 \leq \sin\theta/\lambda \leq 0.7$ Å^{-1}. For the reflections with $\sin\theta/\lambda < 0.5$ Å^{-1} the experimental and the Wakoh and Yamashita form factors clearly disagree. The most striking difference, however, occurs between our experimental form factors and the results due to the recent calculation by Bagayoko et al..[23]

The Cu conduction electrons affect the atomic scattering factors which are accessible in Bragg diffraction experiments only very weakly because of their highly diffuse nature. Therefore one may conclude that the recent band structure calculations predict a d-electron distribution which is contracted too much. On the other hand one may presume that, to a certain extent, the difference between experiment and theory could be caused by an underestimation of the electron transfer from 3d-like localised states to the delocalised conduction states in the band structure calculations. The latter aspect is considered in a strongly simplified model where we assumed that by forming a solid the 4s and one of the 3d electrons of the Cu atom are transferred into free-electron-like conduction states. If we further assume that the atomic d-states are uniformly occupied by the remaining 9d electrons we can calculate a $f(3d^9)$ atomic form factor from the following equation:

$$f(3d^9) = 0.9 \cdot [f(Cu^+) - f(Ar)]$$

where the form factors $f(Cu^+)$ and $f(Ar)$ are from the relativistic Hartree-Fock calculation for the Cu^+ ion and its Ar-core due to Doyle and Turner.[16] In Fig. 5 we have plotted the difference

$$\Delta f = [f(3d^{10}4s^1) - f(3d^9)]$$

which is in very good agreement with the difference between the Doyle and Turner free-atom form factors and our experimental data. This striking agreement between experiment and the simple model may be accidental. On the other hand Fig. 5 certainly demonstrates that an experimental difference charge density map, where again measured and free-atom structure factors are compared, will essentially show holes at the atomic positions and a rather constant charge density in between.

The effect of slight variations in the value of the lattice parameter on the measurable quantities has sometimes been studied in band structure calculations. Fig. 6 shows the difference between the form factors calculated by Koelling[29] for two different lattice parameters: a=3.6032 Å and a=3.6148 Å. The effect of the lattice parameter variation on the calculated form factors seems to be rather large. However, if one takes into account that by varying the lattice parameter the Bragg reflections occur at different values of $\sin\theta/\lambda$, this large effect is very much reduced as demonstrated in Fig. 6b. For a meaningful comparison between theory and experiment or between different theories it is thus important to know the lattice parameters for which the measurements and the calculations are carried out.

In Fig. 6b we have plotted the difference between the form factors due to Bagayoko et al.[23] and four theoretical data sets which we received from D.D. Koelling.[29] As far as the reflections 111, 200, 222, 333 and 511 are concerned Koelling's data correspond to the results given by MacDonald, Daams, Vosko and Koelling[25] for their calculations labelled xo (a=3.6032 Å) and nx (a=3.6148 Å). In the xo calculation only exchange is taken into account, nx corresponds to a non relativistic treatment of exchange and correlation. By comparison with the error band of our experimental data it is obvious that we cannot decide between the four theoretical data sets provided by Koelling.[29] However, there is a net discrepancy between all the modern

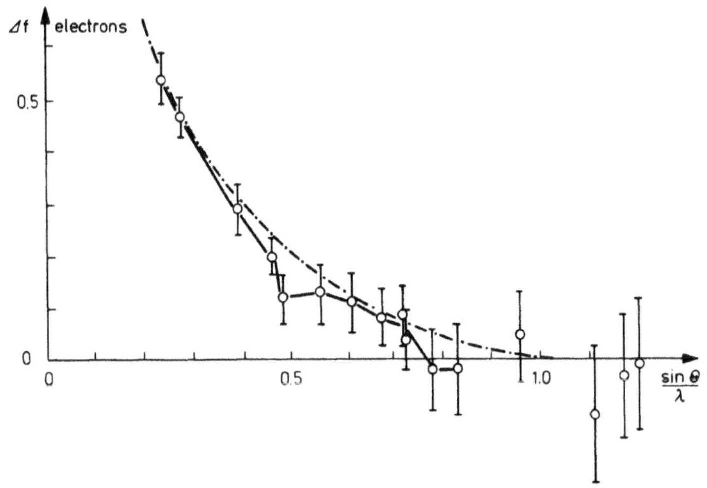

Figure 5. Comparison of the experimental γ-ray scattering factors of Cu due to Schneider et al.[9] with the form factor calculated for an atomic-like model where the 4s and one of the 3d electrons are transferred into plane wave states. The plot shows the difference $\Delta f = f_{DT} - f$, where f_{DT} is the free-atom form factor of Doyle and Turner.[16] f is the result calculated for the atomic-like model again on the basis of the calculation due to Doyle and Turner.

band structure calculations and our experimental results.

In Fig. 6b we have also plotted the difference between the form factors due to Bagayoko et al.[23] and those calculated by Doyle and Turner[16] for the free atom. It is interesting to note that Koelling's form factors[29] agree with Doyle and Turner's values for $\sin\theta/\lambda \geq 1.2$ $Å^{-1}$, whereas there is a net discrepancy between the free-atom values and those of Bagayoko et al. at high values of $\sin\theta/\lambda$. On the other hand the agreement between the free-atom form factors and those due to Bagayoko et al. in the range of $0.55 \leq \sin\theta/\lambda \leq 0.9$ $Å^{-1}$ is rather surprising, and one might conclude that the Gaussian basis set used by Bagayoko et al. gives a rather poor description of the charge density close to the nuclei. Because the number of accessible Bragg reflections is rather high for $\sin\theta/\lambda \geq 0.6$ $Å^{-1}$ the experiment should be able to decide between the theoretical form factors due to Koelling[29] and Bagayoko et al..[23] At present, however, the accuracy of the higher order structure factors as measured by means of γ-ray diffractometry is not sufficient. In order to improve on that the power of our sources should be increased so that a more complete high order data set can be measured in a reasonable time. Furthermore the effect of an harmonici-

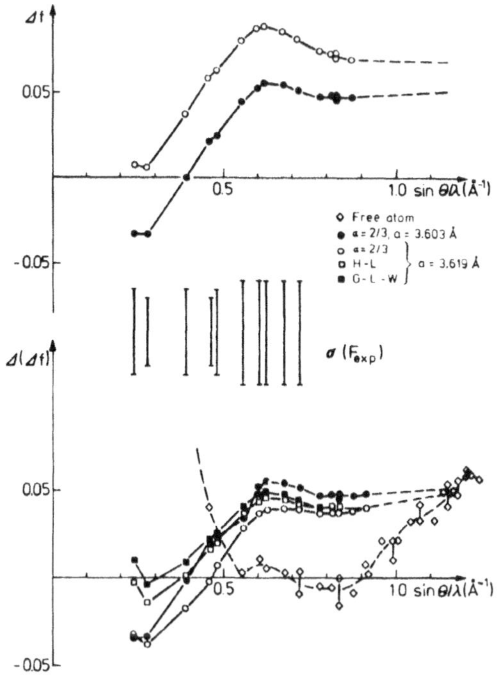

Figure 6. Comparison between different sets of theoretical form factors for Cu. The upper plot shows the difference $\Delta f = f - f_B$, where f_B represents the form factors due to Bagayoko et al..[23-24] f represents theoretical data calculated by Koelling[29] for two lattice parameter values using the Kohn-Sham-Gaspar exchange-correlation potential. The effect that the Bragg reflections occur at different values of $\sin\theta/\lambda$ if one varies the value of the lattice parameter is not considered. The lower plot shows a more extended comparison between the calculation by Bagayoko et al. and four data sets provided by Koelling including calculations for Hedin-Lundqvist and Gunnarson-Lundqvist-Wilkens exchange-correlation potentials. By plotting $\Delta(\Delta f) = [f_{DT} - f_B] - [f_{DT} - f]$ we take into account the effect of the different lattice parameters used in the various calculations because the Doyle and Turner[16] free atoms from factors f_{DT} are always calculated for the appropriate lattice parameter. For comparison we also show the standard deviations σ of the experimental Cu form factors measured by Schneider et al.[9]

ty and thermal diffuse scattering has to be studied explicitly. γ-ray measurements of structure factors on Au single crystals up to momentum transfers of the order of $\sin\theta/\lambda \leq 3$ Å^{-1} are in preparation.[30]

The accuracy of the structure factors measured by means of γ-ray diffractometry allows a study of the deformation density

$$\Delta\rho(\vec{r}) = \frac{1}{V}\sum_{\vec{H}}[F_{obs}(\vec{H}) - F_{theor}(\vec{H})]\, e^{-i2\pi\vec{H}\cdot\vec{r}}$$

where V is the volume of the unit cell and $F_{theor}(\vec{H})$ is the structure factor calculated for the superposition of free-atom charge densities. Because of the high symmetry of the f.c.c. lattice and the small number of reflections available in the range of $\sin\theta/\lambda < 0.8$ Å^{-1}, the error in the difference density is appreciable and therefore the map has to be regarded with considerable caution. This is also obvious from Fig. 5. All possible structure in the difference charge density between the atoms is only due to the small deviation of the measured scattering factors from the $[f(3d^{10}4s^1)-f(3d^9)]$-curve. In Fig. 7 we have plotted the difference density between γ-ray results[9] and the free-atom calculation of Doyle and Turner.[16] The summation only included terms for which $\sin\theta/\lambda \leq 0.8$ Å^{-1}. This cut-off is justified by the good agreement between experiment and free-atom calculation for the measured reflections above $\sin\theta/\lambda = 0.8$ Å^{-1} and, consequently series termination effects may be neglected. At the position of the atoms there is a deficiency of charge of −2.77(25) eÅ^{-1}, which partly appears as a charge pile up of the order of +0.19(8) eÅ^{-3} at the midpoint between nearest-neighbour atoms.

Our deformation density shown in Fig. 7 differs strikingly from another deformation density of copper calculated by Smart and Humphreys[31] which is based on experimental 111, 200 and 220 from factors deduced from critical-voltage electron diffraction measurements and 5 theoretical values, f(311) to f(420), from the band structure calculation of Wakoh and Yamashita.[21] Smart and Humphreys[31] find a concentration of electrons at both the tetrahedral and octahedral interstitial sites; there is no build up of charge in the nearest-neighbour direction in their deformation density. However, we are convinced that these density maps are strongly affected by series termination errors. The actual numbers of the critical-voltage form factors are f(111) = 21.786, f(200) = 20.454 and f(220) = 16.696 electrons.[32] They are about 0.25 electrons higher than our γ-ray diffraction experiments and about 0.1 electrons higher than the theoretical values of Bagayoko et al..[23] At present we cannot offer an explanation for the origin of the marked discrepancy between the two sets of experimental form factors, however, it might be interesting to note, that the form factors of electron diffraction depends on the Fourier transform of the crystal potential rather than the Fourier transform of the charge density.

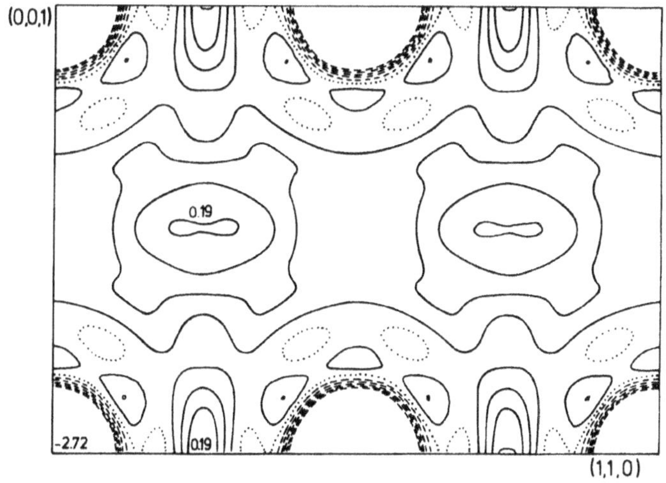

Figure 7. Difference density between the γ-ray diffraction results of Schneider et al.[9] and the free-atom calculation of Doyle and Turner[16] for copper. The contours are at intervals of 0.05 $e\text{Å}^{-3}$. The density at the positions of the Cu atoms is -2.72 $e\text{Å}^{-3}$.

3. MOMENTUM DENSITY - COMPTON PROFILES

The measurement of the energy spectrum of high energy (10 - 500 keV) photons which have been inelastically scattered by a body of electrons can be related to the distribution of electron momenta in the target. If the energy transferred to the electron in the Compton scattering process is much larger than the electron binding energy, the so-called impulse approximation is valid. Under these conditions the scattering cross section is proportional to the Compton profile, $J_{\vec{K}}(p_z)$. Compton scattering is a purely incoherent scattering process and therefore the Compton profile is a single-centred quantity, which implies that there will be a superposition of information from all parts of the unit cell. Although this is not a disadvantage when the structure has a high symmetry, the effect of this superposition complicates the interpretation of data from more complex crystals. In the absence of any evidence to the contrary, it is generally assumed that the core electron distribution is not affected by chemical bonding and is therefore identical to the free atom. Therefore, at present, it is reasonable to subtract an accurate theoretical free-atom core electron Compton profile[33] from the total experimental profile in order to isolate the contribution from the valence electrons.

The valence electron Compton profile displays a directional dependence which is generally only a few percent of the spherical average. This directional variation in the profile is in general demonstrated by plotting the difference $\Delta J(p_z)$ between Compton profiles measured for different orientations of the crystal relative to the scattering vector,

$$\Delta J(p_z) = J_{\vec{K}_1}(p_z) - J_{\vec{K}_2}(p_z)$$

Because of the practical advantages which derive from the fact that many systematic errors in the experimental Compton profiles cancel if one considers this difference profile, the anisotropy $\Delta J(p_z)$ has proved to be the most popular method of presenting Compton profile results for solids. On the other hand, the interpretation of $\Delta J(p_z)$ solely in terms of chemical bonding is very difficult because the observed anisotropy also depends upon the k-space occupancy, i.e. it is partly due to the shape of the Fermi surface, in the case of a metal, or the appropriate Brillouin zones in an insulator or semiconductor. The technique of Compton scattering experiments as well as its applications up to the year 1976 is summarized in a collection of review articles edited by Williams.[34]

The use of high activity γ-ray sources combined with solid state detectors for energy analysis has led to a significant reduction in measuring times in Compton scattering experiments. The Compton spectrometer installed at the Hahn-Meitner-Institut in Berlin[35] operates with 412 keV γ-radiation from sources of radioactive Au of an activity of about 250 Curie. The energy transferred to the recoiling electron is of the order of 250 keV and thus the impulse approximation is justified for probably all elements of the periodic table.[36] On the other hand the momentum resolution of γ-ray experiments is only of the order of 0.4 a.u., i.e. about 20% of the diameter of the Wigner-Seitz sphere of Cu. This is considerably worse than the resolution of 0.1-0.15 a.u. which can obtained with x-ray Compton spectrometers. A detailed comparison of the performance of x-ray and γ-ray Compton spectrometers can be found in the review articles by Reed[37] and Williams.[38] Regarding momentum resolution and intensity of the Compton scattered radiation considerable progress is expected once intense beams of high energy Synchrotron radiation are used in Compton scattering experiments. At present there is only one spectrometer operating at LURE in Paris which, however, uses rather low energy Synchrotron radiation of ~10 keV.[39]

Contact between theory and the results of Compton scattering experiments can be facilitated by reconstructing the momentum density, $\rho(\vec{p})$ from a series of experimental $J_{\vec{K}}(p_z)$ by a double Fourier inversion procedure. This approach has been severely handicapped by the slow rate of data collection in the earlier Compton scattering experiments and has been carried out only in a very limited number of cases (see Mijnarends[40] for a review). At the HMI we measure on average 2 Compton profiles of high statistical accuracy in one week and therefore we have developed an interpolation/inversion procedure based on an expansion of the momentum density in a series of lattice harmonics, with the objective of obtaining 3-D information from a series of 1-D measurements. This is similar to the approach adopted earlier by Mijnarends[41] in the analysis of positron annihilation data. For details of our reconstruction method and its application to experimental Compton data see ref.[42-44].

The interpretation of the results of Compton scattering experiments in direct space reveals some new aspects of how experiments and theory are connected. One can define a function $B(\vec{r})$ which is the Fourier transform of the electron momentum density,

$$B(\vec{r}) = \int \rho(\vec{p}) e^{-i\vec{p}\,\vec{r}} d\vec{p}$$

In the one-electron approximation the band momentum density can be written as a product of momentum space Bloch functions and one obtains

$$B(\vec{r}) = \int \sum_{occ} \chi^*_{\nu\vec{k}}(\vec{p}) \chi_{\nu\vec{k}}(\vec{p}) e^{-i\vec{p}\cdot\vec{r}} d\vec{r}$$

$$\chi_{\nu\vec{k}}(\vec{p}) = \int \psi_{\nu\vec{k}}(\vec{r}) e^{i\vec{p}\cdot\vec{r}} d\vec{r}$$

Using the convolution theorem we get

$$B(\vec{r}) = \sum_{occ} \int \psi_{\nu\vec{k}}(\vec{r}\,') \psi^*_{\nu\vec{k}}(\vec{r}\,'+\vec{r}) d\vec{r}\,'$$

i.e. $B(\vec{r})$ is the autocorrelation function of the corresponding position space wavefunction. Because of the integration over planes in momentum space perpendicular to the scattering vector $\vec{K}$ which occurs in the measurement of Compton profiles, the 1-dimensional Fourier transform of the directional Compton profile $J_{\vec{K}}(p_z)$ yields one component, $B(z)$, of the total autocorrelation function along the direction defined by the scattering vector $\vec{K}$

$$B(z) = \int J_{\vec{K}}(p_z) e^{-ip_z z} dp_z$$

The application of this approach to atoms and simple molecules has been discussed by Weyrich, Pattison and Williams[45] and for solids by Schülke.[46] Schülke has shown that, for a given band, the values of $B(\vec{r})$ at the translational lattice vectors, $\vec{R}$, are the Fourier coefficients of the shape transform of the occupied volume in k-space

$$B(\vec{R}) = \frac{2}{\Omega_{BZ}} \int_{BZ} n_\nu(\vec{k}) e^{-i\vec{k}\cdot\vec{R}} d\vec{k}$$

the integration is over the volume of the Brillouin zone. As a consequence $B(\vec{r})$ for an insulator must have a value of zero at all translational vectors, $\vec{R}$. This property of $B(\vec{r})$ has been confirmed experimentally for silicon and germanium,[47] for example, and also serves as a useful check on theoretical Compton profile calculations (see e.g. the discussion on LiH by Pattison and Weyrich.[48]) In the case of a metal such as copper, an examination of $B(\vec{r})$ in the region of the lattice vectors focusses the attention on the behaviour of those electrons which have an energy close to the Fermi level, i.e. the uppermost occupied band, because these are the electrons whose k-dependence is responsible for the shape of the Fermi surface.

Next the Compton profiles measured by Pattison, Hansen and Schneider[49] on Cu single crystals are compared with the theoretical data due to Bagayoko, Laurent, Singhal and Callaway[23] and Bross.[50] The theory due to Bagayoko et al. was already discussed in the section on charge density where we compared their extensive set of atomic scattering factors with experimental data. In a non self-consistent calculation Bross[50] determined the momentum density and the Compton profiles of Cu using his modified APW method.[51-53] The author justifies the use of the heuristic Chodorow potential[18] to describe exchange and correlation by the very good agreement of the recent angle resolved photoemission data due to Courths[54] with the band structure calculation by Burdick.[17] Another point in favour of the Chodorow potential is given by the work of Janak, Williams and Moruzzi[55] who have shown that the Chodorow potential leads to Fermi surface properties in roughly as good agreement with the experimental results due to Coleridge, Croxon, Scott and Templeton[56] as their self-consistent potential, where the parameter α has been adjusted to reproduce the Fermi surface topology. Stöhr and Bross[57] also calculated atomic scattering factors for Cu by means of the modified APW method and compared their results with those obtained by Arlinghaus[20] using the APW method and the Chodorow potential. Unfortunately this comparison is only in terms of the percentage of deviation of the theoretical from the experimental data due to Jennings et al.[26] Therefore a direct comparison with our experimental scattering factors[9] is not possible. Nevertheless, the 222 value of Stöhr and Bross[57] is significantly larger than the Arlinghaus value and rather close to our experimental data. Using the modified APW method and the Chodorow potential Bross and Stöhr[58] also calculated the anisotropy of the angular correlation of the two photons created by positron annihilation in single crystals of Cu. As shown in Fig. 8 the theory agrees nicely with the experimental data due to Cushner, Erskine and Berko,[59] although the effect of electron-positron correlations, which lead to a momentum dependent "enhancement factor", was neglected. This effect

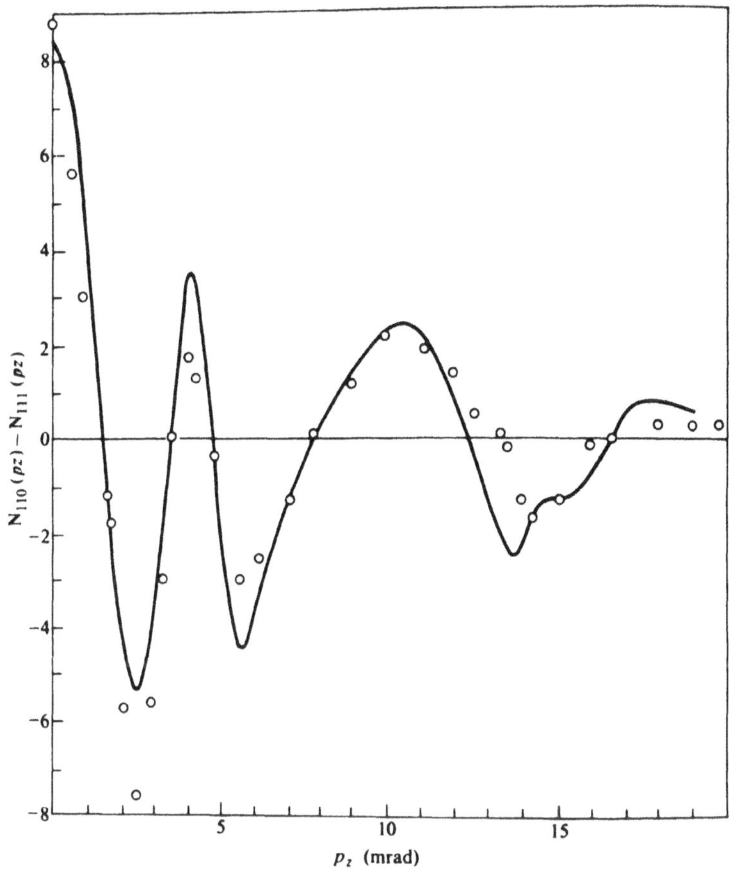

Figure 8. Anisotropy of the angular correlations of positron annihilation radiation in Cu; the circles represent the data of Cushner et al.,[59] the solid line the theoretical results of Bross and Stöhr[58] (from S. Berko Ref. 34). (1 mrad $\hat{=}$ 0.137 atomic units of momentum).

was discussed in detail by Mijnarends and Singru.[60] At this point it may be appropriate to sketch the difference between positron annihilation and Compton scattering experiments which both probe the electron momentum distribution in solids.

Positron annihilation angular correlation experiments show much better momentum space resolution than γ-ray Compton scattering experiments. Particularly if position sensitive area detectors are used, two dimensional intensity distributions are measured, that is, the experiment integrates only over lines through the 3-dimensional momentum distribution instead of the integration over planes in a Compton scattering experiment. On the other hand the interpretation of the information from positron annihilation experiments is complicated by the influence of the positron wavefunction itself, which makes it necessary to calculate crystal wavefunctions for both the electrons and the positron. In addition the effect of electron positron correlations have to be considered. Despite these difficulties, it is still possible to use positron experiments to test band structure calculations as was convincingly demonstrated in the recent results presented by Berko.[61] On the other hand positron annihilation experiments are not very sensitive to the momentum distribution of more localized electrons, like the d-electrons in transition metals. Therefore, the anisotropy in the angular correlations of Cu shown in Fig. 8 vanishes at momenta of the order of 2 a.u., whereas the Compton profile anisotropies observed by means of Compton scattering experiments goes out to momenta of the order of 5 a.u.. To summarize, if one wants to learn something about electron correlation effects or about the more localized electrons in a solid, positron annihilation experiments are not the most suitable ones.

In Fig. 9a the experimental <110> Compton profile due to Pattison et al.[49] is compared with the theoretical one calculated by Bross.[50] For momenta p_z>1 a.u. the agreement between theory and experiment is remarkably good, particularly, the kinks in the experimental Compton profile are very well reproduced by theory. For p_z = 0, however, the experimental value is about 3% larger than predicted by theory. In Fig. 9b we have plotted the difference between the experiment and the theoretical Compton profile due to Bross on one hand and the profile calculated by Bagayoko et al.[23] on the other hand.

At this point one should mention that the experimental data are subject to a number of important corrections before one obtains the Compton profile. The effect of multiple Compton scattering is probably the most important correction to be made, however, because of the effort of various groups over the last ten years, this correction is in rather good shape. We apply a semi-empirical extrapolation procedure developed by Felsteiner and Pattison[62] whenever, as in the case of Cu, Compton measurements can be performed on samples of different thicknesses. Another experimental problem, which we have studied recently,[63] is due to the fact that photons are Compton scattered in the source itself so that our 412 keV γ-line shows a tail of lower energy photons. The amount of low energy contamination in the γ-ray source of our Compton spectrometer was measured by means of the HMI γ-ray diffractometer using the sample crystal as energy analyser. The main effect of the source contamination is a rather strong asymmetry in the measured profile and in the past we have analysed only the high energy side of the experimental Compton profile which is less affected by this problem. From the known low energy contamination of the 412 keV line in our spectrometer we have calculated the asymmetry, which, as shown in Fig. 10, agrees very well with the

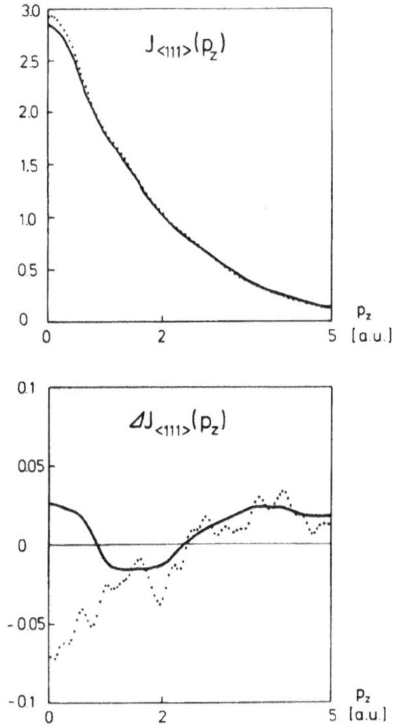

Figure 9. The upper plot shows a comparison of the experimental Compton profile $J^{exp}<111>$ (p_z) due to Pattison et al.[49] (dots) with the theoretical Compton profile $J^{Bross}<111>$ (p_z) calculated by Bross[50] (continuous curve). In the lower part of the figure we compare these two Compton profiles with the theoretical profile $J^{B}<111>$ (p_z) calculated by Bagayoko et al..[23] The continuous curve corresponds to the difference $J^{B}<111> - J^{Bross}<111>$, the (dots) represents $J^{B}<111> - J^{exp}<111>$.

asymmetry in the measured Cu Compton profile. On the other hand this calculation also shows the extent to which the high energy side of the experimental Compton profile is affected by the low energy contamination. The Compton profile obtained from the standard processing of our raw experimental data[64] is in the case of copper about 0.3% too low at $p_z = 0$. Fig. 10 shows that the correction is within the error band reported in our paper[49] but, nevertheless, a proper correction for the effect of source contamination seems worth-while. In the case of Ce this correction would lead to a significant increase of the peak value of the experimental Compton profile of ~1-3% if measured on our Compton spectrometer. Therefore, at least the experi-

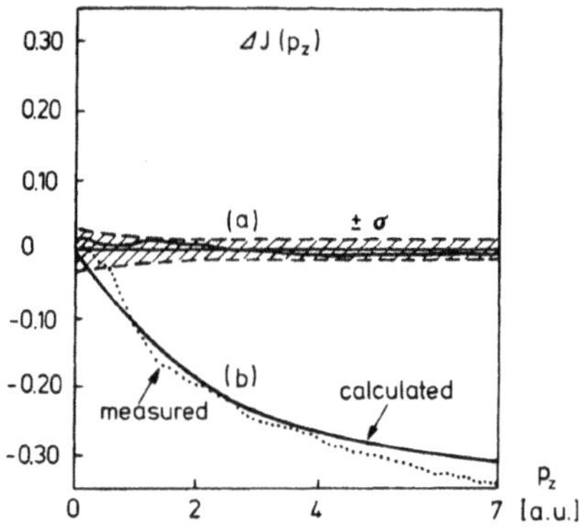

Figure 10. Effect of the low energy contamination of the Au γ-ray source of the HMI Compton spectrometer on the measured copper Compton profile. (a) shows the correction on the high energy side of the measured <111> Compton profile together with the error band given by Pattison et al..[49] In (b) we compare the calculated asymmetry between high and low energy side of the Compton profile with the asymmetry observed in the measurements on Cu.

mental profiles for heavier elements should be carefully corrected for source contamination.

In Fig. 11 our experimental Compton profile anisotropy ΔJ (110-001) is compared with the result of an earlier measurement performed by Eisenberger and Reed[65] by means of a γ-ray Compton spectrometer using 160 keV radiation from ^{123m}Te. Eisenberger and Reed present deconvoluted data which we have convoluted with our resolution function to make the comparison meaningful. The two experiments agree very well within the quoted error. On the other hand the theoretical Compton profile anisotropy calculated by Bagayoko et al.[23] which is also plotted in Fig. 11 does not agree with the experiments. Fig. 12 shows the ΔJ(110-111) and ΔJ(111-001) anisotropies and again the oscillations are reproduced qualitatively by theory, however, the calculated amplitudes are in general much higher than the experimental data. A simple model calculation based on the Seitz approximation, in which higher order Fermi surface volumes all of spherical shape (see Fig. 1) are taken into account in obtaining the momentum density, demonstrates that the major contribution to the Compton profile anisotropy in copper is due to the hybridization which has mixed d-electrons and nearly-free electrons in the top band.[49] Particularly, the oscillations in the Compton profile anisotropy at $p_z \geq 1$ a.u. are mainly caused by the contrast between occupied and unoccupied areas in k-space. Any fine structure in the theoretical anisotropy due to the detailed shape of the Fermi surface cannot be resolved

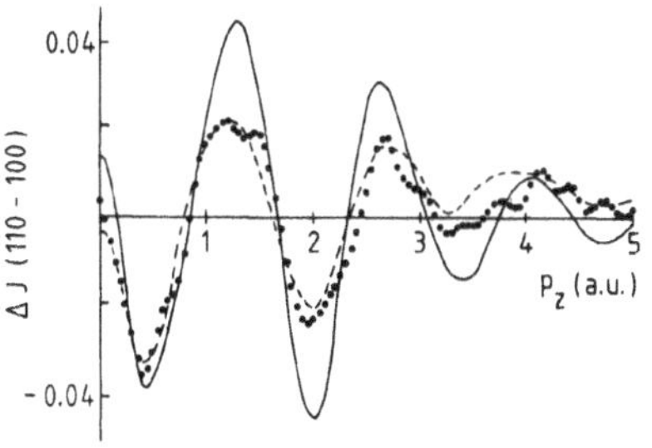

Figure 11. The experimental anisotropy in the Compton profile of copper. The results due to Pattison et al.[49] are shown by the dotted curve, while the earlier experimental results by Eisenberger and Reed[65] are shown by a dashed curve. The continuous curve is the theoretical result due to Bagayoko et al.[23] after convolution with the experimental resolution function. The statistical error in the data of Pattison et al. is ±0.007, ±0.002 and ±0.0015 at p_z = 0.0, 1.0 and 3.0 a.u. respectively.

in present day γ-ray Compton scattering experiments. Fig. 13 shows the theoretical Compton profile anisotropies ΔJ(110-001) and ΔJ(110-111) due to Bagayoko et al.[23] and Bross.[50] Although the two band structure calculations are of rather different nature the calculated Compton profile anisotropies show only small differences, on the other hand, the disagreement between experiment and theory is obvious from Fig. 11 and 12.

In Fig. 14 and Fig. 15 we have plotted the long range behaviour of $B(\vec{r})$ for copper along the three main crystallographic directions as calculated by Bagayoko et al.[23] and by Bross.[50] Again the agreement between the two different theories is surprisingly good. In Fig. 16 the theoretical B(z) function along <110> due to Bagayoko et al. is compared with our experimental data. Because $B(\vec{r})$ is the Fourier coefficient of the shape transform of the occupied volume in k-space for $\vec{r}$ equal to a lattice translational vector, $\vec{R}$, in Table 2 we have compared the accessible experimental with the theoretical values of this quantity. The limited energy resolution of γ-ray Compton scattering experiments, described for our spectrometer by a Gaussian of FWHM of 0.41 a.u., results in an exponential damping of the information in the measured function B(z), which in our case corresponds to a damping factor of 2 at z = 3.56 Å and a factor of 10 at 6.49 Å. Therefore, the better the experimental energy resolution the more $B(\vec{R})$ values become accessible in the experiment. In Table 2 we consider 5 $\vec{R}$-values and find again rather good agreement between the two theories and a net discrepancy with the experiment for 3 $\vec{R}$-values. Because the band-structure calculations reproduce well the shape of the Fermi surface

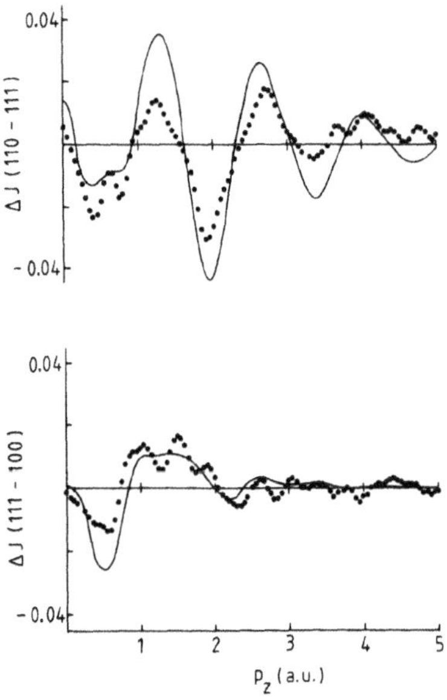

Figure 12. Comparison between the experimental anisotropy in the Compton profile of copper as measured by Pattison et al.[49] and the theoretical anisotropy predicted by Bagayoko et al.[23] in their band structure calculation for the combinations <110> - <111> and <111> - <100>. The theory has been convoluted with the experimental resolution function. The statistical error in the experimental data is ±0.007, ±0.002 and ±0.0015 at p_z = 0.0, 1.0 and 3.0 a.u. respectively.

of Cu, which was measured very accurately, this discrepancy should be related to the hight of the break in the occupation function $n(\vec{k})$ at the Fermi momentum.

Localized secondary peaks in $B(\vec{r})$ centred on the sites of the translational lattice represent another interesting feature in the $B(\vec{r})$ function. They cause most of the anisotropy in the theoretical $B(\vec{r})$ function of Cu. By far the most dominant feature is situated on the nearest neighbour site along <110>. This peak is large enough (~0.2e) and at close enough range (2.55 Å) to be easily observed in the experimental data. Within a simplified picture Pattison et al.[49] discussed the nature of these localized peaks and conclude that they may be directly related to the wavefunctions of the rather localized d-electrons. By making use of the Seitz approximation, one can separate $B(\vec{r})$ for a given band into a product of two components, $B_S(\vec{r}) \cdot B_u(\vec{r})$, where $B_S(\vec{r})$ depends only on the shape of the occupied

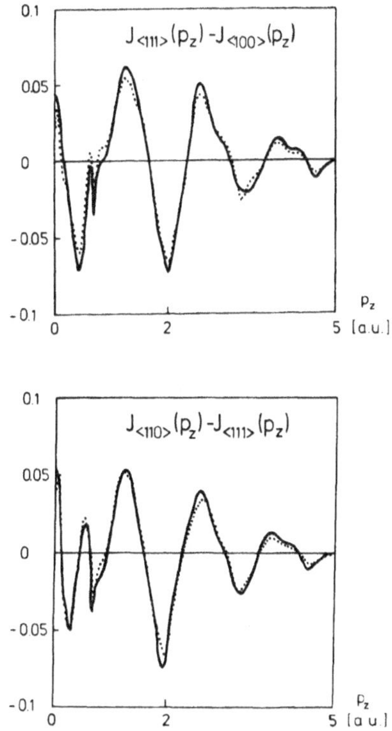

Figure 13. Theoretical Compton profile anisotropies of copper for the combinations <110> - <001> and <110> - <111> as calculated by Bagayoko et al.[23] (continuous curve) and by Bross[50] (dotted curve).

volume in k-space, while $B_u(\vec{r})$ is a wavefunction dependent term which can be written as the autocorrelation function of the band average periodic part of the Bloch wavefunctions. Since the uppermost occupied band in copper is a hybrid of d-electrons and free electrons, the periodic contribution from $B_u(\vec{r})$ will have narrow d-electron peaks around each translational lattice site superimposed on a uniform "background" from the free electrons. Certainly this feature should be studied quantitatively within modern solid state calculations. In their LMTO calculation on Cu Jepsen, Glötzel and Mackintosh[66] have studied the effect of the d-band mixing on the calculated Fermi surface. They showed that by omitting the hybridization of the nearly-free-electron bands with the d-electron bands one obtains an almost spherical Fermi surface. It would be very interesting to see the effect of the degree of hybridization on the local features in $B(\vec{r})$ around the lattice sites.

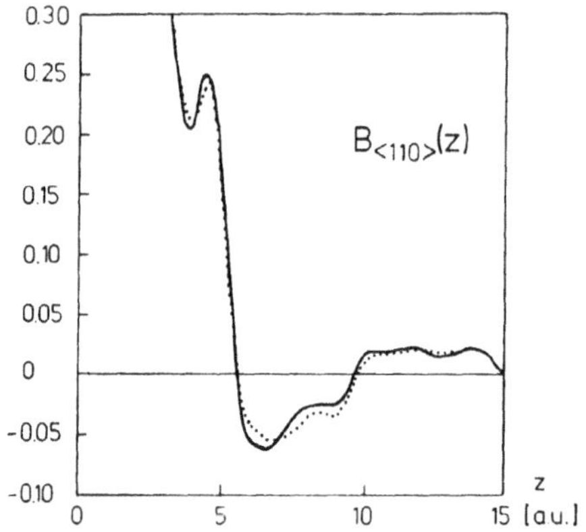

Figure 14. The long range part of the theoretical $B(\vec{r})$ function for copper in the <110> direction as calculated by Bagayoko et al.[23] (continuous curve) and by Bross[50] (dotted curve).

4. DISCUSSION

So far we implicity assumed that the crystal wavefunction can be represented by a Slater determinant constructed from the solutions of the one-particle Schrödinger equations. These one-electron wavefunctions have a physical meaning within the Hartree-Fock scheme. In the density functional formalism one obtains pseudo one-electron wavefunctions $\psi_{\nu\vec{k}}(\vec{r})$ and only the charge density

$$\rho(\vec{r}) = \sum_{occ} |\psi_{\nu\vec{k}}|^2$$

has a physical meaning. Recently, in analogy to the Hohenberg-Kohn theorem,[2] it has been shown by Henderson[67] that the ground state energy is also a unique (though not universal) functional of the momentum density $\rho(\vec{p})$, which in principle opens the possibility of a direct variational determination of $\rho(\vec{p})$. However, due to the lack of appropriate explicit expressions for this functional no actual calculations along these lines have been done, as far as we know.

Lam and Platzman[5] have shown that a correction term has to be added to the results of band structure calculations performed within the Hohenberg-Kohn-Sham density functional formalism[2-4] if momentum densities are calculated. In their original article[5] this correction term is treated only briefly and no explicit calculations are carried out. At this point we want to present the results of such calculations for Cu which demonstrate that the observed discrepancies be-

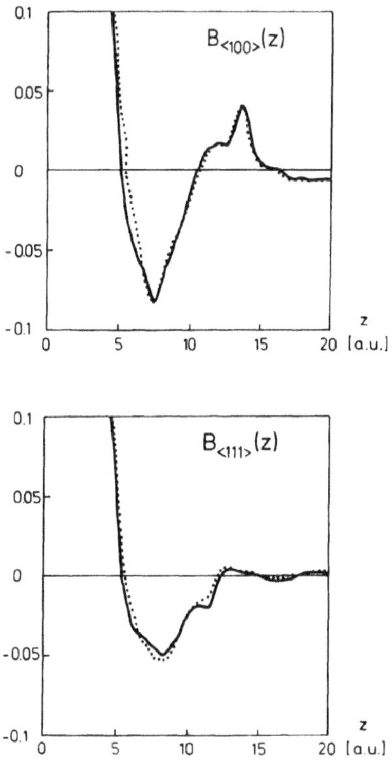

Figure 15. The long range part of the theoretical $B(\vec{r})$ function for copper in the directions <100> and <111> as calculated by Bagayoko et al.[23] (continuous curve) and by Bross[50] (dotted curve).

tween theoretical Compton profiles calculated in the local density approximation and the experimental data cannot be explained by the Lam-Platzman correction term, rather they are due to the inadequacy of the local potentials. The topic will be elaborated formally and in more detail in a separate paper.[68]

The Lam-Platzman correction term for the momentum density in the local density approximation reads

$$\Delta\rho^{LP}(p) = \int_{\substack{\text{unit}\\ \text{cell}}} \rho(\vec{r})\{n^h[p,\rho(\vec{r})] - n^f[p,\rho(\vec{r})]\}d\vec{r}$$

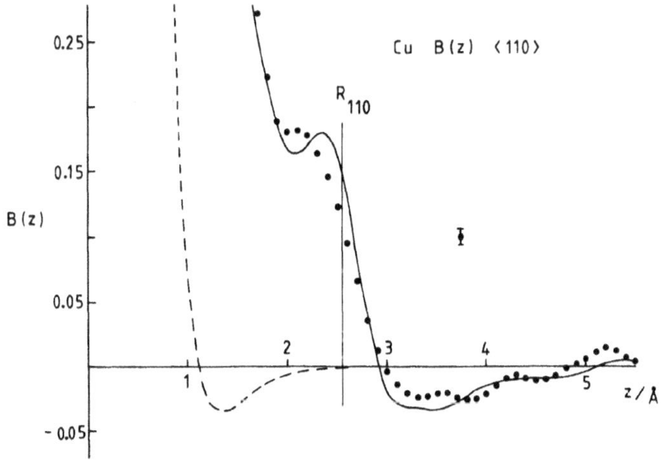

Figure 16. The experimental long range part of $B(\vec{r})$ for the valence electrons in copper in the <110> direction (dotted curve) due to Pattison et al..[49] The statistical error in the experimental curve is ±0.005 e. The theoretical curve for the 11 valence electrons obtained from the band structure calculation of Bagayoko et al.[23] is shown by the continuous curve, and for the 18 core electrons by the broken curve. (Both theoretical curves have been multiplied by the experimental Gaussian function, which, because of the experimental momentum resolution of 0.41 a.u., attenuates the theory by a factor of 2 at 3.56 Å).

$n^f[p,\rho(\vec{r})]$ and $n^h[p,\rho(\vec{r})]$ are the occupation numbers of the free electron gas or the homogeneous interacting electron gas with the (local) charge density $\rho(\vec{r})$. The latter has been calculated by different groups[69–71] and reasonable agreement between different theories beyond the random phase approximation is obtained. Because the uniform electron gas is isotropic, the Lam-Platzman correction term is isotropic as well. This means that the band structure Compton profile anisotropies are not affected at all and can therefore directly be compared with the experiment. $\Delta\rho^{LP}(p)$ plotted in Fig. 17 has been calculated for copper by using the self-consistent muffin-tin charge density as tabulated in Ref.[72] and the occupation numbers of Arponen and Pajanne.[73] As expected, the net effect of the Lam-Platzman correction term is a transfer of electrons from lower to higher momenta. It turns out that the Lam-Platzman correction term is very small for a high density metal like copper. The value of the Fourier transform of $\Delta J^{LP}(p)$ (the Compton profile corresponding to $\Delta\rho^{LP}(p)$) at the first lattice translation $a/\sqrt{2}$ is 0.007, which goes in the right direction, but again, it is much too small to achieve agreement with experiment (see table 2).

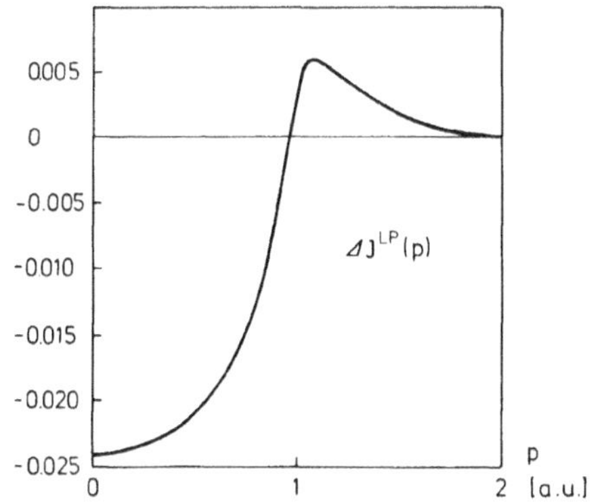

Figure 17. The Lam-Platzman correction term for the momentum density in the local density approximation[5] calculated for copper.

At present we cannot satisfactorily explain the striking difference between the experimental and theoretical Compton profile anisotropies nor the difference in the Fourier transformed Compton profiles B(z). We feel that the disagreement between theoretical and experimental values of B(z) for z equal a lattice translational vector as shown in Table 2 is of special interest. The origin of localized secondary peaks in B(z) centered on the sites of the translational lattice is not yet fully understood and tight binding calculations with the aim to clarify this point are under way.[74] The fact that the two rather different band structure calculations due to Bagayoko et al.[23] and Bross[50] provide nearly identical Compton profile anisotropies, both being in net disagreement with the experimental data, is surprising and needs further discussion. In this context it is interesting to note that within a similar calculation Bross and Stöhr[58] were able to reproduce the anisotropy in the positron annihilation angular correlation data due to Cushner et al..[59] However, we have to keep in mind that these positron annihilation experiments are probably not very suitable to study many body effects because of electron-positron correlations; additionally these experiments are not very sensitive to the momentum distribution of the more localized electrons like d-electrons in transition metals.

Contrary to the Compton profiles there is a net difference between the theoretical structure factors as calculated by Bagayoko et al.[23-24] and by Stöhr and Bross[57], although the comparison is rendered difficult because of the small number of form factors provided by Stöhr et al. and their presentation as percentage deviation from the experimental data due to Jennings et al..[26] Nevertheless, the 111 and 222 form factors calculated by Stöhr and Bross using the modified APW method[51-53] and the Chodorow potential[18] are certainly closer to

Table 2. Values of $B(\vec{R}_i)$ where $\vec{R}_i$ is a "Real Lattice" Vector

$\vec{R}_i$			$\lvert\vec{R}_i\rvert$ [Å]	Bagayoko et al. Ref. 23	Bross Ref. 50	Exp. Ref. 49	ρ(exp)
1/2	1/2	0	2.547	0.214	0.208	0.161	0.005
1	1	0	5.096	-0.003	-0.009	0.033	0.015
1	0	0	3.603	-0.067	-0.066	-0.061	0.007
2	0	0	7.206	0.04	0.04	0.02	0.06
1	1	1	6.241	-0.015	-0.008	0.16	0.03

our experimental data[9] than those due to Bagayoko et al.[23-24] Our experimental data[9] are in very good agreement with the 111 and 200 form factors calculated by Arlinghaus[20] within a non-self-consistent APW calculation again using the heuristic Chodorow potential to describe exchange. However, the 222 value of Arlinghaus is in net disagreement with our experimental value as well as with the result obtained by Stöhr and Bross[57] in their rather similar calculation. In the discussion of Fig. 5 we have seen that the slight charge pile up between next neighbour atoms as observed in our difference charge density map shown in Fig. 7 is mainly due to the experimental value for the 222 structure factor. By calculating the structure factors of Cu Arlinghaus[20] has spherically averaged the density inside the muffin-tin sphere and replaced the density between the spheres by a constant average value. This so-called "double muffin-tin approximation" could be the reason for the discrepancy between the 222 form factors. We may therefore conclude that the non self-consistent APW[20] and the modified APW[57] calculations, both using the heuristic Chodorow potential provide charge densities and structure factors in reasonable agreement with the experimental data due to Jennings et al.[26] and Schneider et al..[9]

On the other hand it is rather worrying that the experimental data do not agree with the theoretical form factors determined within recent self-consistent band structure calculations due to Bagayoko, Laurent, Singhal and Callaway[23-24] and MacDonald, Daams, Vosko and Koelling,[25] both employing exchange-correlation potentials as implied by the local density approximation of the density functional formalism. We follow the statement presented by MacDonald et al.[25] who presume that the phenomenological local potential due to Chodorow[18] might be including exchange-effects more reliably than the standard local density approximation. Both theoretical data sets are higher than the experimental form factors for Bragg reflections of $\sin\theta/\lambda \leq 0.6$ Å^{-1}. Because of the very diffuse nature of the conduction electrons in Cu they affect only very weakly the atomic scattering factors accessible in Bragg diffraction experiments and thus the above result could be interpreted as being due to a theoretical d-electron distribution which is too much contracted. On the other hand more diffuse d-electron distributions should lead to an increase in the theoretical Compton profile anisotropy and thus increase the discrepancy between experimental and theoretical Compton data. As another qualitative explanation for the discrepancy between theoretical and experimental form factors one could assume that the electron transfer from 3-d like localized states to the delocalized may be underestimated in the band structure calculations within the local density approximation. As shown in Fig. 5 we found a remarkably good fit of our experimental data with a simple atomic-like model where the 4s and one of the 3-d electrons are transferred into plane wave states. However, this approach may again lead to an even stronger theoretical Compton profile anisotropy.

The accuracy of our experimental structure factors does not allow for a distinction between the 4 sets of theoretical form factors provided by Koelling[29] extending the information given in the corresponding paper by MacDonald et al..[25] The self-consistent calculations were performed for the Kohn-Sham-Gaspar exchange-correlation potential with $\alpha=2/3$, as well as for the potentials due to Hedin-Lundqvist and Gunnarson-Lundqvist-Wilkens. In the momentum range up to $\sin\theta/\lambda \lesssim 0.45$ Å^{-1} the form factors due to Koelling for the H-L and G-L-W potential agree well with those calculated by Bagayoko et al.. On the other hand there is a clear discrepancy between the results obtained by the two groups for $\sin\theta/\lambda \geq 0.5$ Å^{-1}. By comparison with the free atom form factors due to Doyle and Turner[16] we conclude that the Gaussian basis set used by Bagayoko et al. gives a rather poor description of the charge density close to the nuclei.

At present the accuracy of absolute structure factor measurements in imperfect single crystals by means of γ-ray diffractometry is limited to $\sigma \pm 0.3\%$ for the low order reflections. Further improvements imply a very accurate knowledge of the sample properties which is regarded as the main difficulty. Aldred and Hart[75] performed absolute measurements of X-ray structure factors of Si by means of the Pendellösung method and the internal consistency of the results indicates that probably errors are of the order of ± 0.1%. This is the highest accuracy in structure factor measurements obtained until now. However, the method requires extremely perfect single crystals of large size and unfortunately there are only very few systems for which suitable samples exist. On the other hand there is considerable effort in at least two laboratories to develop techniques which allow the measurement of highly accurate structure factors in small perfect crystals, so that the sample problem should become less severe.[76-78] As far as Compton scattering experiments are concerned the use of high energy Synchrotron radiation beams and crystal analyser systems should lead to a considerable improvement in the momentum resolution, still allowing the measurement of Compton profiles of good statistical accuracy in a reasonable time. Another progress to be expected in Compton scattering experiments using circularly polarised Synchrotron radiation is due to the fact that circularly polarised radiation interacts with unpaired spin electrons through the Compton cross section to yield information about the momentum distribution of magnetic systems.[79]

REFERENCES

1. J.C. Slater, Phys.Rev. 81, 385 (1951).
2. P. Hohenberg and W. Kohn, Phys.Rev. 136, B864 (1964).
3. W. Kohn and L.J. Sham, Phys.Rev. 140, A1133 (1965).
4. L.J. Sham and W. Kohn, Phys.Rev. 145, A561 (1966).
5. L. Lam and P.M. Platzman, Phys.Rev. B9, 5122 (1974).
6. G.E.W. Bauer, in preparation.
7. R.J. Weiss and G. Mazzone, J.Appl.Cryst. 14, 401 (1981).
8. J.R. Schneider, P. Pattison and H.A. Graf, Nucl.Instr.Meth. 166, 1 (1979).
9. J.R. Schneider, N.K. Hansen and H. Kretschmer, Acta Cryst. A37, 711 (1981).
10. J.R. Schneider, Nuclear Science Applications, Vol.1, 227 (1981).
11. R.W. Alkire and W.B. Yelon, J.Appl.Cryst. 14, 362 (1981).
12. W.B. Yelon and F.K. Ross, Nucl.Instr.Meth. 193, 285 (1982).
13. E.C. Snow, Phys.Rev. 171, 785 (1968).
14. E.C. Snow, Phys.Rev. 172, 708 (1968).
15. J.H. Wood, pp. 43-63, in:"Energy Bands in Metals and Alloys", ed. by L.H. Bennet and J.T. Waber, Gordon and Breach (1967).
16. P.A. Doyle and P.S. Turner, Acta Cryst. A24, 390 (1968).
17. G.A. Burdick, Phys.Rev. 129, 138 (1963).
18. M. Chodorow, Phys.Rev. 55, 675 (1939).
19. P. Thiry, D. Chandesris, J. Lecante, C. Guillot, R. Pinchaux and Y. Pétroff, Phys.Rev.Lett. 43, 82 (1979).
20. F.J. Arlinghaus, Phys.Rev. 153, 743 (1967).
21. S. Wakoh and J. Yamashita, J.Phys.Soc. Japan 30, 422 (1971)
22. S. Wakoh, J.Phys.Soc. Japan 20, 1894 (1965).
23. D. Bagayoko, D.G. Laurent, S.P. Singhal and J. Callaway, Phys. Lett. A76, 187 (1980).
24. D. Bagayoko, private communication (1980).
25. A.H. MacDonald, J.M. Daams, S.H. Vosko and D.D. Koelling, Phys. Rev. B25, 713 (1982).
26. L.D. Jennings, D.R. Chipman and J.J. DeMarco, Phys.Rev. 135, 1612 (1964).
27. A. Freund, PhD Thesis, Technische Universität, München (1973) and private communication in which the original experimental structure factors were corrected by about 4%.
28. P. Suortti and L.D. Jennings, Acta Cryst. A33, 1012 (1977).
29. D.D. Koelling, private communication (1980).
30. J.R. Schneider, P. Pattison and H.A. Graf, Phil.Mag. B38, 141 (1978).
31. D.J. Smart and C.J. Humphreys, Inst.Phys.Conf.Ser., No. 52C pp. 211-214 (1980).

32. C.J. Humphreys, private communication (1980).
33. F. Biggs, L.B. Mendelsohn and J.B. Mann, Atom.Data Nucl.Data Tables 16, 202 (1975).
34. B.G. Williams, (ed.): "Compton scattering", McGraw-Hill, New York (1977).
35. P. Pattison and J.R. Schneider, Nucl.Instr.Meth. 158, 145 (1979).
36. P. Pattison and J.R. Schneider, J.Phys. B12, 4013 (1979).
37. W.A. Reed, Acta Cryst. A32, 676 (1976).
38. B.G. Williams, Physica Scripta, Vol.15, 92 (1977).
39. G. Loupias, J. Petiau, A. Issolah and M. Schneider, Phys.Stat. Sol. (b) 102, 79 (1980).
40. P.E. Mijnarends, Topics Curr.Phys. 12, 25 (1979).
41. P.E. Mijnarends, Phys.Rev. 160, 512 (1967).
42. N.K. Hansen, Report No. B 342, Hahn-Meitner-Institut, Berlin (1980).
43. P. Pattison, N.K. Hansen and J.R. Schneider, Chemical Physics 59, 231 (1981).
44. N.K. Hansen, P. Pattison and J.R. Schneider, submitted to Z. Physik B.
45. W. Weyrich, P. Pattison and B.G. Williams, Chem.Phys. 41, 271 (1979).
46. W. Schülke, Phys.Stat.Sol. (b) 82, 181 (1977).
47. P. Pattison and J.R. Schneider, Solid State Communic. 28, 581 (1978).
48. P. Pattison and W. Weyrich, J.Phys.Chem.Solids 40, 213 (1979).
49. P. Pattison, N.K. Hansen and J.R. Schneider, Z. Physik B, in print (1982).
50. H. Bross, private communication (1982), J.Phys. F 12, 2249 (1982).
51. H. Bross, Phys.Kondens.Mat. 3, 119 (1964).
52. H. Bross, Helv.Phys.Acta 41, 717 (1968).
53. H. Bross, J.Phys. F 8, 2631 (1978).
54. R. Courths, Solid State Communic. 40, 529 (1981).
55. J.F. Janak, A.R. Williams and V.L. Moruzzi, Phys.Rev. B11, 1522 (1975).
56. P.T. Coleridge, A.A.M. Croxon, G.B. Scott and I.M. Templeton, J.Phys. E 4, 414, 928 (1971).
57. H. Stöhr and H. Bross, Phys.Stat.Sol. (b) 54, 227 (1972)
58. H. Bross and H. Stöhr, Appl.Phys. 3, 307 (1974).
59. S. Cushner, J.C. Erskine and S. Berko, Phys.Rev. B1, 2852 (1970).
60. P.E. Mijnarends and R.M. Singru, Phys.Rev. B 19, 6038 (1979).
61. S. Berko, in "Positron Annihilation. Proceedings of the V. International Conference on Positron Annihilation", R.R. Hasiguti and K. Fujiware (eds.), p. 65, Japan Institute Met.: Sendai, Japan (1979).
62. J. Felsteiner and P. Pattison, Nucl.Instr.Meth. 173, 323 (1980).
63. G.E.W. Bauer and J.R. Schneider, in preparation.
64. G.E.W. Bauer and P. Pattison, Report No. B 364, Hahn-Meitner-Institut, Berlin (1981).

65. P. Eisenberger and W.A. Reed, Phys.Rev. B 9, 3242 (1974).
66. O. Jepsen, D. Glötzel and A.R. Mackintosh, Phys.Rev. B 23, 2684 (1981).
67. G.A. Henderson, Phys.Rev. A 23, 19 (1981).
68. G.E.W. Bauer, in preparation.
69. E. Daniel and S.H. Vosko, Phys.Rev. 120, 2041 (1960).
70. B.I. Lundqvist, Phys.Kondens.Mat. 7, 117 (1968).
71. J. Arponen and E. Pajanne, J. Phys. C 15, 2683 (1982). and references therein.
72. V.L. Moruzzi, J.F. Janak and A.R. Williams, "Calculated Electronic Properties of Metals", Pergamon (1978).
73. J. Arponen and E. Pajanne, private communication (1982).
74. G.E.W. Bauer, in preparation.
75. P.J.E. Aldred and M. Hart, Proc.R.Soc.Lond. A332, 223, 239 (1973).
76. U. Bonse and R. Teworte, J.Appl.Cryst. 13, 410 (1980).
77. R. Teworte and U. Bonse, Acta Cryst. A37, C-270 (1981).
78. T. Saka, S. Samata and N. Kato, Acta Cryst. A37, C-268 (1981).
79. R.S. Holt and M.J. Cooper, Acta Cryst. A37, C-310 (1981).

MULTIPOLE EXPANSION AS AN ALTERNATIVE

REPRESENTATION OF CHARGE DENSITY

Kaarle Kurki-Suonio and Riitta Sälke

University of Helsinki
Department of Physics
Siltavuorenpenger 20 D
SF-00170 Helsinki 17
Finland

1. INTRODUCTION

In charge density studies the problems of analysis cannot be separated from problems of representation. The aim of the analysis is to obtain chemically and physically meaningful information on the structure, bonding and dynamics of the atoms in a crystal. But in order to detect this kind of information from the charge density we need a clear enough picture of the three-dimensional charge distribution function represented by the diffraction data or by the model fitted to the data. Problems of visualization of the information may seem physically trivial, but in practice they have proved to be essential. The article by Smith, Price and Absar[1] demonstrates these difficulties very clearly while showing different ways by which one has tried to overcome them.

In principle the problem is just the illustration of a real function in three-dimensional space, for which we should need four dimensions. In the study itself one can resort to three-dimensional models and for demonstration purposes time may be useful as the third dimension. However, the journals still are the main forum of scientific communication, which forces us to restrict into two-dimensional illustrations.

Smith et al.[1] demonstrate beautifully, how one can make use of modern computer techniques to produce different kinds of perspective views and projections to visualize e.g. the development of the shape of a molecule, when looked on different charge density values. Still, simple two-dimensional contour maps are the most common means of representation and often the only one which can be used when computing and plotting facilities are restricted.

The multipole expansion offers another completely different possibility of representation. At present multipole expansions are used in many different, often quite sophisticated ways for analysis of diffraction data. However, when first introduced by Atoji[2] the main implication was just their use for representation of the charge density. This original purpose has largely been neglected in spite of its several obvious advantages.

In this article we discuss the representational possibilities offered by multipole expansions as an alternative to more conventional map representations and demonstrate, how they can be used to obtain quantitative information on significant features of the charge density distribution.

2. THE MULTIPOLE EXPANSION

By multipole expansion of the charge density we mean an expansion of the form

$$\rho(\bar{r}) = \sum_{\ell mp} \frac{1}{N_{\ell mp}} \rho_{\ell mp}(r)\, y_{\ell mp}(\theta,\phi) \tag{1}$$

cf. Kurki-Suonio.[3] It is an expansion of $\rho(\bar{r})$ around a given origin in terms of conventional spherical co-ordinates r,θ,ϕ using real spherical harmonics

$$y_{\ell m\pm} \sim P_\ell^m(\cos\theta) \begin{matrix}\cos m\phi \\ \sin m\phi\end{matrix} , \quad \ell = 0,1,\ldots , \quad m = 0,1,\ldots,\ell$$

with the normalization $\int (y_{\ell mp})^2 d\Omega = N_{\ell mp}^2$. We shall refer to ℓ as the order of the term.

The spherical harmonics are well-known standard functions. So, the illustration of the behaviour of each term requires only representation of the radial density $\rho_{\ell mp}(r)$ as a function of the distance r from the origin. Given the function $\rho(\bar{r})$ and the origin, the expansion (1) is unique. Thus, we have a representation of the three-dimensional charge density in terms of a set of one-dimensional functions. This yields a quantitative illustration of the charge density around the origin in all directions simultaneously in terms of a set of curves $\rho_{\ell mp}(r)$.

The perspicuity of this representation decreases rapidly with increasing number of terms required. However, taking the origin at the site of the atomic center few low-order terms are necessary to represent the charge density of the atom and its close surroundings. Moreover, the harmonics can be site-symmetrized i.e. adapted to the point symmetry of the atomic site, which effectively minimizes the number of terms.[3,4]

For the crystal charge density the radial densities around any fixed point $\bar{r}_o$ can be expressed in terms of the crystal structure amplitudes in the form

$$\rho_{\ell mp}(r) = \frac{4\pi(-i)^{\ell}}{VN_{\ell mp}} \sum_{\bar{H}} F_{\bar{H}} \exp(-2\pi i\bar{H}\cdot\bar{r}_o) j_{\ell}(2\pi Hr) y_{\ell mp}(\theta_H, \phi_H) \qquad (2)$$

Taking $\bar{r}_o$ to be the position of each atom of the asymmetric unit in succession we effectively get representations of the atomic charge densities when the radial densities are looked only at small enough distances r. Conventional multipole models used in charge density studies involve a division of the crystal charge density into its atomic constituents; each is assumed to take the form of a multipole expansion. This division and hence the expansions are not unique but depend on the assumed motions of the atoms and on the analytical form chosen for the radial densities. Such use of the multipole expansions always involves some sophisticated interpretation of the charge density. Eq. (2) does not yield any such division. It just gives a pure representation of the crystal charge density as an alternative to contour maps, but is concentrated to the region of each atom separately.

The nature of the representation obtained in this way is certainly very much different in nature from the map representation. The map is more concrete in showing directly the charge density on a plane through the unit cell. Chosen proper planes one gets an immediate picture about the distribution of electrons in various regions. A map is, however, interesting only on a qualitative basis. Many planes are needed to cover the whole unit cell, and from a set of several maps it is not easy to create a clear view about the three-dimensional distribution. The view obtained of the distribution even on a single plane depends on the contour values used, and it is only through considerable experience one can learn to read and to interpret the maps, to discover essential features and to distinguish information from artefacts and noice. Even dramatic changes of single contour shapes may occur as a result of completely insignificant changes of the density. On the other hand, significant features may be left unobserved by occasional improper choice of contour values.

While the map representation gives a picture of the unit cell in slices the multipole representation builds the view of the unit cell of three-dimensional pieces, each showing one atom or a symmetric molecule with its immediate surroundings. The representation of each piece is slightly more abstract since it does not immediately relate density values to certain points of the unit cell. Reading and interpreting the radial density curves requires a good comprehension of the behaviour of the symmetrized spherical harmonics in order to get the three-dimensional mental picture.

On the other hand the representation is quantitative. Large changes of radial density curves indicate significant changes of the density distribution and any significant features are very unlikely to remain unobserved. The information is given in an analysed form, since different low-order multipoles are coupled to different kinds of physical parameters. Also most of the expected interesting observations concern the motions and bonding of atoms i.e. features which are well in the range of the multipole representation and are easily described by low-order multipole components.

Calculation of radial densities through Eq. (2) and plotting the corresponding curves is straightforward and easy. No big computers or sophisticated computational techniques are necessary. So the multipole representation would certainly deserve a much wider use than at present.

3. TWO EXAMPLES

To demonstrate the use of the multipole representation we present two simple examples showing the different steps needed for formation and interpretation of the radial densities and comparing the representation with the use of the contour maps. We discuss the charge densities of the Cl^- ion in NaCl and of the OH^- ion in LiOH as they appear on the basis of the data by Göttlicher[5] and Göttlicher and Kieselbach,[6] respectively. The basic information required is listed in Table 1. In addition to what is needed for calculation of the conventional Fourier series the following basic data must be stated or defined for calculation of the multipole expansion:

1. The center of the expansion $\bar{r}_0$. This is the position of the atom or molecule we wish to study, in this case the atomic site of chlorine in NaCl and that of oxygen in LiOH.

2. The site symmetry of the center. In addition to the point group (m3m and 4mm) one has to state the orientation of the symmetry elements with respect to the crystal axes. It may also be useful to pay attention to possible approximate site symmetry as apparent from the co-ordination of the nearest neighbours.

3. The local co-ordinate system. This defines the meaning of the angular variables (θ,ϕ) and the explicite expressions of the symmetrized harmonics. When adapting the axes properly to the symmetry elements of the site, these are either real spherical

Table 1. Basic data for multipole analysis of NaCl and LiOH

	NaCl	LiOH
Space group	No. 225 Fm3m	No. 129 P4/nmm
Cell parameters	a=b=c=5.64056 Å $\alpha=\beta=\gamma=90^{o}$	a=b=3.549 Å c=4.334 Å $\alpha=\beta=\gamma=90^{o}$
Atomic positions	Na: $0,0,\frac{1}{2}$ Cl: 0,0,0	Li: $\frac{1}{4},\frac{3}{4},0$ O : $\frac{3}{4},\frac{3}{4},0.194$ H : $\frac{3}{4},\frac{3}{4},0.407$
Experimental structure factors	S. Göttlicher[5]	S. Göttlicher and B. Kieselbach[6]
Maximum $\sin\theta/\lambda$	1.0029 $Å^{-1}$	0.9976 $Å^{-1}$
Theoretical model	$Na^{+}Cl^{-}$	$Li^{+}O^{-}$ (!)
Atomic factors	Int. Tables for X-ray Cryst. IV[20]	Int. Tables for X-ray Cryst. IV[20]
Isotropic temperature factors $\langle u^2 \rangle$	Na^{+}: 0.02494 $Å^2$ Cl^{-} : 0.02119 $Å^2$	Li^{+}: 0.02580 $Å^2$ O^{-} : 0.02190 $Å^2$
Site symmetry	Cl^{-} : m3m	O^{-} : 4mm
Matrix of local co-ordinate axes	$\begin{bmatrix}1&0&0\\0&1&0\\0&0&1\end{bmatrix}$	$\begin{bmatrix}1&0&0\\0&1&0\\0&0&1\end{bmatrix}$
Site-symmetrized harmonics, Kara and Kurki-Suonio[4]	even octahedral (cubic ℓ=0,4,6, 8,10,...)	real spherical $y_{\ell mp}$ (ℓ,4μ,+)

harmonics or cubic harmonics with specific selection rules as indicated by Kara and Kurki-Suonio.[4] In both of the present examples the natural local axes are just along the crystal axes i.e. have the directions [100],[010],[001].

4. The multipole components to be calculated. Normally all components allowed by the symmetry up to some not very large order ℓ should be calculated, cf. Kurki-Suonio.[7] No definite rule can be given here for the maximum order. In discussion of approximate symmetries, it is informative to calculate also components allowed by the exact symmetry but not obeying the approximate symmetry because they indicate to which extent the charge density violates this symmetry.[8,9]

The calculation of the radial densities (2) for any components is then straightforward. The expressions (1) and (2) are valid as such even when $y_{\ell mp}$ are replaced by any spherical harmonics of a definite order ℓ, e.g. cubic or icosahedral harmonics. This is useful to note in discussing the approximate symmetries where the harmonics orthogonal to the symmetric ones may have more complicated expressions in terms of $y_{\ell mp}$.

Figures 1 and 2 show the radial difference densities for the low-order components in our examples. They refer to cubic harmonics $K_{\ell j}(\theta,\phi)$ and real spherical harmonics $y_{\ell mp}(\theta,\phi)$, respectively, both normalized to maximum value 1, so that each curve shows directly the relevant contribution to the charge density in the direction (θ,ϕ) where it is largest. They were calculated using Eq. (2) as a

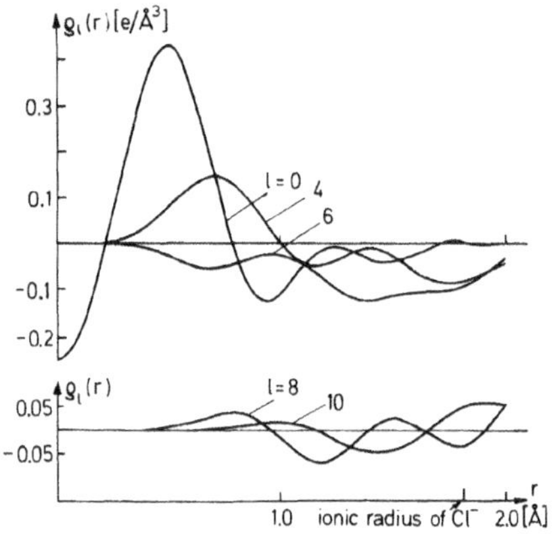

Fig. 1. Low-order radial difference densities of the site-symmetrized multipole expansion centered at the chlorine position in NaCl.

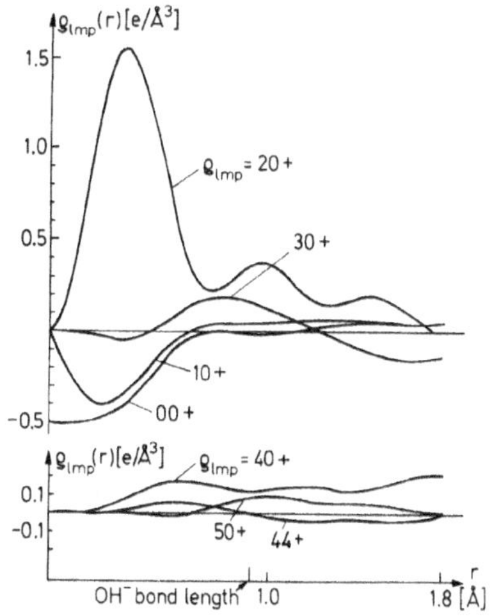

Fig. 2. Low-order radial difference densities of the site-symmetrized multipole expansion centered at the oxygen position in LiOH.

difference series with coefficients $\Delta F = F_{obs} - F_{theor}$. In case of LiOH $\Delta F_{000} = 2$ must be included because there are two electrons missing from the model.

As far as the charge density of Cl^- or OH^- is concerned these are the results to be discussed. One has to learn to visualize these curves as a three-dimensional charge density. Each curve must be understood as multiplied by the corresponding harmonics to yield the three-dimensional contribution to the charge density. Therefore one needs a good acquaintance with the behaviour of low-order Legendre polynomials and associate Legendre functions which define the θ dependence of $y_{\ell mp}$, as well as of the basic symmetry properties of the harmonics. Figures 3 a,b,c show the behaviour of the cubic harmonics K_4, K_6 and K_8, respectively, corresponding to the radial densities of Cl^- in Figure 1. The figure shows one octant of the surface of the unit sphere mapped on an equilateral triangle based on the correspondence $x^2=a$, $y^2=b$, $z^2=c$ where a,b,c are the distances of a point measured from the sides of the triangle. For a good mental picture it is sufficient to note the directions of the main maxima and minima.

So, for instance, we can read from Figure 1 the nature of the

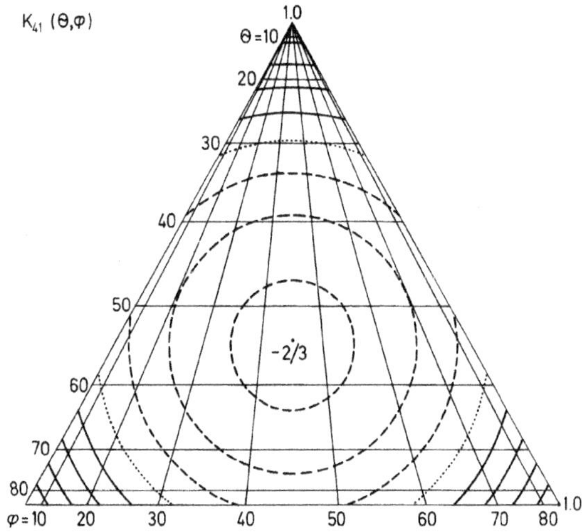
$K_{41}(\Theta,\varphi)$
1.0
Θ = 10
20
30
40
50
60
70
80
-2/3
φ = 10 20 30 40 50 60 70 80
1.0

a

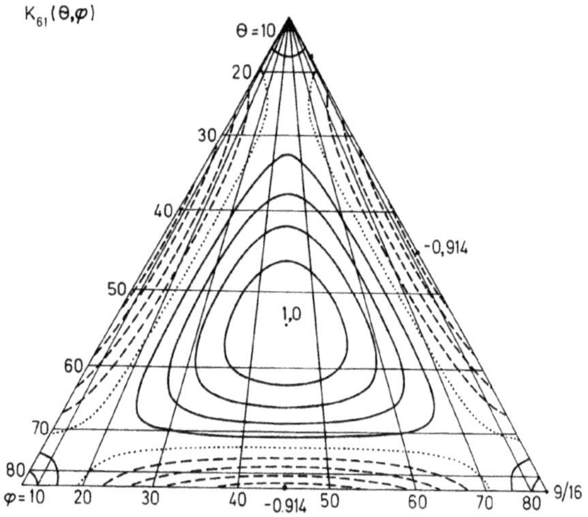
$K_{61}(\Theta,\varphi)$
Θ = 10
20
30
40
50
60
70
80
-0,914
1,0
φ = 10 20 30 40 -0.914 50 60 70 80
9/16

b

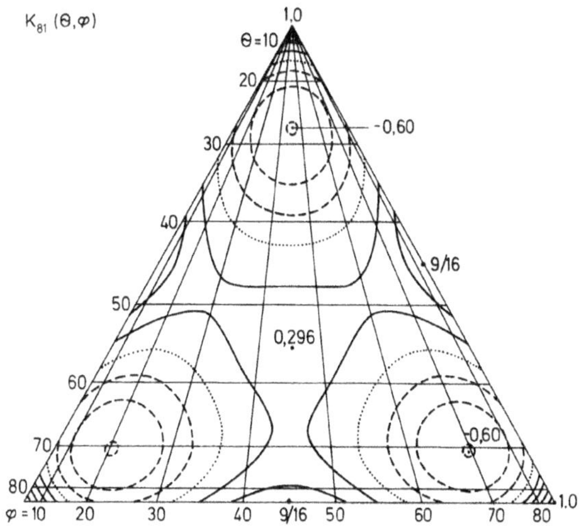

c

Fig. 3 a,b,c. Contour maps of the cubic harmonics $K_4(\theta,\phi)$, $K_6(\theta,\phi)$, $K_8(\theta,\phi)$, respectively, normalized to $Max\{K_\ell\} = 1$ in the first octant. Contour interval 0.2, solid line positive, dashed line negative and dotted line zero.

deformations of chlorine as indicated by the data. The spherical component ρ_0 indicates a slight compression of the ion. In the core region, r<1 Å, ρ_4 shows some transfer of charge from the [111] directions to the [100] directions. In the outer region, r>1...2 Å, which from the point of view of bonding effects ought to be more interesting, the contributions to the density seem to be rather small. The nature of the 4th and 6th order deformations is, however, clear. Both reduce the density in the directions [100], thus effectively decreasing the overlap of the Cl^- and Na^+ as compared to the spherical ion model. They have partly cancelling effects in the [111] directions while the 6th order dominates in the [110] directions allowing an expansion of the chlorine towards the neighbouring chlorines.

Correspondingly, in Figure 2 ρ_{10+} indicates asymmetry of the charge distribution in the tetragonal axis direction. ρ_{20+} shows a symmetric elongation of the charge distribution along the tetragonal axis. The features at small r dominate the behaviour of both of these components, while all contributions look small at larger distances, which should be more interesting from the point of view of the effect of the hydrogen. (Note that there is no hydrogen included in F_{theor}.) At about the hydrogen distance it will be necessary to pay attention also to the main features of both 30+ and 40+. They are similar to those of 10+ and 20+ as described but concentrate to a narrower angular range around the tetragonal axis.

The correspondence between the radial densities and the conventional map representations is demonstrated in Figures 4a and 4b. They show the development of the multipolar charge density of the chlorine ion on the lines [100] and [111], respectively, with increasing number of multipoles as compared to the charge density calculated by Fourier series. The set of Figures 5 a,b,c,d show the corresponding development of multipolar charge density map on the plane x=y through the chlorine as compared to the Fourier map. Similarly Figures 6 and 7 a,b,c,d show the development of multipolar charge density of the oxygen ion in LiOH along the z-axis and on the plane x=0 as compared to the corresponding Fourier representations.

Once the three-dimensional main features have been stated from the radial densities (Figures 1,2), their effect is clearly seen in the corresponding charge densities (Figures 4-7). It is, however, not quite clear how easily one could conclude backwards from the one- and two-dimensional representations, what are the main features three-dimensionally. Also, the shaping of the contour lines in the core region dominates the visual picture, while at the interesting distance r=1...2 Å it is much more difficult to deduce the essential features of the three-dimensional behaviour. For this purpose the multipole representation gives the information in a much more complete and concise form.

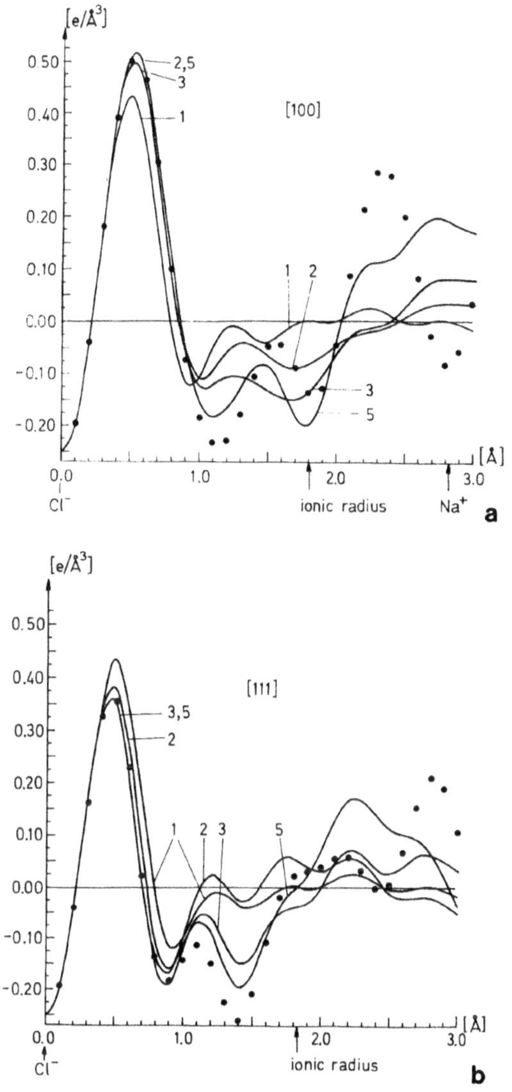

Fig. 4 a,b. Truncated multipole expansions of the difference charge density at the chlorine position in NaCl along the lines a) [100] and b) [111], as compared to the difference Fourier series (•). The number of terms 1,2,3,5 as indicated in the figures corresponds to the maximum order $\ell = 0,4,6,10$, respectively.

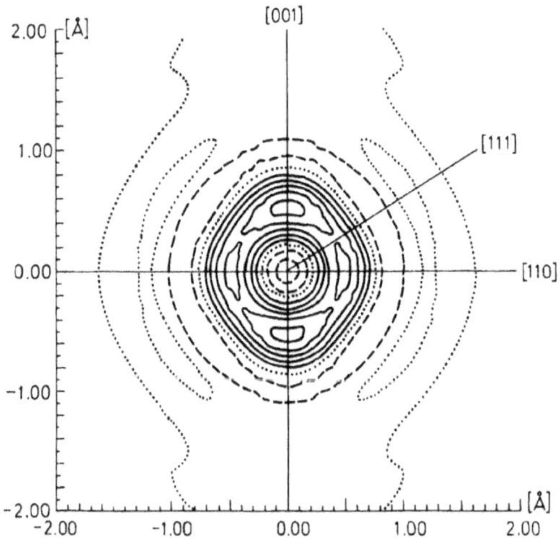
[001]
[111]
[110]
[Å]
2.00
1.00
0.00
-1.00
-2.00
-2.00
-1.00
0.00
1.00
2.00
[Å]

a

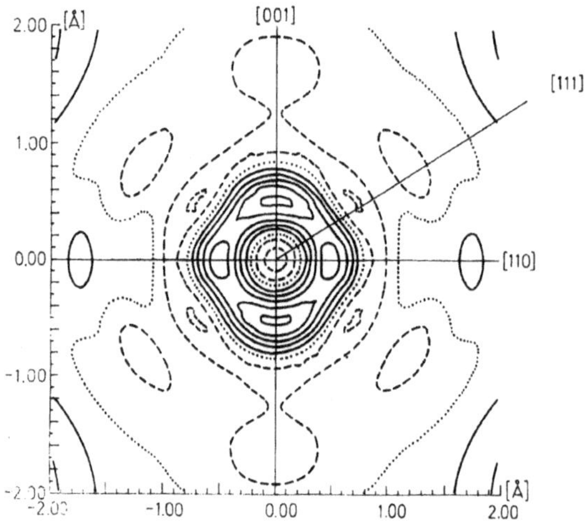
[001]
[111]
[110]
[Å]
2.00
1.00
0.00
-1.00
-1.00
0.00
1.00
2.00
[Å]

b

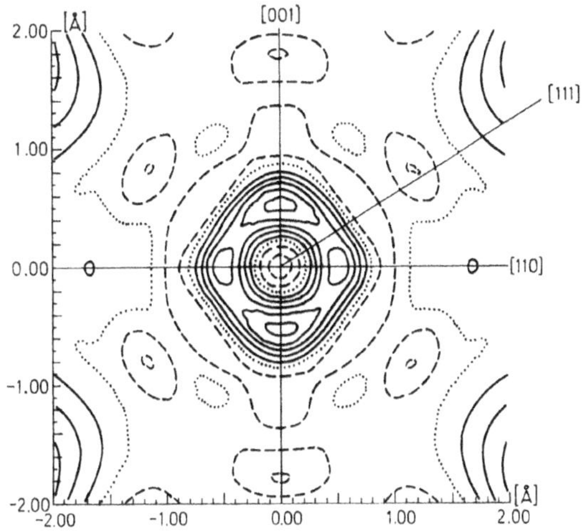

c

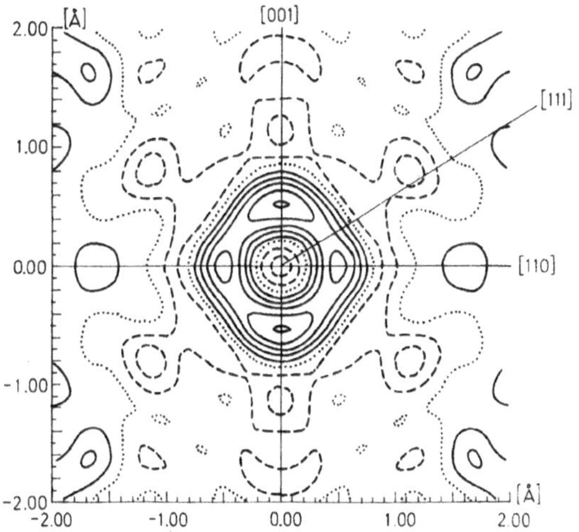

d

Fig. 5 a,b,c,d. Truncated multipole expansions of the difference charge density at the chlorine position in NaCl on the plane x = y up to a) the 4th, b) the 6th and c) the 10th order as compared to d) the difference Fourier series. Contour interval 0.1 e/Å^3.

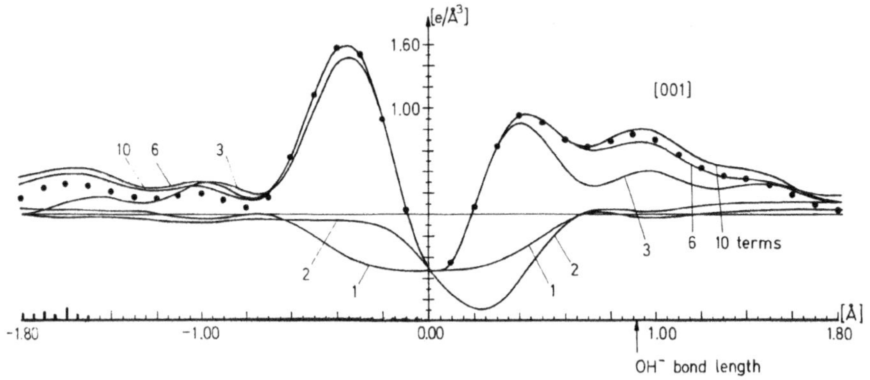

Fig. 6. Truncated multipole expansions of the difference charge density at the oxygen position in LiOH along the tetragonal axis as compared to the difference Fourier series. The number of terms 1,2,3,6,10 as indicated in the figure corresponds to the maximum order $\ell = 0,1,2,4,6$, respectively.

The only question then remains, how far we can rely upon having all important features included in the low order multipoles. To some extent this can be judged on the basis of convergence of the multipole expansion. The nature of the convergence is clearly visible in the Figures 4-7. We see that the convergence is very rapid in the main part of the atom up to about 1 Å distance from the center. In Cl^- three terms (up to the 6th order) and in OH^- three terms (up to the 2nd order) are sufficient to reproduce the shape of the ion. At larger distances higher orders get more important and, as is obvious, close to the distance of the nearest neighbour many more multipoles would be needed for reproduction of the Fourier series value of the density.

In the intermediate region, which is important in discussing the nature of bonding, a few more multipoles seem to give some contribution, and there still remain some deviations from the Fourier value. It is then important to be able to estimate the significance of these less smooth contributions and differences. On general physical arguments one would not expect to have strong high order components in the bonding region. Their presence is rather an indication of inaccuracies of the data. This is certainly true for any very local deviations.

One should remember that charge density is not a very favourable quantity to judge experimental significance of different features because of the reciprocity of the representation space to the space of the experimental data. It is the number of electrons contributing in a feature which determines its significance rather than value of the charge density.[7,10] From this point of view, the radial densities at some distance from the center express more reliably the significant features than charge density values on any single line do, because they present an integrated behaviour over a whole spherical surface. There are, thus, good reasons to think that the smoothing of the Fourier series when presented by a low order multipole expansion is not a defect but an improvement of the picture, filtering away some unnecessary insignificant ripples.

This brings us already close to the question about the ordinary purpose of the charge density representation. If the question is only about getting a clear qualitative over-all picture, then we can hardly convince anyone used to the maps to change over to multipoles. But as soon as we come to the quantitative analysis and interpretation of the data on basis of the representation, then the multipole expansion has some undeniable advantages over the maps.

4. QUESTIONS OF INTERPRETATION

An experienced track-finder is able to recognize also the origin of various features of his maps. He is able to see differences in position and in thermal parameters and many kinds of bonding effects. In the multipole representation different effects are largely concentrated in different components and, thus, even more easily recognizable:

Change of atomic position has a negligible effect on any other component than the first order. Moreover, the ratios of the three first order components express immediately the direction of the displacement, and the magnitude can be derived easily from their values at small r.

Difference in the harmonic temperature factor is visible in the zeroth and in the second order. Each of these six components indicates a difference of its own recognizable nature. The zeroth or *monopole* term refers to the average isotropic temperature parameter $\overline{\langle u^2 \rangle} = (\langle u_x^2 \rangle + \langle u_y^2 \rangle + \langle u_z^2 \rangle)/3$. Significant second order terms indicate anisotropy of the thermal motion. Assuming the local xyz-co-ordinate system to coincide with the thermal axes, 20+ indicates a difference in prolateness or $\langle u_z^2 \rangle - \overline{\langle u^2 \rangle}$, 22+ implies necessity to correct the non-axiality or $\langle u_x^2 \rangle - \langle u_y^2 \rangle$. Rotation of the anisotropic thermal ellipsoid about the local x-, y- and z-axes gives rise to the components 21-, 21+ and 22-, respectively. Each of these effects is visible only in the relevant component with negligible effect on the others.

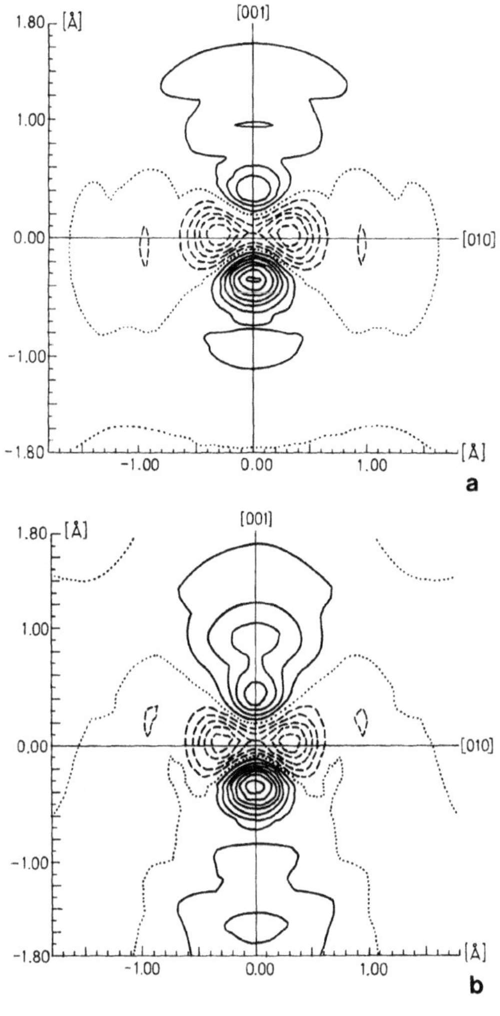
[001]
[Å]
1.80
1.00
0.00
-1.00
-1.80
[010]
-1.00
0.00
1.00
[Å]
a
[001]
[Å]
1.80
1.00
0.00
-1.00
-1.80
[010]
-1.00
0.00
1.00
[Å]
b

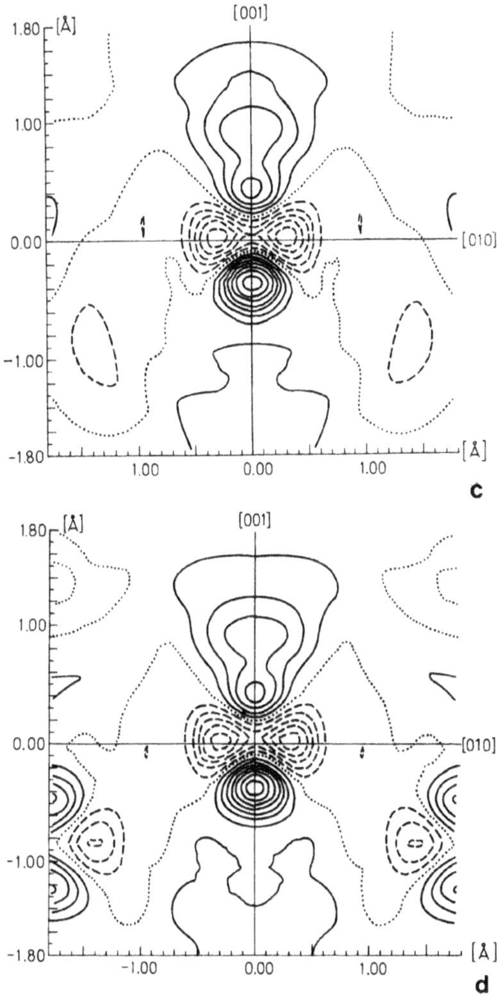

Fig. 7 a,b,c,d. Truncated multipole expansions of the difference charge density at the oxygen position in LiOH on the plane x = 0 up to a) the 2nd, b) the 4th, c) the 6th order as compared to d) the difference Fourier series. Contour interval 0.2 e/$Å^3$.

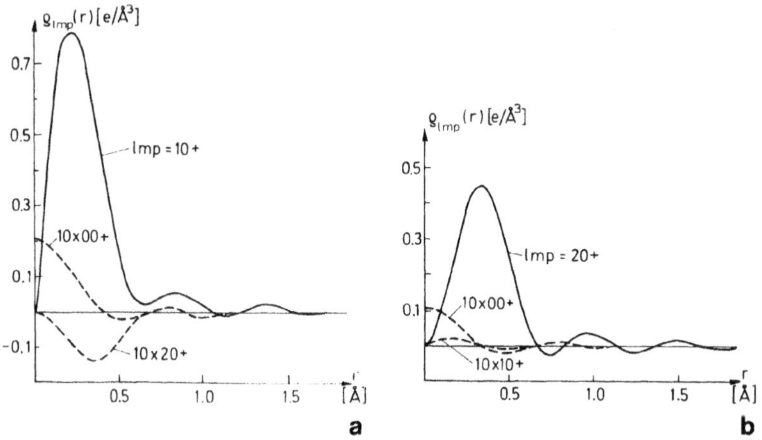

Fig. 8 a,b. The effect of
a) the change $\Delta z = 0.01$ Å of the oxygen position
b) introducing prolateness $\langle u_z^2\rangle - \overline{\langle u2\rangle} = 0.002$ Å^2 of the thermal motion of oxygen
on the radial densities centered at the oxygen in LiOH.

To demonstrate this good separation of information we show in the Figures 8a and 8b the radial densities caused by a change $\Delta z = 0.01$ Å in the position of the oxygen in LiOH and by introducing a prolateness $\langle u_z^2\rangle - \overline{\langle u2\rangle} = 0.002$ Å^2 to its thermal motion, respectively. Both figures are results of a pure model calculation involving no experimental data. They show very clearly that conclusions on the position and on the anisotropy must be based merely on the components 10+ and 20+, respectively. At the same time we recognize that the component 20+ in Figure 2 is mostly due to anisotropy of the thermal motion. Comparison of the magnitudes would give the value 0.0070 Å^2 for the prolateness parameter of the oxygen. Similarly we can conclude that moving the oxygen from the position $z = 0.8408$ Å by 0.0052 Å down the z-axis would make ρ_{10+} vanish without effecting any other components.

Reading quantitative information on other effects, say features which qualitatively are described as electronic deformations or bonding effects is also possible although less straightforward. There are better means than the radial densities themselves to judge the significance of such features, viz. calculation of radial scattering factors corresponding to the different multipoles by spherical volume partition techniques.[7] Also the functions $4\pi r^2\rho_{\ell mp}(r)$ show the significance of different features more clearly, since the area under the curve is directly related to the number of electrons contributing to the deformation described by the component. Quantitative conclusions require then knowledge about the integrals of the harmonics over their lobes. For instance, knowing that the integral of K_4 over its positive lobes is 2.30 we can conclude from ρ_4 in Figure 1 that the angular transfer of electrons involved in the "core deformation" (r<1 Å) is only 0.07, which is at the limit of significance. The corresponding number in the region r = 1...2 Å is 0.3 electrons, which is significant.

In our second example the effect of the hydrogen atom is seen as an electronic deformation of the oxygen. We already saw that visually dominant features can be explained just as small non-interesting corrections to conventional parameters. To discuss the deformation we have to look at the features which at the first sight seem to be small and in fact are small in terms of local density values. However, redrawing the components as $4\pi r^2\rho_{\ell mp}$ would reveal the real relative significance of different features as integrated large-scale features. Taking into account the integrals of $y_{\ell 0+}$ over their main positive lobes we find that the 10+ -asymmetry of OH^-, which extends rather fra from the center, involves the total of only about 0.3 electrons, while the symmetric 20+ deformation, beyond the anisotropy effect, is much larger involving about 0.75 electrons. The 30+ does not add to the integral value of charge asymmetry, but it pushes the excess charge closer to about the hydrogen distance. Also, the 40+ does not essentially add to the charge of the symmetric deformation. It makes it to concentrate to a narrower angular region around the z-axis.

5. SUMMARY

Multipole expansions in the sense they have been discussed in this paper have been applied on both neutron and X-ray data. They have been seen to be most advantageous in case of simple inorganic structures. While providing a simple means of representing the charge density they give an immediate three-dimensional picture and, at the same time, show the information in a well-analyzed form, different components indicating different kinds of effects. Moreover, it is easy to proceed from it into quantitative determination of physically and chemically meaningful parameters. One has been able to see bonding effects, spherical and aspherical deformations of ions,[8-15] anharmonicities of motions,[9,16] librations and preferred orientations of small rigid molecules.[17-19]

The easy adaptation of multipole expansions to any symmetries makes them valuable in studying possible approximate symmetries or deviations from them.[8,9] On the other hand the number of applications is still rather small. So far almost every single application has revealed some interesting new possibilities or problems. A recent study of BeO has proved that even the phase problem of non-centric crystals can be properly treated.[9] It seems that as a means of analysis the multipole expansion would be most profitable when used beside the conventional least squares method to guide the refinement process. As a representation it may similarly find its best use as a simple complimentary method to show quantitatively some interesting three-dimensional features, which can be seen from the charge density maps only incompletely and qualitatively.

REFERENCES

1. V.H. Smith, Jr., P.F. Price, and I. Absar, Isr. J. Chem. 16, 187-197 (1977).
2. M. Atoji, Acta Cryst. 11, 827-829 (1958).
3. K. Kurki-Suonio, Isr. J. Chem. 16, 115-123 (1977).
4. M. Kara and K. Kurki-Suonio, Acta Cryst. A 37, 201-210 (1981).
5. S. Göttlicher, Acta Cryst. B 24, 122-129 (1968).
6. S. Göttlicher and B. Kieselbach, Acta Cryst. A 32, 185-192 (1976).
7. K. Kurki-Suonio, Acta Cryst. A 24, 379-390 (1968).
8. J.P. Vidal, G. Vidal-Valat, M. Galtier, and K. Kurki-Suonio, Acta Cryst. A 37, 826-837 (1981).
9. J.P. Vidal, G. Vidal-Valat, M. Galtier, and K. Kurki-Suonio, (to be published).
10. K. Kurki-Suonio, Italian Crystallographic Association Meeting, Bari, Italy, 1971. Congress Report 31-66 (1972).
11. K. Kurki-Suonio and A. Ruuskanen, Ann. Acad. Sci. Fenn. Ser. A VI 358, 1-29 (1971).
12. K. Kurki-Suonio and V. Meisalo, Ann. Acad. Sci. Fenn. Ser. A VI 241, 1-18 (1967).
13. M. Linkoaho, E. Rantavuori, U. Korhonen, K. Kurki-Suonio, and A. Ruuskanen, Acta Cryst. A 28, 260-264 (1972).
14. A. Ruuskanen and K. Kurki-Suonio, J. Phys. Soc. Japan 34, 715-719 (1973).
15. G. Vidal-Valat, J.P. Vidal, and K. Kurki-Suonio, Acta Cryst. A 34, 594-602 (1978).
16. M. Ahtee, K. Kurki-Suonio, A. Vahvaselkä, A.W. Hewat, J. Harada, and S. Hirotsu, Acta Cryst. B 36, 1023-1028 (1980).
17. K. Kurki-Suonio, M. Merisalo, A. Vahvaselkä, and F.K. Larsen Acta Cryst. A 32, 110-115 (1976).
18. A. Vahvaselkä and K. Kurki-Suonio, Physica Fennica 10, 87-99 (1975).
19. M. Ahtee, K. Kurki-Suonio, B.W. Lucas, and A.W. Hewat, Acta Cryst. A 35, 591-597 (1979).
20. International Tables for X-ray Crystallography, vol. IV, The Kynoch Press, Birmingham, 1974.

MEASURED ELECTRON DENSITY DISTRIBUTIONS IN CRYSTALS AS A BASIS FOR BAND STRUCTURE CALCULATIONS

John Avery and Peter Sommer Larsen

Department of Physical Chemistry
H.C. Ørsted Institute
University of Copenhagen, Denmark

Michael Grodzicki

Institute for Theoretical Physics
University of Hamburg
Jungiusstrasse 9, D-2000 Hamburg 36, FRG

General methods are outlined for constructing Fourier coefficients of the Coulomb and Slater potentials in crystals from X-ray diffraction measurements of the charge density supplemented with atomic Hartree-Fock data. Corrections for temperature effects are discussed. Asymptotic expressions for the Fourier coefficients are also given, as well as an expression for the Coulomb contribution to the binding energy.

1. INTRODUCTION

In recent years, high-quality X-ray crystallographic measurements of electron density distributions have become available for a variety of crystals.[1-8] These measurements, when supplemented with data from atomic Hartree-Fock calculations, can provide a very accurate picture of the charge density distribution $\rho(\underline{x})$ in crystals.[9-15] From the density, the Coulomb potential can be constructed. If a local density approximation is used,[17-19] the exchange potential can also be constructed from the density. Thus X-ray diffraction data can be made the starting point for calculations of the electronic wave functions, band structure and physical properties of crystals. In this paper, we shall describe a general method for determining the Fourier coefficients of the Coulomb and exchange potentials in crystals, using X-ray data supplemented with atomic Hartree-Fock data.

2. CONSTRUCTION OF THE COULOMB POTENTIAL FROM X-RAY DIFFRACTION AND ATOMIC HARTREE-FOCK DATA

Let $\underline{a}_1$, $\underline{a}_2$ and $\underline{a}_3$ be basis vectors for the direct lattice of a crystal, while $\underline{b}_1$, $\underline{b}_2$ and $\underline{b}_3$ are basis vectors of the reciprocal lattice. Then the set of direct lattice vectors will be given by:

$$\underline{\ell} = n_1\underline{a}_1 + n_2\underline{a}_2 + n_3\underline{a}_3; \quad n_1,n_2,n_3 = 0,\pm1,\pm2,\ldots \tag{1}$$

while the reciprocal lattice vectors will be:

$$\underline{g} = 2\pi(m_1\underline{b}_1 + m_2\underline{b}_2 + m_3\underline{b}_3); \quad m_1,m_2,m_3 = 0,\pm1,\pm2,\ldots \tag{2}$$

The vectors $\underline{a}_i$ and $\underline{b}_j$ are related by

$$\underline{a}_i\cdot\underline{b}_j = \delta_{i,j} \tag{3}$$

and the volume of the unit cell in the direct lattice is

$$v = \underline{a}_1\cdot(\underline{a}_2\times\underline{a}_3) \tag{4}$$

Since the Coulomb potential $V(\underline{x})$ in the crystal has the periodicity of the direct lattice, it can be represented by a Fourier series of the form:

$$V(\underline{x}) = \sum_{\underline{g}} (V)_{\underline{g}}\, e^{i\underline{g}\cdot\underline{x}} \tag{5}$$

where

$$(V)_{\underline{g}} = \frac{1}{v}\int_c d^3x\, e^{-i\underline{g}\cdot\underline{x}}\, V(\underline{x}) \tag{6}$$

The subscript c on the integral in equation (6) means that the integral is taken over a unit cell. In a similar way, the charge density (electronic plus nuclear) in the crystal can be represented by a Fourier series of the form:

$$\rho(\underline{x}) = \sum_{\underline{g}} (\rho)_{\underline{g}}\, e^{i\underline{g}\cdot\underline{x}} \tag{7}$$

where

$$(\rho)_{\underline{g}} = \frac{1}{v}\int d^3x\, e^{-i\underline{g}\cdot\underline{x}}\, \rho(\underline{x}) \tag{8}$$

Since the potential and charge density are related by Poisson's equation:

$$\nabla^2 V(\underline{x}) = 4\pi e\rho(\underline{x}) \qquad (9)$$

it follows that for $\underline{g} \neq 0$ the Fourier coefficients are related by

$$(V)_{\underline{g}} = \frac{4\pi e}{\underline{g}^2}(\rho)_{\underline{g}} \qquad (10)$$

(The signs in (9) and (10) are chosen in such a way that V is the potential energy experienced by an electron).

Eq. (10) can be used to relate the Fourier coefficients of the Coulomb potential to the crystallographicly measured values of $(\rho)_{\underline{g}}$. However, the Fourier coefficient of the Coulomb potential corresponding to $\underline{g}=0$ cannot be determined from (10) because, since the crystal is electrically neutral, (8) yields:

$$(\rho)_o = \frac{1}{v}\int_c d^3x\ \rho(\underline{x}) = 0 \qquad (11)$$

and substitution of (11) into (10) gives the indeterminate form 0/0. We shall return later to the question of how to determine $(V)_o$.

3. SUPPLEMENTATION OF X-RAY DATA WITH ATOMIC HARTREE-FOCK DATA

In the preceeding discussion, the charge density $\rho(\underline{x})$ was defined in such a way as to include both the electronic and nuclear charges. Thus, if Z_a is the atomic number of atom a, we can rewrite Eq. (8) in the form:

$$(\rho)_{\underline{g}} = \frac{1}{v}\int_c d^3x\ \rho(\underline{x})e^{-i\underline{g}\cdot\underline{x}} \qquad (12)$$

$$= \frac{1}{v}\int_c d^3x\ [\sum_a Z_a e\delta(\underline{x}-\underline{\delta}_a) + \rho_{el}(\underline{x})]\ e^{-i\underline{g}\cdot\underline{x}}$$

so that

$$(\rho)_{\underline{g}} = \frac{e}{v}\sum_a Z_a e^{-\underline{g}\cdot\underline{\delta}_a} + (\rho_{el})_{\underline{g}} \qquad (13)$$

The Fourier coefficients of the electronic density

$$(\rho_{el})_{\underline{g}} = \frac{1}{v}\int_c d^3x\ \rho_{el}(\underline{x})e^{-i\underline{g}\cdot\underline{x}} \tag{14}$$

are what is measured in an X-ray diffraction experiment.

We would like to use the measured values of $(\rho_{el})_{\underline{g}}$ to construct the Coulomb and exchange potentials in a crystal. However, there are two difficulties which at first sight seem to stand in the way. The first difficulty is that data can only be collected within a limited sphere in reciprocal space; and the second difficulty is that thermal vibrations tend to smear out the sharp peaks of density which would otherwise be observed at the nuclei.

These two difficulties are not as serious as they would appear at first sight, because the charge density near to the nucleus of an atom is dominated by the field of the nucleus, and it is therefore almost unchanged when an atom is incorporated into a crystal. The Fourier coefficients $(\rho_{el})_{\underline{g}}$ corresponding to large values of g describe the charge density near to the nuclei, i.e. in the region where it is almost undeformed. Thus (luckily!) the high-frequency Fourier coefficients which we are unable to measure are just the ones which can be obtained reliably from atomic Hartree-Fock data.

Since the charge density near to the nuclei is undeformed, we know that for large g the T=0 Fourier coefficients must reduce to those which would be calculated for a superposition of neutral atoms (SNA) located at the appropriate positions:

$$[(\rho_{el})_{\underline{g}}]_{T=0} \xrightarrow[\text{large } g]{} [(\rho_{el})_{\underline{g}}]_{SNA} \tag{15}$$

Our method for calculating $[(\rho_{el})_{\underline{g}}]_{SNA}$ is as follows: We first represent the charge density of atom a in the form:

$$\rho_a(r) = -e\sum_j C_j r^{n_j} e^{-\zeta_j r} \tag{16}$$

For example, the coefficients C_j, n_j and ζ_j for carbon and oxygen are shown in Table 1. These coefficients were calculated from Clementi's Hartree-Fock atomic orbitals.[20] We then calculate the atomic form factor, which is the Fourier transform of the radial density distribution of the atom:

$$f_a(k) \equiv \int d^3x\ e^{i\underline{k}\cdot\underline{x}}\rho_a(r) \tag{17}$$

$$= 4\pi\int_0^\infty dr\ r^2 j_o(kr)\rho_a(r)$$

$$= -4\pi e \sum_j C_j \int_0^\infty dr\, r^{n_j+2} j_o(kr) e^{-\zeta_j r}$$

where $j_o(kr) = \sin(kr)/kr$ is the spherical Bessel function of order zero. Integrals of the form

$$J_{n,\ell}(k,\zeta) \equiv \int_0^\infty dr\, r^n j_\ell(kr) e^{-\zeta r} \tag{18}$$

can be expressed in terms of hypergeometric functions:

$$F(a|b|c|x) \equiv 1 + \frac{ab}{c}x + \frac{a(a+1)b(b+1)}{c(c+1)\;2!}x^2 + \ldots \tag{19}$$

and one obtains the result:

$$J_{n,\ell}(k\zeta) = \left(\frac{k}{\zeta}\right)^\ell \frac{(n+\ell)!\; F(\tfrac{1}{2}(\ell-n+1)|\tfrac{1}{2}(\ell-n+2)|\ell+3/2|-(k/\zeta)^2)}{(2\ell+1)!!\zeta^{n+1}\{1+(k/\zeta)^2\}^n} \tag{20}$$

and from (17),

$$f_a(k) = -4\pi e \sum_j C_j J_{n_j+2,0}(k,\zeta_j) \tag{21}$$

The hypergeometric series in Eq. (19) terminates, so that the integrals $J_{n,\ell}$ can be expressed in closed form, the first few with $\ell=0$ being:

$$J_{1,0}(k,\zeta) = \frac{1}{k^2+\zeta^2} \tag{22}$$

$$J_{2,0}(k,\zeta) = \frac{2\zeta}{(k^2+\zeta^2)^2}$$

$$J_{3,0}(k,\zeta) = \frac{2(3\zeta^2-k^2)}{(k^2+\zeta^2)^3}$$

$$J_{4,0}(k,\zeta) = \frac{24\zeta(\zeta^2-k^2)}{(k^2+\zeta^2)^4}$$

etc.

Clementi's Hartree-Fock calculations cover only elements 1-36, because for elements with higher atomic numbers, relativistic effects become important. However, Cromer and Mann[21] have carried out relativistic Hartree-Fock calculations on the heavy elements of the periodic system, and tabulations of atomic form factors based on their results are available for elements 1-92. The atomic form factors of Cromer and Mann are the ones generally used in X-ray crystallographic

Table 1. The coefficients n_j, C_j and ζ_j needed for an analytic representation of the oxygen and carbon radial density distributions (see Eqs. (17-22)).

Oxygen			Carbon		
n_j	C_j	ζ_j	n_j	C_j	ζ_j
0	2.2527	26.6485	0	3.0656	18.5725
0	48.3548	20.9402	0	33.3438	14.6987
1	8.0661	19.2687	1	1.3342	13.5457
1	-2.2242	17.6074	2	0.2590	12.6875
1	-1.1078	15.8869	1	-0.8207	11.8760
2	2.9318	15.8139	0	91.0946	10.8249
0	261.2080	15.2319	1	-0.5081	10.7882
1	-0.3591	15.0824	1	-0.0158	10.3173
1	104.3414	13.5604	1	10.1906	9.6719
1	-34.2968	11.8991	2	1.1897	8.9310
2	52.9635	11.8889	2	5.1920	8.5189
2	16.3259	11.3448	1	-6.6914	8.0022
2	-57.6489	10.2276	2	0.6519	7.7646
1	-18.7412	10.1786	2	0.1035	7.2991
2	5.5550	9.7029	1	-3.9820	6.9144
1	-5.9208	9.3741	2	-7.8426	6.8492
2	0.5220	9.0605	1	-0.1281	6.4435
2	16.2898	8.5663	2	-4.3011	5.7614
2	-37.2501	8.5071	2	-0.1471	5.2905
2	-11.2793	7.7026	2	2.9645	5.1794
2	22.7275	6.8757	2	1.3660	5.1745
2	21.2348	6.8458	2	3.2498	4.0917
2	6.4167	6.0413	2	1.4972	4.0081
2	15.4666	5.2338	2	0.1112	3.6208
2	6.9336	5.1253	2	0.2378	3.5426
2	1.4535	4.5914	2	0.8909	3.0039
2	4.1884	4.3208	2	0.4102	2.8417
2	2.6313	3.5919	2	0.0609	2.5330
2	0.6326	3.5163	2	0.1303	2.3762
2	0.4945	2.9495	2	0.0010	2.0621
2	0.0232	2.3071	2	0.0103	1.9107

[a]The coefficients were derived from Clementi's Hartree-Fock calculations.[20] Coefficients for the other elements up to Z=36 can be obtained by writing to the authors. The Coefficients can also be used to generate the atomic electrostatic potential $V_a(r)$ and the screening function $\chi_a(r) = -rV_a(r)/Z_a$ through the relation:

programs to calculate the R-factor, which expresses the goodness of fit between $[(\rho_{el})_{\underline{g}}]_{T=0}$ and $[(\rho_{el})_{\underline{g}}]_{SNA}$, averaged over all reflections.

In terms of the atomic form factors, we have:

$$[(\rho_{el})_{\underline{g}}]_{SNA} = \frac{1}{v}\sum_a f_a(g)e^{-i\underline{g}\cdot\delta a} \tag{23}$$

where $f_a(g)$ is defined by Eq. (17) and where the sum is taken over the atoms in a unit cell. In Eq. (22), $\underline{\delta}_a$ gives the position of the ath atom with respect to a standard point within the unit cell.

The effect of thermal vibrations of a crystal lattice is to smear out the sharp peaks of density at the atomic nuclei, so that at a finite temperature T the relationship corresponding to Eq. (15) is

$$[(\rho_{el})_{\underline{g}}]_T \xrightarrow[\text{large } g]{} \frac{1}{v}\sum_a f_a(g)e^{-i\underline{g}\cdot\underline{\delta}_a - W_a} \tag{24}$$

where

$$W_a = \frac{1}{2}\sum_{ij} B^a_{ij} g_i g_j \tag{25}$$

Eqs. (24) and (25) are discussed in more detail in an appendix to this paper. By making use of (24) one can find the thermal parameters from the experimental crystallographic data for large g (high scattering angles). Alternatively one can calculate the parameters B^a_{ij} from a knowledge of the force constants of the crystal following the procedure discussed in the appendix.

At all values of g we can write:

$$[(\rho_{el})_{\underline{g}}]_T - \frac{1}{v}\sum_a f_a(g)e^{-i\underline{g}\cdot\underline{\delta}_a - W_a} = [(\Delta\rho_{el})_{\underline{g}}]_T \tag{26}$$

where $\Delta\rho_{el}$ is the "deformation density", i.e. the change in the electronic charge density due to chemical bonding when the atoms are incorporated into a crystal. We can estimate the T=0 Fourier coefficients of the deformation density by writing:

Table 1 continued

$$V_a(r) = -\frac{4\pi}{r}\sum_j \frac{C_j(n_j+1)!}{\zeta_j^{n_j+3}} e^{-\zeta_j r} \sum_{s=0}^{n_j+1} \frac{(n_j-s+2)(\zeta_j r)^s}{s!}$$

$$[(\Delta\rho_{el})_{\underline{g}}]_{T=0} \simeq e^{\overline{W}}[(\Delta\rho_{el})_{\underline{g}}]_T \tag{27}$$

where

$$\overline{W} = \frac{1}{2}\sum_{ij} \overline{B}_{ij} g_i g_j \tag{28}$$

and where $\overline{B}_{ij}$ are thermal parameters averaged over atoms. Thus finally we have for the Fourier coefficients of the electronic charge density:

$$(\rho_{el})_{\underline{g}} = \frac{1}{V}\sum_a f_a(g) e^{-i\underline{g}\cdot\underline{\delta}_a} + [(\Delta\rho_{el})_{\underline{g}}]_{T=0} \tag{29}$$

where the atomic form factors $f_a(g)$ are calculated from atomic Hartree-Fock data, while the Fourier coefficients of the deformation density are determined from X-ray diffraction data.

An alternative approach to the problem of determining the deformation density is by refinement of the crystallographic data with non-spherical atoms. In this approach, the density in the crystal is represented as a superposition of atomic densities, but the atomic densities include contributions from the deformation density, so that one has the relation:

$$[(\rho_{el})_{\underline{g}}]_T = \frac{1}{V}\sum_a \{f_a(g) + \Delta f_a(\underline{g})\} e^{-i\underline{g}\cdot\underline{\delta}_a - W_a} \tag{30}$$

where the values of $[(\rho_{el})_{\underline{g}}]_T$ are taken from experiment, while the values of $f_a(g)$ come from atomic Hartree-Fock calculations. The values of $\Delta f_a(\underline{g})$ are determined by equation (30). The Fourier coefficients of the electronic charge density are then given by:

$$(\rho_{el})_{\underline{g}} = \frac{1}{V}\sum_a \{f_a(g) + \Delta f_a(\underline{g})\} e^{-i\underline{g}\cdot\underline{\delta}_a} \tag{31}$$

4. DETERMINATION OF $(V)_0$

Let us now return to the question of how to determine the Fourier coefficients of the Coulomb potential when $\underline{g}=0$. In order to determine $(V)_0$, we first express the Coulomb potential $V(\underline{x})$ in the form:

$$V(\underline{x}) = -e\sum_{\underline{\ell}} \int \frac{d^3x'\rho(x')}{|\underline{x}-\underline{x}'-\underline{\ell}|} \tag{32}$$

where, as before, the subscript c on the integral means that it is taken over a unit cell. We next introduce the Fourier transform representation of $\frac{1}{r}$ and write:

$$\frac{1}{|\underline{x}-\underline{x}'-\underline{\ell}|} = \frac{1}{2\pi^2}\int \frac{d^3k}{k^2} e^{i\underline{k}\cdot(\underline{x}-\underline{x}'-\underline{\ell})} \tag{33}$$

Substituting (33) into (32) and making use of the identity

$$\sum_{\underline{\ell}} e^{-i\underline{k}\cdot\underline{\ell}} = \frac{(2\pi)^3}{v}\sum_{\underline{g}} \delta(\underline{k}-\underline{g}) \tag{34}$$

we obtain:

$$V(\underline{x}) = -\sum_{\underline{g}\neq 0} \frac{4\pi e}{g^2}(\rho)_{\underline{g}}\, e^{ig\cdot x} + (V)_o \tag{35}$$

where $(\rho)_{\underline{g}}$ is given by Eq. (13), with $(\rho_{el})_{\underline{g}}$ determined by Eq. (29) or by Eq. (31), and where

$$(V)_o = -\frac{4\pi e}{v}\int_c d^3x'\rho(\underline{x}')\int \frac{d^3k}{k^2} e^{i\underline{k}\cdot(\underline{x}-\underline{x}')}\delta(\underline{k}) \tag{36}$$

Let us represent $\delta(\underline{k})$ by

$$\delta(k) = \lim_{k_o\to 0}\begin{cases} \dfrac{1}{(4\pi/3)k_o^3}, & k<k_o \\ 0, & k>k_o \end{cases} \tag{37}$$

Then, making use of (37) and expanding $e^{i\underline{k}\cdot(\underline{x}-\underline{x}')}$ in a Taylor series, we obtain:

$$\int \frac{d^3k}{k^2} e^{i\underline{k}\cdot(\underline{x}-\underline{x}')}\delta(\underline{k}) \tag{38}$$

$$= \lim_{k_o\to 0} \frac{1}{(4\pi/3)k_o^3}\int_{k<k_o} \frac{d^3k}{k^2}\left[1 + i\underline{k}\cdot(\underline{x}-\underline{x}') - \tfrac{1}{2}\{\underline{k}\cdot(\underline{x}-\underline{x}')\}^2 + \ldots\right]$$

so that

$$(V)_o = -\frac{4\pi e}{v} \lim_{k_o \to 0} \frac{1}{(4\pi/3)k_o^3} \int_c d^3x' \rho(\underline{x}') \int_{k<k_o} \frac{d^3k}{k^2} [1 \tag{39}$$

$$+ i\underline{k}\cdot(x-x') - \frac{1}{2}\{k\cdot(x-x')\}^2 + \ldots]$$

The first term in the Taylor series expansion gives zero because of the charge neutrality condition (11). The second term in the Taylor series vanishes when the angular integration is performed. Thus, the first non-vanishing contribution comes from the third term, which yields:

$$(V)_o = \frac{3e}{2v} \lim_{k_o \to 0} \frac{1}{k_o^3} \int_c d^3x' \rho(\underline{x}') \int \frac{d^3k}{k^2} \{\underline{k}\cdot(\underline{x}-\underline{x}')\}^2 \tag{40}$$

Performing the k-integration in (40) we have:

$$\int_{k<k_o} \frac{d^3k}{k^2} \{\underline{k}\cdot(\underline{x}-\underline{x}')\}^2 = \frac{4\pi k_o^3}{9} |\underline{x}-\underline{x}'|^2 \tag{41}$$

so that

$$(V)_o = \frac{2\pi e}{3v} \int_c d^3x' \rho(\underline{x}')(\underline{x}\cdot\underline{x} - 2\underline{x}\cdot\underline{x}' + \underline{x}'\cdot\underline{x}') \tag{42}$$

The first term in (42) vanishes because of charge neutrality (11); and the second term vanishes if we choose the unit cell in such a way that the dipole moment is zero. Thus, provided that the unit cell is electrically neutral and has zero dipole moment, we have the relation:

$$(V_o) = \frac{2\pi e}{3v} \int_c d^3x' \rho(\underline{x}') r'^2 \tag{43}$$

The Fourier series representation of $\rho(\underline{x})$ shown in equation (7) can be used to rewrite (43) in the form:

$$(V)_o = \frac{2\pi e}{3v} \sum_{\underline{g}\neq 0} (\rho)_{\underline{g}} \int_c d^3x\, r^2 e^{i\underline{g}\cdot\underline{x}} \tag{44}$$

where the term corresponding to $\underline{g}=0$ has been omitted because of the charge neutrality condition (11). $(V)_o$ can thus be related to the measured Fourier coefficients of the density.

5. ASYMPTOTIC VALUES OF THE FOURIER COEFFICIENTS

In the near neighborhood of an atomic nucleus, the charge density distribution is dominated by the two 1s core orbitals of the atom. These K-shell electrons see the unscreened nucleus, and therefore, whenever $|\underline{x}-\underline{\ell}-\underline{\delta}_a| \equiv r'$ is very small for one of the values of $\underline{\ell}$ or $\underline{\delta}_a$, it is a good approximation to write:

$$\rho_{el}(\underline{x}) \simeq -\frac{2e}{\pi}\left(\frac{Z_a}{a_o}\right)^3 e^{-\frac{2Z_a r'}{a_o}} \tag{45}$$

The Fourier coefficients $(\rho_{el})_{\underline{g}}$ and $(V)_{\underline{g}}$ corresponding to very high values of $\underline{g}$ are associated with the cusps of charge density in the near neighborhood of nuclei, and therefore we can write:

$$(\rho_{el})_{\underline{g}} = \frac{1}{Nv}\int d^3x\, \rho_{el}(\underline{x}) e^{-i\underline{g}\cdot\underline{x}} \tag{46}$$

$$\xrightarrow[ga_o \gg Z_a]{} \frac{-2e}{\pi v a_o^3}\sum_a Z_a^3 \int d^3x\, e^{-\frac{2Z_a r}{a_o}+i\underline{g}\cdot\underline{x}}$$

$$= -\frac{8e}{v a_o^3}\sum_a Z_a^3 \int_0^\infty dr\, r^2 j_o(gr) e^{-\frac{2Z_a r}{a_o}}$$

Making use of (22) to evaluate the integral, we find the asymptotic expressions:

$$(\rho_{el})_{\underline{g}} \xrightarrow[ga_o \gg Z_a]{} -\frac{2e}{v}\sum_a \frac{(2Z_a)^4 e^{-i\underline{g}\cdot\underline{\delta}_a}}{[(ga_o)^2+(2Z_a)^2]^4} \tag{47}$$

and

$$(V)_{\underline{g}} \xrightarrow[ga_o \gg Z_a]{} \frac{4\pi e^2}{vg^2}\sum_a e^{-i\underline{g}\cdot\underline{\delta}_a}\left\{-Z_a + \frac{2(2Z_a)^4}{[(ga_o)^2+(2Z_a)^2]^2}\right\} \tag{48}$$

The first term in Eq. (48) is produced by the Coulomb potentials of the bare nuclei, while the second term is due to the screening of the nuclei by the K-shell electrons.

6. THE COULOMB CONTRIBUTION TO THE BINDING ENERGY OF CRYSTALS

From a knowledge of the Fourier coefficients of the charge density, it is possible to calculate the Coulomb contribution to the binding energy of crystals. If the Coulomb potential and charge density are given respectively by Eqs. (5) and (7), then making use of (10) and (11) we obtain:

$$-\frac{1}{e}\int d^3x\ \rho(\underline{x})V(\underline{x}) = -\frac{1}{e}\sum_{\underline{g}}\sum_{\underline{g}'\neq 0}(V)_{\underline{g}}(\rho)_{\underline{g}'}\int d^3x\ e^{i(\underline{g}+\underline{g}')\cdot\underline{x}} \tag{49}$$

$$= -\frac{Nv}{e}\sum_{\underline{g}\neq 0}(V)_{\underline{g}}(\rho)_{-\underline{g}} = 4\pi Nv\sum_{\underline{g}\neq 0}\frac{1}{g^2}(\rho)_{\underline{g}}(\rho)_{-\underline{g}}$$

where the coefficients $(\rho)_{\underline{g}}$ are given by (13). In order to calculate the Coulomb contribution to the binding energy, we perform the sum shown in Eq. (49) and subtract from it the Coulomb energies of the isolated atoms:

$$-\frac{N}{e}\sum_a\int d^3x\ V_a(r)\rho_a(r) = 4\pi Ne\sum_a Z_a\int_0^\infty dr\ r\chi_a(r)\rho_a(r) \tag{50}$$

where $\chi_a(r)$ is the screening function of atom a (Table 1).

7. CALCULATION OF FOURIER COEFFICIENTS OF THE SLATER POTENTIAL

The Slater potential, which can be used to approximate the exchange contribution to the effective Hamiltonian, is given by:[23]

$$V_{X\alpha}(\underline{x}) = -3\alpha e^2\left[\frac{3}{8\pi e}\rho(\underline{x})\right]^{1/3} \tag{51}$$

We can see from the discussion given above that the structure factors measured in an X-ray diffraction experiment, supplemented by information from atomic Hartree-Fock calculations, allow us to determine the electronic charge density distribution, $\rho_{el}(\underline{x})$; and having $\rho_{el}(\underline{x})$, we can use Eq. (51) to determine the Fourier coefficients of $V_{X\alpha}$. One possible method is as follows:

We begin by generating $\rho_{el}(\underline{x})$ at a grid of points $\underline{x}_j$ within the unit cell of the direct lattice. To do this, it is convenient to divide the electronic charge density into an undeformed part, $\rho_o(\underline{x})$, calculated from a superposition of the charge densities of the neutral atoms, and the deformation density $\Delta\rho_{el}(\underline{x})$ which is due to the flow of charge which takes place in chemical bonding:

$$\rho_{el}(\underline{x}) = \rho_o(\underline{x}) + \Delta\rho_{el}(\underline{x}) \tag{52}$$

$\rho_o(\underline{x}_j)$ can be calculated from the sum:

$$\rho_o(\underline{x}_j) = \sum_{\underline{\ell}}\sum_a \rho_a(|\underline{x}_j-\underline{\ell}-\underline{\delta}_a|) \tag{53}$$

while $\Delta\rho_{el}(\underline{x}_j)$ is calculated from the Fourier series:

$$\Delta\rho_{el}(\underline{x}_j) = \sum_{\underline{g}} (\Delta\rho_{el})_{\underline{g}} e^{i\underline{g}\cdot\underline{x}_j} \tag{54}$$

The Fourier coefficients of the deformation density are determined from the crystallographic data using Eqs. (26) and (27).

The next step is to take the cube root of $\rho_{el}(\underline{x}_j)$ at each of the grid points. We then define the function

$$\varepsilon(\underline{x}) \equiv \rho_{el}^{1/3}(\underline{x}) - \sum_{\underline{\ell}} \sum_{a} \rho_a^{1/3}(|\underline{x}-\underline{\ell}-\underline{\delta}_a|) \tag{55}$$

In other words, $\varepsilon(\underline{x})$ is the difference between the actual value of $\rho_{el}^{1/3}(\underline{x})$ and the value which would be found by taking the cube root of each neutral atom radial density distribution, and summing over atoms. When $\varepsilon(\underline{x})$ is evaluated at the grid points, it is found to be nearly constant. The large peaks which appear at the atomic nuclei in the function $\rho_{el}^{1/3}(\underline{x})$ do not appear in $\varepsilon(\underline{x})$, because near to a particular nucleus, the electronic charge density is rather well represented by the neutral atom density due to that atom, plus a constant contribution coming from all the other atoms.

We are now in a position to evaluate the Fourier coefficients of

$$\rho_{el}^{1/3}(\underline{x}) = \sum_{\underline{\ell}} \sum_{a} \rho_a^{1/3}(|\underline{x}-\underline{\ell}-\underline{\delta}_a|) + \varepsilon(\underline{x}) \tag{56}$$

To do this, we take the Fourier transform of the first term in Eq. (56) by means of the relations

$$\sum_{\underline{\ell}} \sum_{a} \rho_a^{1/3}(|\underline{x}-\underline{\ell}-\underline{\delta}_a|) = \sum_{\underline{g}} \sum_{a} (\rho_a^{1/3})_{\underline{g}} e^{i\underline{g}\cdot(\underline{x}-\underline{\delta}_a)} \tag{57}$$

and

$$(\rho_a^{1/3})_g = \frac{4\pi}{v} \int_0^\infty dr\, r^2 j_o(gr) \rho_a^{1/3}(r) \tag{58}$$

where j_o is a spherical Bessel function. The remaining term, $\varepsilon(\underline{x})$, can either be approximated by a constant, or else, if higher accuracy is desired, its Fourier transform can be evaluated numerically using the values found at the grid points.

This gives us finally the Fourier coefficients of the Slater potential. If we write $V_{X\alpha}$ in the form:

$$V_{X\alpha}(\underline{x}) = \sum_{\underline{g}} (V_{X\alpha})_{\underline{g}} e^{i\underline{g}\cdot\underline{x}} \tag{59}$$

then the Fourier coefficients $(V_{X\alpha})_{\underline{g}}$ will be given by:

$$(V_{X\alpha})_{\underline{g}} = -3\alpha e^2\left[\frac{3}{8\pi e}\right]^{1/3}\left[\frac{4\pi}{v}\sum_a e^{-i\underline{g}\cdot\underline{\delta}_a}\int_0^\infty dr\, r^2 j_o(gr)\rho_a^{1/3}(r) + \delta_{\underline{g},0}\varepsilon\right] \tag{60}$$

where we have approximated $\varepsilon(\underline{x})$ by a constant $(\varepsilon(\underline{x})\approx\varepsilon)$.

We have evaluated the Fourier coefficients of the Slater potential of a Cu_2O crystal using the method described above. Our experience shows that in the direct lattice sum of Eq. (53), the effects of nearest neigbor cells have to be included. We also found that for the Cu_2O crystal, it was a good approximation to set $\varepsilon(\underline{x})$ equal to a constant.

8. DISCUSSION

The construction of the Coulomb and Slater potentials from X-ray diffraction data using the methods outlined above is a relatively simple procedure. It follows that accurate X-ray diffraction data, supplemented with atomic Hartree-Fock data, can provide a starting point for the calculation of band structures in crystals. Calculations of this type are in progress in our laboratory. Stewart[24-27] has already published Coulomb potentials for a number of crystals. Massa, Clinton and Frishberg[28-32] have suggested that the density matrix could be calculated directly from diffraction data. Their proposal is a very interesting one, although it seems likely that their method could be used only for crystals with narrow bands. For magnetic materials, one might hope to use the data from polarized neutron diffraction experiments[1] to construct the vector potential within a crystal.

We hope that future work in this direction will lead to the development of standard programs for calculating the wave functions and properties of crystals using potentials constructed from accurate diffraction data.

APPENDIX: THE EFFECT OF THERMAL VIBRATIONS ON X-RAY DIFFRACTION INTENSITIES

In this appendix we shall review briefly some of the results of the theory of temperature factors in X-ray crystallography. A more complete treatment can be found in Ref. (33).

The normal modes of a crystal are labeled by two indices, j and $\underline{q}$, where $\underline{q}$ is a wave number in the first Brillouin zone. If there are N unit cells in the crystal, each containing n atoms, then there will be 3nN normal modes and 3n values of j for each value of q. The frequencies and amplitudes of the normal modes are found by solving the 3n secular equations:

$$\sum_{a=1}^{n} \left[\underline{\underline{V}}_{a',a}(\underline{q}) - \underline{\underline{I}}\, \omega^2_{j,\underline{q}} M_a \delta_{a',a} \right] \underline{A}_a(j,\underline{q}) = 0 \qquad \text{(A-1)}$$

The sum in Eq. (A-1) is over the n atoms in the unit cell. M_a is the mass of atom a, $\underline{\underline{I}}$ is the 3×3 unit matrix, and $\underline{u}_a(j,\underline{q})$ is the displacement amplitude of atom a in the $(j,\underline{q})$'th normal mode. Finally,

$$\underline{\underline{V}}_{a',a}(\underline{q}) \equiv \sum_{\underline{\ell}} e^{i\underline{q}\cdot\underline{\ell}}\, \underline{\underline{\kappa}}_{a',a}(\underline{\ell}) \qquad \text{(A-2)}$$

where $\underline{\underline{\kappa}}_{a',a}(\underline{\ell})$ is the force constant tensor between atom a' and atom a in two unit cells separated by the direct lattice vector $\underline{\ell}$, and $\underline{\delta}_a$ is the position of atom a with respect to a standard point within the unit cell.

The classical displacement of atom a in unit cell $\underline{\ell}$ due to the $(j,\underline{q})$'th normal mode is given by:

$$\underline{u}^a_{\underline{\ell}}(j,\underline{q}) = \underline{A}_a(j,\underline{q})\ \mathrm{Re}\left[e^{i(\omega_{j,\underline{q}} t + \underline{q}\cdot\underline{\ell})} \right] \qquad \text{(A-3)}$$

where

$$\underline{A}_a(j,\underline{q}) = \left(\frac{2E_j(\underline{q})}{N M_a \omega^2_{j,\underline{q}}} \right)^{\frac{1}{2}} \underline{e}_a(j,\underline{q}) \qquad \text{(A-4)}$$

$E_j(\underline{q})$ is the average energy in the $(j,\underline{q})$'th mode and $\underline{e}_a(j,\underline{q})$ is a polarization vector which obeys the normalization condition:

$$\sum_a \underline{e}_a(j,\underline{q})\cdot\underline{e}_a(j',\underline{q}) = \delta_{j',j} \qquad \text{(A-5)}$$

At a given instant of time, the displacement of atom a in unit cell $\underline{\ell}$ from its equilibrium position is given by:

$$\underline{R}^a_{\underline{\ell}} = \sum_{j,\underline{q}} \underline{u}^a_{\underline{\ell}}(j,\underline{q}) \qquad \text{(A-6)}$$

In an X-ray scattering experiment, the observed intensity for the scattering vector $\underline{k}$ is given by:

$$I(\underline{k}) = I_e \langle |F(\underline{k})|^2 \rangle \qquad \text{(A-7)}$$

where the brackets denote time averaging and where I_e is a constant. $F(\underline{k})$ is a structure factor given by:

$$F(\underline{k}) = \int d^3x\ \rho_{el}(\underline{x})\, e^{-i\underline{k}\cdot\underline{x}} \qquad \text{(A-8)}$$

If we approximate the electronic density in the crystal by a sum over atoms, we have, at a given instant of time,

$$\rho_{el}(\underline{x}) = \sum_{\underline{\ell}} \sum_{a} \rho_a(\underline{x}-\underline{\ell}-\underline{\delta}_a-\underline{R}^a_{\ell}) \tag{A-9}$$

so that

$$F(\underline{k}) = \sum_a f_a(\underline{k}) e^{-i\underline{k}\cdot\underline{\delta}_a} \sum_{\underline{\ell}} e^{-i\underline{k}\cdot(\underline{\ell}+\underline{R}^a_{\ell})} \tag{A-10}$$

where

$$f_a(\underline{k}) = \int d^3x\ \rho_a(\underline{x}) e^{-i\underline{k}\cdot\underline{x}} \tag{A-11}$$

Substituting (A-10) into (A-7) gives:

$$I(\underline{k}) = I_e \sum_{a,a'} f_a(\underline{k}) f^*_{a'}(\underline{k}) e^{-i\underline{k}\cdot(\underline{\delta}_a-\underline{\delta}_{a'})} \sum_{\underline{\ell},\underline{\ell}'} e^{-i\underline{k}\cdot(\underline{\ell}-\underline{\ell}')} \times \langle e^{-i\underline{k}\cdot(\underline{R}^a_{\ell}-\underline{R}^{a'}_{\ell'})} \rangle \tag{A-12}$$

It can be shown that it follows from (A-12) that the total intensity is composed of an inelastic part called the "thermal diffuse scattering", and an elastic part, which obeys the Bragg condition and which is given by:

$$I_o(\underline{k}) = I_e \frac{(2\pi)^3}{v} N \sum_{\underline{g}} \delta(\underline{k}-\underline{g}) \left| \sum_a f_a(g) e^{-i\underline{g}\cdot\underline{\delta}_a - W_a(\underline{g})} \right|^2 \tag{A-13}$$

where

$$W_a(\underline{g}) = \frac{1}{2} \sum_{\mu,\nu=1}^{3} B^a_{\mu,\nu} g_\mu g_\nu$$

$$= \frac{\hbar}{2NM_a} \sum_{j,\underline{q}} \frac{\{\underline{g}\cdot\underline{e}_a(j,\underline{q})\}^2}{\omega_{j,\underline{q}}} \left(\frac{1}{\exp(\hbar\omega_{j\underline{q}}/kT) - 1} + \frac{1}{2} \right) \tag{A-14}$$

If the force constants of the crystal are known, then $W_a(\underline{g})$ can be found by solving the secular Eqs. (A-1) and substituting the frequencies and eigenvectors into (A-14). Alternatively, the tensor $B^a_{\mu\nu}$ can be determined from X-ray diffraction data by high angle refinement using Eq. (24).

ACKNOWLEDGEMENTS

It is a pleasure to acknowledge stimulating conversations with Profs. R.F. Stewart, E.N. Maslen, P. Coppens, C. Scheringer,

L.J. Massa and S. Trickey. We are also grateful to the Northern European University Computing Center at Lundtofte, Denmark, for the use of their facilities. Finally, we would like to thank lektor Sine Larsen and cand.scient. Michael Gajhede for their advice and help.

REFERENCES

1. P. Becker, Editor, "Electron and Magnetization Densities in Crystals", Plenum, New York (1980).
2. E.N. Maslen, Acta Cryst. B24, 1172 (1968).
3. P. Coppens and A. Vos, Acta Cryst. B27, 146 (1971).
4. F.L. Hirshfeld, Acta Cryst. A32, 239 (1976).
5. R.F. Stewart, J.Chem.Phys. 53, 205 (1970).
6. P. Coppens and E.D. Stephens, Advances in Quantum Chemistry 10, Academic Press, New York (1977).
7. P.J.E. Aldred and M. Hart, Proc.Roy.Soc., (London) A 332, 223 (1973).
8. P. Becker, Phys.Scr. 15, 119 (1977).
9. J. Avery, Int.J.Quantum Chem. 16, 1265 (1979).
10. J. Avery and E. Berg, Int.J.Quantum Chem. 16, 1279 (1979).
11. J. Avery, Int.J.Quantum Chem. Symp. 13, 403 (1979).
12. J. Avery and P.-J. Ørmen, Int.J.Quantum Chem. 18, 953 (1980).
13. J. Avery and P.-J. Ørmen, Int.J.Quantum Chem. Symp. 14, 629 (1980).
14. J. Avery, P.-J. Ørmen and I. Chatterjee, Int.J.Quantum Chem. 21, 515 (1982).
15. J. Avery and P.-J. Ørmen, Int.J.Quantum Chem. Symp. 15, 477 (1981).
16. J. Avery and P.-J. Ørmen, Acta Cryst. A35, 849 (1979).
17. J.C. Slater, Phys.Rev. 81, 385 (1951).
18. E.J. Baerends, D.E. Ellis and P. Ros, Theor.Chim. Acta 27, 339 (1972).
19. O. Gunnarson and P. Johansen, Int.J.Quantum Chem. 10, 307 (1976).
20. E. Clementi, "Tables of Atomic Functions", Supplement to IBM J. Res.Dev. 9, 2 (1965).
21. D.T. Cromer and J.B. Mann, Acta Cryst. A24, 321 (1968).
22. C. Scheringer, Acta Cryst. A34, 905 (1978).
23. J.C. Slater, "Quantum Theory of Molecules and Solids, Vol. 4, The Self-Consistent Field for Molecules and Solids", McGraw-Hill (1974).
24. R.F. Stewart, Acta Cryst. A31, S218 (1975); A32, 182 (1976); A32, 565 (1976); A33, 33 (1977); A34, S27 (1978).
25. R.F. Stewart, Bul.Amer.Phys.Soc. 23, 305 (1977).
26. R.F. Stewart, Chem.Phys.Letters, 65, 335 (1979).
27. R.F. Stewart, J. Bentley and B. Goodman, J.Chem.Phys. 63, 3031 (1975).
28. L.J. Massa and W.L. Clinton, Trans.Amer.Cryst.Assn. 8, 149 (1972.

29. W.L. Clinton and L.J. Massa, Phys.Rev.Letters 29, 1363 (1972).
30. C.A. Frishberg and L.J. Massa, Int.J.Quantum Chem. 18, 801 (1978).
31. C. Frishberg and L.J. Massa, Phys.Rev. B24, 7018 (1981).
32. C. Frishberg and L.J. Massa, Acta Cryst. A38, 93 (1982).
33. B.T.M. Willis and A.W. Pryor, "Thermal Vibrations in Crystallography", Cambridge University Press (1975).

OBSERVATION OF ELECTRON DISTRIBUTIONS ON OUTERMOST LAYERS OF SOLID SURFACES BY PENNING IONIZATION ELECTRON SPECTROSCOPY

Yoshiya Harada and Koichi Ohno

Department of Chemistry, College of General Education
The University of Tokyo, Komaba, Meguro
Tokyo 153, Japan

1. INTRODUCTION

The electronic structure of solid surfaces has been studied by means of various types of electron spectroscopy such as ultraviolet photoelectron spectroscopy (UPS), X-ray photoelectron spectroscopy (XPS or ESCA), Auger electron spectroscopy (AES) and electron impact spectroscopy (EIS). In these methods, however, information on the first layer of the solid surface is mixed with that on inner layers, because photons (XPS, UPS) or electrons (AES, EIS) used for the excitation sources penetrate into the solid and large numbers of electrons within the escape depth are ejected. On the other hand, Penning ionization electron spectroscopy (PIES),[1] which utilizes metastable atoms (He* 2^3S 19.82 eV, Ne* $^3P_2$ 16.62 eV etc.) as the excitation source, is selectively sensitive to the outermost surface layer of the solid, because metastable atoms are physically incapable of surface penetration. Moreover, recent studies have shown that the electron distribution of individual molecular orbitals in the local area outside the solid surface can be probed by PIES.[2,3]

In Penning ionization, electrons are ejected by collisions between target molecules M and metastable atoms A*:

$$M + A^* \longrightarrow M^+ + A + e^- \qquad (1)$$

This process is interpreted in terms of electron exchange, in which an electron in the orbitals of M fills the vacant orbital of A^* and its excited electron is ejected.[4] The transition probability largely depends on the overlapping of the relevant orbital of M and the vacant orbital of A^*. Therefore, an orbital spreading out of the molecule gives a strong band in the Penning spectrum, while a spatially limited and shielded orbital gives a weak band in the spectrum. This characteristic of PIES has been successfully applied to probe electron distributions of various molecules in gas phase.[5-10] When metastable atom beams are introduced onto a solid surface, an orbital "exposed" to the outside of the surface interacts with a metastable atom more effectively than an inner orbital and gives a stronger band in the Penning spectrum. This idea provides the basis for the PIES study of the local electron distribution on the surface.

The observation of the electron distribution of the orbital on the solid surface enables us to study the electronic structure and the orientation of surface molecules. In the following sections recent studies on the surfaces of molecular crystals such as benzene, biphenyl, anthracene, and naphthacene will be reviewed.

2. ORIENTATION OF SURFACE MOLECULES

In this section we describe, taking solid anthracene as an example, how information on the geometrical orientation of surface molecules is drawn from the observation of the electron density by PIES.[2] Fig. 1 shows the Penning electron spectra of vapor and an evaporated film of anthracene by Ar* ($^3P_2$ (1.55 eV) metastable atoms. The photoelectron spectrum of the anthracene film by Ar I (11.7 eV) is also shown for comparison. The spectra give three bands due to the highest three occupied π orbitals (π_7, π_6, π_5,). In the Penning spectra, the second band for the solid is markedly enhanced relative to the other bands in comparison with the spectrum of the vapor. On the other hand, in the photoelectron spectra the intensity ratio of the bands is similar in both solid and vapor.

If we assume that surface molecules of the evaporated film are arranged as in the case of the bulk crystal (cf. the structure in the upper left part of Fig. 1), the intensities of the bands in the Penning spectrum of the film can be explained in terms of the electron densities of the π orbitals. At the top of Fig. 1 the electron densities of each π orbital near the surface are shown.* The density maps show that the π_6 orbital has larger electron density than the π_7 and π_5 orbitals. Therefore, the ionization cross section for the π_6 orbital becomes larger than those for the π_7 and π_5 in line with the general idea described in the introduction. Thus, it is shown that surface molecules are oriented in the evaporated film in a similar way as in the bulk crystal. The present example

*The densities were calculated in a plane parallel to the surface at the distance of about 3 Å from the top of the molecule.

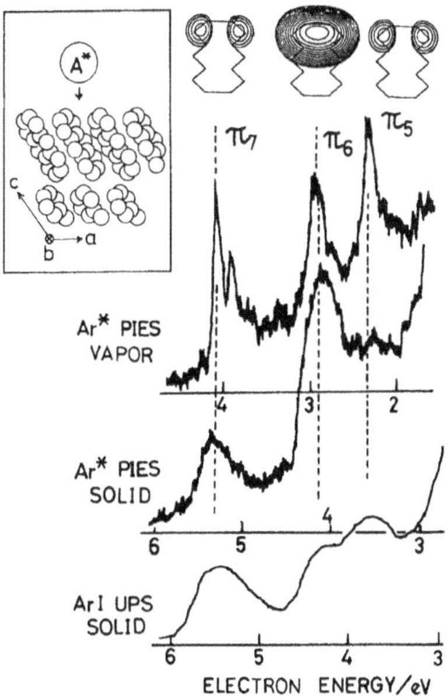

Fig. 1. Ar* ($^3P_2$) Penning ionization electron spectra for anthracene vapor and solid together with Ar I photoelectron spectrum for solid. The orientation of anthracene molecules in the evaporated film is shown on the upper left. Contour lines given at the top of the figure represent the electron densities near the surface (see text).

indicates that the orientation of molecules on the outermost surface layer can be probed much more easily and sensitively by PIES than UPS. In UPS a combination of a careful angle-resolved measurement and theoretical calculations are necessary for this purpose, especially for larger molecules, whereas in PIES a simple angle-unresolved experiment provides useful information on the molecular orientation.

3. PHASE CHANGE ON THE SOLID SURFACE

The observation of the local electron distribution by PIES also enables us to detect the change in the molecular arrangement on the solid surface. Fig. 2 shows the Penning spectra by Ne* ($^3P_2$ 11.62 eV) for biphenyl evaporated films and vapor.[3] In the figure

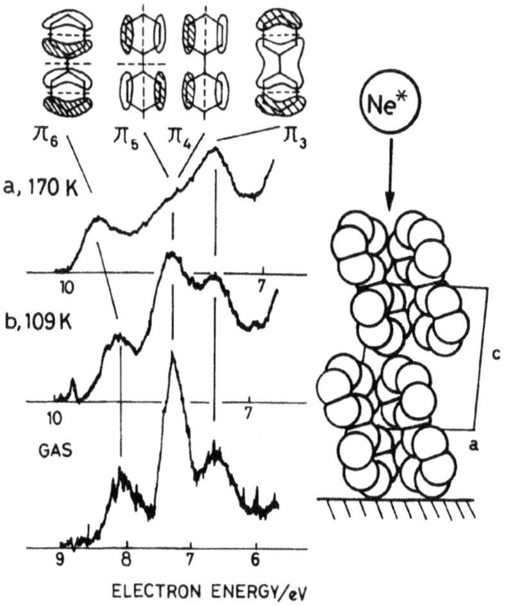

Fig. 2. Ne* ($^3P_2$) Penning spectra for biphenyl films and vapor. Spectra (a) and (b) were obtained from the sample deposited on a substrate at 170 K and 109 K respectively. A schematic diagram of the unit cell of a biphenyl crystal on a metal substrate is given on the right of the figure.

the spectra (a) and (b) were obtained from the samples deposited on a substrate at 170 K (film (a)) and 109 K (film (b)), respectively. The relative intensities of the peaks are significantly different between the spectra (a) and (b). However, when the film (b) was annealed at 170 K for 40 min., the spectrum (b) changed to (a). This change was irreversible; when the temperature of the substrate was lowered back to 109 K, the spectrum (b) was not restored. The photoelectron spectra of the films (a) and (b) were also measured but no difference in the relative intensities was found.

The observed spectral change in the Penning spectra can be interpreted in terms of the change in the molecular arrangement associated with a transition from amorphous to polycrystalline state; the film (b) prepared on the cold substrate at 109 K is amorphous, while the film (a) formed on the hot substrate at 170 K is polycrystalline and the transition from the film (b) to (a) is irreversible. This interpretation is substantiated by the relative intensities of the spectra. Three bands in the spectra in Fig. 2 are assigned to the highest occupied π orbitals, π_6-π_3. The nearly degenerate π_5 and π_4 orbitals have nodal planes along the long axis of the molecule and perpendicular to the molecular plane. Therefore, if molecules are oriented on the polycrystalline surface as shown on the right of Fig. 2 (bulk structure), the electron densities of the π_5 and π_4 orbitals on the surface should be lower, and hence their band intensities should be weaker, than those of the π_6 and π_3 orbitals. This is consistent with the features of the spectrum (a). On the other hand, the Penning spectrum for the film (b) has the intense π_5 and π_4 bands and is close to the gas phase spectrum. This indicates that the molecules on the surface on the film (b) are randomly oriented.

4. ADSORPTION ON THE SOLID SURFACE

Extreme surface sensitivity of PIES offers a promising prospect of the study of the adsorbed surface. In this section we describe the study of benzene molecules adsorbed on a graphite basal plane

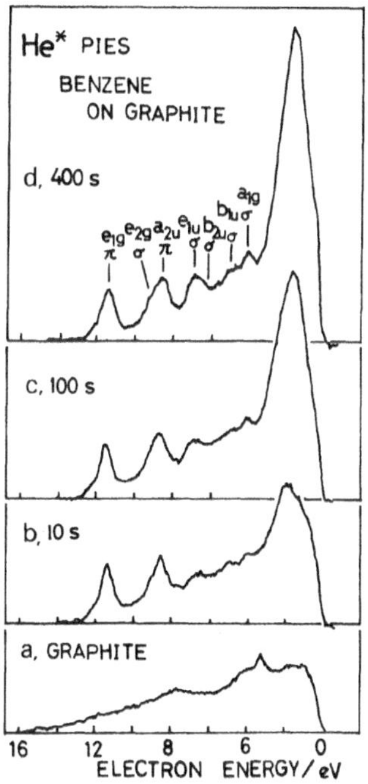

Fig. 3. He* (2^3S) Penning electron spectra for benzene adsorbed on graphite. Spectrum (a) was obtained from clean graphite. Spectra (b), (c), and (d) were obtained from the graphite exposed to benzene vapor of 5×10^{-7} Torr for (b) 10 s, (c) 100 s, and (d) 400 s.

((0001 plane)).[11] The orientation of the adsorbed molecules has been sensitively probed on the basis of the observation of the electron distribution. Fig. 3 shows the He* (2^3S 19.82 eV) Penning spectra measured for graphite exposed to benzene vapor of about 5×10^{-7} Torr for 10, 100 and 400 s in succession at 130 K. The clean graphite spectrum is also shown in Fig. 1. The thickness of the adsorbed layers was estimated to be a few molecular layers at 10 s exposure and several tens of layers or more at 400 s exposure. In the spectra of 10 s exposure in Fig. 3 one can see that the structures in the graphite spectrum have completely disappeared. On the other hand, in the He I photoelectron spectra of the same sample the graphite structures were observed to remain up to 200 s exposure time. These findings together with other earlier work[12,13] indicate that PIES is much more sensitive than UPS.

The bands in the spectra of the adsorbed surface are associated with π and σ orbitals of benzene as shown in Fig. 3. In the figure the relative intensity of the bands is seen to change with increasing exposure time. This observation is accounted for in terms of the change in the molecular orientation on the adsorbed layer surface and the spatial electron distributions of the π and σ orbitals. The π orbitals extend normal to the molecular plane, while the σ orbitals are distributed in the molecular plane. Fig. 4 shows interaction between metastable atoms and benzene molecules adsorbed on a substrate. As shown in Fig. 4 (a), if a metastable atom approaches a benzene molecule adsorbed flat on the substrate, a π orbital exposed outside should give a stronger band in the Penning spectrum than a σ orbital. On the other hand, if a metastable atom approaches a benzene molecule tilted on the substrate, the σ band should also appear with appreciable intensity in the spectrum (Fig 4 (b)). In Fig. 3, all the σ bands are weaker than the π bands in the spectrum for 10 s exposure and their intensities gradually increase with increasing exposure time. This experimental fact points out that benzene molecules are adsorbed flat on the graphite at low coverage and gradually tilted on the surface of the adsorbed multilayer as the layer becomes thick. Thus, in the thin layer the orientation of the adsorbed molecules are determined mainly by the interaction between the molecules and the graphite substrate, whose hexagonal network causes the benzene ring to lie flat on it. With increasing thickness the mutual molecular interaction gradually overcomes the substrate-molecule interaction and molecules become oriented as in the bulk crystal.

In Fig. 3 the b_{2u} σ band scarcely appears even at 400 s exposure, although the other σ bands increase in intensity with increasing exposure time. This can be interpreted in terms of the electron distribution of the orbital. Since the b_{2u} σ orbital is localized along the C-C bonds and shielded by the π orbitals, and other σ orbitals having the hydrogen 1s character, it does not effectively interact with a metastable atom. This character of the b_{2u} σ

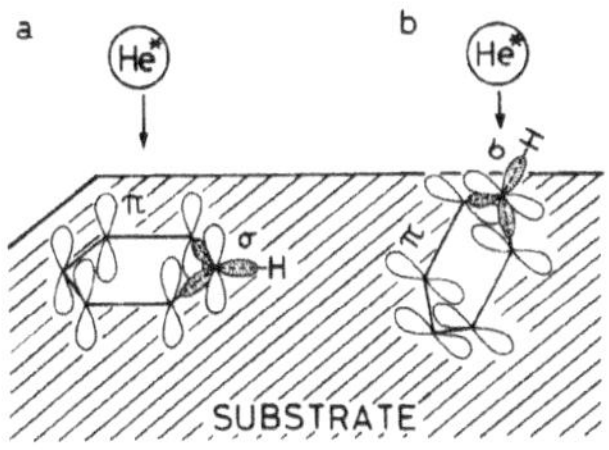

Fig. 4. Interaction between metastable atoms and benzene molecules. (a) benzene molecule adsorbed flat on a substrate, (b) benzene molecule tilted on a substrate.

orbital has also been observed in the Penning spectrum of gas phase molecules.[5]

5. CHEMICAL CHANGE ON THE SURFACE

PIES can be further applied to sensitive detections of chemical change on solid surfaces. Here we describe the example of the photooxidation reaction of tetracene (naphthacene):[14]

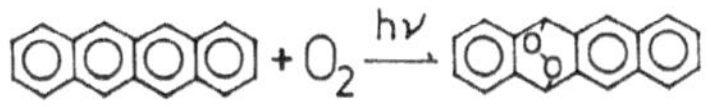

(2)

Fig. 5 shows the Ne 1 photoelectron spectra and Ne* ($^3P_2$ 16.62 eV) Penning spectra for the photooxidation process of a tetracene film. Curves 0 show the spectra for the fresh film and curves 1 those after irradiation for 1 h under oxygen pressure of 2 Torr. In curves 1 a slight change is observed for the Ne* spectrum, whereas almost no change is found for the Ne I spectrum. Further irradiation for 1 h (curves 2) strengthens this tendency to the spectral change.

The change in the Ne* spectrum can be explained in terms of the separation of the conjugated system of tetracene into those of benzene and naphthalene as shown in Eq. (2). In Fig. 5, T, B, and N denote the peak positions corresponding to tetracene, benzene and naphthalene, which are estimated from the photoelectron spectra of gaseous molecules. We can find from the figure that during photooxidation the bands due to tetracene gradually decrease, while those due to the benzene and naphthalene skeletons increase in intensity.

The above results show that in our experimental condition the photooxidation mainly takes place on the outermost layer and that it can be sensitively detected by PIES which probes the local distribution of the electron densities outside the surface.

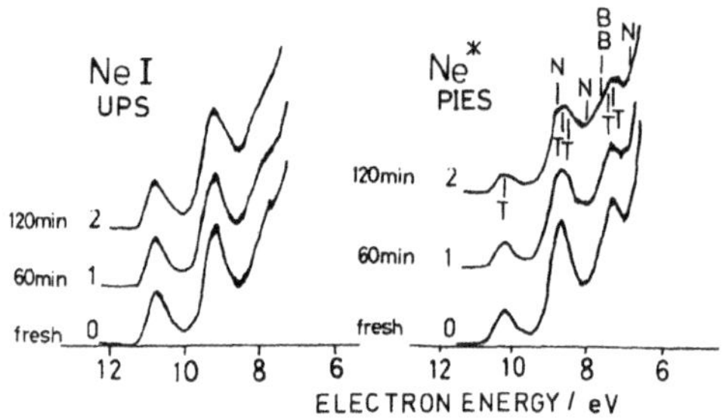

Fig. 5. Ne I photoelectron spectra and Ne* ($^3P_2$) Penning spectra for the photooxidation process of a tetracene (naphthacene) film. Curves 0 show the spectra for the fresh film, curves 1 those after irradiation for 1 h under oxygen pressure of 2 Torr, and curves 2 those after further irradiation for 1 h under the same oxygen atmosphere. In the Ne* spectra, T, B, and N denote the peak positions corresponding to tetracene, benzene, and naphthalene.

6. CONCLUDING REMARKS

We have shown that PIES provides direct information on the electron distribution of the outermost surface layer of the solid and hence is truly surface sensitive in nature. Ion neutralization spectroscopy (INS),[15] in which rare gas ions are used for the excitation source, is another surface sensitive technique that can be compared with PIES. However, a considerable difficulty is associated with the deconvolution of ion neutralization spectra, because the process involves two-step ionization and the energy given by the incident ion is shared by two electrons. PIES is free from this type of problems except for clean metal surfaces, where the resonance tunneling of the excited electron of the metastable atom into one of the continuous vacant levels of metals leads to two-step ionization process (INS process).

Finally we would like to emphasize that PIES of the solid provides valuable information not only on the solid surface itself but also on the electron distribution of individual molecular orbitals at various parts of the molecule, if the orientation of the molecule on the surface in controlled by the selection of an appropriate substrate and of thermal annealing condition.

ACKNOWLEDGEMENT

The authors wish to thank Prof. K. Kuchitsu and his coworkers for their helpful discussions. The studies on biphenyl and adsorbed benzene were performed in collaboration with them.

REFERENCES

1. V. Cermâk, Retarding-potential measurement of the kinetic energy of electrons released in Penning ionization, J. Chem. Phys. 44, 3781 (1966); A. Niehans, Spontaneous ionization in slow collisions, in: "Adv. Chem. Phys, Vol. XLV", J. Wm. McGowan, Ed., Wiley, New York (1981).
2. T. Munakata, K. Ohno, and Y. Harada, Application of Penning ionization electron spectroscopy to the study of the outermost layer of the solid surface, J. Chem. Phys. 72, 2880 (1980).
3. H. Kubota, T. Munakata, T. Hirooka, K. Kuchitsu, and Y. Harada, Observation of molecular reorientation on evaporated biphenyl film surface by Penning ionization electron spectroscopy, Chem. Phys. Lett. 74, 409 (1980).
4. H. Hotop and A. Niehans, Reactions of excited atoms and molecules with atoms and molecules. II Energy analysis of Penning electrons, Z. Phys. 228, 66 (1969).
5. K. Ohno and Y. Harada, Observation of local electron distributions in molecular orbitals, p. 761 in this book.
6. T. Munakata, K. Kuchitsu, and Y. Harada, Comparison of the relative intensities of Penning electron spectrum and photoelectron spectrum. Unsaturated hydrocarbons, Chem. Phys. Lett. 64, 409 (1979).
7. T. Munakata, K. Kuchitsu, and Y. Harada, Comparison of the relative intensities of Penning electron spectra and photoelectron spectra. II. Heterocyclic aromatic molecules, J. Elec. Spectrosc. Relat. Phenom. 20, 235 (1980).
8. T. Munakata, Y. Harada, K. Ohno, and K. Kuchitsu, Penning electron spectrum of ferrocene. Evidence for steric effect on the electronic interaction causing Penning ionization, Chem. Phys. Lett. 84, 6 (1981).
9. T. Munakata, K. Ohno, Y. Harada, and K. Kuchitsu, Assignment of photoelectron bands for naphthalene and anthracene by Penning ionization electron spectroscopy, Chem. Phys. Lett. 83, 243 (1981).
10. K. Ohno, S. Fujisawa, H. Mutoh, and Y. Harada, Application of Penning ionization electron spectroscopy to stereochemistry. Steric shielding effect of methyl groups on Penning ionization in substituted anilines, J. Phys. Chem. 86, 440 (1982).
11. H. Kubota, T. Munakata, T. Hirooka, T. Kondow, K. Kuchitsu, K. Ohno, and Y. Harada, Penning ionization electron spectroscopy of benzene adsorbed on graphite, to be published.
12. H. Conrad, G. Ertl, J. Küppers, and S.W. Wang, Penning ionization electron spectroscopy of chemisorbed CO, Phys. Rev. Lett. 42, 1082 (1979).
13. J. Roussel, C. Boiziau, R. Nuvolone, and C. Reynaud, Electron ejection mechanism in the He 2^3S interaction with a clean and a molecularly covered Ni(111) surface, Surface Sci. 110, L634 (1981).

14. K. Ohno, H. Mutoh, and Y. Harada, Application of Penning ionization electron spectroscopy to the study of chemical reactions on the solid surface; photooxidation of naphthacene and rubrene, Surface Sci. 115, L128 (1982).
15. H.D. Hagstrum, Studies of adsorbate electronic structure using ion neutralization and photoemission spectroscopies, in: "Electron and Ion Spectroscopy of Solids," L. Fiermans, J. Vennik, and W. Dekeyser, eds., Plenum, New York (1978); H.D. Hagstrum, Excited-atom-deexcitation spectroscopy using incident ions, Phys. Rev. Lett. 43, 1050 (1979).

OBSERVATIONS OF LOCAL ELECTRON DISTRIBUTIONS IN MOLECULAR ORBITALS

Koichi Ohno and Yoshiya Harada

Department of Chemistry, College of General Education
The University of Tokyo, Komaba, Meguro
Tokyo 153, Japan

Spatial distributions of electrons in matter are observed by diffraction methods and energy distributions of electrons are measured as ionization potentials by photoelectron spectroscopy. Results of these methods have been compared with total electron densities and orbital energies obtained by theoretical calculations. On the other hand, electron densities of "individual" orbitals, which can also be obtained theoretically, have almost never been confirmed directly by experiments since the birth of the orbital concept in the early days of the quantum theory. Recent studies of Penning ionization electron spectroscopy have shown that the relative intensity of the observed bands depends on the electron distribution of molecular orbitals.[1-3] The present paper reports our recent results to demonstrate that local electron distributions in molecular orbitals can be revealed by Penning ionization electron spectroscopy.

The ionization process of a molecule(M) caused by interactions of the molecule with a metastable atom(A*) has been known as Penning ionization; $A^* + M \rightarrow A + M^+ + e^-$. Analysis of the kinetic energy distribution of the ejected electrons provides a Penning ionization electron spectrum,[4,5] which is similar in many respects to a UV photoelectron spectrum(UPS).[6] Table 1 lists the exciting energy sources for these spectra. Figs. 1 and 2 show He*(2^3S)PIES and He I UPS of molecular ammonia and water. The observed bands are assigned to electrons in molecular orbitals, because electron energies($\frac{1}{2}mv^2$) can be related to ionization potentials(IP) which are equal to the absolute values of the orbital energies in Koopmans' theorem; $\frac{1}{2}mv^2 = E - IP - d$, where E is the energy of the photon or the metastable atom and d is the energy balance arising from relative motions among the particles. For UPS d is negligible, whereas for PIES d is a variable quantity depending on the distance between the metastable atom and the target molecule where the ionization process occurs. The variation of this quantity around the most effective distance may result in obscured vibrational structures. However, since the values of d in PIES are usually not very large and much less than a few hundred meV, UPS and PIES are very similar to each other in their energy scales. On the other hand, as can be seen in Figs. 1 and 2, a considerable difference in the intensities has been observed between UPS and PIES. This must be ascribed to the difference in the mechanisms between photoionization and Penning ionization.

Penning ionization has been considered as an electron transfer process,[7] which is schematically shown in Fig. 3; an electron in a molecular orbital ϕ_M is transferred into the lowest unoccupied orbital χ_a of the metastable atom and followed by removal of an electron from the highest occupied orbital χ_b of the metastable atom to yield an electron in a continuum state ϕ_e. The transition probability of this process is governed by an exchange-type interaction;

$$K = \left| \langle \chi_b(1)\phi_M(2) \left| \frac{1}{r_{12}} \right| \phi_e(1)\chi_a(2) \rangle \right|^2$$

The value of K increases exponentially with the metastable atom approaching to the target molecule. This means that the most favourable distance for the ionization must be in the neighborhood of the molecular surface through which actual penetration of metastable atoms are physically incapable. Furthermore, major contributions

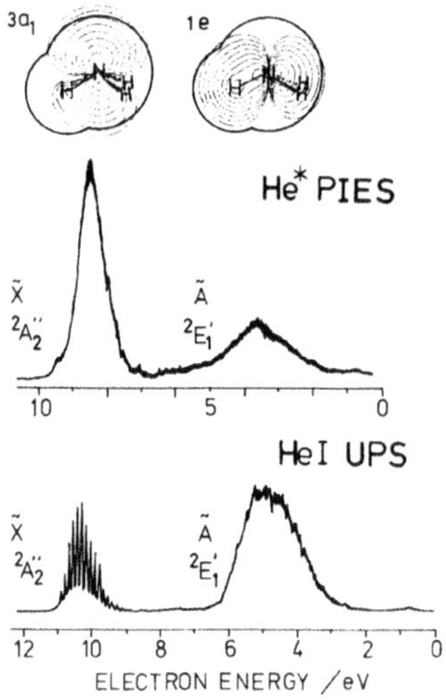

Fig. 1. He*(2^3S)PIES(upper) and He I UPS(lower) of NH_3. The bands are assigned to molecular orbitals for which electron density maps are drawn from the same level of density. The density of the n-th curve is $5\times2^n\times10^{-4}$.

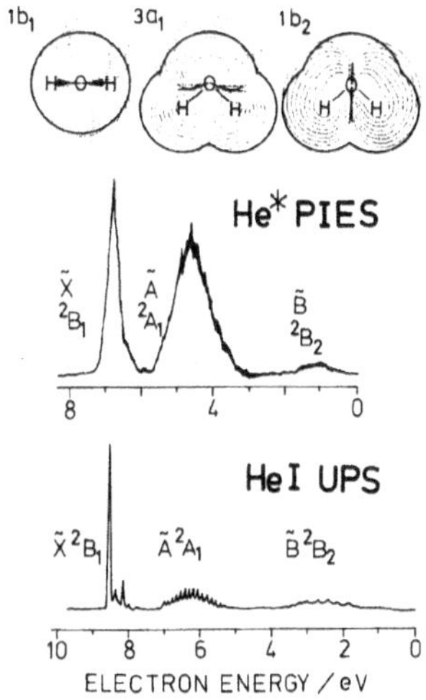

Fig. 2. He*(2^3S)PIES(upper) and He I UPS(lower) of H_2O. Electron density maps are drawn for the relevant orbitals. The density of the n-th curve is $2\times2^n\times10^{-4}$.

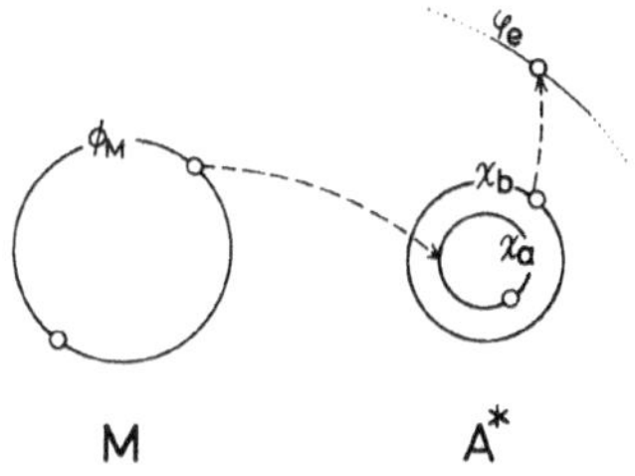

Fig. 3. The electron transfer model of Penning ionization. An electron is transfered from the molecule(M) into an orbital of metastable atom(A*) and an electron in the higher orbital of A* is ejected into a continuum state.

to the integral can be attributed to local regions where ϕ_M and χ_a effectively overlap. Therefore, relative probabilities of individual orbitals are governed by local electron densities of ϕ_M at the molecular surface.

If we imagine a collision of a metastable atom with a target molecule, a clear insight is obtained. The solid curve in Fig. 4 indicates a surface of the molecule up to which metastable atoms are accessible. The dashed curve(a) shows an orbital spreading out of the molecular surface, and the dashed curve(b) shows an orbital enveloped in the surface. Since an orbital exposed to the outside more easily interacts with the vacant orbital of the incoming metastable atom, the outer orbital(a) gives rise to the larger intensity in PIES than does the inner orbital(b). This provides an approximate selection rule for PIES; the outer orbital is active and the inner orbital is inactive. Such an effect may be summarized as a stereoelectronic effect of molecular orbitals on Penning ionization.

In order to verify the selection rule for PIES, experimental results must be compared with theoretical orbital electron densities. In Figs. 1 and 2 electron density maps obtained by ab initio MO calculations with a 4-31G basis set are shown. The contour lines are drawn from the same level of electron density for all the orbitals in each spectrum. The thick solid curves indicate the molecular surface estimated from the van der Waals radii of the atoms.

In the PIES of NH_3(Fig. 1), the $\tilde{X}$ band due to electrons in the nonbonding orbital($3a_1$) is markedly enhanced in comparison with the

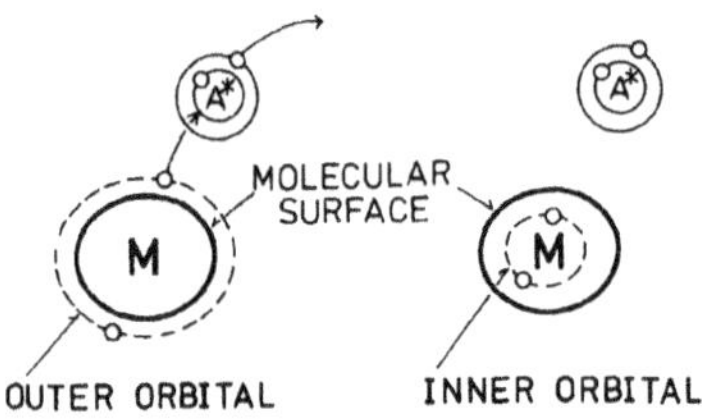

Fig. 4. The selection rule for Penning ionization electron spectra. The matastable atom A* can approach to the molecule up to the molecular surface shown by the solid curve. For the outer orbital electron transfer from M to A* can take place easily, whereas for inner orbital the process is almost forbidden, because penetration of the metastable atom through the molecular surface is physically incapable. Therefore, the outer orbital is active and the inner orbital inactive in Penning ionization.

Table 1. The exciting energy sources for PIES and UPS

PIES			UPS	
He^*	2^1S	20.616 eV	He I	21.218 eV
	2^3S	19.820 eV		
Ne^*	$^3P_0$	16.716 eV	Ne I	16.848 eV
	$^3P_2$	16.619 eV		16.670 eV
Ar^*	$^3P_0$	11.723 eV	Ar I	11.828 eV
	$^3P_2$	11.548 eV		11.623 eV

$\tilde{A}$ band originating from the N-H bonding orbital(1e). In the maps, the nonbonding orbital has a large density in the outside of the surface, whereas for the bonding orbitals only a small density can be seen in the corresponding region. In the PIES of H_2O(Fig. 2), the $\tilde{X}$ and $\tilde{A}$ bands, which are resulting from the $1b_1$ and $3a_1$ orbitals having the nonbonding oxygen 2p character, show much larger intensities than the $\tilde{B}$ band due to the O-H bonding orbital($1b_2$). The electron density maps show that the nonbonding orbitals spread out of the molecular surface and that the bonding orbital is nearly enveloped in the surface. These results clearly give confirmation of the selection rule in PIES.

Figs. 5 and 6 show the PIES and UPS of the nitrogen and acetylene molecules. These molecules are isoelectronic systems showing two sigma($3\sigma_g$ and $2\sigma_u$) and one degenerate pi($1\pi_u$) bands in the energy range concerned. The relative order of the PIES activity is $\sigma>\pi$ for N_2, whereas it is $\pi>\sigma$ for C_2H_2. An intermediate case is shown in Fig. 7. In the PIES of CH3CN, the π orbitals (2e) and the nonbinding orbital ($7a_1$) are active, and the σ orbitals (1e and $6a_1$) are inactive. These examples show that symmetric properties of orbitals cannot necessarily be used for the selection rule of the PIES activity. The electron density maps in Figs. 5-7 show the importance of the local electron distribution in PIES.

In Fig. 8 an example of planar molecule is shown. The π orbitals of benzene can be considered as PIES active orbitals, since the π orbitals spread out of the molecular plane. Electron density maps are drawn for a plane of 1.7 Å above the molecular plane, and they clearly show the strong activity of the π orbitals. In addition to this, it is noteworthy that although the b_{2u} orbital gives a band with considerable intensity in the UPS, the corresponding band is nearly

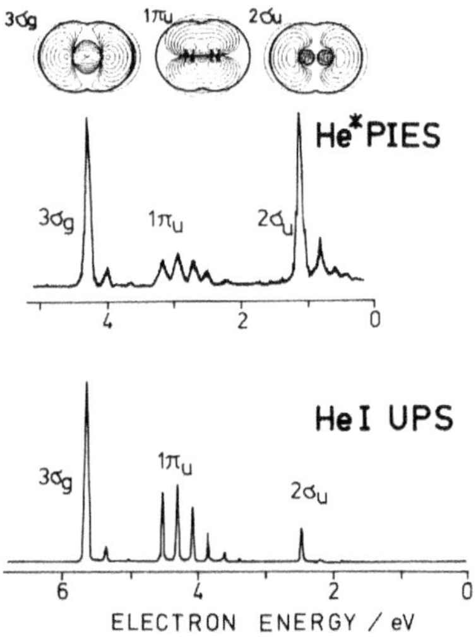

Fig. 5. He*(2^3S)PIES(upper) and He I UPS(lower) of N_2. The density of the n-th curve in the maps is $5\times2^n\times10^{-4}$.

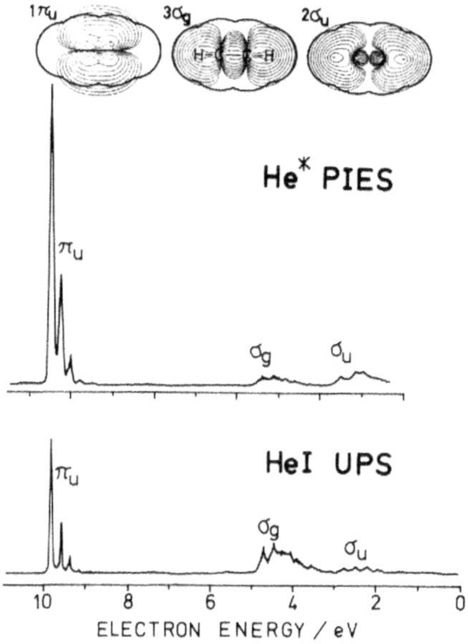

Fig. 6. He*(2^3S)PIES(upper) and He I UPS(lower) of C_2H_2. The density of the n-th curve in the maps is $5\times2^n\times10^{-4}$.

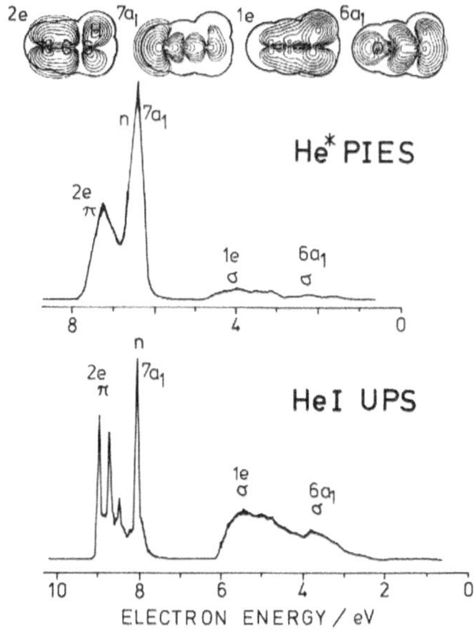

Fig. 7. He*(2^3S)PIES(upper) and He I UPS(lower) of CH_3CN. The density of the n-th curve in the maps is $2{\times}2^n{\times}10^{-4}$.

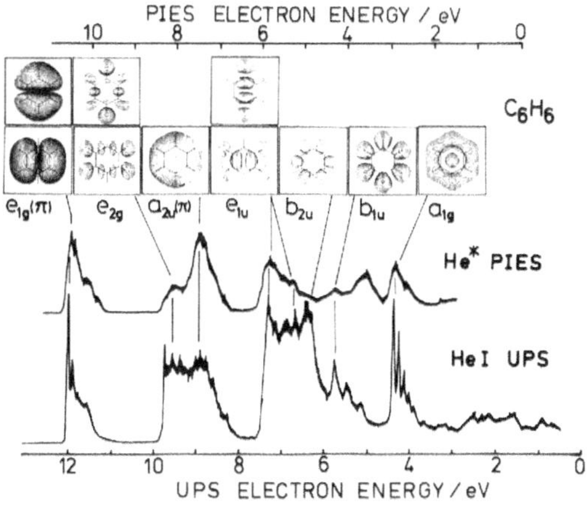

Fig. 8. He*(2^3S)PIES(upper) and He I UPS (lower) of benzene. Electron density maps are drawn for the relevant molecular orbitals at a plane of 1.7 A(van der Waals radii of the carbon atoms) above the molecular plane. The density of the n-th curve is $4{\times}2^{n/2}{\times}10^{-6}$.

missing in the PIES. The striking inactivity of the b_{2u} orbital can be explained by its geometrical distribution; the b_{2u} orbital behaves as an inner orbital because it is located along the carbon skeleton and sandwiched between the π orbitals.

Fig. 9 gives an example indicating that orbitals can be shielded by some bulky groups to become inactive in PIES. The d orbitals(10e' and 3e") in $Fe(CO)_5$ are located at the center of the molecule and considered to be inner orbitals. The carbonyl groups behave as shields protecting the iron orbitals from attacks of metastable atoms. This steric shielding effect leads to very weak intensities of the iron d bands in the PIES. On the other hand, among the orbitals of the carbonyl groups, the 4σ orbitals having large densities on the oxygen atoms spread out to the outer directions. Hence, the striking activity of the 4σ band in the PIES gives a typical example of the behavior of outer orbitals. Furthermore, it is instructive that the 5σ orbitals of the carbonyl groups, which are remarkably active in the PIES of the CO molecule (Fig. 10), are found to become inactive in the complex; a conversion of a nonbonding orbital of CO into a bonding orbital of $Fe(CO)_5$ has thus been observed by this approach.

These results together with some earlier work[1-3,8-10] lead to a conclusion that the electron distribution of the individual molecular orbital can be revealed by Penning ionization electron spectroscopy in which metastable states of rare gas atoms are used to probe electron density maps of molecular orbitals. In this sense this approach may be called metastable atom probe spectroscopy(MAPS). More detailed studies on the electron density maps for a certain part of the molecular surface will be made, when the direction of collisions between metastable atoms and target molecules is controlled, for example by

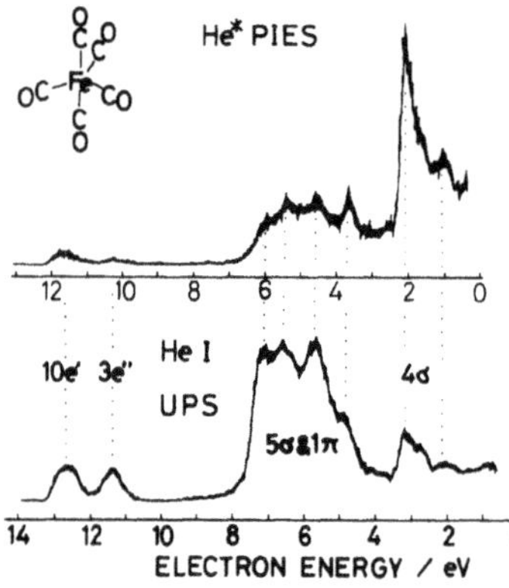

Fig. 9. He*(2^3S)PIES(upper) and He I UPS(lower) of $Fe(CO)_5$. The bands labelled by 10e' and 3" are due to d electrons of the iron atom. The other bands are due to electrons of carbonyl groups.

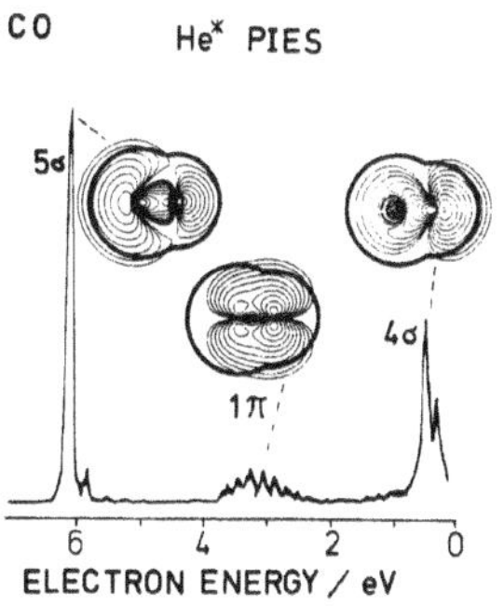

Fig. 10. $He^*(2^3S)$PIES of CO. The density of the n-th curve in the maps is $2.5\times2^n\times10^{-4}$.

introducing metastable atom beams with a definite direction onto a regular array of molecules adsorbed on a solid surface. The variation or the orbital electron density along the direction vertical to the molecular surface will also be observed by controlling the relative velocity between the projectile and the target. Finally, it should be noted that this approach reveals the local electron density of molecular orbitals at the "frontier" of the molecule where the molecule is attacked by the reagent.

REFERENCES

1. T. Munakata, K. Kuchitsu and Y. Harada, Comparison of the relative intensities of Penning electron spectrum and photoelectron spectrum. Unsaturated hydrocarbons, Chem. Phys. Letters 64, 409 (1979).
2. T. Munakata, K. Ohno and Y. Harada, Application of Penning ionization electron spectroscopy to the study of the outermost layer of the solid surface, J. Chem. Phys. 72, 2880 (1980).
3. K. Ohno, S. Fujisawa, H. Mutoh and Y. Harada, Application of Penning ionization electron spectroscopy to stereochemistry. Steric shielding effect of methyl groups on Penning ionization in substituted anilines, J. Phys. Chem. 86, 440 (1982).
4. V.J. Čermák, Retarding-potential measurement of the kinetic energy of electrons released in Penning ionization, J. Chem. Phys. 44, 3781 (1966).
5. A. Niehaus, Spontaneous ionization in slow collisions, in: "Advances in Chemical Physics" Vol. XLV, ed. J. Wm. McGowan, 401, Wiley, New York (1981).
6. D.W. Turner, A.D. Baker, C. Baker and C.R. Brundle, "Molecular Photoelectron Spectroscopy", Wiley, New York (1970).

7. H. Hotop and A. Niehaus, Reactions of excited atoms and molecules with atoms and molecules, II. Energy analysis of Penning electrons, Z. Physik 228, 66 (1969).
8. T. Munakata, K. Kuchitsu and Y. Harada, Comparison of the relative intensities of Penning electron spectra and photoelectron spectra, II. Heterocyclic aromatic molecules, J. Electron Spectry. 20, 235 (1980).
9. T. Munakata, K. Ohno, Y. Harada and K. Kuchitsu, Assignment of photoelectron bands for naphthalene and anthracene by Penning ionization spectroscopy, Chem. Phys. Letters 83, 243 (1981).
10. T. Munakata, Y. Harada, K. Ohno and K. Kuchitsu, Penning electron spectrum of ferrocene. Evidence for steric effect on the electronic interaction causing Penning ionization, Chem. Phys. Letters, 84, 6 (1981).

ACCURATE HELLMANN-FEYNMAN FORCE METHOD FOR THE STUDY OF THE FIRST AND SECOND DERIVATIVES OF POTENTIAL ENERGY HYPERSURFACE

H. Nakatsuji and K. Kanda

Department of Hydrocarbon Chemistry
Faculty of Engineering
Kyoto University, Kyoto, Japan

1. INTRODUCTION

Hellmann-Feynman (H-F) theorem gives an intuitive expression of the force based on an electrostatic interaction between the electron density and nuclei.[1] This simplicity is very valuable in studying complex chemical phenomena such as molecular geometry, chemical reaction and molecular vibration.[2] A past defect of this approach was that the calculation of a wavefunction which satisfies the H-F theorem was believed to be difficult.[3]

However, we recently found a promising method of calculating a reliable H-F force.[4-7] Here, we briefly review the method starting from the underlying theorem and examine the accuracy of the calculated H-F force. We then show some recent applications of the method to geometry optimizations of the molecules in ground and excited states and of the transition state of a chemical reaction. It has also been applied to the analysis of chemical reaction parths.[8]

Further, when we use the H-F theorem, an analytic expression of the second derivative of energy becomes much simpler and more perspective than a straightforward second derivative of the energy.[7,9] We report calculations of force constants (both positive and negative) by this method and explain the electronic origins of the second derivatives.

2. THEOREM

A force acting on nucleus A, $\tilde{F}_A$ is written as

$$\begin{aligned} F_A &= - \langle\Psi|\partial H/\partial R_A|\Psi\rangle - \langle\partial\Psi/\partial R_A|H|\Psi\rangle - \langle\Psi|H|\partial\Psi/\partial R_A\rangle \\ &= - \langle\Psi|\partial H/\partial R_A|\Psi\rangle - \sum_r \Delta_r \, \partial x_r/\partial R_A \end{aligned} \quad (1)$$

where the first term is the H-F force and the rest is an error term which vanishes identically for a correct wavefunction. It is expressed as a sum of the AO error Δ associated with each basis χ_r. (In Eq. (1), R_A is a nuclear coordinate and x_r the center of a basis χ_r.) It is shown that the AO error is expressed as

$$\Delta_r = 2 \sum_i c_{ri} \text{ (SCF requirement projected on } |r'\rangle) \quad (2)$$

where c_{ri} is a mixing coefficient of the basis χ_r in an orbital ϕ_i, and r' is a derivative $\partial\chi_r/\partial x_r$ of the basis r = χ_r. Eq. (2) proves the following theorem:

> A sufficient condition for a general SCF wavefunction to satisfy the Hellmann-Feynman theorem is that the basis set includes the derivative r' for any basis r. The basis set {r, r', r", ...} is such a basis. If the basis is recurrent in the sense $r = r^{(n)}$, then the number of elements can be finite.

This theorem is valid for general SCF theories including closed-shell Hartree-Fock, UHF, RHF for open-shell and excited states, general MC-SCF, and some types of GVB theory. Note that if only the force acting on a nucleus A, F_A, is concerned, the derivative r' may be added only to those bases whose centers are on the nucleus A (see Eq. (1)).

3. A NEW FORCE METHOD

The above theorem gives a basis for a systematic method of improving a wavefunction so that it satisfies the Hellmann-Feynman theorem. As a first stage of such an approach, we have considered an approximation in which only the first derivative AO's {r'} are added to the "parent" AO's {r}. (The set {r, r'} is called a "family".) Then, all of the AO errors of the parent AO's vanish identically as Eq. (2) shows, but the AO errors of the added derivative AO's remain. However, if the parent basis set is already a good basis, the mixing coefficient $c_{r'i}$ of the added derivative AO's r' should be small, so that from Eq. (2) the AO error of the added AO r', $\Delta_{r'}$, may be neglected. This approximation has been confirmed to be very good: test calculations were performed for the closed-shell Hartree-Fock method,[4,6] open-shell RHF[5] and UHF methods, and MC-SCF method.[5] Further, it was also shown that other properties, such as dipole moment, quadrupole moment, etc., are improved at the same time.[6]

Table 1. AO error, Hellmann-Feynman force and energy gradient of CO (a.u.).[a,b]

	Force on Carbon		Force on Oxygen	
	Parent [3s2p]	Family [3s2p] plus first derivatives	Parent [3s2p]	Family [3s2p] plus first derivatives
AO error				
s1	0.4943	0.0062	-0.7986	-0.0079
s2	-0.0932	0.0052	0.0167	0.0
s3	0.0012	0.0002	-0.0096	-0.0008
p1σ	0.3816	0.0117	-0.7835	-0.0248
p2σ	0.0023	0.0003	-0.0036	-0.0002
p1π	0.2574	0.0104	-0.6379	-0.0102
p2π	-0.0001	0.0002	-0.0005	-0.0014
Total error	1.3007	0.0446	-2.8553	-0.0565
H-F force	-3.1998	-1.7094	4.7543	1.7213
Energy gradient	-1.8990	-1.6648	1.8990	1.6648

[a]The CO length is 0.8636 Å (experimental equilibrium length, 1.1283 Å). The CO is on the σ axis in the direction from C to O. The [3s2p] set is due to Dunning and Hay.[9]
[b]SCF energy is -112.3278 a.u. for the parent set and -112.4480 a.u. for the family set.

4. APPLICATION OF THE NEW FORCE METHOD

Table 1 shows the AO error, H-F force, and energy gradient of CO obtained from the parent and family sets of the [3s2p] set.[10] By the addition of the first derivative AO's, the error term decreases dramatically. With the family set, the H-F force agrees well with the direct energy gradient, showing that the H-F theorem is essentially satisfied. The energy lowering due to an addition of the first derivative AO's (a kind of polarization function) is 0.1202 a.u.

Figure 1 shows plots of the transverse forces acting on the proton of H_2O, calculated by several methods, against the HOH angle. When we use the family set, the curves for the H-F force and the energy gradient almost superpose each other; the H-F theorem is satisfied over a wide range of valence angle. Further, by the addition of the first derivative AO's, the calculated equilibrium angle changes from 112.5° to 107.0° and becomes closer to the experimental value, 104.5°.

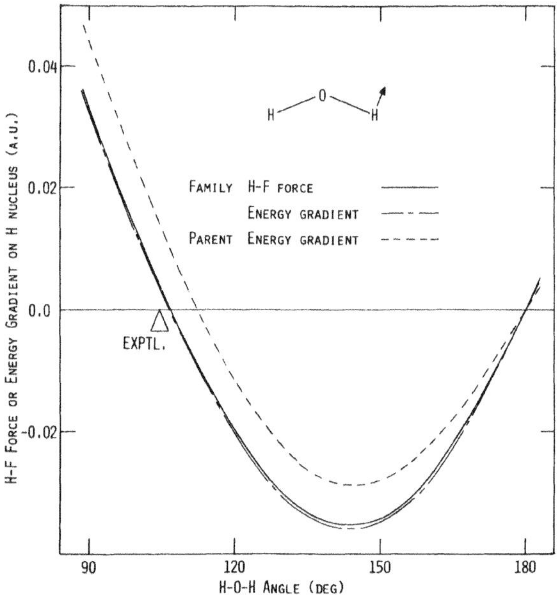

Fig. 1. Hellmann-Feynman force and energy gradient for the transverse force acting on the proton of H_2O vs the bond angle. The basis sets are the parent and family sets of the [3s2p/2s] set.[10]

Table 2 shows the bond lengths and force constants of several diatomic molecules. When the family sets are used, the calculated values are essentially the same independent of the forces used. In further examination, the H-F force method tends to give a bond length a bit longer (~0.01 Å) than the energy gradient values. Comparing the results of the parent and family sets, we see that the force constants are improved considerably by the addition of the first derivative AO's.

Table 3 shows the geometries of the ground state, triplet excited state, and ionized state of HNO determined by the present H-F force method. We have used closed and open-shell RHF method. Since the $\pi \to \pi^*$ transition reduces the electron density from the π bonding region of the N-O bond, the N-O length increases by this transition. The same is true for the ionization from the π MO. Since both $\pi \to \pi^*$ transition and the ionization from π MO do not cause much change in the electron density on the σ plane, the changes in the N-H length and the HNO angle are relatively small.

Table 2. Bond lengths and force constants of diatomic molecules AB.[a]

Molecule AB	Property	Parent: [3s2p/2s] energy gradient	Family: [3s2p/2s] plus first derivatives energy gradient	Family: [3s2p/2s] plus first derivatives H-F force on A	Family: [3s2p/2s] plus first derivatives H-F force on B	Exptl.
LiH	bond length (Å)	1.632	1.604	1.599	1.608	1.596
	force constant (mdyn/Å)	1.203	1.112	1.092	1.112	1.025
BH	bond length (Å)	1.243	1.229	1.235	1.232	1.232
	force constant (mdyn/Å)	3.308	3.196	3.149	3.198	3.046
CO	bong length (Å)	1.138	1.114	1.124	1.127	1.128
	force constant (mdyn/Å)	22.15	20.05	20.47	20.32	19.02
N_2	bond length (Å)	1.100	1.074	1.084	1.084	1.098
	force constant (mdyn/Å)	28.15	25.25	25.50	25.50	22.95
LiF	bond length (Å)	1.606	1.566	1.570	1.583	1.564
	force constant (mdyn/Å)	3.239	2.731	2.717	2.775	2.502

[a]The force constants are calculated at the experimental equilibrium geometry.

Table 3. Geometries of the ground state, triplet excited state, and ionized state of HNO determined by the Hellmann-Feynman force.[a]

State		R(N-O) Å	R(N-H) Å	HNO deg
Ground	Force-optimized	1.193	1.029	108.8
	Exptl.	1.212	1.063	108.6
$^3(\pi\to\pi^*)$ Excited	Force-optimized	1.538	1.015	101.5
$^2(\pi\to\infty)$ Ionized	Force-optimized	1.297	1.011	110.9

[a]The family set of the 4-31G set is used.

5. ANALYTIC SECOND DERIVATIVE OF POTENTIAL ENERGY

When the Hellmann-Feymnan theorem is satisfied for the first derivative of energy, the analytic expression of the second derivative becomes much simpler than the straightforward second derivative of the energy. Further, the intuitive physical meaning of the H-F force is extended to the second derivative. Though these merits have been noted by several authors,[9] it was difficult to realize them, except for some simple systems, because of a lack of a practical method to calculate a wavefunction which satisfies the H-F theorem. However, the present approach has removed this obstacle.

A straightforward second derivative of the Hartree-Fock energy gives an expression

$$\begin{aligned}\frac{\partial^2 E}{\partial X_A \partial Y_B} &= \sum_{r,s} P_{rs} \frac{\partial^2 H_{rs}}{\partial X_A \partial Y_B} + \sum_{r,s} \frac{\partial P_{rs}}{\partial Y_B} \frac{\partial H_{rs}}{\partial X_A} + \frac{\partial^2 V_{nuc}}{\partial X_A \partial Y_B} \\ &- \sum_{r,s} D_{rs} \frac{\partial^2 S_{rs}}{\partial X_A \partial Y_B} - \sum_{r,s} \frac{\partial D_{rs}}{\partial Y_B} \frac{\partial S_{rs}}{\partial X_A} \\ &+ \frac{1}{2} \sum_{rstu} P_{rs} P_{tu} \left[\frac{\partial^2}{\partial X_A \partial Y_B} (rt||su)\right] \\ &+ \sum_{rstu} \frac{\partial P_{rs}}{\partial Y_B} P_{tu} \left[\frac{\partial}{\partial X_A} (rt||su)\right]\end{aligned} \tag{3}$$

where H_{rs}, S_{rs}, and P_{rs} are the core-Hamiltonian matrix overlap matrix, and bond order density matrix, respectively and

$$D_{rs} = \sum_i^{occ} \varepsilon_i c_{ri} c_{si}$$

with ε_i the orbital energy. $(rt||su)$ is an appropriate sum of the Coulomb and exchange repulsion integrals[11] and V_{nuc} is the nuclear repulsion energy. On the other hand, when the H-F theorem is satisfied for the first derivative, only the parts of the first three terms of Eq. (3) remain. Most of the complex terms drop out, and we obtain

$$\begin{aligned}\frac{\partial^2 E}{\partial X_A \partial Y_B} &= \frac{\partial^2 V_{nuc}}{\partial X_A \partial Y_B} + \sum_{r,s} P_{rs}\, \delta_{AB}\, Z_A \langle r|\frac{\partial}{\partial Y_A}\left(\frac{x_A}{r_A^3}\right)|s\rangle \\ &+ \sum_{r,s} P_{rs}\, Z_A \left(\langle\frac{\partial r}{\partial Y_B}|\frac{x_A}{r_A^3}|s\rangle + \langle r|\frac{x_A}{r_A^3}|\frac{\partial s}{\partial Y_B}\rangle\right) \\ &+ \sum_{r,s} \frac{\partial P_{rs}}{\partial Y_B} Z_A \langle r|\frac{x_A}{r_A^3}|s\rangle \end{aligned} \tag{4}$$

where

$$\frac{\partial}{\partial Y_A}\left(\frac{x_A}{r_A^3}\right) = \begin{cases} (r_A^2 - 3x_A^2)/r_A^5 + \frac{4}{3}\pi\delta(A) & X_A = Y_A \\ - 3x_A Y_A/r_A^5 & X_A \neq Y_A \end{cases} \tag{5}$$

Thus equation is much simpler to calculate than Eq. (3) and, furthermore, a simple physical meaning is associated with each term as follows. The first two terms of Eq. (4) show a change in the H-F force when only nucleus A is moved while the electron density surrounding it remains unaltered. They consist of an electric field gradient at nucleus A and a contribution from the density at the nucleus (Fermi term). The third term includes the derivative of the basis AO's and represents the H-F force on A due to the displaced AO's of atom B with keeping the AO coefficients unaltered. Then, the sum of the first three terms shows a net effect when the nucleus and the AO's associated with it are moved simultaneously without changing their AO coefficients (complete following[2]). The last term represents the effect of reorganization of the electron density matrix due to the nuclear motion, $\partial P_{rs}/\partial Y_B$. It arises from the two sources since it is given by a sum of the two terms

$$\begin{aligned}\frac{\partial P_{rs}}{\partial Y_B} &= \sum_{i,j}^{occ} c_{ri} c_{sj} S_{ij}^{(1)} \\ &+ \sum_{i}^{occ} \sum_{a}^{unocc} u_{ai}^{(1)} (c_{ri} c_{sa} + c_{ra} c_{si}) \end{aligned} \tag{6}$$

where $S_{ij}^{(1)}$ is a derivative of the overlap $\langle\phi_i|\phi_i\rangle$ and $u_{ai}^{(1)}$ is a mixing coefficient between occupied and unoccupied MO's. The first term is the renormalization term which arises in order to keep the total wavefunction normalized during the vibration and the second term represents the relaxation of molecular charge distribution during the vibration through a mixing between occupied and unoccupied orbitals.

The role of the reorganization term during a nuclear rearrangement process is generally very important. The renormalization term usually gives a positive contribution to the force constant and is one of the important origins of the electron-cloud incomplete following, which is a density origin of a stable geometry.[2] On the other hand, the relaxation term usually gives a negative contribution to the force constant and works to lower the barrier. It is a origin of the electron-cloud preceding, which is a density origin of many nuclear rearrangement processes.[2] The role of the relaxation term during the course of a chemical reaction is of special interest.

6. CALCULATION OF THE SECOND DERIVATIVES OF STABLE MOLECULE AND TRANSITION STATE

The preceding formula for the second derivatives has been applied to the studies of the force constants of several di- and polyatomic molecules.[7,8] Here, we show the results for CO and CH_5^- as prototypes of stable molecules and transition states. The later is the transition state of the reaction, $CH_4 + H^- \rightarrow H^- + CH_4$, which is a model of S_N2 reaction. We have used the family set of the 4-31G set, so that the H-F theorem is essentially satisfied for the first derivatives as shown in the previous section.

Table 4 shows the analysis of the second derivatives. The force constant of CO was expressed in two different ways, $-\partial F_C/\partial X_C$ and $\partial F_O/\partial X_O$. It is a sum of large cancelling contributions. The nuclear term is positive. The Fermi term and the displaced AO term are very large because of the large inner-shell contribution, but they are cancelling. The sum of the second to fourth terms represents the effect of the AO density completely following the nuclear motion. The sum of the nuclear and completely following terms is negative for CO. In the next three rows the reorganization terms are given. In Figure 2, we have given the contour map of the density differential

$$- \sum_{r,s} \partial P_{rs}/\partial R_{AB}\, \chi_r(\underset{\sim}{r})\chi_s(\underset{\sim}{r})$$

It shows the change in density when the bond distance is shortened. For stretching mode the renormalization term seems always to be positive (Table 4). When two atoms A and B come closer, the overlaps between the AO's become larger. Therefore, in order to keep

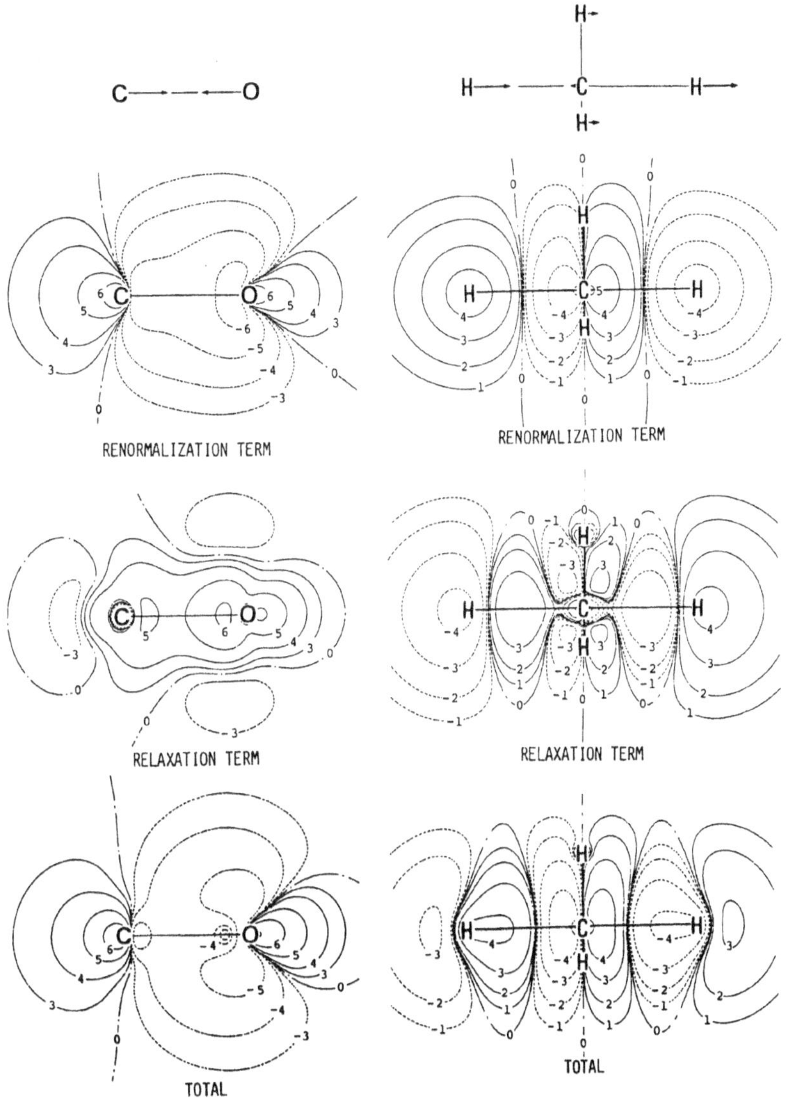

Figure 2. Contour map of the density differential $\sum \partial P_{rs}/\partial Q \times \chi_r(r)\chi_s(r)$ for CO and for the normal mode $Q_2(A_2'')$ of CH_5^-. From the top to the bottom, a sketch of the normal mode, the renormalization term, the relaxation term, and the sum of them are shown. The real lines correspond to an increase in density, and the broken lines correspond to a decrease, with the contour values of 0, ±1, ±2, ±3, ±4, ±5, and ±6 corresponding to 0.0, ±0.001, ±0.003, ±0.01, ±0.03 ±0.1, and ±0.3 a.u., respectively.

Table 4. Force constants of CO and CH_5^- and their analyses (a.u.)

	CO^a		CH_5^{-b}	
	$-\frac{\partial F_C}{\partial X_C}$	$\frac{\partial F_O}{\partial X_O}$	$\frac{\partial F}{\partial Q_1}(A_1')$	$\frac{\partial F}{\partial Q_2}(A_2'')$
Nuclear term	9.906	9.906	0.440	0.142
$Z_A \sum P_{rs} \langle r \vert \frac{3x_A^2-r_A^2}{r_A^5} \vert s \rangle$	-16.522	-14.454	-0.412	-0.089
$Z_A \sum P_{rs} \langle r \vert -\frac{4}{3}\pi\delta(A) \vert s \rangle$	3032.572	9899.184	1.371	79.873
$Z_A \sum P_{rs} \langle r' \frac{x_A}{r_A^3} s \rangle$	-3027.701	-9897.233	-1.337	-79.888
Total	-1.744	-2.596	0.062	0.038
Renormalization term	4.061	5.697	0.036	0.181
Relaxation term	-1.002	-1.785	-0.043	-0.329
Total	3.059	3.911	-0.007	-0.148
Grand total	1.315	1.315	0.055	-0.110
Experimental value	1.222	1.222		

[a]Calculated for CO at the experimental geometry (R_{CO}=1.1283 Å) and for CH_5^- at the transition state geometry determined by the present method (see text).
[b]The normal mode Q_1 is the totally symmetric C-H_a stretching mode and Q_2 is the so-called reaction coordinate. See figures in the text.

the total density normalized, the electron density should flow out of the A-B region as shown at the upper map of Figure 2. When two atoms A and B go apart, the reverse flow of electron density should occur. Thus, the density reorganization due to the renormalization term corresponds to the electron-cloud incomplete following[2] and gives a positive contribution to the force constant. On the other hand, in the relaxation term, the density increases in the bond region as the two atoms come closer. This behavior is the electron-cloud preceding[2] and works to stabilize the system. Then, the relaxation term gives negative contribution to the force constant as seen in Table 4. At the bottom of Figure 2, the total reorganization density is shown. It reflects mainly the renormalization term and then the net effect of reorganization term gives positive contribution to the force constant.

The calculated force constant of CO is 1.315 a.u. It is 7.6% larger than the experimental value. This is a trend of the Hartree-Fock results and an inclusion of electron correlation will reduce the difference. Though different analyses were obtained from $-\partial F_C/\partial X_C$ and $\partial F_O/\partial X_O$, the final results are the same because the H-F theorem is essentially satisfied for the first derivative.

We next show the results for the transition state CH_5^- of the reaction,

$$CH_4 + H^- \rightarrow H^- + CH_4.$$

In the first stage, the geometry of the transition state was determined by the present force method. It was calculated to belong to the D_{3h} symmetry with C-H_a = 1.700 Å and C-H_e = 1.063 Å where H_a and H_e denote axial and equatorial hydrogens, respectively. This may be compared with the results of Dedieu and Veillard,[12] C-H_a = 1.737 Å and C-H_e = 1.062 Å, and of Leforestier,[13], C-H_a = 1.70 Å and C-H_e = 1.07 Å.

The normal modes of the transition state CH_5^- are determined by diagonalizing the Hessian matrix. They consist of the twelve modes; the two A_1', two A_2'', three E', and one E'' symmetry modes. Among these, one normal mode has negative force constant and all others have positive ones. In Table 4, we have shown only the two normal modes of the A_1' and A_2'' symmetries, which may be illustrated as

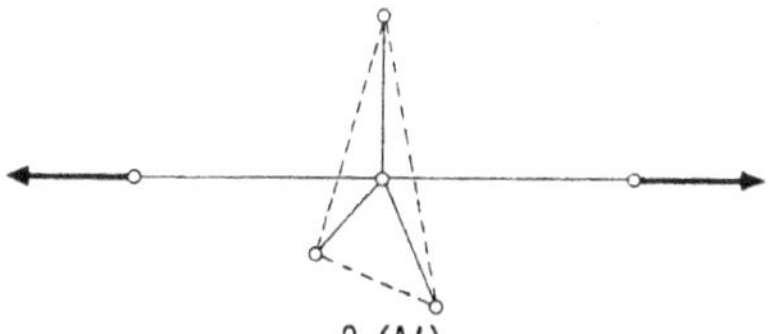
$Q_1(A_1')$

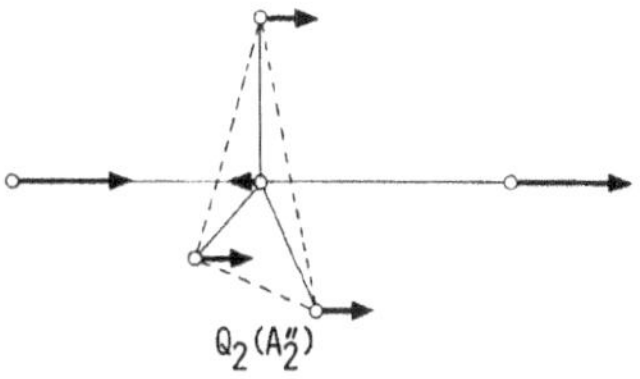
$Q_2(A_2'')$

$Q_1(A_1')$ is the totally symmetric vibration of the axial C-H_a bonds. $Q_2(A_2'')$ is the so-called reaction coordinate involving Walden inversion and has negative force constant near the transition state.

In Table 4, we gave the second derivatives and their analyses. In comparison with CO, the Fermi and displaced AO terms are small because of a large contribution of the coordinates of hydrogens in the normal mode. This is especially so for the Q_1 mode in which carbon does not move during the mode. For CH_5^-, the sum of the nuclear term and the completely following term is positive in contrast to CO. From our experience, this seems to be common to the hydride molecules. A remarkable feature of the reaction coordinate Q_2 is that it has a large negative relaxation term. Because of this contribution, the force constant of the Q_2 mode is negative. At this geometry (transition state) the potential surface of CH_5^- is maximum along the Q_2 coordinate and minimum for all others. For the coordinate Q_1, the force constant is positive, though the reorganization term gives a small negative contribution.

An interesting behavior of electron density along the reaction coordinate is seen from the density differential map shown in the right-hand side of Figure 2. Referring to the relaxation term, we see a typical pattern of the electron-cloud preceding: the density reorganizes itself so as to precede the nuclear motion along the Q_2 coordinate. It occours in all the regions near the moving nuclei, C, H_a, and H_e. In the renormalization term, however, the density near the C and H_e shows a pattern of the electron-cloud imcomplete following. Near the H_a, the renormalization term also exhibits the electron-cloud preceding. In the total sum, the relaxation term surpasses the renormalization term and a beautiful pattern of the electron-cloud preceding is seen. As shown in Table 4, this is an origin of the negative force constant of the reaction coordinate Q_2.

7. CONCLUSION

Here, we have viewed a possibility of an accurate Hellmann-Feynman force approach for the studies of the first and second derivatives of a potential energy hypersurface. Theoretical and physical simplicity of the underlying concept and the reliability of the calculated results would help us to further understand chemical phenomena and to construct a predictive model with a quantum-chemical intuition.

Now that the first obstacle along this approach has been removed, the next step would be (1) to develop a method which fully utilizes the structures of the theory for rapid computations of the desired wavefunction and (2) to apply the present approach to a wider field of chemistry to get a deeper understanding of the phenomena with intuitive concept and quantitative accuracy.

REFERENCES

1. B.M. Deb, ed. "The Force Concept in Chemistry", Van Nostrand Reinhold, New York, 1981.
2. For example, H. Nakatsuji and T. Koga, Force Models for Molecular Geometry, chapter 3 of reference 1.
3. For example, P. Pulay, Calculation of Forces by Non-Hellmann-Feynman Methods, chapter 9 of reference 1.

4. H. Nakatsuji, K. Kanda, and T. Yonezawa, Force in SCF Theories, Chem.Phys. Letters 75, 340 (1980).
5. H. Nakatsuji, T. Hayakawa, and M. Hada, Force in SCF Theories. MC-SCF and open-shell RHF theories, Chem.Phys. Letters 80, 94 (1981).
6. H. Nakatsuji, K. Kanda, M. Hada, and T. Yonezawa, Force in SCF Theories. Test of the New Method. J.Chem.Phys., in press.
7. H. Nakatsuji, K. Kanda, and T. Yonezawa, Force in SCF Theories. Second Derivative of Potential Energy, J.Chem.Phys., in press.
8. H. Nakatsuji, M. Hada, and K. Kanda, Force in SCF Theories. First and Second Derivatives of the Potential Energies of Chemical Reaction Systems, Int.J. Quantum Chem., to be published.
9. For example, J. Goodisman, Calculation and Interpretation of Force Constants, chapter 5 of reference 1.
10. T.H. Dunning, Jr. and P.J. Hay, Gaussian Basis Sets for Molecular Calculations, in: "Methods of Electronic Structure Theory," H.F. Schaefer III, ed., Plenum, New York, 1977.
11. J.A. Pople, R. Krishnan, H.B. Schlegel, and J.S. Binkley, Derivative Studies in Hartree-Fock and Møller-Plesset Theories, Int.J. Quantum Chem.Symp. 13, 225 (1979).
12. A. Dedieu and A. Veillard, A Comparative Study of Some S_N2 Reactions through ab initio Calculations, J.Am.Chem.Soc. 94, 6730 (1972).
13. C. Leforestier, Classical trajectories using the full ab initio potential energy surface $H^- + CH_4 \rightarrow CH_4 + H^-$, J.Chem.Phys. 68, 4406 (1978).

THE CALCULATION OF SPECTROSCOPIC DATA BY THE SCC-Xα METHOD

M. Grodzicki

I. Institut für Theoretische Physik
Jungiusstrasse 9
D-2000 Hamburg 36, FRG

1. INTRODUCTION

A rigorous all-valence electron treatment of large molecules or transition metal clusters is restricted by the computational effort and the high amount of computing time needed. Furthermore, if the electronic structure calculation is performed not for itself, but should serve rather as the basis for the determination and interpretation of spectroscopic data, the use of an approximate molecular orbital theory is both essential and justified. The following four restrictions should be fulfilled by such a method:

(i) The Hamiltonian should be representable as a sum of one-particle operators to avoid the explicit antisymmetrisation of the molecular wavefunction.

(ii) The Hamiltonian should be energy-independent because for large systems the energy levels are often so close together that it becomes time-consuming and difficult to find all the roots of the secular determinant.

(iii) Multidimensional numerical integrations should not occur.

(iv) The restriction to muffin-tin potentials should be avoided since generally large molecules are open structures of low symmetry.

Having in mind these premises, we will develop a semiquantitative MO-method which is the combination of some basic ideas of the iterative extended Hückel theory (IEHT)[1] and the Xα-approximation for the molecular potential.[2] Under this latter assumption the Schrödinger equation (in a.u.) can be written as a set of one-electron equations

$$[-\Delta + V(\vec{r})]\psi_k(\vec{r}) = \varepsilon_k\psi_k(\vec{r}) \tag{1}$$

The molecular orbital ψ_k is given in LCAO approximation by the atomic wavefunctions $\phi_j^{(\nu)}(\vec{r})$ characterising the $(n_j l_j m_j)$th atomic orbital of the νth atom

$$\psi_k(\vec{r}) = \sum_{\nu j} \phi_j^{(\nu)}(\vec{r}-\vec{R}_\nu)c_{jk}^\nu \tag{2}$$

leading to the secular equation

$$\sum_{\nu j} (H_{ij}^{\mu\nu} - \varepsilon_k S_{ij}^{\mu\nu})c_{jk}^\nu = 0 \tag{3}$$

with the matrix elements[3]

$$S_{ij}^{\mu\nu} = \int\phi_i^{(\mu)*}(\vec{r}-\vec{R}_\mu)\phi_j^{(\nu)}(\vec{r}-\vec{R}_\nu)d^3r \tag{4}$$

$$H_{ij}^{\mu\nu} = (\varepsilon_i^\mu + \varepsilon_j^\nu)S_{ij}^{\mu\nu}/2 + (V_{ij}^{\mu\nu} + V_{ji}^{\nu\mu})/2 \tag{5}$$

$$V_{ij}^{\mu\nu} = \int\phi_i^{(\mu)*}(\vec{r}-\vec{R}_\mu)[V(\vec{r}) - V_{at}^{(\nu)}(\vec{r}-\vec{R}_\nu)]\phi_j^{(\nu)}(\vec{r}-\vec{R}_\nu)d^3r \tag{6}$$

where ε_j^ν is the atomic orbital energy of the jth orbital of atom ν.

2. APPROXIMATIONS

Apart from the LCAO representation for the MO, Eq. (3) does not contain any additional approximation so far compared with Eq. (1). On the basis of the matrix Hamiltonian of Eq. (5) in the first step, the various modifications of Hückel-type methods are easily obtained by substituting

$$V_{ij}^{\mu\nu} = (V_{ii}^{\mu\mu} + V_{jj}^{\nu\nu})S_{ij}^{\mu\nu}/2$$

$$\varepsilon_j^\nu + V_{jj}^{\nu\nu} = k\alpha_j^\nu$$

yielding the basic equation of the extended Hückel theory (EHT):[4]

$$H_{ij}^{\mu\nu} = k(\alpha_i^\mu + \alpha_j^\nu)S_{ij}^{\mu\nu}/2 \tag{7}$$

The extended Hückel theory is a nonselfconsistent and purely empirical method since in general the parameter values k and α_j^ν will depend on the molecule under consideration. Therefore, the question arises if an extension of EHT is possible in such a way that a universal set of parameters can be determined in the sense that it depends solely on the atom itself but not on its special bonding situation in a molecule or cluster.

To this end two effects have to be taken into account namely an atomic effect and a molecular effect arising from the influence of the neighbouring atoms. The first effect is considered by choosing the atomic orbital energy charge-dependent:[5]

$$\varepsilon_j^\nu = \varepsilon_{jo}^\nu + \varepsilon_{j1}^\nu Q_\nu + \varepsilon_{j2}^\nu Q_\nu^2 \tag{8}$$

Q_ν is an appropriate defined effective atomic charge and the ε_{jn}^ν are atomic type parameters determined either from experimental data[6] or from atomic Xα-calculations.[7] This modified version of the Hückel method is called iterative extended Hückel theory (IEHT)[1] since it determines the atomic charges Q_ν iteratively in a selfconsistent way. Although the results become significantly better the improvement is not sufficient to achieve the desired universal parameter set unless the influence of the neighbour contributions to the molecular potential is included explicitly.

Therefore, we have to search for a local potential which is incorporable into the selfconsistent charge (SCC) procedure of IEHT and which simultaneously allows for a fast and easy computation of the potential matrix elements $V_{ij}^{\mu\nu}$, Eq. (6). A reasonable local approximation for the atomic potential is Slater's Xα-potential:[2]

$$V(r) = -2Z/r + 8\pi\left[\frac{1}{r}\int_0^r x^2\rho(x)dx + \int_r^\infty x\rho(x)dx\right] - 6\alpha\left[\frac{3}{8\pi}\rho(r)\right]^{1/3} \tag{9}$$

where Z is the nuclear charge and $\rho(r)$ is the atomic charge density represented in the simplest way by a single exponential, i.e.

$$\rho(r) = \frac{N\eta^3}{8\pi} e^{-\eta r} \tag{10}$$

This yields the atomic model potential

$$V_{at}(r) = -2Q/r - N(2e^{-\eta r}/r + \eta e^{-\eta r}) - \alpha'\eta\, N^{1/3} e^{-\eta r/3} \tag{11}$$

N being the number of electrons, Q=Z-N is the effective charge, and $\alpha' = 1.5\alpha(3/\pi^2)^{1/3}$.

The potential parameter η has a simple interpretation. For hydrogen the atomic wavefunction is a single exponential likewise, $\phi \sim e^{-\zeta r}$, and thus we have $\eta=2\zeta$. Owing to the well-known charge and configuration dependence of the orbital exponent[8] the potential parameter is chosen charge dependent, too. For heavier atoms, η may be connected with the Allred-Rochow scale of electro-negativity[9] allowing thus an interpretation in familiar chemical concepts. In conclusion, we have obtained a local model potential as a function of the effective charges of the constituent atoms via N,Q,η and therefore incorporable into the SCC procedure in a natural way.

The molecular potential may be constructed now either from overlapping atomic charge densities[10] or directly by superposition of the atomic potentials, Eq. (11). Then, all angular integrations can be done analytically but in the former case, it remains a numerical radial integration for the exchange potential while in the latter one all integrals can be evaluated analytically.[3] Additionally, to a very good approximation the three-center integrals can be expressed by a sum of two-center integrals[9] wherewith the amount of computation time is not more than about three times that of an IEHT calculation per iteration. Finally it should be mentioned that up to now, as a result of our calculations of various observables of about 250 molecules, the parameter set in fact turns out to be universal, i.e. it depends on the atoms solely but not on the specific molecular bonding.

3. DISCUSSION

Apart from the adiabatic, the Hartree-Fock, the Xα, and the LCAO approximations the following additional simplifications have been introduced in the SCC-Xα method.

(i) The parametrisation of the intraatomic matrix elements

$$\int \phi_i^{(\mu)*}(\vec{r}-\vec{R}_{\mu\nu}) H_{at}^{(\nu)} \phi_j^{(\nu)}(\vec{r}) d^3r = \varepsilon_j^{\nu} S_{ij}^{\mu\nu}$$

where ε_i^{ν} is given in Eq. (8). Although it is a trivial matter to evaluate these matrix elements directly our experience has shown that it is not meaningful to calculate one-center integrals explicitly - especially the kinetic energy integrals - since the changes of intraatomic interactions during molecular formation are not described appropriately in a fixed LCAO basis.

(ii) The simple parametrisation of the atomic charge density, Eq. (10), can be improved by representing $\rho_{at}(r)$ by the sum

$$\rho_{at}(r) = \sum_i a_i \, r^{n_i} e^{-\eta_i r}$$

and fitting the parameters to exact atomic densities. Although the results become slightly better[11] the main drawback consists of a tremendous increase of iteration steps until convergence has been reached, generally by a factor of four compared with the single exponential version.

(iii) The third approximation concerns the fact that $V_{at}(r)$ is kept spherically symmetric during the whole iteration procedure. This may be overcome by replacing the spherical charge density with

$$\rho(\vec{r}) = \sum_{L} \rho_L(r) Y_L(\hat{r})$$

However, it is doubtful whether this possible improvement justifies the large additional amount of computing time expected.

(iv) The last simplification lies in the use of the SCC iteration procedure, which suffers from the arbitrariness of dividing the overlap charge for defining the effective charges Q_ν or the occupation numbers x_{nl} by a population analysis. We have adapted here a weighted form[9] which we have found to be more reliable than the frequently used Mulliken analysis[12] whenever strong polarity occurs or atoms with largely deviating atomic size are involved. The most striking example in this respect is the halogen hydrides where the application of Mulliken population analysis results in Cl being almost neutral in HCl while Br and I in HBr and HI, respectively, become positively charged (Q_{Br} = +0.034e, Q_I = +0.108e). That this is not an artifact of the SCC-Xα parametrisation procedure may be concluded from an earlier ab initio calculation on HBr[13] where Mulliken population analysis again results in a positive charge of +0.0235e on bromine. These unreasonable results are removed by our proposed population analysis yielding only iodine in HI slightly positive. Fortunately, it turns out that the type of population analysis does not affect the final results of the SCC calculation to a major extent, except in this case of small hydrogen containing molecules. It is thus not necessary to perform a full SCF procedure although it would be possible in principle.

The SCC iteration scheme is responsible for the most important advantage of the SCC-Xα method, namely its fast and stable convergence properties; for molecules like IF_5 or IF_7, e.g., less than 10 iterations are needed to obtain occupation numbers with the difference

$$\left| x_{nl}^{(in)} - x_{nl}^{(out)} \right|$$

being less than 0.01. Therefore, it seems possible to handle even larger systems in a selfconsistent manner without serious problems. A further advantage is the possibility to implement the program even on medium-sized computers; a 15 atom and 60 orbital version needs just about 50K words on a TR440 (AEG-Telefunken) machine.

4. APPLICATIONS

The experience from our computations of ionisation potentials and dipole moments on about 250 molecules shows the former to be accurate to 1eV with a few exceptions only while the dipole moments are reproduced generally within 10% of the experimental values (or within 0.1D if $|\mu|<0.5D$). Furthermore, the electric field gradients at the iodine nucleus in several iodine compounds are well described[14] which is remarkable in so far as it is a very sensitive test for the quality of the wavefunctions. We will now describe in detail the calculation of two properties which seem at first sight to arise predominantly from the core states, namely the total electronic charge density $\rho(0)$ at the nucleus and the chemical shift of core electrons.

Let us turn first to the electronic charge density $\rho(0)$ needed e.g. for interpreting the measured Mössbauer isomer shifts δ via the relationship

$$\delta = \alpha\Delta\rho(0) = \alpha[\rho_A(0) - \rho_B(0)] \tag{12}$$

The charge density $\rho_A(0)$ at the atom under consideration in molecule A consists of contributions from the valence and the core electrons

$$\rho(0) = \sum_k^{\text{val}} n_k|\psi_k(0)|^2 + 2\sum_m^{\text{core}} |\psi_m^{(c)}(0)|^2 \tag{13}$$

The core orbital $\psi_m^{(c)}(\vec{r})$, however, does not equal the atomic core orbital $\phi_m^{(c)}(\vec{r})$ but has to be orthogonalised to the molecular orbitals, i.e.

$$\psi_m^{(c)}(\vec{r}) = N_m[\phi_m^{(c)}(\vec{r}) - \delta_m(\vec{r})] \tag{14}$$

where N_m is a normalisation constant and $\delta_m(\vec{r})$ is the orthogonalisation correction given previously.[14] The core wavefunctions $\phi_m^{(c)}(\vec{r})$ have been determined separately from a relativistic atomic Dirac-Slater calculation for various ionic states of the respective atoms to describe correctly the variation of $\phi_m^{(c)}(0)$ due to the electron occupation numbers of the valence orbitals. When this formalism is applied to the five iodine containing compounds I_2, IBr, ICl, IF_5, IF_7, a plot of the computed densities $\rho(0)$ against the experimental isomer shifts[15] shows a good linear relationship (cf. Fig. 1) yielding a value of 0.23 for the calibration constant α of Eq. (12), from which the change $\delta R/R$ of the iodine nuclear radius due to excitation can be calculated as $4.5 \cdot 10^{-4}$ in good agreement with a previous estimate.[16] Furthermore, the contributions to the charge density differences relative to I_2 arising from the various orbitals are listed in Table 1 where the p-orbitals have been omitted, the corresponding $\Delta\rho(0)$ values being constant within $0.6a_o^{-3}$. The results clearly dis-

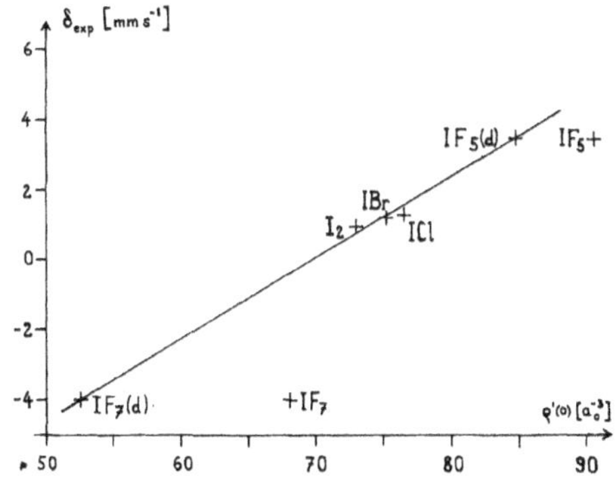

Figure 1. Calculated relativistic electron densities $\rho(0) = 247,900 + \rho'(0)$ in a_o^{-3} vs. experimental isomer shifts δ in mms^{-1} for ^{129}I-containing molecules.

play the importance of the molecular environment for the density at the nucleus of the core electrons and the ability of the SCC-Xα method to describe these changes in a satisfactory manner. Results of similar quality have been obtained recently in the same way for numerous Sn- and Te-compounds.[17,18] The core contributions become even more pronounced in case of iron, where the valence density at the nucleus from the 4s-orbital is almost zero, and therefore the density difference arises exclusively from the indirect change of the core orbitals due to molecular formation. An analogous calculation of spin densities at the nucleus with a spin-polarized SCC-Xα program is in progress now with promising results so far.

As a second example for the applicability of the method, we want to present the results of chemical shift calculations of the $3d_{5/2}$ state of bromine. The binding energy ε_c of a core orbital ψ_c for an atom placed at $\vec{R}_o$ is given as the expectation value of the Hamiltonian

$$\varepsilon_c = \langle\psi_c|H|\psi_c\rangle = -\langle\psi_c\left|\frac{\Delta}{2} + \frac{Z_o}{r}\right|\psi_c\rangle \tag{15}$$

$$- \langle\psi_c\left|\sum_{k\neq o} \frac{Z_k}{|\vec{r}-\vec{R}_{ko}|}\right|\psi_c\rangle + \langle\psi_c\left|V_{el}^{cor}\right|\psi_c\rangle + \langle\psi_c\left|V_{el}^{val}\right|\psi_c\rangle$$

where Z_k is the nuclear charge, $\vec{R}_{ko} = \vec{R}_k - \vec{R}_o$, and V_{el}^{cor} and V_{el}^{val} is the electron-electron interaction potential of the core and valence electrons, respectively. Here we have assumed strict core-valence separability of the electronic molecular system in accordance with the valence-electron-only SCC-Xα method. The evaluation of the matrix elements compatible with its assumptions is easily shown for the Coulomb

Table 1. Difference Density Contributions $\Delta\rho(0) = \rho_X(0) - \rho_{I_2}(0)$ of the Various Iodine ns-Orbitals

	IBr	ICl	IF_5	$IF_5(d)$[a]	IF_7	$IF_7(d)$[a]
5s	3.606	5.751	20.803	9.138	-17.632	-42.139
4s	-2.036	-3.409	-6.043	0.285	14.820	24.102
3s	1.246	1.892	3.407	1.637	-5.117	-7.666
2s	-0.788	-1.174	-1.606	-0.090	3.574	5.824
1s	0.177	0.336	0.905	0.783	-0.699	-0.754
Total	2.205	3.396	17.466	11.753	-5.054	-20.633

[a]With the 5d-orbital on iodine included.

interaction integral

$$\langle\psi_c\left|\frac{Z_k}{|\vec{r}-\vec{R}_{ko}|}\right|\psi_c\rangle = \frac{Z_k}{N_{oc}}\int \frac{\rho_{oc}(\vec{r})d^3r}{|\vec{r}-\vec{R}_{ko}|} \tag{16}$$

$$= \frac{Z_k\,\eta_{oc}^3}{2}\left[\frac{1}{R_{ko}}\int_0^{R_{ko}} e^{-\eta_{oc}r}\,r^2dr + \int_{R_{ko}}^{\infty} e^{-\eta_{oc}r}\,r\,dr\right]$$

when the core charge density is assumed to have again the form of single exponential

$$\rho_{oc}(r) = \frac{N_{oc}\,\eta_{oc}^3}{8\pi}\,e^{-\eta_{oc}r} \tag{17}$$

N_{oc} being the number of core electrons in orbital ψ_c. From Eq. (16) it is now easily seen that the "point charge approximation" does not require the core orbital to collapse to the nucleus[19] but only requires $r^2e^{-\eta_{oc}r}$ to be negligible for $r>R_{ko}$. This is in fact valid for 1s-electrons so that the point charge approximation is indeed justified in this case. It is, however, to a large extent also true for the more diffuse 3d core function of Br, since the neighbour distances R_{ko} are large enough so that this condition is again fulfilled. The electronic Coulomb interaction of ψ_c with the valence electrons is performed using

$$V_{el}^{val}(r') = \int \frac{\rho_{mol}^{val}(\vec{r})d^3r}{|\vec{r}-\vec{r}'|} = \sum_k^{MOs} n_k \int \frac{|\psi_k(\vec{r})|^2 d^3r}{|\vec{r}-\vec{r}'|}$$

where the summation extends over all occupied molecular orbitals. Substituting Eq. (17) and interchanging the order of integration one obtains as the final result in the point charge approximation[20]

$$\langle\psi_c|H|\psi_c\rangle = \varepsilon_{oc} + \varepsilon_{1c}Q_{oc} + \varepsilon_{2c}Q_{oc}^2 - \sum_{k\neq o} Q_{kc}/R_{ko} + \sum_{\substack{\mu\nu\\ ij}} P_{ij}^{\mu\nu} \langle r^{-1}\rangle_{ij}^{\mu\nu} \tag{18}$$

where the core orbital energy is assumed charge dependent (cf. Eq. (8)), Q_{kc} is the core charge, $P_{ij}^{\mu\nu}$ the bond order matrix, and $\langle r^{-1}\rangle_{ij}^{\mu\nu}$ is the expectation value of r^{-1} in the AO-basis.

These approximations are sufficient to reproduce the core binding energy satisfactorily even for the rather high lying Br3$d_{5/2}$ level (Fig. 2) since the second integral in Eq. (16) contributes a nearly constant amount of 2 eV to ε_c except for HBr due to the comparatively short bond distance. Furthermore, we have found that orthogonalisation of ψ_c to the MOs does not yield any significant correction to ε_c ($\leq$0.5 eV) and it may thus be neglected in contrast to $\rho(0)$.

In conclusion, we have shown for two examples how the SCC-Xα method, as a valence-electron-only method, can be utilized even for the description of core electron properties when consistently combined with atomic core data.

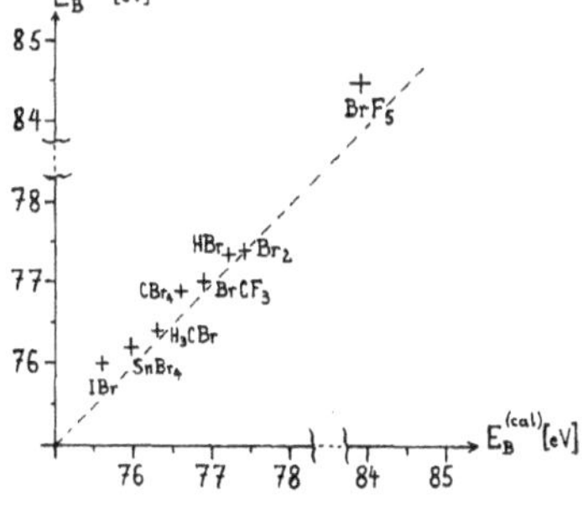

Figure 2. Calculated vs. experimental binding energies of the Br 3$d_{5/2}$ level.

REFERENCES

1. R. Rein, N. Fukuda, H. Win, G.A. Clarke and F.E. Harris, J.Chem. Phys. 45, 4743 (1966).
2. J.C. Slater, Adv.Quant.Chem. 6, 1 (1972).
3. M. Grodzicki, J.Phys. B13, 2683 (1980).
4. M. Wolfsberg and L.J. Helmholtz, J.Chem.Phys. 20, 837 (1952).
5. C.J. Ballhausen and H.B. Gray, Inorg.Chem. 1, 111 (1962).
6. C.E. Moore, Atomic Energy Levels, NBS Circular No. 467, US Govt. Print. Office, Washington, D.C. (1949).
7. C.C. Lu, T.A. Carlson, F.B. Malik, T.C. Tucker and C.W. Nestor, Atom.Nucl.Data Tables 3, 1 (1971).
   K. Huang, M. Aoyagi, M.H. Chen, B. Crasemann and H. Mark, Atom. Nucl.Data Tables 18, 243 (1976).
8. G. Burns, J.Chem.Phys. 41, 1521 (1964).
9. M. Grodzicki, S. Elbel, H. Walther, Z.Naturforsch. (to be published).
10. L.F. Mattheiss, Phys.Rev. 133, A 1399 (1964).
11. K.M. Hasselbach and M. Grodzicki, to be published.
12. R.S. Mulliken, J.Chem.Phys. 23, 1833 (1955).
13. R.H. Findlay, Farad.Trans. II 72, 388 (1975).
14. M. Grodzicki, S. Lauer, A.X. Trautwein and A. Vera in: Mössbauer Spectroscopy and its Chemical Applications," J.G. Stevens and G.K. Shenoy, eds.,Adv.Chem.Series 194, 3, Washington, D.C. (1981).
15. H.deWaard in: Mössbauer Data Index," J.G. Stevens and V.E. Stevens, eds., Plenum Press, New York (1973).
16. W.J.J. Spijkervet and I. Pleifer, Hyperfine Interact. 7, 285 (1979).
17. V. Männing and M. Grodzicki, to be published.
18. K.G. Ladwein, Diplom thesis, Saarbrücken (1982).
19. U. Gelius, Physica Scripta 9, 133 (1974).
20. M. Grodzicki, to be published.

ON CORRELATED ONE-ELECTRON ENERGIES IN THE HFS AND HFG SCHEMES

S. Manoli and M.A. Whitehead

Department of Chemistry, McGill University
Montreal, Quebec
Canada H3A 2K6

It is shown that neither the HFS nor the HFG theories (Hartree-Fock theory modified by Slater and Gopinathan) can give correlated wavefunctions using the Keller and Gasquez form of the Coulomb Hole.

1. INTRODUCTION

The expectation value of the inter-electronic Hamiltonian in terms of 1- and 2-particle density matrices,[1,2] is

$$<\frac{1}{2} \sum_{i,j} \frac{1}{r_{ij}}> = \frac{1}{2} \int \pi(1,2) \frac{1}{r_{12}} d\tau_1 d\tau_2 \tag{1}$$

where $\pi(1,2)$ is the pair probability distribution function, the probability of finding an electron at point 2 when one is at point 1. The 1-particle density matrix is[2,3]

$$\rho(1) = n\int |\Psi(1,2,---N)|^2 d\tau_2 --- d\tau_n dS \tag{2}$$

where n is the number of electrons, dS implies integration over all spin functions, and $d\tau_i$ is the ith volume element. The 2-particle density matrix is,[2,3]

$$\pi(1,2) = n(n-1) \int |\Psi(1,2,---N)|^2 d\tau_3 --- d\tau_n dS \tag{3}$$

For normalized Ψ's

$$\int \rho(1) d\tau_1 = n \tag{4}$$

$$\int \pi(1,2) d\tau_1 d\tau_2 = n(n-1) \tag{5}$$

and

$$\int \pi(1,2)\, d\tau_2 = (n-1)\rho(1) \tag{6}$$

Eq. (4) gives the total number of electrons in the system and Eq. (5) gives the number of 2-particle interactions.

The total density can be written as:

$$\rho(1) = \rho_\uparrow(1) + \rho_\downarrow(1) \tag{7}$$

and the 2-particle density matrix as

$$\pi(1,2) = \pi_{\uparrow\uparrow}(1,2) + \pi_{\uparrow\downarrow}(1,2) + \pi_{\downarrow\uparrow}(1,2) + \pi_{\downarrow\downarrow}(1,2) \tag{8}$$

where $\rho_\uparrow(1)$ is the total up spin density, $\rho_\downarrow(1)$ is the total down spin density, $\pi_{\uparrow\downarrow}(1,2)$ is the interaction of an up spin particle at point 1 with a down spin particle at point 2, $\pi_{\uparrow\uparrow}(1,2)$ is the interaction of 2 up spin particles at points 1 and 2 respectively; similarly for $\pi_{\downarrow\uparrow}(1,2)$ and $\pi_{\downarrow\downarrow}(1,2)$. The 2-particle density matrices in Eq. (8) can be written as[2]

$$\pi_{\uparrow\uparrow}(1,2) = \rho_\uparrow(1)\rho_\uparrow(2) + \rho_\uparrow(1)\rho_\uparrow(2)f_{\uparrow\uparrow}(1,2) \tag{9}$$

and

$$\pi_{\uparrow\downarrow}(1,2) = \rho_\uparrow(1)\rho_\downarrow(2) + \rho_\uparrow(1)\rho_\downarrow(2)f_{\uparrow\downarrow}(1,2) \tag{10}$$

and similarly for $\pi_{\downarrow\uparrow}(1,2)$ and $\pi_{\downarrow\downarrow}(1,2)$. The f's are correlation factors. From Eqs. (9) and (10) and their opposite spin counterparts in Eq. (8), substitution in Eq. (1) gives

$$\begin{aligned}
\left\langle \frac{1}{2} \sum_{i,j} \frac{1}{r_{ij}} \right\rangle = {} & \frac{1}{2} \int \{\rho_\uparrow(1)\rho_\uparrow(2) + \rho_\uparrow(1)\rho_\downarrow(2) + \rho_\downarrow(1)\rho_\uparrow(2) \\
& + \rho_\downarrow(1)\rho_\downarrow(2)\} \frac{1}{r_{12}}\, d\tau_1 d\tau_2 \\
& + \frac{1}{2} \int \{\rho_\uparrow(1)\rho_\uparrow(2)f_{\uparrow\uparrow}(1,2) + \rho_\uparrow(1)\rho_\downarrow(2)f_{\uparrow\downarrow}(1,2) \\
& + \rho_\downarrow(1)\rho_\uparrow(2)f_{\downarrow\uparrow}(1,2) \\
& + \rho_\downarrow(1)\rho_\downarrow(2)f_{\downarrow\downarrow}(1,2)\} \frac{1}{r_{12}}\, d\tau_1 d\tau_2
\end{aligned} \tag{11}$$

In the Self-Interaction Corrected density functional theory[4,5] the total inter-electronic interaction expectation value is

$$\langle \frac{1}{2} \sum_{i,j} \frac{1}{r_{ij}} \rangle = \frac{1}{2} \int \rho(1)\rho(2) \frac{1}{r_{12}} d\tau_1 d\tau_2$$
$$- \frac{1}{2} \sum_i \int \rho_i(1)\rho_i(2) \frac{1}{r_{12}} d\tau_1 d\tau_2 \tag{12}$$
$$+ \frac{1}{2} \int [\rho_\uparrow(1) U_\uparrow^{xc}(1) + \rho_\downarrow(1) U_\downarrow^{xc}(1)] d\tau_1$$

where $\rho_i(1) = n_i u_i^*(1) u_i(1)$, and $U_\uparrow^{xc}(1)$ is the exchange-correlation potential. The first term in Eq. (12) is the Coulomb repulsion energy, the second term is the Self-Interaction Correction energy, and the last term is the exchange-correlation energy. Eqs. (11) and (12) can be equated, and with Eq. (7) give

$$U_\uparrow^{xc}(1) = \int \{\rho_\uparrow(2) f_{\uparrow\uparrow}(1,2) + \rho_\downarrow(2) f_{\uparrow\downarrow}(1,2)$$
$$+ \sum_i^{\uparrow} \frac{\rho_i(1)}{\rho_\uparrow(1)} \rho_i(2)\} \frac{1}{r_{12}} d\tau_2 \tag{13}$$

The first term in Eq. (13) defines the correlation of electrons with parallel spin, the Fermi hole, the second term defines the correlation of electrons with antiparallel spins, the Coulomb hole, and the third term is the Self-Exchange Correlation; similarly for $U_\downarrow^{xc}(1)$.

2. THEORY

The correlation factors $f_{\uparrow\uparrow}(1,2)$ and $f_{\uparrow\downarrow}(1,2)$ remain to be defined. Some of their properties can be found. Integrating Eq. (9) with respect to $d\tau_2$ and equating with Eq. (6) gives

$$\int \rho_\uparrow(2) f_{\uparrow\uparrow}(1,2) d\tau_2 = -1 \tag{14}$$

If particle 1 is far away from particle 2, the independent particle correlation factor for parallel spins is

$$f_{\uparrow\uparrow}^{ind}(1,2) = -1/n_\uparrow \tag{15}$$

satisfying Eq. (5). For the Coulomb hole, the equivalent normalization condition is,

$$\int \pi_{\uparrow\downarrow}(1,2) d\tau_1 d\tau_2 = n_\uparrow n_\downarrow \tag{16}$$

and thus,

$$\int \pi_{\uparrow\downarrow}(1,2) d\tau_2 = n_\downarrow \rho_\uparrow(1) \tag{17}$$

Setting Eq. (17) equal to the integral of Eq. (10) gives,

$$\int \rho_{\downarrow}(2) f_{\uparrow\downarrow}(1,2) = 0 \tag{18}$$

Thus the independent particle correlation factor for antiparallel spins is

$$f^{ind}_{\uparrow\downarrow}(1,2) = 0 \tag{19}$$

for $r_{12} \to \infty$, which satisfies Eq. (16).

As r_{12}, the distance between electrons 1 and 2, approaches zero, both $\pi_{\uparrow\uparrow}(1,2)$ and $\pi_{\uparrow\downarrow}(1,2)$ approach zero. Hence,

$$f_{\uparrow\uparrow}(1,1) = -1 \tag{20}$$

$$f_{\uparrow\downarrow}(1,1) = -1 \tag{21}$$

which indicate that electrons are not allowed to exist together.

Two assumptions are now made: 1/ the Fermi and Coulomb holes are assumed spherically symmetric, and, 2/ the density is assumed to vary slowly around electron 1. Eqs. (14) and (18) are rewritten as:

$$4\pi \int_0^{r_f} \rho_{\uparrow}(1) f_{\uparrow\uparrow}(r) r^2 dr = -1 \tag{22}$$

and

$$4\pi \int_0^{r_c} \rho_{\downarrow}(1) f_{\uparrow\downarrow}(r) r^2 dr = 0 \tag{23}$$

where r_f is the Fermi hole radius and r_c the Coulomb hole radius. The forms of $f_{\uparrow\uparrow}(r)$ and $f_{\uparrow\downarrow}(r)$ are,[3,6]

$$f_{\uparrow\uparrow}(r) = -\{(1-1/n_{\uparrow})(e^{\frac{-br}{r_f}}[1+\frac{br}{r_f}+b(\frac{r}{r_f})^2] + \frac{1}{n_{\uparrow}}\} \tag{24}$$

and[6]

$$f_{\uparrow\downarrow}(r) = -\exp(d\frac{r}{r_f}) \cos(\frac{3\pi}{2}\frac{r}{r_f}) \tag{25}$$

where b and d are parameters to be determined. Eq. (24) obeys the boundary conditions of Eqs. (15) and (20), Eq. (25) obeys the conditions given by Eqs. (19) and (21), and $r_f = r_c$, the Fermi and Coulomb holes have the same size.[6] r_f is obtained in terms of b by substituting Eq. (24) into Eq. (22) and integrating. Then, using r_f, Eq. (23) can be integrated and solved for d, which will depend on b through r_f.

The exchange correlation potential, Eq. (13), can be written as,

$$U_{\uparrow}^{xc}(1) = U_{\uparrow}^{Fer}(1) + U_{\uparrow}^{Cou}(1) \tag{26}$$

where $U_{\uparrow}^{Cou}(1)$ is the exchange potential that correlates electrons with antiparallel spins, and $U_{\uparrow}^{Fer}(1)$ is the exchange potential that correlates electrons of parallel spins. Most multi-electron atom theories in their first approximation, Xα, Ξa, or Hartree-Fock, contain only $U_{\uparrow}^{Fer}(1)$, and only account for the correlation of electrons of the same spin. Consequently, from Eqs. (13) and (26),

$$U_{\uparrow}^{Fer}(1) = \int\left\{\rho_{\uparrow}(2)f_{\uparrow\uparrow}(1,2) + \sum_{i}^{\uparrow} \frac{\rho_i(1)}{\rho_{\uparrow}(1)}\,\rho_i(2)\right\}\frac{1}{r_{12}}\,d\tau_2 \tag{27}$$

and

$$U_{\uparrow}^{Cou}(1) = \int\rho_{\downarrow}(2)f_{\uparrow\downarrow}(1,2)\,\frac{1}{r_{12}}\,d\tau_2 \tag{28}$$

The second term in Eq. (27) can be simplified by using the approximation,[7]

$$\rho_i(1) \cong \frac{\rho_{\uparrow}(1)}{n_{\uparrow}} \tag{29}$$

implying that the ith density is approximately equal to the total density divided by the total number of electrons. Using Eq. (29) and the assumption used to derive Eqs. (22) and (23), Eq. (27) becomes,

$$U_{\uparrow}^{Fer}(1) = 4\pi \int_0^{r_f}\left\{\rho_{\uparrow}(1)f_{\uparrow\uparrow}(r) + \frac{\rho_{\uparrow}(1)}{n_{\uparrow}}\right\} r\,dr \tag{30}$$

Eq. (24) is now substituted in Eq. (30), and integrated. The result is set equal to the Ξa Self-Interaction Corrected exchange potential,[4,5,7]

$$U_{\uparrow}^{Fer}(1) = \frac{-9c}{2}\,a_{\uparrow}^{Fer}\,\rho_{\uparrow}^{\frac{-5}{3}} \sum_{i}^{\uparrow} (\rho_{\uparrow}(1)-\rho_i(1))\,\rho_i(1) \quad \text{(in Hartrees)} \tag{31}$$

and the parameter $a_{\uparrow}^{Fer}$ can be solved

$$a_{\uparrow}^{Fer} = \frac{2}{3}\,(1+7.34751/n_{\uparrow})^{\frac{-2}{3}}\,b = 6.30168 \tag{32}$$

where b is chosen to give a maximum $a_{\uparrow}^{Fer}$ for $n_{\uparrow} = \infty$ of 2/3. With the value of b now defined, r_f can be evaluated and used to integrate

$$U_{\uparrow}^{Cou}(1) = 4\pi \int_0^{r_f} \rho_{\uparrow}(1)f_{\uparrow\downarrow}(r)\,r\,dr \tag{33}$$

with $f\uparrow\downarrow(r)$ given by Eq. (25). This gives,

$$U_\uparrow^{Cou}(1) = \frac{-9c}{2}\, a_\uparrow^{Cou}\, \rho_\uparrow^{\frac{-2}{3}}(1)\, \rho_\downarrow(1) \quad \text{(in Hartrees)} \tag{34}$$

where

$$a_\uparrow^{Cou} = .04000\,(1 + 7.34751/n_\uparrow)^{\frac{-2}{3}} \tag{35}$$

Therefore, the complete, correlated Ξa total energy equation is:

$$\begin{aligned} E = & \sum_i n_i \langle u_i(1)|f_1|u_i(1)\rangle + \frac{1}{2}\sum_{i,j} n_j n_i \langle u_i(1)u_j(2)|\frac{1}{r_{12}}|u_i(1)u_j(2)\rangle \\ & - \frac{1}{2}\sum_i n_i^2 \langle u_i(1)u_i(2)|\frac{1}{r_{12}}|u_i(1)u_i(2)\rangle \\ & - \frac{9c}{4}\, a_\uparrow^{Fer} \sum_i^{\uparrow} n_i \langle u_i(1)|\rho_\uparrow^{\frac{-5}{3}}(1) \sum_j^{\uparrow}(\rho_\uparrow(1) - \rho_j(1))\rho_j(1)|u_i(1)\rangle \\ & - \frac{9c}{4}\, a_\uparrow^{Cou} \sum_i^{\uparrow} n_i \langle u_i(1)|\rho_\uparrow^{\frac{-2}{3}}(1)\rho_\downarrow(1)|u_i(1)\rangle \\ & + \text{opposite spin terms} \end{aligned} \tag{36}$$

where the last term is the correlation energy. Applying the variational theorem to Eq. (36) gives the 1-electron energy equation:

$$\begin{aligned} & \Big\{ f_1 + \sum_i n_i \langle u_i(2)|\frac{1}{r_{12}}|u_i(2)\rangle - n_k \langle u_k(2)|\frac{1}{r_{12}}|u_k(2)\rangle \\ & - \frac{9c}{4}\, a_\uparrow^{Fer}\Big[\frac{4}{3}\rho_\uparrow^{1/3}(1) - 2\rho_k(1)\rho_\uparrow^{-2/3}(1) + \frac{2}{3}\rho_\uparrow^{-5/3}(1)\sum_i^{\uparrow}\rho_i^2(1)\Big] \\ & - \frac{9c}{4}\, a_\uparrow^{Cou}\rho_\uparrow^{-2/3}(1)\rho_\downarrow(1)\Big[\frac{1}{3} + \frac{a_\downarrow^{Cou}}{a_\uparrow^{Cou}}\,\frac{\rho_\uparrow(1)^{2/3}}{\rho_\downarrow(1)}\Big]\Big\}|u_k(1)\rangle \\ & = \varepsilon_k^{\Xi a\,Corr}|u_k(1)\rangle \end{aligned} \tag{37}$$

3. DISCUSSION

Calculations have been carried out on the atoms Be, Ne, Mg, and Ar. The a^{Fer} used in the calculations were not the ones given by Eq. (32), but the ones that gave HF results, a^{HF}. The results are shown in Fig. 1. The values plotted are the differences between the eigenvalues given by different theories and the exact ones obtained from Slater.[8] Energies are given in Rydbergs. The solid line is the HF difference line,[9] the dash-triple-dot line is the correlated Ξa difference line, and the dotted line is the difference between the correlated Xα one-electron energies, calculated from

$$\varepsilon_k^{X\alpha} - \frac{1}{2} \langle u_k(1)u_k(2) | \frac{1}{r_{12}} | u_k(1)u_k(2) \rangle,^{10}$$

and the exact ones, with $\alpha^{Fer} = \alpha^{HF}$ and α^{Cou} given in reference 6. The dash-dot and dashed lines are the one-electron energy difference lines corresponding to the uncorrelated Ξa and Xα theories respectively with $a^{Fer} = a^{HF}$ and $\alpha^{Fer} = \alpha^{HF}$. These results show that there is no correlation in the eigenfunctions.

4. CONCLUSION

The one-electron exchange-correlation potential in the HFS-Xα scheme of Keller and Gazquez[6] is:

$$\begin{aligned} V^{xc}(1) &= -6c\alpha^{Fer}\rho_\uparrow^{1/3}(1) \\ &\quad - \frac{9c}{2}\alpha_\uparrow^{Cou}\rho_\uparrow^{-2/3}(1)\rho_\downarrow(1)\left[\frac{1}{3} + \frac{\alpha_\downarrow^{Cou}}{\alpha_\uparrow^{Cou}}\left(\frac{\rho_\uparrow^{(1)}}{\rho_\downarrow^{(1)}}\right)^{2/3}\right] \end{aligned} \tag{38}$$

for closed shell atoms Eq. (38) reduces to

$$V_\uparrow^{xc}(1) = -6c(\alpha^{Fer} + \alpha^{Cou})\rho_\uparrow^{1/3}(1) \tag{39}$$

and simply scales the value of α to give exact energies. Tseng, Hong, and Whitehead[11] have shown that such scaling does NOT give correlated wavefunctions. Thus, even though this version of the Keller and Gazquez scheme gives good statistical total energies, it cannot give correlated one-electron energies. It had been hoped that the Keller and Gazquez Coulomb hole adapted to the Self-Interaction Corrected Xα theory, the Ξa theory, would give correlated Ξa wavefunctions, because from Eqs. (36) and (37), the parameter a is not scaled, an extra term occurs in the Hamiltonian, which it was hoped would add an effect characteristic of the particular atom and its antiparallel electron interaction. It does not.

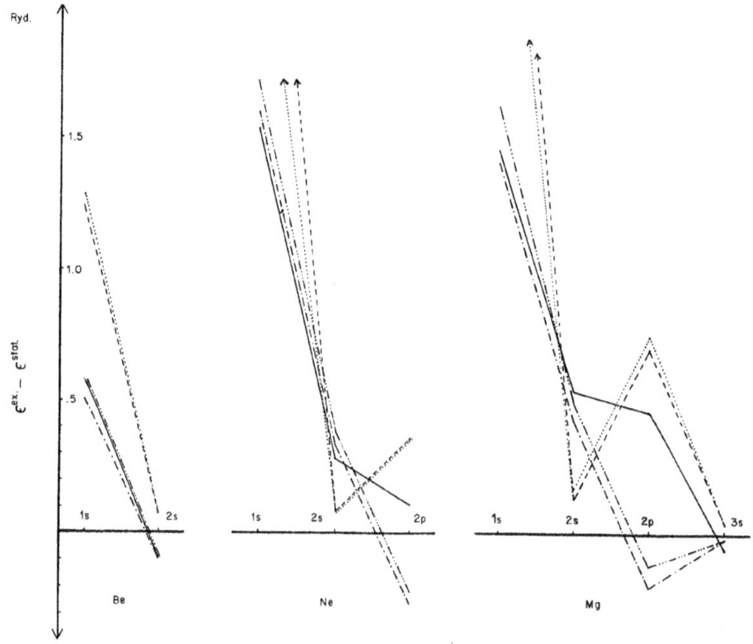

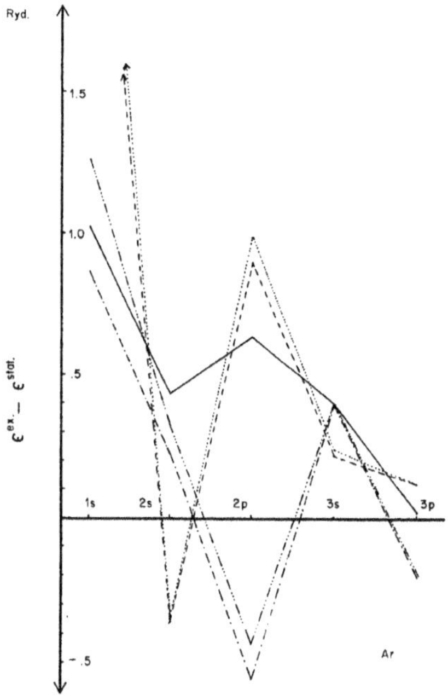

Figure 1. Comparison of correlated and uncorrelated one-electron energies in the Ξα and Xα schemes for Beryllium, Neon, Magnesium and Argon.

——— HF
-·-·- uncorrelated Ξα
-···- correlated Ξα
----- uncorrelated Xα
····· correlated Xα

REFERENCES

1. R. McWeeny, Rev.Mod.Phys. 32, 335 (1960).
2. M.S. Gopinathan, M.A. Whitehead, and R. Bogdanovic, Phys.Rev. A 14, 1 (1976).
3. J.L. Gazquez and J. Keller, Phys.Rev. A 16, 1358 (1977).
4. M.S. Gopinathan, Phys.Rev. A 15, 2135 (1977).
5. T.J. Tseng and M.A. Whitehead, Phys.Rev. A 24, 16 (1981).
6. J. Keller and J.L. Gazquez, Phys.Rev. A 20, 1289 (1979).
7. T.J. Tseng and M.A. Whitehead, Phys.Rev. A 24, 21 (1981).
8. J.C. Slater, Phys.Rev. 98, 1039 (1955).
9. E. Clementi, IBM.J.Res.Dev.Suppl. 9, 2 (1965).
10. M.S. Gopinathan, J.Phys. B: Atom.Molec.Phys 12, 521 (1979).
11. T.J. Tseng, S.H. Hong, and M.A. Whitehead, J.Comp.Chem. 1, 88 (1980).

ON THE ATOMIC WAVEFUNCTIONS GENERATED BY THE HARTREE-FOCK-SLATER METHOD $X\alpha^{HF}$ AND THE MODIFIED HARTREE-FOCK-SLATER METHOD Ξa^{HF} WHICH INCLUDES SELF INTERACTION EXPLICITLY

S. Manoli and M.A. Whitehead

Department of Chemistry, McGill University
801 Sherbrook Street, W., Montreal
Canada H3A 2K6

The Ξa^{HF} wavefunctions better approximate the HF wavefunctions than the $X\alpha^{HF}$ wavefunctions for atoms $Z \leq 20$; the Ξa^{HF} wavefunctions are no better than the $X\alpha^{HF}$ wavefunctions for $Z > 20$.

1. INTRODUCTION

The Total Energy equation in the Hartree-Fock-Slater, HFS or $X\alpha$ method[1] is:

$$E^{HFS} = \sum_i n_i \langle u_i(1)|f_1|u_i(1)\rangle + \frac{1}{2}\sum_{i,j} n_i n_j \langle u_i(1)u_j(2)|g_{12}|u_i(1)u_j(2)|\rangle - \sum_i^{\uparrow} \frac{9C}{2}\alpha\, n_i \langle u_i(1)|\rho_\uparrow^{1/3}(1)|u_i(1)\rangle + \downarrow \text{ spin} \tag{1}$$

where $f_1 = -\nabla_1^2 - 2Z/r$, $g_{12} = 2/r_{12}$, and $C = (3/4\pi)^{1/3}$ in the spin-polarized equations. The exchange potential is approximated by

$$U_\uparrow^{ex}(1) = -9C\,\alpha\,\rho_\uparrow^{1/3}(1) \tag{2}$$

where

$$\rho_\uparrow(1) = \sum_i n_i u_i^*(1) u_i(1)$$

and α is a scaling parameter.[1] The one-electron energy equation is obtained by applying the variational theorem to Eq. (1),

$$\{f_1 + \sum_i n_i \langle u_i(2)|g_{12}|u_i(2)\rangle - 6C\alpha\rho_\uparrow^{1/3}(1)\}|u_i(1)\rangle = \varepsilon_k^{X\alpha}|u_i(1)\rangle \tag{3}$$

The eigenvalue of Eq. (3), $\varepsilon_k^{X\alpha}$, is not directly comparable to the corresponding Hartree-Fock, HF eigenvalue, ε_k^{HF}, and has to be corrected using Slater's Transition State Theory to give a corresponding one-electron energy.[1] Slater evaluated Eq. (3) with the occupation number, n_k, of orbital k equal to 1/2 an electron, to give,

$$\varepsilon_k^{X\alpha}(n_k = \tfrac{1}{2}) = \varepsilon_k^{X\alpha}(n_k = 1) - \tfrac{1}{2}\langle u_k(1)u_k(2)|g_{12}|u_k(1)u_k(2)\rangle \tag{4}$$

$$\varepsilon_k^{X\alpha}(n_k = \tfrac{1}{2})$$

is directly comparable to ε_k^{HF}. Gopinathan[2] confirmed this by calculating

$$E^{HFS}(\text{+ive ion}) - E^{HFS}(\text{neutral atom}) = I_k^{ur} \tag{5}$$

for unrelaxed orbitals (ur), and found that:

$$I_k^{ur} = -\varepsilon_k^{X\alpha}(n_k = 1) + \tfrac{1}{2}\langle u_k(1)u_k(2)|g_{12}|u_k(1)u_k(2)\rangle \tag{6}$$

and since $I_k^{ur} \simeq -\varepsilon_k^{HF}$, a statement of Koopmans' theorem,

$$\varepsilon_k^{HF} = \varepsilon_k^{X\alpha}(n_k = \tfrac{1}{2}) \tag{7}$$

The behaviour of the one-electron exchange potential at large distance is not correct in the Xα theory,[3] because

$$rV_{XHFS\uparrow}(1) = -6C\,\alpha\rho_\uparrow^{1/3}(1)r \rightarrow 0$$

as $r \rightarrow \infty$. The quantity should approach -2 Ryds. and not 0 when r approaches infinity. In the HF theory,

$$rV_{XHF\uparrow}(1) = r\left(V_{HF}^{ex}(1) + V^{SI}(1)\right) \rightarrow -2 \text{ as } r \rightarrow \infty$$

where V^{ex} is the pure exchange potential and V^{SI} the self-interaction or self-exchange potential. Specifically, $V^{SI}(1)$ is that part of the exchange potential which cancels out the self-interaction Coulomb repulsion given by the 2nd term in Eq. (3), and $rV^{SI}(1)$ approaches -2 when r approaches ∞, while $rV_{HF}^{ex}(1)$ approaches 0 when r approaches ∞. Therefore, the Coulomb repulsion of an electron with itself, the self-interaction energy, is not cancelled out exactly by the Xα exchange term.

Gopinathan[3] proposed a scheme to correct the self-interaction energy in the Xα scheme called the Ξa theory.[4,5,6] The total energy equation in the Hartree-Fock Gopinathan, HFG or Ξa scheme is:

$$
\begin{aligned}
E^{HFG} &= \sum_i n_i \langle u_i(1)|f_1|u_i(1)\rangle \\
&+ \frac{1}{2}\sum_{i,j} n_i n_j \langle u_i(1)u_j(2)|g_{12}|u_i(1)u_j(2)\rangle \\
&- \frac{1}{2}\sum_{i}^{\uparrow} n_i^2 \langle u_i(1)u_i(2)|g_{12}|u_i(1)u_i(2)\rangle \\
&+ \frac{1}{2}\sum_i n_i \langle u_i(1)|U_{\uparrow}^{HFG}(1)|u_i(1)\rangle + \downarrow \text{ spin term}
\end{aligned}
\qquad (8)
$$

where

$$U_{\uparrow}^{HFG}(1) = -9C\, a\, \rho_{\uparrow}^{-5/3}(1)\sum_i^{\uparrow} (\rho_{\uparrow}(1) - \rho_i(1))\rho_i(1),$$

$$\rho_i(1) = n_i u_i^*(1) u_i(1)$$

and a is now the scaling parameter.

The 3rd term in Eq. (8) cancels the self-interaction energy present in the 2nd term. The exchange term is now the Pure Exchange, and it is the only term scaled by the parameter a.

Applying the variational theorem to Eq. (8) gives the one-electron energy equation:

$$
\begin{aligned}
\{f_1 &+ \sum_i n_i \langle u_i(2)|g_{12}|u_i(2)\rangle - n_k \langle u_k(2)|g_{12}|u_k(2)\rangle \\
&- \frac{9e}{2}\, a\, [\tfrac{4}{3}\rho_{\uparrow}^{1/3}(1) - 2\rho_k(1)\rho_{\uparrow}^{2/3}(1) \\
&+ \frac{2}{3}\rho_{\uparrow}^{-5/3}(1)\sum_i^{\uparrow} \rho_i^2(1)]\}|u_k(1)\rangle \\
&= \varepsilon_k^{\Xi a}|u_k(1)\rangle
\end{aligned}
\qquad (9)
$$

The second term in Eq. (9) is the Self-Exchange Operator for the electron in the kth orbital. Note that the leading term given in the Pure Exchange is exactly the exchange term given in HFS method. The eigenvalue of Eq. (9), $\varepsilon_k^{\Xi a}$, is directly comparable to the corresponding HF eigenvalues, ε_k^{HF}, and there is no need to invoke Slater's Transition State Theory. Also, the behaviour of the one-electron total exchange potential, including self-exchange, is correct.

$rV_{XHFG\uparrow}(1)$ approaches -2 as r approaches ∞. Tseng, Hong, and Whitehead[4] showed that the $\varepsilon_k^{\Xi a}$'s were closer to the ε_k^{HF}'s than either the $\varepsilon_k^{X\alpha}$'s at $n_k = 1$ or the one-electron energies given by Eq. (4) at $n_k = 1/2$. They also used the Ξa and Xα wavefunctions with the HF one-electron Hamiltonian and concluded that both sets of wavefunctions gave equally good one-electron eigenvalues, $e_k^{HF}(\alpha^{HF}$ or $a^{HF})$.

However, it seemed advisable to test this result over a wider range of atoms before proceeding to molecular problems, as too often preliminary results for a limited selection of atoms prove invalid.

2. METHOD

The wavefunctions from the HFS scheme using α^{HF} and from the HFG scheme using a^{HF} were used with the appropriate HF one-electron operator. The differences between the true HF eigenvalues[7] and those generated with the HF one-electron Hamiltonian on the $X\alpha^{HF}$ or Ξa^{HF} wavefunction are plotted in Figs. 1 to 4 for Li to Kr. Differences between $\varepsilon^{HF}(\alpha,a)$ and ε^{HF} are taken from reference 7. Solid lines correspond to $\varepsilon^{HF}(\Xi a) - \varepsilon^{HF}$, dashed lines correspond to $\varepsilon^{HF}(X\alpha) - \varepsilon^{HF}$.

3. DISCUSSION

The Ξ^{HF} wavefunctions are a better approximation to the HF wavefunctions than the $X\alpha^{HF}$ wavefunctions for atoms between Lithium and Calcium, $Z \leq 20$. The Ξa^{HF} wavefunctions begin to fail for larger systems such as Gallium to Krypton, $Z > 20$.

For helium, Z = 2, the $\varepsilon_{1s}^{HF}(a^{HF}) = -1.8359$ Ryd, exactly equal to the ε^{HF}, whereas $\varepsilon_{1s}^{HF}(\alpha^{HF}) = -1.8432$ Ryd. This is because both the HF and the Ξa method predict the pure-exchange for a paired two electron system to be zero ($a^{HF} = 0$ for He in the Ξa method[5]); in the Xα method, $\alpha = 0.77298$ to give a non-zero exchange, $\alpha\rho_\uparrow^{1/3}(1)$, which only approximately cancels the self-interaction.

In Fig. 1, Lithium, Beryllium, and Boron, the $\varepsilon^{HF}(a^{HF})$'s are generally closer than the corresponding $\varepsilon^{HF}(\alpha^{HF})$'s, but more importantly they "parallel" the baseline of the HF eigenvalues better than the "line" described by the $\varepsilon^{HF}(\alpha^{HF})$'s. Therefore, the Ξa^{HF} wavefunctions reflect the same general trend as the HF wavefunctions. Carbon, Nitrogen, and Oxygen show that the Ξa 1s and 2s wavefunctions still conform to the HF trend better than the Xα wavefunctions. The outermost electrons 2p, however, do not. In

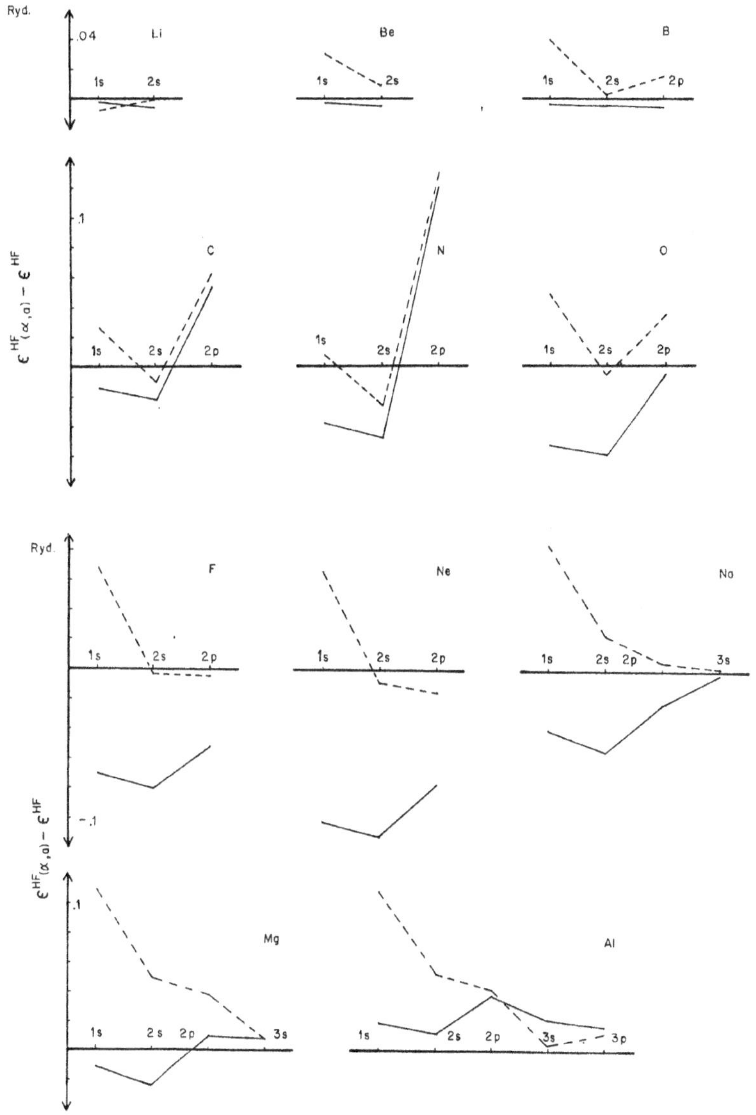

Figure 1. Lithium, Beryllium, Boron, Carbon, Nitrogen, Oxygen, Fluorine, Neon, Sodium, Magnesium, and Aluminium.

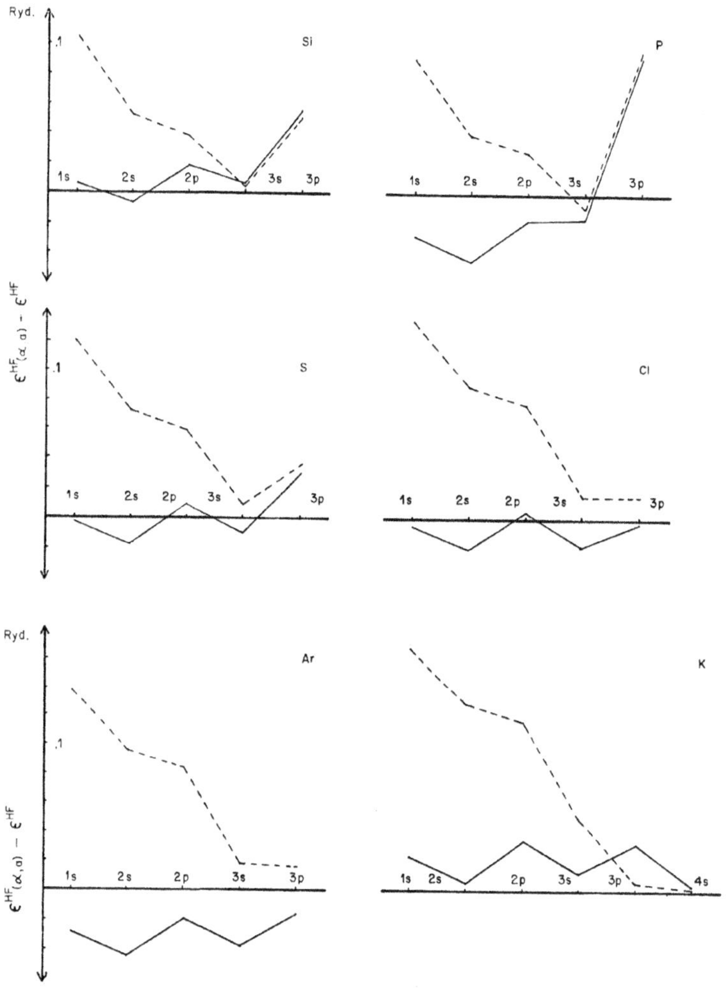

Figure 2. Silicon, Phosphorus, Sulfur, Chlorine, Argon, and Potassium.

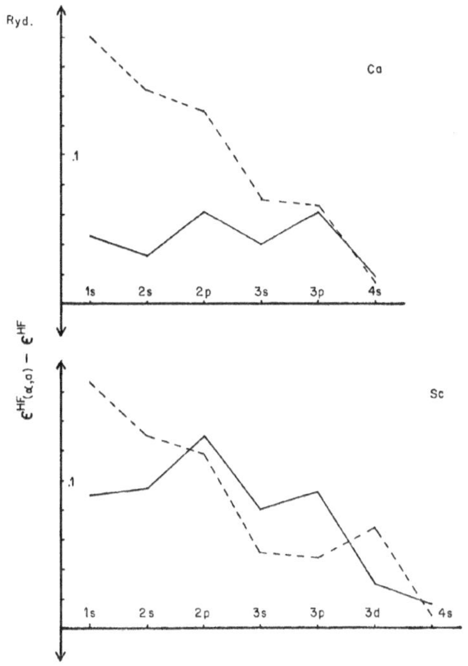

Figure 3. Calcium and Scandium.

fact, Fig. 2, Silicon, Phosphorus, and Sulphur shows the greater the number of electrons in an atom, the worse the agreement with HF, even though the effect is reduced as the system gets larger. This behaviour also occurs for unpaired d electrons of the transition metals, for both the $\Xi\alpha$ and $X\alpha$ wavefunctions, indicating that the greater the difference between $\rho_{\uparrow}(1)$ and $\rho_{\downarrow}(1)$, the worse the $\rho_{\uparrow}{}^{1/3}(1)$ approximation.

Fig. 1, Fluorine and Neon, shows that the $X\alpha$ 2s and 2p wavefunctions are very good approximations to the corresponding HF wavefunctions. However, the $\Xi\alpha$ wavefunctions reflect a better overall "straight line" trend. Figs. 1, 2, and 3, Sodium, Magnesium, Aluminum, Silicon, Phosphorus, Sulphur, Chlorine, Argon, Potassium, and Calcium, clearly show that the $\Xi\alpha$ wavefunctions follow the HF

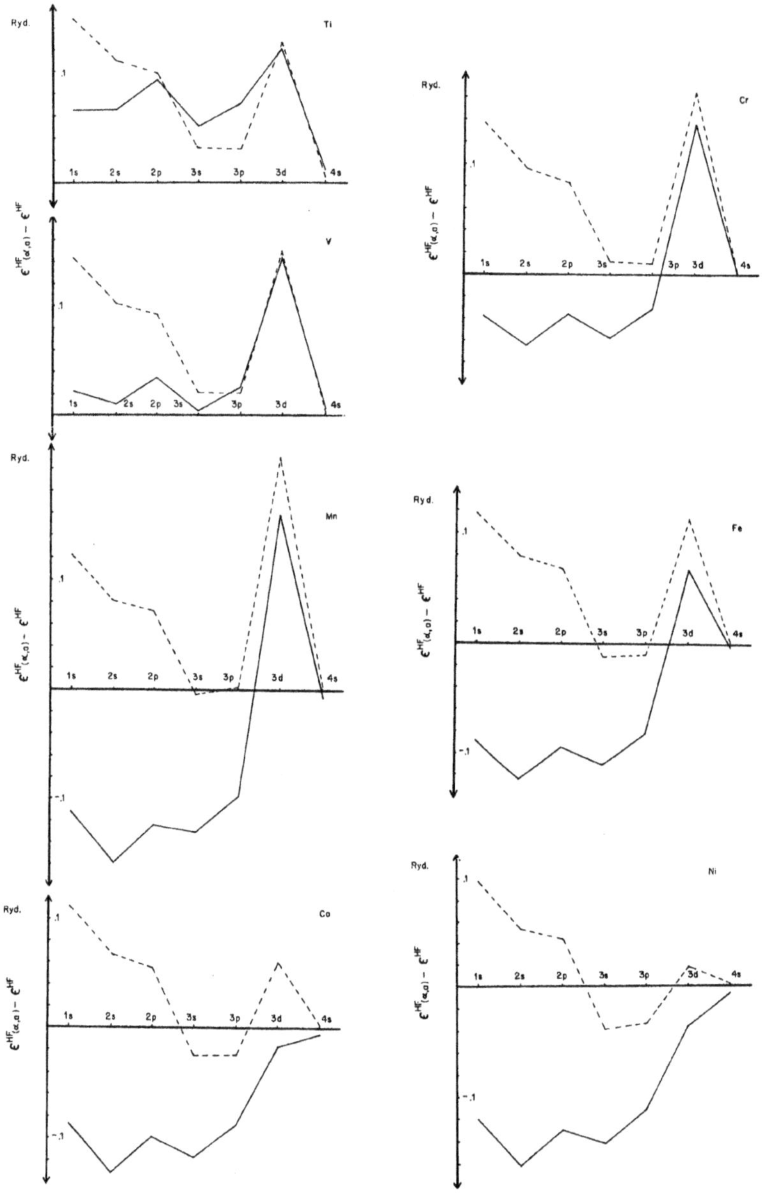
Ryd.
Ti
.1
1s 2s 2p 3s 3p 3d 4s
V
Ryd.
Mn
-.1
Co
Cr
Fe
Ni
$\epsilon^{HF}_{(\alpha,a)} - \epsilon^{HF}$

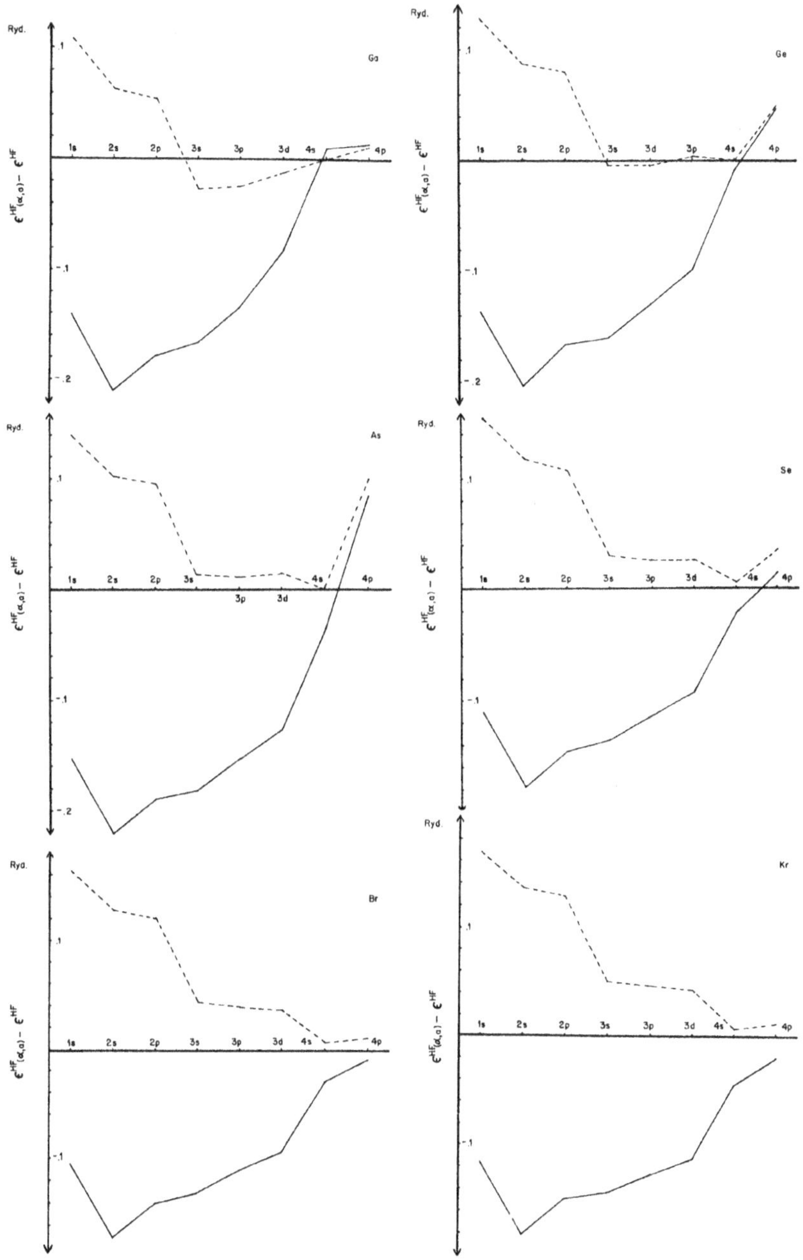

Figure 4. Titanium, Vanadium, Chromium, Manganese, Iron, Cobalt, Nickel, Gallium, Germanium, Arsenic, Selenium, Bromine, and Krypton.

trend very well, while the Xα core wavefunctions are not good.

In Figs. 3 and 4, Scandium through Krypton, except for Cu and Zn due to convergence problems, the Ξα wavefunctions no longer give the HF trend, while the Xα wavefunction maintain the same trend as the atoms with $Z \leq 20$.

For very large systems, when $\rho_k(1) << \rho_\uparrow(1)$, the exchange potentials given in Eqs. (8) and (9) of the HFG Ξα scheme reduce to the HFS Xα ones, because the self-interaction energy becomes less important. For small systems, the dominant term in the Hamiltonian is $-n_k\langle u_k(2)|g_{12}|u_k(2)\rangle$ and the exchange term is small. Thus, it is easy to see why the HFG Ξα scheme works well for small systems, and why it should give results of HFS accuracy in very large systems. But, for atoms in the medium Z range, the HFG exchange potential shows no sign of behaving as the HFS potential, and therefore we are forced to conclude that another form of the exchange potential is needed.

4. CONCLUSION

The energy differences that appear in Figs. 1-4 are quite small, ~ .1 Ryd. = 1.36 eV, but the main concern has been to reproduce the overall HF trend using either the HFS or HFG schemes. The HFG scheme gives better wavefunctions than the HFS scheme for $Z \leq 20$, but the HFG exchange potential fails to behave properly for medium-sized atoms since it approaches the HFS exchange potential for large atoms. Therefore, another form of the pure exchange potential must be found that described the medium-sized atoms better than the HFG potential. The SIC corrected LSD[8] method minimizes with respect to the occupation number which moves the density ρ around; it gives less good values for $-\varepsilon^{stat}$ than the Ξα method[6] for the $-\varepsilon_{ne}$ from He through Ar, but it would be interesting to see how the SIC-LSD wavefunctions perform with the HF Hamiltonian and whether the ε trend parallels the HF or experiment through the whole periodic table.

REFERENCES

1. J.C. Slater, The Self-Consistent Field For Molecules and Solids: Quantum Theory of Molecules and Solids, Vol. 4, (McGraw-Hill, New York) 1974.
2. M.S. Gopinathan, J.Phys B: Atom.Molec.Phys. 12, 521 (1979).
3. M.S. Gopinathan, Phys.Rev. A 15, 2135 (1977).
4. T.J. Tseng, S.H. Hong, M.A. Whitehead, J.Phys.B: Atom.Molec. Phys. 13, 4101 (1980).
5. T.J. Tseng, M.A. Whitehead, Phys.Rev. A 24, 16 (1981).
6. T.J. Tseng, M.A. Whitehead, Phys.Rev. A 24, 21 (1981).
7. E. Clementi, IBM. J.Res.Rev.Suppl. 9, 2 (1965).
8. J.P. Perdew and A. Zunger, Phys.Rev. B 23, 5048 (1981).

AB-INITIO CALCULATIONS INCORPORATING DESIRABLE OPTIONS FOR LARGE MOLECULES AND SOLIDS: A COMPETITIVE COMPLETELY GENERAL ALTERNATIVE TO LOCAL DENSITY METHODS FOR MANY SYSTEMS

Joyce J. Kaufman

Department of Chemistry
The Johns Hopkins University
Baltimore, Maryland 21218, U.S.A.

1. INTRODUCTION

In our group we have a long standing interest in ab-initio quantum chemical calculations on large molecules from our early ab-initio SCF Gaussian calculations in 1963-1964 on diborane[1] (B_2H_6; 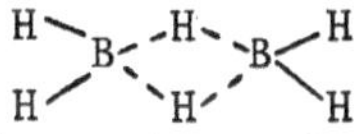 ; with 56 atomic basis functions - a large molecular calculation for those years), through our more recent calculations on the metabolic activation of benzo(a)pyrene[2] [a polycyclic aromatic hydrocarbon precarcinogen; and its proximate carcinogenic oxides and dihydrodiols and its ultimate carcinogen, the 7,8-dihydrodiol-9,10-epoxide (all eight possible stereoisomers), Figure 1a and Figure 1B].

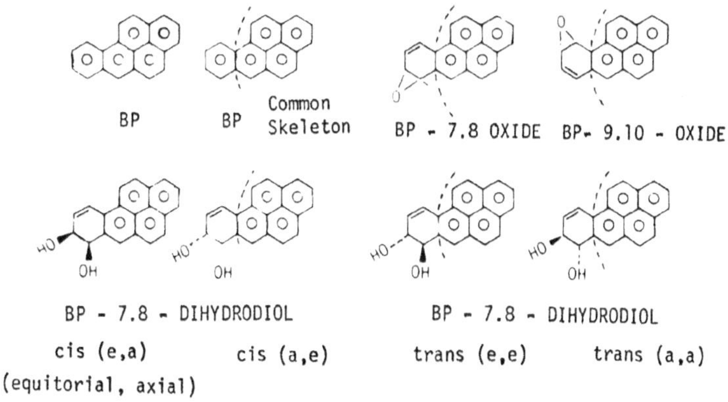

Figure 1a

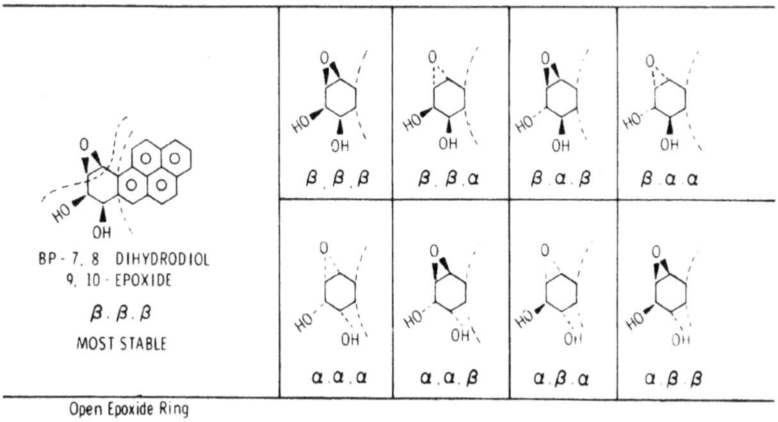

Figure 1b

In the early 1970's with the intent to carry out calculations on large biomedical molecules, we explored the use of the MS-Xα and MS-Xαβ scattered wave methods (both muffin-tin, non-muffin tin and cellular potentials) for molecular calculations.[3-7] The method worked well for us for such blobular systems as the polyhedral boron hydrides,[3] boron trihalides and chlorofluoromethanes,[5] giving good agreement for the transition state calculations with the experimental photoelectron spectra. The transition state calculations also worked well for us for calculation of ionization potentials and electron affinities of atoms and of molecules.[6,7]

However, these methods did not work well for our standard test planar heteroaromatic molecule, pyrrole.

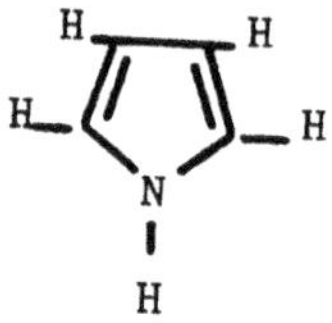

While the calculated transition state orbital energies for pyrrole gave reasonable agreement with experimental photoelectron spectra for the sigma molecular orbitals they gave poor agreement for the pi-molecular orbitals.[8] Examination of the approximations used in the various local density methods and the results for pyrrole indicated quite clearly that the approximations used did not properly represent π-orbitals which physically have a lobe-like structure above and below the aromatic plane.

Moreover, for proper description of molecular states and potential energy surfaces which need configuration interaction calculations and often multi-determinant reference states, the local density methods as then formulated (and even apparently as formulated to date) did not lead to proper spin and symmetry states to use as the reference states.

2. METHODOLOGY

2.1. Molecules

We had earlier (1964-1968) written several versions of a polyatomic large-scale ab-initio Gaussian integral/SCF program, MOSES[9] (Molecular Orbital-SCF Energy System) which were very efficient for calculations on small and even medium size molecules. However, the computational characteristics of large molecules such as morphine (an N-methyl narcotic agonist) and nalorphine (an N-allyl narcotic agonist-antagonist)[10] or promazine and chlorpromazine[11] are quite different than those for small molecules (even using what was at that time a state-of-the-art large molecule program, IBMOL VA of Clementi and Popkie). After carrying out the calculations on these molecules with IBMOL VA we came to the conclusion that there were several new desirable features which should be incorporated into programs for ab-initio calculations on large molecules. Thus in 1974 we embarked on writing several new series of programs.

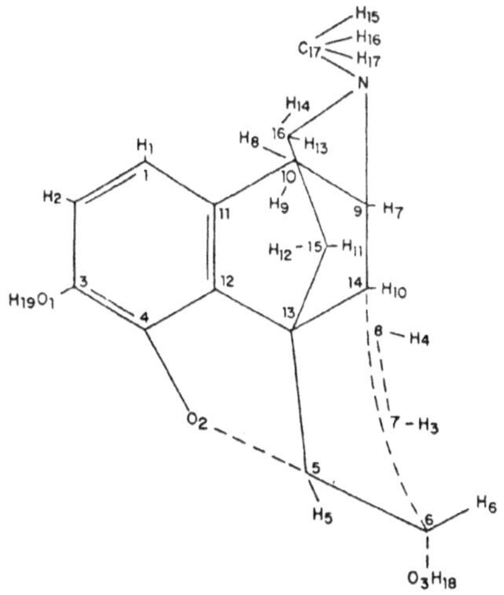

Figure 2

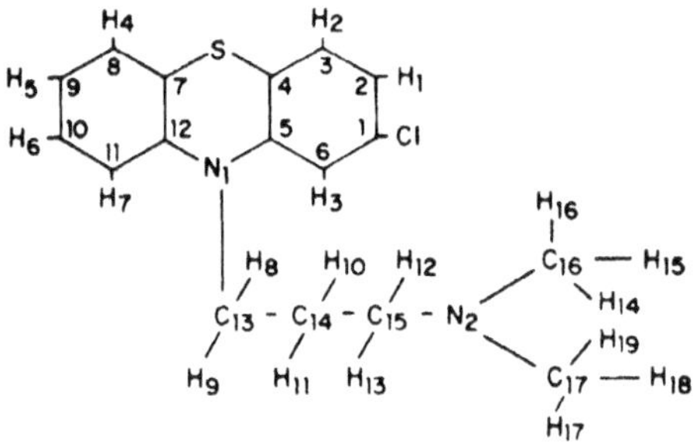

Figure 3

Our new big molecule programs MOLASYS[12] (s and p orbitals) and GIPSY(s,p,d and f orbitals)[13] use fast explicit coding for integrals involving s,p and d integrals.

But even more importantly our programs incorporate several very desirable features as time-saving options to our ab-initio computer programs: ab-initio effective core model potentials (MODPOT) which enable one to treat the valence electrons only explicitly, yet accurately; a charge conserving integral evaluation (which we named VRDDO - variable retention of diatomic differential overlap) especially effective for spatially extended molecules; an efficient MERGE technique which enables us to retain all the integrals for a skeletal fragment which remains unchanged and just recalculate the new integrals for added basis functions, for added atoms or for geometry variations both for intra- and intermolecular calculations; and a new SCF procedure which passes through contributions to Fock matrix elements from invariant skeletal integrals, uses these up until almost convergence then relaxes the conditions and makes up the complete new Fock matrix elements for the final few iterations.

Using these programs we can carry out non-empirical LCAO-MO-SCF plus MC-SCF and CI calculations on large molecules as well as on the more conventional small molecules (LCAO-MO-SCF = linear combination of atomic orbitals to form molecular orbitals by a self-consistent technique[14] to form the best single determinant electronic wave function; MC-SCF = multiconfiguration SCF[15] where both the exponent of the AO's in the MO's and the coefficients of a combination of determinants are varied simultaneously; CI = configuration interaction, a superposition of electronic configurations which a single determinant is insufficient to describe a molecular system satisfactorily[16]).

The ab-initio effective core model potentials are not only desirable for heavy atoms with many inner shell electrons but as we have demenstrated[17], are also very useful for large molecular systems even composed of just first row atoms and hydrogens.

Several years ago Bonifacic and Huzinaga gave a new formulation of the model potential (MODPOT) method for carrying out LCAO-MO-SCF calculations considering only the valence electrons[18]. The effect of the core electrons is taken into account through the use of a modified model potential (MODPOT) Hamiltonian.

Bonifacic and Huzinaga derived model potential expressions for inner s, p and d orbitals. We extended the derivation also to inner f orbitals. This MODPOT procedure of Bonifacic and Huzinaga has been introduced as an option into our MOLASYS and GIPSY computer programs. With the GIPSY computer program we can calculate the usual one-electron overlap, kinetic energy, nuclear attraction and two-electron repulsion integrals between s-, p-, d- and f-contracted Gaussian functions. In addition, the new type of one-electron integrals resulting from the use of the Bonifacic-Huzinaga Hamiltonian with s-, p-, d- and f-type core atomic orbitals can be calculated.

We verified the accuracy of the MODPOT method for a variety of molecules by carrying out the reference calculations with two-electron integrals accurate to 6 decimal places[17a-c]. Comparison of the MODPOT results to the completely ab-initio ones (including inner shell electrons) showed that the valence orbital energies and gross atomic populations are accurate* to more than 2 decimal places. It should be emphasized that the accuracy obtained with the MODPOT approximation is more than acceptable if one wants to reproduce ab-initio results for large molecules. Moreover, for molecules composed even of two second row atoms, such as Cl_2, the savings in time between the MODPOT method compared to the completely ab-initio is a factor of ten. The savings in time also increase dramatically as the size of the molecule goes up.

Our VRDDO (variable retention of diatomic differential overlap) integral prescreening evaluation was inspired in part by a suggestion from solid state physics by Wilhite and Euwema[20]. The approximation consists of neglecting all one-electron integrals (both energy and overlap) and two-electron integrals that involve basis function pairs $\phi_i(1)\phi_j(2)$ whose pseudo-overlap:

$$S_{ij}^* = \int \phi_i(1)\phi_j^*(2)dv_1$$

is less than some threshold τ_1. The VRDDO approximation has been introduced[19] into our MOLASYS[11,12] and GIPSY[13] computer programs.

*Where accurate implies accuracy with respect to the completely ab-initio calculations with the same atomic basis set retaining all integrals 10^6 a.u. or greater.

If both the MODPOT and VRDDO approximations are introduced, the accuracy* with respect to the reference calculations is about the same as that obtained using only the MODPOT approximation.

The relative accuracy of the MODPOT and MODPOT-VRDDO methods along an SCF potential energy surface (such as for molecular conformational changes or interactions between two species, 0.0001 - 0.0002a.u.) appears to be even greater than their absolute accuracy.

For the original testing full basis set calculations (inner and outer shells), VRDDO calculations (with the same full basis set), MODPOT calculations (with the same valence atomic basis set and incorporating the same inner shell basis set into the MODPOT) and MODPOT/VRDDO calculations were carried out on about 75 molecules including:

X = F, Cl, NO_2, CH_3, CF_3

Azoles - 5 member heteroaromatic rings
Pyrrole, Pyrazole, Imidazole
Azines - 6 member heteroaromatic rings
Pyridine, Pyrazine, Pyrimidine

Boron hydrides B_4H_{10}, B_5H_9, $B_6H_6^{-2}$, B_6H_{10}, $B_{10}H_{14}$

and Carboranes $C_2B_4H_6$

Chlorofluoromethanes CX_mY_n $m + n = 4$, X = F Y = CL

H_nX, X_2 X = 1st and 2nd row atoms

Adenine

For one aromatic ring with one substituent (such as NO_2) the VRDDO calculation is 2.5 times faster than the reference calculation (using our own fast ab-initio programs[12,13]), the MODPOT calculation is ≃3.5 times faster and the MODPOT/VRDDO calculation ≃8.5 times faster. Moreover, the relative increase in speed goes up even more dramatically as the molecules get larger (Tables 1 and 2).

As a test of the efficiency of the method, we ran ab-initio MODPOT/VRDDO calculations on the DNA bases (Figure 4), (see Table 3).

We also wrote a new MERGE technique which allows us to retain the integrals for a skeletal fragment which remains unchanged. This is especially useful in studies of large closely related molecules

Table 1. Accuracy Testing Compared to Ab-Initio (Including Inner Shells)[a]

	Orbital Energies(a.u.)	GAP's ($\bar{e}$)	Isomer Energy Differences or Potential Energy Curve Differences (a.u.)
VRDDO	0.001 - 0.002	0.001 - 0.002	0.0001 - 0.001
MODPOT	0.001 - 0.005	0.001 - 0.005	0.0001 - 0.005
MODPOT /VRDDO	0.001 - 0.005	0.001 - 0.005	0.0001 - 0.005

[a]MODPOT/CI
VRDDO/CI
MODPOT/VRDDO/CI

Energy Differences Compared to Ab-Initio (Including Inner Shells)
0.001 - 0.002 eV
Spectra (UV, Visible)
Ionization Potentials
Electron Affinities

Table 2. Nitrobenzene Timings (Sec.)[a] (no symmetry)

	Ab-Initio 6-3G	VRDDO	MODPOT	MODPOT/VRDDO
#2 Integrals	298000	132000	159000	96000
INT 2	2147.4	887.7	486.0	222.1
SCF	543.7	112.7	231.4	98.5
	2691.1	1000.4	717.4	320.6

[a]Times are on a CDC 6600 running under NOS/BE 1.3. Programs were compiled using FTN with OPT = 2.

Table 3.

MOLECULE	NO. ATOMS C,N,O	H	NO. BASIS FNS	TIME (Minutes) Integrals or skeletal integrals	MERGE integrals	SCF
DNA Constituents						
Cytosine	8	5	37	5.1[a]		2.0[b]
Thymine	9	6	38	7.4[a]		3.1[b]
Adenine	10	5	45	9.2[b]		3.5[b]
Guanine	11	5	49	11.2[a]		4.3[b]
Guanine + CH_3^+	12	8	56	6.8[c]	2.3[a]	
				2.9[c]	0.9[d]	1.9[e]

[a] CDC 6600 computer, FTN, OPT = 0. (The integrals run even ~75% faster under OPT = 1 or OPT = 2 on a CDC 6600 computer which supports a full OPT = 2 compiler).
[b] CDC 6600 computer, FTN, OPT = 2.
[c] CDC 6600 computer, FTN, OPT = 1.
[d] CYBER 175, FTN, OPT = 1.
[e] CYBER 175, FTN, OPT = 2.

and their complexes since only an atom or a few atoms are changed and except for the ring which is being attacked, the structure of the rest of the molecule remains invariant. It was not straightforward to incorporate a MERGE technique into our program since we calculate integrals very efficiently by blocks rather than one by one. We are now using this MERGE option for all our calculations since the systems we are investigating at present are quite large. The MERGE option is useful for large molecules (even one and two ring systems), and exceptionally valuable for larger molecules and where the part to be varied is small compared to a large invariant structural fragment. As an example, while the skeletal integrals for benzopyrene-7,8-dihydrodiol (a 5-ring aromatic proximate carcinogen) took 3005 seconds(CDC 6600, SCOPE 3.2, OPT = 1)* to form the ultimate carcinogen benzopyrene 7,8-dihydrodiol-9,10-epoxide, the extra MERGE integrals for the oxygen with the rest of the skeleton took 560 seconds (CDC 6600, SCOPE 3.2,OPT = 0)*.

Quite recently with a new better NOS/BE 1.1 operating system on the CDC 6600, it took only 350 seconds to add both an O and H to the skeleton (CDC 6600 NOS/BE 1.1, FTN, OPT = 2). Thus it was com-

*At that time the particular CDC 6600 computer on which these calculations were run did not support a full OPT = 1 or OPT = 2 compiler. The integrals run ~75% faster under OPT = 1 (and about the same under OPT = 2) than under OPT = 0.

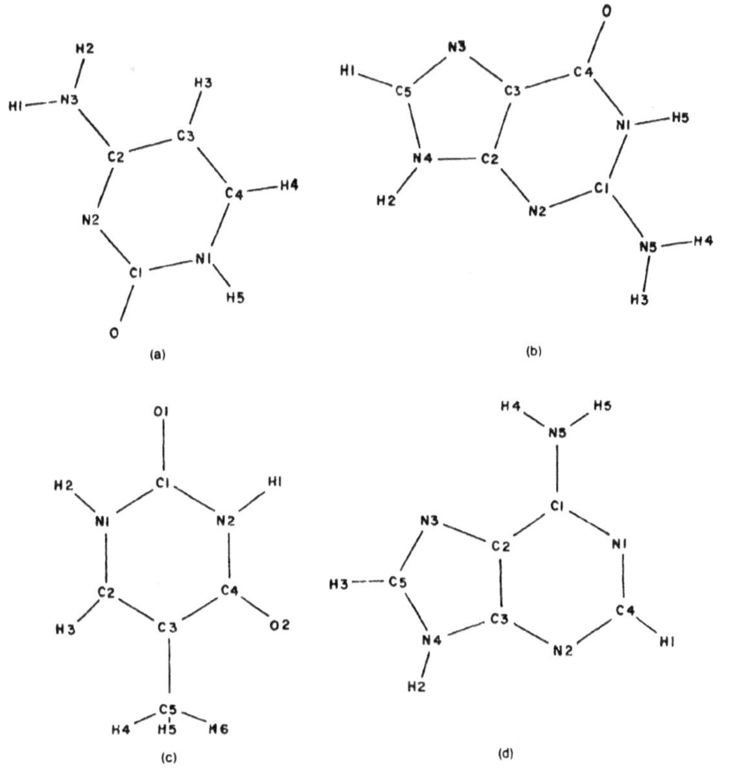

Figure 4

putationally extremely tractable for us to vary the stereochemistry of the epoxide ring and its opening to form an attacking agent.

Most recently we have been investigating such problems as the nitrocubanes [up to octanitrocubane (Figure 5) with 128 basis functions and all the stereoisomers of hexanitrohexazaadamantanes (Figure 6) with 116 basis functions].

We have also derived new SCF procedures which pass through the contributions to the Fock matrix elements from the invariant integrals, converge the calculation with respect to the new integrals and then make up the entire Fock matrix for the last several iterations. We have also derived the analysis for a similar strategy for carrying out the integral transformations for the CI calculations.

Both closed- and open-shell systems can be calculated. The programs can handle several hundred basis functions in a CDC 6600 computer and more in a CDC 7600.

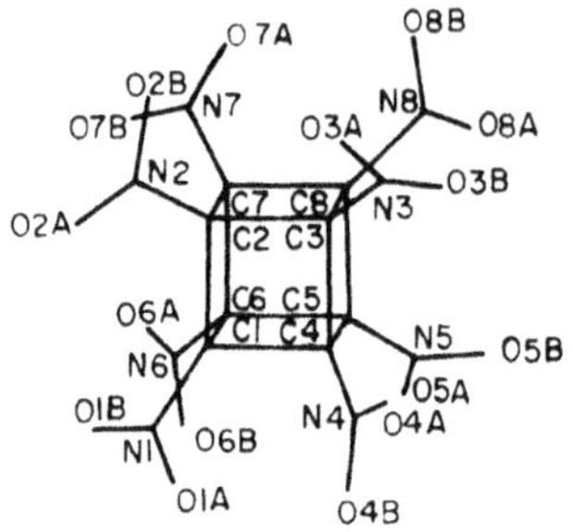

Figure 5

MOLASYS and GIPSY have also been merged with the MCSCF and CI procedures we use. Diffuse and polarization functions can be added to the MODPOT atomic basis sets just as they can to full atomic basis sets which include the inner shells. Using large MODPOT basis sets plus diffuse functions and the MODPOT/CI method we calculated the electron affinities of the alkali atoms with an agreement of 0.001 - 0.002 eV with both experiment and earlier frozen core CI calculations.[17d] In addition we have developed some desirable new strategies for CI procedures for large molecules. For interpreting spectra of large molecules it is not necessary to allow excitations from the lower occupied molecular orbitals or to any of the higher virtual orbitals. It is sufficient in most cases to allow excitations from perhaps the five highest occupied molecular orbitals to the five lowest unoccupied molecular orbitals and to fold into an effective CI Hamiltonian the effect of the other occupied orbitals from which excitations will not be permitted. We have shown that this is numerically equivalent to a frozen core calculation but with the great added advantage that it is not necessary to transform from integrals over atomic orbitals to integrals over molecular orbitals for the molecular orbitals from which no excitations are allowed.

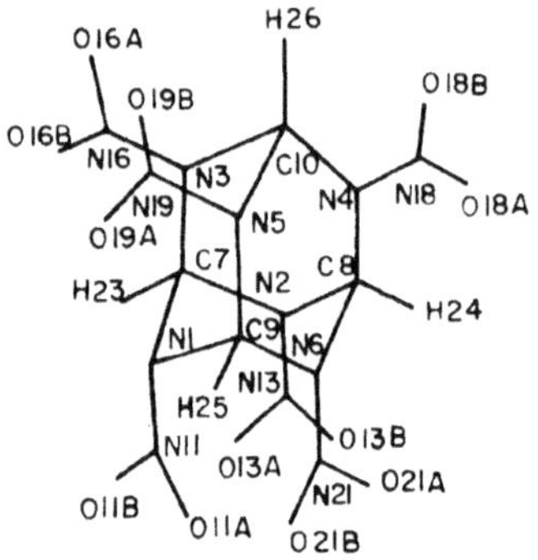

Figure 6

The integral transformation is one of the major bottlenecks in CI calculations on large molecules. For following reaction pathways on big molecules, the same strategy is employed after first transforming to localized orbitals. The effects of the localized orbitals not participating in the reaction are then folded into the effective CI Hamiltonian.

Many more millions of integrals are needed for ab-initio calculations on large molecules than can be held in core. Thus, these integrals are always stored on an external storage device and brought in for each SCF iteration or passed through a number of times for the transformation from integrals over A.O.'s to integrals over M.O.'s.

Based on the increase in integral speed on a vector machine [15 times that of a CDC 7600 (which is 5 times faster than a CDC 6600)] we recently suggested an additional new SCF and CI strategy for ab-initio calculations on large molecules.[21] Using our VRDDO prescreening, calculate the integrals once. Since not all integrals are kept, those integrals retained must be labelled. Use the integrals once for the first SCF iteration, then discard the integrals and keep only the labels packed in core with a counter index. Thus many fewer labels (with a counter index) need be stored than the total number of integrals retained. For each succeeding SCF iteration, go for the label, recalculate the integral and immediately use it to make up its contribution to the Fock matrix element. For efficient transformations random access to the integrals is desirable. A variant of this strategy of packed A.O. labels can be used for transformations for MC-SCF or CI calculations.

These ab-initio SCF wave functions also can be used as the basis for many-body perturbation theory or coupled cluster calculations.

2.2. Intermolecular Interactions and Potential Energy Surfaces

In addition to being capable of extreme accuracy for isolated molecules, ab-initio SCF, MCSCF and CI methods are also capable of describing correctly intermolecular interactions and potential energy surfaces even for large molecules.

Ab-initio SCF calculations of the supermolecule AB gives all the interaction energy between molecule A and molecule B except the dispersion energy. If too small atomic basis sets are used to describe molecule A and molecule B there can be a basis set superposition error in calculating the interaction energy of AB. This is caused in calculation of the supermolecule AB by molecule A using some of the basis set of molecule B and vice versa. This basis set superposition error can be corrected by the "counterpoise correction" suggested by Boys and Bernardi.[22] In that approach isolated molecule A is calculated in its own basis set plus the basis set of molecule B and vice-versa. The original MODPOT basis sets we optimized for intramolecular electronic structural calculations proved to be so well balanced that they have virtually no basis set superposition error[23] and give almost identical interaction energies to those from much larger conventional basis sets[24] (such as 6-31G**[25]).

The SCF interaction energy between molecule A and molecule B can be partitioned into

$$E_{electrostatic} + E_{exchange} + E_{polarization} + E_{C-T}$$

and each of these terms can be calculated separately.[26] There are also basis set superposition errors in the various terms of the partitioned SCF interaction energy. We have derived these.[24]

The various SCF interaction terms can be calculated separately and we have fitted each term separately to analytical functions.[27] Even better, the electrostatic contribution to the interaction energy can be calculated specifically by computing atomic multipole - atomic multipole contributions (based on the charges calculated for the monomer) and combined with the analytic functions for the other terms from the partitioned SCF energy plus the dispersion term.

Potential energy surfaces for dissociation of large molecules and for reactions with large molecules can also be calculated expeditiously with our scheme of carrying out CI calculations on the localized participating localized orbitals.

2.3. Solids, Crystals and Polymers

Since our ab-initio big molecule methods become more and more effective as the systems become more spatially extended they are ideally suited for calculations on solids, crystals and polymers. Moreover, periodicity conditions can be imposed on our methods.

A very recent use we have made has been to calculate the partioned energy contributions to molecular interaction energies, to fit each of these contributions separately to an analytical expression and then to use these expressions to optimize crystal unit cell parameters for various molecular crystals.[27]

3. CONCLUSION

Our disenchantment in 1973 with local density approximations for aromatic systems led us to our derivation and implementation of general strictly ab-initio SCF, MC-SCF and CI methods incorporating desirable options for large molecules.[17,19,21,28] These methods have the virtue of being completely general, reliable and capable of extreme accuracy. We have used our ab-initio MODPOT/VRDDO/MERGE method to carry out ab-initio calculations on hundreds of large molecules and intermolecular complexes in thousands of different conformations.

At this present meeting on Local Density Approximations in Quantum Chemistry and Solid State Theory Ellis[29] mentioned that he had recently carried out SCF local exchange calculations on aromatic compounds and Baerends[30] mentioned that they had also carried out local exchange calculations on aromatic compounds.

However, since 1975, our ab-initio methods also give electronic wave functions which are proper spin and symmetry determinants. Methods using local density approximations do not have this property.

The timings of these ab-initio methods are so rapid that it would seem in many cases that these methods could be competitive timewise with methods using local density approximations.

Moreover, at the recent 4th International Congress of Quantum Chemistry in Uppsala, Sweden, June 1982, Overhauser was extremely critical of results resulting from local density methods as compared to experiment.[31]

In view of the speed, generality, reliability and accuracy of our ab-initio methods incorporating desirable options for large molecules, it would seem that these methods are a competitive and attractive alternative to methods using local density approximations for large molecules, solids, crystals and polymers.

ACKNOWLEDGEMENT

Our current research on ab-initio methods for large molecules, solids, crystals and polymers is supported by the Office of Naval Research under Contract N00014-80-C-0003.

The development of the MERGE technique was supported in part by the National Cancer Institute under Contract N01-CP-75929.

Thanks are due to the late Professor John C. Slater and to Professor John Connolly and Brain Danese, at that time of the University of Florida; To Professor Jamie Keller, National University of Mexico and to Dr. Frank Hermann, IBM, San Jose for their collaboration in allowing us to bring their local density MS-Xα and MS-Xαβ scattered wave programs to our group at the Johns Hopkins University. Professor Keller and Dr. Danese were both visiting scientists with our group at Johns Hopkins University.

REFERENCES

1. L. Burnelle and Joyce J. Kaufman, Molecular orbitals of diborane in terms of a Gaussian basis, J. Chem. Phys. 43, 3540 (1965).
2. Joyce J. Kaufman, H.E. Popkie, S. Palalikit and P.C. Hariharan, Molecular calculations with the ab-initio non-empirical MODPOT, VRDDO and MODPOT/VRDDO procedures. IX. Carcinogenic benzo(a)pyrene and its metabolites using a MERGE technique, Int. J. Quantum Chem. 14, 793 (1978).
3. H.J.T. Preston, Joyce J. Kaufman and W.S. Koski, MS-Xα calculations of polyhedral boron compounds, Int. J. Quantum Chem. S 9, 137 (1975).
4. H.J.T. Preston, Joyce J. Kaufman, J. Keller, J.B. Danese and J.W.D. Connolly, MS-Xα calculations on boron trihalides and comparison with their photoelectron spectra, Chem. Phys. Lett. 37, 55 (1976).
5. H.J.T. Preston and Joyce J. Kaufman, MS-Xα calculations on the chlorofluoromethanes, Chem. Phys. Lett. 50, 157 (1977).
6. Joyce J. Kaufman, Electron affinities - theoretical, invited lecture presented at the International Conference on Heavy Ion Sources, Gatlinburg, Tennessee, October 27-30, 1975. IEEE, Transactions on Nuclear Sciences NS - 23, 936 (1976).
7. H.J.T. Preston and Joyce J. Kaufman, MS-Xα calculations of the electron affinities and ionization potentials of some heavy metals and their hydrides and fluorides, Int. J. Quantum Chem. S 12, 471 (1977).
8. H.J.T. Preston and Joyce J. Kaufman, MS-Xα and MS-Xαβ calculations on pyrrole, The Johns Hopkins University, 1973-1974.
9. (a) L.M. Sachs and M. Geller, MOSES, a FORTRAN IV system for polyatomic molecules, Int. J. Quantum Chem. 1S, 445 (1967).
   (b) I - s and p orbitals.
   (c) II - s and p orbitals; including symmetry option.
   (d) III - s, p and d orbitals; including symmetry option.
10. H.E. Popkie, W.S. Koski and Joyce J. Kaufman, Ab-initio LCAO-MO-SCF calculations on morphine and nalophine and comparison with their measured photoelectron spectra, J. Am. Chem. Soc. 98, 1342 (1976).
11 H.E. Popkie and Joyce J. Kaufman, Ab-initio LCAO-MO-SCF calculations of chlorpromazine and promazine, Int. J. Quantum Chem. 10, 569 (1976).
12. (a) H.E. Popkie, MOLAYSYS: a computer program for molecular orbital calculations on large systems, The Johns Hopkins University, 1974.
    (b) H.E. Popkie, MOLASYS-MERGE, The Johns Hopkins University, 1978.
13. (a) H.E. Popkie, GIPSY (Gaussian Integral Program System): a system of computer programs for the evaluation of one- and two-electron integrals involving s-, p- and d-type contracted cartesian Gaussian functions, The Johns Hopkins University, 1975.

(b) H.E. Popkie, GIPSY (Gaussian Integral Program System): a system of computer programs for the evaluation of one- and two-electron integrals involving s-, p-, d- and f-type contracted cartesian Gaussian functions, The Johns Hopkins University, 1976.

14. C.C.J. Roothaan, New developments in molecular orbital theory, Rev. Mod. Phys. 23, 69 (1951).

15. J. Hinze, MC-SCF. I. The multi-configuration self-consistent-field method, J. Chem. Phys. 59, 6424 (1973).

16. I. Shavitt, The method of configuration interaction, in: "Modern Theoretical Chemistry, Volume II., Electronic Structure Ab-Initio Methods", H.F. Schaefer III, ed., Plenum Press, New York (1976), pp. 189-275.

17. (a) Joyce J. Kaufman, Quantum chemical calculations on small, medium and large molecules, presented at the Summer Conference in Theoretical Chemistry, Boulder, Colorado, June 1975.

(b) H.E. Popkie and Joyce J. Kaufman, Molecular calculations with the VRDDO, MODPOT and MODPOT/VRDDO procedures, I. HF, F_2, HCl, Cl_2, formamide, pyrrole, pyridine and nitrobenzene, Int. J. Quantum Chem. S10, 47 (1976).

(c) H.E. Popkie and Joyce J. Kaufman, Molecular calculations with the VRDDO, MODPOT and MODPOT/VRDDO procedures, II. Cyclopentadiene, benzene, diazoles diazines and benzonitrile, J. Chem. Phys. 66, 4827 (1977).

(d) H.E. Popkie and Joyce J. Kaufman, Molecular calculations with the VRDDO, MODPOT and MODPOT/VRDDO procedures, III. MODPOT/SCF + CI calculations to determine electron affinities of alkali metal atoms, Chem. Phys. Lett. 47, 55 (1977).

(e) Joyce J. Kaufman, H.E. Popkie and H.J.T. Preston, Ab-initio and approximately rigorous calculations on small, medium and large molecules, Int. J. Quantum Chem. 11, 1005 (1977).

(f) H.E. Popkie and Loyce J. Kaufman, Molecular calculations with the MODPOT, VRDDO and MODPOT/VRDDO procedures, IV. Boron hydrides and carboranes, Int. J. Quantum Chem. 12, 937 (1977).

(g) H.E. Popkie and Joyce J. Kaufman, Molecular calculations with the MODPOT, VRDDO and MODPOT/VRDDO procedures, V. Ab-initio and MODPOT LCAO-MO-SCF Calculations on the chlorofluoromethanes, Int. J. Quantum Chem. S11, 433 (1977).

(h) Joyce J. Kaufman and H.E. Popkie, Molecular calculations with non-empirical ab-initio MODPOT, VRDDO and MODPOT/VRDDO procedures, VI. Nucleic acid constituents: A, G, C, T, presented at the Sixth Canadian Conference on Theoretical Chemstry, Fredericton, N.B., Canada, June 1977.

(i) Joyce J. Kaufman, H.E. Popkie and H.J.T. Preston, Molecular calculations with the non-empirical ab-initio MODPOT, VRDDO and MODPOT/VRDDO procedures, VII. Prototype normal neurotransmitters and their metabolites, presented at the Sixth Canadian Conference on Theoretical Chemistry, Fredericton, N.B., Canada, June 1977.

(j) Joyce J. Kaufman, H. E. Popkie and H.J.T. Preston, Molecular calculations with the ab-initio non-empirical MODPOT, VRDDO and MODPOT/VRDDO procedures. VIII. Charge delocalization in the anions of aromatic carboxylic acids and phenolic compounds, Int. J. Quantum Chem. S12, 283 (1978).

(k) Joyce J. Kaufman, H.E. Popkie, S. Palalikit and P.C. Hariharan, Molecular calculations with the ab-initio non-empirical MODPOT, VRDDO and MODPOT/VRDDO procedures, IX. Carcinogenic benzo(a)pyrene and its metabolites using a MERGE technique, Int. J. Quantum Chem. 14, 793 (1978).

(ℓ) P.C. Hariharan, H.E. Popkie and Joyce J. Kaufman, Non-empirical ab-initio MODPOT, VRDDO and MODPOT/VRDDO calculations, X. The attack of the simplest ultimate carcinogen CH_3^+, on guanine by a MERGE technique and the fundamental difference between methylating versus ethylating ultimate carcinogens, Int. J. Quantum Chem. S13, 255 (1979).

(m) W.A. Sokalski, P.C. Hariharan, H.E. Popkie, Joyce J. Kaufman and C. Petrongolo, Molecular calculations with the non-empirical ab-initio MODPOT, VRDDO and MODPOT/VRDDO procedures, XI. Theoretical study of the $[C_6H_5OH...OC_6H_5]^-$ molecular complex: ab-initio MODPOT/VRDDO calculations and electrostatic molecular potential contour maps, Int. J. Quantum Chem. 18, 173 (1980).

(n) P.C. Hariharan, H.E. Popkie, and Joyce J. Kaufman, Molecular calculations with the non-empirical ab-initio MODPOT, VRDDO, and MODPOT/VRDDO procedures, XII. Carcinogenic 3-methylcholanthrene and its metabolites using a MERGE technique, Int. J. Quantum Chem. 20, 645 (1981).

(o) F.L. Tobin, P.C. Hariharan, Joyce J. Kaufman and R.S. Miller, Ab-initio MODPOT/VRDDO/MERGE calculations on energetic compounds, I. Mechanism of initiation of cationic polymerization from electrostatic molecular potential contour maps,* Int. J. Quantum Chem. S15, 203 (1981).
*Paper XIII in the series "Molecular Calculations with the Non-Empirical Ab-Initio MODPOT/VRDDO/MERGE Procedures".

(p) A.H. Lowrey, P.C. Hariharan and Joyce J. Kaufman, Molecular calculations with the non-empirical ab-initio MODPOT/VRDDO/MERGE procedures, XIV. 2,6-dimethyl-N-nitroso-morpholine and its α-OH isomers: conformations and electrostatic molecular potential contour maps, Int. J. Quantum Chem. QBS8, 149 (1981).

(q) P.C. Hariharan, V. Lewchenko, W.S. Koski and Joyce J. Kaufman, Quantum chemical calculations for understanding and predicting toxicity, I. Ab-initio MODPOT/VRDDO calculations and electrostatic molecular potential contour maps of organophosphorus anticholinesterases and relation to mechanism of action,* an invited plenary lecture presented at the International Sanibel Symposium on Quantum Biology and Quantum Pharmacology, Palm Coast, Florida, March 1982. In press, Int. J. Quantum Chem., Quantum Biology Symposium Issue.

*Paper XV in the series "Molecular Calculations with the Non-Empirical Ab-Initio MODPOT/VRDDO/MERGE Procedures".

(r) P.C. Hariharan, W.S. Koski, Joyce J. Kaufman, R.S. Miller and A.H. Lowrey, Ab-initio MODPOT/VRDDO/MERGE calculations on energetic compounds, II. Nitroexplosives: RDX and α-, β- and δ-HMX,* presented at the International Sanibel Symposium on Quantum Chemistry and the Quantum Theory of Condensed Matter, Palm Coast, Florida, March 1982. In press, Int. J. Quantum Chem., Symposium Issue.
*Paper XVI in the series "Molecular Calculations with the Non-Empirical Ab-Initio MODPOT/VRDDO/MERGE Procedures".

(s) P.C. Hariharan, W.S. Koski, Joyce J. Kaufman and R.S. Miller, Ab-initio MODPOT/VRDDO/MERGE calculations on energetic compounds, III. Nitroexplosives: polyaminopolynitrobenzenes (including DATB, TATB and tetryl).* In press, Int. J. Quantum Chem., Congress Issue, 4th International Congress of Quantum Chemistry.
*Paper XVII in the series "Molecular Calculations with the Non-Empirical Ab-Initio MODPOT/VRDDO/MERGE Procedures".

18. (a) V. Bonifacic and S. Huzinaga, Atomic and molecular calculations with the model potential method, I., J. Chem. Phys. 60, 2779 (1974).
(b) V. Bonifacic and S. Huzinaga, Atomic and molecular calculations with the model potential method, II., J.Chem. Phys. 62, 1507 (1975).
(c) V. Bonifacic and S. Huzinaga, Atomic and molecular calculations with the model potential method, III., J.Chem. Phys. 62, 1509 (1975).
(d) D. McWilliams and S. Huzinaga, Gaussian basis sets suitable for accurate valence-shell calculations using the model potential method, J. Chem. Phys. 63, 4678 (1975).
(e) S. Huzinaga, personal discussion, August 1975.
(f) S. Huzinaga, Atomic and molecular calculations with model potential methods, presented at the Quantum Chemical Symposium, First North American Chemical Congress, Mexico City, December 1975.
(g) V. Bonifacic and S. Huzinaga, Atomic and molecular calculations with the model potential method, IV., J. Chem. Phys. 64, 956 (1976).
(h) S. Huzinaga, private communication, July 1977.
(i) O. Gropen, S. Huzinaga and A.D. McLean, Model potential SCF calculations on Cl_2, Br_2, and I_2, J. Chem. Phys. 73, 402 (1980).

19. H.E. Popkie and Joyce J. Kaufman, Test of charge conserving integral approximations for a variable retention of diatomic differential overlap (VRDDO) procedure for semi-ab-initio molecular orbital calculations on large molecules, Int. J. Quantum Chem., QBS2, 279 (1975).

20. (a) D.L. Wilhite and R.N. Euwema, Charge-conserving integral approximations for ab-initio quantum chemistry, Chem. Phys. Lett. 20, 610 (1973).

(b) D.L. Wilhite and R.N. Euwema, Charge-conserving integral approximations for ab-initio quantum chemistry, J. Chem. Phys. 61, 375 (1974).
21. Joyce J. Kaufman, P.C. Hariharan and H.E. Popkie, Symposium note: Additional new computational strategies for ab-initio calculations on large molecules, Int. J. Quantum Chem. S15, 199 (1981).
22. S.F. Boys and F. Bernardi, The calculations of small molecular interactions by the differences of separate total energies. Some procedures with reduced errors, Mol. Phys. 19, 553 (1970).
23. W.A. Sokalski, P.C. Hariharan and Joyce J. Kaufman, The Johns Hopkins University, 1981.
24. W.A. Sokalski, S. Roszak, P.C. Hariharan and Joyce J. Kaufman, An improved SCF interaction energy decomposition scheme corrected for basis set superposition effect, presented as part of an invited plenary lecture at the 4th International Congress on Quantum Chemistry, Satellite Symposium on Quantum Chemistry in Biology, Stockholm, Sweden, June 1982. In press, Int. J. Quantum Chem., Congress Issue.
25. R. Ditchfield, W.J. Hehre and J.A. Pople, Self-consistent molecular-orbital methods, IX. An extended Gaussian-type basis for molecular-orbital studies of organic molecules, J. Chem. Phys. 54, 724 (1971).
26. (a) K. Morokuma, Molecular orbital studies of hydrogen bonds, III. C=O...H-O hydrogen bond in $H_2CO...H_2O$ and $H_2CO...2H_2$),, J. Chem. Phys. 55, 1236 (1971).
(b) K. Kitaura and K. Morokuma, A new energy decomposition scheme for molecular interactions within the Hartree-Fock approximation, Int. J. Quantum Chem. 10, 325 (1976).
27. P.C. Hariharan, V. Lewchenko, A.H. Lowrey and Joyce J. Kaufman, Research in progress, The Johns Hopkins University.
28. Joyce J. Kaufman, H.E. Popkie and P.C. Hariharan, New optimal strategies for ab-initio quantum chemical calculations on large drugs, carcinogens, teratogens and biomolecules, an invited lecture presented at the Symposium on Computer Assisted Drug Design, Division of Computers in Chemistry at the American Chemical Society National Meeting, Honolulu, Hawaii, April 1979. In: "Computer Assisted Drug Design", E.C. Olson and R.E. Christoffersen, eds., ACS Symposium Series 112, Am. Chem. Soc., Washington D.C. (1979).
29. D.M. Ellis, Comment at this meeting. Unpublished.
30. (a) E.S. Baerends, Comment at this meeting.
(b) T. Ziegler, A. Rauk and E.J. Baerends, On the calculation of multiplet energies by the Hartree-Fock-Slater method, Theoret. Chim. Acta (Berl.) 43, 261 (1977).

INDEX

GPSR Compliance

The European Union's (EU) General Product Safety Regulation (GPSR) is a set of rules that requires consumer products to be safe and our obligations to ensure this.

If you have any concerns about our products, you can contact us on ProductSafety@springernature.com

In case Publisher is established outside the EU, the EU authorized representative is:

Springer Nature Customer Service Center GmbH
Europaplatz 3
69115 Heidelberg, Germany

Batch number: 10762183

Printed by Printforce, the Netherlands